Graduate Texts in Mathematics 151

Springer

New York
Berlin
Heidelberg
Barcelona
Hong Kong
London
Milan
Paris
Singapore
Tokyo

Graduate Texts in Mathematics

(continued after index)

Joseph H. Silverman

Advanced Topics in the Arithmetic of Elliptic Curves

With 17 Illustrations

 Springer

Joseph H. Silverman
Mathematics Department
Brown University
Providence, RI 02912
USA
jhs@math.brown.edu

Mathematics Subject Classifications (1991): 14-01, 11Gxx, 14Gxx, 14H52

Library of Congress Cataloging-in-Publication Data
Silverman, Joseph H., 1955–
    Advanced topics in the arithmetic of elliptic curves / Joseph H. Silverman.
        p.   cm. — (Graduate texts in mathematics ; v. 151)
    Includes bibliographical references and index.
    ISBN 0-387-94325-0 (New York) — ISBN 3-540-94325-0 (Berlin) (hardcover)
    ISBN 0-387-94328-5 (New York) — ISBN 3-540-94328-5 (Berlin) (softcover)
    1. Curves, Elliptic.   2. Curves, Algebraic.   3. Arithmetic.
I. Title.   II. Series.
QA567.S442   1994
516.3′52—dc20   94-21787

Printed on acid-free paper.

Production managed by Hal Henglein; manufacturing supervised by Vincent Scelta.
Photocomposed copy prepared from the author's TeX file.
Printed and bound by R.R. Donnelley & Sons, Harrisonburg, VA.
Printed in the United States of America.

9 8 7 6 5 4 3 2 (Corrected second printing, 1999)

ISBN 0-387-94325-0   Springer-Verlag   New York   Berlin   Heidelberg (hardcover)
ISBN 3-540-94325-0   Springer-Verlag   Berlin   Heidelberg   New York
ISBN 0-387-94328-5   Springer-Verlag   New York   Berlin   Heidelberg (softcover)   SPIN 10727882
ISBN 3-540-94328-5   Springer-Verlag   Berlin   Heidelberg   New York

For Susan

Preface

In the introduction to the first volume of *The Arithmetic of Elliptic Curves* (Springer-Verlag, 1986), I observed that "the theory of elliptic curves is rich, varied, and amazingly vast," and as a consequence, "many important topics had to be omitted." I included a brief introduction to ten additional topics as an appendix to the first volume, with the tacit understanding that eventually there might be a second volume containing the details. You are now holding that second volume.

Unfortunately, it turned out that even those ten topics would not fit into a single book, so I was forced to make some choices. The following material is covered in this book:

    I. Elliptic and modular functions for the full modular group.

   II. Elliptic curves with complex multiplication.

  III. Elliptic surfaces and specialization theorems.

  IV. Néron models, Kodaira-Néron classification of special fibers,
          Tate's algorithm, and Ogg's conductor-discriminant formula.

   V. Tate's theory of q-curves over p-adic fields.

  VI. Néron's theory of canonical local height functions.

So what's still missing? First and foremost is the theory of modular curves of higher level and the associated modular parametrizations of elliptic curves. There is little question that this is currently the hottest topic in the theory of elliptic curves, but any adequate treatment would seem to require (at least) an entire book of its own. (For a nice introduction, see Knapp [1].) Other topics that I have left out in order to keep this book at a manageable size include the description of the image of the ℓ-adic representation attached to an elliptic curve and local and global duality theory. Thus, at best, this book covers approximately half of the material described in the appendix to the first volume. I apologize to those who may feel disappointed, either at the incompleteness or at the choice of particular topics.

In addition to the complete areas which have been omitted, there are several topics which might have been naturally included if space had been available. These include a description of Iwasawa theory in Chapter II,

the analytic theory of p-adic functions (rigid analysis) in Chapter V, and
Arakelov intersection theory in Chapter VI.

It has now been almost a decade since the first volume was written.
During that decade the already vast mathematical literature on elliptic
curves has continued to explode, with exciting new results appearing with
astonishing rapidity. Despite the many omissions detailed above, I am
hopeful that this book will prove useful, both for those who want to learn
about elliptic curves and for those who hope to advance the frontiers of our
knowledge. I offer all of you the best of luck in your explorations!

Computer Packages

There are several computer packages now available for performing computations on elliptic curves. PARI and SIMATH have many built-in elliptic
curve functions, there are packages available for commercial programs such
as Mathematica and Maple, and the author has written a small stand-alone
program which runs on Macintosh computers. Listed below are addresses,
current as of March 1994, where these packages may be acquired via anony-
mous ftp.

>   PARI (includes many elliptic curve functions)
>   math.ucla.edu   128.97.4.254
>   megrez.ceremab.u-bordeaux.fr   147.210.16.17
>   (directory pub/pari)
>   (unix, mac, msdos, amiga versions available)
>   SIMATH (includes many elliptic curve functions)
>   ftp.math.orst.edu
>   ftp.math.uni-sb.de
>   apecs (arithmetic of plane elliptic curves, Maple package)
>   math.mcgill.ca   132.206.1.20
>   (directory pub/apecs)
>   Elliptic Curve Calculator (Mathematica package)
>   Elliptic Curve Calculator (stand-alone Macintosh program)
>   gauss.math.brown.edu   128.148.194.40
>   (directory dist/EllipticCurve)

A description of many of the algorithms used for doing computations on
elliptic curves can be found in H. Cohen [1, Ch. 7] and Cremona [1].

Acknowledgments

I would like to thank Peter Landweber and David Rohrlich for their careful reading of much of the original draft of this book. My thanks also go
to the many people who offered corrections, suggestions, and encourage-
ment, including Michael Artin, Ian Connell, Rob Gross, Marc Hindry, Paul
Lockhart, Jonathan Lubin, Masato Kuwata, Elisabetta Manduchi, Michael
Rosen, Glenn Stevens, Felipé Voloch, and Siman Wong.

As in the first volume, I have consulted a great many sources while
writing this book. Citations have been included for major theorems, but

many results which are now considered "standard" have been presented as such. In any case, I claim no originality for any of the unlabeled theorems in this book, and apologize in advance to anyone who may feel slighted. Sources which I found especially useful included the following:

Chapter I	Apostol [1], Lang [1,2,3], Serre [3], Shimura [1]
Chapter II	Lang [1], Serre [6], Shimura [1]
Chapter IV	Artin [1], Bosch-Lütkebohmert-Raynaud [1], Tate [2]
Chapter V	Robert [1], Tate [9]
Chapter VI	Lang [3,4], Tate [3]

I would like to thank John Tate for providing me with a copy of his unpublished manuscript (Tate [9]) containing the theory of q-curves over complete fields. This material, some of which is taken verbatim from Professor Tate's manuscript, forms the bulk of Chapter V, Section 3. In addition, the description of Tate's algorithm in Chapter IV, Section 9, follows very closely Tate's original exposition in [2], and I appreciate his allowing me to include this material.

Portions of this book were written while I was visiting the University of Paris VII (1992), IHES (1992), Boston University (1993), and Harvard (1994). I would like to thank everyone at these institutions for their hospitality during my stay.

Finally, and most importantly, I would like to thank my wife Susan for her constant love and understanding, and Debby, Danny, and Jonathan for providing all of those wonderful distractions so necessary for a truly happy life.

Joseph H. Silverman
March 27, 1994

Acknowledgments for the Second Printing

I would like to thank the following people who kindly provided corrections which have been incorporated in this second revised printing: Andrew Baker, Brian Conrad, Guy Diaz, Darrin Doud, Lisa Fastenberg, Benji Fisher, Boris Iskra, Steve Harding, Sharon Kineke, Joan-C. Lario, Yihsiang Liow, Ken Ono, Michael Reid, Ottavio Rizzo, David Rohrlich, Samir Siksek, Tonghai Yang, Horst Zimmer.

Providence, Rhode Island   *February, 1999*

Contents

Contents

CHAPTER VI

Local Height Functions

APPENDIX A

Some Useful Tables

Introduction

In the first volume of *The Arithmetic of Elliptic Curves*, we presented the basic theory culminating in two fundamental global results, the Mordell-Weil theorem on the finite generation of the group of rational points and Siegel's theorem on the finiteness of the set of integral points. This second volume continues our study of elliptic curves by presenting six important, but somewhat more specialized, topics.

We begin in Chapter I with the theory of elliptic functions and modular functions for the full modular group $\Gamma(1) = \mathrm{SL}_2(\mathbb{Z})/\{\pm 1\}$. We develop this material in some detail, including the theory of Hecke operators and the L-series associated to cusp forms for $\Gamma(1)$. Chapter II is devoted to the study of elliptic curves with complex multiplication. The main theorem here states that if $K/\mathbb{Q}$ is a quadratic imaginary field and if $E/\mathbb{C}$ is an elliptic curve whose endomorphism ring is isomorphic to the ring of integers of K, then $K(j(E))$ is the Hilbert class field of K; and further, the maximal abelian extension of K is generated by $j(E)$ and the x-coordinates[†] of the torsion points in $E(\mathbb{C})$. This is analogous to the cyclotomic theory, where the maximal abelian extension of $\mathbb{Q}$ is generated by the points of finite order in the multiplicative group $\mathbb{C}^*$. At the end of Chapter II we show that the L-series of an elliptic curve with complex multiplication is the product of two Hecke L-series with Grössencharacter, thereby obtaining at one stroke the analytic continuation and functional equation.

The common theme of Chapters III and IV is one-parameter families of elliptic curves. Chapter III deals with the classical geometric case, where the family is parametrized by a projective curve over a field of characteristic zero. Such families are called elliptic surfaces. Thus an elliptic surface consists of a curve C, a surface $\mathcal{E}$, and a morphism $\pi : \mathcal{E} \to C$ such that almost every fiber $\pi^{-1}(t)$ is an elliptic curve. The set of sections

$$\{\text{maps } \sigma : C \to \mathcal{E} \text{ such that } \pi \circ \sigma(t) = t\}$$

[†] If $j(E) = 1728$ or $j(E) = 0$, one has to use x^2 or x^3 instead of x.

to an elliptic surface forms a group, and we prove an analogue of the Mordell-Weil theorem which asserts that this group is (usually) finitely generated. In the latter part of Chapter III we study canonical heights and intersection theory on $\mathcal{E}$ and prove specialization theorems for both the canonical height and the group of sections.

Chapter IV continues our study of one-parameter families of elliptic curves in a more general setting. We replace the base curve C by a scheme $S = \operatorname{Spec} R$, where R is a discrete valuation ring. The generic fiber of the arithmetic surface $\mathcal{E} \to S$ is an elliptic curve E defined over the fraction field K of R, and its special fiber is a curve $\tilde{\mathcal{E}}$ (possibly singular, reducible, or even non-reduced) defined over the residue field k of R. We prove that if $\mathcal{C} \to S$ is a minimal proper regular arithmetic surface whose generic fiber is E, and if we write $\mathcal{E}$ for the part of $\mathcal{C}$ that is smooth over S, then $\mathcal{E}$ is a group scheme over S and satisfies Néron's universal mapping property. In particular, $E(K) \cong \mathcal{E}(R)$; that is, every K-rational point on the generic fiber E extends to an R-valued point of $\mathcal{E}$. We also describe the Kodaira-Néron classification of the possible configurations for the special fiber $\tilde{\mathcal{C}}$ and give Tate's algorithm for computing the special fiber. At the end of Chapter IV we discuss the conductor of an elliptic curve and prove (some cases of) Ogg's formula relating the conductor, minimal discriminant, and number of components of $\tilde{\mathcal{C}}$.

In Chapter V we return to the analytic theory of elliptic curves. We begin with a brief review of the theory over $\mathbb{C}$, which we then use to analyze elliptic curves defined over $\mathbb{R}$. But the main emphasis of Chapter V is on elliptic curves defined over p-adic fields. Every elliptic curve E defined over $\mathbb{C}$ is analytically isomorphic to $\mathbb{C}^*/q^{\mathbb{Z}}$ for some $q \in \mathbb{C}^*$. Similarly, Tate has shown that if E is defined over a p-adic field K and if the j-invariant of E is non-integral, then E is analytically isomorphic to $K^*/q^{\mathbb{Z}}$ for some $q \in K^*$. (It may be necessary to replace K by a quadratic extension.) Further, the isomorphism $E(\bar{K}) \cong \bar{K}^*/q^{\mathbb{Z}}$ respects the action of the Galois group $G_{\bar{K}/K}$, a fact which is extremely important for the study of arithmetic questions. In Chapter V we describe Tate's theory of q-curves and give some applications.

The final chapter of this volume contains a brief exposition of the theory of canonical local height functions. These local heights can be used to decompose the global canonical height described in the first volume [AEC, VIII §9]. We prove the existence of canonical local heights and give explicit formulas for them. Local heights are useful in studying some of the more refined properties of the global height.

As with the first volume, this book is meant to be an introductory text, albeit at an upper graduate level. For this reason we have occasionally made simplifying assumptions. We mention in particular that in Chapter II we restrict attention to elliptic curves whose ring of complex multiplications is integrally closed; in Chapter III we only consider elliptic surfaces over fields of characteristic 0; and in Chapter IV we assume that all Dedekind domains and discrete valuation rings have perfect residue fields. Possibly it would be preferable not to make these assumptions, but we feel that the loss of generality is more than made up for by the concomitant clarity of the exposition.

Prerequisites

The main prerequisite for reading this book is some familiarity with the basic theory of elliptic curves as described, for example, in the first volume. Beyond this, the prerequisites vary enormously from chapter to chapter. Chapter I requires little more than a first course in complex analysis. Chapter II uses class field theory in an essential way, so a brief summary of class field theory has been included in (II §3). Chapter III requires various classical results from algebraic geometry, such as the theory of surfaces and the theory of divisors on varieties. As always, summaries, references, and examples are supplied as needed.

Chapter IV is technically the most demanding chapter of the book. The reader will need some acquaintance with the theory of schemes, such as given in Hartshorne [1, Ch. II] or Eisenbud-Harris [1]. But beyond that, there are portions of Chapter IV, especially IV §6, which use advanced techniques and concepts from modern algebraic geometry. We have attempted to explain all of the main points, with varying degrees of precision and reliance on intuition, but the reader who wants to fill in every detail will face a non-trivial task. Finally, Chapters V and VI are basically self-contained, although they do refer to earlier chapters. More precisely, the interdependence of the chapters of this book is illustrated by the following guide:

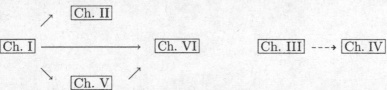

The dashed line connecting Chapter III to Chapter IV is meant to indicate that although there are few explicit cross-references, mastery of the subject matter of Chapter III will certainly help to illuminate the more difficult material covered in Chapter IV.

References and Exercises

The first volume of *The Arithmetic of Elliptic Curves* (Springer-Verlag, 1986) is denoted by [AEC], so for example [AEC, VIII.6.7] is Theorem 6.7 in Chapter VIII of [AEC]. All other bibliographic references are given by the author's name followed by a reference number in square brackets, for example Tate [7, theorem 5.1]. Cross-references within the same chapter are given by number in parentheses, such as (3.7) or (4.5a). References from within one chapter to another chapter or appendix are preceded by the appropriate Roman numeral or letter, as in (IV.6.1) or (A §3). Exercises appear at the end of each chapter and are numbered consecutively, so, for example, exercise 4.23 is the 23rd exercise at the end of Chapter IV.

Just as in the first volume, numerous exercises have been included at the end of each chapter. The reader desiring to gain a real understanding of the subject is urged to attempt as many as possible. Some of these exercises are (special cases of) results which have appeared in the literature. A list of comments and citations for the exercises will be found at the end of the book. Exercises marked with a single asterisk are somewhat more difficult, and two asterisks signal an unsolved problem.

Standard Notation

Throughout this book, we use the symbols

$$\mathbb{Z}, \ \mathbb{Q}, \ \mathbb{R}, \ \mathbb{C}, \ \mathbb{F}_q, \ \text{and} \ \mathbb{Z}_p$$

to represent the integers, rational numbers, real numbers, complex numbers, field with q elements, and p-adic integers respectively. Further, if R is any ring, then R^* denotes the group of invertible elements of R; and if A is an abelian group, then $A[m]$ denotes the subgroup of A consisting of all elements with order dividing m. A more complete list of notation will be found at the end of the book.

CHAPTER I

Elliptic and Modular Functions

In most of our previous work in [AEC], the major theorems have been of the form *"Let E/K be an elliptic curve. Then E/K has such-and-such a property."* In this chapter we will change our perspective and consider the set of elliptic curves as a whole. We will take the collection of all (isomorphism classes of) elliptic curves and make it into an algebraic curve, a so-called modular curve. Then by studying functions and differential forms on this modular curve, we will be able to make deductions about elliptic curves. Further, the Fourier coefficients of these modular functions and modular forms turn out to be extremely interesting in their own right, especially from a number-theoretic viewpoint. We will be able to prove some of their properties in the last part of the chapter.

This chapter thus has two main themes, each of which provides a paradigm for major areas of current research in number theory and algebraic geometry. First, when studying a collection of algebraic varieties or algebraic structures, one can often match the objects being studied (up to isomorphism) with the points of some other algebraic variety, called a moduli space. Then one can use techniques from algebraic geometry to study the moduli space as a variety and thereby deduce facts about the original collection of objects. A subtheme of this first main theme is that the moduli space itself need not be a projective variety, so a first task is to find a "natural" way to complete the moduli space.

Our second theme centers around the properties of functions and differential forms on a moduli space. Using techniques from algebraic geometry and complex analysis, one studies the dimensions of these spaces of modular functions and forms and also gives explicit Laurent, Fourier, and product expansions. Next one uses the geometry of the objects to define linear operators (called Hecke operators) on the space of modular forms, and one shows that the Hecke operators satisfy certain relations. One then takes a modular form which is a eigenfunction for the Hecke operators and deduces that the Fourier coefficients of the modular form satisfy the same relations. Finally, one reinterprets all of these results by associating an L-series to a modular form and showing that the L-series has an Euler product expansion and analytic continuation and that it satisfies a functional equation.

§1. The Modular Group

Recall [AEC VI.3.6] that a lattice $\Lambda \subseteq \mathbb{C}$ defines an elliptic curve $E/\mathbb{C}$ via the complex analytic map

$$\mathbb{C}/\Lambda \longrightarrow E_\Lambda(\mathbb{C}) : y^2 = 4x^3 - g_2 x - g_3$$
$$z \longmapsto \big(\wp(z; \Lambda), \wp'(z; \Lambda)\big).$$

Here

$$\wp(z; \Lambda) = \frac{1}{z^2} + \sum_{\substack{\omega \in \Lambda \\ \omega \neq 0}} \left(\frac{1}{(z - \omega)^2} - \frac{1}{\omega^2} \right)$$

is the Weierstrass $\wp$-function relative to the lattice Λ. (See [AEC VI,§3].) Further, if Λ_1 and Λ_2 are two lattices, then we have

$$E_{\Lambda_1} \cong_{/\mathbb{C}} E_{\Lambda_2} \qquad \text{if and only if} \qquad \Lambda_1 \text{ and } \Lambda_2 \text{ are homothetic.}$$

(See [AEC VI.4.1.1]. Recall Λ_1 and Λ_2 are *homothetic* if there is a number $c \in \mathbb{C}^*$ such that $\Lambda_1 = c\Lambda_2$.)

Thus the set of elliptic curves over $\mathbb{C}$ is intimately related to the set of lattices in $\mathbb{C}$, which we denote by $\mathcal{L}$:

$$\mathcal{L} = \{\text{lattices in } \mathbb{C}\}.$$

We let $\mathbb{C}^*$ act on $\mathcal{L}$ by multiplication,

$$c\Lambda = \{c\omega : \omega \in \Lambda\}.$$

Then the above discussion may be summarized by saying that there is an injection

$$\mathcal{L}/\mathbb{C}^* \hookrightarrow \frac{\{\text{elliptic curves defined over } \mathbb{C}\}}{\mathbb{C}\text{-isomorphism}}.$$

According to the Uniformization Theorem for Elliptic Curves (stated but not proven in [AEC VI.5.1]), this map is a bijection. One of our goals in this chapter is to prove this fact (4.3). But first we will need to describe the set $\mathcal{L}/\mathbb{C}^*$ more precisely. We will put a complex structure on $\mathcal{L}/\mathbb{C}^*$, and ultimately we will show that $\mathcal{L}/\mathbb{C}^*$ is isomorphic to $\mathbb{C}$.

Let $\Lambda \in \mathcal{L}$. We can describe Λ by choosing a basis, say

$$\Lambda = \mathbb{Z}\omega_1 + \mathbb{Z}\omega_2.$$

Switching ω_1 and ω_2 if necessary, we always assume that the pair (ω_2, ω_1) gives a positive orientation. (That is, the angle from ω_2 to ω_1 is positive and between $0°$ and $180°$. See Figure 1.1.)

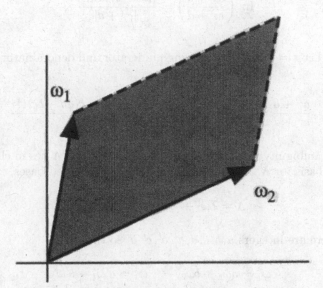

An Oriented Basis for the Lattice Λ

Figure 1.1

Since we only care about Λ up to homothety, we can normalize our basis by looking instead at

$$\frac{1}{\omega_2}\Lambda = \mathbb{Z}\frac{\omega_1}{\omega_2} + \mathbb{Z}.$$

Our choice of orientation implies that the imaginary part of ω_1/ω_2 satisfies

$$\mathrm{Im}(\omega_1/\omega_2) > 0,$$

which suggests looking at the upper half-plane

$$\mathbf{H} = \{\tau \in \mathbb{C} : \mathrm{Im}(\tau) > 0\}.$$

We have just shown that the natural map

$$\mathbf{H} \longrightarrow \mathcal{L}/\mathbb{C}^*,$$
$$\tau \longmapsto \Lambda_\tau = \mathbb{Z}\tau + \mathbb{Z}$$

is surjective. It is not, however, injective. When do two τ's give the same lattice? We start with an easy calculation.

Lemma 1.1. *Let $a, b, c, d \in \mathbb{R}$, $\tau \in \mathbb{C}$, $\tau \notin \mathbb{R}$. Then*

$$\operatorname{Im}\left(\frac{a\tau + b}{c\tau + d}\right) = \frac{(ad - bc)\operatorname{Im}(\tau)}{|c\tau + d|^2}.$$

PROOF. Let $\tau = s + it$. Multiplying numerator and denominator by $c\bar{\tau} + d$, we find

$$\frac{a\tau + b}{c\tau + d} = \frac{\{ac|\tau|^2 + (ad + bc)s + bd\} + \{(ad - bc)t\}i}{|c\tau + d|^2}.$$

$\square$

The ambiguity in associating a $\tau \in \mathsf{H}$ to a lattice Λ lies in choosing an oriented basis for Λ. Suppose that we take two oriented bases,

$$\Lambda = \mathbb{Z}\omega_1 + \mathbb{Z}\omega_2 = \mathbb{Z}\omega_1' + \mathbb{Z}\omega_2'.$$

Then there are integers $a, b, c, d, a', b', c', d'$ so that

$$\omega_1' = a\omega_1 + b\omega_2, \qquad \omega_1 = a'\omega_1' + b'\omega_2',$$
$$\omega_2' = c\omega_1 + d\omega_2, \qquad \omega_2 = c'\omega_1' + d'\omega_2'.$$

Substituting the left-hand expressions into the right-hand ones and using the fact that ω_1 and ω_2 are $\mathbb{R}$-linearly independent, we see that

$$\begin{pmatrix} a & b \\ c & d \end{pmatrix}\begin{pmatrix} a' & b' \\ c' & d' \end{pmatrix} = \begin{pmatrix} 1 & 0 \\ 0 & 1 \end{pmatrix}.$$

Further, using Lemma 1.1 (with $\tau = \omega_1/\omega_2$) and the fact that our bases are oriented, we find that

$$0 < \operatorname{Im}\left(\frac{\omega_1'}{\omega_2'}\right) = \operatorname{Im}\left(\frac{a\omega_1 + b\omega_2}{c\omega_1 + d\omega_2}\right) = \frac{(ad - bc)\operatorname{Im}(\omega_1/\omega_2)}{|c(\omega_1/\omega_2) + d|^2};$$

and so

$$ad - bc > 0.$$

In other words, the matrix $\begin{pmatrix} a & b \\ c & d \end{pmatrix}$ is in the special linear group over $\mathbb{Z}$,

$$\begin{pmatrix} a & b \\ c & d \end{pmatrix} \in \mathrm{SL}_2(\mathbb{Z}) = \left\{\begin{pmatrix} \alpha & \beta \\ \gamma & \delta \end{pmatrix} : \alpha, \beta, \gamma, \delta \in \mathbb{Z}, \alpha\delta - \beta\gamma = 1\right\}.$$

This proves the first half of the following lemma.

Lemma 1.2. (a) *Let $\Lambda \subset \mathbb{C}$ be a lattice, and let ω_1, ω_2 and ω_1', ω_2' be two oriented bases for Λ. Then*

$$\begin{aligned} \omega_1' &= a\omega_1 + b\omega_2 \\ \omega_2' &= c\omega_1 + d\omega_2 \end{aligned} \quad \text{for some matrix} \quad \begin{pmatrix} a & b \\ c & d \end{pmatrix} \in \mathrm{SL}_2(\mathbb{Z}).$$

(b) *Let $\tau_1, \tau_2 \in \mathbf{H}$. Then Λ_{τ_1} is homothetic to Λ_{τ_2} if and only if there is a matrix*

$$\begin{pmatrix} a & b \\ c & d \end{pmatrix} \in \mathrm{SL}_2(\mathbb{Z}) \quad \text{such that} \quad \tau_2 = \frac{a\tau_1 + b}{c\tau_1 + d}.$$

(c) *Let $\Lambda \subset \mathbb{C}$ be a lattice. Then there is a $\tau \in \mathbf{H}$ such that Λ is homothetic to $\Lambda_\tau = \mathbb{Z}\tau + \mathbb{Z}$.*

PROOF. (a) This was done above.
(b) Using (a), we find that

$$\Lambda_{\tau_1} \text{ is homothetic to } \Lambda_{\tau_2}$$
$$\Longleftrightarrow \mathbb{Z}\tau_2 + \mathbb{Z} = \mathbb{Z}\alpha\tau_1 + \mathbb{Z}\alpha \quad \text{for some } \alpha \in \mathbb{C}^*,$$
$$\Longleftrightarrow \begin{cases} \tau_2 = a\alpha\tau_1 + b\alpha \\ 1 = c\alpha\tau_1 + d\alpha \end{cases} \quad \text{for some } \begin{pmatrix} a & b \\ c & d \end{pmatrix} \in \mathrm{SL}_2(\mathbb{Z}),$$
$$\Longrightarrow \tau_2 = \frac{a\tau_1 + b}{c\tau_1 + d}.$$

Conversely, if $\tau_2 = (a\tau_1 + b)/(c\tau_1 + d)$, let $\alpha = c\tau_1 + d$. Then again using (a), we find

$$\alpha\Lambda_{\tau_2} = \mathbb{Z}(a\tau_1 + b) + \mathbb{Z}(c\tau_1 + d) = \mathbb{Z}\tau_1 + \mathbb{Z} = \Lambda_{\tau_1}.$$

Hence Λ_{τ_1} and Λ_{τ_2} are homothetic.
(c) Write $\Lambda = \omega_1\mathbb{Z} + \omega_2\mathbb{Z}$ with an oriented basis and take $\tau = \omega_1/\omega_2$. $\qquad \square$

In view of Lemma 1.2(b), it is natural to define an action of $\mathrm{SL}_2(\mathbb{Z})$ on $\mathbf{H}$ as follows:

$$\gamma\tau = \frac{a\tau + b}{c\tau + d} \quad \text{for } \gamma = \begin{pmatrix} a & b \\ c & d \end{pmatrix} \in \mathrm{SL}_2(\mathbb{Z}) \text{ and } \tau \in \mathbf{H}.$$

The fact that $\gamma\tau$ is in $\mathbf{H}$ follows from Lemma 1.1, and the fact that this defines a group action is an easy calculation. This action gives an equivalence relation on the points of $\mathbf{H}$, and Lemma 1.2(b) tells us what the cosets are. There is a bijection

$$\mathrm{SL}_2(\mathbb{Z}) \backslash \mathbf{H} \quad \overset{\text{one--to--one}}{\longleftrightarrow} \quad \mathcal{L}/\mathbb{C}^*,$$
$$\tau \quad \longmapsto \quad \Lambda_\tau.$$

We can actually do a little bit better, since the matrix

$$-1 = \begin{pmatrix} -1 & 0 \\ 0 & -1 \end{pmatrix}$$

acts trivially on $\mathbf{H}$.

Definition. The *modular group,* denoted $\Gamma(1)$, is the quotient group

$$\Gamma(1) = \mathrm{SL}_2(\mathbb{Z})/\{\pm 1\}.$$

Although $\Gamma(1)$ is the quotient $\mathrm{SL}_2(\mathbb{Z})/\{\pm 1\}$, we will generally just write down matrices and leave it to the reader to remember that $\left(\begin{smallmatrix} -1 & 0 \\ 0 & -1 \end{smallmatrix}\right)$ is equal to $\left(\begin{smallmatrix} 1 & 0 \\ 0 & 1 \end{smallmatrix}\right)$. For an explanation of the notation $\Gamma(1)$, see exercise 1.6 where we define groups $\Gamma(N)$ for all integers $N \geq 1$.

Remark 1.3. Note that ± 1 are the only elements of $\mathrm{SL}_2(\mathbb{Z})$ which fix **H**. For suppose that $\gamma = \left(\begin{smallmatrix} a & b \\ c & d \end{smallmatrix}\right)$ satisfies $\gamma\tau = \tau$ for all $\tau \in \mathbf{H}$. This means that

$$c\tau^2 - (d-a)\tau - b = 0 \qquad \text{for all } \tau \in \mathbf{H},$$

from which we conclude that $c = b = 0$ and $a = d$. Hence $\gamma = \pm 1$.

Remark 1.4. The group $\Gamma(1)$ contains two particularly important elements, which we will denote

$$S = \begin{pmatrix} 0 & -1 \\ 1 & 0 \end{pmatrix}, \qquad T = \begin{pmatrix} 1 & 1 \\ 0 & 1 \end{pmatrix}.$$

Their action on **H** is given by

$$S(\tau) = -\frac{1}{\tau}, \qquad T(\tau) = \tau + 1.$$

Notice also that the elements S and $ST = \left(\begin{smallmatrix} 0 & -1 \\ 1 & 1 \end{smallmatrix}\right)$ have finite order,

$$S^2 = \begin{pmatrix} 0 & -1 \\ 1 & 0 \end{pmatrix}^2 = 1 \quad \text{and} \quad (ST)^3 = \begin{pmatrix} 0 & -1 \\ 1 & 1 \end{pmatrix}^3 = 1,$$

so $\Gamma(1)$ contains finite subgroups of order 2 and 3.

The next proposition provides us with a good description of the quotient space $\Gamma(1)\backslash\mathbf{H}$.

Proposition 1.5. *Let $\mathcal{F} \subset \mathbf{H}$ be the set*

$$\mathcal{F} = \left\{\tau \in \mathbf{H} : |\tau| \geq 1 \quad \text{and} \quad |\operatorname{Re}(\tau)| \leq \tfrac{1}{2}\right\}.$$

(See Figure 1.2 for a picture of $\mathcal{F}$ and some of its translates by elements of $\Gamma(1)$.)
(a) Let $\tau \in \mathbf{H}$. Then there is a $\gamma \in \Gamma(1)$ such that $\gamma\tau \in \mathcal{F}$.
(b) Suppose that both τ and $\gamma\tau$ are in $\mathcal{F}$ for some $\gamma \in \Gamma(1)$, $\gamma \neq 1$. Then one of the following is true:

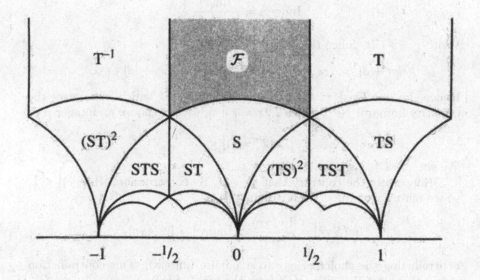

$\mathcal{F}$ *and Some of Its* $\Gamma(1)$-*Translates*

Figure 1.2

(i) $\mathrm{Re}(\tau) = -\frac{1}{2}$ and $\gamma\tau = \tau + 1$;

(ii) $\mathrm{Re}(\tau) = \frac{1}{2}$ and $\gamma\tau = \tau - 1$;

(iii) $|\tau| = 1$ and $\gamma\tau = -1/\tau$.

(c) *Let* $\tau \in \mathcal{F}$, *and let*

$$I(\tau) = \{\gamma \in \Gamma(1) : \gamma\tau = \tau\}$$

be the stabilizer of τ. *Then*

$$I(\tau) = \begin{cases} \{1, S\} & \text{if } \tau = i; \\ \{1, ST, (ST)^2\} & \text{if } \tau = \rho = e^{2\pi i/3}; \\ \{1, TS, (TS)^2\} & \text{if } \tau = -\bar{\rho} = e^{2\pi i/6}; \\ \{1\} & \text{otherwise.} \end{cases}$$

PROOF. (a) We prove something stronger. Let Γ' be the subgroup of $\Gamma(1)$ generated by $S = \begin{pmatrix} 0 & -1 \\ 1 & 0 \end{pmatrix}$ and $T = \begin{pmatrix} 1 & 1 \\ 0 & 1 \end{pmatrix}$, and let $\tau \in \mathbf{H}$. We will prove that there is a $\gamma \in \Gamma'$ such that $\gamma\tau \in \mathcal{F}$.

For any $\gamma = \begin{pmatrix} a & b \\ c & d \end{pmatrix} \in \Gamma(1)$, Lemma 1.1 says that

$$\operatorname{Im}(\gamma\tau) = \frac{\operatorname{Im}(\tau)}{|c\tau + d|^2}.$$

Write $\tau = s + it$. Since $t > 0$, it is clear that

$$|c\tau + d|^2 = (cs + d)^2 + (ct)^2 \to \infty \quad \text{as} \quad |c| + |d| \to \infty.$$

Hence, for our fixed τ, there is a matrix $\gamma_0 \in \Gamma'$ which *maximizes* the quantity $\operatorname{Im}(\gamma_0\tau)$. Next, since $T^n\tau = \tau + n$, we can choose an integer n so that

$$\left|\operatorname{Re}(T^n\gamma_0\tau)\right| \leq \tfrac{1}{2}.$$

We set $\gamma = T^n\gamma_0$ and claim that $\gamma\tau \in \mathcal{F}$.

Suppose to the contrary that $\gamma\tau \notin \mathcal{F}$. By construction, $\left|\operatorname{Re}(\gamma\tau)\right| \leq \tfrac{1}{2}$, so we must have $|\gamma\tau| < 1$. But then

$$\operatorname{Im}(S\gamma\tau) = \frac{\operatorname{Im}(\gamma\tau)}{|\gamma\tau|^2} > \operatorname{Im}(\gamma\tau) = \operatorname{Im}(\gamma_0\tau),$$

contradicting the choice of $\gamma_0\tau$ to maximize $\operatorname{Im}(\gamma_0\tau)$. This contradiction shows that $\gamma\tau \in \mathcal{F}$, which completes the proof of (a).

(b,c)   We may assume that $\operatorname{Im}(\gamma\tau) \geq \operatorname{Im}(\tau)$, since otherwise we replace the pair $\tau, \gamma\tau$ by the pair $\gamma\tau, \gamma^{-1}(\gamma\tau)$. Writing $\gamma = \begin{pmatrix} a & b \\ c & d \end{pmatrix}$ as usual, we have

$$\operatorname{Im}(\tau) \leq \operatorname{Im}(\gamma\tau) = \frac{\operatorname{Im}(\tau)}{|c\tau + d|^2}, \quad \text{so} \quad |c\tau + d| \leq 1.$$

Since $\operatorname{Im}(\tau) \geq \tfrac{1}{2}\sqrt{3}$, we must have $|c| \leq 2/\sqrt{3}$, so $|c| \leq 1$. Replacing γ by $-\gamma$ if necessary, it suffices to consider the cases $c = 0$ and $c = 1$.

$\boxed{c = 0}$

Then $a = d = 1$ and $\gamma\tau = \tau + b$. Since

$$|\operatorname{Re}(\tau)| \leq \tfrac{1}{2} \quad \text{and} \quad |\operatorname{Re}(\gamma\tau)| = |\operatorname{Re}(\tau + b)| \leq \tfrac{1}{2},$$

it follows that

$$b = \pm 1 \quad \text{and} \quad \operatorname{Re}(\tau) = \mp\tfrac{1}{2}.$$

$\boxed{c = 1}$

By assumption, $|\tau| \geq 1$ and $|\tau + d| \leq 1$. Writing $\tau = s + it$, this means that

$$1 \leq s^2 + t^2 \quad \text{and} \quad (s + d)^2 + t^2 \leq 1;$$

so

$$1 \leq s^2 + t^2 \leq 1 - 2ds - d^2 = 1 - d(d \pm 1) - d(2s \mp 1).$$

Since $d \in \mathbb{Z}$, the quantity $d(d \pm 1)$ is non-negative. Similarly since $|s| \leq \frac{1}{2}$, the quantity $d(2s \mp 1)$ is non-negative for one of the choices of $+/-$ sign. We conclude that

$$|\tau| = s^2 + t^2 = 1 \qquad \text{and} \qquad d(2s + d) = 0.$$

We now look at several subcases.

$\boxed{c = 1, d = 0}$

Then $\gamma = \begin{pmatrix} a & -1 \\ 1 & 0 \end{pmatrix}$, and since $|\tau| = 1$, we have

$$\tfrac{1}{2} \geq |\operatorname{Re}(\gamma\tau)| = |\operatorname{Re}(a - \tau^{-1})| = |a - s|.$$

Hence one of the following three cases holds:

$$a = 0, \quad |s| \leq \tfrac{1}{2}, \quad |\tau| = 1, \quad \gamma = S, \qquad \gamma\tau = -1/\tau;$$
$$a = 1, \quad s = \tfrac{1}{2}, \quad \tau = -\bar{\rho}, \quad \gamma = TS, \quad \gamma(-\bar{\rho}) = -\bar{\rho};$$
$$a = -1, \quad s = -\tfrac{1}{2}, \quad \tau = \rho, \quad \gamma = (ST)^2, \quad \gamma\rho = \rho.$$

$\boxed{c = 1, d = 1, s = -\tfrac{1}{2}}$

Then $\tau = \rho$ and $\gamma = \begin{pmatrix} a & a-1 \\ 1 & 1 \end{pmatrix}$, so

$$\gamma\tau = a - \frac{1}{\rho + 1} = a + \rho.$$

Since $\gamma\tau \in \mathcal{F}$, this leads to two cases:

$$a = 0, \quad \gamma = ST, \qquad \gamma\rho = \rho;$$
$$a = 1, \quad \gamma = \begin{pmatrix} 1 & 0 \\ 1 & 1 \end{pmatrix}, \quad \gamma\rho = -\bar{\rho}.$$

$\boxed{c = 1, d = -1, s = \tfrac{1}{2}}$

Then $\tau = -\bar{\rho}$, $\gamma = \begin{pmatrix} a & -a-1 \\ 1 & -1 \end{pmatrix}$, and $\gamma\tau = a + \tau$, so just as in the previous case there are two possibilities:

$$a = 0, \qquad \gamma = (TS)^2, \qquad \gamma(-\bar{\rho}) = -\bar{\rho};$$
$$a = -1, \quad \gamma = \begin{pmatrix} -1 & 0 \\ 1 & -1 \end{pmatrix}, \quad \gamma(-\bar{\rho}) = \rho.$$

$\square$

The geometric description of the quotient space $\Gamma(1)\backslash \mathbf{H}$ provided by Proposition 1.5 can be used to give a quick proof of the following purely algebraic fact.

Corollary 1.6. *The modular group* $\Gamma(1)$ *is generated by the matrices*

$$S = \begin{pmatrix} 0 & -1 \\ 1 & 0 \end{pmatrix} \quad \text{and} \quad T = \begin{pmatrix} 1 & 1 \\ 0 & 1 \end{pmatrix}.$$

PROOF. As in the proof of Proposition 1.5(a), we let Γ' be the subgroup of $\Gamma(1)$ generated by S and T. Fix some τ in the interior $\mathcal{F}$, such as $\tau = 2i$. Let $\gamma \in \Gamma(1)$. From the proof of (1.5a) there is a $\gamma' \in \Gamma'$ such that $\gamma'(\gamma\tau) \in \mathcal{F}$. Thus τ is in the interior of $\mathcal{F}$, and $(\gamma'\gamma)\tau$ is in $\mathcal{F}$. We conclude from (1.5b) that $\gamma\gamma' = 1$. Therefore $\gamma = \gamma'^{-1} \in \Gamma'$, which proves that $\Gamma' = \Gamma(1)$. $\qquad\square$

Remark 1.6.1. It is in fact true that $\Gamma(1)$ is the free product of its subgroups $\langle S \rangle$ and $\langle ST \rangle$ of orders 2 and 3. See exercise 1.1.

§2. The Modular Curve $X(1)$

The quotient space $\Gamma(1)\backslash\mathsf{H}$ classifies the set of lattices in $\mathbb{C}$ up to homothety. Proposition 1.5 provides a nice geometric description of $\Gamma(1)\backslash\mathsf{H}$. The vertical sides of the fundamental domain $\mathcal{F}$ are identified by T, and the two arcs of the circle $|\tau| = 1$ are identified by S, as shown in Figure 1.3. Making these identifications, we see that as a topological space, $\Gamma(1)\backslash\mathsf{H}$ looks like a 2-sphere with one point missing. Our next tasks are to supply that missing point, define a topology, and make the resulting surface into a Riemann surface.

Rather than adding a single point to $\Gamma(1)\backslash\mathsf{H}$, we will give a more general construction which is useful for generalizing the results of this chapter.

Definition. The *extended upper half-plane* H^* is the union of the upper half-plane H and the $\mathbb{Q}$-rational points of the projective line,

$$\mathsf{H}^* = \mathsf{H} \cup \mathbb{P}^1(\mathbb{Q}) = \mathsf{H} \cup \mathbb{Q} \cup \{\infty\}.$$

One should think of $\mathbb{P}^1(\mathbb{Q})$ as consisting of the rational points on the real axis together with a point at infinity. The points in $\mathbb{P}^1(\mathbb{Q})$ are called the *cusps of* H^*.

There is a natural action of $\Gamma(1)$ on $\mathbb{P}^1(\mathbb{Q})$ defined by

$$\begin{pmatrix} a & b \\ c & d \end{pmatrix} \begin{bmatrix} x \\ y \end{bmatrix} = \begin{bmatrix} ax + by \\ cx + dy \end{bmatrix}.$$

(Here we use $\begin{bmatrix} x \\ y \end{bmatrix}$ to denote homogeneous coordinates for a point in $\mathbb{P}^1(\mathbb{Q})$.) Thus $\Gamma(1)$ acts on the extended upper half-plane H^*. We define

$$Y(1) = \Gamma(1)\backslash\mathsf{H} \quad \text{and} \quad X(1) = \Gamma(1)\backslash\mathsf{H}^*.$$

The points in the complement $X(1) \smallsetminus Y(1)$ are called the *cusps of $X(1)$*. We now show that $X(1)$ has only one cusp and calculate its stabilizer.

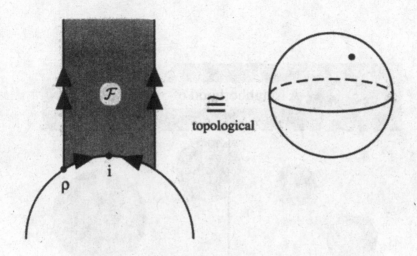

The Geometry of $\Gamma(1)\backslash \mathbf{H}$

Figure 1.3

Lemma 2.1. (a)

$$X(1) \smallsetminus Y(1) = \{\infty\}.$$

(b) *The stabilizer in $\Gamma(1)$ of $\infty \in \mathbf{H}^*$ is*

$$I(\infty) = \left\{ \begin{pmatrix} 1 & b \\ 0 & 1 \end{pmatrix} \in \Gamma(1) \right\} = \langle \textit{the subgroup of } \Gamma(1) \textit{ generated by } T \rangle.$$

PROOF. (a) Let $\begin{bmatrix} x \\ y \end{bmatrix} \in \mathbb{P}^1(\mathbb{Q})$ be any point in $\mathbf{H}^* \smallsetminus \mathbf{H}$. Since x and y are homogeneous coordinates, we may assume that $x, y \in \mathbb{Z}$ and $\gcd(x, y) = 1$. Choose $a, b \in \mathbb{Z}$ so that $ax + by = 1$. Then

$$\gamma = \begin{pmatrix} a & b \\ -y & x \end{pmatrix} \in \Gamma(1) \qquad \text{and} \qquad \gamma \begin{bmatrix} x \\ y \end{bmatrix} = \begin{bmatrix} 1 \\ 0 \end{bmatrix}.$$

Therefore every point in $\mathbf{H}^* \smallsetminus \mathbf{H}$ is equivalent (under the action of $\Gamma(1)$) to ∞.

(b) We have $\begin{pmatrix} a & b \\ c & d \end{pmatrix} \begin{bmatrix} 1 \\ 0 \end{bmatrix} = \begin{bmatrix} 1 \\ 0 \end{bmatrix}$ if and only if $c = 0$. Hence $\begin{pmatrix} a & b \\ c & d \end{pmatrix}$ has the form $\begin{pmatrix} 1 & b \\ 0 & 1 \end{pmatrix}$. $\qquad \square$

Topologically, $X(1)$ looks like a 2-sphere. To make this precise, we need to describe a topology on $X(1)$. We start by giving a topology for $\mathbf{H}^*$.

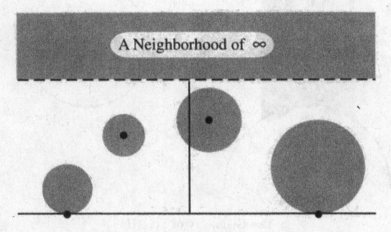

Some Open Sets in $\mathbf{H}^*$

Figure 1.4

Definition. The *topology of* $\mathbf{H}^*$ is defined as follows. For $\tau \in \mathbf{H}$, we take the usual open neighborhoods of τ contained in $\mathbf{H}$. For the cusp ∞, we take as a basis of open neighborhoods the sets

$$\{\tau \in \mathbf{H} : \mathrm{Im}(\tau) > \kappa\} \cup \{\infty\} \qquad \text{for every } \kappa > 0.$$

For a cusp $\tau \neq \infty$, we take as a basis of open neighborhoods the sets

$$\{\text{the interior of a circle in } \mathbf{H} \text{ tangent to the real axis at } \tau\} \cup \{\tau\}.$$

(See Figure 1.4.)

Remark 2.2.1. For any cusp $\tau_0 \neq \infty$, Lemma 2.1(a) says that there is a transformation $\gamma \in \Gamma(1)$ with $\gamma\infty = \tau_0$. Then one easily checks that γ sends a set of the form $\{\mathrm{Im}(\tau) > \kappa\}$ to the interior of a circle in $\mathbf{H}$ tangent to the real axis at τ_0. (See exercise 1.2.) In other words, the fundamental neighborhoods of ∞ and of the finite cusps are sent one-to-another by the elements of $\Gamma(1)$.

Remark 2.2.2. From the definition, it is clear that distinct points of $\mathbf{H}^*$ have disjoint neighborhoods. Hence $\mathbf{H}^*$ is a Hausdorff space. It is also clear from (2.2.1) that the elements of $\Gamma(1)$ define homeomorphisms of $\mathbf{H}^*$.

The next lemma will help us describe the topology on the quotient space $X(1) = \Gamma(1)\backslash\mathbf{H}^*$. It will also be used later to define a complex structure on $X(1)$.

Lemma 2.3. *For any two points $\tau_1', \tau_2 \in \mathbf{H}^*$, let*

$$I(\tau_1, \tau_2) = \{\gamma \in \Gamma(1) : \gamma\tau_1 = \tau_2\},$$

and similarly, for any two subsets $U_1, U_2 \subseteq \mathbf{H}^*$, let

$$I(U_1, U_2) = \{\gamma \in \Gamma(1) : \gamma U_1 \cap U_2 \neq \emptyset\}.$$

Then, for all $\tau_1, \tau_2 \in \mathbf{H}^*$, there exist open neighborhoods $U_1, U_2 \subseteq \mathbf{H}^*$ of τ_1, τ_2 respectively such that

$$I(U_1, U_2) = I(\tau_1, \tau_2).$$

(In other words, if γU_1 and U_2 have a point in common, then necessarily $\gamma\tau_1 = \tau_2$.)

PROOF. For any $\alpha, \beta \in \Gamma(1)$ we have

$$I(\alpha\tau_1, \beta\tau_2) = \beta I(\tau_1, \tau_2)\alpha^{-1} \qquad \text{and} \qquad I(\alpha U_1, \beta U_2) = \beta I(U_1, U_2)\alpha^{-1}.$$

It thus suffices to prove the lemma for any $\Gamma(1)$-translates of τ_1 and τ_2. Using (1.5a) and (2.1a), we may assume that

$$\tau_1, \tau_2 \in \mathcal{F}^* = \mathcal{F} \cup \{\infty\}.$$

From (1.5) and (2.1), we have a good description of how $\Gamma(1)$ acts on $\mathbf{H}^*$ and $\mathcal{F}^*$, as illustrated in Figure 1.2. We consider three cases, depending on whether or not our points are at ∞.

$\boxed{\tau_1, \tau_2 \in \mathcal{F}}$

From (1.5) (or Figure 1.2) we see that $I(\mathcal{F}, \mathcal{F})$ is finite; explicitly,

$$I(\mathcal{F}, \mathcal{F}) = \{1, T, TS, TST, (TS)^2, S, ST, STS, (ST)^2, T^{-1}\}.$$

Let

$$\mathcal{G} = \text{Interior}\Big(\bigcup_{\gamma \in I(\mathcal{F}, \mathcal{F})} \gamma\mathcal{F}\Big).$$

Then $\mathcal{G}$ is an open subset of $\mathbf{H}$ containing $\mathcal{F}$. Further, $I(\mathcal{G}, \mathcal{G})$ is finite, since

$$I(\mathcal{G}, \mathcal{G}) \subseteq \bigcup_{\gamma_1, \gamma_2 \in I(\mathcal{F}, \mathcal{F})} I(\gamma_1\mathcal{F}, \gamma_2\mathcal{F}) = \bigcup_{\gamma_1, \gamma_2 \in I(\mathcal{F}, \mathcal{F})} \gamma_2 I(\mathcal{F}, \mathcal{F})\gamma_1^{-1}.$$

Next we observe that if $\gamma \in I(\mathcal{G}, \mathcal{G}) \smallsetminus I(\tau_1, \tau_2)$, so $\gamma\tau_1 \neq \tau_2$, then we can find open sets V_γ, W_γ in $\mathbf{H}$ satisfying

$$\gamma\tau_1 \in V_\gamma, \quad \tau_2 \in W_\gamma, \quad \text{and} \quad V_\gamma \cap W_\gamma = \emptyset.$$

Let

$$U_1 = \mathcal{G} \cap \bigcap_{\substack{\gamma \in I(\mathcal{G},\mathcal{G}) \\ \gamma \notin I(\tau_1,\tau_2)}} \gamma^{-1} V_\gamma, \qquad U_2 = \mathcal{G} \cap \bigcap_{\substack{\gamma \in I(\mathcal{G},\mathcal{G}) \\ \gamma \notin I(\tau_1,\tau_2)}} W_\gamma.$$

By construction, $\tau_1 \in U_1$ and $\tau_2 \in U_2$, so

$$I(\tau_1, \tau_2) \subseteq I(U_1, U_2).$$

Suppose that they are not equal, say $\gamma \in I(U_1, U_2) \smallsetminus I(\tau_1, \tau_2)$. Then

$$\gamma \in I(\mathcal{G}, \mathcal{G}) \smallsetminus I(\tau_1, \tau_2), \qquad \text{and so} \qquad \gamma \in I(\gamma^{-1} V_\gamma, W_\gamma) = I(V_\gamma, W_\gamma)\gamma.$$

But $V_\gamma \cap W_\gamma = \emptyset$, so $1 \notin I(V_\gamma, W_\gamma)$. This contradiction shows the other inclusion and completes the proof that $I(\tau_1, \tau_2) = I(U_1, U_2)$.

$\boxed{\tau_1 \in \mathcal{F},\ \tau_2 = \infty}$

Let U_1 be an open disk centered at τ_1. As in the proof of Proposition 1.5, we observe that the quantity

$$\kappa = \kappa(U_1) = \sup_{\substack{\tau \in U_1 \\ \gamma \in \Gamma(1)}} \operatorname{Im}(\gamma\tau) = \sup_{\substack{\tau \in U_1 \\ \left(\begin{smallmatrix} a & b \\ c & d \end{smallmatrix}\right) \in \Gamma(1)}} \frac{\operatorname{Im}(\tau)}{|c\tau + d|^2}$$

is finite. (Note that if $\tau = s + it \in U_1$, then s and t are bounded, so

$$|c\tau + d|^2 = (cs + d)^2 + (ct)^2 \to \infty \quad \text{as} \quad |c| + |d| \to \infty \quad \textit{uniformly in } \tau \in U_1.)$$

Now

$$U_2 = \{\tau \in \mathbf{H} : \operatorname{Im}(\tau) > \kappa\} \cup \{\infty\}$$

will be a neighborhood of ∞ satisfying

$$\gamma U_1 \cap U_2 = \emptyset \qquad \text{for all } \gamma \in \Gamma(1).$$

Hence

$$I(U_1, U_2) = \emptyset = I(\tau_1, \tau_2).$$

$\boxed{\tau_1 = \tau_2 = \infty}$

Let

$$U_\infty = \{\tau \in \mathbf{H} : \operatorname{Im}(\tau) > 2\} \cup \{\infty\}.$$

From (1.5) (or Figure 1.2) we see that the only elements of $\Gamma(1)$ which take some point in U_∞ to another point in U_∞ are powers of T. Hence from (2.1b) we conclude that

$$I(U_\infty, U_\infty) = \{T^k \in \Gamma(1) : k \in \mathbb{Z}\} = I(\infty, \infty).$$

$\qquad\qquad\qquad\qquad\qquad\qquad\qquad\qquad\qquad\qquad\qquad\qquad\qquad\qquad\qquad\qquad\square$

Next we define a topology on $X(1)$ and use Lemma 2.3 to show that $X(1)$ is a Hausdorff space. Note that this fact requires proof; it is not immediate from the fact that $\mathbf{H}^*$ is Hausdorff. (See exercise 1.3.)

Definition. Let

$$\phi : \mathsf{H}^* \longrightarrow \Gamma(1)\backslash \mathsf{H}^* = X(1)$$

be the natural projection. The *quotient topology on* $X(1)$ is defined by the condition that $U \subseteq X(1)$ is open if and only if $\phi^{-1}(U)$ is open. Equivalently, it is the weakest topology for which ϕ is continuous. Note that ϕ is also an open map, that is, it takes open sets to open sets. For if $W \subset \mathsf{H}^*$ is open, then so is

$$\phi^{-1}(\phi W) = \bigcup_{\gamma \in \Gamma(1)} \gamma W.$$

Proposition 2.4. $X(1)$ *with its quotient topology is a compact Hausdorff space.*

PROOF. We start by checking that $X(1)$ is compact. Let $\{U_i\}_{i \in I}$ be an open cover of $X(1)$. Then $\{\phi^{-1}(U_i)\}_{i \in I}$ is an open cover of H^*. In particular, some $\phi^{-1}(U_i)$ contains ∞, say $\infty \in \phi^{-1}(U_{i_1})$. By definition of the topology on H^*, there is a constant $\kappa > 0$ so that

$$\phi^{-1}(U_{i_1}) \supseteq \{\tau \in \mathsf{H} : \mathrm{Im}(\tau) > \kappa\} \cup \{\infty\}.$$

Hence the set $\mathcal{F} \smallsetminus \phi^{-1}(U_{i_1})$ is compact (it is closed and bounded), so there is a finite subcover

$$\mathcal{F} \smallsetminus \phi^{-1}(U_{i_1}) \subseteq \phi^{-1}(U_{i_2}) \cup \cdots \cup \phi^{-1}(U_{i_n}).$$

Then $U_{i_1} \cup \cdots \cup U_{i_n}$ covers $X(1)$.

Next we verify that $X(1)$ is Hausdorff. Let $x_1, x_2 \in X(1)$ be distinct points, and let $\tau_1, \tau_2 \in \mathsf{H}^*$ be points with $\phi(\tau_i) = x_i$. Then $\gamma\tau_1 \neq \tau_2$ for all $\gamma \in \Gamma(1)$, so in the notation of (2.3), $I(\tau_1, \tau_2) = \emptyset$. From (2.3), there are open neighborhoods $U_1, U_2 \subseteq \mathsf{H}^*$ of τ_1, τ_2 satisfying $I(U_1, U_2) = \emptyset$. Then $\phi(U_1), \phi(U_2)$ are disjoint neighborhoods of x_1, x_2. $\square$

Making $X(1)$ into a compact Hausdorff space is a good start, but recall that our ultimate goal is to give $X(1)$ a complex structure. We recall what this means.

Definition. Let X be a topological space. A *complex structure* on X is an open covering $\{U_i\}_{i \in I}$ of X and homeomorphisms

$$\psi_i : U_i \xrightarrow{\sim} \psi_i(U_i) \subset \mathbb{C}$$

such that each $\psi_i(U_i)$ is an open subset of $\mathbb{C}$ and such that for all $i, j \in I$ with $U_i \cap U_j \neq \emptyset$, the map

$$\psi_j \circ \psi_i^{-1} : \psi_i(U_i \cap U_j) \longrightarrow \psi_j(U_i \cap U_j)$$

is holomorphic. The map ψ_i is called a *local parameter* for the points in U_i. A *Riemann surface* is a connected Hausdorff space which has a complex structure defined on it.

Theorem 2.5.   *The following defines a complex structure on $X(1)$ which gives it the structure of a compact Riemann surface of genus 0:*

Let $x \in X(1)$, choose $\tau_x \in \mathbf{H}^*$ with $\phi(\tau_x) = x$, and let $U_x \subset \mathbf{H}^*$ be a neighborhood of τ_x satisfying

$$I(U_x, U_x) = I(\tau_x).$$

(Such a U_x exists from Lemma 2.3 with $\tau_1 = \tau_2 = \tau_x$ and $U_x = U_1 \cap U_2$.) Then

$$I(\tau_x)\backslash U_x \subset X(1)$$

is a neighborhood of x, so $\{I(\tau_x)\backslash U_x\}_{x \in X(1)}$ is an open cover of $X(1)$.

$\boxed{x \neq \infty}$

Let $r = \#I(\tau_x)$, and let g_x be the holomorphic isomorphism

$$g_x : \mathbf{H} \longrightarrow \{z \in \mathbb{C} : |z| < 1\}, \qquad g_x(\tau) = \frac{\tau - \tau_x}{\tau - \bar{\tau}_x}.$$

Then the map

$$\psi_x : I(\tau_x)\backslash U_x \longrightarrow \mathbb{C}, \qquad \psi_x(\phi(\tau)) = g_x(\tau)^r$$

is well defined and gives a local parameter at x.

$\boxed{x = \infty}$

We may take $\tau_x = \infty$, so $I(\tau_x) = \{T^k\}$. Then

$$\psi_x : I(\tau_x)\backslash U_x \longrightarrow \mathbb{C}, \qquad \psi_x(\phi(\tau)) = \begin{cases} e^{2\pi i \tau} & \text{if } \phi(\tau) \neq \infty, \\ 0 & \text{if } \phi(\tau) = \infty \end{cases}$$

is well defined and gives a local parameter at x.

Remark 2.5.1. If $I(\tau_x) = \{1\}$, then the natural map

$$\phi : U_x \xrightarrow{\sim} I(\tau_x)\backslash U_x \subset X(1)$$

is already a homeomorphism, so

$$\psi_x = \phi^{-1} : I(\tau_x)\backslash U_x \longrightarrow U_x$$

is a local parameter at x. Thus the only real complication occurs when x equals $\phi(i)$, $\phi(\rho)$, or $\phi(\infty)$. (See also exercise 1.4.)

Remark 2.5.2. The following commutative diagrams illustrate the definitions of the local parameters $\psi_x : I(\tau_x)\backslash U_x \hookrightarrow \mathbb{C}$.

$$
\begin{array}{ccc}
U_x & \xrightarrow{\phi} & I(\tau_x)\backslash U_x \\
\downarrow{\scriptstyle g_x} & & \downarrow{\scriptstyle \psi_x} \\
\mathbb{C} & \xrightarrow{z \mapsto z^r} & \mathbb{C}
\end{array}
\qquad\qquad
\begin{array}{ccc}
U_x & \xrightarrow{\phi} & I(\tau_x)\backslash U_x \\
{\scriptstyle g_\infty}\searrow & & \downarrow{\scriptstyle \psi_x} \\
& & \mathbb{C}
\end{array}
$$

$$
x \neq \infty, \quad g_x(\tau) = \frac{\tau - \tau_x}{\tau - \bar{\tau}_x}
\qquad\qquad
x = \infty, \quad g_\infty(\tau) = e^{2\pi i \tau}
$$

PROOF (of Theorem 2.5). We already know that $X(1)$ is a compact Hausdorff space (2.4), and it is clearly connected due to the continuous surjection $\phi : \mathbf{H}^* \to X(1)$. Further, an inspection of Figure 1.2 shows that $X(1)$ has genus 0. (For those who dislike such a visual argument, we will later give an explicit map $j : X(1) \to \mathbb{P}^1(\mathbb{C})$. See (4.1) below. The interested reader can check that our proof that j is analytic does not depend on the a priori knowledge that $X(1)$ has genus 0. Then the elementary argument described in exercise 1.11 shows that j is bijective, hence an isomorphism.)

By construction, the set

$$
\phi(U_x) = I(U_x, U_x)\backslash U_x = I(\tau_x)\backslash U_x
$$

is a neighborhood of x. We must verify that the maps

$$
\psi_x : I(\tau_x)\backslash U_x \longrightarrow \mathbb{C}
$$

are well-defined homeomorphisms (onto their images) and that they satisfy the compatibility conditions for a complex structure.

We begin with a lemma which shows that the function $g_x(\tau)$ behaves nicely with respect to the transformations in $I(\tau_x)$.

Lemma 2.6. Let $a \in \mathbf{H}$, let $R : \mathbf{H} \to \mathbf{H}$ be a holomorphic map with $R(a) = a$, and let $g(\tau) = (\tau - a)/(\tau - \bar{a})$. Suppose further that

$$
\overbrace{R \circ \cdots \circ R}^{r \text{ times}}(\tau) = \tau
$$

and that $r \geq 1$ is the smallest integer with this property. Then there is a primitive r^{th}-root of unity ζ such that

$$
g(R\tau) = \zeta g(\tau) \quad \text{for all } \tau \in \mathbf{H}.
$$

PROOF. Note that g is an isomorphism

$$
g : \mathbf{H} \xrightarrow{\sim} \{z \in \mathbb{C} : |z| < 1\}
$$

with $g(a) = 0$, so the map

$$G = g \circ R \circ g^{-1} : \{z \in \mathbb{C} : |z| < 1\} \longrightarrow \{z \in \mathbb{C} : |z| < 1\}$$

is a holomorphic automorphism of the unit disk with $G(0) = 0$. It follows that $G(z) = cz$ for some constant $c \in \mathbb{C}$. (See, e.g., Ahlfors [1].) Since the r-fold composition $G \circ \cdots \circ G(z) = z$ and r is chosen minimally, we conclude that c is a primitive r^{th}-root of unity. $\qquad\qquad\square$

We resume the proof of Theorem 2.5. Suppose first that $x \neq \infty$. Note that from (1.5), $I(\tau_x)$ is cyclic, say generated by R. Then (2.6) implies that

$$g_x(R\tau) = \zeta g(\tau) \qquad \text{for all } \tau \in \mathbf{H},$$

where ζ is a primitive r^{th}-root of unity. Hence

$$\psi_x\big(\phi(R\tau)\big) = g_x(R\tau)^r = \zeta^r g_x(\tau)^r = \psi_x\big(\phi(\tau)\big),$$

so ψ_x is well defined on the quotient $I(\tau_x)\backslash U_x$.

Next we check that ψ_x is injective. Let $\tau_1, \tau_2 \in U_x$. Then

$$
\begin{aligned}
\psi_x\big(\phi(\tau_1)\big) = \psi_x\big(\phi(\tau_2)\big) &\iff g_x(\tau_1)^r = g_x(\tau_2)^r \\
&\iff g_x(\tau_1) = \zeta^i g_x(\tau_2) \quad \text{for some } 0 \le i < r, \\
&\iff g_x(\tau_1) = g_x(R^i \tau_2) \quad \text{for some } 0 \le i < r, \\
&\iff \tau_1 = R^i \tau_2 \quad \text{for some } 0 \le i < r, \\
&\iff \phi(\tau_1) = \phi(\tau_2).
\end{aligned}
$$

Hence ψ_x is injective. Finally, it is clear from the commutative diagram given in (2.5.2) that both ψ_x and ψ_x^{-1} are continuous, since the maps ϕ, g_x, and $z \mapsto z^r$ are all continuous and open. Therefore ψ_x is a homeomorphism.

The case $x = \infty$ is similar. From (2.1b) we know that $I(\infty) = \{T^k\}$ consists of the translations $\tau \mapsto \tau + k$ for $k \in \mathbb{Z}$. Hence $\psi_x\big(\phi(\tau)\big) = e^{2\pi i \tau}$ is well defined and injective on the quotient $I(\infty)\backslash U_\infty$. And, as above, ψ_x and ψ_x^{-1} are continuous, since both ϕ and $\tau \mapsto e^{2\pi i \tau}$ are continuous and open. Hence ψ_x is a homeomorphism.

It remains to check compatibility. First let $x, y \in X(1)$ with $x, y \neq \infty$. Then

$$\psi_y \circ \psi_x^{-1}(z) = \psi_y \circ \phi \circ (\psi_x \circ \phi)^{-1}(z) = g_y^{r_y} \circ g_x^{-1}\left(z^{1/r_x}\right).$$

Now g_y and g_x^{-1} are holomorphic, so the only possible problem would be the appearance of fractional powers of z. Let ζ be the primitive $r_x{}^{\text{th}}$-root of unity such that $g_x(R_x\tau) = \zeta g_x(\tau)$. Then using the fact that $\phi \circ \gamma = \phi$ for any $\gamma \in \Gamma(1)$, we find

$$g_y^{r_y} \circ g_x^{-1}(\zeta z) = \psi_y \circ \phi \circ R_x \circ g_x^{-1}(z) = \psi_y \circ \phi \circ g_x^{-1}(z) = g_y^{r_y} \circ g_x^{-1}(z).$$

It follows that $g_y^{T_y} \circ g_x^{-1}(z)$ is a power series in z^{r_x}, which proves that the composition $\psi_y \circ \psi_x^{-1}(z)$ is holomorphic. (Note the importance of knowing that ζ is a *primitive* $r_x{}^{\text{th}}$-root of unity.)

By exactly the same computation, taking $g_\infty(\tau) = \exp(2\pi i \tau)$, the function

$$\psi_\infty \circ \psi_x^{-1}(z) = \exp\left(2\pi i g_x^{-1}(z^{1/r_x})\right)$$

is holomorphic.

Finally, we note that

$$g_y^{T_y}(\tau + 1) = \psi_y \circ \phi \circ T(\tau) = \psi_y \circ \phi(\tau) = g_y^{T_y}(\tau),$$

so $g_y^{T_y}(\tau)$ is a holomorphic function in the variable $q = e^{2\pi i \tau}$. (Note τ is restricted to $U_y \cap U_\infty$; it is not allowed to tend toward $i\infty$.) Hence the transition map

$$\psi_y \circ \psi_\infty^{-1}(z) = g_y^{T_y}\left(\frac{1}{2\pi i} \log z\right)$$

is holomorphic.

This completes the proof that the open sets $I(\tau_x)\backslash U_x$ and the maps

$$\psi_x : I(\tau_x)\backslash U_x \to \mathbb{C}$$

define a complex structure on $X(1)$. $\qquad\square$

§3. Modular Functions

In the previous section we showed that the quotient space $X(1) = \Gamma(1)\backslash \mathbf{H}^*$ has the structure of a Riemann surface of genus 0. It is natural to look at the meromorphic functions on this Riemann surface.

Example 3.1. Recall that to each $\tau \in \mathbf{H}$ we have associated a lattice $\Lambda_\tau = \mathbb{Z}\tau + \mathbb{Z}$ and an elliptic curve $\mathbb{C}/\Lambda_\tau$. From Lemma 1.2(b) there is a well-defined map (of sets)

$$\begin{array}{ccc} \Gamma(1)\backslash \mathbf{H} & \longrightarrow & \mathbb{C} \\ \tau & \longmapsto & j\left(\mathbb{C}/\Lambda_\tau\right). \end{array}$$

We will show later (4.1) that with the complex structure described in (2.5), the j function is a meromorphic function on $X(1)$ which gives a complex analytic isomorphism

$$j : X(1) \overset{\sim}{\longrightarrow} \mathbb{P}^1(\mathbb{C}).$$

Every meromorphic function f on $X(1)$ is thus a rational function of j, that is, $f \in \mathbb{C}(j)$. In order to have a richer source of functions, we will study functions on $\mathbf{H}$ that have "nice" transformation properties relative to the action of $\Gamma(1)$ on $\mathbf{H}$. Although these transformation properties may look somewhat artificial at first, the corresponding functions actually define differential forms on $X(1)$, so they are in fact natural objects to study. (See (3.5) below for further details.)

Definition. Let $k \in \mathbb{Z}$, and let $f(\tau)$ be a function on $\mathbf{H}$. We say that f is *weakly modular of weight* $2k$ *(for* $\Gamma(1)$*)* if the following two conditions are satisfied:

(i) f is meromorphic on $\mathbf{H}$;

(ii) $f(\gamma\tau) = (c\tau + d)^{2k} f(\tau)$ for all $\gamma = \begin{pmatrix} a & b \\ c & d \end{pmatrix} \in \Gamma(1)$, $\tau \in \mathbf{H}$.

Remark 3.2. Note that a function satisfying $f(\gamma\tau) = (c\tau + d)^{\kappa} f(\tau)$ for an odd integer κ is necessarily the zero function, since taking $\gamma = \begin{pmatrix} -1 & 0 \\ 0 & -1 \end{pmatrix}$ yields $f(\tau) = -f(\tau)$. This explains why we restrict attention to even weights.

Remark 3.3. Since (1.6) says that $\Gamma(1)$ is generated by the two matrices $S = \begin{pmatrix} 0 & -1 \\ 1 & 0 \end{pmatrix}$ and $T = \begin{pmatrix} 1 & 1 \\ 0 & 1 \end{pmatrix}$, a meromorphic function f on $\mathbf{H}$ is weakly modular of weight $2k$ if it satisfies the two identities

$$f(\tau + 1) = f(\tau) \qquad \text{and} \qquad f\left(\frac{-1}{\tau}\right) = \tau^{2k} f(\tau).$$

From the first it follows that we can express f as a function of

$$q = e^{2\pi i \tau},$$

and f will be meromorphic in the punctured disk

$$\{q : 0 < |q| < 1\}.$$

Thus f has a Laurent expansion $\tilde{f}$ in the variable q, or in other words, f has a Fourier expansion:

$$\tilde{f}(q) = \sum_{n=-\infty}^{\infty} a_n q^n.$$

Definition. With notation as in (3.3), f is said to be

$$\textit{meromorphic at } \infty \text{ if } \tilde{f} = \sum_{n=-n_0}^{\infty} a_n q^n \quad \text{for some integer } n_0,$$

$$\textit{holomorphic at } \infty \text{ if } \tilde{f} = \sum_{n=0}^{\infty} a_n q^n.$$

If f is meromorphic at ∞, say $\tilde{f} = a_{-n_0} q^{-n_0} + \cdots$ with $a_{-n_0} \neq 0$, then the *order of* f *at* ∞ is

$$\mathrm{ord}_{\infty}(f) = \mathrm{ord}_{q=0}(\tilde{f}) = -n_0.$$

If f is holomorphic at ∞, its *value at* ∞ is defined to be

$$f(\infty) = \tilde{f}(0) = a_0.$$

Definition. A weakly modular function that is meromorphic at ∞ is called a *modular function*.

Definition. A modular function that is everywhere holomorphic (i.e., everywhere on $\mathbf{H}$ and at ∞) is called a *modular form*. If in addition $f(\infty) = 0$, then f is called a *cusp form*.

Example 3.4.1. Let Λ be a lattice. The *Eisenstein series*

$$G_{2k}(\Lambda) = \sum_{\substack{\omega \in \Lambda \\ \omega \neq 0}} \frac{1}{\omega^{2k}}$$

is absolutely convergent for all integers $k \geq 2$. (See [AEC VI.3.1].) For $\tau \in \mathbf{H}$ we let

$$G_{2k}(\tau) = G_{2k}(\Lambda_\tau) = \sum_{\substack{m,n \in \mathbb{Z} \\ (m,n) \neq (0,0)}} \frac{1}{(m\tau + n)^{2k}}.$$

By inspection,

$$G_{2k}(c\Lambda) = c^{-2k} G_{2k}(\Lambda) \qquad \text{for any } c \in \mathbb{C}^*,$$

whereas

$$\Lambda_{\gamma\tau} = \mathbb{Z}\frac{a\tau + b}{c\tau + d} + \mathbb{Z} = \frac{1}{c\tau + d}\big(\mathbb{Z}(a\tau + b) + \mathbb{Z}(c\tau + d)\big) = \frac{1}{c\tau + d}\Lambda_\tau.$$

Hence

$$\begin{aligned}
G_{2k}(\gamma\tau) = G_{2k}(\Lambda_{\gamma\tau}) &= G_{2k}\big((c\tau + d)^{-1}\Lambda_\tau\big) \\
&= (c\tau + d)^{2k} G_{2k}(\Lambda_\tau) = (c\tau + d)^{2k} G_{2k}(\tau).
\end{aligned}$$

Thus G_{2k} is weakly modular of weight $2k$.

Proposition 3.4.2. *Let $k \geq 2$ be an integer. The Eisenstein series G_{2k} is a modular form of weight $2k$. Its value at ∞ is given by $G_{2k}(\infty) = 2\zeta(2k)$, where $\zeta(s)$ is the Riemann zeta function. (For the complete Fourier expansion of G_{2k}, see (7.1).)*

PROOF. We have just shown that G_{2k} is weakly modular, so it remains to show that G_{2k} is holomorphic on $\mathbf{H}$ and at ∞ and to compute its value at ∞. Note that if τ is in the fundamental domain $\mathcal{F}$ described in Proposition 1.5, then

$$|m\tau + n|^2 = m^2|\tau|^2 + 2mn\,\mathrm{Re}(\tau) + n^2 \geq m^2 - mn + n^2 = |m\rho - n|^2.$$

Hence the series obtained from $G_{2k}(\tau)$ by putting in absolute values is dominated, term-by-term, by the series obtained from $G_{2k}(\rho)$ by putting in absolute values. Therefore G_{2k} is holomorphic on $\mathcal{F}$. But $\mathbf{H}$ is covered by the $\Gamma(1)$-translates of $\mathcal{F}$, and $G_{2k}(\gamma\tau) = (c\tau + d)^{2k}G_{2k}(\tau)$, so G_{2k} is holomorphic on all of $\mathbf{H}$.

Next we look at the behavior of $G_{2k}(\tau)$ as $\tau \to i\infty$. Since the series for G_{2k} converges uniformly, we can take the limit term-by-term. Terms of the form $(m\tau + n)^{-2k}$ with $m \neq 0$ will tend to zero, whereas the others give n^{-2k}. Hence

$$\lim_{\tau \to i\infty} G_{2k}(\tau) = \sum_{\substack{n=-\infty \\ n \neq 0}}^{\infty} \frac{1}{n^{2k}} = 2\zeta(2k).$$

This shows that G_{2k} is holomorphic at ∞ and gives its value.   □

Example 3.4.3. It is customary to let

$$g_2(\tau) = 60G_4(\tau) \qquad \text{and} \qquad g_3(\tau) = 140G_6(\tau).$$

(See [AEC VI.3.5.1].) The *(modular) discriminant* is the function

$$\Delta(\tau) = g_2(\tau)^3 - 27g_3(\tau)^2.$$

It is a modular form of weight 12, since from (3.4.2) we know that $G_4(\tau)$ and $G_6(\tau)$ are modular forms of weights 4 and 6 respectively.

Using the well-known values (see (7.2) and (7.3.2))

$$\zeta(4) = \frac{\pi^4}{90} \qquad \text{and} \qquad \zeta(6) = \frac{\pi^6}{945},$$

we find that

$$g_2(\infty) = 120\zeta(4) = \frac{4\pi^4}{3}, \qquad g_3(\infty) = 280\zeta(6) = \frac{8\pi^6}{27}, \qquad \Delta(\infty) = 0.$$

Hence $\Delta(\tau)$ is a cusp form of weight 12. We will see below (3.10.2) that it is essentially the only one.

Remark 3.5. Let $\gamma = \begin{pmatrix} a & b \\ c & d \end{pmatrix} \in \mathrm{SL}_2(\mathbb{Z})$, and let $d\tau$ be the usual differential form on **H**. Then

$$d(\gamma\tau) = d\left(\frac{a\tau + b}{c\tau + d}\right) = \frac{ad - bc}{(c\tau + d)^2}\, d\tau = (c\tau + d)^{-2}d\tau.$$

Thus $d\tau$ has "weight -2." In particular, if $f(\tau)$ is a modular function of weight $2k$, then the k-form

$$f(\tau)\,(d\tau)^k$$

is $\Gamma(1)$-invariant. It thus defines a k-form on the quotient space $\Gamma(1)\backslash\mathbf{H}$, at least away from the orbits of i and ρ, where the complex structure is a bit more complicated.

We will soon show that $f(\tau)\,(d\tau)^k$ actually defines a meromorphic k-form on $X(1)$. We begin with a brief digression concerning differential forms on arbitrary Riemann surfaces. In particular, formula (3.6b) below will be crucial in our determination of the space of modular forms of a given weight.

Definition. Let $X/\mathbb{C}$ be a smooth projective curve, or, equivalently, a compact Riemann surface. Recall that Ω_X is the $\mathbb{C}(X)$-vector space of differential 1-forms on X. (See [AEC II §4].) The *space of (meromorphic) k-forms on X* is the k-fold tensor product

$$\Omega_X^k = \Omega_X^{\otimes k} = \Omega_X \otimes_{\mathbb{C}(X)} \cdots \otimes_{\mathbb{C}(X)} \Omega_X.$$

Ω_X^k is a 1-dimensional $\mathbb{C}(X)$-vector space [AEC II.4.2a]. Notice that if we set $\Omega_X^0 = \mathbb{C}(X)$, then $\bigoplus_{k=0}^{\infty} \Omega_X^k$ has a natural structure as a graded $\mathbb{C}(X)$-algebra.

Let $\omega \in \Omega_X^k$, $x \in X$, and choose a uniformizer $t \in \mathbb{C}(X)$ at x. Then

$$\omega = g(dt)^k$$

for some function $g \in \mathbb{C}(X)$. We define the *order of ω at x* to be

$$\operatorname{ord}_x(\omega) = \operatorname{ord}_x(g).$$

It is independent of the choice of t. (If t' is another uniformizer, then applying [AEC II.4.3b] we find that dt/dt' is holomorphic and non-vanishing at x.) Just as with 1-forms, we define the *divisor of ω* by

$$\operatorname{div}(\omega) = \sum_{x \in X} \operatorname{ord}_x(\omega)(x) \in \operatorname{Div}(X);$$

we say that ω is *regular* (or *holomorphic*) if

$$\operatorname{ord}_x(\omega) \geq 0 \qquad \text{for all } x \in X.$$

Proposition 3.6. *Let $X/\mathbb{C}$ be a smooth projective curve of genus g, let $k \geq 1$ be an integer, and let $\omega \in \Omega_X^k$.*
(a) *Let K_X be a canonical divisor on X [AEC II §4]. Then $\operatorname{div}(\omega)$ is linearly equivalent to kK_X.*
(b)
$$\deg(\operatorname{div}\omega) = k(2g - 2).$$

PROOF. (a) Let $\eta \in \Omega_X^1$ be a non-zero 1-form with divisor $\operatorname{div}(\eta) = K_X$. Then

$$F = \omega/\eta^k \in \Omega_X^0 = \mathbb{C}(X)$$

is a function on X, so

$$\text{div}(\omega) = k \, \text{div}(\eta) + \text{div}(\omega/\eta^k) = kK_X + \text{div}(F)$$

is linearly equivalent to kK_X.
(b) From (a), $\deg(\text{div}\,\omega) = k \deg(K_X)$. Now apply the Riemann-Roch theorem [AEC II.5.4b], which says that $\deg(K_X) = 2g - 2$. $\square$

The next proposition gives the precise relationship between a modular function f of weight $2k$ and the corresponding k-form $f(\tau)\,(d\tau)^k$.

Proposition 3.7. *Let f be a non-zero modular function of weight $2k$.*
(a) *The k-form $f(\tau)\,(d\tau)^k$ on H descends to give a meromorphic k-form ω_f on the Riemann surface $X(1)$. In other words, there is a k-form $\omega_f \in \Omega^k_{X(1)}$ such that*

$$\phi^*(\omega_f) = f(\tau)\,(d\tau)^k,$$

where $\phi : \mathsf{H} \to X(1)$ is the usual projection.
(b) *Let $x \in X(1)$, and let $\tau_x \in \mathsf{H}^*$ with $\phi(\tau_x) = x$. Then*

$$\text{ord}_x(\omega_f) = \begin{cases} \text{ord}_{\tau_x}(f) & \text{if } x \neq \phi(i), \phi(\rho), \phi(\infty); \\ \frac{1}{2}\,\text{ord}_i(f) - \frac{1}{2}k & \text{if } x = \phi(i); \\ \frac{1}{3}\,\text{ord}_\rho(f) - \frac{2}{3}k & \text{if } x = \phi(\rho); \\ \text{ord}_\infty(f) - k & \text{if } x = \phi(\infty). \end{cases}$$

Remark 3.7.1. If f is a modular function, then it is easy to see that the order of vanishing of f at $\tau \in \mathsf{H}$ depends only on the $\Gamma(1)$-equivalence class of τ. The point is that since $f(\gamma\tau) = (c\tau + d)^{2k} f(\tau)$ and $c\tau + d \neq 0$, we have

$$\text{ord}_\tau(f) = \text{ord}_\tau\left(f \circ \gamma^{-1}\right) = \text{ord}_{\gamma\tau}(f).$$

Thus the expression in (3.7b) really does not depend on the choice of the representative τ_x.

PROOF. (a) As we have seen, the k-form $f(\tau)\,(d\tau)^k$ is invariant for the action of $\Gamma(1)$ on H. We must show that for each $x = \phi(\tau_x) \in X(1)$, the k-form $f(\tau)\,(d\tau)^k$ descends locally around x to a meromorphic k-form on $X(1)$, and that it vanishes to the indicated order. Clearly, we will need to use the description of the complex structure on $X(1)$ provided by Theorem 2.5. We consider two cases.

$\boxed{x \neq \infty}$

Using the notation from (2.5), there is a commutative diagram

$$\begin{array}{ccc} U_x & \xrightarrow{\phi} & I(\tau_x)\backslash U_x \\ \downarrow{\scriptstyle g_x} & & \downarrow{\scriptstyle \psi_x} \\ \mathbb{C} & \xrightarrow[z \mapsto w = z^r]{} & \mathbb{C} \end{array}$$

which defines a local parameter

$$\psi_x\big(\phi(\tau)\big) = g_x(\tau)^r$$

at x. We write

$$z = g_x(\tau) = \frac{\tau - \tau_x}{\tau - \bar{\tau}_x}, \qquad w = z^r, \qquad \text{and} \qquad \tau = g_x^{-1}(z) = \frac{\bar{\tau}_x z - \tau_x}{z - 1},$$

so $w = z^r$ is our local parameter.

Let R be a generator of $I(\tau_x)$. Then from (2.6),

$$g_x(R\tau) = \zeta g(\tau), \qquad \text{and so} \qquad R\tau = R \circ g_x^{-1}(z) = g_x^{-1}(\zeta z).$$

Here ζ is some primitive r^{th}-root of unity.

Now

$$\begin{aligned} f(\tau)\,(d\tau)^k &= f\big(g_x^{-1}(z)\big)\,\big(dg_x^{-1}(z)\big)^k \\ &= f\big(g_x^{-1}(z)\big)g_x^{-1\prime}(z)^k\,(dz)^k = F(z)\,(dz)^k, \end{aligned}$$

where $F(z) = f\big(g_x^{-1}(z)\big)g_x^{-1\prime}(z)^k$ is a meromorphic function of z. Note further that since g_x is a local isomorphism, we have

$$\operatorname{ord}_{\tau=\tau_x}(f) = \operatorname{ord}_{z=0}(F).$$

We must show that $F(z)\,(dz)^k$ is a meromorphic function of $w = z^r$.

To do this, we use the fact (3.5) that $f(\tau)\,(d\tau)^k$ is $\Gamma(1)$-invariant. This implies

$$\begin{aligned} F(z)\,(dz)^k &= f(\tau)\,(d\tau)^k = f(R\tau)\,(dR\tau)^k \\ &= f\big(g_x^{-1}(\zeta z)\big)\,\big(dg_x^{-1}(\zeta z)\big)^k = F(\zeta z)\,(d\zeta z)^k = F(\zeta z)\zeta^k\,(dz)^k. \end{aligned}$$

In particular, the function $z^k F(z)$ is invariant under the substitution $z \mapsto \zeta z$. Since ζ is a *primitive* r^{th}-root of unity, it follows that

$$z^k F(z) = F_1(z^r)$$

for some meromorphic function $F_1(w)$. Hence

$$\begin{aligned} F(z)\,(dz)^k &= r^{-k} z^{k(1-r)} F(z)\,\big(d(z^r)\big)^k \\ &= r^{-k} z^{-rk} F_1(z^r)\,\big(d(z^r)\big)^k = r^{-k} w^{-k} F_1(w)\,(dw)^k, \end{aligned}$$

which proves that $f(\tau)\,(d\tau)^k$ descends to a meromorphic k-form ω_f in a neighborhood of x.

Finally, we compute

$$\operatorname{ord}_{\tau=\tau_x} f(\tau) = \operatorname{ord}_{z=0} F(z) = \operatorname{ord}_{z=0} z^{-k} F_1(z^r) = -k + r \operatorname{ord}_{w=0} F_1(w);$$

$$\operatorname{ord}_x \omega_f = \operatorname{ord}_{w=0} r^{-k} w^{-k} F_1(w) = -k + \operatorname{ord}_{w=0} F_1(w).$$

Eliminating $\operatorname{ord}_{w=0} F_1(w)$ from these two equations yields

$$\operatorname{ord}_x \omega_f = \frac{1}{r} \operatorname{ord}_{\tau_x} f - \left(1 - \frac{1}{r}\right) k.$$

It only remains to note that from (1.5),

$$r = \begin{cases} 1 & \text{if } x \neq \phi(i), \phi(\rho), \\ 2 & \text{if } x = \phi(i), \\ 3 & \text{if } x = \phi(\rho). \end{cases}$$

$\boxed{x = \infty}$

Again using (2.5), we have a local parameter

$$\psi_x : I(\infty)\backslash U_\infty \longrightarrow \mathbb{C}, \qquad \psi_x(\phi(\tau)) = e^{2\pi i \tau}.$$

Let $q = e^{2\pi i \tau}$ be the local parameter at ∞, and write $f(\tau) = \tilde{f}(q)$ as in (3.3). Since $d\tau = (2\pi i q)^{-1} dq$, we have

$$f(\tau)(d\tau)^k = \tilde{f}(q)(2\pi i q)^{-k}(dq)^k.$$

By definition, $\tilde{f}$ is meromorphic at $q = 0$, so $f(\tau)(d\tau)^k$ descends to a meromorphic k-form ω_f in a neighborhood of ∞. Finally,

$$\operatorname{ord}_\infty \omega_f = \operatorname{ord}_{q=0} \tilde{f}(q)(2\pi i q)^{-k} = \operatorname{ord}_\infty(f) - k.$$

$\square$

Proposition 3.7 describes the local behavior of the k-form $\omega_f \in \Omega^k_{X(1)}$. The Riemann-Roch theorem, specifically Proposition 3.6(b), gives a global description of its degree. Combining these results, we obtain the following important formula.

Corollary 3.8. *Let f be a non-zero modular function of weight $2k$. Then*

$$\frac{1}{2} \operatorname{ord}_i(f) + \frac{1}{3} \operatorname{ord}_\rho(f) + \operatorname{ord}_\infty(f) + \sum_{\substack{\tau \in \Gamma(1)\backslash \mathbf{H}^* \\ \tau \neq i, \rho, \infty}} \operatorname{ord}_\tau(f) = \frac{k}{6}.$$

(Here the sum is over any set of representatives for $\Gamma(1)\backslash\mathbf{H}^*$ excluding the equivalence classes containing i, ρ, and ∞.)

PROOF. First note that from (3.7.1), the sum is independent of the choice of representatives for $\Gamma(1)\backslash\mathbf{H}^*$. Let $\omega_f \in \Omega^k_{X(1)}$ be the k-form corresponding to $f(\tau)\,(d\tau)^k$ as in Proposition 3.7. By the Riemann-Roch theorem (3.6b) and the fact that $X(1)$ has genus 0 (2.5), we find that

$$\deg(\operatorname{div}\omega_f) = -2k.$$

On the other hand, (3.7) gives

$$\deg(\operatorname{div}\omega_f) = \left(\tfrac{1}{2}\operatorname{ord}_i f - \tfrac{1}{2}k\right) + \left(\tfrac{1}{3}\operatorname{ord}_\rho f - \tfrac{2}{3}k\right)$$
$$+ (\operatorname{ord}_\infty f - k) + \sum_{\substack{\tau \in \Gamma(1)\backslash \mathbf{H}^* \\ \tau \neq i,\rho,\infty}} \operatorname{ord}_\tau(f).$$

Equating these two expressions for $\deg(\operatorname{div}\omega_f)$ gives the desired formula.
□

Using Corollary 3.8, we can give a good description of the space of all modular forms of a given weight. We set the notation

$$M_{2k} = \{\text{modular forms of weight } 2k \text{ for } \Gamma(1)\},$$

$$M_{2k}^0 = \{\text{cusp forms of weight } 2k \text{ for } \Gamma(1)\}.$$

Note that both M_{2k} and M_{2k}^0 are $\mathbb{C}$-vector spaces.

Example 3.9. For all $k \geq 2$, the Eisenstein series $G_{2k}(\tau)$ is in M_{2k} but is not in M_{2k}^0. The modular discriminant $\Delta(\tau)$ is in M_{12}^0. See (3.4.2) and (3.4.3).

Theorem 3.10. (a) *For all integers* $k \geq 2$,

$$M_{2k} \cong M_{2k}^0 + \mathbb{C}G_{2k}.$$

(b) *For all integers* k, *the map*

$$M_{2k-12} \longrightarrow M_{2k}^0, \qquad f \longmapsto f\Delta$$

is an isomorphism of $\mathbb{C}$-*vector spaces.*
(c) *The dimension of* M_{2k} *as a* $\mathbb{C}$-*vector space is given by*

$$\dim M_{2k} = \begin{cases} 0 & \text{if } k < 0; \\ [k/6] & \text{if } k \geq 0,\ k \equiv 1 \pmod 6; \\ [k/6+1] & \text{if } k \geq 0,\ k \not\equiv 1 \pmod 6. \end{cases}$$

(The square brackets denote greatest integer. For an alternative proof of (c) using the Riemann-Roch theorem, see exercises 1.8 and 1.9.)

PROOF. (a) By definition, M_{2k}^0 is the kernel of the map

$$M_{2k} \longrightarrow \mathbb{C}, \qquad f \longmapsto f(\infty),$$

so M_{2k}/M_{2k}^0 has dimension at most 1. On the other hand, for $k \geq 2$, the Eisenstein series G_{2k} is in M_{2k} and is not in M_{2k}^0. (See (3.4.2).) Hence

$$M_{2k} = M_{2k}^0 + \mathbb{C}G_{2k} \qquad \text{for all } k \geq 2.$$

(b) First we note that

$$G_4(\rho) = (\rho+1)^4 G_4(ST\rho) = \rho^2 G_4(\rho) \quad \text{and} \quad G_6(i) = i^6 G_6(Si) = -G_6(i),$$

which implies that $G_4(\rho) = 0$ and $G_6(i) = 0$. Since G_4 and G_6 are modular forms of weight 4 and 6 respectively (3.4.2), it follows from (3.8) that they have no other zeros in $\Gamma(1)\backslash\mathbf{H}$. In particular,

$$\Delta(\rho) = \big(60G_4(\rho)\big)^3 - 27\big(140G_6(\rho)\big)^2 = -2^4 3^3 5^2 7^2 G_6(\rho)^2 \neq 0,$$

so $\Delta(\tau)$ is not identically zero.

Thus $\Delta(\tau)$ is a non-zero modular form of weight 12 with $\Delta(\infty) = 0$. It follows from (3.8) that

$$\mathrm{ord}_\infty(\Delta) = 1$$

and that $\Delta(\tau) \neq 0$ for all $\tau \in \mathbf{H}$. (For an alternative proof that $\Delta(\tau) \neq 0$ for all $\tau \in \mathbf{H}$, see [AEC VI.3.6a].) Therefore $1/\Delta$ has a simple pole at ∞ and no other poles, so the map

$$M_{2k}^0 \longrightarrow M_{2k-12},$$
$$f \longmapsto f/\Delta$$

is well-defined. (The main point is that as long as f vanishes at ∞, then f/Δ will still be holomorphic at ∞.) This gives an inverse to the map in (b), so $M_{2k}^0 \cong M_{2k-12}$.

(c) If $k < 0$, then (3.8) implies immediately that $M_{2k} = 0$. (Note that all of the terms in the left-hand sum are non-negative.) Similarly, if $f \in M_0$, then (3.8) says that f has no zeros on $\mathbf{H}^*$. Thus f gives a holomorphic non-vanishing function on $X(1)$. But $X(1)$ is a compact Riemann surface, so an analytic map $[f,1] : X(1) \to \mathbb{P}^1(\mathbb{C})$ is necessarily either constant or surjective. Hence f is constant and $M_0 \cong \mathbb{C}$.

Next we use (3.8) to describe all functions $f \in M_{2k}$ for small values of k. Note that for small values of k, the equation

$$\frac{1}{2}a + \frac{1}{3}b + c = \frac{k}{6}$$

will have very few solutions in non-negative integers a, b, c. For example, if $k = 1$, there are no solutions. We compile the results in Table 1.1.

Table 1.1

k	$\text{ord}_i f$	$\text{ord}_r f$	$\text{ord}_\tau f$ $\tau \neq i, \rho$	basis for M_{2k}
1	—	—	—	$\emptyset$
2	0	1	0	G_4
3	1	0	0	G_6
4	0	2	0	G_4^2
5	1	1	0	$G_4 G_6$

Everything in Table 1.1 is clear except that the functions in the final column actually form a basis. They are in M_{2k} from (3.4.2), so we need to show that M_{2k} has dimension 1. But if $f_1, f_2 \in M_{2k}$ with $2 \leq k \leq 5$, then Table 1.1 shows that f_1 and f_2 have exactly the same zeros. Hence $f_1/f_2 \in M_0 = \mathbb{C}$, which proves that $\dim(M_{2k}) = 1$ for $2 \leq k \leq 5$.

We have now verified (c) for all integers $k \leq 5$. On the other hand, if $k \geq 0$, then using (a) and (b) we find that

$$\dim M_{2k+12} = \dim M_{2k+12}^0 + 1 \qquad \text{from (a)}$$
$$= \dim M_{2k} + 1 \qquad \text{from (b).}$$

Thus the left-hand side of (c) increases by 1 when k is replaced by $k + 6$. Since the same is true of the right-hand side, an easy induction argument completes the proof. □

Example 3.10.1. Each of the vector spaces

$$M_0, M_4, M_6, M_8, M_{10}, M_{14}$$

has dimension 1. For example, since $G_4^2 \in M_8$ and $G_8 \in M_8$, it follows immediately that

$$G_8 = c G_4^2$$

for some constant $c \in \mathbb{C}$. Letting $\tau \to i\infty$ and using (3.4.2), we can even compute

$$c = \frac{2\zeta(8)}{4\zeta(4)^2} = \frac{3}{7}.$$

(See (7.2) for the calculation of $\zeta(8)$.) Similarly, $G_{10} = \frac{5}{11} G_4 G_6$ and $G_{14} = \frac{30}{143} G_4^2 G_6$. More generally, M_{2k} has a basis consisting of functions of the form $G_4^a G_6^b$. (See exercise 1.10.) To appreciate the subtlety of identities such as these, the reader might try to give a proof that $G_8 = \frac{3}{7} G_4^2$ directly from the series definition (3.4.1) of the G_{2k}'s.

Example 3.10.2. Since $M_{2k} \cong M^0_{2k+12}$ from (3.10b), the spaces

$$M^0_{12}, M^0_{16}, M^0_{18}, M^0_{20}, M^0_{22}, M^0_{26}$$

also have dimension 1. In particular, up to multiplication by a constant, there is only one cusp form of weight 12, namely $\Delta(\tau)$.

§4. Uniformization and Fields of Moduli

We begin by proving the Uniformization Theorem for elliptic curves, which was stated but not proved in [AEC VI.5.1]. This theorem says that every elliptic curve over $\mathbb{C}$ is parametrized by Weierstrass elliptic functions. Our main tool will be Theorem 3.7(a), which says in particular that every modular function of weight 0 defines a meromorphic function on the Riemann surface $X(1)$. For a more elementary, but less intrinsic, proof of the Uniformization Theorem, see exercise 1.11.

Definition. The *modular j-invariant* $j(\tau)$ is the function

$$j(\tau) = 1728 \frac{g_2(\tau)^3}{\Delta(\tau)}.$$

Thus $j(\tau)$ is the j-invariant of the elliptic curve

$$E_{\Lambda_\tau} : y^2 = 4x^3 - g_2(\tau)x - g_3(\tau),$$

and $E_{\Lambda_\tau}(\mathbb{C})$ has a parametrization using the Weierstrass $\wp$-function,

$$\begin{array}{ccc} \mathbb{C}/\Lambda_\tau & \longrightarrow & E_{\Lambda_\tau}(\mathbb{C}), \\ z & \longmapsto & \big(\wp(z;\Lambda_\tau), \wp'(z;\Lambda_\tau)\big). \end{array}$$

(For details, see [AEC VI.3.6].)

Theorem 4.1. $j(\tau)$ *is a modular function of weight 0. It induces a (complex analytic) isomorphism*

$$j : X(1) \xrightarrow{\sim} \mathbb{P}^1(\mathbb{C}).$$

PROOF. From (2.4.2) and (2.4.3), both $\Delta(\tau)$ and $g_2(\tau)^3 = 2^6 3^3 5^3 G_4(\tau)^3$ are modular forms, and both have weight 12, so their quotient is a modular function of weight 0. By (3.7a) with $k = 0$, j defines a meromorphic function on $X(1)$. (N.B. This means that j is meromorphic relative to the complex structure on $X(1)$ described by (2.5).) Hence j gives a finite complex-analytic map

$$j : X(1) \longrightarrow \mathbb{P}^1(\mathbb{C}).$$

Finally, we note that $g_2(i\infty) = 120\zeta(4) \neq 0$ (3.4.2) and $\Delta(i\infty) = 0$ (3.4.3). Since Δ has weight 12, (3.8) implies that

$$\mathrm{ord}_\infty \Delta = 1.$$

Thus j has a simple pole at the cusp $\infty \in X(1)$ and no other poles on $X(1)$, so the map $j : X(1) \to \mathbb{P}^1(\mathbb{C})$ is an analytic map of degree 1 between compact Riemann surfaces. It is therefore an isomorphism.   □

Corollary 4.2.   *Let f be a modular function of weight 0.*
(a) *The function f is a rational function of j, that is, $f \in \mathbb{C}(j)$.*
(b) *If in addition f is holomorphic on $\mathbf{H}$, then f is a polynomial function of j, that is, $f \in \mathbb{C}[j]$.*

PROOF. (a) From (3.7a), f defines a meromorphic function on $X(1)$, and so by (4.1), $f \circ j^{-1}$ is a meromorphic function on $\mathbb{P}^1(\mathbb{C})$. But the only meromorphic functions on $\mathbb{P}^1(\mathbb{C})$ are rational functions, so

$$f \circ j^{-1}(t) = P(t) \qquad \text{for some } P(T) \in \mathbb{C}(T).$$

Substituting $t = j(x)$ with $x \in X(1)$ gives $f(x) = P(j(x))$.
(b) From (a), we know that $f = P(j)$ for some rational function $P(T) \in \mathbb{C}(T)$. Suppose P is not a polynomial. Then there is a $t_0 \in \mathbb{C}$ such that $P(t_0) = \infty$. The isomorphism $j : X(1) \overset{\sim}{\to} \mathbb{P}^1(\mathbb{C})$ from (4.1) sends $\mathbf{H}$ to $\mathbb{C} \subset \mathbb{P}^1(\mathbb{C})$, so we can find a $\tau_0 \in \mathbf{H}$ with $j(\tau_0) = t_0$. But then $f(\tau_0) = P(j(\tau_0)) = P(t_0) = \infty$, contradicting the assumption that f is holomorphic on $\mathbf{H}$. Hence $P(T)$ must be a polynomial.   □

Corollary 4.3.   (Uniformization Theorem For Elliptic Curves over $\mathbb{C}$) *Let $A, B \in \mathbb{C}$ satisfy $4A^3 + 27B^2 \neq 0$. Then there is a unique lattice $\Lambda \subset \mathbb{C}$ such that*

$$g_2(\Lambda) = 60G_4(\Lambda) = -4A \qquad \text{and} \qquad g_3(\Lambda) = 140G_6(\Lambda) = -4B.$$

The map
$$\begin{array}{ccc} \mathbb{C}/\Lambda & \longrightarrow & E : y^2 = x^3 + Ax + B, \\ z & \longmapsto & \left(\wp(z; \Lambda), \tfrac{1}{2}\wp'(z; \Lambda)\right) \end{array}$$
is a complex analytic isomorphism.

PROOF. Using Theorem 4.1, we can choose a $\tau \in \mathbf{H}$ such that

$$j(\tau) = 1728\frac{4A^3}{4A^3 + 27B^2}.$$

Assume first that $AB \neq 0$. It follows from this and the definition of $j(\tau)$ that

$$\frac{27B^2}{4A^3} = \frac{1728}{j(\tau)} - 1 = -\frac{27g_3(\tau)^2}{g_2(\tau)^3}, \qquad \text{so} \qquad \left(\frac{B}{g_3(\tau)}\right)^2 \left(\frac{g_2(\tau)}{A}\right)^3 = -4.$$

Let

$$\alpha = \sqrt{\frac{Ag_3(\tau)}{Bg_2(\tau)}} \qquad \text{and} \qquad \Lambda = \alpha\Lambda_\tau = \mathbb{Z}\alpha\tau + \mathbb{Z}\alpha.$$

Then

$$g_2(\Lambda) = \alpha^{-4} g_2(\Lambda_\tau) = \frac{B^2 g_2(\tau)^3}{A^2 g_3(\tau)^2} = -4A,$$

$$g_3(\Lambda) = \alpha^{-6} g_3(\Lambda_\tau) = \frac{B^3 g_2(\tau)^3}{A^3 g_3(\tau)^2} = -4B.$$

Similarly, if $A = 0$, then $j(\tau) = 0$ and $g_2(\tau) = 0$, whereas if $B = 0$, then $j(\tau) = 1728$ and $g_3(\tau) = 0$. Hence in these two cases it suffices to take $\Lambda = \alpha\Lambda_\tau$ with

$$\alpha = \sqrt[6]{\frac{g_3(\tau)}{-4B}} \quad \text{if } A = 0, \text{ and} \qquad \alpha = \sqrt[4]{\frac{g_2(\tau)}{-4A}} \quad \text{if } B = 0.$$

This gives the existence of Λ. Since we will not need the uniqueness of Λ in our subsequent work, we will leave this fact to the reader. (See exercise 1.12.) Finally, we note that the second part of Corollary 4.3 is essentially a restatement of [AEC VI.3.6b]. $\square$

We are now ready to relate the function $j(\tau)$, defined as a meromorphic function on the Riemann surface $X(1)$, to the j-invariant defined in [AEC III §1] which classifies isomorphism classes of elliptic curves. We let

$$\mathcal{ELL}_\mathbb{C} = \frac{\{\text{elliptic curves defined over } \mathbb{C}\}}{\mathbb{C}\text{-isomorphism}}.$$

Thus an element of $\mathcal{ELL}_\mathbb{C}$ is a $\mathbb{C}$-isomorphism class of elliptic curves. We also recall the notation

$$\mathcal{L} = \{\text{lattices in } \mathbb{C}\}$$

from §1. Much of our preceding discussion is summarized in the following proposition.

Proposition 4.4. *There are one-to-one correspondences between the following four sets, given by the indicated maps:*

$$\begin{array}{ccccccc}
\mathcal{ELL}_\mathbb{C} & \longleftarrow & \mathcal{L}/\mathbb{C}^* & \longleftarrow & \Gamma(1)\backslash \mathbf{H} & \longrightarrow & \mathbb{C}, \\
\{E_\Lambda\} & \longleftarrow & \{\Lambda\} = \{\Lambda_\tau\} & \longleftarrow & \tau & \longrightarrow & j(\tau).
\end{array}$$

Here $\Lambda_\tau = \mathbb{Z}\tau + \mathbb{Z}$, $\{E_\Lambda\}$ denotes the $\mathbb{C}$-isomorphism class of the elliptic curve $E_\Lambda : y^2 = 4x^3 - g_2(\Lambda)x - g_3(\Lambda)$, and $\{\Lambda\}$ is the homothety class of the lattice Λ.

PROOF. Since $j(i\infty) = \infty$, the bijectivity of $\Gamma(1)\backslash\mathbf{H} \overset{j}{\to} \mathbb{C}$ is (4.1). The bijectivity of $\Gamma(1)\backslash\mathbf{H} \to \mathcal{L}/\mathbb{C}^*$ is (1.2bc). Finally, the injectivity of $\mathcal{L}/\mathbb{C}^* \to \mathcal{ELL}_\mathbb{C}$ is [AEC 4.1.1] and the surjectivity is (4.3). $\qquad\square$

Let us describe in a bit more detail the bijective map

$$\mathcal{ELL}_\mathbb{C} \longrightarrow \mathbb{C}$$

given in Proposition 4.4. Let $\{E\} \in \mathcal{ELL}_\mathbb{C}$ be an isomorphism class of elliptic curves, and choose a Weierstrass equation

$$E : y^2 = x^3 + Ax + B$$

for some curve E in this class. Now take a basis γ_1, γ_2 for the homology group $H_1(E(\mathbb{C}), \mathbb{Z})$, and compute the periods

$$\omega_1 = \int_{\gamma_1} \frac{dx}{y} \quad \text{and} \quad \omega_2 = \int_{\gamma_2} \frac{dx}{y}.$$

(See [AEC VI §1].) Switching ω_1 and ω_2 if necessary, we may assume that

$$\tau_E = \frac{\omega_1}{\omega_2} \in \mathbf{H}.$$

Then evaluate the holomorphic function $j(\tau)$ at $\tau = \tau_E$.

Thus the map

$$j : \mathcal{ELL}_\mathbb{C} \longrightarrow \mathbb{C}, \qquad \{E\} \longmapsto j(\tau_E)$$

involves two transcendental (i.e., non-algebraic) operations, namely the computation of the periods ω_1, ω_2 and the evaluation of the function $j(\tau)$. From this perspective, it seems unlikely that rationality properties of $j(\tau_E)$ should have anything to do with rationality properties of E. To describe the relationship that does exist, we make the following two definitions.

Definition. Let $\{E\} \in \mathcal{ELL}_\mathbb{C}$, and let $K \subseteq \mathbb{C}$. We say that K is a *field of definition* for $\{E\}$ if there is an elliptic curve E_0 in the isomorphism class $\{E\}$ such that E_0 is defined over K. We say that K is a *field of moduli* for $\{E\}$ if for all automorphisms $\sigma \in \text{Aut}(\mathbb{C}/\mathbb{Q})$,

$$E^\sigma \in \{E\} \qquad \text{if and only if} \qquad \sigma \text{ acts trivially on } K.$$

Note that the field of moduli exists and is unique, since by Galois theory an equivalent definition is that the field of moduli is the fixed field of the group

$$\{\sigma \in \text{Aut}(\mathbb{C}/\mathbb{Q}) : E^\sigma \in \{E\}\}.$$

From the complex analytic viewpoint described above, it is not clear that the number $j(\{E\})$ should have any relationship to fields of definition and moduli for $\{E\}$. Note that there are lots of bijections $\mathcal{ELL}_\mathbb{C} \to \mathbb{C}$. For example, $j'(\{E\}) = e^\pi j(\{E\}) + e^{-\pi}$ is also a bijection. But clearly, it is not possible for both j and j' to have good rationality properties.

Proposition 4.5.   *Let $\{E\} \in \mathcal{ELL}_{\mathbb{C}}$.*
(a) $\mathbb{Q}\big(j(\{E\})\big)$ *is the field of moduli for* $\{E\}$.
(b) $\mathbb{Q}\big(j(\{E\})\big)$ *is the minimal field of definition for* $\{E\}$.

PROOF. The j-invariant $j(E)$ of the elliptic curve

$$E \; : \; y^2 = 4x^3 - g_2(\tau)x - g_3(\tau)$$

is

$$j(E) = 1728\frac{g_2(\tau_E)^3}{g(\tau_E)^3 - 27g_3(\tau_E)^2} = j(\tau_E) = j(\{E\}),$$

so for any $\sigma \in \mathrm{Aut}(\mathbb{C}/\mathbb{Q})$,

$$j(E^\sigma) = j(E)^\sigma.$$

(a) From [AEC III.1.4b] we have

$$E^\sigma \in \{E\} \qquad \text{if and only if} \qquad j(E^\sigma) = j(E).$$

Since $j(E^\sigma) = j(E)^\sigma$, this shows that $\mathbb{Q}(j(E))$ is the field of moduli for $\{E\}$.
(b) We know from [AEC III.1.4bc] that there exists an elliptic curve E_0 defined over $\mathbb{Q}(j(E))$ with $j(E_0) = j(E)$, and so satisfying $E_0 \cong_{/\mathbb{C}} E$. This shows that $\mathbb{Q}(j(E))$ is a field of definition for $\{E\}$.

On the other hand, if K is any field of definition for $\{E\}$, let E_0/K be a curve in $\{E\}$ given by an equation

$$E_0 \; : \; y^2 = x^3 + Ax + B \qquad \text{with } A, B \in K.$$

Then

$$j(E) = j(E_0) = 1728\frac{4A^3}{4A^3 + 27B^2} \in K,$$

so $\mathbb{Q}(j(E)) \subseteq K$.   □

Remark 4.6. The reader should note that the proof of Proposition 4.5 is very elementary because we have explicit Weierstrass equations with which to work. (This is how [AEC III.1.4bc] was proven.) For modular curves of higher level the problem becomes considerably more difficult, since one cannot rely on explicit equations. (See Shimura [1, §6.7].) Finally, we should mention that an analogous statement is false for abelian varieties of higher dimension; the field of moduli for an isomorphism class of abelian varieties need not be a field of definition.

§5. Elliptic Functions Revisited

Let $\Lambda \subset \mathbb{C}$ be a lattice. Our fundamental elliptic function is the Weierstrass $\wp$-function,

$$\wp(z;\Lambda) = \frac{1}{z^2} + \sum_{\substack{\omega \in \Lambda \\ \omega \neq 0}} \left(\frac{1}{(z-\omega)^2} - \frac{1}{\omega^2} \right).$$

As we have seen [AEC VI §3], $\wp$ defines a meromorphic function on the elliptic curve $\mathbb{C}/\Lambda$. It has a pole of order 2 at $0 \in \mathbb{C}/\Lambda$ and no other poles. We have also computed the Laurent series of $\wp$ around $z = 0$ [AEC VI.3.5a],

$$\wp(z;\Lambda) = \frac{1}{z^2} + \sum_{k=1}^{\infty} (2k+1)G_{2k+2}(\Lambda)z^{2k},$$

valid for $|z|$ less than the smallest non-zero vector in Λ.

Since $\wp(z;\Lambda)$ has no residues, we can integrate it to find a new function which will almost be periodic for the lattice Λ. Note, however, that when we integrate the series for $\wp(z;\Lambda)$ term-by-term, it is necessary to adjust the constant of integration in each term so as to ensure convergence.

Proposition 5.1. (a) *The series*

$$\zeta(z;\Lambda) = \frac{1}{z} + \sum_{\substack{\omega \in \Lambda \\ \omega \neq 0}} \left(\frac{1}{z-\omega} + \frac{1}{\omega} + \frac{z}{\omega^2} \right)$$

is absolutely and uniformly convergent on compact subsets of $\mathbb{C} \smallsetminus \Lambda$. It defines a meromorphic function on $\mathbb{C}$ with simple poles on Λ and no other poles. $\zeta(z;\Lambda)$ is called the Weierstrass ζ-function (associated to the lattice Λ).
(b) *The Laurent series for ζ around $z = 0$ is*

$$\zeta(z;\Lambda) = \frac{1}{z} - \sum_{k=1}^{\infty} G_{2k+2}(\Lambda)z^{2k+1}.$$

PROOF. (a) Let $C \subset \mathbb{C} \smallsetminus \Lambda$ be a compact set, and let

$$\varepsilon = \inf\{|z-\omega| : z \in C, \ \omega \in \Lambda\} \qquad \text{and} \qquad M = \sup\{|z| : z \in C\}.$$

Since C is compact, we have $\varepsilon > 0$ and $M < \infty$.

Let $z \in C$, and let $\omega \in \Lambda$ satisfy $|\omega| > 2M$. Then

$$\left| \frac{1}{z-\omega} + \frac{1}{\omega} + \frac{z}{\omega^2} \right| = \left| \frac{z^2}{\omega^3} \cdot \frac{1}{1-\dfrac{z}{\omega}} \right| \leq \frac{2M^2}{|\omega|^3}.$$

On the other hand, there are only finitely many terms in the sum with $0 < |\omega| \le 2M$, and for $z \in C$ those terms all satisfy

$$\left| \frac{1}{z - \omega} + \frac{1}{\omega} + \frac{z}{\omega^2} \right| \le \frac{1}{|z - \omega|} + \frac{1}{|\omega|} + \frac{|z|}{|\omega|^2} \le \frac{1}{\varepsilon} + \frac{1}{|\omega|} + \frac{M}{|\omega|^2}.$$

Hence

$$\left| \frac{1}{z} \right| + \sum_{\substack{\omega \in \Lambda \\ \omega \ne 0}} \left| \frac{1}{z - \omega} + \frac{1}{\omega} + \frac{z}{\omega^2} \right| \le \frac{1}{\varepsilon} + \sum_{\substack{\omega \in \Lambda \\ 0 < |\omega| \le 2M}} \left(\frac{1}{\varepsilon} + \frac{1}{|\omega|} + \frac{M}{|\omega|^2} \right) + \sum_{\substack{\omega \in \Lambda \\ \omega > 2M}} \frac{2M^2}{|\omega|^3}.$$

We know [AEC VI.3.1a] that the last series converges, which proves the series defining $\zeta(z; \Lambda)$ converges absolutely and uniformly on C.

It follows that $\zeta(z; \Lambda)$ is holomorphic on $\mathbb{C} \smallsetminus \Lambda$, and an inspection of the series defining ζ shows immediately that it has simple poles at each point of Λ.

(b) Let z be a complex number such that $|z| < |\omega|$ for all non-zero $\omega \in \Lambda$. Then

$$\frac{1}{z - \omega} + \frac{1}{\omega} + \frac{z}{\omega^2} = -\frac{1}{\omega} \left\{ \frac{1}{1 - \dfrac{z}{\omega}} - 1 - \frac{z}{\omega} \right\}$$

$$= -\frac{1}{\omega} \sum_{k=2}^{\infty} \left(\frac{z}{\omega} \right)^k,$$

so

$$\zeta(z; \Lambda) = \frac{1}{z} + \sum_{\substack{\omega \in \Lambda \\ \omega \ne 0}} -\frac{1}{\omega} \sum_{k=2}^{\infty} \left(\frac{z}{\omega} \right)^k$$

$$= \frac{1}{z} - \sum_{k=2}^{\infty} G_{k+1}(\Lambda) z^k.$$

This is the desired series once one notes that $G_k(\Lambda) = 0$ for odd k. $\square$

Differentiating the series (5.1a), we see that $\zeta'(z; \Lambda) = -\wp(z; \Lambda)$. Thus the derivative of $\zeta(z; \Lambda)$ is periodic for the lattice Λ, so ζ itself will have some sort of "quasi-periodicity" property as explained in the following proposition.

Proposition 5.2. (a) For all $z \in \mathbb{C}$,

$$\frac{d}{dz} \zeta(z; \Lambda) = -\wp(z; \Lambda).$$

(b) For all $\omega \in \Lambda$ and all $z \in \mathbb{C}$,

$$\zeta(z + \omega; \Lambda) = \zeta(z; \Lambda) + \eta(\omega),$$

where the number $\eta(\omega)$ is independent of z. The map

$$\eta : \Lambda \longrightarrow \mathbb{C}$$

is called the quasi-period map associated to Λ. If $\omega \in \Lambda$ and $\omega \notin 2\Lambda$, then $\eta(\omega)$ is given by the formula

$$\eta(\omega) = 2\zeta(\tfrac{1}{2}\omega; \Lambda).$$

(c) *The quasi-period map is a homomorphism of Λ into $\mathbb{C}$.*
(d) *(Legendre Relation) Let $\Lambda = \mathbb{Z}\omega_1 + \mathbb{Z}\omega_2$ be a lattice with basis satisfying $\mathrm{Im}(\omega_1/\omega_2) > 0$. Then*

$$\omega_1\eta(\omega_2) - \omega_2\eta(\omega_1) = 2\pi i.$$

PROOF. (a) The series (5.1a) defining ζ converges absolutely and uniformly, so it can be differentiated term-by-term. The result is the defining series for $-\wp$.
(b)

$$\frac{d}{dz}\zeta(z + \omega; \Lambda) = -\wp(z + \omega; \Lambda) = -\wp(z; \Lambda) = \frac{d}{dz}\zeta(z; \Lambda).$$

Integrating, we find that the quantity

$$\eta(\omega) = \zeta(z + \omega; \Lambda) - \zeta(z; \Lambda)$$

is independent of z. If, further, $\omega \notin 2\Lambda$, then ζ does not have a pole at $\pm\tfrac{1}{2}\omega$. Putting $z = -\tfrac{1}{2}\omega$ and using the fact (evident from the defining series) that $\zeta(-z; \Lambda) = -\zeta(z; \Lambda)$, we find in this case that $\eta(\omega) = 2\zeta(\tfrac{1}{2}\omega; \Lambda)$.
(c) We compute

$$
\begin{aligned}
\eta(\omega + \omega') &= \zeta(z + \omega + \omega'; \Lambda) - \zeta(z; \Lambda) \\
&= \{\zeta(z + \omega + \omega'; \Lambda) - \zeta(z + \omega; \Lambda)\} + \{\zeta(z + \omega; \Lambda) - \zeta(z; \Lambda)\} \\
&= \eta(\omega') + \eta(\omega).
\end{aligned}
$$

(d) We integrate $\zeta(z; \Lambda)$ around a fundamental parallelogram offset slightly so as not to contain points of Λ on its boundary. Thus let D be the region

$$D = \{a + t_1\omega_1 + t_2\omega_2 : 0 \le t_1, t_2 \le 1\},$$

and let

$$\partial D = L_1 + L_2 + L_3 + L_4$$

be its boundary as illustrated in Figure 1.5.

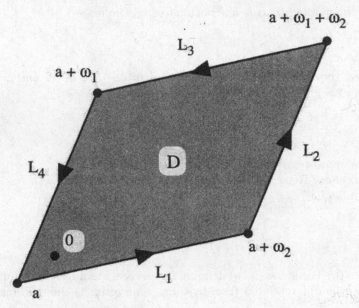

An Offset Fundamental Domain for $\mathbb{C}/\Lambda$

Figure 1.5

The only pole of ζ in D is a simple pole of residue 1 at $z = 0$. (Look at the series (5.1a) defining ζ.) Hence

$$\int_{\partial D} \zeta(z;\Lambda)\,dz = 2\pi i.$$

On the other hand, using (b) we get some cancellation when computing the line integrals over opposite sides. Thus

$$\int_{L_1+L_3} \zeta(z;\Lambda)\,dz = \int_0^1 \zeta(a + t\omega_2;\Lambda)\,\omega_2 dt + \int_1^0 \zeta(a + \omega_1 + t\omega_2;\Lambda)\,\omega_2 dt$$

$$= \int_0^1 \zeta(a + t\omega_2;\Lambda)\,\omega_2 dt - \int_0^1 \big(\zeta(a + t\omega_2) + \eta(\omega_1)\big)\,\omega_2 dt$$

$$= -\eta(\omega_1)\omega_2.$$

Similarly,

$$\int_{L_2+L_4} \zeta(z;\Lambda)\,dz = \eta(\omega_2)\omega_1.$$

Therefore

$$2\pi i = \int_{\partial D} \zeta(z;\Lambda)\,dz = \int_{L_1+L_2+L_3+L_4} \zeta(z;\Lambda)\,dz = \eta(\omega_2)\omega_1 - \eta(\omega_1)\omega_2.$$

$\square$

Remark 5.3. Let $E/\mathbb{C}$ be an elliptic curve given by a Weierstrass equation, and let

$$\omega_E = \frac{dx}{2y + a_1 x + a_3},$$

be the associated invariant differential. The lattice Λ for E is the set of periods

$$\int_\gamma \omega_E,$$

where γ runs over all closed paths on $E(\mathbb{C})$. (Equivalently, γ runs through the cycles in $H_1(E(\mathbb{C}), \mathbb{Z})$. See [AEC VI §1].) The classical name for an everywhere holomorphic differential such as ω_E on a Riemann surface such as $E(\mathbb{C})$ is a *differential of the first kind*.

Similarly, a *differential of the second kind* is a meromorphic differential with no residues (i.e., with no simple poles), and a *differential of the third kind* is a meromorphic differential with at worst simple poles.

The differential

$$\wp(z; \Lambda) \, dz = x \, \omega_E$$

is thus a differential of the second kind on $\mathbb{C}/\Lambda$, and its indefinite integral is the multi-valued function $-\zeta(z; \Lambda)$. The indeterminacy in ζ is given by the numbers

$$\int_\gamma x \, \omega_E = \int_a^{a+\omega} \wp(z; \Lambda) \, dz = -\zeta(a + \omega; \Lambda) + \zeta(a; \Lambda) = -\eta(\omega),$$

where $\omega = \int_\gamma \omega_E$ is the period associated to the closed path γ.

In terms of our original Weierstrass equation, there is the period map

$$H_1(E(\mathbb{C}), \mathbb{Z}) \longrightarrow \mathbb{C}, \qquad \gamma \longmapsto \int_\gamma \omega_E,$$

whose image is the lattice Λ. Using this to identify Λ with the first homology of $E(\mathbb{C})$, we see that the quasi-period map associates to a path the negative of the corresponding period for the differential $x\omega_E$:

$$\eta : H_1(E(\mathbb{C}), \mathbb{Z}) \longrightarrow \mathbb{C}, \qquad \gamma \longmapsto -\int_\gamma x \, \omega_E.$$

The last function we want to examine is essentially the integral of ζ. To eliminate the indeterminacy caused by the simple poles of ζ, we take the exponential of the integral. This leads to a familiar function which we used in [AEC VI §3] to construct elliptic functions with a given divisor.

Proposition 5.4. (a) *The infinite product*

$$\sigma(z; \Lambda) = z \prod_{\substack{\omega \in \Lambda \\ \omega \neq 0}} \left(1 - \frac{z}{\omega}\right) e^{z/\omega + (1/2)(z/\omega)^2}$$

defines a holomorphic function on $\mathbb{C}$ *with simple zeros on* Λ *and no other zeros. It is called the Weierstrass* σ*-function (associated to the lattice* Λ*).*
(b)

$$\frac{d}{dz} \log \sigma(z; \Lambda) = \zeta(z; \Lambda), \qquad \frac{d^2}{dz^2} \log \sigma(z; \Lambda) = -\wp(z; \Lambda).$$

(c) *For all* $z \in \mathbb{C}$ *and* $\omega \in \Lambda$,

$$\sigma(z + \omega; \Lambda) = \psi(\omega) e^{\eta(\omega)(z + \frac{1}{2}\omega)} \sigma(z; \Lambda),$$

where $\eta : \Lambda \to \mathbb{C}$ *is the quasi-period map for* Λ*, and* ψ *is defined by*

$$\psi : \Lambda \longrightarrow \{\pm 1\}, \qquad \psi(\omega) = \begin{cases} 1 & \text{if } \omega \in 2\Lambda; \\ -1 & \text{if } \omega \notin 2\Lambda. \end{cases}$$

PROOF. (a) This is a restatement of [AEC VI.3.3a].
(b) Taking the derivative of

$$\log \sigma(z; \Lambda) = \log(z) + \sum_{\substack{\omega \in \Lambda \\ \omega \neq 0}} \left\{ \log \left(1 - \frac{z}{\omega}\right) + \frac{z}{\omega} + \frac{1}{2} \left(\frac{z}{\omega}\right)^2 \right\}$$

gives the defining series (5.1a) for ζ, and then from (5.2a) we see that the second derivative is $-\wp$. Note that the logarithms are locally well defined up to the addition of a constant which disappears when we differentiate and also that we must take the principal branch of $\log \left(1 - \dfrac{z}{\omega}\right)$ for almost all ω in order to ensure the convergence of the series.
(c) From (b) and (5.2b),

$$\frac{d}{dz} \log \frac{\sigma(z + \omega; \Lambda)}{\sigma(z; \Lambda)} = \zeta(z + \omega; \Lambda) - \zeta(z; \Lambda) = \eta(\omega),$$

so

$$\sigma(z + \omega; \Lambda) = C e^{\eta(\omega)z} \sigma(z; \Lambda)$$

for some constant C not depending on z. Note also that σ is an odd function, a fact that is clear from the product defining σ.

We consider two cases. First, if $\omega \notin 2\Lambda$, then σ does not vanish at $\pm \frac{1}{2}\omega$. Hence putting $z = -\frac{1}{2}\omega$ gives

$$\sigma \left(\tfrac{1}{2}\omega; \Lambda\right) = C e^{-\frac{1}{2}\eta(\omega)\omega} \sigma \left(-\tfrac{1}{2}\omega; \Lambda\right) = -C e^{-\frac{1}{2}\eta(\omega)\omega} \sigma \left(\tfrac{1}{2}\omega; \Lambda\right),$$

so

$$C = -e^{\frac{1}{2}\eta(\omega)\omega}.$$

Next, if $\omega \in 2\Lambda$, then σ has a simple zero at $\pm\frac{1}{2}\omega$. Using L'Hôpital's rule yields

$$Ce^{-\frac{1}{2}\eta(\omega)\omega} = \lim_{z \to -\omega/2} \frac{\sigma(z + \omega; \Lambda)}{\sigma(z; \Lambda)} = \frac{\sigma'\left(\frac{1}{2}\omega; \Lambda\right)}{\sigma'\left(-\frac{1}{2}\omega; \Lambda\right)} = 1.$$

(Note that σ' is an even function, since σ is odd.) Hence in this case we find that

$$C = e^{\frac{1}{2}\eta(\omega)\omega},$$

which completes the proof of (c). $\qquad\square$

Any elliptic function can be factored as a product of Weierstrass σ-functions reflecting its zeros and poles. We give a general result and two important examples. To ease notation, since the lattice Λ is fixed, we will write $\sigma(z)$ and $\wp(z)$ instead of $\sigma(z; \Lambda)$ and $\wp(z; \Lambda)$.

Proposition 5.5. *Let $f(z)$ be a non-zero elliptic function for the lattice Λ. Write the divisor of f as*

$$\mathrm{div}(f) = \sum_{i=1}^{r} n_i(a_i)$$

for some $a_i \in \mathbb{C}$, and let

$$b = \sum_{i=1}^{r} n_i a_i.$$

(See [AEC VI §2] for the definition of the divisor of an elliptic function.) Then there is a constant $c \in \mathbb{C}^*$ so that

$$f(z) = c \frac{\sigma(z)}{\sigma(z - b)} \prod_{i=1}^{r} \sigma(z - a_i)^{n_i}.$$

Corollary 5.6.

(a)
$$\wp(z) - \wp(a) = -\frac{\sigma(z + a)\sigma(z - a)}{\sigma(z)^2 \sigma(a)^2}.$$

(b)
$$\wp'(z) = -\frac{\sigma(2z)}{\sigma(z)^4}.$$

PROOF (of Proposition 5.5). Let

$$g(z) = \frac{\sigma(z)}{\sigma(z-b)} \prod_{i=1}^{r} \sigma(z-a_i)^{n_i}.$$

From [AEC VI.2.2c] we know that $b \in \Lambda$, so using (5.4c), we find that

$$\frac{\sigma(z)}{\sigma(z-b)} = \pm e^{\eta(b)(z-\frac{1}{2}b)}$$

is holomorphic and non-vanishing on all of $\mathbb{C}$. Since $\sigma(z)$ has simple zeros on Λ and no other zeros, it follows that g has exactly the same zeros and poles as f. Hence $f(z)/g(z)$ is everywhere holomorphic.

Next we verify that g is an elliptic function. Let $\omega \in \Lambda$, and use (5.4c) to write

$$\frac{\sigma(z+\omega)}{\sigma(z)} = Ae^{Bz}$$

for certain constants A and B which depend on ω but not on z. Then

$$\frac{g(z+\omega)}{g(z)} = \frac{\sigma(z-b)}{\sigma(z-b+\omega)} \frac{\sigma(z+\omega)}{\sigma(z)} \prod_{i=1}^{r} \left(\frac{\sigma(z+\omega-a_i)}{\sigma(z-a_i)} \right)^{n_i}$$

$$= e^{-B(z-b)} e^{Bz} \prod_{i=1}^{r} \left(Ae^{B(z-a_i)} \right)^{n_i}$$

$$= e^{B(b-\Sigma n_i a_i)} \left(Ae^{Bz} \right)^{\Sigma n_i} = 1.$$

The last equality follows from the definition of b and the fact [AEC VI.2.2b] that the divisor of an elliptic function has degree 0.

This proves that $g(z)$ is an elliptic function, and so $f(z)/g(z)$ is an everywhere holomorphic elliptic function. From [AEC VI.2.1] we conclude that it is constant.   □

PROOF (of Corollary 5.6). (a) Since $\wp(z)$ is an even function of order 2, we see immediately that the zeros of $\wp(z) - \wp(a)$ are a and $-a$. Thus

$$\mathrm{div}\big(\wp(z) - \wp(a)\big) = (-a) + (a) - 2(0).$$

Applying (5.5) we find

$$\wp(z) - \wp(a) = C \frac{\sigma(z+a)\sigma(z-a)}{\sigma(z)^2}$$

for some constant C. Multiplying by z^2 and using

$$\lim_{z \to 0} z^2 \wp(z) = 1 \qquad \text{and} \qquad \lim_{z \to 0} \frac{\sigma(z)}{z} = 1$$

gives the value of C,

$$1 = C\sigma(a)\sigma(-a) = -C\sigma(a)^2.$$

(b) Divide (a) by $z - a$ and let $z \to a$. This yields

$$\wp'(a) = -\lim_{z \to a} \frac{\sigma(z - a)}{z - a} \frac{\sigma(z + a)}{\sigma(z)^2 \sigma(a)^2} = -\sigma'(0) \frac{\sigma(2a)}{\sigma(a)^4}.$$

Since $\sigma'(0) = 1$, this is the desired result. $\qquad\qquad\qquad\qquad\square$

§6. q-Expansions of Elliptic Functions

As we have seen in §§1–4, it is often convenient to use normalized lattices

$$\Lambda_\tau = \mathbb{Z}\tau + \mathbb{Z} \qquad \text{with } \tau \in \mathbf{H}.$$

We then use the obvious notation

$$\wp(z; \tau), \zeta(z; \tau), \sigma(z; \tau) \qquad \text{for} \qquad \wp(z; \Lambda_\tau), \zeta(z; \Lambda_\tau), \sigma(z; \Lambda_\tau).$$

We will soon see that $\wp$, ζ, and σ are quite well behaved when considered as functions of two variables $(z; \tau) \in \mathbb{C} \times \mathbf{H}$.

Note that since $1 \in \Lambda_\tau$, the $\wp$ function satisfies the relation

$$\wp(z + 1; \tau) = \wp(z; \tau).$$

This means that it is possible to expand $\wp$ as a Fourier series in the variable $u = e^{2\pi i z}$. Similarly, since $\Lambda_{\tau+1} = \Lambda_\tau$, the $\wp$ function satisfies

$$\wp(z; \tau + 1) = \wp(z; \tau).$$

Thus, as a function of τ, the $\wp$ function should have a Fourier expansion in terms of $q = e^{2\pi i \tau}$.

This idea can be formulated more intrinsically as follows. Let

$$u = u_z = e^{2\pi i z} \qquad \text{and} \qquad q = q_\tau = e^{2\pi i \tau},$$

and let

$$q^{\mathbb{Z}} = \{q^k : k \in \mathbb{Z}\}$$

be the cyclic subgroup of $\mathbb{C}^*$ generated by q. Then there is a complex-analytic isomorphism

$$\mathbb{C}/\Lambda_\tau \xrightarrow{\sim} \mathbb{C}^*/q^{\mathbb{Z}}, \qquad z \longmapsto u = e^{2\pi i z}.$$

Note that this is an isomorphism of complex Lie groups, since it is clearly a homomorphism.

Our first step is to express $\wp(z; \tau)$ as a power series in the variables $u = e^{2\pi i z}$ and $q = e^{2\pi i \tau}$. The quickest way to do this is to write down (by magic?!?) the correct expression, and then verify that it gives the same function as $\wp(z; \tau)$. We opt instead for a somewhat lengthier, but hopefully more perspicuous, derivation.

Consider first the series

$$\wp(z; \Lambda) = \frac{1}{z^2} + \sum_{\substack{\omega \in \Lambda \\ \omega \neq 0}} \frac{1}{(z - \omega)^2} - \frac{1}{\omega^2}$$

defining $\wp$. How does it arise? From [AEC VI.2.3] we know that any non-constant elliptic function must have at least two poles, so we look for a meromorphic function $F(z)$ satisfying

(i) $F(z + \omega) = F(z)$ for all $z \in \mathbb{C}$, $\omega \in \Lambda$;

(ii) $F(z)$ has a double pole at each point in Λ and no other poles.

The simplest function with a double pole at ω is $(z - \omega)^{-2}$. By averaging over $\omega \in \Lambda$, we find a series

$$F(z) = \sum_{\omega \in \Lambda} \frac{1}{(z - \omega)^2}$$

which *formally* satisfies (i) and (ii). The problem is that this series is not absolutely convergent. However, by subtracting an appropriate constant from each term, we can create a series which does converge and has the desired properties. This is how we "discovered" $\wp(z; \Lambda)$ in [AEC VI §3].

We apply the same principle to express $\wp(z; \tau)$ as a function of u and q. Exponentiating the conditions (i) and (ii), we look for a function $F(u; q)$ satisfying

(iii) $F(q^k u; q) = F(u; q)$ for all $u \in \mathbb{C}^*$, $k \in \mathbb{Z}$;

(iv) $F(u; q)$ has a double pole at each $u \in q^{\mathbb{Z}}$ and no other poles.

As above, we look for F to be an average

$$F(u; q) = \sum_{n \in \mathbb{Z}} f(q^n u)$$

for some elementary function f. Such an F will clearly satisfy the periodicity condition (iii):

To obtain (iv), we need $f(T)$ to have a double pole at $T = 1$. For example, we might use $f(T) = (1 - T)^{-2}$. But the series

$$\sum_{n \in \mathbb{Z}} \frac{1}{(1 - q^n u)^2}$$

does not converge, since $|q| < 1$. The terms with $n \to -\infty$ are all right, since then $q^n \to \infty$. But as $n \to \infty$, the nth term goes to 1.

In order to get convergence, we want $f(T)$ to have a double pole at $T = 1$ and also to satisfy

$$\lim_{T \to 0} f(T) = 0 \quad \text{and} \quad \lim_{T \to \infty} f(T) = 0.$$

The simplest such function is $f(T) = T(1-T)^{-2}$, which leads us to consider the function $F(u; q)$ in the following lemma.

Lemma 6.1.   *Let*

$$F(u; q) = \sum_{n \in \mathbb{Z}} \frac{q^n u}{(1 - q^n u)^2}.$$

(a) *The series defining F, considered as a function of z, converges absolutely and uniformly on compact subsets of $\mathbb{C} \smallsetminus \Lambda_\tau$.*
(b) *F is an elliptic function for the lattice Λ_τ. It has a double pole at each $z \in \Lambda_\tau$ and no other poles.*
(c) *The Laurent series for F around $z = 0$ begins*

$$F(u; q) = \frac{1}{(2\pi i)^2 z^2} - \left\{ \frac{1}{12} - 2 \sum_{n \geq 1} \frac{q^n}{(1 - q^n)^2} \right\} + (\text{powers of } z).$$

PROOF. (a) Note that

$$\frac{q^n u}{(1 - q^n u)^2} = \frac{q^{-n} u^{-1}}{(1 - q^{-n} u^{-1})^2}.$$

We use this identity to rewrite the terms in F having $n < 0$. This gives an alternative expression for F,

$$F(u; q) = \frac{u}{(1 - u)^2} + \sum_{n \geq 1} \left\{ \frac{q^n u}{(1 - q^n u)^2} + \frac{q^n u^{-1}}{(1 - q^n u^{-1})^2} \right\}.$$

Now let $C \subset \mathbb{C} \smallsetminus \Lambda_\tau$ be a compact set. Then $u = e^{2\pi i z}$ is bounded away from 0 and ∞ uniformly for $z \in C$. Since $q^n \to 0$ as $n \to \infty$, it follows that there are constants c_1 and c_2 so that

$$\left| \frac{q^n u}{(1 - q^n u)^2} \right| + \left| \frac{q^n u^{-1}}{(1 - q^n u^{-1})^2} \right| \leq c_1 |q^n| \quad \text{for all } z \in C, n \geq c_2.$$

This shows that except possibly for the terms with $n \leq c_2$, the series is absolutely and uniformly convergent on C.

Consider now one of the finitely many terms with $n \leq c_2$. We know that $u \neq q^n$ for $z \in C$, so the compactness of C ensures that

$$\inf_{z \in C} |1 - q^n u^{\pm 1}| > 0.$$

Hence the terms with $n \leq c_2$ are also uniformly bounded.

(b) From (a), F is a holomorphic function on $\mathbb{C} \smallsetminus \Lambda_\tau$, and looking at the series defining F, it is clear that F has a double pole at each point in Λ_τ. Finally, since the transformations $z \mapsto z + 1$ and $z \mapsto z + \tau$ correspond to $u \mapsto u$ and $u \mapsto qu$ respectively, it is clear again from the series that F is an elliptic function for the lattice Λ_τ.

(c) Note that $u = e^{2\pi i z} \to 1$ as $z \to 0$. Hence the pole at $z = 0$ in the series for F comes from the term with $n = 0$. Now a little freshman calculus yields

$$\frac{u}{(1 - u)^2} = \frac{e^{2\pi i z}}{(1 - e^{2\pi i z})^2} = \frac{1}{(2\pi i z)^2} - \frac{1}{12} + \text{(powers of } z\text{)}.$$

Hence using the alternative series for F given above, we find

$$\lim_{z \to 0} \left\{ F(u; q) - \frac{1}{(2\pi i z)^2} + \frac{1}{12} \right\} = \lim_{z \to 0} \left\{ F(u; q) - \frac{u}{(1 - u)^2} \right\}$$

$$= \lim_{u \to 1} \sum_{n \geq 1} \left\{ \frac{q^n u}{(1 - q^n u)^2} + \frac{q^n u^{-1}}{(1 - q^n u^{-1})^2} \right\}$$

$$= 2 \sum_{n \geq 1} \frac{q^n}{(1 - q^n)^2}.$$

$\square$

Theorem 6.2.   Let $u = e^{2\pi i z}$ and $q = e^{2\pi i \tau}$.

(a) $\quad \dfrac{1}{(2\pi i)^2} \wp(z; \tau) = \displaystyle\sum_{n \in \mathbb{Z}} \frac{q^n u}{(1 - q^n u)^2} + \frac{1}{12} - 2 \sum_{n \geq 1} \frac{q^n}{(1 - q^n)^2}$

$$= \sum_{n \geq 0} \frac{q^n u}{(1 - q^n u)^2} + \sum_{n \geq 1} \frac{q^n u^{-1}}{(1 - q^n u^{-1})^2}$$

$$+ \frac{1}{12} - 2 \sum_{n \geq 1} \frac{q^n}{(1 - q^n)^2}.$$

(b) $\quad \dfrac{1}{(2\pi i)^3} \wp'(z; \tau) = \displaystyle\sum_{n \in \mathbb{Z}} \frac{q^n u (1 + q^n u)}{(1 - q^n u)^3}$

$$= \sum_{n \geq 0} \frac{q^n u (1 + q^n u)}{(1 - q^n u)^3} - \sum_{n \geq 1} \frac{q^n u^{-1} (1 + q^n u^{-1})}{(1 - q^n u^{-1})^3}.$$

Remark 6.2.1. For $|X| < 1$ there is the elementary identity

$$\frac{X}{(1-X)^2} = X\frac{d}{dX}\left(\frac{1}{1-X}\right) = \sum_{m \geq 1} mX^m.$$

This is sometimes used to rewrite the final sum in (6.2a) as

$$\sum_{n \geq 1} \frac{q^n}{(1-q^n)^2} = \sum_{n \geq 1}\sum_{m \geq 1} mq^{mn} = \sum_{m \geq 1} \frac{mq^m}{1-q^m}.$$

PROOF (of Theorem 6.2). (a) Let $F(u; q)$ be as in (6.1). Consider the function

$$\frac{1}{(2\pi i)^2}\wp(z; \tau) - F(u; q) + \frac{1}{12} - 2\sum_{n \geq 1}\frac{q^n}{(1-q^n)^2}.$$

From (6.1b) we see that this expression is an elliptic function for the lattice Λ_τ which is holomorphic on $\mathbb{C} \setminus \Lambda_\tau$. Further, comparing the Laurent series for F given in (6.1c) with the known Laurent series for $\wp$, we see that it is also holomorphic at $z = 0$, and in fact it vanishes there. It thus represents an everywhere holomorphic elliptic function which vanishes at $z = 0$. Applying [AEC VI.2.1], we conclude that it is identically zero. This proves the first equality in (a), and the second is an easy rearrangement of the terms in the initial sum. (See the proof of (6.1a).)

(b) Apply $\dfrac{d}{dz} = 2\pi i u\dfrac{d}{du}$ to (a). $\qquad\qquad\square$

The next step is to find a q-expansion for $\zeta(z; \tau)$ analogous to (6.2). By construction,

$$\frac{d}{dz}\zeta(z; \tau) = -\wp(z; \tau),$$

so we try integrating the series (6.2a) for $\wp$ term-by-term. Proceeding blindly, we find

$$\int \frac{q^n u}{(1-q^n u)^2}dz = \int \frac{q^n u}{(1-q^n u)^2}\frac{du}{2\pi i u} = \frac{1}{2\pi i}\cdot\frac{1}{(1-q^n u)}.$$

Unfortunately, the series

$$\sum_{n \in \mathbb{Z}}\frac{1}{1-q^n u}$$

is clearly divergent, the nth term goes to 1 as $n \to \infty$. But just as in the original definition of $\wp$, we can improve the convergence by adding a constant onto each term.

The second expression for $\wp$ in (6.2a) has the form

$$\frac{1}{(2\pi i)^2}\wp(z; \tau) = \sum_{n \geq 0}\frac{q^n u}{(1-q^n u)^2} + \sum_{n \geq 1}\frac{q^n u^{-1}}{(1-q^n u^{-1})^2} + C_1,$$

where

$$C_1 = C_1(q) = \frac{1}{12} - 2\sum_{n\geq 1}\frac{q^n}{(1-q^n)^2}.$$

(We will soon identify C_1 more precisely.) Now integrate:

$$\frac{1}{2\pi i}\zeta(z;\tau) = \frac{1}{2\pi i}\int -\wp(z;\tau)\,dz = -\int\frac{1}{(2\pi i)^2}\wp(z;\tau)\frac{du}{u}$$

$$= -\int\left\{\sum_{n\geq 0}\frac{q^n u}{(1-q^n u)^2} + \sum_{n\geq 1}\frac{q^n u^{-1}}{(1-q^n u^{-1})^2} + C_1\right\}\frac{du}{u}$$

$$= -\sum_{n\geq 0}\left\{\frac{1}{1-q^n u} - 1\right\} - \sum_{n\geq 1}\frac{-q^n u^{-1}}{1-q^n u^{-1}} - 2\pi i C_1 z + C_2$$

$$= \sum_{n\geq 0}\frac{-q^n u}{1-q^n u} + \sum_{n\geq 1}\frac{q^n u^{-1}}{1-q^n u^{-1}} - 2\pi i C_1 z + C_2$$

for some constant of integration $C_2 = C_2(q)$. Note that the last series is absolutely and uniformly convergent on compact subsets of $\mathbb{C}\smallsetminus\Lambda_\tau$, so it defines a meromorphic function on $\mathbb{C}$. (The proof is identical to the proof of (6.1a).) Further, the (d/dz)-derivative of this series can be computed term-by-term and agrees with the series (6.2a) for $\wp$. This proves that it equals ζ for some choice of C_2.

To find C_2, we compute the first few terms of the Laurent series around $z = 0$. We already know (5.1b) that

$$\zeta(z;\tau) = \frac{1}{z} - G_4(\tau)z^3 + \text{higher powers of } z.$$

On the other hand, the pole at $z = 0$ (i.e., at $u = 1$) in the above q-series comes from the $n = 0$ term, so we find

$$\frac{-u}{1-u} + \underbrace{\sum_{n\geq 1}\left\{\frac{-q^n u}{1-q^n u} + \frac{q^n u^{-1}}{1-q^n u^{-1}}\right\}}_{\text{vanishes at } z = 0\ (u = 1)} - 2\pi i C_1 z + C_2$$

$$= -\frac{e^{2\pi i z}}{1-e^{2\pi i z}} + C_2 + (\text{powers of } z)$$

$$= \frac{1}{2\pi i z} + \frac{1}{2} + C_2 + (\text{powers of } z).$$

Since the Laurent series for ζ has no constant term, we see that $C_2 = -\frac{1}{2}$. This proves part of the following theorem.

Theorem 6.3. Let $\zeta(z;\tau)$ be the Weierstrass ζ-function and $\eta : \Lambda_\tau \to \mathbb{C}$ the quasi-period homomorphism associated to Λ.

(a) $\dfrac{1}{2\pi i}\zeta(z;\tau) = \displaystyle\sum_{n\geq 0}\frac{-q^n u}{1-q^n u} + \sum_{n\geq 1}\frac{q^n u^{-1}}{1-q^n u^{-1}} + \frac{1}{2\pi i}\eta(1)z - \frac{1}{2}.$

(b) $\dfrac{1}{(2\pi i)^2}\eta(1) = \dfrac{1}{12}\left\{-1 + 24\displaystyle\sum_{n\geq 1}\dfrac{q^n}{(1-q^n)^2}\right\}.$

PROOF. Let

$$G(z;\tau) = \sum_{n\geq 0}\frac{-q^n u}{1-q^n u} + \sum_{n\geq 1}\frac{q^n u^{-1}}{1-q^n u^{-1}} \quad \text{and}$$

$$C_1(q) = \frac{1}{12} - 2\sum_{n\geq 1}\frac{q^n}{(1-q^n)^2}.$$

We proved above that

$$\frac{1}{2\pi i}\zeta(z;\tau) = G(z;\tau) - 2\pi i C_1(q) - \frac{1}{2}.$$

Now evaluate at $z = \frac{1}{2}$. From (5.2b), $\zeta(\frac{1}{2};\tau) = \eta(1)$. Further, $z = \frac{1}{2}$ corresponds to $u = e^{\pi i} = -1$, so all of the terms in $G(\frac{1}{2};\tau)$ cancel except the $n = 0$ term. Thus

$$G\left(\frac{1}{2};\tau\right) = \sum_{n\geq 0}\frac{q^n}{1+q^n} + \sum_{n\geq 1}\frac{-q^n}{1+q^n} = \frac{1}{2}.$$

Hence

$$\frac{1}{2\pi i}\eta(1) = \frac{1}{2\pi i}\zeta\left(\frac{1}{2};\tau\right) = G\left(\frac{1}{2};\tau\right) - 2\pi i C_1(q) - \frac{1}{2} = -2\pi i C_1(q),$$

so

$$C_1(q) = -\frac{1}{(2\pi i)^2}\eta(1).$$

This completes the proof of both parts (a) and (b).   □

Finally, we integrate the series for $\zeta(z;\tau)$ and exponentiate to obtain an important q-product expansion for $\sigma(z;\tau)$.

Theorem 6.4. *The Weierstrass σ-function has the product expansion*

$$\sigma(z;\tau) = -\frac{1}{2\pi i}e^{\frac{1}{2}\eta(1)z^2}e^{-\pi i z}(1-u)\prod_{n\geq 1}\frac{(1-q^n u)(1-q^n u^{-1})}{(1-q^n)^2},$$

where $u = e^{2\pi i z}$ and $q = e^{2\pi i\tau}$ as usual, and $\eta(1)$ is the quasi-period associated to the period $1 \in \Lambda_\tau$.

PROOF. By construction,

$$\frac{\sigma'(z;\tau)}{\sigma(z;\tau)} = \frac{d}{dz}\log\sigma(z;\tau) = \zeta(z;\tau).$$

Using (6.3a) and integrating gives

$$
\begin{aligned}
\log \sigma(z; \tau) &= \int \zeta(z; \tau) \, dz = \int \frac{1}{2\pi i} \zeta(z; \tau) \frac{du}{u} \\
&= \sum_{n \geq 0} \int \frac{-q^n u}{1 - q^n u} \frac{du}{u} + \sum_{n \geq 1} \int \frac{q^n u^{-1}}{1 - q^n u^{-1}} \frac{du}{u} + \int \left(\eta(1) z - \pi i \right) dz \\
&= \sum_{n \geq 0} \log(1 - q^n u) + \sum_{n \geq 1} \log(1 - q^n u^{-1}) + \tfrac{1}{2} \eta(1) z^2 - \pi i z + C_3.
\end{aligned}
$$

We claim that the series will converge provided we use the principal branch of the logarithm when evaluating $\log(1 - q^n u)$ and $\log(1 - q^n u^{-1})$. To see this, note that for n sufficiently large we have $|q^n u^{\pm 1}| \leq \frac{1}{2}$. So, for all but finitely many n,

$$
\left| \log(1 - q^n u^{\pm 1}) \right| = \left| \sum_{k=1}^{\infty} \frac{1}{k} (q^n u^{\pm 1})^k \right| \leq 2 \left| q^n u^{\pm 1} \right|.
$$

Hence the series will converge.

Exponentiating, we eliminate any ambiguity arising from the choice of a branch of the logarithm and obtain the product representation

$$
\sigma(z; \tau) = e^{\frac{1}{2} \eta(1) z^2 - \pi i z + C_3} \prod_{n \geq 0} (1 - q^n u) \prod_{n \geq 1} (1 - q^n u^{-1}).
$$

It remains to find C_3. Recall that σ was normalized by the condition that $\sigma(z; \tau)/z \to 1$ as $z \to 0$. It is the $n = 0$ term in the product which vanishes at $z = 0$, so we find

$$
\begin{aligned}
1 &= \lim_{z \to 0} \frac{\sigma(z; \tau)}{z} \\
&= \lim_{\substack{z \to 0 \\ u \to 1}} e^{\frac{1}{2} \eta(1) z^2 - \pi i z + C_3} \left(\frac{1 - u}{z} \right) \prod_{n \geq 1} (1 - q^n u)(1 - q^n u^{-1}) \\
&= e^{C_3} (-2\pi i) \prod_{n \geq 1} (1 - q^n)^2.
\end{aligned}
$$

Hence

$$
e^{C_3} = -\frac{1}{2\pi i} \prod_{n \geq 1} \frac{1}{(1 - q^n)^2},
$$

which gives the desired product formula for $\sigma(z; \tau)$. $\square$

§7. q-**Expansions of Modular Functions**

The Eisenstein series $G_{2k}(\tau)$ is a modular function of weight $2k$. It satisfies $G_{2k}(\tau + 1) = G_{2k}(\tau)$, so it has a Fourier expansion in terms of the variable $q = e^{2\pi i \tau}$. In this section we will compute the Fourier series of G_{2k} and use it to deduce various properties of the Fourier expansions for $\Delta(\tau)$ and $j(\tau)$.

Proposition 7.1. *Let $k \geq 2$. Then*

$$G_{2k}(\tau) = 2\zeta(2k) + 2\frac{(2\pi i)^{2k}}{(2k-1)!} \sum_{n \geq 1} \sigma_{2k-1}(n)q^n,$$

where

$$\zeta(s) = \sum_{n \geq 1} \frac{1}{n^s} \qquad \text{and} \qquad \sigma_k(n) = \sum_{d|n} d^k$$

are respectively the Riemann ζ-function and the kth-power divisor function.

PROOF.

$$G_{2k}(\tau) = \sum_{\substack{m,n \in \mathbb{Z} \\ (m,n) \neq (0,0)}} \frac{1}{(m\tau + n)^{2k}}$$

$$= \sum_{\substack{n \in \mathbb{Z} \\ n \neq 0}} \frac{1}{n^{2k}} + 2 \sum_{m=1}^{\infty} \sum_{n \in \mathbb{Z}} \frac{1}{(m\tau + n)^{2k}}.$$

The first sum is just $2\zeta(2k)$. Notice that the rightmost (inner) sum is clearly invariant under $\tau \mapsto \tau + 1$. We now compute its Fourier expansion.

Lemma 7.1.1. *Let $k \geq 1$ be an integer. Then for all $\tau \in \mathbf{H}$,*

$$\sum_{n \in \mathbb{Z}} \frac{1}{(\tau + n)^{2k}} = \frac{(2\pi i)^{2k}}{(2k-1)!} \sum_{r=1}^{\infty} r^{2k-1} e^{2\pi i r \tau}.$$

PROOF. Ignoring questions of convergence, we have a formal identity

$$\sum_{n \in \mathbb{Z}} \frac{1}{(\tau + n)^{2k}} = \sum_{n \in \mathbb{Z}} \frac{1}{(2k-1)!} \frac{d^{2k}}{d^{2k}\tau} \log(\tau + n)$$

$$= \frac{1}{(2k-1)!} \frac{d^{2k}}{d^{2k}\tau} \log \prod_{n \in \mathbb{Z}} (\tau + n).$$

Of course, this product does not converge. But we do get convergence if we factor an n out of each term. (Remember this is just a formal manipulation

to see what the answer should be. Otherwise one might rightly object that
dividing by $(\infty!)^2$ is a highly dubious procedure.) The product

$$\tau \prod_{\substack{n\in\mathbb{Z} \\ n\neq 0}} \left(1+\frac{\tau}{n}\right) = \tau \prod_{n=1}^{\infty} \left(1-\frac{\tau^2}{n^2}\right)$$

converges to give a function that is holomorphic on $\mathbb{C}$, has simple zeros
at each integer, and no other zeros. With this description, the reader will
undoubtedly recognize the usual product expansion for the sine function.
(See Ahlfors [1].)

$$\sin(\pi\tau) = \pi\tau \prod_{n=1}^{\infty} \left(1-\frac{\tau^2}{n^2}\right).$$

We now reverse our formal argument to produce a rigorous proof.
Starting with the product expansion of the sine function, we take the logarithmic derivative, yielding

$$\frac{d}{d\tau}\log(\sin\pi\tau) = \frac{1}{\tau} + \sum_{n=1}^{\infty} \frac{-2\tau}{n^2-\tau^2}$$

$$= \frac{1}{\tau} + \sum_{n=1}^{\infty} \left(\frac{-1}{n+\tau} + \frac{1}{n-\tau}\right).$$

Now taking $(2k-1)$ more derivatives, we find

$$\frac{d^{2k}}{d^{2k}\tau}\log(\sin\pi\tau) = -(2k-1)!\left\{\frac{1}{\tau^{2k}} + \sum_{n=1}^{\infty}\left(\frac{1}{(n+\tau)^{2k}} + \frac{1}{(n-\tau)^{2k}}\right)\right\}$$

$$= -(2k-1)! \sum_{n\in\mathbb{Z}} \frac{1}{(n+\tau)^{2k}}.$$

Next we compute the Fourier series of (a branch of) $\log(\sin\pi\tau)$. Writing

$$\sin(\pi\tau) = \frac{1}{2i}\left(e^{\pi i\tau} - e^{-\pi i\tau}\right) = -\frac{1}{2i}e^{-\pi i\tau}\left(1-e^{2\pi i\tau}\right),$$

we find (for $\tau \in \mathbf{H}$)

$$\log(\sin\pi\tau) = -\log(-2i) - \pi i\tau + \log\left(1-e^{2\pi i\tau}\right)$$

$$= -\log(-2i) - \pi i\tau - \sum_{r=1}^{\infty}\frac{1}{r}e^{2\pi ir\tau}.$$

Differentiating $2k$ times (with $k \geq 1$) yields

$$\frac{d^{2k}}{d^{2k}\tau}\log(\sin\pi\tau) = \sum_{r=1}^{\infty}(2\pi i)^{2k}r^{2k-1}e^{2\pi ir\tau}.$$

Equating this expression for $d^{2k} \log(\sin \pi \tau)/d\tau^{2k}$ with the expression obtained above gives the desired result.

$\square$

We resume the proof of Proposition 7.1. Applying (7.1.1) with $m\tau$ in place of τ, we find

$$G_{2k}(\tau) = 2\zeta(2k) + 2 \sum_{m=1}^{\infty} \sum_{n \in \mathbb{Z}} \frac{1}{(m\tau + n)^{2k}}$$

$$= 2\zeta(2k) + 2 \frac{(2\pi i)^{2k}}{(2k-1)!} \sum_{m=1}^{\infty} \sum_{r=1}^{\infty} r^{2k-1} e^{2\pi i r m \tau}$$

$$= 2\zeta(2k) + 2 \frac{(2\pi i)^{2k}}{(2k-1)!} \sum_{n=1}^{\infty} \sum_{r|n} r^{2k-1} e^{2\pi i n \tau}. \qquad \square$$

As is well known, $\zeta(2k)$ is a rational multiple of π^{2k}. It is frequently convenient to factor a π^{2k} out of the Fourier series for G_{2k}, yielding a series with rational coefficients. We briefly recall the details concerning special values of the Riemann ζ-function at even integers.

Definition. The *Bernoulli numbers* B_k are defined by the power series expansion

$$\frac{x}{e^x - 1} = \sum_{k=0}^{\infty} B_k \frac{x^k}{k!}.$$

For example, one easily computes

$$B_0 = 1, \qquad B_1 = -\frac{1}{2}, \qquad B_2 = \frac{1}{6}, \qquad B_4 = -\frac{1}{30},$$

$$B_6 = \frac{1}{42}, \qquad B_8 = -\frac{1}{30}, \qquad \text{and } B_{2k+1} = 0 \text{ for all } k \geq 1.$$

For a longer table of B_k's and the corresponding values of $\zeta(2k)$, see (A §1).

Proposition 7.2. *For all integers $k \geq 1$,*

$$\zeta(2k) = \sum_{n=1}^{\infty} \frac{1}{n^{2k}} = -\frac{(2\pi i)^{2k}}{2(2k)!} B_{2k}.$$

PROOF. First we use the definition of the B_k's to write

$$\pi x \cot(\pi x) = \pi i x \frac{e^{\pi i x} + e^{-\pi i x}}{e^{\pi i x} - e^{-\pi i x}} = \pi i x \left(1 + \frac{2}{e^{2\pi i x} - 1}\right)$$

$$= \pi i x + \sum_{k=0}^{\infty} B_k \frac{(2\pi i x)^k}{k!} = \sum_{k=0}^{\infty} B_{2k} \frac{(2\pi i x)^{2k}}{(2k)!}.$$

Next we use the product expansion for $\sin(\pi x)$ already considered in the proof of (7.1),

$$\sin(\pi x) = \pi x \prod_{n=1}^{\infty} \left(1 - \frac{x^2}{n^2}\right).$$

Taking the logarithmic derivative yields

$$\pi \cot(\pi x) = \frac{1}{x} + \sum_{n=1}^{\infty} -\frac{2x}{n^2} \cdot \frac{1}{1 - (x^2/n^2)}$$

$$= \frac{1}{x} + \sum_{n=1}^{\infty} \left\{ -\frac{2}{x} \sum_{k=1}^{\infty} \left(\frac{x^2}{n^2}\right)^k \right\}$$

$$= \frac{1}{x} \left\{ 1 - 2 \sum_{k=1}^{\infty} \zeta(2k) x^{2k} \right\}.$$

Comparing the two Laurent series for $\pi x \cot(\pi x)$ gives the desired result.

$\square$

Remark 7.3.1. We can now define a *normalized Eisenstein series* $E_{2k}(\tau)$ as the series

$$E_{2k}(\tau) = 1 - \frac{4k}{B_{2k}} \sum_{n \geq 1} \sigma_{2k-1}(n) q^n.$$

Using (7.1) and (7.2) we see that

$$G_{2k}(\tau) = 2\zeta(2k) E_{2k}(\tau).$$

The fact that the E_{2k}'s have leading coefficient 1 makes them particularly easy to compare. For example, E_4^2 and E_8 are both modular forms of weight 8. Since M_8 has dimension 1 from (3.10.1), we know that they are multiples of one another. But since they are normalized, we see on comparing their constant terms that $E_4^2 = E_8$. Equating Fourier coefficients gives the identity

$$\sigma_7(n) = \sigma_3(n) + 120 \sum_{m=1}^{n-1} \sigma_3(m) \sigma_3(n - m).$$

The reader will be able to construct many more identities of this sort.

Remark 7.3.2. We can also write $g_2(\tau)$ and $g_3(\tau)$ in terms of normalized Eisenstein series:

$$g_2(\tau) = 60 G_4(\tau) = 120 \zeta(4) E_4(\tau) = (2\pi)^4 \frac{1}{2^2 3} E_4(\tau);$$

$$g_3(\tau) = 140 G_6(\tau) = 280 \zeta(6) E_6(\tau) = (2\pi)^6 \frac{1}{2^3 3^3} E_6(\tau).$$

These expressions are useful for computing the Fourier expansion of $\Delta(\tau)$ and $j(\tau)$, as explained in the next proposition.

Proposition 7.4. (a) *The modular discriminant has the Fourier expansion*

$$\Delta(\tau) = (2\pi)^{12} \sum_{n \geq 1} \tau(n)q^n,$$

where $\tau(1) = 1$ and $\tau(n) \in \mathbb{Z}$ for all n. The arithmetic function $n \mapsto \tau(n)$ is called the Ramanujan τ-function.
(b) *The modular j-function has the Fourier expansion*

$$j(\tau) = \frac{1}{q} + \sum_{n \geq 0} c(n)q^n,$$

where $c(n) \in \mathbb{Z}$ for all n.

PROOF. (a) Using (7.3.2) we compute

$$\Delta(\tau) = g_2(\tau)^3 - 27g_3(\tau)^2 = \frac{(2\pi)^{12}}{2^6 3^3}\left(E_4(\tau)^3 - E_6(\tau)^2\right).$$

We must show that every coefficient of $E_4^3 - E_6^2$ is divisible by $2^6 3^3 = 12^3$. From (7.3.1) we have

$$E_4(\tau) = 1 + 240 \sum_{n \geq 1} \sigma_3(n)q^n \quad \text{and} \quad E_6(\tau) = 1 - 504 \sum_{n \geq 1} \sigma_5(n)q^n.$$

To ease notation, let us write

$$E_4(\tau) = 1 + 240A \quad \text{and} \quad E_6(\tau) = 1 - 504B.$$

Then

$$E_4(\tau)^3 - E_6(\tau)^2 = (1 + 240A)^3 - (1 - 504B)^2$$
$$= 12^2(5A + 7B) + 12^3(100A^2 - 147B^2 + 8000A^3).$$

It remains to show that every coefficient of $5A + 7B$ is divisible by 12. We have

$$5A + 7B = \sum_{n \geq 1}\left(5\sigma_3(n) + 7\sigma_5(n)\right)q^n = \sum_{n \geq 1}\sum_{d \mid n}(5d^3 + 7d^5)q^n,$$

and for any integer d,

$$5d^3 + 7d^5 = d^3(5 + 7d^2) \equiv \begin{cases} d^3(1 - d^2) \equiv 0 \,(\mathrm{mod}\ 4), \\ d^3(-1 + d^2) \equiv 0 \,(\mathrm{mod}\ 3). \end{cases}$$

Hence $5d^3 + 7d^5 \equiv 0\,(\mathrm{mod}\ 12)$. This proves that

$$\Delta(\tau) = (2\pi)^{12} \sum_{n \geq 1} \tau(n)q^n$$

for integers $\tau(n)$.

Finally, the coefficient of q is

$$\frac{(2\pi)^{12}}{12^3} \cdot 12^2 \cdot \big(5\sigma_3(1) + 7\sigma_5(1)\big) = (2\pi)^{12},$$

so $\tau(1) = 1$.

(b) We use (a), (7.3.1), (7.3.2), and the definition of $j(\tau)$ to compute

$$j(\tau) = 1728\frac{g_2(\tau)^3}{\Delta(\tau)} = 1728\frac{\dfrac{(2\pi)^{12}}{12^3}E_4(\tau)^3}{(2\pi)^{12}\displaystyle\sum_{n \ge 1}\tau(n)q^n}$$

$$= \frac{\left(1 + 240\displaystyle\sum_{n \ge 1}\sigma_3(n)q^n\right)^3}{q + \displaystyle\sum_{n \ge 2}\tau(n)q^n}.$$

Since the $\sigma_3(n)$'s and the $\tau(n)$'s are all integers, this last expression gives a Laurent series of the form $q^{-1} + \sum c(n)q^n$ with integer coefficients. (Note the reciprocal of a power series with integer coefficient and leading term 1 will again have integer coefficients.) ☐

Remark 7.4.1. Using the formulas developed in the proof of (7.4), it is easy to compute the first few values of $\tau(n)$ and $c(n)$. Thus

$$(2\pi)^{-12}\Delta(\tau) = q - 24q^2 + 252q^3 - 1472q^4 + 4830q^5 + \cdots,$$

$$j(\tau) = q^{-1} + 744 + 196884q + 21493760q^2 + \cdots.$$

For a more extensive list, see (A §2).

Remark 7.4.2. In the next section we will prove that $\Delta(\tau)$ has the product expansion

$$\Delta(\tau) = (2\pi)^{12}q\prod_{n \ge 1}(1 - q^n)^{24}.$$

This gives an alternative (but less elementary) proof of (7.4a)

Remark 7.4.3. The $c(n)$ coefficients of $j(\tau)$ have many interesting arithmetical properties. For example, Lehner [1,2] proved that they satisfy the following divisibility conditions. (See also Apostol [1, Ch. 4].)

$$\begin{array}{lll}
n \equiv 0 \pmod{2^e} & \Longrightarrow & c(n) \equiv 0 \pmod{2^{3e+8}}, \\
n \equiv 0 \pmod{3^e} & \Longrightarrow & c(n) \equiv 0 \pmod{3^{2e+3}}, \\
n \equiv 0 \pmod{5^e} & \Longrightarrow & c(n) \equiv 0 \pmod{5^{e+1}}, \\
n \equiv 0 \pmod{7^e} & \Longrightarrow & c(n) \equiv 0 \pmod{7^e}, \\
n \equiv 0 \pmod{11^e} & \Longrightarrow & c(n) \equiv 0 \pmod{11^e}.
\end{array}$$

Remark 7.4.4. The values in (7.4.1) and (A §2) suggest that the $c(n)$'s grow quite rapidly. This is indeed the case, as is clear from the following asymptotic formula proven by Petersson [1] using the circle method of Hardy, Ramanujan, and Littlewood:

$$c(n) \sim \frac{e^{4\pi\sqrt{n}}}{\sqrt{2}n^{3/4}} \qquad \text{as } n \to \infty.$$

It also turns out that the $c(n)$'s are intimately connected with representations of the largest sporadic groups, in particular with the Fischer-Griess monster group. See Conway [1] and Conway-Norton [1] for an interesting account of this surprising connection.

Remark 7.4.5. Ramanujan's τ-function also has many interesting properties. For example, we will later prove (10.7) that it satisfies the identities

$$\tau(mn) = \tau(m)\tau(n) \qquad \text{if } (m,n) = 1,$$
$$\tau(p^{e+1}) = \tau(p)\tau(p^e) - p^{11}\tau(p^{e-1}) \qquad \text{for } p \text{ prime and } e \geq 1.$$

These identities were conjectured by Ramanujan; the first proof was given by Mordell.

We will also prove (11.2) that the $\tau(n)$'s grow much more slowly than the $c(n)$'s. Precisely, we will show that there is a constant c such that $|\tau(n)| \leq cn^6$ for all $n \geq 1$. Another conjecture of Ramanujan, proven by Deligne as a consequence of his proof of the Riemann hypothesis for varieties over finite fields, says that one can do better.

Theorem 7.5. (Deligne [1,2])

$$|\tau(n)| \leq \sigma_0(n)n^{11/2} \qquad \text{for all } n \geq 1.$$

(Here $\sigma_0(n)$ is the number of divisors of n. For example, if n is prime, then $\sigma_0(n) = 2$.)

In the other direction, there is the following open conjecture of Lehmer.

Conjecture 7.6. (Lehmer [1])

$$\tau(n) \neq 0 \qquad \text{for all } n \geq 1.$$

§8. Jacobi's Product Formula for $\Delta(\tau)$

In this section we will prove Jacobi's beautiful product expansion for the modular discriminant $\Delta(\tau)$.

Theorem 8.1. (Jacobi)

$$\Delta(\tau) = (2\pi)^{12} q \prod_{n \geq 1} (1 - q^n)^{24}.$$

Remark 8.2. We will derive the product (8.1) directly from the definition of $\Delta(\tau)$ and the product representation (6.4) for the Weierstrass σ-function. There are other methods which can be used to prove (8.1). For example, see Serre [3, Ch. 7, Thm. 6] for a proof based on rearrangement of conditionally convergent double series, and Apostol [1, Ch. 3, §2] or Siegel [1] for an exposition of Siegel's clever proof using residue calculations. The heart of both of these proofs lies in first proving that the function

$$F(\tau) = q \prod_{n \geq 1} (1 - q^n)^{24}$$

satisfies

$$F(-1/\tau) = \tau^{12} F(\tau).$$

Since F visibly satisfies

$$F(\tau + 1) = F(\tau) \qquad \text{and} \qquad \lim_{\tau \to i\infty} F(\tau) = 0,$$

and since S and T generate the modular group $\Gamma(1)$, it follows that F is a cusp form of weight 12. Hence $F(\tau)/\Delta(\tau)$ is a holomorphic modular function of weight 0, so it is constant. Finally, letting $\tau \to i\infty$, one easily checks that this constant is $(2\pi)^{-12}$.

PROOF (of Theorem 8.1). By definition,

$$\Delta(\tau) = g_2(\tau)^3 - 27 g_3(\tau)^2$$

is the discriminant of the cubic polynomial

$$4X^3 - g_2(\tau)X - g_3(\tau) = 4(x - e_1)(x - e_2)(x - e_3).$$

But we know the roots of this polynomial from [AEC, VI.3.6], namely

$$e_1 = \wp\left(\frac{1}{2}, \tau\right), \qquad e_2 = \wp\left(\frac{\tau}{2}, \tau\right), \qquad e_3 = \wp\left(\frac{\tau + 1}{2}, \tau\right).$$

Thus

$$\Delta(\tau) = 16(e_1 - e_2)^2(e_1 - e_3)^2(e_2 - e_3)^2.$$

The idea of our proof is to express $\Delta(\tau)$ in terms of special values of the Weierstrass σ-function and then use the product expansion (6.4) for σ.

If we differentiate the equation

$$\wp'^2 = 4(\wp - e_1)(\wp - e_2)(\wp - e_3)$$

and divide by $2\wp'$, we find

$$\wp'' = 2(\wp - e_1)(\wp - e_2) + 2(\wp - e_1)(\wp - e_3) + 2(\wp - e_2)(\wp - e_3).$$

Now if we evaluate successively at $z = \dfrac{1}{2}$, $z = \dfrac{\tau}{2}$, and $z = \dfrac{\tau+1}{2}$, we see in each case that only one of the three terms survives:

$$\wp''\left(\frac{1}{2}, \tau\right) = 2(e_1 - e_2)(e_1 - e_3),$$

$$\wp''\left(\frac{\tau}{2}, \tau\right) = 2(e_2 - e_1)(e_2 - e_3),$$

$$\wp''\left(\frac{\tau+1}{2}, \tau\right) = 2(e_3 - e_1)(e_3 - e_2).$$

Comparing these formulas with the expression for $\Delta(\tau)$, we write $\Delta(\tau)$ in terms of values of $\wp''$,

$$\Delta(\tau) = -2\wp''\left(\frac{1}{2}, \tau\right) \wp''\left(\frac{\tau}{2}, \tau\right) \wp''\left(\frac{\tau+1}{2}, \tau\right).$$

Recall (5.6b) that we have expressed $\wp'$ in terms of the Weierstrass σ-function,

$$\wp'(z, \tau) = -\frac{\sigma(2z, \tau)}{\sigma(z, \tau)^4}.$$

Taking derivatives gives

$$\wp''(z, \tau) = -2\frac{\sigma'(2z, \tau)}{\sigma(z, \tau)^4} + 4\frac{\sigma(2z, \tau)\sigma'(z, \tau)}{\sigma(z, \tau)^5}.$$

If we evaluate $\wp''$ successively at $z = \dfrac{1}{2}, \dfrac{\tau}{2}, \dfrac{\tau+1}{2}$, the second term will vanish, since $\sigma(z, \tau)$ has zeros at points in the lattice $\mathbb{Z}\tau + \mathbb{Z}$. We obtain

$$\wp''\left(\frac{\omega}{2}, \tau\right) = -2\frac{\sigma'(\omega, \tau)}{\sigma\left(\frac{\omega}{2}, \tau\right)^4} \qquad \text{for } \omega = 1, \tau, \tau + 1.$$

Combining this with the above formula expressing Δ in terms of $\wp''$, we find

$$\Delta(\tau) = 16 \frac{\sigma'(1,\tau)\sigma'(\tau,\tau)\sigma'(\tau+1,\tau)}{\left[\sigma\left(\frac{1}{2},\tau\right)\sigma\left(\frac{\tau}{2},\tau\right)\sigma\left(\frac{\tau+1}{2},\tau\right)\right]^4}.$$

In order to compute the values of the derivatives in this expression, we take the transformation formula (5.4c) for $\sigma(z,\tau)$ and differentiate it with respect to z. This gives

$$\sigma'(z+\omega,\tau) = \psi(\omega)\eta(\omega)e^{\eta(\omega)(z+\frac{1}{2}\omega)}\sigma(z) + \psi(\omega)e^{\eta(\omega)(z+\frac{1}{2}\omega)}\sigma'(z)$$

$$\text{for all } \omega \in \mathbb{Z}\tau + \mathbb{Z}.$$

Now put $z = 0$ and use the fact that $\sigma(0) = 0$ and $\sigma'(0) = 1$ to get $\sigma'(\omega) = \psi(\omega)e^{\omega\eta(\omega)/2}$. Taking $\omega = 1$, $\omega = \tau$, and $\omega = \tau + 1$ in succession yields

$$\sigma'(1) = -e^{\eta(1)/2}, \quad \sigma'(\tau) = -e^{\tau\eta(\tau)/2}, \quad \text{and} \quad \sigma'(\tau+1) = -e^{(\eta+1)\eta(\tau+1)/2}.$$

Next we use Legendre's relation (5.2d), which in our situation reads $\tau\eta(1) - \eta(\tau) = 2\pi i$, to eliminate $\eta(\tau)$. After some algebra we obtain

$$\sigma'(1) = -e^{\frac{1}{2}\eta}, \quad \sigma'(\tau) = -e^{\frac{1}{2}\eta\tau^2}q^{-\frac{1}{2}}, \quad \text{and} \quad \sigma'(\tau+1) = -e^{\frac{1}{2}\eta(\tau+1)^2},$$

where to ease notation we write $\eta = \eta(1)$.

The next step is to use the product expansion (6.4) for σ to compute σ at the half periods. Thus

$$\sigma\left(\frac{1}{2},\tau\right)^4 = \frac{1}{(2\pi i)^4}e^{\frac{1}{2}\eta}\cdot 2^4\left(\prod_{n\geq 1}\frac{1+q^n}{1-q^n}\right)^8,$$

$$\sigma\left(\frac{\tau}{2},\tau\right)^4 = \frac{1}{(2\pi i)^4}e^{\frac{1}{2}\eta\tau^2}q^{-1}(1-q^{\frac{1}{2}})\prod_{n\geq 1}\frac{\left(1-q^{n+\frac{1}{2}}\right)^4\left(1-q^{n-\frac{1}{2}}\right)^4}{(1-q^n)^8}$$

$$= \frac{1}{(2\pi i)^4}e^{\frac{1}{2}\eta\tau^2}q^{-1}\left(\prod_{n\geq 1}\frac{1-q^{n-\frac{1}{2}}}{1-q^n}\right)^8,$$

$$\sigma\left(\frac{\tau+1}{2},\tau\right)^4 = \frac{1}{(2\pi i)^4}e^{\frac{1}{2}\eta(\tau+1)^2}q^{-1}(1+q^{\frac{1}{2}})$$

$$\times \prod_{n\geq 1}\frac{\left(1+q^{n+\frac{1}{2}}\right)^4\left(1+q^{n-\frac{1}{2}}\right)^4}{(1-q^n)^8}$$

$$= \frac{1}{(2\pi i)^4}e^{\frac{1}{2}\eta(\tau+1)^2}q^{-1}\left(\prod_{n\geq 1}\frac{1+q^{n-\frac{1}{2}}}{1-q^n}\right)^8.$$

$$= (2\pi)^{12}q \left(\prod_{n \geq 1} \frac{(1-q^n)^3}{(1+q^n)(1-q^{2n-1})} \right)^8$$

$$= (2\pi)^{12}q \left(\prod_{n \geq 1} \frac{(1-q^n)^2(1-q^{2n})}{1+q^n} \right)^8$$

$$\text{since } \left(\prod 1 - q^{2n-1}\right)\left(\prod 1 - q^{2n}\right) = \prod 1 - q^n$$

$$= (2\pi)^{12}q \prod_{n \geq 1} (1-q^n)^{24} \qquad \text{since } 1 - q^{2n} = (1+q^n)(1-q^n).$$

$\square$

In view of the exponent appearing in the product expansion for $\Delta(\tau)$, it is natural to study the function obtained by taking 24th-roots.

Definition. The *Dedekind η-function* $\eta(\tau)$ is defined by the product

$$\eta(\tau) = e^{2\pi i \tau/24} \prod_{n \geq 1} (1-q^n) \qquad \text{for } \tau \in \mathbf{H}, \, q = e^{2\pi i \tau}.$$

Warning. Do not confuse the Dedekind η-function with the quasi-period map (5.2b) $\eta : \Lambda \longrightarrow \mathbb{C}$. This may be especially confusing when $\Lambda = \Lambda_\tau$, since then the symbol $\eta(\tau)$ has two meanings, and it is quite possible for both to appear in a single formula. For example,

$$(2\pi)^{12} \left(\text{Dedekind } \eta(\tau) \right)^{24} = \Delta(\tau) = \text{product of values of } \sigma(\tau),$$

and using the product expansion (6.4) for σ will give a formula involving the quasi-period $\eta(1)$. Why, you may ask, do we continue to use this confusing notation? Tradition!

Proposition 8.3. (a) *The Dedekind η-function satisfies the identities*

$$\eta(\tau + 1) = e^{2\pi i/24}\eta(\tau), \qquad \text{and} \qquad \eta\left(-\frac{1}{\tau}\right) = \sqrt{-i\tau}\, \eta(\tau).$$

Here we take the branch of $\sqrt{}$ which is positive on the positive real axis.
(b)
$$\Delta(\tau) = (2\pi)^{12}\eta(\tau)^{24}.$$

PROOF. Note first that (b) is immediate from the definition of $\eta(\tau)$ and Jacobi's product formula (8.1) for $\Delta(\tau)$. Next, since the transformation $\tau \mapsto \tau + 1$ does not change q, we see from the definition of $\eta(\tau)$ that

$$\eta(\tau + 1) = e^{2\pi i(\tau+1)/24} \prod_{n \geq 1} (1-q^n) = e^{2\pi i/24}\eta(\tau).$$

Finally, we know that $\Delta(\tau)$ is a modular form of weight 12, so

$$\Delta\left(-\frac{1}{\tau}\right) = \tau^{12}\Delta(\tau).$$

Using (b) and taking 24th-roots shows that

$$\eta\left(-\frac{1}{\tau}\right) = \varepsilon\sqrt{-i\tau}\,\eta(\tau)$$

for some 24th-root of unity ε. Now evaluate at $\tau = i$. Since $-1/i = i$, we find that $\varepsilon = 1$. $\square$

Remark 8.4. More generally, let

$$\gamma = \begin{pmatrix} a & b \\ c & d \end{pmatrix} \in \mathrm{SL}_2(\mathbb{Z}) \qquad \text{with } c \geq 0.$$

Taking the 24th-root of (8.3b) and using the known transformation property of $\Delta(\tau)$ shows that

$$\eta(\gamma\tau) = e^{2\pi i \Phi(\gamma)/24}\sqrt{-i(c\tau + d)}\eta(\tau)$$

for some integer $\Phi(\gamma)$ depending on γ. For example, (8.3a) says that $\Phi(S) = 0$ and $\Phi(T) = 1$. Note that although $\Phi(\gamma)$ is only defined modulo 24, we can pin down a particular value for $\Phi(\gamma)$ by fixing a branch of $\log\eta(\tau)$, setting

$$\frac{2\pi i}{24}\Phi(\gamma) = \log\eta(\gamma\tau) - \log\eta(\tau) + \tfrac{1}{2}\log\{-i(c\tau + d)\} \qquad \text{if } c \geq 0$$

and requiring $\Phi(-\gamma) = \Phi(\gamma)$ if $c < 0$.

For many purposes it is important to know precisely how η transforms. The following theorem of Dedekind supplies the answer. First we need one definition.

Definition. Let x and y be relatively prime integers with $y > 0$. The *Dedekind sum* $s(x, y)$ is defined to be

$$s(x, y) = \sum_{j=1}^{y-1} \frac{j}{y}\left(\frac{jx}{y} - \left[\frac{jx}{y}\right] - \frac{1}{2}\right).$$

(The square brackets denote the greatest integer function.)

Theorem 8.5. (Dedekind) *Let $\gamma = \begin{pmatrix} a & b \\ c & d \end{pmatrix} \in \mathrm{SL}_2(\mathbb{Z})$ with $c > 0$. The Dedekind η-function satisfies the transformation formula*

$$\eta(\gamma\tau) = e^{2\pi i \Phi(\gamma)/24} \sqrt{-i(c\tau + d)}\, \eta(\tau),$$

where $\sqrt{}$ is the branch of the square root which is positive on the positive real axis, $\Phi(\gamma)$ is given by the formula

$$\Phi(\gamma) = \frac{1}{c} + \frac{d}{c} - 12 s(d, c),$$

and $s(x, y)$ is the Dedekind sum defined above.

PROOF. Since we will not need this result, we omit the lengthy proof. The interested reader might consult Apostol [1, Thm. 3.4] or Lang [2, Ch. IX]. □

Remark 8.6. Dedekind sums $s(x, y)$ satisfy many interesting relations. Of particular importance is *Dedekind's reciprocity law:* Let $x, y > 0$ be integers with $\gcd(x, y) = 1$. Then

$$12 s(x, y) + 12 s(y, x) = \frac{x}{y} + \frac{y}{x} + \frac{1}{xy} - 3.$$

See Apostol [1, Thm. 3.7] or exercise 1.17. A good source for information about Dedekind sums is Grosswald-Rademacher [1].

§9. Hecke Operators

Let $E/\mathbb{C}$ be an elliptic curve. We have seen amply demonstrated in [AEC] the importance of studying isogenies connecting our given elliptic curve E with other elliptic curves. If $E(\mathbb{C}) \cong \mathbb{C}/\Lambda$ for some lattice $\Lambda \in \mathcal{L}$, then an isogeny $E' \to E$ of degree n corresponds to a sublattice $\Lambda' \subset \Lambda$ of index n by the natural map

$$\mathbb{C}/\Lambda' \longrightarrow \mathbb{C}/\Lambda, \qquad z \longmapsto z.$$

In keeping with our general philosophy in this chapter, rather than focusing on a single isogeny, we instead consider the set of all isogenies to E of degree n. Equivalently, we look at all sublattices of Λ of index n. This is the same as studying degree n maps from E to other elliptic curves, since we can always take the dual isogeny. In our situation, the dual isogeny $\mathbb{C}/\Lambda \to \mathbb{C}/\Lambda'$ is induced by the map $z \mapsto nz$. This leads to the notion of a Hecke operator.

Definition. For any set S, let $\mathrm{Div}(S)$ denote the divisor group of the set S, that is, the free abelian group generated by the elements of S,

$$\mathrm{Div}(S) = \bigoplus_{s \in S} \mathbb{Z} \cdot s.$$

A homomorphism $T : \mathrm{Div}(S) \to \mathrm{Div}(S)$ is called a *correspondence on S*. Notice that a correspondence is determined by linearity once its values are known on the elements of S.

Definition. Let $n \geq 1$ be an integer. The nth *Hecke operator* $T(n)$ is the correspondence on the set of lattices $\mathcal{L}$ whose value at a lattice $\Lambda \in \mathcal{L}$ is

$$T(n)\Lambda = \sum_{\substack{\Lambda' \subset \Lambda \\ [\Lambda : \Lambda'] = n}} (\Lambda').$$

If two lattices are homothetic, then they give the same elliptic curve. This suggests that we should also look at the following homothety operator.

Definition. Let $\lambda \in \mathbb{C}^*$. The *homothety operator* R_λ is the correspondence on $\mathcal{L}$ whose value at a lattice $\Lambda \in \mathcal{L}$ is

$$R_\lambda \Lambda = \lambda \Lambda.$$

Since the $T(n)$'s and the R_λ's are homomorphisms which map the group $\mathrm{Div}(\mathcal{L})$ to itself, they can be composed with one another. The following fundamental calculation describes the algebra that they generate.

Theorem 9.1.

(a) $R_\lambda R_\mu = R_{\lambda\mu}$ for all $\lambda, \mu \in \mathbb{C}^*$.

(b) $R_\lambda T(n) = T(n) R_\lambda$ for all $\lambda \in \mathbb{C}^*$, $n \geq 1$.

(c) $T(mn) = T(m)T(n)$ for all $m, n \geq 1$ with $\gcd(m, n) = 1$.

(d) $T(p^e)T(p) = T(p^{e+1}) + p T(p^{e-1}) R_p$ for p prime, $e \geq 1$.

Proof. (a)
$$R_\lambda R_\mu(\Lambda) = R_\lambda(\mu\Lambda) = \lambda\mu\Lambda = R_{\lambda\mu}(\Lambda).$$

(b) This follows immediately from the definitions and from the fact that Λ' is a sublattice of Λ of index n if and only if $\lambda\Lambda'$ is a sublattice of $\lambda\Lambda$ of index n.

(c) Let $\Lambda'' \overset{mn}{\subset} \Lambda$, where the superscript mn denotes the index. Since m and n are relatively prime, the quotient Λ/Λ'' has a unique decomposition

$$\Lambda/\Lambda'' = \Phi_m \times \Phi_n, \qquad \text{with } |\Phi_m| = m \text{ and } |\Phi_n| = n.$$

It follows that there is a *unique* intermediate lattice Λ' satisfying

$$\Lambda'' \overset{m}{\subset} \Lambda' \overset{n}{\subset} \Lambda,$$

namely

$$\Lambda' = \{x \in \Lambda : mx \in \Lambda''\}.$$

Using this fact, it is now easy to verify (c).

$$T(mn)\Lambda = \sum_{\Lambda'' \overset{mn}{\subset} \Lambda} (\Lambda'') = \sum_{\Lambda' \overset{n}{\subset} \Lambda} \sum_{\Lambda'' \overset{m}{\subset} \Lambda'} (\Lambda'')$$

$$= \sum_{\Lambda' \overset{n}{\subset} \Lambda} T(m)(\Lambda') = T(m)\Big(\sum_{\Lambda' \overset{n}{\subset} \Lambda} (\Lambda') \Big) = T(m)T(n)\Lambda.$$

(d) Let $\Lambda \in \mathcal{L}$. For a given sublattice $\Lambda' \overset{p^{e+1}}{\subset} \Lambda$, let $a(\Lambda')$ and $b(\Lambda')$ be the integers defined by

$$a(\Lambda') = \#\{\Gamma : \Lambda' \subset \Gamma \overset{p}{\subset} \Lambda\} \quad \text{and} \quad b(\Lambda') = \begin{cases} 1 & \text{if } \Lambda' \subset p\Lambda, \\ 0 & \text{if } \Lambda' \not\subset p\Lambda. \end{cases}$$

Then

$$T(p^e)T(p)\Lambda = \sum_{\Gamma \overset{p}{\subset} \Lambda} \sum_{\Lambda' \overset{p^e}{\subset} \Gamma} (\Lambda') = \sum_{\Lambda' \overset{p^{e+1}}{\subset} \Lambda} a(\Lambda')(\Lambda'),$$

$$T(p^{e+1})\Lambda = \sum_{\Lambda' \overset{p^{e+1}}{\subset} \Lambda} (\Lambda'),$$

$$T(p^{e-1})R_p\Lambda = \sum_{\Lambda'' \overset{p^{e-1}}{\subset} p\Lambda} (\Lambda'') = \sum_{\Lambda' \overset{p^{e+1}}{\subset} \Lambda} b(\Lambda')(\Lambda').$$

(Note that $p\Lambda$ has index p^2 in Λ.) The identity (d) we are trying to prove is thus reduced to verifying

$$a(\Lambda') = 1 + pb(\Lambda') \quad \text{for all } \Lambda' \overset{p^{e+1}}{\subset} \Lambda.$$

We consider two cases.

| Case 1. $\Lambda' \subset p\Lambda, \quad b(\Lambda') = 1$ |

Let $\Gamma \overset{p}{\subset} \Lambda$. Then $\Gamma \supset p\Lambda \supset \Lambda'$, so $a(\Lambda')$ is increased by one for each such Γ. Hence

$$a(\Lambda') = \#\{\Gamma : \Gamma \overset{p}{\subset} \Lambda\} = p + 1 = 1 + pb(\Lambda').$$

[Quick proof of the middle equality: The set of $\Gamma \overset{p}{\subset} \Lambda$ corresponds to subgroups of $\Lambda/p\Lambda \cong (\mathbb{Z}/p\mathbb{Z})^2$ of index p (or equivalently of order p). These are the lines in $\mathbb{A}^2(\mathbb{Z}/p\mathbb{Z})$, so there are $\#\mathbb{P}^1(\mathbb{Z}/p\mathbb{Z}) = p+1$ of them. For a proof of a more general result, see (9.3) below.]

Case 2. $\Lambda' \not\subset p\Lambda, \quad b(\Lambda') = 0$

Let Γ satisfy $\Lambda' \subset \Gamma \overset{p}{\subset} \Lambda$. Note that $p\Lambda \subset \Gamma$. We have inclusions

$$0 \subset \frac{\Lambda'}{\Lambda' \cap p\Lambda} \subseteq \frac{\Gamma}{p\Lambda} \overset{p}{\subset} \frac{\Lambda}{p\Lambda}.$$

↑
Not equal,
since $\Lambda' \not\subset p\Lambda$.

↑
Index p,
since $\Gamma \overset{p}{\subset} \Lambda$.

But $\Lambda/p\Lambda$ has order p^2, so we conclude that the middle inclusion must be an equality. Therefore

$$\Gamma = \Lambda' + p\Lambda.$$

Thus for a given $\Lambda' \not\subset p\Lambda$ there is exactly one Γ satisfying $\Lambda' \subset \Gamma \overset{p}{\subset} \Lambda$. Hence

$$a(\Lambda') = 1 = 1 + pb(\Lambda').$$

□

Corollary 9.1.1.   *Every $T(n)$ is a polynomial in the $T(p)$'s and R_p's for primes p. More precisely, the rings*

$$\mathbb{Z}[T(n), R_n : n \in \mathbb{Z}, n \geq 1] \qquad and \qquad \mathbb{Z}[T(p), R_p : p \text{ prime}]$$

are the same. This ring is called the Hecke algebra (of $\Gamma(1)$). (Notice that the Hecke algebra is a subring of the ring of correspondences

$$\mathrm{End}(\mathrm{Div}(\mathcal{L})) = \{ homomorphisms \ \mathrm{Div}(\mathcal{L}) \to \mathrm{Div}(\mathcal{L}) \}. \ )$$

PROOF.   Factor $n = p_1^{e_1} \cdots p_r^{e_r}$. From (9.1a) and (9.1c) we find

$$R_n = \prod_{i=1}^{r} R_{p_i}^{e_i} \qquad and \qquad T(n) = \prod_{i=1}^{r} T(p_i^{e_i}).$$

Finally, (9.1d) and an easy induction on e shows that $T(p^e)$ is a polynomial in $T(p)$ and R_p.

□

Corollary 9.1.2.   *The Hecke algebra $\mathbb{Z}[T(n), R_n : n \in \mathbb{Z}, n \geq 1]$ is commutative. In particular,*

$$T(m)T(n) = T(n)T(m) \qquad for \ all \ m, n \geq 1.$$

(Note that $\text{End}(\text{Div}(\mathcal{L}))$ *is definitely not commutative.)*

PROOF. From (9.1a,b,c), we are reduced to showing that $T(p^e)$ commutes with $T(p^f)$. This follows from (9.1.1), since both $T(p^e)$ and $T(p^f)$ are polynomials in $T(p)$ and R_p, which commute from (9.1b). $\square$

Example 9.2. Using (9.1d), it is easy to illustrate (9.1.1) for small powers $T(p^e)$. For example,

$$T(p^2) = T(p)^2 - pR_p,$$
$$T(p^3) = T(p)^3 - 2pR_pT(p),$$
$$T(p^4) = T(p)^4 - 3pR_pT(p)^2 + p^2R_p^2.$$

For a general recursion, see exercise 1.19.

The Hecke operator $T(n)$ sends a lattice Λ to the sum of its sublattices of index n. We now describe these sublattices more precisely. Let $\Lambda \in \mathcal{L}$, and fix an oriented basis $\mathbb{Z}\omega_1 + \mathbb{Z}\omega_2$ for Λ. For any $\Lambda' \overset{n}{\subset} \Lambda$, we choose an oriented basis ω_1', ω_2' for Λ' and write

$$\omega_1' = a\omega_1 + b\omega_2, \qquad \omega_2' = c\omega_1 + d\omega_2,$$

with integers a, b, c, d. Then one easily checks that

$$n = [\Lambda : \Lambda'] = \det \begin{pmatrix} a & b \\ c & d \end{pmatrix} = ad - bc.$$

Here's a quick geometric proof of this fact. The linear transformation $\alpha = \begin{pmatrix} a & b \\ c & d \end{pmatrix}$ acting on the vector space $\mathbb{R}^2 \cong \mathbb{R}\omega_1 + \mathbb{R}\omega_2 = \mathbb{C}$ sends a fundamental parallelogram D for $\mathbb{C}/\Lambda$ to a fundamental parallelogram for $\mathbb{C}/\Lambda'$. Hence

$$[\Lambda : \Lambda'] = \frac{\text{Area of } \alpha D}{\text{Area of } D} = \det(\alpha).$$

Conversely, if $ad - bc = n$, then

$$\Lambda' = \mathbb{Z}(a\omega_1 + b\omega_2) + \mathbb{Z}(c\omega_1 + d\omega_2)$$

is a sublattice of Λ of index n. We thus obtain a map

$$\{\alpha \in M_2(\mathbb{Z}) : \det(\alpha) = n\} \longrightarrow \{\Lambda' : \Lambda' \overset{n}{\subset} \Lambda\}$$
$$\alpha = \begin{pmatrix} a & b \\ c & d \end{pmatrix} \longmapsto \alpha(\Lambda) = \mathbb{Z}(a\omega_1 + b\omega_2) + \mathbb{Z}(c\omega_1 + d\omega_2).$$

(Here $M_2(\mathbb{Z})$ is the ring of 2×2 matrices with integral coefficients.) Note that $\alpha(\Lambda)$ depends on the choice of basis for Λ, although our notation does not reflect this dependence.

It is possible, of course, for different α's to give the same sublattice. According to (1.2), we have $\alpha(\Lambda) = \alpha'(\Lambda)$ if and only if $\alpha = \gamma\alpha'$ for some $\gamma \in \mathrm{SL}_2(\mathbb{Z})$. Note that the basis for $\alpha(\Lambda)$ will be oriented if the basis for Λ is, since (1.1) gives

$$\mathrm{Im}\left(\frac{a\omega_1 + b\omega_2}{c\omega_1 + d\omega_2}\right) = \frac{(\det\alpha)\,\mathrm{Im}(\omega_1/\omega_2)}{|c(\omega_1/\omega_2) + d|^2},$$

and $\det(\alpha) = n \geq 1$. This proves the first half of the next lemma, which we state after setting some notation.

Notation. Let $n \geq 1$ be an integer. We define

$$\mathcal{D}_n = \left\{ \begin{pmatrix} a & b \\ c & d \end{pmatrix} \in M_2(\mathbb{Z}) : ad - bc = n \right\},$$

$$\mathcal{S}_n = \left\{ \begin{pmatrix} a & b \\ 0 & d \end{pmatrix} \in M_2(\mathbb{Z}) : ad = n,\, a, d > 0,\, 0 \leq b < d \right\}.$$

Note that $\mathcal{S}_n$ is a finite subset of $\mathcal{D}_n$ having order

$$\#\mathcal{S}_n = \sum_{d|n}\sum_{b=0}^{d-1} 1 = \sigma_1(n).$$

Note also that $\mathrm{SL}_2(\mathbb{Z})$ acts on $\mathcal{D}_n$ via multiplication: if $\gamma \in \mathrm{SL}_2(\mathbb{Z})$ and $\alpha \in \mathcal{D}_n$, then $\det(\gamma\alpha) = n$, so $\gamma\alpha \in \mathcal{D}_n$.

Lemma 9.3. *Let $\Lambda \in \mathcal{L}$ be a lattice given with a fixed oriented basis $\Lambda = \mathbb{Z}\omega_1 + \mathbb{Z}\omega_2$.*
(a) There is a one-to-one correspondence

$$\mathrm{SL}_2(\mathbb{Z})\backslash\mathcal{D}_n \overset{1\text{-}1}{\longleftrightarrow} \left\{ \Lambda' : \Lambda' \overset{n}{\subset} \Lambda \right\}$$

$$\alpha = \begin{pmatrix} a & b \\ c & d \end{pmatrix} \longmapsto \alpha(\Lambda) = \mathbb{Z}(a\omega_1 + b\omega_2) + \mathbb{Z}(c\omega_1 + d\omega_2).$$

(b) The natural inclusion $\mathcal{S}_n \subset \mathcal{D}_n$ induces a one-to-one correspondence

$$\mathcal{S}_n \overset{1\text{-}1}{\longleftrightarrow} \mathrm{SL}_2(\mathbb{Z})\backslash\mathcal{D}_n.$$

PROOF. (a) This was proven during the discussion above.
(b) Let $\alpha = \begin{pmatrix} a & b \\ c & d \end{pmatrix} \in \mathcal{D}_n$. We construct a $\gamma \in \mathrm{SL}_2(\mathbb{Z})$ such that $\gamma\alpha \in \mathcal{S}_n$. Suppose first that $c \neq 0$. Write the fraction $-a/c$ in lowest terms, say $-a/c = s/r$. Since r and s are relatively prime, we can find integers p and q so that $ps - qr = 1$. Then

$$\begin{pmatrix} p & q \\ r & s \end{pmatrix}\begin{pmatrix} a & b \\ c & d \end{pmatrix} = \begin{pmatrix} * & * \\ 0 & * \end{pmatrix} \quad \text{and} \quad \begin{pmatrix} p & q \\ r & s \end{pmatrix} \in \mathrm{SL}_2(\mathbb{Z}),$$

so we are reduced to the case that $c = 0$. Replacing α by $-\alpha$ if necessary, we may also assume that $a, d > 0$. Finally, for an appropriate choice of $t \in \mathbb{Z}$, the matrix

$$\begin{pmatrix} 1 & t \\ 0 & 1 \end{pmatrix} \begin{pmatrix} a & b \\ 0 & d \end{pmatrix} = \begin{pmatrix} a & b + td \\ 0 & d \end{pmatrix}$$

satisfies $0 \le b + td < d$, so it is in $\mathcal{S}_n$. This proves that $\mathcal{S}_n$ surjects onto $\mathrm{SL}_2(\mathbb{Z}) \backslash \mathcal{D}_n$.

Suppose now that $\alpha, \alpha' \in \mathcal{S}_n$ have the same image. Thus there is a $\gamma = \begin{pmatrix} p & q \\ r & s \end{pmatrix}$ in $\mathrm{SL}_2(\mathbb{Z})$ such that

$$\begin{pmatrix} a & b \\ 0 & d \end{pmatrix} = \begin{pmatrix} p & q \\ r & s \end{pmatrix} \begin{pmatrix} a' & b' \\ 0 & d' \end{pmatrix} = \begin{pmatrix} a'p & b'p + d'q \\ a'r & b'r + d's \end{pmatrix}.$$

Since $a' \ne 0$, the lower left-hand entry gives $r = 0$. Next, comparing diagonal entries, we find

$$a = a'p, \qquad d = d's, \qquad ps \frac{a'd'}{ad} = 1, \qquad a, d, a', d' > 0.$$

It follows that $p = s = 1$, and so $a = a'$ and $d = d'$. Finally, we have

$$b = b' + d'q \qquad \text{and (by assumption)} \qquad 0 \le b, b' < d' = d.$$

Hence $|d'q| = |b - b'| < d'$, from which we conclude that $q = 0$ and $b = b'$. Therefore $\alpha = \alpha'$. $\square$

Proposition 9.4. Let $\Lambda \in \mathcal{L}$ be a lattice, and let $\Lambda = \mathbb{Z}\omega_1 + \mathbb{Z}\omega_2$ be an oriented basis for Λ. Then the Hecke operator $T(n)$ is given explicitly by the formulas

$$T(n)\Lambda = \sum_{\substack{ad=n, a \ge 1 \\ 0 \le b < d}} \left(\mathbb{Z}(a\omega_1 + b\omega_2) + \mathbb{Z}d\omega_2 \right) = \sum_{\alpha \in \mathcal{S}_n} (\alpha(\Lambda)).$$

(The notation $\alpha(\Lambda)$ is as in (9.3a).)

PROOF. Immediate from (9.3), which says that the sublattices of Λ of index n are precisely the lattices $\alpha(\Lambda)$ with $\alpha \in \mathcal{S}_n$. $\square$

Example 9.4.1. For primes p, (9.4) gives the formula

$$T(p)\Lambda = (\mathbb{Z}p\omega_1 + \mathbb{Z}\omega_2) + \sum_{b=0}^{p-1} (\mathbb{Z}(\omega_1 + b\omega_2) + \mathbb{Z}p\omega_2).$$

§10. Hecke Operators Acting on Modular Forms

In the last section we described Hecke operators $T(n)$ which assign to a lattice $\Lambda \in \mathcal{L}$ a formal sum of lattices

$$T(n)\Lambda = \sum_{\Lambda' \overset{n}{\subset} \Lambda} (\Lambda'),$$

and we also gave homothety operators R_λ defined by $R_\lambda(\Lambda) = \lambda\Lambda$. Letting $F : \mathcal{L} \longrightarrow \mathbb{C}$ be any function on the space of lattices, we define new functions $T(n)F$ and $R_\lambda F$ on $\mathcal{L}$ in the natural way,

$$(T(n)F)(\Lambda) = \sum_{\Lambda' \overset{n}{\subset} \Lambda} F(\Lambda') \qquad \text{and} \qquad (R_\lambda F)(\Lambda) = F(\lambda\Lambda).$$

We would like to define an action of $T(n)$ on the space of modular functions f of weight $2k$. Unfortunately, a modular function f is not a well-defined function on the space of lattices $\mathcal{L}$; it is only a function on the space of lattices with given bases:

$$\Lambda = \mathbb{Z}\omega_1 + \mathbb{Z}\omega_2 \longmapsto f\left(\frac{\omega_1}{\omega_2}\right).$$

However, we can use the fact that f is modular to construct a function on $\mathcal{L}$ having a certain homogeneity property, as described in the following proposition.

Proposition 10.1. *There is a one-to-one correspondence*

$$\left\{ \begin{array}{c} \text{weakly modular functions} \\ f : \mathbf{H} \to \mathbb{C} \text{ of weight } 2k \end{array} \right\} \overset{1\text{-}1}{\longleftrightarrow} \left\{ \begin{array}{c} \text{lattice functions } F : \mathcal{L} \to \mathbb{C} \\ \text{satisfying } F(\lambda\Lambda) = \lambda^{-2k} F(\Lambda) \\ \text{for all } \lambda \in \mathbb{C}^* \end{array} \right\}$$

$$\begin{array}{ccc} f & \longrightarrow & F_f(\mathbb{Z}\omega_1 + \mathbb{Z}\omega_2) = \omega_2^{-2k} f(\omega_1/\omega_2), \\ f_F(\tau) = F(\Lambda_\tau) & \longleftarrow & F. \end{array}$$

PROOF. First we check that $F_f(\Lambda)$ depends only on Λ, and not on the choice of an (oriented) basis for Λ. From (1.2a), any other oriented basis has the form

$$(a\omega_1 + b\omega_2, c\omega_1 + d\omega_2) \qquad \text{for some} \quad \begin{pmatrix} a & b \\ c & d \end{pmatrix} \in \mathrm{SL}_2(\mathbb{Z});$$

so

$$F_f\big((a\omega_1 + b\omega_2)\mathbb{Z} + (c\omega_1 + d\omega_2)\mathbb{Z}\big) = (c\omega_1 + d\omega_2)^{-2k} f\left(\frac{a\omega_1 + b\omega_2}{c\omega_1 + d\omega_2}\right)$$

$$= (c\omega_1 + d\omega_2)^{-2k} \left(c\frac{\omega_1}{\omega_2} + d\right)^{2k} f\left(\frac{\omega_1}{\omega_2}\right)$$

$$= \omega_2^{-2k} f\left(\frac{\omega_1}{\omega_2}\right) = F_f(\mathbb{Z}\omega_1 + \mathbb{Z}\omega_2).$$

Next, it is clear that

$$F_f(\lambda\Lambda) = F_f(\mathbb{Z}\lambda\omega_1 + \mathbb{Z}\lambda\omega_2) = \lambda^{-2k}F_f(\Lambda).$$

Similarly, if $\gamma = \begin{pmatrix} a & b \\ c & d \end{pmatrix} \in \mathrm{SL}_2(\mathbb{Z})$, then

$$\Lambda_{\gamma\tau} = (c\tau+d)^{-1}\big(\mathbb{Z}(a\tau+b) + \mathbb{Z}(c\tau+d)\big) = (c\tau+d)^{-1}\Lambda_\tau,$$

which implies that

$$f_F(\gamma\tau) = F(\Lambda_{\gamma\tau}) = F\big((c\tau+d)^{-1}\Lambda_\tau\big) = (c\tau+d)^{2k}F(\Lambda_\tau) = (c\tau+d)^{2k}f_F(\tau).$$

This shows that the indicated maps are well defined.

Finally, we check that they are inverse to one another, which will prove that they give one-to-one correspondences between the indicated sets.

$$F_{f_F}(\Lambda) = \omega_2^{-2k}f_F\left(\frac{\omega_1}{\omega_2}\right) = \omega_2^{-2k}F\left(\Lambda_{\omega_1/\omega_2}\right)$$

$$= \omega_2^{-2k}F\left(\mathbb{Z}\frac{\omega_1}{\omega_2} + \mathbb{Z}\right) = F(\mathbb{Z}\omega_1 + \mathbb{Z}\omega_2) = F(\Lambda).$$

$$f_{F_f}(\tau) = F_f(\Lambda_\tau) = F_f(\mathbb{Z}\tau + \mathbb{Z}) = f(\tau). \qquad \square$$

Using (10.1), we can define Hecke operators on the space of modular functions of weight $2k$. It turns out to be convenient to multiply by the scalar factor n^{2k-1}, which will prevent the appearance of denominators in (10.3) below.

Definition. The nth *Hecke operator* $T_{2k}(n)$ on the space of (weakly) modular functions of weight $2k$ is defined by the formula

$$(T_{2k}(n)f)(\tau) = n^{2k-1}\sum_{\Lambda' \overset{n}{\subset} \Lambda_\tau} F_f(\Lambda') = n^{2k-1}\sum_{\substack{ad=n,\, a\geq 1 \\ 0\leq b<d}} d^{-2k}f\left(\frac{a\tau+b}{d}\right).$$

Here $F_f(\mathbb{Z}\omega_1 + \mathbb{Z}\omega_2) = \omega_2^{-2k}f(\omega_1/\omega_2)$ is as in (10.1). The equality of the last two expressions is immediate from (9.3), which says that

$$\{\Lambda' : \Lambda' \overset{n}{\subset} \Lambda_\tau\} = \{\mathbb{Z}(a\tau+b) + \mathbb{Z}d : ad = n,\ a \geq 1,\ 0 \leq b < d\}.$$

Theorem 10.2. *Let f be a modular function (respectively modular form, respectively cusp form) of weight $2k$. Then so is $T_{2k}(n)f$.*

PROOF. First, we verify that $T_{2k}(n)f$ has weight $2k$. By definition, $T_{2k}(n)f$ is associated as in (10.1) to the lattice function $n^{2k-1}T(n)F_f$. The scalar factor n^{2k-1} is immaterial, and we have

$$(T(n)F)(\lambda\Lambda) = \sum_{\Lambda' \overset{n}{\subset} \lambda\Lambda} F(\Lambda') = \sum_{\Lambda' \overset{n}{\subset} \Lambda} F(\lambda\Lambda')$$

$$= \lambda^{-2k}\sum_{\Lambda' \overset{n}{\subset} \Lambda} F(\Lambda') = \lambda^{-2k}(T(n)F)(\Lambda).$$

Again invoking (10.1), it follows that $T(n)F_f$ corresponds to a weakly modular function of weight $2k$, and so $T_{2k}(n)f$ is weakly modular of weight $2k$.

Next we observe that if f is meromorphic (respectively holomorphic) on $\mathbf{H}$, then the formula

$$\left(T_{2k}(n)f\right)(\tau) = n^{2k-1} \sum_{\substack{ad=n,\,a\geq 1 \\ 0\leq b<d}} d^{-2k} f\left(\frac{a\tau+b}{d}\right)$$

shows that the same is true of $T(n)f$.

It remains to check the behavior of $T_{2k}(n)f$ at ∞. The next proposition gives an explicit formula for the Fourier coefficients of $T_{2k}(n)f$, from which it follows by inspection (10.3.2) that $T_{2k}(n)f$ is meromorphic (respectively holomorphic, respectively zero) at ∞ if f is. This completes the proof of (10.2).   □

Proposition 10.3.   Let $f(\tau) = \sum c(m)q^m$ be a modular function of weight $2k$. Then the Fourier series for $T_{2k}(n)f$ is

$$\left(T_{2k}(n)f\right)(\tau) = \sum_{m\in\mathbb{Z}} \gamma(m)q^m, \quad \text{where} \quad \gamma(m) = \sum_{a\mid\gcd(m,n)} a^{2k-1}c\left(\frac{mn}{a^2}\right).$$

As a special case of (10.3), we list the values of $\gamma(0)$, $\gamma(1)$, and $\gamma(p)$ for primes p. Notice in particular that $\gamma(1) = c(n)$. Thus in some sense $T_{2k}(n)$ acts as a shifting operator on the Fourier coefficients of f.

Corollary 10.3.1.   With notation as in (10.3),

(a)
$$\gamma(0) = c(0)\sigma_{2k-1}(n) \qquad \text{and} \qquad \gamma(1) = c(n).$$

(b) For primes p,

$$\gamma(p) = \begin{cases} c(pn) + p^{2k-1}c(n/p) & \text{if } p\mid n, \\ c(pn) & \text{if } p\nmid n. \end{cases}$$

Remark 10.3.2. Notice that if $c(m) = 0$ for $m \leq -m_0 \leq 0$, then $\gamma(m) = 0$ for $m \leq -m_0 n$. This is clear because $\gamma(m)$ is a sum of terms of the form $c(mn/a^2)$ with $a\mid\gcd(m,n)$, so $mn/a^2 \leq -m_0$. Thus $T_{2k}(n)f$ will be meromorphic (respectively holomorphic, respectively zero) at ∞ if f is.

PROOF (of Proposition 10.3). We use the formula defining $T_{2k}(n)f$ and compute

$$\left(T_{2k}(n)f\right)(\tau) = n^{2k-1} \sum_{\substack{ad=n,\, a\geq 1 \\ 0\leq b<d}} d^{-2k} f\left(\frac{a\tau+b}{d}\right)$$

$$= n^{2k-1} \sum_{\substack{ad=n,\, a\geq 1 \\ 0\leq b<d}} d^{-2k} \sum_{m\in\mathbb{Z}} c(m)e^{2\pi im(a\tau+b)/d}$$

$$= n^{2k-1} \sum_{m\in\mathbb{Z}} \sum_{\substack{ad=n,\, a\geq 1}} c(m)d^{-2k}e^{2\pi ima\tau/d} \sum_{0\leq b<d} e^{2\pi imb/d}.$$

The innermost sum is

$$\sum_{0\leq b<d} e^{2\pi imb/d} = \begin{cases} d & \text{if } d\mid m, \\ 0 & \text{if } d\nmid m. \end{cases}$$

Replacing m by $md = mn/a$ and using $n/d = a$, we find

$$\left(T_{2k}(n)f\right)(\tau) = \sum_{m\in\mathbb{Z}} \sum_{ad=n,\, a\geq 1} a^{2k-1}e^{2\pi ima\tau}c\left(\frac{mn}{a}\right),$$

and collecting equal powers of $q = e^{2\pi i\tau}$ (let $M = ma$) yields

$$= \sum_{M\in\mathbb{Z}} \sum_{a\mid \gcd(M,n)} a^{2k-1}c\left(\frac{Mn}{a^2}\right) e^{2\pi iM\tau}. \qquad \square$$

Suppose that a modular function f is an *eigenfunction* for the Hecke operator $T_{2k}(n)$. This means that there is a constant $\lambda(n) \in \mathbb{C}$ so that

$$\left(T_{2k}(n)f\right)(\tau) = \lambda(n)f(\tau) \qquad \text{for all } \tau \in \mathbf{H}.$$

Using (10.3) to compare the Fourier coefficients of $\lambda(n)f$ and $T_{2k}(n)f$, it is clear that the eigenvalue $\lambda(n)$ is related in some way to the Fourier coefficients of f.

Of particular importance are those modular forms which are simultaneous eigenfunctions for every $T_{2k}(n)$. Although it may seem unlikely, a priori, that there are any such functions, we will later observe (10.9) that in fact M_{2k}^0 has a basis of such functions. In any case, we can already construct the following examples.

Example 10.4. The modular discriminant $\Delta(\tau)$ is an eigenfunction for every Hecke operator $T_{12}(n)$, $n \geq 1$. To see this, note that (10.2) says that $T_{12}(n)\Delta$ is also a cusp form of weight 12. But from (3.10.2) the space of weight 12 cusp forms M_{12}^0 has dimension 1. It follows that $T_{12}(n)\Delta$ is a constant multiple of Δ.

Similarly, $G_4(\tau)$ and $G_6(\tau)$ are eigenfunctions for $T_4(n)$ and $T_6(n)$ respectively, since the spaces M_4 and M_6 have dimension 1 (3.10.1). In fact, it is not hard to show that $G_{2k}(\tau)$ is an eigenfunction for $T_{2k}(n)$ for all $k \geq 2$ and $n \geq 1$. See exercise 1.25.

We now describe the relationship between the eigenvalues and the Fourier coefficients of simultaneous eigenfunctions.

Theorem 10.5. *Let $f(\tau) = \sum c(m)q^m \neq 0$ be a cusp form of weight $2k$, and suppose that f is an eigenfunction for all Hecke operators $T_{2k}(n)$, say*

$$T_{2k}(n)f = \lambda(n)f.$$

Then

$$c(1) \neq 0, \quad \text{and} \quad c(n) = \lambda(n)c(1) \quad \text{for all } n \geq 1.$$

PROOF. Comparing the leading coefficient in

$$\lambda(n)f = \lambda(n)c(1)q + \cdots \quad \text{and} \quad T_{2k}(n)f = c(n)q + \cdots,$$

(see (10.3.1b)), we find that $c(n) = \lambda(n)c(1)$. This proves the second part of the theorem.

Suppose now that $c(1) = 0$. Then what we have proven implies that $c(n) = \lambda(n)c(1) = 0$ for all $n \geq 1$, so $f = 0$. This contradicts our original assumption that $f \neq 0$. Hence $c(1) \neq 0$. □

Definition. A simultaneous eigenfunction as in (10.5) is called *normalized* if $c(1) = 1$. In view of (10.5), every simultaneous eigenfunction is a constant multiple of a normalized eigenfunction.

In the last section we proved several identities (9.1) for Hecke operators $T(n)$ acting on the space of lattices $\mathcal{L}$. These give us the following identities for the action of Hecke operators on modular functions, which in turn give us relations on the Fourier coefficients of simultaneous eigenfunctions.

Proposition 10.6. *Let f be a (weakly) modular function of weight $2k$.*

(a) $T_{2k}(mn)f = T_{2k}(m)T_{2k}(n)f \quad$ *for all $m, n \in \mathbb{Z}$ with $\gcd(m,n) = 1$.*

(b) $T_{2k}(p^e)T_{2k}(p)f = T_{2k}(p^{e+1})f + p^{2k-1}T_{2k}(p^{e-1})f$

$$\text{for all primes } p \text{ and all } e \geq 1.$$

PROOF. (a) This is immediate from (9.1c).

(b) We apply the identity (9.1d) to the lattice function F_f described in (10.1). Since

$$R_\lambda F_f(\Lambda) = F_f(\lambda\Lambda) = \lambda^{-2k}F_f(\Lambda),$$

we find

$$T(p^e)T(p)F_f = T(p^{e+1})F_f + p^{1-2k}T(p^{e-1})F_f.$$

By definition, $T_{2k}(n)f = n^{2k-1}T(n)F_f$, so multiplying by $p^{(e+1)(2k-1)}$ gives the desired result. □

Corollary 10.6.1. *Let $f(\tau) = \sum c(n)q^n \neq 0$ be a cusp form of weight $2k$ that is a normalized eigenfunction for every Hecke operator $T_{2k}(n)$.*

(a) $c(mn) = c(m)c(n)$ *for all $m, n \in \mathbb{Z}$ with $\gcd(m, n) = 1$.*

(b) $c(p^e)c(p) = c(p^{e+1}) + p^{2k-1}c(p^{e-1})$ *for all primes p and $e \geq 1$.*

PROOF. (a) We combine (10.5) and (10.6a) to find (note that $c(1) = 1$)

$$c(mn)f = \lambda(mn)f = T_{2k}(mn)f$$
$$= T_{2k}(m)T_{2k}(n)f = \lambda(m)\lambda(n)f = c(m)c(n)f.$$

(b) This follows similarly from (10.5) and (10.6b).   □

Example 10.7. Let $\tau(n)$ be Ramanujan's τ-function defined by

$$(2\pi)^{-12}\Delta(\tau) = \sum_{n \geq 1} \tau(n)q^n = q \prod_{n \geq 1}(1 - q^n)^{24}.$$

(See (7.4a) and (8.1).) Then (10.4) says that $\sum \tau(n)q^n$ is a normalized simultaneous eigenfunction, so (10.6.1) gives the relations

$$\tau(mn) = \tau(m)\tau(n) \qquad \text{for all } m, n \in \mathbb{Z} \text{ with } \gcd(m, n) = 1.$$

$$\tau(p^e)\tau(p) = \tau(p^{e+1}) + p^{11}\tau(p^{e-1}) \qquad \text{for all primes } p \text{ and all } e \geq 1.$$

These identities, conjectured by Ramanujan, were first proven by Mordell.

Remark 10.8. It is clear from (10.6.1) that modular forms that are simultaneous eigenfunctions have many interesting arithmetical properties. (We will see some additional ones in the next section.) We have given examples of such functions (e.g., $\Delta(\tau)$), but so far we only know finitely many such examples. The following theorem of Petersson shows that there are many functions to which (10.6.1) applies. We will not give the proof, which requires additional machinery involving subgroups of $\mathrm{SL}_2(\mathbb{Z})$.

Theorem 10.9. (Petersson [2]) *The set*

$$\{f \in M_{2k}^0 : f \text{ is a normalized eigenfunction for all } T_{2k}(n), \, n \geq 1\}$$

is a basis for the space M_{2k}^0 of cusp forms of weight $2k$.

PROOF. See Lang [2, Ch. III §4], Ogg [1], Shimura [1, Ch. 3 §§4,5], or exercise 1.22.   □

§11. L-Series Attached to Modular Forms

Let $f(\tau) = \sum c(n)q^n \neq 0$ be a cusp form of weight $2k$ which is a normalized eigenfunction for all Hecke operators $T_{2k}(n)$. Then the Fourier coefficients of f satisfy the identities given in (10.6.1). We now show that these identities equivalent to an Euler product decomposition for a certain Dirichlet series attached to f.

Definition. For any power series

$$f = \sum_{n \geq 1} c(n)q^n \in \mathbb{C}[\![q]\!],$$

the *L-series attached to f* is the (formal) Dirichlet series

$$L(f, s) = \sum_{n \geq 1} c(n)n^{-s}.$$

Proposition 11.1. *Let $f = \sum\limits_{n \geq 1} c(n)q^n$ be a power series with $c(1) = 1$. Then the coefficients of f satisfy the identities*

(i) $c(mn) = c(m)c(n)$ *for all m, n with $\gcd(m, n) = 1$,*

(ii) $c(p^e)c(p) = c(p^{e+1}) + p^{2k-1}c(p^{e-1})$ *for primes p and $e \geq 1$,*

if and only if the associated L-series $L(f, s)$ has the Euler product expansion

(iii) $$L(f, s) = \prod_p \frac{1}{1 - c(p)p^{-s} + p^{2k-1-2s}}.$$

(Note that this is an equality of formal Dirichlet series. We have said nothing yet about convergence properties.)

PROOF. Suppose first that f satisfies (i) and (ii). The multiplicativity relation (i) implies that we can decompose $L(f, s)$ into a product over primes,

$$L(f, s) = \sum_{n \geq 1} c(n)n^{-s} = \prod_p \sum_{e \geq 0} c(p^e)p^{-es}.$$

If we multiply the inner sum by $1 - c(p)p^{-s} + p^{2k-1-2s}$, we find

$$(1 - c(p)p^{-s} + p^{2k-1-2s})\left(\sum_{e \geq 0} c(p^e)p^{-es}\right)$$

$$= \sum_{e \geq 0} c(p^e)p^{-es} - \sum_{e \geq 0} c(p)c(p^e)p^{-(e+1)s} + \sum_{e \geq 0} c(p^e)p^{2k-1-(e+2)s}$$

$$= \{c(1) + c(p)p^{-s}\} - \{c(p)c(1)p^{-s}\}$$

$$+ \sum_{e \geq 2}\left(c(p^e) - c(p)c(p^{e-1}) + c(p^{e-2})p^{2k-1}\right)p^{-es}$$

$$= 1 \qquad \text{using (ii) and } c(1) = 1.$$

Hence

$$\sum_{e \geq 0} c(p^e) p^{-es} = \frac{1}{1 - c(p)p^{-s} + p^{2k-1-2s}},$$

which proves that $L(f, s)$ has the Euler product expansion (iii).

We leave the converse as an exercise, since it will not be needed in the sequel. See exercise 1.23.   □

In order to prove that the formal Dirichlet series $L(f, s)$ converges in some half-plane $\mathrm{Re}(s) > s_0$, we need an estimate for the size of the Fourier coefficients of f.

Theorem 11.2. (Hecke) Let $f(\tau)$ be a cusp form of weight $2k$ with Fourier expansion $\sum c(n)q^n$. There is a constant κ, depending only on f, such that

$$|c(n)| \leq \kappa n^k \quad \text{for all } n \geq 1.$$

Remark 11.2.1. Let $f(\tau) = \sum c(n)q^n$ be a normalized cusp form of weight $2k$ which is a simultaneous eigenfunction for all Hecke operators $T_{2k}(n)$. Then the Fourier coefficients of f actually satisfy the stronger estimate

$$|c(n)| \leq \sigma_0(n) n^{k - \frac{1}{2}},$$

where $\sigma_0(n)$ is the number of positive divisors of n. This is the generalized Ramanujan conjecture (for $\Gamma(1)$), which was proven by Deligne [1,2] as a consequence of his proof of the Riemann hypothesis for varieties over finite fields.

Remark 11.2.2. If $f(\tau)$ is a modular form of weight $2k$ which is not a cusp form, then the Fourier coefficients of f grow at the faster rate

$$\kappa_1 n^{2k-1} \leq |c(n)| \leq \kappa_2 n^{2k-1}.$$

See exercise 1.24.

PROOF (of Theorem 11.2). For any $y > 0$, we can extract the nth Fourier coefficient of f by integrating

$$c(n) = \int_0^1 e^{-2\pi i n(x+iy)} f(x + iy) \, dx.$$

Hence

$$|c(n)| \leq e^{2\pi n y} \sup_{0 \leq x \leq 1} |f(x + iy)|.$$

Next consider the (non-negative) real-valued function

$$\phi(\tau) = |f(\tau)| (\mathrm{Im}\, \tau)^k.$$

Using (1.1) and the fact that f is modular of weight $2k$, we see that

$$\phi(\gamma\tau) = \phi(\tau) \qquad \text{for all } \gamma \in \Gamma(1).$$

Hence

$$\sup_{\tau\in\mathbf{H}} \phi(\tau) = \sup_{\tau\in\mathcal{F}} \phi(\tau),$$

where $\mathcal{F}$ is the usual fundamental domain (1.5) for $\Gamma(1)\backslash\mathbf{H}$. Further, ϕ is continuous on $\mathcal{F}$, and

$$\lim_{\tau\to i\infty} \phi(\tau) = \lim_{\tau\to i\infty} \left|\sum_{n\geq 1} c(n)e^{2\pi i n\tau}\right|(\operatorname{Im}\tau)^k = 0.$$

Note the importance of knowing that f is a cusp form, since if $c(0) \neq 0$, the limit would not exist. It follows that ϕ is bounded on $\mathcal{F}$, and so it is bounded on all of $\mathbf{H}$.

Let

$$C = \sup_{\tau\in\mathbf{H}} \phi(\tau).$$

Then

$$\left|f(x+iy)\right| = \phi(x+iy)y^{-k} \leq Cy^{-k} \qquad \text{for all } x+iy \in \mathbf{H}.$$

Substituting this estimate for f into the above inequality for $\left|c(n)\right|$ yields

$$\left|c(n)\right| \leq Cy^{-k}e^{2\pi n y}.$$

This inequality is valid for all $y > 0$. In particular, putting $y = 1/n$ gives the desired result. $\qquad\qquad\qquad\square$

Corollary 11.3. *Let f be a cusp form of weight $2k$. Then the associated L-series $L(f,s)$ converges to give a holomorphic function in the half-plane*

$$\operatorname{Re}(s) > k + 1.$$

PROOF. From (11.2) we have

$$\left|c(n)n^{-s}\right| = \left|c(n)\right|\left|n^{-\operatorname{Re}(s)}\right| \leq \kappa n^{k-\operatorname{Re}(s)}.$$

Hence $\sum c(n)n^{-s}$ is absolutely convergent provided $\operatorname{Re}(s) > k + 1$. $\qquad\square$

Our next goal is to show that the L-series $L(f,s)$ attached to a cusp form f has an analytic continuation to all of $\mathbb{C}$ and that it satisfies a functional equation similar to the functional equation satisfied by the Riemann ζ-function.

Theorem 11.4.   (Hecke) *Let $f(\tau)$ be a cusp form of weight $2k$.*
(a) *$L(f, s)$ has an analytic continuation to all of $\mathbb{C}$.*
(b) *Let*

$$R(f, s) = (2\pi)^{-s}\Gamma(s)L(f, s),$$

where $\Gamma(s)$ is the usual Γ-function. Then

$$R(f, 2k - s) = (-1)^k R(f, s) \qquad \text{for all } s \in \mathbb{C}.$$

PROOF. The Γ-function is given by the integral

$$\Gamma(s) = \int_0^\infty t^{s-1}e^{-t}\, dt \qquad \text{for Re}(s) > 0.$$

(For basic facts about Γ, see Ahlfors [1].) Replacing t by $2\pi n t$ in the integral, we obtain the useful formula

$$n^{-s} = (2\pi)^s\Gamma(s)^{-1}\int_0^\infty t^{s-1}e^{-2\pi nt}\, dt.$$

Write $f(\tau) = \sum c(n)q^n$. Multiplying our formula for n^{-s} by $c(n)$ and summing over all $n \geq 1$ gives

$$L(f, s) = \sum_{n \geq 1} c(n)n^{-s} = \sum_{n \geq 1}\left\{c(n)(2\pi)^s\Gamma(s)^{-1}\int_0^\infty t^{s-1}e^{-2\pi nt}\, dt\right\}$$

$$= (2\pi)^s\Gamma(s)^{-1}\int_0^\infty t^{s-1}\sum_{n \geq 1}c(n)e^{-2\pi nt}\, dt$$

$$= (2\pi)^s\Gamma(s)^{-1}\int_0^\infty t^{s-1}f(it)\, dt.$$

Note that since $|c(n)| \leq \kappa n^k$ from (11.2), the quantity

$$\sum_{n \geq 1} c(n)\int_0^\infty t^{s-1}e^{-2\pi nt}\, dt$$

is absolutely convergent for Re$(s) > k + 1$, so it is permissible for us to reverse the order of the sum and the integral.

We split the above integral for $L(f, s)$ into two parts. For large t the integral will converge for all $s \in \mathbb{C}$. For small t we replace t by $1/t$ and use the fact that f satisfies

$$f\left(\frac{i}{t}\right) = f\big(S(it)\big) = (it)^{2k}f(it).$$

Thus

$$(2\pi)^{-s}\Gamma(s)L(f,s) = \int_0^\infty t^{s-1}f(it)\,dt \qquad \text{from above,}$$

$$= \int_0^1 t^{s-1}f(it)\,dt + \int_1^\infty t^{s-1}f(it)\,dt$$

$$= \int_\infty^1 \left(\frac{1}{t}\right)^{s-1} f\left(\frac{i}{t}\right) d\left(\frac{1}{t}\right) + \int_1^\infty t^{s-1}f(it)\,dt$$

$$= \int_1^\infty (-1)^k t^{2k-s-1}f(it)\,dt + \int_1^\infty t^{s-1}f(it)\,dt.$$

This gives us the integral representation

$$L(f,s) = (2\pi)^s\Gamma(s)^{-1}\int_1^\infty \left\{t^{s-1} + (-1)^k t^{2k-s-1}\right\}f(it)\,dt,$$

valid a priori for $\operatorname{Re}(s) > k + 1$.

But $\Gamma(s)^{-1}$ is holomorphic on $\mathbb{C}$, and by inspection the integral is absolutely and uniformly convergent for s in any compact subset of $\mathbb{C}$. (Note that since f is a cusp form, $|f(it)|$ goes to 0 like a multiple of $e^{-2\pi t}$ as $t \to \infty$.) Hence this integral gives the analytic continuation of $L(f,s)$ to $\mathbb{C}$. Finally, we observe that the expression

$$\varepsilon(s,t) = t^{s-1} + (-1)^k t^{2k-s-1} \qquad \text{satisfies} \quad \varepsilon(2k-s,t) = (-1)^k\varepsilon(s,t).$$

It follows immediately that

$$R(f,s) = (2\pi)^{-s}\Gamma(s)L(f,s) = \int_1^\infty \varepsilon(s,t)f(it)\,dt$$

has the same functional equation, $R(f, 2k-s) = (-1)^k R(f,s)$. $\qquad\square$

We record as a corollary the useful integral expression for $L(f,s)$ derived during the course of proving (11.4).

Corollary 11.4.1. *Let $f(\tau)$ be a cusp form of weight $2k$. Then*

$$L(f,s) = (2\pi)^s\Gamma(s)^{-1}\int_1^\infty \left\{t^{s-1} + (-1)^k t^{2k-s-1}\right\}f(it)\,dt.$$

EXERCISES

1.1. Prove that the modular group $\Gamma(1)$ is the free product of its subgroups $\langle S\rangle$ and $\langle ST\rangle$ of orders 2 and 3.

1.2. Let $\tau_0 \in \mathbb{Q}$, and let $\gamma \in \Gamma(1)$ satisfy $\gamma \infty = \tau_0$. Prove that γ sends the set

$$\{\tau \in \mathbf{H} : \mathrm{Im}(\tau) > \kappa\}$$

to the interior of a circle in $\mathbf{H}$ which is tangent to the real axis at τ_0. Prove that the radius of the circle goes to 0 as $\kappa \to \infty$.

1.3. Give an example of a Hausdorff space X and a topological group Γ acting continuously on X such that the quotient space $\Gamma \backslash X$, taken with the quotient topology, is not Hausdorff. (By definition, the action of Γ on X is continuous if the map $\Gamma \times X \to X$, $(\gamma, x) \mapsto \gamma x$, is continuous.)

1.4. For any $a \in \mathbb{C}$, let $g_a(\tau) = (\tau - a)/(\tau - \bar{a})$.
(a) Prove directly that

$$g_i(S\tau) = -g_i(\tau) \qquad \text{and} \qquad g_\rho(ST\tau) = \rho^2 g_\rho(\tau).$$

(As usual, $i = e^{\pi i/2}$ and $\rho = e^{2\pi i/3}$.)
(b) Find the largest disk $U \subset \mathbf{H}$ centered at i such that the map

$$\{1, S\} \backslash U \hookrightarrow \mathbb{C}, \qquad \tau \longmapsto g_i(\tau)^2$$

is injective. Compute its image and its inverse.
(c) Same as (b) for U centered at ρ with

$$\{1, ST, (ST)^2\} \backslash U \hookrightarrow \mathbb{C}, \qquad \tau \longmapsto g_\rho(\tau)^3.$$

1.5. Let $\tau \in \mathbf{H}$ be a point satisfying a quadratic equation

$$\tau^2 - a\tau + b, \qquad a, b \in \mathbb{Z}, \qquad a^2 - 4b < 0.$$

Suppose further that $\mathbb{Z}[\tau]$ is the ring of integers of the quadratic imaginary field $\mathbb{Q}(\tau)$.
(a) Prove that the fractional ideals of $\mathbb{Q}(\tau)$ are in one-to-one correspondence with the lattices L contained in $\mathbb{Q}(\tau)$ which satisfy $\tau L \subseteq L$. (In this context, a lattice is a free $\mathbb{Z}$-module of rank 2.)
(b) Prove that every ideal class is represented by a fractional ideal of the form

$$\mathbb{Z} + \frac{x + \tau}{y} \mathbb{Z}$$

satisfying the following conditions:
(i) $x, y \in \mathbb{Z}$ with $y > 0$,
(ii) $4y^2 - (4b - a^2) \le (2x + a)^2 \le y^2$,
(iii) $y | x^2 + ax + b$.
Conclude that the class number of $\mathbb{Q}(\tau)$ is finite.
(c) Prove that the ideal classes in (b) are distinct provided that we discard all pairs (x, y) satisfying either of the following conditions:
(iv) $2x + a = -y$,
(v) $x^2 + ax + b = y^2$ with $2x + a < 0$.
(d) Use the above algorithm to compute the class number of the following quadratic fields:

$$\mathbb{Q}(\sqrt{-3}), \quad \mathbb{Q}(\sqrt{-5}), \quad \mathbb{Q}(\sqrt{-23}), \quad \mathbb{Q}(\sqrt{-29}), \quad \mathbb{Q}(\sqrt{-47}).$$

1.6.   (a) Prove that the natural reduction map

$$SL_2(\mathbb{Z}) \longrightarrow SL_2(\mathbb{Z}/N\mathbb{Z})$$

is surjective.

(b) Define $\Gamma(N)$ to be the subgroup of $\Gamma(1)$ consisting of matrices congruent to $1 \pmod{N}$, that is,

$$\Gamma(N) = \left\{ \begin{pmatrix} a & b \\ c & d \end{pmatrix} \in \Gamma(1) : \begin{array}{l} a \equiv d \equiv 1 \pmod{N} \\ b \equiv c \equiv 0 \pmod{N} \end{array} \right\}.$$

Prove that

$$[\Gamma(1) : \Gamma(N)] = \begin{cases} 6 & \text{if } N = 2, \\ \frac{1}{2}N^3 \prod_{p|N}(1 - p^{-2}) & \text{if } N \geq 3. \end{cases}$$

(c) Prove that $\Gamma(N)$ is a normal subgroup of $\Gamma(1)$ and that

$$\Gamma(1)/\Gamma(N) \cong SL_2(\mathbb{Z}/N\mathbb{Z})/\{\pm 1\}.$$

1.7.   Define subgroups $\Gamma_0(N)$ and $\Gamma_1(N)$ of $\Gamma(1)$ by

$$\Gamma_0(N) = \left\{ \begin{pmatrix} a & b \\ c & d \end{pmatrix} \in \Gamma(1) : c \equiv 0 \pmod{N} \right\},$$

$$\Gamma_1(N) = \left\{ \begin{pmatrix} a & b \\ c & d \end{pmatrix} \in \Gamma(1) : a \equiv d \equiv 1 \pmod{N}, \ c \equiv 0 \pmod{N} \right\}.$$

(a) Prove that $\Gamma_1(N)$ is a normal subgroup of $\Gamma_0(N)$, and show that

$$\Gamma_0(N)/\Gamma_1(N) \cong (\mathbb{Z}/N\mathbb{Z})^* /\{\pm 1\}.$$

(b) Prove that

$$\Gamma_1(N)/\Gamma(N) \cong \mathbb{Z}/N\mathbb{Z} \quad \text{(taken additively)}.$$

(c) Prove the following two formulas:

$$[\Gamma(1) : \Gamma_0(N)] = N \prod_{p|N}(1 + p^{-1}).$$

$$[\Gamma(1) : \Gamma_1(N)] = \begin{cases} 3 & \text{if } N = 2, \\ \frac{1}{2}N^2 \prod_{p|N}(1 - p^{-2}) & \text{if } N \geq 3. \end{cases}$$

1.8. Let $X/\mathbb{C}$ be a smooth projective curve of genus g. For any divisor $D = \sum n_x(x)$ with real coefficients $n_x \in \mathbb{R}$, let

$$[D] = \sum_{x \in X} [n_x](x) \in \mathrm{Div}(X)$$

be the *integer part of D,* where $[n_x]$ denotes the greatest integer in n_x. Also let

$$\mathcal{L}(D) = \{f \in \mathbb{C}(X)^* : \mathrm{div}(f) \geq -D\} \cup \{0\}.$$

(a) Prove that $\mathcal{L}(D) \cong \mathcal{L}([D])$.

(b) Let $k \geq 1$ be an integer, and let K_X be a canonical divisor on X. Prove that

$$\{\omega \in \Omega_X^k : \mathrm{div}(\omega) \geq -D\} \cong \mathcal{L}(kK_X + [D]).$$

1.9. Let $\phi : \mathbf{H}^* \to X(1)$ be the usual projection, and let

$$D_0 = \tfrac{1}{2}(\phi(i)) + \tfrac{2}{3}(\phi(\rho)) + (\phi(\infty)) \in \mathrm{Div}(X(1)) \otimes \mathbb{Q}.$$

(a) Prove that the map

$$M_{2k} \longrightarrow \{\omega \in \Omega_{X(1)}^k : \mathrm{div}(\omega) \geq -kD_0\}, \qquad f \longmapsto \omega_f$$

is an isomorphism. Here ω_f is the k-form described in (3.7a) having the property $\phi^* \omega_f = f(\tau)\,(d\tau)^k$.

(b) Conclude that $M_{2k} \cong \mathcal{L}(kK_{X(1)} + [kD_0])$. Use the Riemann-Roch theorem [AEC II.5.4] to calculate the dimension of M_{2k}, thereby giving an alternative proof of (3.10c).

1.10. (a) Prove that the set

$$\{G_4^a G_6^b : a, b \in \mathbb{Z}, a, b \geq 0, 2a + 3b = k\}$$

is a basis for M_{2k}.

(b) Conclude that the map

$$\mathbb{C}[X, Y] \longrightarrow \bigoplus_{k=0}^{\infty} M_k, \qquad P(X, Y) \longmapsto P(G_4, G_6),$$

is an isomorphism of graded $\mathbb{C}$-algebras, where we grade $\mathbb{C}[X, Y]$ by assigning weights $\mathrm{wt}(X) = 2$ and $\mathrm{wt}(Y) = 3$. In particular, the functions $G_4(\tau)$ and $G_6(\tau)$ are algebraically independent over $\mathbb{C}$.

1.11. This exercise outlines an elementary proof that the modular j-invariant defines a *bijective* map

$$j : \Gamma(1)\backslash \mathbf{H} \longrightarrow \mathbb{C}.$$

Fix some $j_0 \in \mathbb{C}$, let H be a large real number, and let $\mathcal{F}(H) \subset \mathbf{H}$ be the region bounded by the curves

$$|\tau| = 1, \quad \mathrm{Re}(\tau) = \tfrac{1}{2}, \quad \mathrm{Re}(\tau) = -\tfrac{1}{2}, \quad \mathrm{Im}(\tau) = H.$$

Let $\partial\mathcal{F}(H)$ be the boundary of $\mathcal{F}(H)$, which we take with a counter-clockwise orientation. Assume for now that $j(\tau) \neq j_0$ for all $\tau \in \partial\mathcal{F}(H)$.
(a) Prove that

$$\#\{\tau \in \mathcal{F}(H) : j(\tau) = j_0\} = \frac{1}{2\pi i} \int_{\partial\mathcal{F}(H)} \frac{j'(\tau)}{j(\tau) - j_0} \, d\tau.$$

(b) Prove that

$$\lim_{H \to \infty} \frac{1}{2\pi i} \int_{\partial\mathcal{F}(H)} \frac{j'(\tau)}{j(\tau) - j_0} \, d\tau = 1.$$

(*Hint.* Use $j(\tau) = j(\tau+1) = j(-1/\tau)$ to cancel out most of the line integral, and use $j(\tau) = q^{-1} +$ (power series in q) to evaluate the remaining piece.)
(c) Conclude that $j(\mathcal{F}) = \mathbb{C}$ and that j is injective on the interior of $\mathcal{F}$.
(d) If $j(\tau_0) = j_0$ for some $\tau \in \partial\mathcal{F}$, use a slightly modified region to show that j is still injective. Conclude that j maps the quotient $\Gamma(1)\backslash\mathbf{H}$ bijectively to $\mathbb{C}$.
(e) Use the bijectivity from (d) to prove the Uniformization Theorem (4.3).

1.12. Let $\Lambda, \Lambda' \subset \mathbb{C}$ be lattices satisfying

$$G_4(\Lambda) = G_4(\Lambda') \quad \text{and} \quad G_6(\Lambda) = G_6(\Lambda').$$

Prove that $\Lambda = \Lambda'$.

1.13. *Let $\Lambda = \mathbb{Z}\omega_1 + \mathbb{Z}\omega_2$ be a lattice given with an oriented basis, and let $\eta : \Lambda \to \mathbb{C}$ be the associated quasi-period map. Prove that

$$\frac{\eta(\omega_1)}{\omega_1} = \sum_{\substack{n \in \mathbb{Z} \\ n \neq 0}} \sum_{\substack{m \in \mathbb{Z} \\ m \neq 0}} \frac{1}{(m\omega_1 + n\omega_2)^2}, \qquad \frac{\eta(\omega_2)}{\omega_2} = \sum_{\substack{m \in \mathbb{Z} \\ m \neq 0}} \sum_{\substack{n \in \mathbb{Z} \\ n \neq 0}} \frac{1}{(m\omega_1 + n\omega_2)^2}.$$

N.B. These double series are not absolutely convergent; the order of summation really does matter.

1.14. (a) Prove that

$$\prod_{\substack{u,v \in \frac{1}{N}\Lambda/\Lambda \\ u \neq v}} (\wp(z+u;\Lambda) - \wp(z+v;\Lambda))$$

$$= \pm N^{N^2} \wp'(Nz;\Lambda)^{N^2-1} \Delta(\Lambda)^{\frac{(2N^2-3)(N^2-1)}{12}}.$$

(b) Prove that

$$\prod_{\substack{u,v \in \frac{1}{N}\Lambda/\Lambda \\ u \not\equiv \pm v \pmod{\Lambda} \\ u,v \not\equiv 0 \pmod{\Lambda}}} (\wp(u;\Lambda) - \wp(v;\Lambda)) = \pm N^{-2(N^2-3)} \Delta(\Lambda)^{\frac{(N^2-1)(N^2-3)}{6}}.$$

1.15. Let $E/\mathbb{C}$ be the elliptic curve associated to the oriented lattice $\Lambda = \mathbb{Z}\omega_1 + \mathbb{Z}\omega_2$. Recall the Weil pairing

$$e_m : E[m] \times E[m] \longrightarrow \mu_m$$

defined in [AEC III §8]. Prove that on

$$E[m] = m^{-1}\Lambda/\Lambda \subset \mathbb{C}/\Lambda,$$

the Weil pairing is given by the formula

$$e_m\left(\frac{a\omega_1 + b\omega_2}{m}, \frac{c\omega_1 + d\omega_2}{m}\right) = e^{2\pi i(ad-bc)/m}.$$

(*Hint.* Use (5.5) to write the elliptic functions appearing in the definition of e_m as products of σ functions.)

1.16. Let $s(x, y)$ be the Dedekind sum defined in §8.
(a) Prove that

$$s(1, y) = \frac{(y-1)(y-2)}{12y}.$$

(b) Derive a similar formula for $s(2, y)$. (The answer will depend on the parity of y.)
(c) Prove that $s(y^2 + 1, y) = 0$ for all integers $y > 0$.

1.17. Let $\gamma = \begin{pmatrix} a & b \\ c & d \end{pmatrix} \in \mathrm{SL}_2(\mathbb{Z})$ with $c > 0$, and let Φ be as in (8.4) and (8.5).
(a) Prove that $\Phi(\gamma T) = \Phi(\gamma) + 1$.
(b) * Prove that $\Phi(\gamma S) = \Phi(\gamma) - 3$ provided that $d > 0$. (*Hint.* Use the definition of Φ (8.5) and (8.3a) to show that

$$2\big(\Phi(\gamma S) - \Phi(\gamma)\big)\frac{2\pi i}{24} = \log(cS\tau + d) + \log(\tau) - \log(d\tau - c) - \frac{\pi i}{2}.$$

Now evaluate at $\tau = i$.)
(c) Use (b) and (8.5) to deduce Dedekind's reciprocity law (8.6),

$$12s(x, y) + 12s(y, x) = \frac{x}{y} + \frac{y}{x} + \frac{1}{xy} - 3.$$

1.18. Let

$$P_0 = \prod_{n \geq 1}(1 - q^n), \qquad P_1 = \prod_{n \geq 1}(1 - q^{n-\frac{1}{2}}),$$

$$P_2 = \prod_{n \geq 1}(1 + q^n), \qquad P_3 = \prod_{n \geq 1}(1 + q^{n-\frac{1}{2}}).$$

(a) Prove that

$$\left\{\wp\left(\frac{1}{2}\right) - \wp\left(\frac{\tau}{2}\right)\right\}^{\frac{1}{4}} = P_0 P_3^2,$$

$$\left\{\wp\left(\frac{1}{2}\right) - \wp\left(\frac{\tau+1}{2}\right)\right\}^{\frac{1}{4}} = P_0 P_1^2,$$

$$\left\{\wp\left(\frac{\tau+1}{2}\right) - \wp\left(\frac{\tau}{2}\right)\right\}^{\frac{1}{4}} = 2q^{\frac{1}{8}} P_0 P_2^2.$$

(*Hint.* Use (5.6a) and the product expansion (6.4) of σ.)
(b) Prove that
$$P_1 P_2 P_3 = 1.$$

(*Hint.* It's easier to show that $P_0 P_1 P_2 P_3 = P_0$.)
(c) Use (a) and (b) to prove Jacobi's formula

$$\Delta(\tau) = (2\pi)^{12} q \prod_{n \geq 1} (1 - q^n)^{24}.$$

1.19. Verify the following identities for the Hecke and homothety operators acting as correspondences on the space of lattices $\mathcal{L}$. (Note that there are similar identities for the operators $T_{2k}(n)$ acting on the space of modular forms of weight $2k$ which will differ from these identities by various scalar factors.)

(a) $$T(p)^e = \sum_{0 \leq r \leq e/2} \left[\binom{e}{r} - \binom{e}{r-1}\right] p^r R_p^r T(p^{e-2r}).$$

(b) $$T(p^r)T(p^s) = \sum_{i=0}^{r} p^i R_p^i T(p^{r+s-2i}) \qquad \text{for } 0 \leq r \leq s.$$

(c) $$T(m)T(n) = \sum_{d \mid \gcd(m,n)} d R_d T\left(\frac{mn}{d^2}\right).$$

1.20. (a) Let $f(\tau)$ be a modular function of weight $2k$. Prove that

$$g = (2k+1)\left(\frac{df}{d\tau}\right)^2 - 2k \cdot f \cdot \frac{d^2 f}{d\tau^2}$$

is a modular function of weight $4k+4$.
(b) If f is a modular form, prove that g is a cusp form.
(c) If f is the Eisenstein series $G_4(\tau)$, prove that

$$g = \frac{1}{2^4 3^3 5^2 \pi^2} \Delta(\tau).$$

Similarly, if $f = G_6(\tau)$, prove that $g = cG_4(\tau)\Delta(\tau)$, and find the value of the constant c.

1.21. For any matrix $\alpha = \begin{pmatrix} a & b \\ c & d \end{pmatrix}$ with real coefficients and $\det(\alpha) > 0$, define

$$\mu(\alpha, \tau) = c\tau + d.$$

For any function $f : \mathbf{H} \to \mathbb{C}$, define a new function $f|[\alpha]_{2k}$ by

$$\left(f|[\alpha]_{2k}\right)(\tau) = (\det \alpha)^k \mu(\alpha, \tau)^{-2k} f(\alpha \tau).$$

(a) Prove that f is weakly modular of weight $2k$ if and only if

$$f|[\gamma]_{2k} = f \qquad \text{for all } \gamma \in \mathrm{SL}_2(\mathbb{Z}).$$

(b) Prove that
$$\omega_f \circ \alpha = \omega_{f|[\alpha]_{2k}},$$
where ω_f is the differential form described in (3.7).

(c) Verify the identities

$$\mu(\alpha\beta, \tau) = \mu(\alpha, \beta\tau)\mu(\beta, \tau) \quad \text{and} \quad \left(f|[\alpha]_{2k}\right)|[\beta]_{2k} = f|[\alpha\beta]_{2k}.$$

(d) Prove that the action of $T_{2k}(n)$ on a weight $2k$ modular function f is given by the formula

$$T_{2k}(n)f = n^{2k-1} \sum_{\alpha \in \mathrm{SL}_2(\mathbb{Z}) \backslash \mathcal{D}_n} f|[\alpha]_{2k}.$$

(See §9 for the definition of $\mathcal{D}_n$.)

1.22. Let $f, g \in M_{2k}^0$. The *Petersson inner product of f and g* is defined by the integral

$$\langle f, g \rangle = \int_{\mathcal{F}} f(\tau)\overline{g(\tau)}(\mathrm{Im}\,\tau)^k \frac{d\tau \wedge \overline{d\tau}}{-2i(\mathrm{Im}\,\tau)^2}.$$

(Here $\mathcal{F}$ is the usual fundamental domain for $\Gamma(1)\backslash\mathbf{H}$. See (1.5).)

(a) Prove that the integral converges. (Note that f and g are assumed to be cusp forms.)

(b) Prove that $\langle\ ,\ \rangle$ is a positive definite Hermitian inner product on the complex vector space M_{2k}^0.

(c) Let $\omega(f, g)$ be the integrand

$$\omega(f, g) = f(\tau)\overline{g(\tau)}(\mathrm{Im}\,\tau)^k \frac{d\tau \wedge \overline{d\tau}}{-2i(\mathrm{Im}\,\tau)^2}.$$

Prove that for any matrix α with real coefficients and $\det(\alpha) > 0$, and any functions f, g on $\mathbf{H}$,

$$\omega(f, g) \circ \alpha = \omega\left(f|[\alpha]_{2k}, g|[\alpha]_{2k}\right).$$

In particular, if $f, g \in M_{2k}$ and $\gamma \in \mathrm{SL}_2(\mathbb{Z})$, then

$$\omega(f, g) \circ \gamma = \omega(f, g).$$

(See exercise 1.21 for the notation $f\|[\alpha]_{2k}$.)

(d) * Prove that $T_{2k}(n)$ is self-adjoint with respect to the Petersson inner product:

$$\langle T_{2k}(n)f, g \rangle = \langle f, T_{2k}(n)g \rangle \qquad \text{for all } f, g \in M_{2k}^0, \, n \geq 1.$$

(e) If $f, g \in M_{2k}^0$ are normalized eigenfunctions for every $T_{2k}(n)$, prove that either

$$\langle f, g \rangle = 0 \qquad \text{or} \qquad f = g.$$

(f) Prove that

$$\{ f \in M_{2k}^0 : f \text{ is a normalized eigenfunction for all } T_{2k}(n), \, n \geq 1 \}$$

is a basis for M_{2k}^0.

1.23. Prove that if $L(f, s)$ has an Euler product expansion as in (iii) of (11.1), then the coefficients of f satisfy the identities (i) and (ii) of (11.1).

1.24. (a) Let $G_{2k}(\tau) = \sum c(n)q^n$ be the Fourier expansion of the Eisenstein series G_{2k}. Prove that there are constants $\kappa_1, \kappa_2 > 0$, depending only on k, such that

$$\kappa_1 n^{2k-1} \leq |c(n)| \leq \kappa_2 n^{2k-1} \qquad \text{for all } n \geq 1.$$

(b) Let $f(\tau) = \sum c(n)q^n$ be a modular form of weight $2k$ which is not a cusp form (i.e. $c(0) \neq 0$). Prove that there are constants $\kappa_1, \kappa_2 > 0$, depending on f, such that

$$\kappa_1 n^{2k-1} \leq |c(n)| \leq \kappa_2 n^{2k-1} \qquad \text{for all } n \geq 1.$$

1.25. (a) Prove that the normalized Eisenstein series E_{2k} is a normalized eigenfunction for every Hecke operator $T_{2k}(n)$. See (7.3.1) for the definition of E_{2k}.

(b) Let $f \in M_{2k}$ be a modular form of weight $2k \geq 4$ which is not a cusp form, and suppose that f is a normalized eigenfunction for every Hecke operator $T_{2k}(n)$. Prove that $f = E_{2k}$.

1.26. Let $f \in M_{2k}$ be a modular form of weight $2k \geq 4$ which is not a cusp form, say f has the Fourier expansion $f = c(0) + c(1)q + \cdots$ with $c(0) \neq 0$. Let $L(f, s)$ be the L-series attached to f as described in §11.

(a) Prove that $L(f, s)$ can be analytically continued to $\mathbb{C} \smallsetminus \{2k\}$ and that it has a simple pole at $s = 2k$ with residue

$$\mathrm{res}_{s=2k} L(f, s) = \frac{(-1)^k c(0)(2\pi)^{2k}}{\Gamma(2k)}.$$

(b) Let $R(f, s) = (2\pi)^{-s}\Gamma(s)L(f, s)$. Prove that $L(f, s)$ satisfies the functional equation $R(f, 2k - s) = (-1)^k R(f, s)$.

1.27. Let $f(\tau)$ be a cusp form of weight $2k$ with k an even integer.

(a) Prove that

$$L(f,k) = \frac{2}{(k-1)!} \sum_{n\geq 1} \frac{c(n)}{n^k} \Gamma(k, 2\pi n),$$

where $\Gamma(s, x)$ is the incomplete Γ-function

$$\Gamma(s, x) = \int_x^\infty t^{s-1} e^{-t}\, dt \qquad \text{for } s \in \mathbb{C},\ x > 0.$$

(b) Prove that

$$L(f,k) = 2(2\pi)^k \sum_{n\geq 1} c(n) e^{-2\pi n} \sum_{m=1}^k \frac{1}{(2\pi n)^m (k-m)!}.$$

(Note that this series converges quite rapidly, since from (11.2), $|c(n)|$ grows no faster than n^k.)

1.28. Let $f(\tau) = \sum c(n) q^n$ be a cusp form of weight $2k$, let p be a prime, and let

$$\chi : (\mathbb{Z}/p\mathbb{Z})^* \longrightarrow \mathbb{C}^*$$

be a primitive Dirichlet character and extend χ to $\mathbb{Z}$ be setting $\chi(p) = 0$. The Gauss sum $g(\chi)$ associated to χ is given by the formula

$$g(\chi) = \sum_{b=0}^{p-1} \chi(b) e^{2\pi i b/p}.$$

We define the *the twist of f by χ* to be the function

$$f(\chi, \tau) = \sum_{n=1}^\infty c(n) \chi(n) q^n.$$

(As usual, we set $\chi(n) = 0$ if $\gcd(p, n) > 1$.) We will denote the associated twisted L-series by

$$L(f, \chi, s) = L(f(\chi, \cdot), s) = \sum_{n\geq 1} c(n) \chi(n) n^{-s}.$$

(a) Prove that

$$\chi(n) = \frac{1}{p} g(\chi) \sum_{a=0}^{p-1} \overline{\chi(-a)}\, e^{2\pi i a n/p}.$$

(b) Let $R(f, \chi, s)$ be the function

$$R(f, \chi, s) = \left(\frac{p}{2\pi} \right)^s \frac{\Gamma(s)}{g(\chi)} L(f, \chi, s).$$

Prove that R has the integral representation

$$R(f, \chi, s) = \int_{1/p}^\infty \left\{ (pt)^s \frac{f(\chi, it)}{g(\chi)} + (-1)^k (pt)^{2k-s} \frac{f(\bar\chi, it)}{g(\bar\chi)} \right\} \frac{dt}{t}.$$

(c) Prove that $L(f, \chi, s)$ has an analytic continuation to all of $\mathbb{C}$ and that it satisfies the functional equation

$$R(f, \chi, s) = (-1)^k \chi(-1) R(f, \bar\chi, 2k - s).$$

1.29. Let $a_1, a_2, \ldots$ be a sequence of complex numbers, and suppose that there is a constant $c > 0$ such that $|a_n| \leq n^c$ for all n. Let $\lambda > 0$ be a constant and $k > 0$ an integer, and define functions

$$\phi(s) = \sum_{n \geq 1} a_n n^{-s}, \quad \Phi(s) = \left(\frac{2\pi}{\lambda}\right)^{-s} \Gamma(s)\phi(s), \quad f(\tau) = \sum_{n \geq 1} a_n e^{2\pi i n \tau / \lambda}.$$

(a) Prove that $\phi(s)$ is absolutely convergent provided $\mathrm{Re}(s)$ is sufficiently large.

(b) Prove that $f(\tau)$ is holomorphic on $\mathbf{H}$.

(c) *Prove that the following two facts are equivalent:

(I) $\Phi(s)$ has an analytic continuation to all of $\mathbb{C}$, is bounded on every vertical strip, and satisfies the functional equation

$$\Phi(k - s) = \pm\Phi(s).$$

(A vertical strip is a region of the form $c_1 \leq \mathrm{Re}(s) \leq c_2$.)

(II) $f(\tau)$ satisfies the functional equation

$$f\left(-\frac{1}{\tau}\right) = \pm\left(\frac{\tau}{i}\right)^k f(\tau).$$

CHAPTER II

Complex Multiplication

Most elliptic curves over $\mathbb{C}$ have only the multiplication-by-m endomorphisms. An elliptic curve that possesses extra endomorphisms is said to have *complex multiplication*, or CM for short. Such curves have many special properties. For example, the endomorphism ring of a CM curve E is an order in a quadratic imaginary field K, and the j-invariant and torsion points of E generate abelian extensions of K. This is analogous to the way in which the torsion points of $\mathbb{G}_m(\mathbb{C}) = \mathbb{C}^*$ generate abelian extensions of $\mathbb{Q}$. An important result in the cyclotomic theory is the Kronecker-Weber Theorem, which says that every abelian extension of $\mathbb{Q}$ is contained in a cyclotomic extension. We will prove corresponding results for a quadratic imaginary field K. For example, we will show how to construct an elliptic curve E such that $K\bigl(j(E)\bigr)$ is the Hilbert class field of K, and we will explain how to use the torsion points of E to generate the maximal abelian extension of K.

We have generally not tried to assign credit for the results described in this chapter but will content ourselves with mentioning Kronecker, Weber, Fricke, Hasse, Deuring, and Shimura, who are largely responsible for that part of the theory of complex multiplication that we will cover. In particular, the algebraic proofs in §§4 and 5 are essentially due to Deuring, and the idelic description of complex multiplication in §§8 and 9 is mainly due to Shimura.

The material included in this chapter barely scratches the surface of the theory of complex multiplication; a complete treatment of even the basics would fill (at least) an entire volume. The reader desiring further information might profitably consult the following sources, as well as the references they contain. We must especially acknowledge Lang [1], Serre [6], and Shimura [1], whose expositions strongly influenced our organization of this chapter.

Borel et al. [1]:   A development of the basic theory of CM using an analytic approach, together with some useful computational methods.

Cassou-Noguès-Taylor [1]:   The basic theory of CM is developed in the first few chapters, followed by the use of CM to generate rings of integers.

Coates [1]: The basic theory of CM is described, followed by an introduction to the Iwasawa theory of CM elliptic curves.

Lang [1]: Part II develops the theory of CM much as we do, with additional material on the arithmetic properties of special values of elliptic and modular functions.

Perrin-Riou [1]: Iwasawa theory for CM elliptic curves.

Serre [6]: A very brief, but beautifully written, summary of the main theorems of CM for elliptic curves.

Shimura [1]: The idelic formulation of CM for elliptic curves is covered in Chapter 5 and is extended to abelian varieties in §§5.5 and 7.8. For a more complete treatment of the theory of complex multiplication on abelian varieties, see Shimura-Taniyama [1].

Vlǎduţ [1]: A nice historical account of Kronecker's Jugendtraum, the theory of complex multiplication, and the relationship with the theory of modular forms.

The main prerequisite for this chapter is some familiarity with the basic theorems of class field theory. We have provided in §3 a resumé (without proof) of the results we will need. We also assume that the reader is familiar with basic properties of elliptic curves over the complex numbers.

§1. Complex Multiplication over $\mathbb{C}$

In this section we are going to discuss elliptic curves with complex multiplication from the viewpoint of complex analysis. Although interesting in its own right, this should be viewed mainly as the preparation needed to study arithmetic questions.

Let $E/\mathbb{C}$ be an elliptic curve with complex multiplication. We know from [AEC VI.5.5] that $\operatorname{End}(E) \otimes \mathbb{Q}$ is isomorphic to a quadratic imaginary field and that $\operatorname{End}(E)$ is an order in that field. If $\operatorname{End}(E) \cong R \subset \mathbb{C}$ and $K = R \otimes \mathbb{Q}$, then we will say that that "E has complex multiplication by R" or that "E has complex multiplication by K." We also let

$$R_K = \text{ring of integers (maximal order) of } K.$$

Much of the theory becomes easier if one restricts attention to elliptic curves with complex multiplication by R_K, so we will usually take this course. For the general theory, see Lang [1] or Shimura [1, Ch. 5].

The uniformization theorem for elliptic curves [AEC VI.5.1] says that for every elliptic curve $E/\mathbb{C}$ there is a lattice $\Lambda \subset \mathbb{C}$ and an isomorphism

$$
\begin{array}{rcl}
f: & \mathbb{C}/\Lambda & \xrightarrow{\sim} & E(\mathbb{C}) \\
& z & \longmapsto & \left(\wp(z,\Lambda), \wp'(z,\Lambda)\right).
\end{array}
$$

We will denote the elliptic curve corresponding to a lattice Λ by E_Λ; it is given by the usual Weierstrass equation

$$E_\Lambda : y^2 = 4x^3 - g_2(\Lambda)x - g_3(\Lambda).$$

If E has complex multiplication, then there are two ways to embed the order $\mathrm{End}(E)$ into $\mathbb{C}$. It is important to pin down one of these embeddings. This is done by the following proposition, which also provides an important tool for studying arithmetic properties of various analytically defined maps. The reader might compare Proposition 1.1 with [AEC III.5.3], which gives the case that $\alpha \in \mathbb{Z}$. We will use Proposition 1.1 to make deductions in a manner similar to the way we used [AEC III.5.3] to deduce [AEC III.5.4] and [AEC III.5.5].

Proposition 1.1.   *Let $E/\mathbb{C}$ be an elliptic curve with complex multiplication by the ring $R \subset \mathbb{C}$. There is a unique isomorphism*

$$[\,\cdot\,] : R \xrightarrow{\sim} \mathrm{End}(E)$$

such that for any invariant differential $\omega \in \Omega_E$ on E (see [AEC III §5]),

$$[\alpha]^*\omega = \alpha\omega \qquad \text{for all } \alpha \in R.$$

We say in this case that the pair $\big(E, [\,\cdot\,]\big)$ is normalized.

PROOF. Choosing a lattice Λ and an isomorphism $E \cong E_\Lambda$, it suffices to prove the proposition for E_Λ. (Note that [AEC III §1, Table 1.2] says an isomorphism has the effect of multiplying an invariant differential by a constant.)

Next we recall [AEC VI.5.3] that the endomorphism ring of E_Λ is isomorphic to

$$\{\alpha \in \mathbb{C} : \alpha\Lambda \subset \Lambda\} = R \subset \mathbb{C}.$$

More precisely, each $\alpha \in R$ gives an endomorphism $[\alpha] : E_\Lambda \to E_\Lambda$ determined by the commutativity of the following diagram:

$$
\begin{array}{ccc}
\mathbb{C}/\Lambda & \xrightarrow[\;z \mapsto \alpha z\;]{\;\phi_\alpha\;} & \mathbb{C}/\Lambda \\[2pt]
\Big\downarrow{\scriptstyle f} & & \Big\downarrow{\scriptstyle f} \\[2pt]
E_\Lambda & \xrightarrow{\;[\alpha]\;} & E_\Lambda
\end{array}
$$

We claim that this map $[\,\cdot\,] : R \xrightarrow{\sim} \mathrm{End}(E)$ satisfies $[\alpha]^*\omega = \alpha\omega$.

To verify our claim, we first note that any two non-zero invariant differentials on E_Λ are scalar multiples of one another. This follows trivially from the fact that their quotient would be a translation invariant function, hence would be constant. So if we take any invariant differential $\omega \in \Omega_E$

and pull back via the isomorphism $f : \mathbb{C}/\Lambda \to E_\Lambda(\mathbb{C})$, we obtain a multiple of the invariant differential dz on $\mathbb{C}/\Lambda$, say

$$f^*\omega = c\, dz.$$

Now tracing around the commutative diagram shown above gives the desired result:

$$[\alpha]^*\omega = (f^{-1})^* \circ \phi_\alpha^* \circ f^*(\omega)$$
$$= (f^{-1})^* \circ \phi_\alpha^*(c\, dz) = (f^{-1})^*(c\alpha\, dz) = \alpha\omega.$$

$\square$

Corollary 1.1.1. *Let* $(E_1, [\,\cdot\,]_{E_1})$ *and* $(E_2, [\,\cdot\,]_{E_2})$ *be normalized elliptic curves with complex multiplication by* R, *and let* $\phi : E_1 \to E_2$ *be an isogeny. Then*

$$\phi \circ [\alpha]_{E_1} = [\alpha]_{E_2} \circ \phi \qquad \text{for all } \alpha \in R.$$

PROOF. Let $0 \neq \omega \in \Omega_{E_2}$ be an invariant differential. Then

$$(\phi \circ [\alpha]_{E_1})^* \omega = [\alpha]_{E_1}^* (\phi^*\omega)$$
$$= \alpha\phi^*\omega \quad \text{since } \phi^*\omega \text{ is an invariant differential on } E_1$$
$$= \phi^*\alpha\omega$$
$$= \phi^* ([\alpha]_{E_2}^* \omega)$$
$$= ([\alpha]_{E_2} \circ \phi)^* \omega.$$

Every non-zero isogeny $E_1 \to E_2$ is separable (we're working in characteristic 0), so [AEC II.4.2c] says that the map

$$\mathrm{Hom}(E_1, E_2) \longrightarrow \mathrm{Hom}(\Omega_{E_2}, \Omega_{E_1}), \qquad \psi \longmapsto \psi^*,$$

is injective. Therefore $\phi \circ [\alpha]_{E_1} = [\alpha]_{E_2} \circ \phi$. $\square$

We have seen in Chapter I that in order to understand particular elliptic curves, it is often useful to study the set of all elliptic curves. Similarly, in order to study a particular elliptic curve with complex multiplication, it turns out that one should look at the set of all elliptic curves with the same endomorphism ring. Of course, by "elliptic curves" we really mean isomorphism classes of elliptic curves, which leads us to define the following set:

$$\mathcal{ELL}(R) = \frac{\{\text{elliptic curves } E/\mathbb{C} \text{ with } \mathrm{End}(E) \cong R\}}{\text{isomorphism over } \mathbb{C}}$$

$$= \frac{\{\text{lattices } \Lambda \text{ with } \mathrm{End}(E_\Lambda) \cong R\}}{\text{homothety}}.$$

If we start with a quadratic imaginary field K, how might we construct an elliptic curve with complex multiplication by R_K? If $\mathfrak{a}$ is a non-zero ideal of R_K, or more generally if it is a non-zero fractional ideal of K, then using the embedding $\mathfrak{a} \subset K \subset \mathbb{C}$ we see that $\mathfrak{a}$ is a lattice in $\mathbb{C}$. (This is clear from the definition of fractional ideal, which for quadratic imaginary fields implies that $\mathfrak{a}$ is a $\mathbb{Z}$-module of rank 2 which is not contained in $\mathbb{R}$.) Hence we can form an elliptic curve $E_{\mathfrak{a}}$ whose endomorphism ring is

$$
\begin{aligned}
\text{End}(E_{\mathfrak{a}}) &\cong \{\alpha \in \mathbb{C} : \alpha\mathfrak{a} \subset \mathfrak{a}\} \\
&= \{\alpha \in K : \alpha\mathfrak{a} \subset \mathfrak{a}\} \qquad \text{since } \mathfrak{a} \subset K \\
&= R_K \qquad \text{since } \mathfrak{a} \text{ is a fractional ideal.}
\end{aligned}
$$

Thus each non-zero fractional ideal $\mathfrak{a}$ of K will give an elliptic curve with complex multiplication by R_K. On the other hand, since homothetic lattices give isomorphic elliptic curves, we see that $\mathfrak{a}$ and $c\mathfrak{a}$ give the same elliptic curve in $\mathcal{ELL}(R_K)$. This suggests that we look at the group of fractional ideals modulo principal ideals, which the reader will recognize as one of the fundamental objects of study in algebraic number theory:

$$
\begin{aligned}
\mathcal{CL}(R_K) &= \text{ideal class group of } R_K \\
&= \frac{\{\text{non-zero fractional ideals of } K\}}{\{\text{non-zero principal ideals of } K\}}.
\end{aligned}
$$

If $\mathfrak{a}$ is a fractional ideal of K, we denote by $\bar{\mathfrak{a}}$ its ideal class in $\mathcal{CL}(R_K)$. We have seen that there is a map

$$
\mathcal{CL}(R_K) \longrightarrow \mathcal{ELL}(R_K), \qquad \bar{\mathfrak{a}} \longmapsto E_{\mathfrak{a}}.
$$

More generally, if Λ is any lattice with $E_\Lambda \in \mathcal{ELL}(R_K)$ and $\mathfrak{a}$ is any non-zero fractional ideal of K, we can form the product

$$
\mathfrak{a}\Lambda = \{\alpha_1\lambda_1 + \cdots + \alpha_r\lambda_r : \alpha_i \in \mathfrak{a}, \ \lambda_i \in \Lambda\}.
$$

We will now prove the elementary, but crucial, fact that this induces a simply transitive action of the ideal class group $\mathcal{CL}(R_K)$ on the set of elliptic curves $\mathcal{ELL}(R_K)$. This proposition forms the basis for all of our subsequent work on complex multiplication.

Proposition 1.2. (a) *Let Λ be a lattice with $E_\Lambda \in \mathcal{ELL}(R_K)$, and let $\mathfrak{a}$ and $\mathfrak{b}$ be non-zero fractional ideals of K.*

(i) $\mathfrak{a}\Lambda$ *is a lattice in $\mathbb{C}$.*

(ii)   *The elliptic curve $E_{\mathfrak{a}\Lambda}$ satisfies $\text{End}(E_{\mathfrak{a}\Lambda}) \cong R_K$.*

(iii) $E_{\mathfrak{a}\Lambda} \cong E_{\mathfrak{b}\Lambda}$ *if and only if* $\bar{\mathfrak{a}} = \bar{\mathfrak{b}}$ *in $\mathcal{CL}(R_K)$.*

Hence there is a well-defined action of $\mathcal{CL}(R_K)$ on $\mathcal{ELL}(R_K)$ determined by

$$
\bar{\mathfrak{a}} * E_\Lambda = E_{\mathfrak{a}^{-1}\Lambda}.
$$

(The reason for using $\mathfrak{a}^{-1}$ instead of $\mathfrak{a}$ will become apparent below.)
(b) *The action of $\mathcal{CL}(R_K)$ on $\mathcal{ELL}(R_K)$ described in (a) is simply transitive. In particular,*

$$\# \, \mathcal{CL}(R_K) = \# \, \mathcal{ELL}(R_K).$$

PROOF. (a) (i) By assumption, $\text{End}(E_\Lambda) = R_K$, so $R_K \Lambda = \Lambda$. Choose a non-zero integer $d \in \mathbb{Z}$ so that $d\mathfrak{a} \subset R_K$, which is possible by the definition of fractional ideal. Then $\mathfrak{a}\Lambda \subset \dfrac{1}{d}\Lambda$, so $\mathfrak{a}\Lambda$ is a discrete subgroup of $\mathbb{C}$. Similarly, choosing a non-zero integer d so that $dR_K \subset \mathfrak{a}$, we find that $d\Lambda \subset \mathfrak{a}\Lambda$, hence $\mathfrak{a}\Lambda$ spans $\mathbb{C}$. This proves that $\mathfrak{a}\Lambda$ is a lattice.
(ii) For any $\alpha \in \mathbb{C}$ and any fractional ideal $\mathfrak{a} \neq 0$, we have

$$\alpha\mathfrak{a}\Lambda \subset \mathfrak{a}\Lambda \iff \mathfrak{a}^{-1}\alpha\mathfrak{a}\Lambda \subset \mathfrak{a}^{-1}\mathfrak{a}\Lambda \iff \alpha\Lambda \subset \Lambda.$$

Hence

$$\begin{aligned}
\text{End}(E_{\mathfrak{a}\Lambda}) &= \{\alpha \in \mathbb{C} \, : \, \alpha\mathfrak{a}\Lambda \subset \mathfrak{a}\Lambda\} \\
&= \{\alpha \in \mathbb{C} \, : \, \alpha\Lambda \subset \Lambda\} = \text{End}(E_\Lambda) = R_K.
\end{aligned}$$

(iii) From [AEC VI.4.1.1], the isomorphism class of $E_{\mathfrak{a}\Lambda}$ is exactly determined by the homothety class of $\mathfrak{a}\Lambda$. In other words, $E_{\mathfrak{a}\Lambda} \cong E_{\mathfrak{b}\Lambda}$ if and only if there is a $c \in \mathbb{C}^*$ such that $\mathfrak{a}\Lambda = c\mathfrak{b}\Lambda$. Multiplying by $\mathfrak{a}^{-1}$ and using the fact that $R_K \Lambda = \Lambda$, we see that

$$E_{\mathfrak{a}\Lambda} \cong E_{\mathfrak{b}\Lambda} \iff \Lambda = c\mathfrak{a}^{-1}\mathfrak{b}\Lambda.$$

Similarly, multiplying by $c^{-1}\mathfrak{b}^{-1}$ gives

$$E_{\mathfrak{a}\Lambda} \cong E_{\mathfrak{b}\Lambda} \iff \Lambda = c^{-1}\mathfrak{a}\mathfrak{b}^{-1}\Lambda.$$

Hence if $E_{\mathfrak{a}\Lambda} \cong E_{\mathfrak{b}\Lambda}$, then both $c\mathfrak{a}^{-1}\mathfrak{b}$ and $c^{-1}\mathfrak{a}\mathfrak{b}^{-1}$ take Λ to itself, so they are both contained in R_K, and hence are equal to R_K. Therefore

$$\mathfrak{a} = c\mathfrak{b},$$

from which we see immediately that $c \in K$ and $\bar{\mathfrak{a}} = \bar{\mathfrak{b}}$. This completes the proof of (iii).

Finally, the trivial observation

$$\bar{\mathfrak{a}} * (\bar{\mathfrak{b}} * E_\Lambda) = \bar{\mathfrak{a}} * E_{\mathfrak{b}^{-1}\Lambda} = E_{\mathfrak{a}^{-1}(\mathfrak{b}^{-1}\Lambda)} = E_{(\mathfrak{a}\mathfrak{b})^{-1}\Lambda} = (\bar{\mathfrak{a}}\bar{\mathfrak{b}}) * E_\Lambda$$

shows that the definition $\bar{\mathfrak{a}} * E_\Lambda = E_{\mathfrak{a}^{-1}\Lambda}$ gives a group action of $\mathcal{CL}(R_K)$ on $\mathcal{ELL}(R_K)$.

(b) Let E_{Λ_1} and E_{Λ_2} be two elliptic curves in $\mathcal{ELL}(R_K)$. To show that the class group $\mathcal{CL}(R_K)$ acts transitively on $\mathcal{ELL}(R_K)$, we must find a fractional ideal $\mathfrak{a}$ with the property $\bar{\mathfrak{a}} * E_{\Lambda_1} = E_{\Lambda_2}$. Choose any non-zero element $\lambda_1 \in \Lambda_1$, and consider the lattice $\mathfrak{a}_1 = \dfrac{1}{\lambda_1}\Lambda_1$. From [AEC VI.5.5] we see that $\mathfrak{a}_1$ is contained in K, and by assumption it is a finitely generated R_K-module, hence it is a fractional ideal of K. Similarly, choosing a non-zero $\lambda_2 \in \Lambda_2$, we obtain a second fractional ideal $\mathfrak{a}_2 = \dfrac{1}{\lambda_2}\Lambda_2$ of K. Then

$$\frac{\lambda_2}{\lambda_1}\mathfrak{a}_2\mathfrak{a}_1^{-1}\Lambda_1 = \Lambda_2.$$

So if we let $\mathfrak{a} = \mathfrak{a}_2^{-1}\mathfrak{a}_1$, then

$$\bar{\mathfrak{a}} * E_{\Lambda_1} = E_{\mathfrak{a}^{-1}\Lambda_1} = E_{\frac{\lambda_1}{\lambda_2}\Lambda_2} \cong E_{\Lambda_2}.$$

Note the last equality follows from the fact that homothetic lattices give isomorphic elliptic curves. This shows that the action of $\mathcal{CL}(R_K)$ on $\mathcal{ELL}(R_K)$ is transitive.

To prove that the action is simply transitive, we must show that if $\mathfrak{a} * E_\Lambda = \mathfrak{b} * E_\Lambda$, then $\bar{\mathfrak{a}} = \bar{\mathfrak{b}}$. But this is immediate from part (ii) of (a). $\qquad\square$

We have already seen two sorts of elliptic curves which have complex multiplication, namely the curves with $j = 0$ and $j = 1728$ whose automorphism groups are strictly larger than $\{\pm 1\}$. (See [AEC III.10.1].) Now we'll look at these curves from a complex analytic viewpoint.

Example 1.3.1. Let $\Lambda = \mathbb{Z}[i]$ be the lattice of Gaussian integers. Then the endomorphism ring of E_Λ is $\mathbb{Z}[i]$. In particular, $\text{Aut}(E_\Lambda) \cong \{\pm 1, \pm i\}$, so our general theory [AEC III.10.1] tells us that $j(E_\Lambda) = 1728$. But we can see this directly in the following way. The lattice Λ satisfies $i\Lambda = \Lambda$. Hence

$$g_3(\Lambda) = g_3(i\Lambda) = i^6 g_3(\Lambda) = -g_3(\Lambda),$$

so $g_3(\Lambda) = 0$. Therefore E_Λ is given by the Weierstrass equation

$$E_\Lambda : y^2 = 4x^3 - g_2(\Lambda)x,$$

from which we see immediately that $j(E_\Lambda) = 1728$.

Since $j(E_\Lambda)$ is rational, we know that E_Λ is isomorphic over $\mathbb{C}$ to an elliptic curve defined over $\mathbb{Q}$; for example, it is isomorphic to the curve $y^2 = x^3 + x$. But it does not follow that $g_2(\Lambda)$ itself is in $\mathbb{Q}$. In fact, a theorem of Hurwitz [1] says that

$$g_2(\mathbb{Z}[i]) = 64 \left(\int_0^1 \frac{dt}{\sqrt{1-t^4}} \right)^4.$$

Example 1.3.2. Similarly, let $\rho = e^{2\pi i/3}$ be a primitive cube root of unity, and let $\Lambda = \mathbb{Z}[\rho]$ be the associated lattice. Then $\rho\Lambda = \Lambda$, so

$$g_2(\Lambda) = g_2(\rho\Lambda) = \rho^4 g_2(\Lambda) = \rho g_2(\Lambda),$$

and hence $g_2(\Lambda) = 0$. Thus E_Λ is given by the equation

$$E_\Lambda : y^2 = 4x^3 - g_3(\Lambda),$$

so $j(E_\Lambda) = 0$. This confirms [AEC III.10.1], since $\mathrm{Aut}(E_\Lambda) = \mathbb{Z}[\rho]^* = \{\pm 1, \pm\rho, \pm\rho^2\}$. Further, we see that E_Λ is $\mathbb{C}$-isomorphic to the curve $y^2 = x^3 + 1$, which is defined over $\mathbb{Q}$.

If E has complex multiplication by K, we will eventually use torsion points of E to generate abelian extensions of K. We could restrict ourselves to studying points of order m for various integers m, but because E has complex multiplication, there are other natural finite subgroups to look at. In general, if $\mathfrak{a}$ is any integral ideal of R_K, we define

$$E[\mathfrak{a}] = \{P \in E : [\alpha]P = 0 \text{ for all } \alpha \in \mathfrak{a}\}.$$

We call $E[\mathfrak{a}]$ the *group of $\mathfrak{a}$-torsion points of E*. For example, if $\mathfrak{a} = mR_K$, then $E[\mathfrak{a}]$ is just $E[m]$. Notice that the definition of $E[\mathfrak{a}]$ depends on choosing a particular isomorphism $[\cdot] : R_K \xrightarrow{\sim} \mathrm{End}(E)$; we always choose the normalized isomorphism described in (1.1).

If $\mathfrak{a}$ is an integral ideal of R_K, then $\Lambda \subset \mathfrak{a}^{-1}\Lambda$. This means that there is a natural homomorphism

$$\mathbb{C}/\Lambda \longrightarrow \mathbb{C}/\mathfrak{a}^{-1}\Lambda, \qquad z \longmapsto z,$$

which in turn induces a natural isogeny

$$E_\Lambda \longrightarrow \bar{\mathfrak{a}} * E_\Lambda.$$

The following useful proposition gives a precise description of this isogeny and of $E[\mathfrak{a}]$.

Proposition 1.4. *Let $E \in \mathcal{ELL}(R_K)$, and let $\mathfrak{a}$ be an integral ideal of R_K.*
(a) *$E[\mathfrak{a}]$ is the kernel of the natural map $E \to \bar{\mathfrak{a}} * E$.*
(b) *$E[\mathfrak{a}]$ is a free $R_K/\mathfrak{a}$-module of rank 1.*

PROOF. Let Λ be a lattice corresponding to E. Fixing an analytic isomorphism $\mathbb{C}/\Lambda \cong E(\mathbb{C})$, we find that

$$\begin{aligned}
E[\mathfrak{a}] &\cong \{z \in \mathbb{C}/\Lambda : \alpha z = 0 \text{ for all } \alpha \in \mathfrak{a}\} \\
&= \{z \in \mathbb{C} : \alpha z \in \Lambda \text{ for all } \alpha \in \mathfrak{a}\}/\Lambda \\
&= \{z \in \mathbb{C} : z\mathfrak{a} \subset \Lambda\}/\Lambda \\
&= \mathfrak{a}^{-1}\Lambda/\Lambda \\
&= \ker\left(\mathbb{C}/\Lambda \xrightarrow{z \mapsto z} \mathbb{C}/\mathfrak{a}^{-1}\Lambda\right) \\
&= \ker(E \to \bar{\mathfrak{a}} * E).
\end{aligned}$$

(b) Continuing with the notation from (a), we choose a non-zero lattice element $\lambda \in \Lambda$. Then [AEC VI.5.5] says that the lattice $(1/\lambda)\Lambda$ is contained in K, and it is a finitely generated R_K-module, so it is a fractional ideal of K. Since homothetic lattices give isomorphic elliptic curves, we may assume that Λ is a fractional ideal of K.

From (a) we know that $E[\mathfrak{a}] \cong \mathfrak{a}^{-1}\Lambda/\Lambda$ as $R_K/\mathfrak{a}$-modules. Note that if $\mathfrak{q}$ is any integral ideal dividing $\mathfrak{a}$, then the fact that $R_K\Lambda = \Lambda$ implies

$$(\mathfrak{a}^{-1}\Lambda/\Lambda) \otimes_{R_K} (R_K/\mathfrak{q}) \cong \mathfrak{a}^{-1}\Lambda/(\Lambda + \mathfrak{q}\mathfrak{a}^{-1}\Lambda) = \mathfrak{a}^{-1}\Lambda/\mathfrak{q}\mathfrak{a}^{-1}\Lambda.$$

Hence if we use the Chinese Remainder Theorem to write

$$R_K/\mathfrak{a} \cong \prod_{\mathfrak{p} \text{ prime}} R_K/\mathfrak{p}^{e(\mathfrak{p})}, \qquad \text{then} \qquad E[\mathfrak{a}] \cong \prod_{\mathfrak{p} \text{ prime}} \mathfrak{a}^{-1}\Lambda/\mathfrak{p}^{e(\mathfrak{p})}\mathfrak{a}^{-1}\Lambda.$$

So it suffices to prove that if $\mathfrak{b}$ is a fractional ideal of R_K (such as $\mathfrak{b} = \mathfrak{a}^{-1}\Lambda$) and if $\mathfrak{p}^e$ is a power of a prime ideal, then $\mathfrak{b}/\mathfrak{p}^e\mathfrak{b}$ is a free $R_K/\mathfrak{p}^e$-module of rank one.

To ease notation, we momentarily write

$$R' = R_K/\mathfrak{p}^e, \qquad \mathfrak{p}' = \mathfrak{p}/\mathfrak{p}^e, \qquad \text{and} \qquad \mathfrak{b}' = \mathfrak{b}/\mathfrak{p}^e\mathfrak{b}.$$

Notice that R' is a local ring with maximal ideal $\mathfrak{p}'$. (In fact, the only ideals in R' are $(0), \mathfrak{p}'^{e-1}, \cdots, \mathfrak{p}', (1)$.) Consider the quotient

$$\mathfrak{b}'/\mathfrak{p}'\mathfrak{b}' \cong \mathfrak{b}/\mathfrak{p}\mathfrak{b} \quad \text{as a vector space over the field } R'/\mathfrak{p}' \cong R_K/\mathfrak{p}.$$

We claim that it is a one-dimensional vector space.

First we observe that any two elements of $\mathfrak{b}$ are R_K-linearly dependent, so the dimension of $\mathfrak{b}/\mathfrak{p}\mathfrak{b}$ over $R_K/\mathfrak{p}$ is at most one. On the other hand, if the dimension were zero, then we would have $\mathfrak{b} = \mathfrak{p}\mathfrak{b}$, which is absurd. Hence the dimension is one. By Nakayama's lemma (Atiyah-MacDonald [1, Prop 2.8]) applied to the local ring R' and the R'-module $\mathfrak{b}'$, it follows that $\mathfrak{b}'$ is a free R'-module of rank one. This completes the proof of Proposition 1.4.

$$\square$$

We can use (1.4) to compute the degree of the isogeny $E \to \bar{\mathfrak{a}} * E$, as well as the degree of an endomorphism $[\alpha] : E \to E$.

Corollary 1.5. *Let $E \in \mathcal{ELL}(R_K)$.*
(a) *For all integral ideals $\mathfrak{a} \subset R_K$, the natural map $E \to \bar{\mathfrak{a}} * E$ has degree $\mathrm{N}_{\mathbb{Q}}^K \mathfrak{a}$.*
(b) *For all $\alpha \in R_K$, the endomorphism $[\alpha] : E \to E$ defined in (1.1) has degree $\left| \mathrm{N}_{\mathbb{Q}}^K \alpha \right|$.*

PROOF. Both parts are immediate from (1.4). For example,

$$\deg(E \to \bar{\mathfrak{a}} * E) = \#E[\mathfrak{a}] \qquad \text{from (1.4a)}$$
$$= \mathrm{N}_{\mathbb{Q}}^K \mathfrak{a} \qquad \text{from (1.4b)}.$$

Similarly,

$$\deg[\alpha] = \#\ker[\alpha] = \#E[\alpha R_K] = \mathrm{N}^K_{\mathbb{Q}}(\alpha R_K) = \left|\mathrm{N}^K_{\mathbb{Q}}\alpha\right|.$$

□

Remark 1.6. Before going on to arithmetic questions, we want to make one brief remark about terminology. The classical name for the j-invariant of an elliptic curve with complex multiplication is a *singular j-invariant*. This terminology, meant to single out such j-invariants as being unusual, is somewhat unfortunate, since it suggests that the elliptic curve itself has singularities. We will not use the word "singular" in this sense, but the reader should be aware of this usage, since it is still fairly common.

Notice that an elliptic curve defined over a finite field always has a "singular" j-invariant, since its endomorphism ring is always larger than $\mathbb{Z}$ [AEC, V.3.1]. In those rare cases that the endomorphism ring is a quaternion algebra, the singularity is especially exceptional, which explains the origin of the term "supersingular" to describe such curves.

§2. Rationality Questions

In this section we will study the field of definition for complex multiplication elliptic curves and their endomorphisms. We begin by showing that every elliptic curve with complex multiplication is defined over an algebraic extension of $\mathbb{Q}$.

Proposition 2.1. (a) *Let $E/\mathbb{C}$ be an elliptic curve, and let $\sigma : \mathbb{C} \to \mathbb{C}$ be any field automorphism of $\mathbb{C}$. Then*

$$\mathrm{End}(E^\sigma) \cong \mathrm{End}(E).$$

(b) *Let $E/\mathbb{C}$ be an elliptic curve with complex multiplication by the ring of integers R_K of a quadratic imaginary field K. Then $j(E) \in \bar{\mathbb{Q}}$. (Later we will show that $j(E)$ is an algebraic integer. See (II §6) and (V.6.3).)*
(c)

$$\mathcal{ELL}(R_K) \cong \frac{\{\text{elliptic curves } E/\bar{\mathbb{Q}} \text{ with } \mathrm{End}(E) \cong R_K\}}{\text{isomorphism over } \bar{\mathbb{Q}}}.$$

(Note that the original definition of $\mathcal{ELL}(R_K)$ is in terms of isomorphism classes of elliptic curves over $\mathbb{C}$, not over $\bar{\mathbb{Q}}$.)

PROOF. (a) This is clear, since if $\phi : E \to E$ is an endomorphism of E, then $\phi^\sigma : E^\sigma \to E^\sigma$ is an endomorphism of E^σ.

(b) Let $\sigma \in \mathrm{Aut}(\mathbb{C})$ be as in (a). Now E^σ is obtained from E by letting σ act on the coefficients of a Weierstrass equation for E, and $j(E)$ is a rational combination of those coefficients, so it is clear that

$$j(E^\sigma) = j(E)^\sigma.$$

On the other hand, (a) implies that $\mathrm{End}(E^\sigma) \cong R_K$, so (1.2b) implies that E^σ is in one of only finitely many $\mathbb{C}$-isomorphism classes of elliptic curves. Since the isomorphism class of an elliptic curve is determined by its j-invariant [AEC III.1.4b], it follows that $j(E)^\sigma$ takes on only finitely many values as σ ranges over $\mathrm{Aut}(\mathbb{C})$. Therefore $[\mathbb{Q}(j(E)) : \mathbb{Q}]$ is finite, so $j(E)$ is an algebraic number.

(c) For any subfield F of $\mathbb{C}$, let us momentarily denote by $\mathcal{ELL}_F(R_K)$ the set

$$\mathcal{ELL}_F(R_K) \cong \frac{\{\text{elliptic curves } E/F \text{ with } \mathrm{End}(E) \cong R_K\}}{\text{isomorphism over } F}.$$

If we fix an embedding $\bar{\mathbb{Q}} \subset \mathbb{C}$, then there is a natural map

$$\varepsilon : \mathcal{ELL}_{\bar{\mathbb{Q}}}(R_K) \longrightarrow \mathcal{ELL}_{\mathbb{C}}(R_K).$$

We need to show that this map is a bijection.

Let $E/\mathbb{C}$ represent an element of $\mathcal{ELL}_{\mathbb{C}}(R_K)$. Then we have:

  (i) $j(E) \in \bar{\mathbb{Q}}$, from (b);
  (ii) there is an elliptic curve $E'/\mathbb{Q}(j(E))$ with $j(E') = j(E)$, from [AEC III.1.4c];
  (iii) E' is isomorphic to E over $\mathbb{C}$, from [AEC III.1.4b].

These three facts imply that $\varepsilon(E') = E$, which proves that ε is surjective.

Next let $E_1/\bar{\mathbb{Q}}$ and $E_2/\bar{\mathbb{Q}}$ represent elements of $\mathcal{ELL}_{\bar{\mathbb{Q}}}(R_K)$, and suppose that $\varepsilon(E_1) = \varepsilon(E_2)$. Then $j(E_1) = j(E_2)$ from [AEC III.1.4b], and another application of [AEC III.1.4b] says that E_1 and E_2 are isomorphic over $\bar{\mathbb{Q}}$. Hence E_1 and E_2 represent the same element of $\mathcal{ELL}_{\bar{\mathbb{Q}}}(R_K)$, which shows that ε is also injective. $\square$

Next we study the effect that field automorphisms have on the maps $[\alpha] : E \to E$ described in (1.1). In particular, we will find a field of definition for these maps. Note that if ϕ is an endomorphism of E and σ is any automorphism of $\mathbb{C}$, then ϕ^σ will be an endomorphism of E^σ.

Theorem 2.2. (a) *Let $E/\mathbb{C}$ be an elliptic curve with complex multiplication by the ring $R \subset \mathbb{C}$. Then*

$$[\alpha]_E{}^\sigma = [\alpha^\sigma]_{E^\sigma} \qquad \text{for all } \alpha \in R \text{ and all } \sigma \in \mathrm{Aut}(\mathbb{C}),$$

where the isomorphisms $[\,\cdot\,]_E : R \xrightarrow{\sim} \mathrm{End}(E)$ and $[\,\cdot\,]_{E^\sigma} : R \xrightarrow{\sim} \mathrm{End}(E^\sigma)$ are normalized as in (1.1).

(b) *Let E be an elliptic curve defined over a field $L \subset \mathbb{C}$ and with complex multiplication by the quadratic imaginary field $K \subset \mathbb{C}$. Then every endomorphism of E is defined over the compositum LK.*

(c) *Let E_1/L and E_2/L be elliptic curves defined over a field $L \subset \mathbb{C}$. Then there is a finite extension L'/L such that every isogeny from E_1 to E_2 is defined over L'.*

PROOF. Let $\omega \in \Omega_E$ be a non-zero invariant differential on E. Then the normalization described in (1.1) says that

$$[\alpha]_E^* \omega = \alpha\omega \qquad \text{for all } \alpha \in R.$$

Further, ω^σ is an invariant differential on E^σ, so again from (1.1) we get

$$[\beta]_{E^\sigma}^* \omega^\sigma = \beta\omega^\sigma \qquad \text{for all } \beta \in R.$$

Now for any $\alpha \in R$ and any $\sigma \in \text{Aut}(\mathbb{C})$, we compute

$$\left([\alpha]_E{}^\sigma\right)^* (\omega^\sigma) = \left([\alpha]_E^* \omega\right)^\sigma = (\alpha\omega)^\sigma = \alpha^\sigma \omega^\sigma = [\alpha^\sigma]_{E^\sigma}^* (\omega^\sigma).$$

Thus $[\alpha]_E{}^\sigma$ and $[\alpha^\sigma]_{E^\sigma}$ have the same effect on the invariant differential ω^σ.

Now we use [AEC II.4.2c], which says that the natural map

$$\text{End}(E^\sigma) \longrightarrow \text{End}(\Omega_{E^\sigma}), \qquad \psi \longmapsto \psi^*,$$

is injective. (Note we are working in characteristic 0, so all finite maps are separable.) This proves that $[\alpha]_E{}^\sigma = [\alpha^\sigma]_{E^\sigma}$.

(b) Let $\sigma \in \text{Aut}(\mathbb{C})$ be an automorphism of $\mathbb{C}$ that fixes L. Since E is defined over L, we can take a Weierstrass equation for E with coefficients in L, so $E^\sigma = E$. Then (a) says that for all $\alpha \in R$,

$$[\alpha]_E{}^\sigma = [\alpha^\sigma]_{E^\sigma} = [\alpha^\sigma]_E.$$

If in addition σ fixes K, then $\alpha^\sigma = \alpha$. This proves that

$$[\alpha]_E{}^\sigma = [\alpha]_E \qquad \text{for all } \sigma \in \text{Aut}(\mathbb{C}) \text{ such that } \sigma \text{ fixes } LK.$$

Hence the endomorphism $[\alpha]$ is defined over LK.

(c) As in (b), we take Weierstrass equations for E_1 and E_2 with coefficients in L. Let $\phi \in \text{Hom}(E_1, E_2)$ be an isogeny. Then for any $\sigma \in \text{Aut}(\mathbb{C})$ such that σ fixes L, we have $\phi^\sigma \in \text{Hom}(E_1, E_2)$. Note that $\deg \phi^\sigma = \deg \phi$. From [AEC III.4.11], we see that an isogeny $\phi \in \text{Hom}(E_1, E_2)$ is determined by its kernel, at least up to an automorphism of E_1 and E_2. Since E_1 has only finitely many subgroups of any given finite order, and since $\text{Aut}(E_1)$ and $\text{Aut}(E_2)$ are finite, it follows that $\text{Hom}(E_1, E_2)$ contains only finitely many isogenies of a given degree. Therefore the set

$$\{\phi^\sigma : \sigma \in \text{Aut}(\mathbb{C}), \ \sigma \text{ fixes } L\}$$

is finite, which implies that ϕ is defined over a finite extension of L. Finally, we observe from [AEC III.7.5] that $\mathrm{Hom}(E_1, E_2)$ is a finitely generated group, so it suffices to take a field of definition for some finite set of generators. □

Remark 2.2.1. Notice that the proof of (2.1b) together with the estimate in (1.2b) shows that if $\mathrm{End}(E) \cong R_K$, then

$$[\mathbb{Q}(j(E)) : \mathbb{Q}] \le h_K,$$

where $h_K = \# \mathcal{CL}(R_K)$ is the class number of K. We will prove later (4.3) that this is an equality. In particular, $j(E)$ is in $\mathbb{Q}$ if and only if K has class number 1. For a complete list of these $\mathbb{Q}$-rational j-invariants, see Appendix A §3.

Remark 2.2.2. In view of (2.2.1), we see that if R_K has class number 1, then E has a model defined over $\mathbb{Q}$. We have already seen examples of this in (1.3.1) and (1.3.2), where we looked at curves with complex multiplication by $\mathbb{Z}[i]$ and $\mathbb{Z}[\rho]$. (Here $\rho = e^{2\pi i/3}$.) We can also illustrate (2.2) for these curves. For example, to normalize the curve

$$E : y^2 = x^3 + x,$$

we use the isomorphism $[\,\cdot\,] : \mathbb{Z}[i] \to \mathrm{End}(E)$ determined by

$$[i](x, y) = (-x, iy).$$

To see that this is the correct normalization, we compute

$$[i]^* \frac{dx}{y} = \frac{d(-x)}{iy} = i \frac{dx}{y}.$$

If $\sigma \in \mathrm{Aut}(\mathbb{C})$ is complex conjugation, then

$$
\begin{aligned}
\left([i](x, y)\right)^\sigma = (-x, iy)^\sigma &= (-x^\sigma, i^\sigma y^\sigma) \\
&= (-x^\sigma, -iy^\sigma) = [-i](x^\sigma, y^\sigma) = [i^\sigma](x^\sigma, y^\sigma).
\end{aligned}
$$

Hence $[i]^\sigma$ equals $[i^\sigma]$, as it should by (2.2).

Similarly, for the curve

$$E : y^2 = x^3 + 1$$

we take the isomorphism $[\,\cdot\,] : \mathbb{Z}[\rho] \to \mathrm{End}(E)$ determined by

$$[\rho](x, y) = (\rho x, y).$$

Remark 2.2.3. There is an interesting converse to (2.1b). Suppose that $\Lambda = \mathbb{Z}\omega_1 + \mathbb{Z}\omega_2$ is a lattice satisfying

$$3 \leq [\mathbb{Q}(\omega_1/\omega_2) : \mathbb{Q}] < \infty;$$

that is, ω_1/ω_2 is an algebraic number of degree at least 3 over $\mathbb{Q}$. Then one can show that $j(E_\Lambda)$ is a transcendental number. Notice the analogy with the Gel'fond-Schneider theorem, which says that if $\alpha \in \bar{\mathbb{Q}}$ with $\alpha \neq 0, 1$ and if β satisfies $2 \leq [\mathbb{Q}(\beta) : \mathbb{Q}] < \infty$, then α^β is transcendental. The transcendence of $j(E_\Lambda)$ was first proven by Schneider. The interested reader will find a proof of this fact in Schneider [1, Thm. 17] or Waldschmidt [1, Cor. 3.2.4]. For a general account of the transcendence properties of elliptic and modular functions, see for example Waldschmidt [1, Ch. 3].

It is an immediate consequence of (1.4b) and (2.2b) that the torsion points of E generate abelian extensions of $K(j(E))$. Before giving the proof, we remind the reader of the analogous result for cyclotomic fields. Thus let $\zeta \in \mathbb{C}^*$ be a primitive N^{th}-root of unity and let $\sigma \in \text{Gal}(\mathbb{Q}(\zeta)/\mathbb{Q})$. Then ζ^σ is another primitive N^{th}-root of unity, say $\zeta^\sigma = \zeta^{\rho(\sigma)}$, and it is an easy matter to check that the map

$$\rho : \text{Gal}(\mathbb{Q}(\zeta)/\mathbb{Q}) \longrightarrow \text{Aut}(\mu_N) \cong (\mathbb{Z}/N\mathbb{Z})^*$$

is an injective homomorphism. (Here $\mu_N = \zeta^{\mathbb{Z}}$ is the group of Nth-roots of unity.) Hence $\mathbb{Q}(\zeta)/\mathbb{Q}$ is an abelian extension. We now prove the same thing for elliptic curves. (For another proof, see exercise 2.6.)

Theorem 2.3. *Let $E/\mathbb{C}$ be an elliptic curve with complex multiplication by the ring of integers R_K of the quadratic imaginary field K, and let*

$$L = K(j(E), E_{\text{tors}})$$

be the field generated by the j-invariant of E and the coordinates of all of the torsion points of E. Then L is an abelian extension of $K(j(E))$. (N.B. In general, L will not be an abelian extension of K.)

PROOF. To ease notation, let $H = K(j(E))$. Further let

$$L_m = K(j(E), E[m]) = H(E[m])$$

be the extension of H generated by the m-torsion points of E. Since L is the compositum of all of the L_m's, it suffices to show that L_m is an abelian extension of H.

As usual, there is a representation

$$\rho : \text{Gal}(\bar{K}/H) \longrightarrow \text{Aut}(E[m])$$

determined by the condition

$$\rho(\sigma)(T) = T^\sigma \qquad \text{for all } \sigma \in \text{Gal}(\bar{K}/H) \text{ and all } T \in E[m].$$

(See [AEC III §7].) For an arbitrary elliptic curve, all we would be able to deduce from this is that $\text{Gal}(L_m/H)$ injects into the automorphism group of the abelian group $E[m]$, so we would find that $\text{Gal}(L_m/H)$ is isomorphic to a subgroup of $\text{GL}_2(\mathbb{Z}/m\mathbb{Z})$.

But the fact that our elliptic curve has complex multiplication gives us additional information. We take a model for E defined over $H = K(j(E))$, and then (2.2b) says that every endomorphism of E is also defined over H. So elements of $\text{Gal}(L_m/H)$ will commute with elements of R_K in their action on $E[m]$:

$$([\alpha]T)^\sigma = [\alpha](T^\sigma) \qquad \text{for all } \sigma \in \text{Gal}(L_m/H), \ T \in E[m], \text{ and } \alpha \in R_K.$$

In other words, ρ is actually a homomorphism from $\text{Gal}(\bar{K}/H)$ to the group of R_K/mR_K-module automorphisms of $E[m]$. Hence ρ induces an injection

$$\phi : \text{Gal}(L_m/H) \hookrightarrow \text{Aut}_{R_K/mR_K}(E[m]).$$

Now we use (1.4b), which says that $E[m]$ is a free R_K/mR_K-module of rank one. This implies that

$$\text{Aut}_{R_K/mR_K}(E[m]) \cong (R_K/mR_K)^*,$$

and hence $\text{Gal}(L_m/H)$ is abelian. $\qquad\qquad\qquad\qquad\qquad\qquad \square$

Before proceeding with the general theory, we will pause to construct a few more examples of elliptic curves having complex multiplication. More precisely, we will find all elliptic curves that possess an endomorphism of degree 2. We already know one such curve, namely the curve $y^2 = x^3 + x$ with complex multiplication by $\mathbb{Z}[i]$, since the map $[1 + i]$ has degree 2. From (1.5b), we need to find all quadratic imaginary fields K that have an element $\alpha \in R_K$ satisfying $|N_{\mathbb{Q}}^K \alpha| = 2$. This is an easy exercise (which we leave to the reader), the answer being that there are three such fields:

$$K = \mathbb{Q}(\sqrt{-1}), \qquad R_K = \mathbb{Z}[\sqrt{-1}], \qquad \alpha = 1 + \sqrt{-1};$$

$$K = \mathbb{Q}(\sqrt{-2}), \qquad R_K = \mathbb{Z}[\sqrt{-2}], \qquad \alpha = \sqrt{-2};$$

$$K = \mathbb{Q}(\sqrt{-7}), \qquad R_K = \mathbb{Z}\left[\frac{1 + \sqrt{-7}}{2}\right], \qquad \alpha = \frac{1 + \sqrt{-7}}{2}.$$

Since all three of these rings R_K have class number 1, we know from (2.2.1) that the corresponding elliptic curves have j-invariants in $\mathbb{Q}$.

How can we find equations for these curves and their endomorphisms? It is possible to proceed analytically, but we will take another approach. If $\phi : E \to E$ has degree 2, then its kernel $E[\phi]$ consists of two points, O and a point of order 2. If we move the point of order 2 to $(0,0)$, then E will have a Weierstrass equation of the form

$$E : y^2 = x^3 + ax^2 + bx.$$

For this elliptic curve E, we have already determined an elliptic curve E' and an isogeny $\phi : E \to E'$ whose kernel is $\{O, (0,0)\}$, namely

$$E' : Y^2 = X^3 - 2aX^2 + (a^2 - 4b)X, \quad \phi(x,y) = \left(x + a + \frac{b}{x}, y\left(1 - \frac{b}{x^2}\right) \right).$$

(See [AEC III.4.5], although note we have taken the negative of the isogeny defined there and have substituted $x^3 + ax^2 + bx$ for y^2.) Hence E will possess an endomorphism of degree 2 if and only if this E' is isomorphic to E.

To see when E and E' are isomorphic, we set their j-invariants to be equal and solve for a and b, or more precisely for the ratio a^2/b. Now

$$j(E) = \frac{256(a^2 - 3b)^3}{b^2(a^2 - 4b)} \quad \text{and} \quad j(E') = \frac{16(a^2 + 12b)^3}{b(a^2 - 4b)^2}.$$

Setting $j(E) = j(E')$, we find after some calculation that

$$16b(a^2 - 4b)a^2(a^2 - 8b)(16a^4 - 81a^2b + 324b^2) = 0.$$

The first two cases, $b = 0$ and $a^2 - 4b = 0$, give singular curves, so we discard them. The third case, $a = 0$, gives the curve $y^2 = x^3 + bx$ with $j(E) = 1728$ and complex multiplication by $\mathbb{Z}[i]$.

Next consider the case $a^2 - 8b = 0$. Taking $b = 2$ and $a = 4$ gives the curve

$$E : y^2 = x^3 + 4x^2 + 2x$$

with $j(E) = 8000 = 2^6 5^3$. Similarly, E' is given by the equation

$$E' : Y^2 = X^3 - 8X^2 + 8X$$

with j-invariant $j(E') = 8000$. Hence E and E' are isomorphic, and we easily find an isomorphism

$$E' \longrightarrow E, \quad (X, Y) \longmapsto \left(-\frac{1}{2}X, -\frac{1}{2\sqrt{-2}}Y \right).$$

Composing the isogeny $E \to E'$ with this isomorphism, we obtain the desired endomorphism of E of degree 2:

$$E : y^2 = x^3 + 4x^2 + 2x \overset{[\sqrt{-2}]}{\longmapsto} \qquad E$$

$$(x, y) \longmapsto \left(-\frac{1}{2} \left(x + 4 + \frac{2}{x} \right), -\frac{y}{2\sqrt{-2}} \left(1 - \frac{2}{x^2} \right) \right).$$

There remains the case $16a^4 - 81a^2b + 324b^2$, which (taking $a = 36$) leads to the elliptic curves

$$E : y^2 = x^3 + 36x^2 + 18 \left(9 + 5\sqrt{-7} \right) x,$$

$$E' : Y^2 = X^3 - 72X^2 + 72 \left(9 - 5\sqrt{-7} \right) X$$

with $j(E) = j(E') = -3375 = -3^3 5^3$. We will leave it to the reader to make the appropriate variable changes which lead to models for E and E' over $\mathbb{Q}$ and to an explicit formula for the corresponding endomorphism of degree 2. The final answer is given in the following summary of our calculations.

Proposition 2.3.1. *There are exactly three isomorphism classes of elliptic curves over $\mathbb{C}$ which possess an endomorphism of degree 2. The following are representatives for these curves and endomorphisms.*

(i) $E : y^2 = x^3 + x,$ $j = 1728,$ $\alpha = 1 + \sqrt{-1},$

$$[\alpha](x, y) = \left(\alpha^{-2} \left(x + \frac{1}{x} \right), \alpha^{-3} y \left(1 - \frac{1}{x^2} \right) \right);$$

(ii) $E : y^2 = x^3 + 4x^2 + 2x,$ $j = 8000,$ $\alpha = \sqrt{-2},$

$$[\alpha](x, y) = \left(\alpha^{-2} \left(x + 4 + \frac{2}{x} \right), \alpha^{-3} y \left(1 - \frac{2}{x^2} \right) \right);$$

(iii) $E : y^2 = x^3 - 35x + 98,$ $j = -3375,$ $\alpha = \dfrac{1 + \sqrt{-7}}{2},$

$$[\alpha](x, y) = \left(\alpha^{-2} \left(x - \frac{7(1 - \alpha)^4}{x + \alpha^2 - 2} \right), \alpha^{-3} y \left(1 + \frac{7(1 - \alpha)^4}{(x + \alpha^2 - 2)^2} \right) \right).$$

We now resume our development of the general theory of complex multiplication. From here on we will use (2.1) to identify $\mathcal{ELL}(R_K)$ with the $\bar{\mathbb{Q}}$-isomorphism classes of elliptic curves having complex multiplication by R_K. Then there is a natural action of $\mathrm{Gal}(\bar{K}/K)$ on $\mathcal{ELL}(R_K)$ defined by the property that $\sigma \in \mathrm{Gal}(\bar{K}/K)$ sends the isomorphism class of E to the isomorphism class of E^σ. On the other hand, (1.2b) says that the action of the class group $\mathcal{CL}(R_K)$ on $\mathcal{ELL}(R_K)$ is simply transitive, so there is a unique $\bar{\mathfrak{a}} \in \mathcal{CL}(R_K)$, depending on σ, such that $\bar{\mathfrak{a}} * E = E^\sigma$. In other words, there is a well defined map

$$F : \mathrm{Gal}(\bar{K}/K) \longrightarrow \mathcal{CL}(R_K)$$

characterized by the property

$$E^\sigma = F(\sigma) * E \qquad \text{for all } \sigma \in \mathrm{Gal}(\bar{K}/K).$$

It is by studying this map F that we will be able to precisely describe the field $K\big(j(E)\big)$. An easy property of F, which we will prove below, is that F is a homomorphism. A much deeper property, which we will also prove, is that F is independent of the choice of the curve $E \in \mathcal{ELL}(R_K)$. The astute reader will have noticed that F is actually well defined on the larger group $\mathrm{Gal}(\bar{\mathbb{Q}}/\mathbb{Q})$. However, it is only on the smaller group $\mathrm{Gal}(\bar{K}/K)$ that F will be independent of E.

Before proving these basic properties about F, we want to stress that the definition of F has an essential analytic component, since $F(\sigma)$ depends on the way in which the lattice of an elliptic curves changes when the lattice is multiplied by an ideal. Thus if we denote by $j(\Lambda)$ the j-invariant of the elliptic curve E_Λ, then as described in Chapter I, $j(\Lambda)$ is an analytic function of Λ. The map F is then characterized by the formula

$$j(\Lambda)^\sigma = j\big(F(\sigma)^{-1}\Lambda\big),$$

so F converts the algebraic action of σ into the analytic action of multiplication by $F(\sigma)^{-1}$.

Proposition 2.4. *Let $K/\mathbb{Q}$ be a quadratic imaginary field. There exists a homomorphism*

$$F : \mathrm{Gal}(\bar{K}/K) \longrightarrow \mathcal{CL}(R_K)$$

uniquely characterized by the condition

$$E^\sigma = F(\sigma) * E \qquad \text{for all } \sigma \in \mathrm{Gal}(\bar{K}/K) \text{ and all } E \in \mathcal{ELL}(R_K).$$

PROOF. As described above, (2.1) and (1.2b) ensure that for any element $\sigma \in \mathrm{Gal}(\bar{K}/K)$ and any $E \in \mathcal{ELL}(R_K)$, there is a unique $\bar{\mathfrak{a}} \in \mathcal{CL}(R_K)$ with $E^\sigma = \bar{\mathfrak{a}} * E$. So for a fixed E, we get a well-defined map

$$F : \mathrm{Gal}(\bar{K}/K) \to \mathcal{CL}(R_K)$$

determined by the property $E^\sigma = F(\sigma) * E$ for all $\sigma \in \mathrm{Gal}(\bar{K}/K)$. It is easy to check that F is a homomorphism, since

$$F(\sigma\tau) * E = E^{\sigma\tau} = (E^\tau)^\sigma = \big(F(\tau) * E\big)^\sigma$$
$$= F(\sigma) * \big(F(\tau) * E\big) = \big(F(\sigma)F(\tau)\big) * E.$$

(Note that $\text{Gal}(\bar{K}/K)$ acts on the left.)

It remains to show that the definition of F is independent of the choice of a particular elliptic curve in $\mathcal{ELL}(R_K)$. So let $E_1, E_2 \in \mathcal{ELL}(R_K)$, let $\sigma \in \text{Gal}(\bar{K}/K)$, and write $E_1^\sigma = \bar{\mathfrak{a}}_1 * E_1$ and $E_2^\sigma = \bar{\mathfrak{a}}_2 * E_2$. We need to show that $\bar{\mathfrak{a}}_1 = \bar{\mathfrak{a}}_2$. Since $\mathcal{CL}(R_K)$ acts transitively on $\mathcal{ELL}(R_K)$, we can find some $\bar{\mathfrak{b}}$ with $E_2 = \bar{\mathfrak{b}} * E_1$. Then

$$(\bar{\mathfrak{b}} * E_1)^\sigma = E_2^\sigma = \bar{\mathfrak{a}}_2 * E_2 = \bar{\mathfrak{a}}_2 * (\bar{\mathfrak{b}} * E_1) = (\bar{\mathfrak{a}}_2 \bar{\mathfrak{b}} \bar{\mathfrak{a}}_1^{-1}) * E_1^\sigma.$$

So if we can prove that $(\bar{\mathfrak{b}} * E_1)^\sigma$ is equal to $\bar{\mathfrak{b}} * E_1^\sigma$, then we can cancel $\bar{\mathfrak{b}}$ from both sides to conclude that $E_1^\sigma = (\bar{\mathfrak{a}}_2 \bar{\mathfrak{a}}_1^{-1}) * E_1^\sigma$; and then (1.2(iii)) will give $\bar{\mathfrak{a}}_1 = \bar{\mathfrak{a}}_2$. Hence the following proposition completes the proof of Proposition 2.4. (Note that $\bar{\mathfrak{b}}^\sigma = \bar{\mathfrak{b}}$, since $\mathfrak{b} \subset K$ and $\sigma \in \text{Gal}(\bar{K}/K)$.)

Proposition 2.5. *Let $E/\bar{\mathbb{Q}}$ be an elliptic curve representing an element of $\mathcal{ELL}(R_K)$, let $\bar{\mathfrak{a}} \in \mathcal{CL}(R_K)$, and let $\sigma \in \text{Gal}(\bar{\mathbb{Q}}/\mathbb{Q})$. Then*

$$(\bar{\mathfrak{a}} * E)^\sigma = \bar{\mathfrak{a}}^\sigma * E^\sigma.$$

Although the statement of Proposition 2.5 looks relatively innocuous, it is giving a relationship between the algebraic action of σ and the analytic action of multiplication by $\bar{\mathfrak{a}}$. This suggests that the proof may not be entirely straightforward. The main idea is to find an algebraic description of $\bar{\mathfrak{a}} * E$. One of the tools we will need is the following lemma from commutative algebra, whose proof we leave as an exercise.

Lemma 2.5.1. *Let R be a Dedekind domain, let $\mathfrak{a}$ be a fractional ideal of R, and let M be a torsion-free R-module. Then the natural map*

$$\phi: \quad \mathfrak{a}^{-1}M \quad \longrightarrow \quad \text{Hom}_R(\mathfrak{a}, M)$$
$$x \quad \longmapsto \quad (\phi_x : \alpha \mapsto \alpha x)$$

is an isomorphism.

PROOF (of Proposition 2.5). Choose a lattice Λ so that $E \cong E_\Lambda$. Also fix a resolution (i.e., an exact sequence)

$$R_K^m \xrightarrow{A} R_K^n \longrightarrow \mathfrak{a} \longrightarrow 0, \tag{i}$$

where A is an $m \times n$ matrix with coefficients in R_K. The idea underlying the proof of Proposition 2.5 is that we should have

$$\mathbb{C}/\mathfrak{a}^{-1}\Lambda \cong \bar{\mathfrak{a}} * E \cong \text{Hom}(\mathfrak{a}, E),$$

where we want to describe $\text{Hom}(\mathfrak{a}, E)$ as an algebraic variety and not just as an R_K-module. (Here and in the following, Hom means homomorphisms of R_K-modules.) We begin by applying Hom to the "product" of the exact sequence (i) and the exact sequence of R_K-modules

$$0 \longrightarrow \Lambda \longrightarrow \mathbb{C} \longrightarrow E \longrightarrow 0. \qquad (ii)$$

This gives us the following commutative diagram:

$$
\begin{array}{ccccc}
0 & & 0 & & 0 \\
\downarrow & & \downarrow & & \downarrow \\
0 \longrightarrow \operatorname{Hom}(\mathfrak{a},\Lambda) & \longrightarrow & \operatorname{Hom}(\mathfrak{a},\mathbb{C}) & \longrightarrow & \operatorname{Hom}(\mathfrak{a},E) \\
\downarrow & & \downarrow & & \downarrow \\
0 \longrightarrow \operatorname{Hom}(R_K^n,\Lambda) & \longrightarrow & \operatorname{Hom}(R_K^n,\mathbb{C}) & \longrightarrow & \operatorname{Hom}(R_K^n,E) \\
\downarrow{\scriptstyle A} & & \downarrow{\scriptstyle A} & & \downarrow{\scriptstyle A} \\
0 \longrightarrow \operatorname{Hom}(R_K^m,\Lambda) & \longrightarrow & \operatorname{Hom}(R_K^m,\mathbb{C}) & \longrightarrow & \operatorname{Hom}(R_K^m,E)
\end{array}
\qquad (iii)
$$

For any R_K-module M, we have $\operatorname{Hom}(R_K^n, M) \cong M^n$, and applying Lemma 2.5.1, first with $M = \Lambda$ and then with $M = \mathbb{C}$, we get

$$\operatorname{Hom}(\mathfrak{a},\Lambda) = \mathfrak{a}^{-1}\Lambda \qquad \text{and} \qquad \operatorname{Hom}(\mathfrak{a},\mathbb{C}) = \mathfrak{a}^{-1}\mathbb{C} = \mathbb{C}.$$

Using these isomorphisms, we can rewrite the diagram (iii) as

$$
\begin{array}{ccccc}
0 & & 0 & & 0 \\
\downarrow & & \downarrow & & \downarrow \\
0 \longrightarrow \mathfrak{a}^{-1}\Lambda & \longrightarrow & \mathbb{C} & \longrightarrow & \operatorname{Hom}(\mathfrak{a},E) \\
\downarrow & & \downarrow & & \downarrow \\
0 \longrightarrow \Lambda^n & \longrightarrow & \mathbb{C}^n & \longrightarrow & E^n \longrightarrow 0 \\
\downarrow{\scriptstyle {}^tA} & & \downarrow{\scriptstyle {}^tA} & & \downarrow{\scriptstyle {}^tA} \\
0 \longrightarrow \Lambda^m & \longrightarrow & \mathbb{C}^m & \longrightarrow & E^m \longrightarrow 0
\end{array}
\qquad (iv)
$$

Here ${}^tA$ is the transpose of the matrix A, and the bottom two rows are clearly exact on the right, since they are just a number of copies of the exact sequence (ii).

Applying the snake lemma to the bottom two rows of (iv) gives the exact sequence

$$0 \longrightarrow \mathfrak{a}^{-1}\Lambda \longrightarrow \mathbb{C} \longrightarrow \left(\ker E^n \overset{{}^tA}{\to} E^m\right) \longrightarrow \Lambda^n / {}^tA\Lambda^m. \qquad (v)$$

Notice that $E^n \overset{{}^tA}{\to} E^m$ is an algebraic map of algebraic varieties, since ${}^tA$ is an $m \times n$ matrix whose coefficients are elements of $\mathrm{End}(E) = R_K$. Hence the inverse image of the point $(0, 0, \ldots, 0) \in E^m$ is an algebraic subvariety of E^n. Of course, E^n and E^m are group varieties, so what we are saying is that the kernel of $E^n \overset{{}^tA}{\to} E^m$ is an algebraic group variety. Further, (2.2a) says that for any $\sigma \in \mathrm{Aut}(\mathbb{C})$, the corresponding map from $E^{\sigma n} \to E^{\sigma m}$ is obtained by applying σ to the entries of ${}^tA$, treating those entries as elements of $R_K \subset \mathbb{C}$.

On the other hand, looking at the complex topology for one more moment, we note that $\Lambda^n / {}^t A \Lambda^m$ is discrete and $\mathbb{C}/\mathfrak{a}^{-1}\Lambda$ is connected. Hence the exact sequence (v) gives

$$(\mathfrak{a} * E)(\mathbb{C}) = \mathbb{C}/\mathfrak{a}^{-1}\Lambda \cong \text{identity component of } \ker(E^n \overset{{}^tA}{\to} E^m).$$

We have thus described $\mathfrak{a} * E$ algebraically in terms of the algebraic map $E^n \overset{{}^tA}{\to} E^m$, and it now easy to finish the proof of Proposition 2.5. For any $\sigma \in \mathrm{Gal}(\bar{\mathbb{Q}}/\mathbb{Q})$, we apply our characterization first to E and then to E^σ to deduce that

$$(\mathfrak{a} * E)^\sigma = \Big(\text{identity component of } \ker(E^n \overset{{}^tA}{\to} E^m)\Big)^\sigma$$
$$= \text{identity component of } \ker\Big((E^\sigma)^n \overset{{}^tA^\sigma}{\to} (E^\sigma)^m\Big)$$
$$= \mathfrak{a}^\sigma * E^\sigma.$$

This completes the proof of Proposition 2.5, and with it the proof of Proposition 2.4.

$$\square$$

§3. Class Field Theory — A Brief Review

Class field theory describes the abelian extensions of a number field K in terms of the arithmetic of K. The theory of complex multiplication provides an analytic realization of class field theory for quadratic imaginary fields, much as cyclotomic theory gives a realization of class field theory for $\mathbb{Q}$. In this section we will briefly review, without proof, the basic facts from class field theory which will be used in the sequel. We will begin with the classical version using ideals and ideal class groups. Afterwards we will present the more modern idelic version. For proofs of the theorems stated in this section and for additional material on (global) class field theory, the reader might consult Lang [5], Tate [7], or Neukirch [1]. We will mostly restrict attention to totally imaginary fields, that is, fields with no real embeddings, since (except for §7) that is the only case we will use in the sequel.

Let K be a totally imaginary number field and let L be a finite abelian extension of K; that is, L/K is Galois with abelian Galois group. As usual, we write R_K and R_L for the rings of integers of K and L respectively. Let $\mathfrak{p}$ be a prime of K which does not ramify in L, and let $\mathfrak{P}$ be a prime of L lying over $\mathfrak{p}$. Thus the picture is

L	$\mathfrak{P}$	$R_L/\mathfrak{P}$
finite abelian extension	unramified prime	extension of finite fields
K	$\mathfrak{p}$	$R_K/\mathfrak{p}$

By restriction, we get a homomorphism from the decomposition group of $\mathfrak{P}$ to the Galois group of the residue fields,

$$\{\sigma \in \mathrm{Gal}(L/K) : \mathfrak{P}^\sigma = \mathfrak{P}\} \longrightarrow \begin{pmatrix} \text{Galois group of} \\ R_L/\mathfrak{P} \text{ over } R_K/\mathfrak{p} \end{pmatrix}.$$

The right-hand Galois group is cyclic, generated by the Frobenius automorphism

$$x \longmapsto x^{\mathrm{N}_\mathbb{Q}^K \mathfrak{p}}.$$

Further, since $\mathfrak{p}$ is unramified, there is a unique element $\sigma_\mathfrak{p} \in \mathrm{Gal}(L/K)$ which maps to Frobenius. Our notation reflects the fact that $\sigma_\mathfrak{p}$ is determined by the prime ideal $\mathfrak{p}$ in K. For a general Galois extension L/K, $\mathfrak{p}$ will only determine the conjugacy class of $\sigma_\mathfrak{p}$, and making a new choice for $\mathfrak{P}$ will change $\sigma_\mathfrak{p}$ by conjugation. But in our situation $\sigma_\mathfrak{p}$ will not change, since we have assumed that L/K is abelian. Thus $\sigma_\mathfrak{p} \in \mathrm{Gal}(L/K)$ is uniquely determined by the condition

$$\sigma_\mathfrak{p}(x) \equiv x^{\mathrm{N}_\mathbb{Q}^K \mathfrak{p}} \pmod{\mathfrak{P}} \qquad \text{for all } x \in R_L.$$

Let $\mathfrak{c}$ be an integral ideal of K that is divisible by all primes that ramify in L/K, and let

$I(\mathfrak{c}) = $ group of fractional ideals of K which are relatively prime to $\mathfrak{c}$.

Then the *Artin map* is defined using the $\sigma_\mathfrak{p}$'s and linearity:

$$(\,\cdot\,, L/K) : I(\mathfrak{c}) \longrightarrow \mathrm{Gal}(L/K),$$

$$(\mathfrak{a}, L/K) = \left(\prod_\mathfrak{p} \mathfrak{p}^{n_\mathfrak{p}}, L/K \right) \overset{\mathrm{def}}{=} \prod_\mathfrak{p} \sigma_\mathfrak{p}^{n_\mathfrak{p}}.$$

Notice that the Artin map is defined by piecing together local information, one prime at a time. The following theorem, which is a weak version of Artin's reciprocity law, provides important global information.

Proposition 3.1. (Artin Reciprocity) *Let L/K be a finite abelian extension of number fields. There exists an integral ideal $\mathfrak{c} \subset R_K$, divisible by precisely the primes of K that ramify in L, such that*

$$\big((\alpha), L/K\big) = 1 \qquad \text{for all } \alpha \in K^* \text{ satisfying } \alpha \equiv 1 \,(\mathrm{mod}\; \mathfrak{c}).$$

If (3.1) is true for the ideals $\mathfrak{c}_1$ and $\mathfrak{c}_2$, then it also true for $\mathfrak{c}_1 + \mathfrak{c}_2$. There is thus a largest ideal for which (3.1) is true. We call this ideal the *conductor of L/K* and denote it by $\mathfrak{c}_{L/K}$.

In view of (3.1), it is natural to define the group of principal ideals congruent to 1 modulo $\mathfrak{c}$:

$$P(\mathfrak{c}) = \big\{ (\alpha) : \alpha \in K^*, \; \alpha \equiv 1 \,(\mathrm{mod}\; \mathfrak{c}) \big\}.$$

Artin reciprocity says that the kernel of the Artin map contains $P(\mathfrak{c})$ for an appropriate choice of $\mathfrak{c}$. More precisely,

$$\mathfrak{a} \in P(\mathfrak{c}_{L/K}) \Longrightarrow (\mathfrak{a}, L/K) = 1.$$

It is important to observe that a principal ideal (α) may be in $P(\mathfrak{c})$ even if $\alpha \not\equiv 1 \,(\mathrm{mod}\; \mathfrak{c})$; all that is necessary is that there exist a unit $\xi \in R_K^*$ such that $\xi\alpha \equiv 1 \,(\mathrm{mod}\; \mathfrak{c})$.

Let $\mathfrak{p}$ be a prime of K which is unramified in L. Then $\mathfrak{p}$ splits completely in L if and only if the extension of residue fields has degree 1, or equivalently if and only if $(\mathfrak{p}, L/K) = 1$. Thus the unramified prime ideals in the kernel of the Artin map are precisely the primes of K that split completely in L.

Definition. Let $\mathfrak{c}$ be an integral ideal of K. A *ray class field of K (modulo $\mathfrak{c}$)* is a finite abelian extension $K_\mathfrak{c}/K$ with the property that for any finite abelian extension L/K,

$$\mathfrak{c}_{L/K} \,|\, \mathfrak{c} \Longrightarrow L \subset K_\mathfrak{c}.$$

Intuitively, one can think of the ray class field as the "largest" field with a given conductor. However, it is important to note that the conductor of $K_\mathfrak{c}$ need not actually equal $\mathfrak{c}$. For example, the ray class field of $\mathbb{Q}(i)$ modulo the ideal (2) is just $\mathbb{Q}(i)$ itself, so $\mathbb{Q}(i)_{(2)}$ has conductor (1).

Theorem 3.2. (Class Field Theory) *Let L/K be a finite abelian extension of number fields, and let $\mathfrak{c}$ be an integral ideal of K.*
(a) *The Artin map*

$$(\,\cdot\,, L/K) : I(\mathfrak{c}_{L/K}) \longrightarrow \mathrm{Gal}(L/K)$$

is a surjective homomorphism.

(b) *The kernel of the Artin map is* $(\mathrm{N}_K^L I_L)P(\mathfrak{c}_{L/K})$, *where* I_L *is the group of non-zero fractional ideals of* L.

(c) *There exists a unique ray class field* $K_\mathfrak{c}$ *of* K *(modulo* $\mathfrak{c}$*). The conductor of* $K_\mathfrak{c}/K$ *divides* $\mathfrak{c}$.

(d) *The ray class field* $K_\mathfrak{c}$ *is characterized by the property that it is an abelian extension of* K *and satisfies*

$$\left\{ \begin{array}{c} \text{primes of } K \text{ that} \\ \text{split completely in } K_\mathfrak{c} \end{array} \right\} = \{ \text{prime ideals in } P(\mathfrak{c}) \}.$$

Example 3.3. Consider the ray class field of K modulo the unit ideal $\mathfrak{c} = (1)$. It is the maximal abelian extension of K which is unramified at all primes. We call $K_{(1)}$ the *Hilbert class field of* K and denote it by H or H_K. Notice that

$$I(\mathfrak{c}_{H/K}) = I\big((1)\big) = \{ \text{all non-zero fractional ideals of } K \},$$

$$P(\mathfrak{c}_{H/K}) = P\big((1)\big) = \{ \text{all non-zero principal ideals of } K \},$$

so the Artin map induces an isomorphism between the ideal class group of K and the Galois group of the Hilbert class field of K:

$$(\cdot, H/K) : \mathcal{CL}(R_K) \xrightarrow{\sim} \mathrm{Gal}(H/K).$$

We will also need the following version of Dirichlet's theorem on primes in arithmetic progressions.

Theorem 3.4. *Let* K *be a number field and* $\mathfrak{c}$ *an integral ideal of* K. *Then every ideal class in* $I(\mathfrak{c})/P(\mathfrak{c})$ *contains infinitely many degree 1 primes of* K.

The Idelic Formulation of Class Field Theory

We will now briefly recall how class field theory is formulated using ideles. This material will not be used until §7, so the reader may wish to omit the rest of this section until arriving at that point.

Let K be a number field, and for each absolute value v on K, let K_v be the completion of K at v. Further, let R_v be the ring of integers of K_v if v is non-archimedean, and let $R_v = K_v$ otherwise. The idele group of K is the group

$$\mathbf{A}_K^* = \prod_v{}' K_v^*,$$

where prime indicates that the product is restricted relative to the R_v's. This means that an element $s \in \prod K_v^*$ in the unrestricted product is in $\mathbf{A}_K^*$

if and only if $x_v \in R_v^*$ for all but finitely many v. In particular, we can embed K^* into $\mathbf{A}_K^*$ by using the natural diagonal embedding

$$K \lhook\joinrel\longrightarrow \mathbf{A}_K^*, \qquad \alpha \longmapsto (\cdots, \alpha, \alpha, \alpha, \cdots),$$

since any $\alpha \in K^*$ is in R_v^* for all but finitely many K. Similarly, for any given v we embed K_v^* as a subgroup of $\mathbf{A}_K^*$ via

$$K_v^* \lhook\joinrel\longrightarrow \mathbf{A}_K^*, \qquad t \longmapsto (\ldots, 1, 1, t, 1, 1, \ldots).$$
$$\underset{v\text{-component}}{\uparrow}$$

If v is a non-archimedean absolute value corresponding to a prime ideal $\mathfrak{p}$, we will often write $K_{\mathfrak{p}}$ and $R_{\mathfrak{p}}$ in place of K_v and R_v. We will also write $\mathrm{ord}_{\mathfrak{p}}$ for the corresponding normalized valuation.

Let $s \in \mathbf{A}_K^*$ be an idele. We define the *ideal of s* to be the fractional ideal of K given by

$$(s) = \prod_{\mathfrak{p}} \mathfrak{p}^{\mathrm{ord}_{\mathfrak{p}} s_{\mathfrak{p}}},$$

where the product is over all prime ideal of K. Note that (s) is well defined, since $s_{\mathfrak{p}}$ is a $\mathfrak{p}$-adic unit for all but finitely many $\mathfrak{p}$.

One makes $\mathbf{A}_K^*$ into a topological group in the usual way; we will not need the precise definition of the topology. For any integral ideal $\mathfrak{c}$ of K, let $U_{\mathfrak{c}}$ be the subgroup of $\mathbf{A}_K^*$ defined by

$$U_{\mathfrak{c}} = \left\{ s \in \mathbf{A}_K^* : s_{\mathfrak{p}} \in R_{\mathfrak{p}}^* \text{ and } s_{\mathfrak{p}} \equiv 1 \,(\mathrm{mod}\ \mathfrak{c}R_{\mathfrak{p}}) \text{ for all primes } \mathfrak{p} \right\}.$$

Then $U_{\mathfrak{c}}$ is an open subgroup of $\mathbf{A}_K^*$, and one proves that $K^*U_{\mathfrak{c}}$ is a subgroup of finite index in $\mathbf{A}_K^*$.

If L/K is a finite extension, then there is a natural norm map from $\mathbf{A}_L^*$ to $\mathbf{A}_K^*$. This is a continuous homomorphism

$$\mathrm{N}_K^L : \mathbf{A}_L^* \longrightarrow \mathbf{A}_K^*$$

defined by the prescription that the v-component of $\mathrm{N}_K^L x$ is

$$\prod_{w|v} \mathrm{N}_{K_v}^{L_w} x_w.$$

The idelic formulation of class field theory is given in terms of the reciprocity map described in the following theorem.

Theorem 3.5. *Let K be a number field, and let K^{ab} be the maximal abelian extension of K. There exists a unique continuous homomorphism*

$$\mathbf{A}_K^* \longrightarrow \mathrm{Gal}(K^{\mathrm{ab}}/K), \qquad s \longmapsto [s, K],$$

with the following property:

> *Let L/K be a finite abelian extension, and let $s \in \mathbf{A}_K^*$ be an idele whose ideal (s) is not divisible by any primes that ramify in L. Then*

$$[s, K]\big|_L = \big((s), L/K\big).$$

Here $(\,\cdot\,, L/K)$ is the Artin map, and $\mathrm{Gal}(K^{\mathrm{ab}}/K)$ is given the usual profinite topology. The homomorphism $[\,\cdot\,, K]$ is called the *reciprocity map for K*.

The reciprocity map has the following additional properties:

(a) The reciprocity map is surjective, and K^* is contained in its kernel.

(b) The reciprocity map is compatible with the norm map,

$$[x, L]\big|_{K^{\mathrm{ab}}} = \big[\mathrm{N}_K^L x, K\big] \qquad \text{for all } x \in \mathbf{A}_L^*.$$

(c) Let $\mathfrak{p}$ be a prime ideal of K, let $I_{\mathfrak{p}}^{\mathrm{ab}} \subset \mathrm{Gal}(K^{\mathrm{ab}}/K)$ be the inertia group of $\mathfrak{p}$ for the extension K^{ab}/K, let $\pi_{\mathfrak{p}} \in K_{\mathfrak{p}}^*$ be a uniformizer at $\mathfrak{p}$, and let L/K be any abelian extension that is unramified at $\mathfrak{p}$. Then

$$[\pi_{\mathfrak{p}}, K]\big|_L = (\mathfrak{p}, L/K) = \text{Frobenius for } L/K \text{ at } \mathfrak{p},$$

and

$$[R_{\mathfrak{p}}^*, K] = I_{\mathfrak{p}}^{\mathrm{ab}}.$$

There is, of course, much more to class field theory that we have not mentioned. For example, one often wants to know the exact kernel of the reciprocity map and the correspondence between subgroups of $\mathbf{A}_K^*$ and subfields of K^{ab}. However, the only additional fact that we will need in this chapter is the following idelic characterization of ray class fields.

Theorem 3.6. *Let K be a number field, let $\mathfrak{c}$ be an integral ideal of K, let $K_{\mathfrak{c}}$ be the ray class field of K modulo $\mathfrak{c}$, and let $U_{\mathfrak{c}}$ be the subgroup of $\mathbf{A}_K^*$ described above. Then the reciprocity map induces an isomorphism*

$$[\,\cdot\,, K] : \mathbf{A}_K^*/K^* U_{\mathfrak{c}} \xrightarrow{\sim} \mathrm{Gal}(K_{\mathfrak{c}}/K).$$

In other words, $[s, K]$ acts trivially on the ray class field $K_{\mathfrak{c}}$ if and only if s can be written as $s = \alpha u$ with $\alpha \in K^*$ and $u \in U_{\mathfrak{c}}$.

§4. The Hilbert Class Field

Our goal in this section is to prove the following theorem.

Theorem 4.1. *Let $K/\mathbb{Q}$ be a quadratic imaginary field with ring of integers R_K, and let $E/\mathbb{C}$ be an elliptic curve with $\mathrm{End}(E) \cong R_K$. Then $K(j(E))$ is the Hilbert class field H of K.*

Remark 4.1.1. Note that it is easy to produce an elliptic curve with endomorphism ring equal to R_K. For example, we could take E to be the curve corresponding to the lattice R_K. Then

$$j(E) = j(R_K) = 1728 \frac{g_2(R_K)^3}{g_2(R_K)^3 - 27g_3(R_K)^2}$$

is given in terms of series $g_2(R_K)$ and $g_3(R_K)$ involving the elements of R_K. Alternatively, if we write $R_K = \mathbb{Z}\tau + \mathbb{Z}$, then

$$j(E) = j(R_K) = \frac{1}{e^{2\pi i \tau}} + \sum_{n=0}^{\infty} c(n)e^{2\pi i n \tau},$$

where the $c(n) \in \mathbb{Z}$ are the coefficients in the q-series expansion of j (I.7.4b). So Theorem 4.1 says that the Hilbert class field of a quadratic imaginary field K is generated by the value of a certain holomorphic function $j(\tau)$ evaluated at a generator for the ring of integers of K.

We will actually prove much more than the mere statement of Theorem 4.1. We will give an explicit description of how the Galois group of H/K acts on $j(E)$. To do this, we recall the homomorphism

$$F : \mathrm{Gal}(\bar{K}/K) \longrightarrow \mathcal{CL}(R_K)$$

from §2 characterized by the condition

$$E^{\sigma} = F(\sigma) * E \qquad \text{for all } \sigma \in \mathrm{Gal}(\bar{K}/K) \text{ and all } E \in \mathcal{ELL}(R_K).$$

Note that the kernel of F is actually a finite quotient of $\mathrm{Gal}(\bar{K}/K)$, since any E will be defined over some finite extension L/K, and then $F(\sigma) = 1$ for $\sigma \in \mathrm{Gal}(\bar{K}/L)$. Since $\mathcal{CL}(R_K)$ is an abelian group, F factors through

$$F : \mathrm{Gal}(K^{\mathrm{ab}}/K) \longrightarrow \mathcal{CL}(R_K),$$

where K^{ab} is the maximal abelian extension of K. Recall also the Frobenius element $\sigma_{\mathfrak{p}} \in \mathrm{Gal}(K^{\mathrm{ab}}/K)$ corresponding to a prime $\mathfrak{p}$ in K. The following proposition, together with basic class field theory, will serve to completely determine F.

Proposition 4.2. *There is a finite set of rational primes $S \subset \mathbb{Z}$ such that if $p \notin S$ is a prime which splits in K, say as $pR_K = \mathfrak{p}\mathfrak{p}'$, then*

$$F(\sigma_\mathfrak{p}) = \bar{\mathfrak{p}} \in \mathcal{CL}(R_K).$$

Proposition 4.2 does not look very strong, since it determines F on fewer than half of all Frobenius elements. But we will be able to use it to get complete information about F. Before proceeding with the proof of Proposition 4.2, we will derive some of its consequences, including a proof of Theorem 4.1.

Theorem 4.3. *Let E be an elliptic curve representing an isomorphism class in $\mathcal{ELL}(R_K)$.*
(a) *$K(j(E))$ is the Hilbert class field H of K.*
(b) *$[\mathbb{Q}(j(E)) : \mathbb{Q}] = [K(j(E)) : K] = h_K$,*
where $h_K = \# \mathcal{CL}(R_K) = \# \operatorname{Gal}(H/K)$ is the class number of K.
(c) *Let $E_1, \ldots, E_h$ be a complete set of representatives for $\mathcal{ELL}(R_K)$. Then $j(E_1), \ldots, j(E_h)$ is a complete set of $\operatorname{Gal}(\bar{K}/K)$ conjugates for $j(E)$.*
(d) *For every prime ideal $\mathfrak{p}$ of K,*

$$j(E)^{\sigma_\mathfrak{p}} = j(\bar{\mathfrak{p}} * E).$$

More generally, for every non-zero fractional ideal $\mathfrak{a}$ of K,

$$j(E)^{(\mathfrak{a}, H/K)} = j(\bar{\mathfrak{a}} * E).$$

Remark 4.3.1. It is now clear why we took the inverse when we defined the action of an ideal class $\bar{\mathfrak{a}}$ on an elliptic curve E_Λ. If we had used the more natural definition $\bar{\mathfrak{a}} * E_\Lambda = E_{\mathfrak{a}\Lambda}$, then the action of the Artin symbol on $j(E)$ in (4.3d) would instead have been $j(E)^{(\mathfrak{a}, H/K)} = j(\bar{\mathfrak{a}}^{-1} * E)$. Thus we put the inverse into the action of $\mathcal{CL}(R_K)$ on $\mathcal{ELL}(R_K)$ so that the Artin symbol would act without an inverse.

PROOF (of Theorem 4.3). Let L/K be the finite extension corresponding to the homomorphism $F : \operatorname{Gal}(\bar{K}/K) \to \mathcal{CL}(R_K)$, by which we mean that L is the fixed field of the kernel of F. Then

$$\operatorname{Gal}(\bar{K}/L) = \ker F$$
$$= \{\sigma \in \operatorname{Gal}(\bar{K}/K) : F(\sigma) = 1\}$$
$$= \{\sigma \in \operatorname{Gal}(\bar{K}/K) : F(\sigma) * E = E\} \quad \text{since by (1.2), } \mathcal{CL}(R_K)$$
$$\qquad\qquad\qquad\qquad\qquad\qquad\quad \text{acts simply transitively on } \mathcal{ELL}(R_K)$$
$$= \{\sigma \in \operatorname{Gal}(\bar{K}/K) : E^\sigma = E\} \quad \text{from the definition of } F$$
$$= \{\sigma \in \operatorname{Gal}(\bar{K}/K) : j(E^\sigma) = j(E)\}$$
$$= \{\sigma \in \operatorname{Gal}(\bar{K}/K) : j(E)^\sigma = j(E)\}$$
$$= \operatorname{Gal}(\bar{K}/K(j(E))).$$

Hence $L = K\big(j(E)\big)$. Further, since F maps $\mathrm{Gal}(L/K)$ injectively into $\mathcal{CL}(R_K)$, we see that L/K is an abelian extension. So we have shown that $L = K\big(j(E)\big)$ is an abelian extension of K.

Let $\mathfrak{c}_{L/K}$ be the conductor of L/K, and consider the composition of the Artin map with F,

$$I(\mathfrak{c}_{L/K}) \xrightarrow{\ (\,\cdot\,,L/K)\ } G_{L/K} \xrightarrow{\ F\ } \mathcal{CL}(R_K).$$

We claim that this composition is just the natural projection of $I(\mathfrak{c}_{L/K})$ onto $\mathcal{CL}(R_K)$. In other words, we wish to establish the

$\qquad Claim:\quad F\big((\mathfrak{a}, L/K)\big) = \bar{\mathfrak{a}} \qquad$ for all $\mathfrak{a} \in I(\mathfrak{c}_{L/K})$.

Let $\mathfrak{a} \in I(\mathfrak{c}_{L/K})$, and let S be the finite set of primes described in (4.2). From Dirichlet's theorem (3.4) there exists a degree 1 prime $\mathfrak{p} \in I(\mathfrak{c}_{L/K})$ in the same $P(\mathfrak{c}_{L/K})$-ideal class as $\mathfrak{a}$ and not lying over a prime in S. In other words, there is an $\alpha \in K^*$ satisfying

$$\alpha \equiv 1 \ (\mathrm{mod}\ \mathfrak{c}_{L/K}) \qquad \text{and} \qquad \mathfrak{a} = (\alpha)\mathfrak{p}.$$

We compute

$$
\begin{aligned}
F\big((\mathfrak{a}, L/K)\big) &= F\big(((\alpha)\mathfrak{p}, L/K)\big) && \text{since } \mathfrak{a} = (\alpha)\mathfrak{p} \\
&= F\big((\mathfrak{p}, L/K)\big) && \text{since } \alpha \equiv 1 \ (\mathrm{mod}\ \mathfrak{c}_{L/K}) \\
&= \bar{\mathfrak{p}} && \text{from (4.2), since } \mathrm{N}^K_{\mathbb{Q}}\mathfrak{p} \notin S \\
&= \bar{\mathfrak{a}} && \text{since } \mathfrak{a} = (\alpha)\mathfrak{p}.
\end{aligned}
$$

This completes the proof of the claim.

Notice that as an immediate consequence we find that

$$F\big(((\alpha), L/K)\big) = 1 \qquad \text{for } all \text{ principal ideals } (\alpha) \in I(\mathfrak{c}_{L/K}),$$

and not just for those that are congruent to 1 modulo $\mathfrak{c}_{L/K}$. We also know that the map $F : \mathrm{Gal}(L/K) \to \mathcal{CL}(R_K)$ is injective, so this implies that

$$((\alpha), L/K) = 1 \qquad \text{for all } (\alpha) \in I(\mathfrak{c}_{L/K}).$$

But the conductor of L/K is the smallest integral ideal $\mathfrak{c}$ with the property that

$$\alpha \equiv 1 \ (\mathrm{mod}\ \mathfrak{c}) \implies ((\alpha), L/K) = 1.$$

(See §3.) It follows that $\mathfrak{c}_{L/K} = (1)$. The conductor is divisible by every prime that ramifies (3.1), from which we conclude that the extension L/K is everywhere unramified. Therefore L is contained in the Hilbert class field H of K.

On the other hand, the natural map $I(\mathfrak{c}_{L/K}) = I((1)) \to \mathcal{CL}(R_K)$ is clearly surjective, so the claim implies that $F : \mathrm{Gal}(L/K) \to \mathcal{CL}(R_K)$ is surjective, hence an isomorphism. Therefore

$$[L : K] = \# \mathrm{Gal}(L/K) = \# \mathcal{CL}(R_K) = \# \mathrm{Gal}(H/K) = [H : K].$$

This combined with the inclusion $L \subset H$ proves that $L = H$. Since $L = K(j(E))$, this completes the proof of (a), as well as the second equality in (b).

To prove the first equality in (b), we use the observation (2.2.1) that

$$[\mathbb{Q}(j(E)) : \mathbb{Q}] \le h_K.$$

This inequality combined with $[K(j(E)) : K] = h_K$ and $[K : \mathbb{Q}] = 2$ implies that $[\mathbb{Q}(j(E)) : \mathbb{Q}] = h_K$, which completes the proof of (b).

Next, from (1.2b) we know that $\mathcal{CL}(R_K)$ acts transitively on the set of j-invariants

$$\mathcal{J} = \{ j(E_1), \ldots, j(E_h) \},$$

since by [AEC III.1.4b] the set $\mathcal{ELL}(R_K)$ may be identified with the j-invariants of its elements. The map $F : \mathrm{Gal}(\bar{K}/K) \to \mathcal{CL}(R_K)$ is defined by identifying the action of $\mathrm{Gal}(\bar{K}/K)$ on $\mathcal{J}$ with the action of $\mathcal{CL}(R_K)$ on $\mathcal{J}$, so $\mathrm{Gal}(\bar{K}/K)$ also acts transitively on $\mathcal{J}$. Therefore $\mathcal{J}$ is a complete set of $\mathrm{Gal}(\bar{K}/K)$ conjugates of $j(E)$, which proves (c).

Finally, we see that the claim proven above gives (d) for all ideals in $I(\mathfrak{c}_{L/K})$. But $\mathfrak{c}_{L/K} = (1)$, so $I(\mathfrak{c}_{L/K})$ is the set of all non-zero fractional ideals of K. $\square$

It remains to prove Proposition 4.2. For that purpose we will need the following result which says that isogenies behave nicely under reduction.

Proposition 4.4. *Let L be a number field, $\mathfrak{P}$ a maximal ideal of L, E_1/L and E_2/L elliptic curves with good reduction at $\mathfrak{P}$, and $\tilde{E}_1$ and $\tilde{E}_2$ their reductions modulo $\mathfrak{P}$. Then the natural reduction map*

$$\mathrm{Hom}(E_1, E_2) \longrightarrow \mathrm{Hom}(\tilde{E}_1, \tilde{E}_2), \qquad \phi \longmapsto \tilde{\phi},$$

is injective. Further, it preserves degrees,

$$\deg(\phi) = \deg(\tilde{\phi}).$$

PROOF. Since the degree of a non-zero isogeny is non-zero, the injectivity follows from the preservation of the degree. However, the proof of injectivity is more elementary, so we will give a separate proof.

Let $\phi : E_1 \to E_2$ be an isogeny satisfying $\tilde{\phi} = [0]$. For any integer m prime to $\mathfrak{P}$, [AEC VII.3.1b] says that $E_2[m]$ injects into $\tilde{E}_2$. On the other hand, if $T \in E_1[m]$, then by assumption

$$\widetilde{\phi(T)} = \tilde{\phi}(\tilde{T}) = \tilde{O}.$$

Since $\phi(T) \in E_2[m]$, it follows that $\phi(T) = O$. Therefore $E_1[m] \subset \ker(\phi)$. This holds for arbitrarily large m, so we must have $\phi = [0]$.

Now we begin the proof that $\deg(\phi) = \deg(\tilde{\phi})$. Choose a rational prime ℓ relatively prime to $\mathfrak{P}$. Our idea is to use the Weil pairing and calculate everything on the Tate modules. (See [AEC III.8.3] for the properties of the Weil pairing $e_E : T_\ell(E) \times T_\ell(E) \to T_\ell(\boldsymbol{\mu})$ that we will need.) For any $x, y \in T_\ell(E_1)$ we have

$$e_{E_1}(x,y)^{\deg \phi} = e_{E_1}\big((\deg \phi)x, y\big) = e_{E_1}(\hat{\phi}\phi x, y) = e_{E_2}(\phi x, \phi y), \qquad \text{(i)}$$

and a similar calculation on $\tilde{E}_1$ gives

$$e_{\tilde{E}_1}(\tilde{x}, \tilde{y})^{\deg \tilde{\phi}} = e_{\tilde{E}_2}(\tilde{\phi}\tilde{x}, \tilde{\phi}\tilde{y}). \qquad \text{(ii)}$$

Next we observe that if E/L is any elliptic curve with good reduction at $\mathfrak{P}$, then $T_\ell(E) \cong T_\ell(\tilde{E})$. This crucial equality is a consequence of [AEC VII.3.1b], which says that $E[\ell^n] \cong \tilde{E}[\ell^n]$ for all n. Looking at the definition of the Weil pairing [AEC III §8], we see that

$$\widetilde{e_E(x,y)} = e_{\tilde{E}}(\tilde{x}, \tilde{y}) \qquad \text{for all } x, y \in T_\ell(E). \qquad \text{(iii)}$$

We now take $x, y \in T_\ell(E_1)$ and compute

$$
\begin{aligned}
e_{\tilde{E}_1}(\tilde{x}, \tilde{y})^{\deg \phi} &= \widetilde{e_{E_1}(x,y)}^{\deg \phi} && \text{from (iii)} \\
&= \widetilde{e_{E_2}(\phi x, \phi y)} && \text{from (i)} \\
&= e_{\tilde{E}_2}(\widetilde{\phi x}, \widetilde{\phi y}) && \text{from (iii)} \\
&= e_{\tilde{E}_2}(\tilde{\phi}\tilde{x}, \tilde{\phi}\tilde{y}) && \\
&= e_{\tilde{E}_1}(\tilde{x}, \tilde{y})^{\deg \tilde{\phi}} && \text{from (ii).}
\end{aligned}
$$

This equality holds for all $x, y \in T_\ell(E_1)$, hence for all $\tilde{x}, \tilde{y} \in T_\ell(\tilde{E}_1)$. The non-degeneracy of the Weil pairing on $T_\ell(\tilde{E}_1)$ now implies that $\deg \phi = \deg \tilde{\phi}$. $\qquad \square$

PROOF (of Proposition 4.2). We know that $\mathcal{ELL}(R_K)$ is finite from (1.2b) and that every curve in $\mathcal{ELL}(R_K)$ can be defined over $\bar{\mathbb{Q}}$ from (2.1c), so we can choose a finite extension field L/K and representatives $E_1, \ldots, E_n$

defined over L for the distinct $\bar{K}$ isomorphism classes in $\mathcal{ELL}(R)$. Further, using (2.2c) we may replace L by a finite extension so that every isogeny connecting every pair of E_i's is defined over L. We now let S be the finite set of rational primes satisfying any one of the following three conditions:

(i) p ramifies in L,

(ii) some E_i has bad reduction at some prime of L lying over p,

(iii) p divides either the numerator or the denominator of one of the numbers $N_{\mathbb{Q}}^{L}\big(j(E_i) - j(E_k)\big)$ for some $i \neq k$.

Notice that condition (iii) means that if $p \notin S$ and if $\mathfrak{P}$ is a prime of L dividing p, then $\bar{E}_i \not\cong \bar{E}_k \pmod{\mathfrak{P}}$, since their j-invariants are not the same modulo $\mathfrak{P}$.

Now let $p \notin S$ be a prime which splits as $pR_K = \mathfrak{p}\mathfrak{p}'$ in K, and let $\mathfrak{P}$ be a prime of L lying over $\mathfrak{p}$. Also let Λ be a lattice for E, so $E(\mathbb{C}) \cong \mathbb{C}/\Lambda$. Choose some integral ideal $\mathfrak{a} \subset R_K$ relatively prime to p such that $\mathfrak{a}\mathfrak{p}$ is principal, say

$$\mathfrak{a}\mathfrak{p} = (\alpha).$$

From [AEC VI.4.1b] there are isogenies connecting E, $\bar{\mathfrak{p}} * E$, and $\bar{\mathfrak{a}} * \bar{\mathfrak{p}} * E$ corresponding to the natural analytic maps as indicated in the following diagram:

$$
\begin{array}{ccccccc}
\mathbb{C}/\Lambda & \xrightarrow{z \mapsto z} & \mathbb{C}/\mathfrak{p}^{-1}\Lambda & \xrightarrow{z \mapsto z} & \mathbb{C}/\mathfrak{a}^{-1}\mathfrak{p}^{-1}\Lambda = \mathbb{C}/(\alpha^{-1})\Lambda & \xrightarrow[\sim]{z \mapsto \alpha z} & \mathbb{C}/\Lambda \\
\downarrow\wr & & \downarrow\wr & & \downarrow\wr & & \downarrow\wr \\
E & \xrightarrow{\phi} & \bar{\mathfrak{p}} * E & \xrightarrow{\psi} & \bar{\mathfrak{a}} * \bar{\mathfrak{p}} * E = (\alpha) * E & \xrightarrow[\sim]{\lambda} & E
\end{array}
$$

Next we choose a Weierstrass equation for E/L which is minimal at $\mathfrak{P}$ (see [AEC VII §§1,2]) and let

$$\omega = \frac{dx}{2y + a_1 x + a_2}$$

be the associated invariant differential on E. The pull-back of ω to $\mathbb{C}/\Lambda$ will be some multiple of dz. Since the map along the top row of our diagram is simply $z \to \alpha z$, we see that dz pulls back to $d(\alpha z) = \alpha dz$. Tracing around the commutative diagram, we conclude that

$$(\lambda \circ \psi \circ \phi)^* \omega = \alpha \omega.$$

As usual, we will use a tilde to denote reduction modulo $\mathfrak{P}$. Since the equation for E/L is minimal at $\mathfrak{P}$, we obtain an equation for $\tilde{E}$ by reducing the coefficients modulo $\mathfrak{P}$, and so the reduced differential

$$\tilde{\omega} = \frac{dx}{2y + \tilde{a}_1 x + \tilde{a}_3}$$

is a non-zero invariant differential on $\tilde{E}$. Further, since $(\alpha) = \mathfrak{a}\mathfrak{p}$ and since $\mathfrak{P}$ divides $\mathfrak{p}$, we find

$$(\tilde{\lambda} \circ \tilde{\psi} \circ \tilde{\phi})^* \tilde{\omega} = \widetilde{(\lambda \circ \psi \circ \phi)^*} \omega = \tilde{\alpha}\tilde{\omega} = \tilde{0}.$$

It follows from [AEC II.4.2c] that

$$\tilde{\lambda} \circ \tilde{\psi} \circ \tilde{\phi} \text{ is inseparable.}$$

On the other hand, using (4.4) and (1.5a), we see that

$$\deg \tilde{\phi} = \deg \phi = \mathrm{N}_{\mathbb{Q}}^K \mathfrak{p} = p,$$
$$\deg \tilde{\psi} = \deg \psi = \mathrm{N}_{\mathbb{Q}}^K \mathfrak{a},$$
$$\deg \tilde{\lambda} = \deg \lambda = 1.$$

Since $\mathrm{N}_{\mathbb{Q}}^K \mathfrak{a}$ is prime to p by assumption, both $\tilde{\psi}$ and $\tilde{\lambda}$ are separable, so we conclude that

$$\tilde{\phi} : \tilde{E} \longrightarrow \widetilde{\bar{\mathfrak{p}} * E}$$

must be inseparable. Now any map (such as $\tilde{\phi}$) factors as a qth-power Frobenius map followed by a separable map [AEC II.2.12], so the fact that $\tilde{\phi}$ has degree p and is inseparable implies that $\tilde{\phi}$ must "be" the pth-power Frobenius map. More precisely, there is an isomorphism from $\tilde{E}^{(p)}$ to $\widetilde{\bar{\mathfrak{p}} * E}$ so that the composition

$$\tilde{E} \xrightarrow[\text{Frobenius}]{p^{\text{th}}\text{-power}} \tilde{E}^{(p)} \xrightarrow{\sim} \widetilde{\bar{\mathfrak{p}} * E}$$

equals $\tilde{\phi}$.

In particular, we find that

$$j\left(\widetilde{\bar{\mathfrak{p}} * E}\right) = j\left(\tilde{E}^{(p)}\right) = j(\tilde{E})^p,$$

so

$$j(\bar{\mathfrak{p}}*E) \equiv j(E)^p = j(E)^{\mathrm{N}_{\mathbb{Q}}^K \mathfrak{p}} \equiv j(E)^{\sigma_{\mathfrak{p}}} = j\left(E^{\sigma_{\mathfrak{p}}}\right) = j\left(F(\sigma_{\mathfrak{p}})*E\right) \pmod{\mathfrak{P}}.$$

But from the original choice of excluded primes S, we have

$$j(E_i) \equiv j(E_k) \pmod{\mathfrak{P}} \quad \text{if and only if} \quad E_i \cong E_k.$$

Hence $\bar{\mathfrak{p}} * E \cong F(\sigma_{\mathfrak{p}}) * E$, and the simplicity of the action of $\mathcal{CL}(R_K)$ on $\mathcal{ELL}(R_K)$ (1.2b) gives the desired conclusion

$$F(\sigma_{\mathfrak{p}}) = \bar{\mathfrak{p}}.$$

$\square$

We also record for later use the following fact which we proved during the course of proving Proposition 4.2.

Lemma 4.5. *Let E be an elliptic curve with complex multiplication by R_K, and suppose that E is defined over a number field L. Then for all but finitely many degree 1 primes $\mathfrak{p}$ of K, the natural map*

$$E \longrightarrow \bar{\mathfrak{p}} * E$$

has degree p, and its reduction

$$\tilde{E} \longrightarrow \widetilde{\bar{\mathfrak{p}} * E}$$

is purely inseparable. (Here we reduce modulo some prime $\mathfrak{P}$ of L lying above $\mathfrak{p}$.)

§5. The Maximal Abelian Extension

Let E be an elliptic curve with complex multiplication by a quadratic imaginary field K. In this section we are going to describe the field generated by the points in $E(\mathbb{C})_{\text{tors}}$, much as we described the field generated by $j(E)$ in the last section. Our goal is to use the torsion points of E to generate abelian extensions of K.

Before beginning, we briefly recall the analogous (but simpler) case of cyclotomic extensions. In this case the elliptic curve $E(\mathbb{C})$ is replaced by the multiplicative group $\mathbb{G}_m(\mathbb{C}) = \mathbb{C}^*$. Let

$$\boldsymbol{\mu}_N = \ker\left(\mathbb{G}_m(\mathbb{C}^*) \underset{z \to z^N}{\longrightarrow} \mathbb{G}_m(\mathbb{C}^*)\right)$$

be the group of N-torsion points of $\mathbb{G}_m$ as usual; that is, $\boldsymbol{\mu}_N$ is the group of Nth-roots of unity. As is well known, the extension $\mathbb{Q}(\boldsymbol{\mu}_N)/\mathbb{Q}$ is an abelian extension that is ramified only at primes dividing N. Let p be a prime with $p \nmid N$, choose a generator ζ for $\boldsymbol{\mu}_N$, and let $\sigma_p \in \text{Gal}(\mathbb{Q}(\zeta)/\mathbb{Q})$ be the Frobenius element associated to p. Also let $\mathfrak{P}$ be a prime of $\mathbb{Q}(\zeta)$ lying above p. Then by the definition of σ_p we have

$$\zeta^{\sigma_p} \equiv \zeta^p \pmod{\mathfrak{P}}.$$

But $1, \zeta, \zeta^2, \ldots, \zeta^{N-1}$ are distinct modulo $\mathfrak{P}$, since our assumption $\mathfrak{P} \nmid N$ implies that $X^N - 1$ is separable in characteristic p. Hence the congruence is an equality,

$$\zeta^{\sigma_p} = \zeta^p,$$

and so we conclude that

$$\sigma_p = 1 \iff p \equiv 1 \pmod{N}.$$

Therefore $\mathbb{Q}(\zeta) = \mathbb{Q}(\mu_N)$ is the ray class field of $\mathbb{Q}$ of conductor N. (Actually, it is the ray class field of conductor $N\infty$. Since the base field $\mathbb{Q}$ is not totally imaginary, we have to also consider ramification of the infinite place.)

Now let $L/\mathbb{Q}$ be any abelian extension, and let N be the conductor of L. Then class field theory (3.2) says that L is contained in the ray class field of conductor N, so we recover the following famous result.

Theorem 5.1. (Kronecker-Weber Theorem) *Every abelian extension of $\mathbb{Q}$ is contained in a cyclotomic extension; that is, given any finite abelian extension $L/\mathbb{Q}$, there is a root of unity ζ such that $L \subset \mathbb{Q}(\zeta)$.*

Thus the ray class fields of $\mathbb{Q}$ are generated by the values of the analytic function

$$e^{2\pi i z} = \sum_{n \geq 0} \frac{(2\pi i z)^n}{n!}$$

evaluated at points of finite order in the circle group $\mathbb{R}/\mathbb{Z}$; that is, they are generated by numbers of the form $e^{2\pi i a/N}$ with $a, N \in \mathbb{Z}$. Further, the action of a Frobenius element σ_p on the value $e^{2\pi i a/N}$ is given explicitly by the formula

$$\left(e^{2\pi i a/N} \right)^{\sigma_p} = e^{2\pi i a p/N} \qquad \text{provided } p \nmid N.$$

Thus the Galois action of σ_p is transformed into a multiplication action on the circle group. The reader who has a good understanding of this cyclotomic theory will have no trouble seeing how all of its main elements are reproduced in the theory of complex multiplication as described in this section and in §8, albeit with a number of additional technical complications.

As usual, we let R_K be the ring of integers in a quadratic imaginary field K, and let E be an elliptic curve with complex multiplication by R_K. We will always assume that the isomorphism $[\cdot] : R_K \xrightarrow{\sim} \text{End}(E)$ is normalized as in (1.1).

We begin with an important lemma which tells us when an endomorphism of the reduced curve $\tilde{E}$ (mod $\mathfrak{P}$) actually comes from an endomorphism of E.

Lemma 5.2. *Suppose that E is defined over the number field L, let $\mathfrak{P}$ be a prime of L at which E has good reduction, and let $\tilde{E}$ be the reduction of E modulo $\mathfrak{P}$. Let*

$$\theta : \text{End}(E) \longrightarrow \text{End}(\tilde{E})$$

be the natural map which takes an endomorphism to its reduction modulo $\mathfrak{P}$. Then for any $\gamma \in \text{End}(\tilde{E})$,

$$\gamma \in \text{Image}(\theta) \iff \gamma \text{ commutes with every element in Image}(\theta).$$

In other words, $\mathrm{Image}(\theta)$ *is its own commutator inside* $\mathrm{End}(\tilde{E})$.

PROOF. One direction is trivial, since if $\gamma \in \mathrm{Image}(\theta)$, then γ certainly commutes with elements of $\mathrm{Image}(\theta)$. This is immediate from the fact that $\mathrm{End}(E) \cong R_K$, which implies that $\mathrm{Image}(\theta)$ is a commutative ring.

For the other direction, we first note that θ is injective from (4.4). From [AEC III.9.4], $\mathrm{End}(\tilde{E})$ is an order in either a quadratic imaginary field or in a quaternion algebra. If it is an order in a quadratic imaginary field, then θ is an isomorphism, since, by assumption, $\mathrm{End}(E) \cong R_K$ is the maximal order in K. So in this case we are done.

Next we consider the case that $\mathrm{End}(\tilde{E})$ is an order in a quaternion algebra $\mathcal{H}$. Then $\mathrm{Image}(\theta) \otimes \mathbb{Q}$ is a quadratic subfield of $\mathcal{H}$, call it $\mathcal{K}$. (Note $\mathcal{K} \cong K$, but it is possible for $\mathcal{H}$ to contain several distinct subfields each isomorphic to K.) We start by choosing a $\mathbb{Q}$-basis $\{1, \alpha\}$ for $\mathcal{K}$ such that $\alpha^2 \in \mathbb{Q}$; and then we extend it to a $\mathbb{Q}$-basis for $\mathcal{H}$ of the form

$$\mathcal{H} = \mathbb{Q} + \mathbb{Q}\alpha + \mathbb{Q}\beta + \mathbb{Q}\alpha\beta$$

satisfying

$$\alpha^2, \beta^2, (\alpha\beta)^2 \in \mathbb{Q} \qquad \text{and} \qquad \alpha\beta = -\beta\alpha.$$

(See the proof [AEC III.9.3].) Now it is easy to find the commutator of $\mathcal{K}$ in $\mathcal{H}$. For any $\gamma \in \mathcal{H}$, we write $\gamma = d + a\alpha + b\beta + c\alpha\beta$ with $a, b, c, d \in \mathbb{Q}$ and compute:

γ commutes with $\mathcal{K}$

$$\Longleftrightarrow \gamma\alpha = \alpha\gamma$$
$$\Longleftrightarrow (d + a\alpha + b\beta + c\alpha\beta)\alpha = \alpha(d + a\alpha + b\beta + c\alpha\beta)$$
$$\Longleftrightarrow d\alpha + a\alpha^2 + b\beta\alpha + c\alpha\beta\alpha = d\alpha + a\alpha^2 + b\alpha\beta + c\alpha^2\beta$$
$$\Longleftrightarrow -b\alpha\beta - c\alpha^2\beta = b\alpha\beta + c\alpha^2\beta \quad \text{since } \alpha\beta = -\beta\alpha$$
$$\Longleftrightarrow b = c = 0$$
$$\qquad \text{since } \alpha^2 \in \mathbb{Q} \text{ and } \{1, \alpha, \beta, \alpha\beta\} \text{ is a } \mathbb{Q}\text{-basis for } \mathcal{H}$$
$$\Longleftrightarrow \gamma = d + a\alpha \in \mathbb{Q} + \mathbb{Q}\alpha = \mathcal{K}.$$

Finally, let $\delta \in \mathrm{End}(\tilde{E})$ commute with $\mathrm{Image}(\theta)$. Then δ commutes with $\mathcal{K}$, so from what we have just done, δ is in $\mathcal{K}$. But we also know that δ is integral over $\mathbb{Z}$ and that $\mathrm{Image}(\theta) \cong R_K$ is the maximal order in $\mathcal{K} \cong K$, hence $\delta \in \mathrm{Image}(\theta)$. This completes the proof of Lemma 5.2. $\qquad\square$

As usual, let E be an elliptic curve with complex multiplication by R_K. In the last section (4.3a) we proved that

$$H = K\big(j(E)\big)$$

is the Hilbert class field of K. Since $j(E) \in H$, this means we can find an equation for E with coefficients in H, so we may as well assume that E is defined over H. The next proposition says that we can lift the pth-power Frobenius map $\tilde{E} \to \tilde{E}^{(p)}$ to a map in characteristic 0.

Proposition 5.3. *Let K be a quadratic imaginary field, H the Hilbert class field of K, and E/H an elliptic curve with complex multiplication by R_K. Let $\sigma_{\mathfrak{p}} \in G_{H/K}$ be the Frobenius element associated to a prime $\mathfrak{p}$ of R_K, and let $\mathfrak{P}$ be a prime of H lying over $\mathfrak{p}$. Assume that $\mathfrak{p}$ has degree 1 and is not in the finite set of primes specified in (4.5), so in particular E has good reduction at $\mathfrak{P}$. Then there exists an isogeny*

$$\lambda : E \longrightarrow E^{\sigma_{\mathfrak{p}}}$$

whose reduction modulo $\mathfrak{P}$,

$$\tilde{\lambda} : \tilde{E} \longrightarrow \tilde{E}^{(p)},$$

is the pth-power Frobenius map.

Remark 5.3.1. In general, there is no reason to expect an elliptic curve to be isogenous to one of its Galois conjugates. Of course, there are always maps

$$
\begin{array}{ccc}
E & & E^{\sigma_{\mathfrak{p}}} \\
\downarrow & & \downarrow \\
\tilde{E} & \xrightarrow[\text{Frobenius}]{p^{\text{th}} \text{ power}} & \tilde{E}^{(p)}
\end{array}
$$

where the vertical maps are "reduction modulo $\mathfrak{P}$." The content of Proposition 5.3 is that there is an isogeny $\lambda : E \to E^{\sigma_{\mathfrak{p}}}$ which makes this picture into a commutative square. Thus λ lifts the Frobenius map from characteristic p to characteristic 0.

PROOF (of Proposition 5.3). To ease notation, we will write σ in place of $\sigma_{\mathfrak{p}}$. From (4.5) there is an isogeny $E \to \bar{\mathfrak{p}} * E$ whose reduction $\tilde{E} \to \widetilde{\bar{\mathfrak{p}} * E}$ is purely inseparable of degree p. Composing this isogeny with the isomorphism $\bar{\mathfrak{p}} * E \cong E^{\sigma}$ provided by (4.3), we get an isogeny $\tilde{\lambda} : \tilde{E} \to \widetilde{E^{\sigma}}$ which is purely inseparable of degree p. It follows from [AEC III.4.6] that $\tilde{\lambda}$ factors as

$$\tilde{E} \xrightarrow{\phi} \tilde{E}^{(p)} \xrightarrow{\varepsilon} \widetilde{E^{\sigma}},$$

where ϕ is the pth-power Frobenius map and $\deg \varepsilon = 1$. But, by definition, the reduction of E^{σ} is precisely $\tilde{E}^{(p)}$, so ε is an automorphism of $\widetilde{E^{\sigma}}$. If we can show that ε is the reduction modulo $\mathfrak{P}$ of some $\varepsilon_0 \in \operatorname{Aut}(E^{\sigma})$, then we can replace λ by $\varepsilon_0^{-1} \circ \lambda$ and be done. So we need to prove the

Claim: ε lies in the image of $\operatorname{Aut}(E^{\sigma})$ inside $\operatorname{Aut}(\widetilde{E^{\sigma}})$.

From (5.2), it suffices to show that ε commutes with the image of $\operatorname{End}(E^{\sigma})$ inside $\operatorname{End}(\widetilde{E^{\sigma}})$. (This will allow us to lift ε to an ε_0 in $\operatorname{End}(E^{\sigma})$, and then (4.4) will imply that ε_0 has degree 1, so it is in $\operatorname{Aut}(E^{\sigma})$.) Recall that we have normalized isomorphisms

$$[\,\cdot\,]_E : R_K \xrightarrow{\sim} \operatorname{End}(E) \qquad \text{and} \qquad [\,\cdot\,]_{E^{\sigma}} : R_K \xrightarrow{\sim} \operatorname{End}(E^{\sigma}),$$

and that from (1.1.1) these isomorphisms satisfy

$$\lambda \circ [\alpha]_E = [\alpha]_{E^\sigma} \circ \lambda \qquad \text{for all } \alpha \in R_K.$$

Next we look at the reduction of $[\alpha]$ modulo $\mathfrak{P}$. In general, suppose that $f : V \to W$ is any rational map of algebraic varieties over a field k of characteristic p, let $\phi_V : V \to V^{(p)}$ and $\phi_W : W \to W^{(p)}$ be the pth-power Frobenius maps, and let $\sigma \in \text{Aut}(k)$ be the pth-power Frobenius automorphism of k. Then $f^\sigma : V^{(p)} \to W^{(p)}$ is a rational map and

$$\phi_W \circ f = f^\sigma \circ \phi_V.$$

To see that this is true, write $f = [f_0, \ldots, f_n]$ (locally) as a map given by homogeneous polynomials. The desired result then follows from the observation that for a polynomial $f(\mathbf{x}) = f(x_1, \ldots, x_m) = \sum a_i \mathbf{x}^i$, we have

$$f^\sigma\big(\phi(\mathbf{x})\big) = \sum a_i^p \mathbf{x}^{ip} = \left(\sum a_i \mathbf{x}^i\right)^p = \phi(f(\mathbf{x})).$$

We now apply this general fact to the map $\widetilde{[\alpha]}_E : \tilde{E} \to \tilde{E}$. Note that

$$[\alpha]_E^\sigma = [\alpha]_{E^\sigma}$$

from (2.2a), since $\sigma \in G_{H/K}$ fixes $\alpha \in K$. Thus we get

$$\phi \circ \widetilde{[\alpha]}_E = \widetilde{[\alpha]}_E^\sigma \circ \phi = \widetilde{[\alpha]}_{E^\sigma} \circ \phi.$$

Using this, we compute

$$
\begin{aligned}
\widetilde{[\alpha]}_{E^\sigma} \circ \varepsilon \circ \phi &= \widetilde{[\alpha]}_{E^\sigma} \circ \tilde{\lambda} && \text{since } \varepsilon \circ \phi = \tilde{\lambda} \\
&= \tilde{\lambda} \circ \widetilde{[\alpha]}_E && \text{from above,} \\
&= \varepsilon \circ \phi \circ \widetilde{[\alpha]}_E \\
&= \varepsilon \circ \widetilde{[\alpha]}_{E^\sigma} \circ \phi && \text{from above.}
\end{aligned}
$$

Therefore $\widetilde{[\alpha]}_{E^\sigma} \circ \varepsilon = \varepsilon \circ \widetilde{[\alpha]}_{E^\sigma}$, which completes the proof of our claim and with it the proof of Proposition 5.3. $\square$

An important special case of Proposition 5.3 occurs when the ideal $\mathfrak{p}$ is principal, in which case $\sigma_{\mathfrak{p}} = (\mathfrak{p}, H/K) = 1$. Then λ is an endomorphism of E. We can identify that endomorphism quite precisely as follows.

Corollary 5.4. *Let K be a quadratic imaginary field, H the Hilbert class field of K, and E/H an elliptic curve with complex multiplication by R_K. For all but finitely many degree 1 prime ideals $\mathfrak{p}$ of K that satisfy*

$$(\mathfrak{p}, H/K) = 1,$$

there is a unique $\pi = \pi_{\mathfrak{p}} \in R_K$ such that

$$
\begin{array}{ccc}
E & \xrightarrow{\;[\pi]\;} & E \\
\downarrow & & \downarrow \\
\tilde{E} & \xrightarrow[\text{Frobenius}]{p^{\text{th}} \text{ power}} & \tilde{E}
\end{array}
$$

$$\mathfrak{p} = \pi R_K, \qquad \text{and}$$

is a commutative diagram. (Note that the condition $(\mathfrak{p}, H/K) = 1$ is equivalent to $\mathfrak{p}$ being a principal ideal.)

PROOF. Let $\mathfrak{P}$ be a prime of H lying over $\mathfrak{p}$. Having excluded finitely many $\mathfrak{p}$'s, including those for which $\tilde{E}$ (mod $\mathfrak{P}$) is singular, we may use (5.3) to obtain a commutative diagram

$$
\begin{array}{ccc}
E & \xrightarrow{\;\lambda\;} & E^{\sigma_{\mathfrak{p}}} \\
\downarrow & & \downarrow \\
\tilde{E} & \xrightarrow{\;\phi\;} & \tilde{E}^{(p)}.
\end{array}
$$

Here $\sigma_{\mathfrak{p}} = (\mathfrak{p}, H/K)$, λ is an isogeny, ϕ is the pth-power Frobenius map, and the vertical maps are reduction modulo $\mathfrak{P}$.

Our assumption that $(\mathfrak{p}, H/K) = 1$ means that $E^{\sigma_{\mathfrak{p}}} = E$, so λ is really an endomorphism of E, say $\lambda = [\pi]$. It also implies that $\tilde{E}^{(p)} = \tilde{E}$. Thus we have a commutative diagram

$$
\begin{array}{ccc}
E & \xrightarrow{\;[\pi]\;} & E \\
\downarrow & & \downarrow \\
\tilde{E} & \xrightarrow{\;\phi\;} & \tilde{E}.
\end{array}
$$

Now we compute

$$
\begin{aligned}
\mathrm{N}^K_{\mathbb{Q}}\mathfrak{p} &= p & &\text{since } \mathfrak{p} \text{ has degree 1} \\
&= \deg \phi & &\text{since } \phi \text{ is } p^{\text{th}} \text{ power Frobenius} \\
&= \deg [\pi] & &\text{from (4.4), since } \widetilde{[\pi]} = \phi \\
&= |\mathrm{N}^K_{\mathbb{Q}}\pi| & &\text{from (1.4b).}
\end{aligned}
$$

Since $\mathfrak{p}$ is a prime ideal in the quadratic field K, this means that either

$$\mathfrak{p} = \pi R_K \qquad \text{or} \qquad \mathfrak{p} = \pi' R_K,$$

where π' is the $\mathrm{Gal}(K/\mathbb{Q})$-conjugate of π. We can use the fact that $(E, [\,\cdot\,])$ is normalized to check which one it is.

We take an equation for E/H with good reduction at $\mathfrak{P}$ and let $\omega \in \Omega_E$ be a non-zero invariant differential whose reduction $\tilde{\omega}$ is a non-zero invariant differential on $\tilde{E}$. Then the normalization (1.1) says that $[\pi]^*\omega = \pi\omega$, so

$$\tilde{\pi}\tilde{\omega} = \widetilde{\pi\omega} = \widetilde{[\pi]^*\omega} = \widetilde{[\pi]}^*\tilde{\omega} = \phi^*\tilde{\omega} = 0.$$

The last equality follows from [AEC II.4.2c], since the Frobenius map ϕ is inseparable. Now $\Omega_{\tilde{E}}$ is a one-dimensional vector space generated by $\tilde{\omega}$, so $\tilde{\pi} = 0$. In other words,

$$\pi \equiv 0 \pmod{\mathfrak{P}},$$

so $\pi \in \mathfrak{P} \cap K = \mathfrak{p}$. Since we saw above that $\mathfrak{p}$ equals either πR_K or $\pi' R_K$, we conclude that $\mathfrak{p} = \pi R_K$. This finishes the existence half of (5.4).

To see that π is uniquely determined, we need merely observe that the composition

$$R_K \xrightarrow{[\,\cdot\,]} \mathrm{End}(E) \longrightarrow \mathrm{End}(\tilde{E})$$

is injective. Since π is required to satisfy $\widetilde{[\pi]} = \phi \in \mathrm{End}(\tilde{E})$, there is at most one such π. $\square$

Our goal is to show that the torsion points of an elliptic curve E with complex multiplication by R_K can be used to generate abelian extensions of K. It would be nice if the torsion points themselves should generate abelian extensions of K, but unfortunately it turns out that they only generate abelian extensions of the Hilbert class field H of K. In order to pick out the correct subfield, we take a model for E defined over H and fix a (finite) map

$$h : E \longrightarrow E/\mathrm{Aut}(E) \cong \mathbb{P}^1$$

also defined over H. Such a map h is called a *Weber function for E/H*.

Example 5.5.1. If we take a Weierstrass equation for E of the form

$$y^2 = x^3 + Ax + B \qquad \text{with } A, B \in H,$$

then the following is a Weber function for E/H:

$$h(P) = h(x, y) = \begin{cases} x & \text{if } AB \neq 0, \\ x^2 & \text{if } B = 0, \\ x^3 & \text{if } A = 0. \end{cases}$$

So in essence, except for the two exceptional cases $j = 0$ and $j = 1728$, a Weber function is just an x-coordinate for the curve.

Example 5.5.2. It is also possible to define a Weber function analytically in such a way that we don't have to worry about fields of definition. For example, if we choose a lattice Λ and an isomorphism

$$f : \mathbb{C}/\Lambda \xrightarrow{\sim} E(\mathbb{C}), \qquad z \longmapsto \big(\wp(z,\Lambda),\, \wp'(z,\Lambda)\big),$$

then the following is a Weber function for E:

$$h\big(f(z)\big) = \begin{cases} \dfrac{g_2(\Lambda)g_3(\Lambda)}{\Delta(\Lambda)}\wp(z,\Lambda) & \text{if } j(E) \neq 0, 1728, \\[2mm] \dfrac{g_2(\Lambda)^2}{\Delta(\Lambda)}\wp(z,\Lambda)^2 & \text{if } j(E) = 1728, \\[2mm] \dfrac{g_3(\Lambda)}{\Delta(\Lambda)}\wp(z,\Lambda)^3 & \text{if } j(E) = 0. \end{cases}$$

Here $\Delta(\Lambda) = g_2(\Lambda)^2 - 27g_3(\Lambda)^3 \neq 0$ is the usual modular discriminant. The reader may easily verify that this Weber function is model independent; that is, it does not change if we take a new lattice for E, or equivalently a new Weierstrass equation for E. Since we know from (4.3a) that it is possible to find an equation for E defined over H, it follows from the model independence that this Weber function $h : E \to \mathbb{P}^1$ is defined over H.

To generate abelian extensions of K, we will use the values of a Weber function on torsion points, which essentially means we will take the x-coordinates of the torsion points. Recall from §1 that for any integral ideal $\mathfrak{c}$ of R_K we defined the group of $\mathfrak{c}$-torsion points of E to be

$$E[\mathfrak{c}] = \big\{ P \in E \,:\, [\gamma]P = 0 \text{ for all } \gamma \in \mathfrak{c} \big\}.$$

The reader is advised to compare the following theorem with the cyclotomic theory discussed at the beginning of this section.

Theorem 5.6. *Let K be a quadratic imaginary field, let E be an elliptic curve with complex multiplication by R_K, and let $h : E \to \mathbb{P}^1$ be a Weber function for E/H as described above. Let $\mathfrak{c}$ be an integral ideal of R_K. Then the field*

$$K\big(j(E), h\big(E[\mathfrak{c}]\big)\big)$$

is the ray class field of K modulo $\mathfrak{c}$.

Corollary 5.7. *With notation as in Theorem 5.6,*

$$K^{\mathrm{ab}} = K\big(j(E), h(E_{\mathrm{tors}})\big).$$

In particular, if $j(E) \neq 0, 1728$ and if we take an equation for E with coefficients in $K\big(j(E)\big)$, then the maximal abelian extension of K is generated by $j(E)$ and the x-coordinates of the torsion points of E.

PROOF (of Theorem 5.6). Let

$$L = K\big(j(E), h\big(E[\mathfrak{c}]\big)\big).$$

Then $L \supset K\big(j(E)\big)$, and from (4.3a) we know that $K\big(j(E)\big) = H$ is the Hilbert class field of K. In order to show that L is the ray class field of K modulo $\mathfrak{c}$, we need to prove that

$$(\mathfrak{p}, L/K) = 1 \iff \mathfrak{p} \in P(\mathfrak{c}).$$

As usual, it suffices to prove this for all but finitely many degree 1 primes in K.

Suppose first that $\mathfrak{p}$ is a degree 1 prime of K with $\mathfrak{p} \in P(\mathfrak{c})$. This means that

$$\mathfrak{p} = \mu R_K \qquad \text{for some } \mu \in R_K \text{ with } \mu \equiv 1 \,(\mathrm{mod}\, \mathfrak{c}).$$

In particular, $\mathfrak{p}$ is principal, so $(\mathfrak{p}, H/K) = 1$. Hence we can apply (5.4) (after excluding finitely many $\mathfrak{p}$'s) to get some $\pi \in R_K$ such that

$$\mathfrak{p} = \pi R_K \qquad \text{and} \qquad
\begin{array}{ccc}
E & \xrightarrow{[\pi]} & E \\
\downarrow & & \downarrow \\
\tilde{E} & \xrightarrow{\phi} & \tilde{E}
\end{array}
\qquad \text{commutes.}$$

Since $\pi R_K = \mathfrak{p} = \mu R_K$, there is a unit $\xi \in R_K^*$ such that $\pi = \xi \mu$. Notice that $[\xi] \in \mathrm{Aut}(E)$, so $[\pi]$ and $[\mu]$ differ by an automorphism of E.

We already know that $(\mathfrak{p}, L/K)$ fixes $H = K\big(j(E)\big)$, so in order to show that it fixes all of L, we must show that it fixes $h\big(E[\mathfrak{c}]\big)$. Let $T \in E[\mathfrak{c}]$ be any $\mathfrak{c}$-torsion point. Then the commutative diagram gives

$$\widetilde{T^{(\mathfrak{p}, L/K)}} = \phi(\tilde{T}) = \widetilde{[\pi]T}.$$

On the other hand, [AEC VII.3.1b] tells us that the reduction map $E \to \tilde{E}$ is injective on torsion points whose order is prime to $\mathfrak{p}$. So if we exclude from consideration the finitely many $\mathfrak{p}$'s which divide $\#E[\mathfrak{c}]$, then the reduction map

$$E[\mathfrak{c}] \longrightarrow \tilde{E}[\mathfrak{c}]$$

is injective. Therefore

$$T^{(\mathfrak{p}, L/K)} = [\pi]T.$$

Now we compute

$$\begin{aligned}
h(T)^{(\mathfrak{p}, L/K)} &= h\left(T^{(\mathfrak{p}, L/K)}\right) && \text{since } (\mathfrak{p}, H/K) = 1 \text{ and} \\
& && h : E \to \mathbb{P}^1 \text{ is defined over } H \\
&= h\big([\pi]T\big) && \text{from above} \\
&= h\big([\xi] \circ [\mu]T\big) && \text{since } \pi = \xi\mu \\
&= h\big([\mu]T\big) && \text{since } h \text{ is } \mathrm{Aut}(E)\text{-invariant and} \\
& && [\xi] \in \mathrm{Aut}(E) \\
&= h(T) && \text{since } T \in E[\mathfrak{c}] \text{ and } \mu \equiv 1 \,(\mathrm{mod}\, \mathfrak{c}).
\end{aligned}$$

This completes the proof that

$$\mathfrak{p} \in P(\mathfrak{c}) \Longrightarrow (\mathfrak{p}, L/K) = 1.$$

In order to prove the converse, we take a prime of degree 1 satisfying $(\mathfrak{p}, L/K) = 1$. Then

$$(\mathfrak{p}, H/K) = (\mathfrak{p}, L/K)\big|_H = 1,$$

so (excluding finitely many $\mathfrak{p}$'s) we can apply (5.5) as usual to get a $\pi \in R_K$ such that

$$\mathfrak{p} = \pi R_K \qquad \text{and} \qquad
\begin{array}{ccc}
E & \xrightarrow{[\pi]} & E \\
\downarrow & & \downarrow \\
\tilde{E} & \xrightarrow{\phi} & \tilde{E}
\end{array}
\qquad \text{commutes.}$$

We also choose some $\sigma \in G_{\bar{K}/K}$ whose restriction to K^{ab} is $(\mathfrak{p}, K^{\mathrm{ab}}/K)$. Then in particular $\sigma|_L = (\mathfrak{p}, L/K) = 1$, and also $\sigma|_H = 1$ since $H \subset L$.

Now let $T \in E[\mathfrak{c}]$ be any $\mathfrak{c}$-torsion point. We compute

$$
\begin{aligned}
\tilde{h}(\widetilde{[\pi]}\tilde{T}) &= \tilde{h}(\widetilde{[\pi]T}) \\
&= \tilde{h}(\phi(\tilde{T})) && \text{from the commutative diagram} \\
&= \tilde{h}\left(\tilde{T}^\sigma\right) && \text{since } \sigma \text{ reduces to } p^{\mathrm{th}} \text{ power Frobenius} \\
&= \widetilde{h(T^\sigma)} \\
&= \widetilde{h(T)^\sigma} && \text{since } \sigma|_H = 1 \text{ and } h \text{ is defined over } H \\
&= \widetilde{h(T)} && \text{since } h(T) \in L \text{ and } \sigma|_L = 1 \\
&= \tilde{h}(\tilde{T}).
\end{aligned}
$$

Next we observe that the reduction of h modulo $\mathfrak{P}$ is the map

$$\tilde{h} : \tilde{E} \longrightarrow \widetilde{E/\operatorname{Aut} E} \cong \tilde{E}/\widetilde{\operatorname{Aut} E}.$$

(N.B. The image is not $\tilde{E}/\operatorname{Aut}\tilde{E}$, since $\operatorname{Aut}\tilde{E}$ may be larger than $\widetilde{\operatorname{Aut}(E)}$.) It follows from this and the equality $\tilde{h}(\widetilde{[\pi]}\tilde{T}) = \tilde{h}(\tilde{T})$ proven above that there is an automorphism $[\xi] \in \operatorname{Aut}(E)$ such that

$$\widetilde{[\pi]}\tilde{T} = \widetilde{[\xi]}\tilde{T}.$$

Again using the injectivity of the torsion $E[\mathfrak{c}] \hookrightarrow \tilde{E}[\mathfrak{c}]$ from [AEC VII.3.1b], we find that $[\pi - \xi]T = O$.

A priori, the particular ξ for which $[\pi - \xi]T = O$ might depend on T. But from (1.4b) we know that $E[\mathfrak{c}]$ is a free $R_K/\mathfrak{c}$-module of rank one.

Hence there is a single $\xi \in R_K^*$ such that $[\pi - \xi]$ annihilates all of $E[\mathfrak{c}]$, which implies that $\pi \equiv \xi \pmod{\mathfrak{c}}$. Therefore

$$\xi^{-1}\pi \equiv 1 \pmod{\mathfrak{c}},$$

and of course we have $\mathfrak{p} = \pi R_K = (\xi^{-1}\pi)R_K$ since ξ is a unit. This proves that $\mathfrak{p} \in P(\mathfrak{c})$, which completes the proof of Theorem 5.6. $\square$

PROOF (of Corollary 5.7). Let L/K be any finite abelian extension and let $\mathfrak{c}_{L/K}$ be the conductor of L/K. By class field theory (3.2c), L is contained in the ray class field of K modulo $\mathfrak{c}_{L/K}$. Using (5.6), this means that

$$L \subset K\big(j(E), h(E[\mathfrak{c}_{L/K}])\big).$$

Taking the compositum over all conductors gives $L \subset K\big(j(E), h(E_{\text{tors}})\big)$, and then taking the union over all L's gives $K^{\text{ab}} \subset K\big(j(E), h(E_{\text{tors}})\big)$. But (5.6) says that $K\big(j(E), h(E_{\text{tors}})\big)$ is a compositum of abelian extensions, hence it is abelian, hence it equals K^{ab}. This completes the first part of (5.7).

The second part of (5.7) is then immediate from (5.5.1), which says that if $j(E) \neq 0, 1728$, then the x-coordinate on a Weierstrass equation for $E/\mathbb{Q}(j(E))$ is a Weber function for E. $\square$

Example 5.8. Corollary 5.7 raises the obvious question of what happens if we adjoin all of E_{tors} to K, rather than just the values of a Weber function. In general one does not get an abelian extension of K, although we have seen (2.3) that E_{tors} generates an abelian extension of H. (The reader might try to use (5.4) to construct another proof of this fact.) Suppose now we look at the special case that K has class number 1, so $H = K$. Then we have inclusions

$$K^{\text{ab}} = H\big(h(E_{\text{tors}})\big) \subset H(E_{\text{tors}}) \subset H^{\text{ab}} = K^{\text{ab}}.$$

Thus

$$K \text{ has class number } 1 \quad \Longrightarrow \quad K^{\text{ab}} = K\big(h(E_{\text{tors}})\big) = K(E_{\text{tors}}).$$

The j-invariants of these curves will be in $\mathbb{Q}$. For a complete list of all CM j-invariants in $\mathbb{Q}$, together with representative Weierstrass equations, see Appendix A §3.

Example 5.8.1. We will illustrate (5.6) and (5.8) with the curve

$$E : y^2 = x^3 + x$$

which has complex multiplication by the ring of Gaussian integers $\mathbb{Z}[i]$ in the field $K = \mathbb{Q}(i)$. Clearly

$$E[2] = \big\{O, \, (0,0), \, (\pm i, 0)\big\},$$

so $K(E[2]) = K$. One can easily check that the ray class field of K modulo 2 is K, so this confirms (5.6).

Next we look at points of order 3. Letting $T = (x, y) \in E$, the duplication formula reads

$$2T = \left(\frac{x^4 - 2x^2 + 1}{4y^2}, \frac{x^6 + 5x^4 - 5x^2 - 1}{8y^3} \right).$$

So setting $x(2T) = x(T)$, we find (after some algebra) that

$$3T = O \iff 3x^4 + 6x^2 - 1 = 0.$$

The four roots of this equation are

$$\alpha, \; -\alpha, \; \frac{1}{\sqrt{3}\,\alpha}, \; -\frac{1}{\sqrt{3}\,\alpha}, \quad \text{where} \quad \alpha = \sqrt{\frac{2\sqrt{3} - 3}{3}}.$$

Since the Weber function on E is $h(x, y) = x^2$, this gives $K(h(E[3])) = K(\sqrt{3})$, which the reader may verify is indeed the ray class field of K modulo 3.

Substituting these four values for x into $y^2 = x^3 + x$ and solving for y, we find the y-coordinates of the points in $E[3]$. If we let

$$\beta = \sqrt[4]{\frac{8\sqrt{3} - 12}{9}} = \sqrt{\frac{2\alpha}{\sqrt{3}}},$$

then the nine points in $E[3]$ are

$$E[3] = \left\{ O, \; (\alpha, \pm\beta), \; (-\alpha, \pm i\beta), \; \left(\frac{1}{\sqrt{3}\alpha}, \frac{\pm 2}{\sqrt[4]{27}\beta} \right), \; \left(\frac{-1}{\sqrt{3}\alpha}, \frac{\pm 2i}{\sqrt[4]{27}\beta} \right) \right\}.$$

Since $K = \mathbb{Q}(i)$ has class number 1, (5.8) says that the field $K(\beta)$ is an abelian extension of K, but it is not necessarily a ray class field. We leave it for the reader to check directly that $K(\beta)/K$ is abelian.

Next, $T = (x, y)$ is a point of exact order 4 if and only if $y(2P) = 0$. Using the duplication formula given above, if we let $\gamma = (\sqrt{2} - 1)\,i$, then the x-coordinates of the points of order 4 satisfy

$$0 = x^6 + 5x^4 - 5x^2 - 1 = (x - 1)(x + 1)(x - \gamma)(x + \gamma)(x - \gamma^{-1})(x + \gamma^{-1}).$$

Hence $K(h(E[4])) = K(\gamma^2) = K(\sqrt{2})$, which is the ray class field of K modulo 4. Finally, if we let $\delta = (1 + i)(\sqrt{2} - 1)$, then we find that

$$E[4] = \left\{ O, \; (0, 0), \; (\pm i, 0), \; (1, \pm\sqrt{2}), \; (\gamma, \pm\delta), \right.$$

$$\left. (-\gamma, \pm i\delta), \; (\gamma^{-1}, \pm\gamma^{-2}\delta), \; (-\gamma^{-1}, \pm i\gamma^{-2}\delta) \right\}.$$

So in this case $K(E[4]) = K(\gamma, \delta) = K(\sqrt{2})$ is equal to $K(h(E[4]))$.

§6. Integrality of j

We have seen (2.1b) that the j-invariant of an elliptic curve E with complex multiplication is an algebraic number. In this section we are going to prove that $j(E)$ is in fact an algebraic integer or, equivalently, that E has everywhere potential good reduction. The results of this section will not be needed until §10, so the reader primarily interested in the relationship between complex multiplication and class field theory may wish to skip directly to §7.

Theorem 6.1. *Let $E/\mathbb{C}$ be an elliptic curve with complex multiplication. Then $j(E)$ is an algebraic integer.*

We are going to give three proofs of this important fact, two in this section and a third in (V.6.3). In order to help the reader understand the different approaches used in these three proofs, we will start with a brief description of each.

The Complex Analytic Proof

Let Λ_1 and Λ_2 be lattices corresponding to elliptic curves $E_1/\mathbb{C}$ and $E_2/\mathbb{C}$, and suppose that E_1 and E_2 are isogenous. Then we will show that $j(E_1)$ and $j(E_2)$ are algebraically dependent over $\mathbb{Q}$ by explicitly constructing a polynomial $F(X,Y) \in \mathbb{Z}[X,Y]$ with $F\big(j(E_1), j(E_2)\big) = 0$. If E has complex multiplication, then by taking $E_1 = E_2 = E$ we will obtain a monic polynomial with $j(E)$ as a root. Thus we show that $j(E)$ is integral over $\mathbb{Z}$ by explicitly constructing a monic polynomial with $j(E)$ as a root. This proof has the advantage of being very explicit, and the disadvantage that it does not generalize to higher dimensions.

The ℓ-adic (Good Reduction) Proof

This proof, which is due to Serre and Tate [1], readily generalizes to abelian varieties of arbitrary dimension. The idea is to use the criterion of Néron-Ogg-Shafarevich [AEC VII.7.3] to prove directly that E has potential good reduction at all primes, which implies by [AEC VII.5.5] that $j(E)$ is integral at all primes. Thus let L be a local field and E/L an elliptic curve with complex multiplication. We have seen (2.3) that the action of $\mathrm{Gal}(\bar{L}/L)$ on the Tate module $T_\ell(E)$ is abelian. (For another proof of this fact that uses nothing more than a little linear algebra, see exercise 2.6.) In other words, $\mathrm{Gal}(L^{\mathrm{ab}}/L)$ acts on $T_\ell(E)$. Next we use the description of $\mathrm{Gal}(L^{\mathrm{ab}}/L)$ provided by local class field theory to show that the action must factor through a finite quotient of $\mathrm{Gal}(L^{\mathrm{ab}}/L)$, which allows us to apply [AEC VII.7.3].

The p-adic (Bad Reduction) Proof

For this proof, which is due to Serre, we assume that $j(E)$ is not integral at some prime $\mathfrak{p}$ and prove that E has no non-trivial endomorphisms. Let L

be a complete local field with maximal ideal $\mathfrak{p}$, and let E/L be an elliptic curve whose j-invariant is non-integral at $\mathfrak{p}$. We will show (V §§3,5) that after replacing L by a quadratic extension, there is an element $q \in L^*$ and a $\mathfrak{p}$-adic analytic isomorphism of groups

$$\bar{L}^*/q^{\mathbb{Z}} \xrightarrow{\sim} E(\bar{L}).$$

Using this isomorphism, we construct (V.6.1) an element of $\mathrm{Gal}(\bar{L}/L)$ which acts on the Tate module of E via the matrix $\left(\begin{smallmatrix} 1 & 1 \\ 0 & 1 \end{smallmatrix}\right)$ (relative to a suitable basis). Then the fact that endomorphisms commute with the action of Galois will allow us to conclude that there are no non-trivial endomorphisms, so E does not have complex multiplication.

It is worth remarking that the second and third proofs are local; one shows that $j(E)$ is integral by working one prime at a time. The first proof, on the other hand, is more global in nature. We are going to give the first two proofs in this section. For proof number three, see (V.6.3).

Example 6.2.1. Note that the three elliptic curves in (2.3.1) possessing an endomorphism of degree 2 all have j-invariants in $\mathbb{Z}$, as they should from (6.1), since the corresponding quadratic imaginary fields have class number 1. More generally, if K has class number 1, then $j(R_K)$ will be a rational integer.

As is well known, there are only nine quadratic imaginary fields of class number 1, a fact conjectured by Gauss and proven by Heegner [1]. (See also Baker [1] and Stark [1].) These fields are

$$\mathbb{Q}\left(\sqrt{-1}\right), \quad \mathbb{Q}\left(\sqrt{-2}\right), \quad \mathbb{Q}\left(\sqrt{-3}\right), \quad \mathbb{Q}\left(\sqrt{-7}\right), \quad \mathbb{Q}\left(\sqrt{-11}\right),$$
$$\mathbb{Q}\left(\sqrt{-19}\right), \quad \mathbb{Q}\left(\sqrt{-43}\right), \quad \mathbb{Q}\left(\sqrt{-67}\right), \quad \mathbb{Q}\left(\sqrt{-163}\right).$$

A list of the corresponding j-invariants is given in Appendix A §3. It follows for example that

$$j\left(\frac{1+\sqrt{-163}}{2}\right) \in \mathbb{Z}.$$

Recall (I.7.4b) that $j(\tau)$ has the q-expansion

$$j(q) = \frac{1}{q} + 744 + 196884q + 21493760q^2 + \cdots,$$

where $q = e^{2\pi i \tau}$. If we substitute $\tau = \left(1 + \sqrt{-163}\right)/2$, then

$$q = -e^{-\pi\sqrt{163}} \approx -3.809 \cdot 10^{-18}$$

is very small. Thus the main term in $j(q)$ will be $1/q$, which means that $1/q$ should be "almost" an integer. Computing $1/q$ to 40 significant digits we find that

$$e^{\pi\sqrt{163}} = 262537412640768743.999999999999250072597\ldots,$$

so $e^{\pi\sqrt{163}}$ is an integer to 12 decimal places. Of course, we know a priori that $e^{\pi\sqrt{163}}$ is not an integer and in fact is not even an algebraic number, since the Gel'fond-Schneider theorem says that $e^{\pi\alpha} = (-1)^{-i\alpha}$ is transcendental whenever $i\alpha$ is algebraic of degree at least 2 over $\mathbb{Q}$.

Example 6.2.2. Now let's look at an example with class number larger than 1. For example, consider the field $K = \mathbb{Q}\left(\sqrt{-15}\right)$ and its ring of integers $R_K = \mathbb{Z}[\alpha]$, where to ease notation we will write $\alpha = \dfrac{1+\sqrt{-15}}{2}$. It is not hard to check that R_K has class number 2 and that a non-trivial ideal class is given by $\mathfrak{a} = 2\mathbb{Z} + \alpha\mathbb{Z}$. Further, one can check that the field $H = K\left(\sqrt{5}, \sqrt{-3}\right)$ is everywhere unramified over K, so it is the Hilbert class field of K. (See exercise 2.11.) It follows from (4.3) that $H = K\bigl(j(R_K)\bigr)$ and that $\mathbb{Q}\bigl(j(R_K)\bigr)$ is a quadratic extension of $\mathbb{Q}$ contained in H and disjoint from K. Hence $\mathbb{Q}\bigl(j(R_K)\bigr)$ is either $\mathbb{Q}\left(\sqrt{5}\right)$ or $\mathbb{Q}\left(\sqrt{-3}\right)$. We will see in a moment that $j(R_K) \in \mathbb{R}$, so we must have $\mathbb{Q}\bigl(j(R_K)\bigr) = \mathbb{Q}\left(\sqrt{5}\right)$. (This also follows from exercise 2.9, which says in general that $j(\mathfrak{c}) \in \mathbb{R}$ if and only if $\bar{\mathfrak{c}}^2 = 1$ in $\mathcal{CL}(R_K)$.)

It remains to compute $j(R_K)$ explicitly as an element of $\mathbb{Q}\left(\sqrt{5}\right)$. Let A and B be rational numbers so that

$$j(R_K) = A + B\sqrt{5}.$$

From (4.3c) we see that $j(\mathfrak{a})$ is the $\mathrm{Gal}(\bar{K}/K)$-conjugate of $j(R_K)$, so

$$j(\mathfrak{a}) = A - B\sqrt{5}.$$

Solving these two equations for A and B gives

$$A = \frac{j(R_K) + j(\mathfrak{a})}{2} \qquad \text{and} \qquad B = \frac{j(R_K) - j(\mathfrak{a})}{2\sqrt{5}}.$$

In order to compute the two values of j numerically, we can use the q-series (I.7.4b), where $q = e^{2\pi i\tau}$ is the parameter for the normalized lattice $\mathbb{Z} + \mathbb{Z}\tau$:

$$j(q) = \frac{1}{q} + 744 + 196884q + 21493760q^2 + 864299970q^3 + 20245856256q^4$$
$$+ 333202640600q^5 + 4252023300096q^6 + 44656994071935q^7 + \cdots.$$

Thus for R_K we find that

$$j(R_K) = j(\mathbb{Z} + \alpha\mathbb{Z}) = j\left(e^{2\pi i\alpha}\right) = j(-e^{-\sqrt{15}\pi})$$
$$\approx j\left(-5.19748331238 \cdot 10^{-6}\right) \approx -191657.832863.$$

(Notice in this case that $q = -e^{-\sqrt{15}\pi} \in \mathbb{R}$, so $j(R_K) \in \mathbb{R}$.) Similarly for $\mathfrak{a}$ we calculate

$$j(\mathfrak{a}) = j(2\mathbb{Z} + \alpha\mathbb{Z}) = j\left(\mathbb{Z} + \frac{1}{2}\alpha\mathbb{Z}\right) = j(e^{-\sqrt{15}\pi/2}i)$$
$$\approx j\left(2.27979896315 \cdot 10^{-3}i\right) \approx 632.83286254.$$

Using these values gives (to 12 significant digits) $A = -95512.5000002$ and $B = -42997.5000001$, so

$$j(R_K) \approx -95512.5 - 42997.5\sqrt{5} = -52515 - 85995\frac{1+\sqrt{5}}{2} \in \mathbb{Z}\left[\frac{1+\sqrt{5}}{2}\right].$$

Thus $j(R_K)$ is (at least approximately) integral over $\mathbb{Z}$, which gives a numerical verification of Theorem 6.1 for this example.

The Analytic Proof of Theorem 6.1

Before beginning, we give a few words of motivation. It is not hard to see that an elliptic curve E has complex multiplication if and only if there is an endomorphism $E \to E$ whose degree is not a square. This suggests that we take an arbitrary elliptic curve E and a positive integer n and study the set of all elliptic curves E' for which there is an isogeny $E \to E'$ of degree n. We took this point of view in (I §9,10) when we studied Hecke operators. What we are going to do is show that in this situation $j(E')$ is integral over $\mathbb{Z}[j(E)]$. We will do this by explicitly constructing a monic polynomial $F_n(j(E), X)$ with coefficients in $\mathbb{Z}[j(E)]$ having $j(E')$ as a root. Finally, if E has complex multiplication, then for an appropriate choice of n we can take $E' = E$. This means that $F_n(j(E), j(E)) = 0$, which we will show implies that $j(E)$ is integral over $\mathbb{Z}$.

We now begin the analytic proof of Theorem 6.1. We fix a positive integer n and recall the sets of matrices $\mathcal{D}_n$ and $\mathcal{S}_n$ defined in (I §9):

$$\mathcal{D}_n = \left\{ \begin{pmatrix} a & b \\ c & d \end{pmatrix} \in M_2(\mathbb{Z}) : ad - bc = n \right\},$$

$$\mathcal{S}_n = \left\{ \begin{pmatrix} a & b \\ 0 & d \end{pmatrix} \in M_2(\mathbb{Z}) : ad = n, \ d > 0, \ 0 \le b < d \right\}.$$

We also recall (I.9.2), which says that $\mathcal{S}_n = \mathrm{SL}_2(\mathbb{Z}) \backslash \mathcal{D}_n$. For any matrix $\alpha = \begin{pmatrix} a & b \\ c & d \end{pmatrix} \in M_2(\mathbb{R})$ with $\det \alpha > 0$, we define the function $j \circ \alpha$ as usual by the formula

$$j \circ \alpha(\tau) = j\left(\frac{a\tau + b}{c\tau + d}\right).$$

Of course, if $\alpha \in \mathrm{SL}_2(\mathbb{Z})$, then $j \circ \alpha = j$. We are going to study the polynomial

$$F_n(X) = \prod_{\alpha \in \mathcal{S}_n} (X - j \circ \alpha) = \sum_m s_m X^m$$

whose coefficients $s_m = s_m(\tau)$ are holomorphic functions on the upper half-plane $\mathbf{H}$. More precisely, s_m is the mth elementary symmetric function in the $j \circ \alpha$'s. We are going to prove several claims concerning the s_m's.

Claim 1: $s_m(\gamma\tau) = s_m(\tau)$ for all $\gamma \in \mathrm{SL}_2(\mathbb{Z})$ and all $\tau \in \mathbf{H}$.

Let $\gamma \in \mathrm{SL}_2(\mathbb{Z})$. For any $\alpha \in \mathcal{S}_n$ we have $\alpha\gamma \in \mathcal{D}_n$, so (I.9.2) says that there is a (unique) $\delta_\alpha \in \mathrm{SL}_2(\mathbb{Z})$ such that $\delta_\alpha \alpha\gamma$ is back in $\mathcal{S}_n$. Further, if $\delta_\alpha \alpha\gamma = \delta_\beta \beta\gamma$ for some $\beta \in \mathcal{S}_n$, then $\beta = (\delta_\beta^{-1}\delta_a)\alpha$, so (I.9.2) implies that $\alpha = \beta$. In other words, the map

$$\mathcal{S}_n \longrightarrow \mathcal{S}_n, \qquad \alpha \longmapsto \delta_\alpha \alpha\gamma,$$

is one-to-one, hence is a bijection since $\mathcal{S}_n$ is a finite set.

Now we observe that

$$\begin{aligned}
\{j \circ (\alpha\gamma) : \alpha \in \mathcal{S}_n\} &= \{j \circ \delta_\alpha^{-1} \circ (\delta_a \alpha\gamma) : \alpha \in \mathcal{S}_n\} \\
&= \{j \circ (\delta_a \alpha\gamma) : \alpha \in \mathcal{S}_n\} \quad \text{since } j \text{ is } \mathrm{SL}_2(\mathbb{Z})\text{-invariant} \\
&= \{j \circ \alpha : \alpha \in \mathcal{S}_n\} \quad \text{since } \mathcal{S}_n = \{\delta_a \alpha\gamma : \alpha \in \mathcal{S}_n\}.
\end{aligned}$$

Hence any symmetric function on the set $\{j \circ \alpha : \alpha \in \mathcal{S}_n\}$ will be invariant under $\tau \mapsto \gamma\tau$ for $\gamma \in \mathrm{SL}_2(\mathbb{Z})$. In particular, this applies to the $s_m(\tau)$'s, which completes the proof of Claim 1.

Claim 2: $s_m \in \mathbb{C}[j]$.

In other words, we are claiming that there is a polynomial $f_m(X) \in \mathbb{C}[X]$ such that $s_m(\tau) = f_m(j(\tau))$ for all $\tau \in \mathbf{H}$. From Claim 1 we know that s_m is holomorphic on $\mathbf{H}$ and is $\mathrm{SL}_2(\mathbb{Z})$-invariant. In particular, $s_m(\tau+1) = s_m(\tau)$, so s_m has a Fourier expansion in $q = e^{2\pi i \tau}$. We want to study what happens as $\tau \to i\infty$, or equivalently as $q \to 0$. Recall (I.7.4b) that j has the Fourier expansion $j = q^{-1} + \sum_{k \geq 0} c_k q^k$, so j has a pole of order 1 at $q = 0$. Now if $\alpha = \left(\begin{smallmatrix} a & b \\ 0 & d \end{smallmatrix}\right) \in \mathcal{S}_n$, then

$$j \circ \alpha(\tau) = e^{-2\pi i \frac{a\tau+b}{d}} + \sum_{k=0}^{\infty} c_k e^{2\pi i k \frac{a\tau+b}{d}},$$

so in particular $q^{n+1}(j \circ \alpha)(\tau) \to 0$ as $q \to 0$. It follows from the definition of the s_m's that there is an integer N such that $q^N s_m(\tau) \to 0$ as $q \to 0$. This means that each $s_m(\tau)$ is meromorphic at ∞ (see I §3), so s_m is a modular function of weight 0 which is holomorphic on $\mathbf{H}$. Now (I.4.2b) says that $s_m \in \mathbb{C}[j]$, which completes the proof of Claim 2.

Claim 3: The Fourier expansion of s_m has coefficients in $\mathbb{Z}$.

To ease notation, let $\zeta = e^{2\pi i/n}$ and $Q = q^{1/n} = e^{2\pi i \tau/n}$. For any $\alpha = \begin{pmatrix} a & b \\ 0 & d \end{pmatrix} \in \mathcal{S}_n$ we have

$$q \circ \alpha(\tau) = e^{2\pi i \frac{a\tau + b}{d}} = \zeta^{ab} Q^{a^2}.$$

(Note that $ad = n$.) Using the q-expansion of $j(\tau)$ (I.7.4b), we find as above that $j \circ \alpha$ has the Q-expansion

$$j \circ \alpha(t) = \zeta^{-ab} Q^{-a^2} + \sum_{k=0}^{\infty} c_k \zeta^{abk} Q^{a^2 k},$$

where $c_0, c_1, \ldots$ are integers. In particular, the Fourier coefficients of $j \circ \alpha$ lie in $\mathbb{Z}[\zeta]$, and so the same is true of the s_m's.

Let $\sigma \in \mathrm{Gal}\big(\mathbb{Q}(\zeta)/\mathbb{Q}\big)$, and write $\zeta^\sigma = \zeta^{r(\sigma)}$ for some integer $r(\sigma)$ relatively prime to n. If we apply σ to the Q-Fourier coefficients of $j \circ \alpha$, we get the series

$$(j \circ \alpha)^\sigma = \zeta^{-r(\sigma)ab} Q^{-a^2} + \sum_{k=0}^{\infty} c_k \zeta^{r(\sigma)abk} Q^{a^2 k}.$$

Comparing the series for $j \circ \alpha$ and $(j \circ \alpha)^\sigma$, we see that

$$\left(j \circ \begin{pmatrix} a & b \\ 0 & d \end{pmatrix} \right)^\sigma = j \circ \begin{pmatrix} a & r(\sigma)b \\ 0 & d \end{pmatrix}.$$

In general, the value of $j \circ \begin{pmatrix} a & b \\ 0 & d \end{pmatrix}$ only depends on $b \,(\mathrm{mod}\, d)$, since

$$\begin{pmatrix} 1 & k \\ 0 & 1 \end{pmatrix} \begin{pmatrix} a & b \\ 0 & d \end{pmatrix} = \begin{pmatrix} a & b + kd \\ 0 & d \end{pmatrix}$$

and j is $\mathrm{SL}_2(\mathbb{Z})$-invariant. Further, if r is any integer prime to $n = ad$, then the set $\{rb : 0 \le b < d\}$ is a complete set of residue classes modulo d. It follows that for any integer r relatively prime to n we have

$$\left\{ j \circ \begin{pmatrix} a & rb \\ 0 & d \end{pmatrix} : \begin{pmatrix} a & b \\ 0 & d \end{pmatrix} \in \mathcal{S}_n \right\} = \{ j \circ \alpha : \alpha \in \mathcal{S}_n \}.$$

Applying this with $r = r(\sigma)$ for $\sigma \in \mathrm{Gal}\big(\mathbb{Q}(\zeta)/\mathbb{Q}\big)$, it follows that

$$\{ (j \circ \alpha)^\sigma : \alpha \in \mathcal{S}_n \} = \{ j \circ \alpha : \alpha \in \mathcal{S}_n \}.$$

Now consider the Q-Fourier coefficients of the $s_m(\tau)$'s, which we know from above lie in $\mathbb{Z}[\zeta]$. Since $s_m(\tau)$ is a symmetric polynomial in the functions $\{ j \circ \alpha : \alpha \in \mathcal{S}_n \}$, we see that its Q-Fourier coefficients are fixed by $\mathrm{Gal}\big(\mathbb{Q}(\zeta)/\mathbb{Q}\big)$ and so lie in $\mathbb{Q}$. Hence the Fourier coefficients of $s_m(\tau)$ are in $\mathbb{Z}[\zeta] \cap \mathbb{Q} = \mathbb{Z}$. Finally, we note that $\sigma_m(\tau+1) = \sigma_m(\tau)$ from Claim 1, so σ_m is in fact represented by a Fourier series in $q = Q^n$. This completes the proof of Claim 3.

Claim 4: $s_m(\tau) \in \mathbb{Z}[j]$.

We already know from Claim 2 that $s_m \in \mathbb{C}[j]$ and from Claim 3 that $s_m \in \mathbb{Z}[\![q, q^{-1}]\!]$. We will show that

$$\mathbb{C}[j] \cap \mathbb{Z}[\![q, q^{-1}]\!] = \mathbb{Z}[j],$$

which will give the desired result. Let $f(j) \in \mathbb{C}[j] \cap \mathbb{Z}[\![q, q^{-1}]\!]$ be a polynomial of degree d, and write $f(j) = a_0 j^d + a_1 j^{d-1} + \cdots + a_d$ with $a_i \in \mathbb{C}$. Substituting in the q-expansion of j (I.7.4b) gives

$$f = \frac{a_0}{q^d} + \frac{a_1 + 744 d a_0}{q^{d-1}} + \cdots,$$

so the fact that $f \in \mathbb{Z}[\![q, q^{-1}]\!]$ implies that $a_0 \in \mathbb{Z}$. Now

$$f - a_0 j^d = a_1 j^{d-1} + \cdots + a_d \in \mathbb{C}[j] \cap \mathbb{Z}[\![q, q^{-1}]\!],$$

so repeating the above argument gives $a_1 \in \mathbb{Z}$. Continuing in this way, we find that every coefficient of f in $\mathbb{Z}$, which completes the proof of Claim 4.

Combining Claims 1, 2, 3, and 4, we have completed the proof of the first half of the following important result.

Theorem 6.3. (a) *There is a polynomial $F_n(Y, X) \in \mathbb{Z}[Y, X]$ so that*

$$\prod_{\alpha \in \mathcal{S}_n} (X - j \circ \alpha) = F_n(j, X).$$

(b) *Let $\beta \in M_2(\mathbb{Z})$ be a matrix with integer coefficients and $\det \beta > 0$. Then the function $j \circ \beta$ is integral over the ring $\mathbb{Z}[j]$.*
(c) *If n is not a perfect square, then the polynomial $H_n(X) = F_n(X, X)$ is non-constant and has leading coefficient ± 1.*

PROOF. (a) The four claims proven above say that

$$\prod_{\alpha \in \mathcal{S}_n} (X - j \circ \alpha) = \sum_m s_m X^m \qquad \text{with } s_m \in \mathbb{Z}[j].$$

(b) Let $n = \det \beta$, so $\beta \in \mathcal{D}_n$. Using (I.9.2), we can find a matrix $\gamma \in SL_2(\mathbb{Z})$ such that $\gamma \beta \in \mathcal{S}_n$. The $SL_2(\mathbb{Z})$-invariance of j says that $j \circ \beta = j \circ (\gamma \beta)$, while the definition of F_n shows that $X = j \circ (\gamma \beta)$ is a root of $F_n(j, X)$. Since F_n is monic by definition and has coefficients in $\mathbb{Z}[j]$ from (a), it follows that $j \circ \beta$ is integral over $\mathbb{Z}[j]$.
(c) Let $\alpha = \begin{pmatrix} a & b \\ 0 & d \end{pmatrix} \in \mathcal{S}_n$. Then using the Q-expansion of $j \circ \alpha$ described above during the proof of Claim 3, we see that the Q-expansion of $j - j \circ \alpha$ is

$$j - j \circ \alpha = \left(\frac{1}{Q^n} + \sum_{k=0}^{\infty} c_k Q^{nk} \right) - \left(\frac{1}{\zeta^{ab} Q^{a^2}} + \sum_{k=0}^{\infty} c_k \zeta^{abk} Q^{a^2 k} \right).$$

(Here we are again writing $\zeta = e^{2\pi i/n}$ and $Q = q^{1/n}$.) Since n is not a square, the leading terms cannot cancel, so $j - j \circ \alpha$ has a pole as $Q \to 0$ and the coefficient of the leading term is necessarily a root of unity. (Precisely, the coefficient is 1 if $n > a^2$, and it is $-\zeta^{-ab}$ otherwise.) It follows that $F_n(j, j)$ has a pole as $Q \to 0$ and that the leading Q-coefficient is a root of unity. But the Q-expansion of $F_n(j, j)$ has integer coefficients, so the leading coefficient is a root of unity in $\mathbb{Z}$; hence it must be ± 1. Further, $F_n(j, j)$ is actually a series in $q = Q^n$, so we have proven that

$$F_n(j, j) = \pm\frac{1}{q^m} + \cdots \in q^{-m}\mathbb{Z}[\![q]\!]$$

for some $m \geq 1$. But we also know that $F_n(j, j) \in \mathbb{Z}[j]$ and that j has a simple pole at $q = 0$. Hence $F_n(j, j) = \pm j^m + \cdots \in \mathbb{Z}[j]$, which proves that $F_n(X, X)$ is a non-constant polynomial with leading coefficient ± 1.

$\square$

It is now a simple matter to complete the proof of Theorem 6.1.

Corollary 6.3.1. (Theorem 6.1). *Let $E/\mathbb{C}$ be an elliptic curve with complex multiplication. Then $j(E)$ is an algebraic integer.*

PROOF. Let $R \cong \mathrm{End}(E)$ be an order in a quadratic imaginary field K. We consider first the case that $R = R_K$ is the ring of integers of K. Choose some element $\rho \in R$ such that $n = |\mathrm{N}_\mathbb{Q}^K \rho|$ is not a perfect square. For example, if $K = \mathbb{Q}(i)$, take $\rho = 1 + i$, and if $K = \mathbb{Q}(\sqrt{-D})$ with square-free $D \geq 2$, take $\rho = \sqrt{-D}$. Then (1.5b) says that the isogeny $[\rho] : E \to E$ has degree n. Fix a $\tau \in \mathsf{H}$ with $j(\tau) = j(E)$. Then multiplication by ρ sends the lattice $\mathbb{Z}\tau + \mathbb{Z}$ to a sublattice of index n, say

$$\begin{aligned} \rho\tau &= a\tau + b \\ \rho &= c\tau + d \end{aligned} \qquad \text{for some } a, b, c, d \in \mathbb{Z} \text{ with } ad - bc = n.$$

So if we let $\alpha = \begin{pmatrix} a & b \\ c & d \end{pmatrix} \in \mathcal{D}_n$, then

$$j(\alpha\tau) = j\left(\frac{a\tau + b}{c\tau + d}\right) = j(\tau) = j(E).$$

By definition, $j \circ \alpha$ is a root of $F_n(j, X)$, so if we substitute $X = j \circ \alpha$ and evaluate at τ, we get

$$0 = F_n\big(j(\tau), j(\alpha\tau)\big) = F_n\big(j(E), j(E)\big) = H_n\big(j(E)\big).$$

From (6.3c), the polynomial $H_n(X)$ has integer coefficients and leading coefficient ± 1. This proves that $j(E)$ is integral over $\mathbb{Z}$.

Now we deal with the case that R is an arbitrary order in K. Let $\Lambda = \mathbb{Z}\omega_1 + \mathbb{Z}\omega_2$ be a lattice for E. From [AEC VI.5.5] we know that $K = \mathbb{Q}(\omega_1/\omega_2)$. Hence replacing Λ by $\lambda\Lambda$ for an appropriate $\lambda \in \mathbb{C}^*$, we may assume that $\Lambda \subset R_K$. We also choose a $\tau \in \mathsf{H}$ so that $R_K = \mathbb{Z}\tau + \mathbb{Z}$. Then we can write

$$\begin{aligned}\omega_1 &= a\tau + b \\ \omega_2 &= c\tau + d\end{aligned} \qquad \text{for some } a, b, c, d \in \mathbb{Z}.$$

Let $n = ad - bc$. Switching ω_1 and ω_2 if necessary, we may assume that $n \geq 1$. The matrix $\alpha = \begin{pmatrix} a & b \\ c & d \end{pmatrix}$ is in $\mathcal{D}_n$, so (6.3b) says that the function $j \circ \alpha$ is integral over the ring $\mathbb{Z}[j]$. Taking the equation $F_n(j, X) = 0$ which gives that integrality and evaluating it at τ, we find that $j(\alpha\tau)$ is integral over $\mathbb{Z}[j(\tau)]$. But $j(\alpha\tau) = j(E)$, and we already know that $j(\tau)$ is integral over $\mathbb{Z}$ because it is the j-invariant of an elliptic curve with complex multiplication by R_K. Therefore $j(E)$ is integral over $\mathbb{Z}$. $\square$

Example 6.3.2. The polynomials $F_n(Y, X) \in \mathbb{Z}[Y, X]$ and $H_n(X) \in \mathbb{Z}[X]$ described in (6.3) can be extremely complicated. For example,

$$\begin{aligned}F_2(Y, X) = &-(XY)^2 + X^3 + Y^3 + 2^4 \cdot 3 \cdot 31 \cdot XY(X + Y) \\ &+ 3^4 \cdot 5^3 \cdot 4027 \cdot XY - 2^4 \cdot 3^4 \cdot 5^3 (X^2 + Y^2) \\ &+ 2^8 \cdot 3^7 \cdot 5^6 (X + Y) - 2^{12} \cdot 3^9 \cdot 5^9,\end{aligned}$$
$$\begin{aligned}H_2(X) = &-X^4 + 2 \cdot 1489 \cdot X^3 + 3^4 \cdot 5^4 \cdot 17 \cdot 47 \cdot X^2 \\ &+ 2^9 \cdot 3^7 \cdot 5^6 \cdot X - 2^{12} \cdot 3^9 \cdot 5^9.\end{aligned}$$

According to [AEC III.4.5], the elliptic curves $E : y^2 = x^3 + ax^2 + bx$ and $E' : y^2 = x^3 - 2ax^2 + (a^2 - 4b)x$ are connected by an isogeny $E \to E'$ of degree 2. It follows from exercise 2.19 that

$$F_2\big(j(E), j(E')\big) = F_2^*\left(\frac{256(a^2 - 3b)^3}{b^2(a^2 - 4b)}, \frac{16(a^2 + 12b)^3}{b(a^2 - 4b)^2} \right) = 0,$$

a fact that the interested reader can check by a direct computation (preferably with the assistance of a symbolic calculator).

The ℓ-adic Proof of Theorem 6.1

We now begin the ℓ-adic proof of Theorem 6.1. If E/L is an elliptic curve with complex multiplication, then we know from (2.3) that the action of $\mathrm{Gal}(\bar{L}/L)$ on E_{tors} is abelian, so in particular the action on the Tate module $T_\ell(E)$ is abelian. (For another proof of this result which uses only a little linear algebra and is valid even if $\mathrm{End}(E)$ is not a maximal order, see exercise 2.6.) We now use Proposition 2.3, local class field theory, and the criterion of Néron-Ogg-Shafarevich to prove that E has everywhere potential good reduction. Although somewhat involved, this proof has the advantage that it generalizes to abelian varieties of arbitrary dimension (see Serre-Tate [1]).

Theorem 6.4. *Let L be a number field and E/L an elliptic curve with complex multiplication. Then E has potential good reduction at every prime of L.*

PROOF. (Serre-Tate) Every endomorphism of R is defined over a finite extension of L (2.2b), so replacing L by a finite extension, we may assume that $\mathrm{End}_L(E)$ is strictly larger than $\mathbb{Z}$. Fix a prime v of L. We set the following notation:

$$L_v = \text{the completion of } L \text{ at } v,$$

$$R_v = \text{the ring of integers of } L_v,$$

$$\mathfrak{M}_v = \text{the maximal ideal of } R_v,$$

$$p = \mathrm{char}\, R_v/\mathfrak{M}_v = \text{the residue characteristic of } R_v,$$

$$\ell = \text{a rational prime not equal to 2 or } p,$$

$$I_v = \text{the inertia subgroup of } \mathrm{Gal}(\bar{L}_v/L_v),$$

$$L_v^{\mathrm{ab}} = \text{the maximal abelian extension of } L_v,$$

$$I_v^{\mathrm{ab}} = \text{the inertia subgroup of } \mathrm{Gal}(L_v^{\mathrm{ab}}/L_v).$$

By assumption, $\mathrm{End}_L(E) \neq \mathbb{Z}$, so certainly $\mathrm{End}_{L_v}(E)$ is strictly larger than $\mathbb{Z}$. Applying (2.3), we see that the action of $\mathrm{Gal}(\bar{L}_v/L_v)$ on $T_\ell(E)$ is abelian. In particular, I_v acts through the quotient I_v^{ab}.

Local class field theory says that there is an isomorphism

$$I_v^{\mathrm{ab}} \cong R_v^*.$$

(See, e.g., Lang [5], Serre [4, XIV §6, Cor. 2(ii) to Thm. 1], Serre [5].) This gives us a very good picture of I_v^{ab}, since we can decompose R_v^* using the exact sequence

$$1 \longrightarrow \underbrace{R_{v,1}^*}_{\text{pro}-p \text{ group}} \longrightarrow \underset{\|\wr}{R_v^*} \longrightarrow \underbrace{(R_v/\mathfrak{M}_v)^*}_{\text{finite}} \longrightarrow 1.$$
$$I_v^{\mathrm{ab}}$$

Here $R_{v,1}^*$ is the group of 1-units,

$$R_{v,1}^* = \big\{u \in R_v^* : u \equiv 1 \,(\mathrm{mod}\,\mathfrak{M}_v)\big\}.$$

There is an isomorphism from the formal multiplicative group $\hat{\mathbb{G}}_{\mathrm{m}}(\mathfrak{M}_v)$ to $R_{v,1}^*$ given by

$$\hat{\mathbb{G}}_{\mathrm{m}}(\mathfrak{M}_v) \xrightarrow{\sim} R_{v,1}^*, \qquad t \longmapsto 1+t.$$

Hence $R_{v,1}^*$ is a pro-p group; that is, it is the inverse limit of finite groups of p-power order (see [AEC IV.3.1.2] and [AEC IV.3.2]).

Similarly, if we fix an isomorphism $\operatorname{Aut} T_\ell(E) \cong \operatorname{GL}_2(\mathbb{Z}_\ell)$ corresponding to some basis for $T_\ell(E)$, then there is an exact sequence

$$1 \longrightarrow \underbrace{\operatorname{GL}_2(\mathbb{Z}_\ell)_1}_{\text{pro-}\ell \text{ group}} \longrightarrow \operatorname{GL}_2(\mathbb{Z}_\ell) \longrightarrow \operatorname{GL}_2(\mathbb{Z}/\ell\mathbb{Z}) \longrightarrow 1.$$

$$\qquad\qquad\qquad\qquad\quad \| \wr \qquad\qquad\qquad \| \wr$$

$$\qquad\qquad\qquad \operatorname{Aut} T_\ell(E) \longrightarrow \underbrace{\operatorname{Aut} E[\ell]}_{\text{finite}}$$

Here $\operatorname{GL}_2(\mathbb{Z}_\ell)_1$ is the group of matrices congruent to the identity matrix modulo ℓ, and it is not hard to see that this is a pro-ℓ group. More precisely, the logarithm map gives an isomorphism

$$\operatorname{GL}_2(\mathbb{Z}_\ell)_1 \xrightarrow{\sim} M_2(\ell\mathbb{Z}_\ell), \qquad 1 + \ell A \longmapsto \log(1 + \ell A) = \sum_{n=1}^{\infty} \frac{(-1)^{n+1}\ell^n A^n}{n},$$

where $M_2(\mathbb{Z}_\ell)$ is the group of 2×2 matrices with coefficients in $\ell\mathbb{Z}_\ell$ under addition. This isomorphism is the GL_2 analogue of [AEC IV.6.4b]. (See also exercises 2.22 and 2.23.)

It follows from the above discussion that the map

$$I_v \longrightarrow \operatorname{Aut} T_\ell(E)$$

fits into the following diagram:

$$I_v$$
$$\downarrow$$
$$I_v^{\text{ab}}$$
$$\| \wr$$

$$1 \longrightarrow R_{v,1}^* \longrightarrow R_v^* \longrightarrow (R_v/\mathfrak{M}_v)^* \longrightarrow 1$$
$$\downarrow$$
$$1 \longrightarrow \operatorname{GL}_2(\mathbb{Z}_\ell)_1 \longrightarrow \operatorname{Aut} T_\ell(E) \longrightarrow \operatorname{GL}_2(\mathbb{Z}/\ell\mathbb{Z}) \longrightarrow 1$$

Next we observe that since $\ell \neq p$, then there can be no non-trivial homomorphisms from a pro-p group to a pro-ℓ group, so the images of $R_{v,1}^*$ and $\operatorname{GL}_2(\mathbb{Z}_\ell)_1$ in $\operatorname{Aut} T_\ell(E)$ have trivial intersection. Therefore there is an injection

$$\operatorname{Image}\bigl(R_{v,1}^* \longrightarrow \operatorname{Aut} T_\ell(E)\bigr) \hookrightarrow \operatorname{GL}_2(\mathbb{Z}/\ell\mathbb{Z}).$$

Since also $(R_v/\mathfrak{M}_v)^*$ is finite, it follows that

$$\operatorname{Image}\bigl(R_v^* \longrightarrow \operatorname{Aut} T_\ell(E)\bigr) \quad \text{is finite,}$$

since it consists of finitely many cosets of $\text{Image}\big(R_{v,1}^* \longrightarrow \text{Aut}\, T_\ell(E)\big)$.

This proves that the image of I_v in $\text{Aut}\, T_\ell(E)$ is finite. Now the criterion of Néron-Ogg-Shafarevich (specifically [AEC VII.7.3]) says that E has potential good reduction at v, which concludes the proof of Theorem 6.4.

$\square$

It is now a simple matter to deduce (6.1) from (6.4).

Corollary 6.4.1.   (Theorem 6.1). *Let $E/\mathbb{C}$ be an elliptic curve with complex multiplication. Then $j(E)$ is an algebraic integer.*

PROOF. The elementary result (2.1b) says that $j(E)$ is an algebraic number, so we may take an equation for E with coefficients in the number field $L = \mathbb{Q}\big(j(E)\big)$. Then (6.4) says that E has potential good reduction at every prime of L, so [AEC VII.5.5] implies that $j(E)$ is integral at every prime of L. $\square$

§7. Cyclotomic Class Field Theory

In this section we are going to formulate the class field theory of $\mathbb{Q}$ in terms of special values of analytic functions, specifically special values of the exponential function. This is analogous to the way we will later be describing the class field theory of quadratic imaginary fields via the theory of complex multiplication. We hope that studying the simpler cyclotomic case first will aid the reader in understanding the more intricate proofs required in the complex multiplication case. However, the results in this section will not be used later, so the reader who already feels comfortable with class field theory may wish to skip directly to §8.

We begin with the multiplicative group

$$\mathbb{G}_m(\mathbb{C}) \cong \mathbb{C}^*.$$

The exponential map provides a complex analytic parametrization of the multiplicative group,

$$\begin{array}{ccc} f: & \mathbb{C}/\mathbb{Z} & \overset{\sim}{\longrightarrow} & \mathbb{G}_m(\mathbb{C}) \\ & t & \longmapsto & e^{2\pi i t}. \end{array}$$

Sitting inside of $\mathbb{G}_m(\mathbb{C})$ is its torsion subgroup

$$\mathbb{G}_m(\mathbb{C})_{\text{tors}} = f\left(\mathbb{Q}/\mathbb{Z}\right).$$

The elements of $\mathbb{G}_m(\mathbb{C})_{\text{tors}}$ are roots of unity, so they generate abelian extensions of $\mathbb{Q}$. Our aim is to give an analytic description of the action of $\text{Gal}(\mathbb{Q}^{\text{ab}}/\mathbb{Q})$ on $\mathbb{G}_m(\mathbb{C})_{\text{tors}}$.

From class field theory (see §3) we have the reciprocity map, which is a surjective homomorphism

$$\mathbf{A}_{\mathbb{Q}}^* \longrightarrow \mathrm{Gal}(\mathbb{Q}^{\mathrm{ab}}/\mathbb{Q}), \qquad s \longmapsto [s, \mathbb{Q}].$$

Each idele s thus defines an isomorphism

$$\mathbb{G}_{\mathrm{m}}(\mathbb{C})_{\mathrm{tors}} \longrightarrow \mathbb{G}_{\mathrm{m}}(\mathbb{C})_{\mathrm{tors}},$$
$$\zeta \longmapsto \zeta^{[s,\mathbb{Q}]}.$$

The algebraic action of $\mathrm{Gal}(\mathbb{Q}^{\mathrm{ab}}/\mathbb{Q})$ on $\mathbb{G}_{\mathrm{m}}(\mathbb{C})_{\mathrm{tors}}$ is determined by these isomorphisms.

In general, if $x \in \mathbf{A}_{\mathbb{Q}}^*$ is any idele, we want to define a subgroup $x\mathbb{Z} \subset \mathbb{Q}$ and a multiplication-by-x map

$$\mathbb{Q}/\mathbb{Z} \xrightarrow{\ x\ } \mathbb{Q}/x\mathbb{Z}.$$

The definition of $x\mathbb{Z}$ is easy; it is just the ideal of x, which we recall is the fractional ideal of $\mathbb{Q}$ given by

$$x\mathbb{Z} = (x) = \prod_p p^{\mathrm{ord}_p x_p} \cdot \mathbb{Z} = N_x \mathbb{Z}.$$

For convenience, we will write N_x as indicated for a rational number generating the ideal $x\mathbb{Z}$. Later we will pin down N_x precisely by requiring that $\mathrm{sign}(N_x) = \mathrm{sign}(x_\infty)$.

In order to define the multiplication-by-x map, we decompose $\mathbb{Q}/\mathbb{Z}$ into its p-primary components and multiply the p-component by x_p. Note that

$$(p\text{-primary component of } \mathbb{Q}/\mathbb{Z}) = \mathbb{Z}[p^{-1}]/\mathbb{Z} \cong \mathbb{Q}_p/\mathbb{Z}_p.$$

The first equality is immediate, since if $t = a/n \in \mathbb{Q}/\mathbb{Z}$ has p-power order, then n must be a power of p. For the second equality, we clearly have an injection

$$\mathbb{Z}[p^{-1}]/\mathbb{Z} \hookrightarrow \mathbb{Q}_p/\mathbb{Z}_p.$$

To check surjectivity, let $\xi \in \mathbb{Q}_p/\mathbb{Z}_p$. We can write $\xi = \alpha/p^e$ for some $\alpha \in \mathbb{Z}_p$ and some integer $e \geq 0$. Choose an integer $a \in \mathbb{Z}$ with $a \equiv \alpha \pmod{p^e \mathbb{Z}_p}$. Then

$$\frac{a}{p^e} \in \mathbb{Z}[p^{-1}]/\mathbb{Z} \qquad \text{and} \qquad \frac{a}{p^e} \equiv \frac{\alpha}{p^e} = \xi \text{ in } \mathbb{Q}_p/\mathbb{Z}_p.$$

Similarly, we observe that for any $N \in \mathbb{Q}^*$, the p-primary part of $\mathbb{Q}/N\mathbb{Z}$ is isomorphic to $\mathbb{Q}_p/N\mathbb{Z}_p$.

It is a general fact that an abelian group whose elements all have finite order is the direct sum of its p-primary components. (See (8.1) and (8.1.1) for something stronger.) Hence

$$\mathbb{Q}/\mathbb{Z} \cong \bigoplus_p \mathbb{Q}_p/\mathbb{Z}_p,$$

and for any idele $x \in \mathbf{A}_{\mathbb{Q}}^*$,

$$\mathbb{Q}/x\mathbb{Z} = \mathbb{Q}/N_x\mathbb{Z} \cong \bigoplus_p \mathbb{Q}_p/N_x\mathbb{Z}_p = \bigoplus_p \mathbb{Q}_p/x_p\mathbb{Z}_p.$$

The last equality follows from the fact that $\mathrm{ord}_p(N_x) = \mathrm{ord}_p(x_p)$, so the ideals $N_x\mathbb{Z}_p$ and $x_p\mathbb{Z}_p$ are the same. Now we can define the multiplication-by-x map to be multiplication of the p-component by x_p; in other words, multiplication-by-x is defined by the commutativity of the following diagram:

$$
\begin{array}{ccc}
\mathbb{Q}/\mathbb{Z} & \overset{x}{\longrightarrow} & \mathbb{Q}/x\mathbb{Z} \\
\downarrow{\wr} & & \downarrow{\wr} \\
\bigoplus_p \mathbb{Q}_p/\mathbb{Z}_p & \longrightarrow & \bigoplus_p \mathbb{Q}_p/x_p\mathbb{Z}_p \\
(t_p) & \longmapsto & (x_p t_p)
\end{array}
$$

We are now ready for the main theorem of this section. The reader should compare this cyclotomic result (7.1) with the corresponding complex multiplication theorem (8.2).

Theorem 7.1. *Fix the following quantities:*

$\sigma \in \mathrm{Aut}(\mathbb{C})$, *an automorphism of the complex numbers,*

$s \in \mathbf{A}_{\mathbb{Q}}^*$, *an idele of $\mathbb{Q}$ satisfying $[s, \mathbb{Q}] = \sigma|_{\mathbb{Q}^{\mathrm{ab}}}$.*

Further, fix the complex analytic isomorphism

$$f : \mathbb{C}/\mathbb{Z} \xrightarrow{\sim} \mathbb{G}_{\mathrm{m}}(\mathbb{C}), \qquad f(t) = e^{2\pi i t}.$$

Then there exists a unique complex analytic isomorphism

$$f' : \mathbb{C}/s^{-1}\mathbb{Z} \xrightarrow{\sim} \mathbb{G}_{\mathrm{m}}(\mathbb{C})$$

so that the following diagram commutes:

$$
\begin{array}{ccc}
\mathbb{Q}/\mathbb{Z} & \overset{s^{-1}}{\longrightarrow} & \mathbb{Q}/s^{-1}\mathbb{Z} \\
\downarrow{f} & & \downarrow{f'} \\
\mathbb{G}_{\mathrm{m}}(\mathbb{C}) & \overset{\sigma}{\longrightarrow} & \mathbb{G}_{\mathrm{m}}(\mathbb{C}).
\end{array}
$$

Remark 7.1.1. Theorem 7.1 says that

$$f(t)^{[s,\mathbb{Q}]} = f'(s^{-1}t) \qquad \text{for all } t \in \mathbb{Q}/\mathbb{Z}.$$

Of course, f' depends on s. We will see during the proof of (7.1) that

$$f'(t) = e^{2\pi i N_s t},$$

where N_s is a certain non-zero rational number. Thus written out explicitly, (7.1) says that

$$\left(e^{2\pi i t}\right)^{[s,\mathbb{Q}]} = e^{2\pi i N_s(s^{-1}t)} \qquad \text{for all } t \in \mathbb{Q}/\mathbb{Z}.$$

This formula makes it very clear how the algebraic (Galois) action of $[s, \mathbb{Q}]$ is transformed into the analytic (multiplication) action $t \to N_s s^{-1} t$. Later we will have more to say about N_s, see (7.2).

PROOF (of Theorem 7.1). Let $t \in \mathbb{Q}/\mathbb{Z}$, say $t = a/n \,(\mathrm{mod}\ \mathbb{Z})$ as a fraction in lowest terms. To ease notation, let $\zeta = f(t)$ be the corresponding primitive n^{th}-root of unity. Suppose first that our idele s has the property

$$s_p \equiv 1 \quad (\mathrm{mod}\ n\mathbb{Z}_p) \quad \text{for all primes } p, \text{ and further that} \quad s_\infty > 0. \qquad (*)$$

In particular, s_p is a unit for all primes dividing n, and we know that $\mathbb{Q}(\zeta)$ is ramified only at these primes, so by (3.5) the action of $[s, \mathbb{Q}]$ on $\mathbb{Q}(\zeta)$ is given by the Artin symbol

$$[s, \mathbb{Q}]\big|_{\mathbb{Q}(\zeta)} = ((s), \mathbb{Q}(\zeta)/\mathbb{Q}).$$

For any idele s, we will write $N_s \in \mathbb{Q}^*$ for the unique rational number satisfying

$$N_s \mathbb{Z} = (s) = s\mathbb{Z} \qquad \text{and} \qquad \mathrm{sign}(N_s) = \mathrm{sign}(s_\infty).$$

Then $((s), \mathbb{Q}(\zeta)/\mathbb{Q}) = (N_s \mathbb{Z}, \mathbb{Q}(\zeta)/\mathbb{Q})$, from which it follows that $\zeta^{[s, \mathbb{Q}]} = \zeta^{N_s}$, or equivalently

$$f(t)^{[s, \mathbb{Q}]} = f(t)^{N_s}.$$

Next we decompose $t \,(\mathrm{mod}\ \mathbb{Z})$ into p-primary components,

$$t \equiv \frac{a}{n} \equiv \sum_p \frac{a_p}{p^{e_p}} \in \bigoplus_p \mathbb{Q}_p/\mathbb{Z}_p.$$

Then

$$s^{-1} t \equiv \sum_p \frac{s_p^{-1} a_p}{p^{e_p}} \in \bigoplus_p \mathbb{Q}_p/\mathbb{Z}_p \quad \text{by definition of multiplication by } s^{-1}$$

$$\equiv \sum_p \frac{a_p}{p^{e_p}} \quad \text{from } (*), \text{ which says that } s_p^{-1} - 1 \in n\mathbb{Z}_p = p^{e_p}\mathbb{Z}_p$$

$$\equiv t.$$

This suggests that we should take f' to be the map f_s defined by

$$f_s : \mathbb{C}/s^{-1}\mathbb{Z} \longrightarrow \mathbb{G}_{\mathrm{m}}(\mathbb{Z}), \qquad f_s(t) = e^{2\pi i N_s t}.$$

Then for any idele s satisfying $(*)$, we have

$$f_s(s^{-1} t) = f_s(t) = e^{2\pi i N_s t} = f(t)^{N_s} = f(t)^{[s, \mathbb{Q}]}.$$

To recapitulate, we have proven that

$$f(t)^{[s, \mathbb{Q}]} = f_s(s^{-1} t) \qquad \text{provided } t \equiv \frac{a}{n} \in \mathbb{Q}/\mathbb{Z} \text{ and } s \text{ satisfies } (*).$$

Now let $s \in \mathbf{A}_{\mathbb{Q}}^*$ and $t \in \mathbb{Q}/\mathbb{Z}$ be arbitrary, and as usual write $t \equiv a/n \pmod{\mathbb{Z}}$. Using the weak approximation theorem, we can find a rational number $r \in \mathbb{Q}^*$ so that rs satisfies $(*)$. From the definitions, it is easy to check that

$$N_{rs} = rN_s \quad \text{and} \quad (rs)^{-1}t = r^{-1}(s^{-1}t).$$

Notice that the requirement $\text{sign}(N_s) = \text{sign}(s_\infty)$ ensures that $N_{rs} = rN_s$, even when r is negative. Using these equalities, we compute

$$
\begin{aligned}
f(t)^{[s,\mathbb{Q}]} &= f(t)^{[rs,\mathbb{Q}]} && \text{since } [rs,\mathbb{Q}] = [s,\mathbb{Q}] \\
&= f_{rs}\big((rs)^{-s}t\big) && \text{from above, since } rs \text{ satisfies } (*) \\
&= e^{2\pi i N_{rs}\cdot(rs)^{-1}t} && \text{from the definition of } f_{rs} \\
&= e^{2\pi i N_s \cdot s^{-1}t} && \text{since } N_{rs} = rN_s \text{ and } (rs)^{-1}t = r^{-1}(s^{-1}t) \\
&= f_s(s^{-1}t) && \text{from the definition of } f_s.
\end{aligned}
$$

This completes the proof of the existence half of (7.1), with the additional information that f' is given by the map $f'(t) = f_s(t) = e^{2\pi i N_s t}$. As for uniqueness, we need merely observe that the commutative diagram determines f' on $\mathbb{Q}/s^{-1}\mathbb{Z}$, which is a dense subset of $\mathbb{G}_m(\mathbb{C})$, so there is at most one possibility for f'.   □

As an alternative version of (7.1), we could use only the single analytic parametrization f and replace the multiplication-by-s^{-1} map so as to make the following diagram commute:

$$
\begin{array}{ccc}
\mathbb{Q}/\mathbb{Z} & \dashrightarrow & \mathbb{Q}/\mathbb{Z} \\
\downarrow{\scriptstyle f} & & \downarrow{\scriptstyle f} \\
\mathbb{G}_m(\mathbb{C}) & \xrightarrow{\;\sigma\;} & \mathbb{G}_m(\mathbb{C}).
\end{array}
$$

We can (try to) do this because every $\sigma \in \text{Aut}(\mathbb{C})$ maps $\mathbb{G}_m(\mathbb{C})$ to itself. In the case of an elliptic curve E, this will only be possible for those σ's such that $E^\sigma \cong E$. For the elliptic analogues of our next two results (7.2) and (7.3), see (9.1) and (9.2).

Theorem 7.2. *Let* $s \in \mathbf{A}_{\mathbb{Q}}^*$ *be an idele. With notation as in (7.1), there is a unique rational number* $N_s \in \mathbb{Q}^*$ *such that the following diagram commutes:*

$$
\begin{array}{ccc}
\mathbb{Q}/\mathbb{Z} & \xrightarrow{N_s s^{-1}} & \mathbb{Q}/\mathbb{Z} \\
\downarrow{\scriptstyle f} & & \downarrow{\scriptstyle f} \\
\mathbb{G}_m(\mathbb{C}) & \xrightarrow{\;\sigma\;} & \mathbb{G}_m(\mathbb{C}).
\end{array}
$$

More precisely, N_s *is the unique rational number satisfying*

$$N_s\mathbb{Z} = (s) = s\mathbb{Z} \quad \text{and} \quad \text{sign}(N_s) = \text{sign}(s_\infty). \tag{$*$}$$

PROOF. During the proof of (7.1) we showed that f' is the map $f'(t) = e^{2\pi i N_s t}$, where N_s is as specified in $(*)$. The commutative square in (7.1) then says that

$$f(t)^{[s,\mathbb{Q}]} = f'(s^{-1}t) = e^{2\pi i N_s s^{-1} t} = f(N_s s^{-1} t).$$

This proves that the square in (7.2) is commutative with N_s chosen to satisfy $(*)$. It remains to prove that this commutative diagram uniquely determines N_s. But if N_s' also makes the diagram commute, then we find that multiplication by $N_s^{-1} N_s'$ induces the identity map on $\mathbb{Q}/\mathbb{Z}$. Hence $N_s = N_s'$. $\square$

From (7.2) we have a well-defined map

$$\mathbf{A}_{\mathbb{Q}}^* \longrightarrow \mathbb{Q}^* \subset \mathbb{C}, \qquad s \longmapsto N_s,$$

and it is clear that this map is a homomorphism. Further, the explicit description of N_s given by $(*)$ in (7.2) shows that the map is continuous. Recall that for any number field L, a homomorphism

$$\chi : \mathbf{A}_L^* \longrightarrow \mathbb{C}^*$$

is called a Grössencharacter of L if it is continuous and satisfies $\chi(L^*) = 1$; that is, the kernel of χ must contain the image of L^* in $\mathbf{A}_L^*$. It is easy to see that our map $s \mapsto N_s$ does not have this property. In fact, if s is the image of some $a \in L^*$, then clearly $N_s = a$. We can get a Grössencharacter by making a small modification to N_s.

Theorem 7.3.   *For any idele $s \in \mathbf{A}_{\mathbb{Q}}^*$, let s_∞ be the archimedean component of s. Define a map*

$$\chi : \mathbf{A}_{\mathbb{Q}}^* \longrightarrow \mathbb{R}^* \subset \mathbb{C}^*, \qquad \chi(s) = N_s s_\infty^{-1},$$

where $N_s \in \mathbb{Q}^*$ is the unique rational number satisfying

$$N_s \mathbb{Z} = (s) = s\mathbb{Z} \qquad and \qquad \mathrm{sign}(N_s) = \mathrm{sign}(s_\infty).$$

Then χ is a Grössencharacter of $\mathbb{Q}$.

PROOF. It is clear that both of the maps

$$s \longmapsto N_s \qquad and \qquad s \longmapsto s_\infty$$

are continuous homomorphisms from $\mathbf{A}_{\mathbb{Q}}^*$ to $\mathbb{C}^*$. Further, they clearly take the same value on the image of $\mathbb{Q}^*$ in $\mathbf{A}_{\mathbb{Q}}^*$. Hence χ is a continuous homomorphism that is trivial on $\mathbb{Q}^*$; that is, it is a Grössencharacter.

$\square$

§8. The Main Theorem of Complex Multiplication

Let K be a quadratic imaginary field with ring of integers R_K as usual. For each prime ideal $\mathfrak{p}$ of K, let $K_\mathfrak{p}$ be the completion of K at $\mathfrak{p}$ and let $R_\mathfrak{p}$ be the ring of integers of $K_\mathfrak{p}$. Similarly, if $\mathfrak{a}$ is any fractional ideal of K, let $\mathfrak{a}_\mathfrak{p} = \mathfrak{a} R_\mathfrak{p}$ be the fractional ideal of $K_\mathfrak{p}$ generated by $\mathfrak{a}$.

Let M be an R_K-module. The $\mathfrak{p}$-primary component of M, which by definition is that part of M annihilated by some power of $\mathfrak{p}$, is denoted by

$$M[\mathfrak{p}^\infty] = \{m \in M : \mathfrak{p}^e m = (0) \text{ for some } e \geq 0\}.$$

We begin with an elementary lemma about $\mathfrak{p}$-primary decompositions.

Lemma 8.1. (a) *Let M be a torsion R_K-module; that is, for every $m \in M$ there is a non-zero $\alpha \in R_K$ such that $\alpha m = 0$. Then the natural summation map*

$$S : \bigoplus_\mathfrak{p} M[\mathfrak{p}^\infty] \xrightarrow{\sim} M, \qquad S(\mu) = \sum_\mathfrak{p} \mu_\mathfrak{p},$$

is an isomorphism. Here the sums are over all prime ideals of R_K, and $\mu_\mathfrak{p}$ denotes the $\mathfrak{p}$-component of μ.
(b) *Let $\mathfrak{a}$ be a fractional ideal of K. Then for each prime ideal $\mathfrak{p}$ of K, the inclusion $K \hookrightarrow K_\mathfrak{p}$ induces an isomorphism*

$$T : (K/\mathfrak{a})[\mathfrak{p}^\infty] \xrightarrow{\sim} K_\mathfrak{p}/\mathfrak{a}_\mathfrak{p}.$$

(c) *Again let $\mathfrak{a}$ be a fractional ideal of K. Then there is an isomorphism*

$$K/\mathfrak{a} \cong \bigoplus_\mathfrak{p} K_\mathfrak{p}/\mathfrak{a}_\mathfrak{p}.$$

Remark 8.1.1. As our proof will show, Lemma 8.1 is true more generally for any Dedekind domain R with fraction field K. For example, taking $R = \mathbb{Z}$ gives the decomposition of a torsion abelian group into p-primary components as discussed in §7.

PROOF (of Lemma 8.1). (a) Suppose first that $\mu \in \ker(S)$. For each prime $\mathfrak{p}$, let $e(\mathfrak{p}) \geq 0$ be the smallest integer such that $\mathfrak{p}^{e(\mathfrak{p})} \mu_\mathfrak{p} = (0)$. Note that $e(\mathfrak{p})$ exists, since $\mu_\mathfrak{p} \in M[\mathfrak{p}^\infty]$, and that all but finitely many of the $e(\mathfrak{p})$'s are zero since μ has only finitely many non-zero components. Now fix a prime ideal $\mathfrak{q}$ and let

$$\mathfrak{d} = \prod_{\mathfrak{p} \neq \mathfrak{q}} \mathfrak{p}^{e(\mathfrak{p})}.$$

By construction, we have $\partial \mu_{\mathfrak{p}} = (0)$ for all $\mathfrak{p} \neq \mathfrak{q}$. On the other hand, since $S(\mu) = 0$ by assumption, we have

$$(0) = \partial S(\mu) = \partial \sum_{\mathfrak{p}} \mu_{\mathfrak{p}} = \sum_{\mathfrak{p}} \partial \mu_{\mathfrak{p}} = \partial \mu_{\mathfrak{q}}.$$

But ∂ is relatively prime to $\mathfrak{q}$, so $\partial + \mathfrak{q}^{e(\mathfrak{q})} = (1)$. Hence

$$\left(\mu_{\mathfrak{q}}\right) = \left(\partial + \mathfrak{q}^{e(\mathfrak{q})}\right) \mu_{\mathfrak{q}} = \partial \mu_{\mathfrak{q}} + \mathfrak{q}^{e(\mathfrak{q})} \mu_{\mathfrak{q}} = (0) + (0) = (0),$$

which proves that $\mu_{\mathfrak{q}} = 0$. Since $\mathfrak{q}$ was arbitrary, we have proven that $\mu = 0$ and hence that S is injective.

Next we check surjectivity. Take any element $m \in M$, and choose a non-zero $\alpha \in R_K$ with $\alpha m = 0$. Factor the ideal αR_K as

$$\alpha R_K = \mathfrak{p}_1^{e_1} \mathfrak{p}_2^{e_2} \cdots \mathfrak{p}_r^{e_r}.$$

Then we can find $\varepsilon_1, \ldots, \varepsilon_r \in R_K$ satisfying

$$\varepsilon_1 + \cdots + \varepsilon_r = 1 \quad \text{and} \quad \varepsilon_i \equiv \begin{cases} 1 \,(\mathrm{mod}\ \mathfrak{p}_i^{e_i}), \\ 0 \,(\mathrm{mod}\ \mathfrak{p}_j^{e_j}) & \text{for } j \neq i. \end{cases}$$

[Proof: Let $\mathfrak{e}_i = \alpha \mathfrak{p}^{-e_i}$. Then $\mathfrak{e}_1 + \cdots + \mathfrak{e}_r = (1)$, so it suffices to take $\varepsilon_i \in \mathfrak{e}_i$ with $\varepsilon_1 + \cdots + \varepsilon_r = 1$.] Notice that $\mathfrak{p}^{e_i} \varepsilon_i \subset \alpha R_K$, so $\mathfrak{p}^{e_i} \varepsilon_i m \subset \alpha m R_K = (0)$. Hence $\varepsilon_i m \in M[\mathfrak{p}_i^\infty]$, so if we set

$$\mu_{\mathfrak{p}} = \begin{cases} \varepsilon_i m & \text{if } \mathfrak{p} = \mathfrak{p}_i \text{ for some } 1 \leq i \leq r, \\ 0 & \text{otherwise}, \end{cases}$$

then $\mu \in \oplus M[\mathfrak{p}^\infty]$ and $S(\mu) = \sum \mu_{\mathfrak{p}} = \varepsilon_1 m + \cdots + \varepsilon_r m = m$. This completes the proof that S is surjective.

(b) First, suppose that $\bar{\alpha} \in (K/\mathfrak{a})[\mathfrak{p}^\infty]$ is in the kernel of T. Choosing a representative $\alpha \in K$ for $\bar{\alpha}$, this means that $\mathfrak{p}^e \alpha \subset \mathfrak{a}$ for some integer $e \geq 0$ and that $\alpha \in \mathfrak{a}_{\mathfrak{p}} = \mathfrak{a} R_{\mathfrak{p}}$. These two inclusions imply respectively that

$$\mathrm{ord}_{\mathfrak{q}}(\alpha) \geq \mathrm{ord}_{\mathfrak{q}}(\mathfrak{a}) \quad \text{for all } \mathfrak{q} \neq \mathfrak{p} \quad \text{and} \quad \mathrm{ord}_{\mathfrak{p}}(\alpha) \geq \mathrm{ord}_{\mathfrak{p}}(\mathfrak{a}).$$

Therefore $\alpha R_K \subset \mathfrak{a}$, so $\alpha \in \mathfrak{a}$, which means that $\bar{\alpha} = 0$. This proves that T is injective.

Next, let $\bar{\beta} \in K_{\mathfrak{p}}/\mathfrak{a}_{\mathfrak{p}}$ and choose a representative $\beta \in K_{\mathfrak{p}}$ for $\bar{\beta}$. By the weak approximation theorem (essentially the Chinese Remainder Theorem) we can find an $\alpha \in K$ satisfying

$$\mathrm{ord}_{\mathfrak{p}}(\alpha - \beta) \geq \mathrm{ord}_{\mathfrak{p}}(\mathfrak{a}) \quad \text{and} \quad \mathrm{ord}_{\mathfrak{q}}(\alpha) \geq \mathrm{ord}_{\mathfrak{q}}(\mathfrak{a}) \quad \text{for all } \mathfrak{q} \neq \mathfrak{p}.$$

The first inequality says that $\alpha \equiv \beta \pmod{\mathfrak{a}_\mathfrak{p}}$, so $T(\bar\alpha) = \bar\beta$. Let e be a non-negative integer greater than $\mathrm{ord}_\mathfrak{p}(\mathfrak{a}) - \mathrm{ord}_\mathfrak{p}(\alpha)$. Then $\mathrm{ord}_\mathfrak{q}(\mathfrak{p}^e \alpha) \geq \mathrm{ord}_\mathfrak{q}(\mathfrak{a})$ for all primes $\mathfrak{q}$, including $\mathfrak{q} = \mathfrak{p}$, so $\mathfrak{p}^e \alpha \in \mathfrak{a}$. Hence $\bar\alpha \in (K/\mathfrak{a})\,[\mathfrak{p}^\infty]$, which completes the proof that T is surjective.

(c) This is immediate from (a) and (b).   □

Let $x \in \mathbf{A}_K^*$ be an idele. Recall that the ideal of x is the fractional ideal

$$(x) = \prod_\mathfrak{p} \mathfrak{p}^{\mathrm{ord}_\mathfrak{p}(x_\mathfrak{p})}.$$

If $\mathfrak{a}$ is any fractional ideal of K, we define $x\mathfrak{a}$ to be the product $(x)\mathfrak{a}$. Using the equality $(x)_\mathfrak{p} = (x)R_\mathfrak{p} = x_\mathfrak{p} R_\mathfrak{p}$, we see that

$$(x\mathfrak{a})_\mathfrak{p} = (x)\mathfrak{a}R_\mathfrak{p} = x_\mathfrak{p}\mathfrak{a}R_\mathfrak{p} = x_\mathfrak{p}\mathfrak{a}_\mathfrak{p}.$$

Now (8.1c) gives natural isomorphisms

$$K/\mathfrak{a} \cong \bigoplus_\mathfrak{p} K_\mathfrak{p}/\mathfrak{a}_\mathfrak{p} \qquad \text{and} \qquad K/x\mathfrak{a} \cong \bigoplus_\mathfrak{p} K_\mathfrak{p}/x_\mathfrak{p}\mathfrak{a}_\mathfrak{p}.$$

We define the *multiplication-by-x map on $K/\mathfrak{a}$* to be multiplication of the $\mathfrak{p}$-primary component by $x_\mathfrak{p}$. In other words, multiplication-by-x is defined by the commutativity of the following diagram:

$$
\begin{array}{ccc}
K/\mathfrak{a} & \xrightarrow{\;x\;} & K/x\mathfrak{a} \\
\Big\downarrow{\wr} & & \Big\downarrow{\wr} \\
\bigoplus_\mathfrak{p} K_\mathfrak{p}/\mathfrak{a}_\mathfrak{p} & \longrightarrow & \bigoplus_\mathfrak{p} K_\mathfrak{p}/x_\mathfrak{p}\mathfrak{a}_\mathfrak{p} \\
(t_\mathfrak{p}) & \longmapsto & (x_\mathfrak{p} t_\mathfrak{p})
\end{array}
$$

We also recall from §3 that the reciprocity map for K,

$$\mathbf{A}_K^* \longrightarrow \mathrm{Gal}(K^{\mathrm{ab}}/K), \qquad s \longmapsto [s, K],$$

is surjective and its kernel contains K^*.

We are now ready to state and prove the main theorem of complex multiplication in its adelic formulation. At the risk of making the statement overly long, we include a summary of our notation and assumptions.

Theorem 8.2. (The Main Theorem of Complex Multiplication) *Fix the following quantities:*

$K/\mathbb{Q}$ *a quadratic imaginary field with ring of integers R_K,*

$E/\mathbb{C}$ *an elliptic curve with $\mathrm{End}(E) \cong R_K$,*

σ $\in \text{Aut}(\mathbb{C})$, *an automorphism of the complex numbers,*

s $\in \mathbf{A}_K^*$, *an idele of K satisfying $[s, K] = \sigma|_{K^{ab}}$.*

Further, fix a complex analytic isomorphism

$$f : \mathbb{C}/\mathfrak{a} \xrightarrow{\sim} E(\mathbb{C}),$$

where $\mathfrak{a}$ is a fractional ideal of K. Then there exists a unique complex analytic isomorphism

$$f' : \mathbb{C}/s^{-1}\mathfrak{a} \xrightarrow{\sim} E^\sigma(\mathbb{C})$$

(depending on f and σ) so that the following diagram commutes:

$$
\begin{array}{ccc}
K/\mathfrak{a} & \xrightarrow{s^{-1}} & K/s^{-1}\mathfrak{a} \\
\downarrow{\scriptstyle f} & & \downarrow{\scriptstyle f'} \\
E(\mathbb{C}) & \xrightarrow{\sigma} & E^\sigma(\mathbb{C}).
\end{array}
$$

Remark 8.2.1. The statement of Theorem 8.2 remains true for elliptic curves whose endomorphism ring is a non-maximal order of K. Of course, one first must explain how to multiply $K/\mathfrak{a}$ by an idele x when $\mathfrak{a}$ is an arbitrary lattice in K. For details, see Shimura [1] or Lang [1, Ch. 8, 10].

Remark 8.2.2. Notice how Theorem 8.2 transforms the algebraic action of σ on the torsion subgroup $f(K/\mathfrak{a}) = E_{\text{tors}}$ into the analytic action of multiplication by s^{-1}:

$$f(t)^{[s, K]} = f'(s^{-1}t) \qquad \text{for } t \in K/\mathfrak{a} \text{ and } s \in \mathbf{A}_K^*.$$

Compare with (7.1.1).

PROOF (of Theorem 8.2). Clearly, there is at most one f', since the commutative diagram determines f' on $K/s^{-1}\mathfrak{a}$, which is a dense subset of $\mathbb{C}/s^{-1}\mathfrak{a}$.

Suppose that $E_1/\mathbb{C}$ is an elliptic curve that is isomorphic to E and that $f_1 : \mathbb{C}/\mathfrak{a}_1 \to E_1(\mathbb{C})$ is an analytic isomorphism. We are going to begin by proving that if Theorem 8.2 is true for (E_1, f_1), then it is also true for (E, f). This will allow us to reduce to the case that E is defined over $\mathbb{Q}(j(E))$ and $\mathfrak{a}$ is an integral ideal.

So we are assuming that there is an analytic isomorphism

$$f_1' : \mathbb{C}/s^{-1}\mathfrak{a}_1 \xrightarrow{\sim} E_1^\sigma(\mathbb{C})$$

and a commutative diagram

$$
\begin{array}{ccc}
K/\mathfrak{a}_1 & \xrightarrow{s^{-1}} & K/s^{-1}\mathfrak{a}_1 \\
\downarrow{\scriptstyle f_1} & & \downarrow{\scriptstyle f_1'} \\
E_1(\mathbb{C}) & \xrightarrow{\sigma} & E_1^\sigma(\mathbb{C}).
\end{array}
$$

Since E and E_1 are isomorphic, we may fix an isomorphism $i : E_1 \xrightarrow{\sim} E$. Further, the lattices for E and E_1 are homothetic [AEC VI.4.1.1], so we have $\mathfrak{a}_1 = \gamma\mathfrak{a}$ for some $\gamma \in K^*$. Then each of the squares in the following diagram is commutative:

$$
\begin{array}{ccc}
K/\mathfrak{a} & \xrightarrow{s^{-1}} & K/s^{-1}\mathfrak{a} \\
\downarrow{\scriptstyle\gamma} & & \downarrow{\scriptstyle\gamma} \\
K/\mathfrak{a}_1 & \xrightarrow{s^{-1}} & K/s^{-1}\mathfrak{a}_1 \\
\downarrow{\scriptstyle f_1} & & \downarrow{\scriptstyle f_1'} \\
E_1(\mathbb{C}) & \xrightarrow{\sigma} & E_1^\sigma(\mathbb{C}) \\
\downarrow{\scriptstyle i} & & \downarrow{\scriptstyle i^\sigma} \\
E(\mathbb{C}) & \xrightarrow{\sigma} & E^\sigma(\mathbb{C}).
\end{array}
$$

Hence (8.2) is true for (E, f) if we take for f' the map

$$
f' : \mathbb{C}/s^{-1}\mathfrak{a} \xrightarrow{\sim} E^\sigma(\mathbb{C}), \qquad f'(z) = i^\sigma \circ f_1'(\gamma z).
$$

We are now reduced to proving (8.2) under the assumptions that E is defined over $\mathbb{Q}(j(E))$ and that $\mathfrak{a} \subset R_K$ is an integral ideal. Fix an integer $m \geq 3$ and let L/K be a finite Galois extension satisfying

$$
j(E) \in L \qquad \text{and} \qquad E[m] \subset E(L).
$$

We note that (5.6) implies that L contains $K_{(m)}$, the ray class field of K modulo m. We are going to begin by proving that (8.2) is true on the m-torsion points of E. As usual, our main tool will be reduction modulo a suitable prime.

Let $\mathfrak{P}$ be a prime ideal of L satisfying the following five conditions:

(i) $\sigma|_L = (\mathfrak{P}, L/K)$; that is, the restriction of σ to L is a Frobenius element for $\mathfrak{P}$.

(ii) $\mathfrak{p} = \mathfrak{P} \cap K$ is a prime of degree 1; that is, $p = \mathrm{N}_\mathbb{Q}^K \mathfrak{p}$ is a rational prime.

(iii) $\mathfrak{p}$ is unramified in L.

(iv) $\mathfrak{p}$ is not one of the finitely many primes excluded in (4.5).

(v) $\mathfrak{P}$ does not divide m.

Such an ideal always exists, since the Tchebotarev Density Theorem (Lang [5, Ch. VIII, Thm. 10]) says that there are infinitely many primes satisfying (i) and (ii), whereas each of (iii), (iv), and (v) excludes only finitely many primes.

Using the fact that L contains $K_{(m)}$, we find

$$\begin{aligned} [s, K]\big|_{K_{(m)}} &= \sigma\big|_{K_{(m)}} \qquad &&\text{since } [s, K] = \sigma\big|_{K^{\mathrm{ab}}} \text{ by assumption} \\ &= \big(\mathfrak{p}, K_{(m)}/K\big) \qquad &&\text{from (i) and (ii).} \end{aligned}$$

This last map is the Frobenius element associated to $\mathfrak{p}$. Let $\pi \in \mathbf{A}_K^*$ be an idele with a uniformizer at the $\mathfrak{p}$-component and 1's elsewhere. Then (3.5c) says that $[\pi, K]$ also equals $\big(\mathfrak{p}, K_{(m)}/K\big)$. Hence $[s\pi^{-1}, K]$ acts trivially on $K_{(m)}$, so the idelic characterization (3.6) of the ray class field $K_{(m)}$ says that the idele $s\pi^{-1}$ factors as

$$s\pi^{-1} = \alpha u.$$

Here $\alpha \in K^*$, and for each prime $\mathfrak{q}$, $u \in \mathbf{A}_K^*$ satisfies

$$u_{\mathfrak{q}} \in R_{\mathfrak{q}}^* \qquad \text{and} \qquad u_{\mathfrak{q}} \equiv 1 \,(\mathrm{mod}\, mR_{\mathfrak{q}}).$$

Next we use (5.3) to find an isogeny

$$\lambda : E \longrightarrow E^{\sigma}$$

whose reduction modulo $\mathfrak{P}$ is the pth-power Frobenius map. Note that since L contains $K\big(j(E)\big)$, (i) implies that

$$\sigma\big|_{K(j(E))} = \big(\mathfrak{P}, L/K\big)\big|_{K(j(E))} = \big(\mathfrak{p}, K\big(j(E)\big)/K\big).$$

Since E is defined over $\mathbb{Q}\big(j(E)\big)$, we see that the isogeny described in (5.3) is indeed from E to E^{σ}.

We claim that on m-torsion points, λ acts like σ. To see this, let $T \in E[m]$ and use tilde's to denote reduction modulo $\mathfrak{P}$. Then

$$\widetilde{\lambda(T)} = \tilde{\lambda}(\tilde{T}) = \widetilde{T^{\sigma}},$$

since both λ and $\sigma|_L = (\mathfrak{P}, L/K)$ act on the residue field modulo $\mathfrak{P}$ as the pth-power map. Now $\mathfrak{P} \nmid m$ from (v), so [AEC VII.3.1b] says that on m-torsion points the reduction map $E^{\sigma}[m] \to \widetilde{E^{\sigma}}[m]$ is injective. Hence

$$\lambda(T) = T^{\sigma} \qquad \text{for all } T \in E[m].$$

In other words, we have produced a commutative diagram

$$\begin{array}{ccc} E[m] & \xrightarrow{\ \lambda\ } & E^{\sigma}[m] \\ \downarrow & & \downarrow \\ E(\mathbb{C}) & \xrightarrow{\ \sigma\ } & E^{\sigma}(\mathbb{C}). \end{array}$$

(N.B. This diagram describes a deep mathematical relationship, because it transforms the algebraic action of σ on m-torsion points into the geometric action of λ. Notice that for general points $P \in E(\bar{K})$, $\lambda(P)$ and P^σ will not be equal.)

We also note from (4.3d) that E^σ is isomorphic to $\bar{\mathfrak{p}} * E$, where $\bar{\mathfrak{p}} * E$ is an elliptic curve associated to the lattice $\mathfrak{p}^{-1}\mathfrak{a}$. Using the given analytic isomorphism $f : \mathbb{C}/\mathfrak{a} \to E(\mathbb{C})$, this means there is an analytic isomorphism $f'' : \mathbb{C}/\mathfrak{p}^{-1}\mathfrak{a} \to E^\sigma(\mathbb{C})$ so that we have a commutative diagram

$$
\begin{array}{ccc}
\mathbb{C} & \xrightarrow{\ \mathrm{id}\ } & \mathbb{C} \\
\downarrow & & \downarrow \\
\mathbb{C}/\mathfrak{a} & \longrightarrow & \mathbb{C}/\mathfrak{p}^{-1}\mathfrak{a} \\
\Big\downarrow{\scriptstyle f} & & \Big\downarrow{\scriptstyle f''} \\
E(\mathbb{C}) & \xrightarrow{\ \lambda\ } & E^\sigma(\mathbb{C}).
\end{array}
\qquad (*)
$$

Recall that we factored the idele s as $s = \alpha \pi u$. Since every component of u is a unit, and every component of π is 1 except for a uniformizer in the $\mathfrak{p}$-component, we have

$$(s) = (\alpha)(\pi) = (\alpha)\mathfrak{p}, \qquad \text{and so} \qquad s^{-1}\mathfrak{a} = \alpha^{-1}\mathfrak{p}^{-1}\mathfrak{a}.$$

Thus multiplication by α^{-1} gives an isomorphism

$$\mathbb{C}/\mathfrak{p}^{-1}\mathfrak{a} \xrightarrow{\ \alpha^{-1}\ } \mathbb{C}/s^{-1}\mathfrak{a},$$

so we can extend $(*)$ to form the larger commutative diagram

$$
\begin{array}{ccccc}
\mathbb{C} & \xrightarrow{\ \mathrm{id}\ } & \mathbb{C} & \xrightarrow{\ \alpha^{-1}\ } & \mathbb{C} \\
\downarrow & & \downarrow & & \downarrow \\
\mathbb{C}/\mathfrak{a} & \longrightarrow & \mathbb{C}/\mathfrak{p}^{-1}\mathfrak{a} & \longrightarrow & \mathbb{C}/s^{-1}\mathfrak{a} \\
\Big\downarrow{\scriptstyle f} & & \Big\downarrow{\scriptstyle f''} & & \Big\downarrow{\scriptstyle f'} \\
E(\mathbb{C}) & \xrightarrow{\ \lambda\ } & E^\sigma(\mathbb{C}) & \xrightarrow{\ \mathrm{id}\ } & E^\sigma(\mathbb{C}).
\end{array}
\qquad (**) \ .
$$

Here $f' : \mathbb{C}/s^{-1}\mathfrak{a} \xrightarrow{\sim} E^\sigma(\mathbb{C})$ is the unique analytic isomorphism making $(**)$ commute. We claim that f' satisfies

$$f(t)^\sigma \stackrel{?}{=} f'(s^{-1}t) \qquad \text{for all } t \in m^{-1}\mathfrak{a}/\mathfrak{a}.$$

To verify this claim, we note from above that

$$f(t)^\sigma = \lambda\big(f(t)\big).$$

Combining this with the commutativity of $(**)$, we must check that

$$f'(\alpha^{-1}t) \overset{?}{=} f'(s^{-1}t) \qquad \text{for all } t \in m^{-1}\mathfrak{a}/\mathfrak{a};$$

and since f' is bijective, this is equivalent to showing that

$$\alpha^{-1}t - s^{-1}t \overset{?}{\in} s^{-1}\mathfrak{a} \qquad \text{for all } t \in m^{-1}\mathfrak{a}.$$

Recalling that $s^{-1}t \in K/s^{-1}\mathfrak{a}$ is defined by multiplying the q-primary component by $s_{\mathfrak{q}}$, we must prove for each prime $\mathfrak{q}$ of K that

$$\alpha^{-1}t - s_{\mathfrak{q}}^{-1}t \overset{?}{\in} s_{\mathfrak{q}}^{-1}\mathfrak{a}_{\mathfrak{q}} \qquad \text{for all } t \in m^{-1}\mathfrak{a}_{\mathfrak{q}}.$$

Now multiplying through by $s_{\mathfrak{q}}$ and using the decomposition $s_{\mathfrak{q}} = \alpha\pi_{\mathfrak{q}}u_{\mathfrak{q}}$ from above, we must check that

$$\pi_{\mathfrak{q}}u_{\mathfrak{q}}t - t \overset{?}{\in} \mathfrak{a}_{\mathfrak{q}} \qquad \text{for all } t \in m^{-1}\mathfrak{a}_{\mathfrak{q}};$$

or equivalently that

$$(\pi_{\mathfrak{q}}u_{\mathfrak{q}} - 1)\mathfrak{a}_{\mathfrak{q}} \overset{?}{\subset} m\mathfrak{a}_{\mathfrak{q}}.$$

By construction, $u_{\mathfrak{q}}$ is in $R_{\mathfrak{q}}^*$ and satisfies $u_{\mathfrak{q}} \equiv 1 \,(\mathrm{mod}\, mR_{\mathfrak{q}})$, so we are reduced to proving

$$(\pi_{\mathfrak{q}} - 1)\mathfrak{a}_{\mathfrak{q}} \overset{?}{\subset} m\mathfrak{a}_{\mathfrak{q}}.$$

There are two cases to consider. First, if $\mathfrak{q} \neq \mathfrak{p}$, then $\pi_{\mathfrak{q}} = 1$, so we are done. Second, if $\mathfrak{q} = \mathfrak{p}$, we know that $\pi_{\mathfrak{p}}$ is a uniformizer, so $(\pi_{\mathfrak{p}}-1)\mathfrak{a}_{\mathfrak{p}} = \mathfrak{a}_{\mathfrak{p}}$. Further, we know from (v) that $\mathfrak{p} \nmid m$, so m is a p-adic unit and hence $m\mathfrak{a}_{\mathfrak{p}} = \mathfrak{a}_{\mathfrak{p}}$. This proves the desired inclusion for all $\mathfrak{q}$, thereby completing the proof of our claim.

To recapitulate, for each integer $m \geq 3$ we have produced an analytic isomorphism

$$f'_m : \mathbb{C}/s^{-1}\mathfrak{a} \overset{\sim}{\longrightarrow} E^\sigma(\mathbb{C})$$

and a commutative diagram

$$
\begin{array}{ccc}
m^{-1}\mathfrak{a}/\mathfrak{a} & \overset{s^{-1}}{\longrightarrow} & m^{-1}s^{-1}\mathfrak{a}/s^{-1}\mathfrak{a} \\
\downarrow{\scriptstyle f} & & \downarrow{\scriptstyle f'_m} \\
E(\mathbb{C}) & \overset{\sigma}{\longrightarrow} & E^\sigma(\mathbb{C}).
\end{array}
$$

To complete the proof of (8.2) it suffices to show that all of the f'_m maps are the same, since then these commutative diagrams will fit together to give the desired result on all of $K/\dot{\mathfrak{a}}$.

So let $n \geq 1$ be an integer, and let $f'_{mn} : \mathbb{C}/s^{-1}\mathfrak{a} \xrightarrow{\sim} E^{\sigma}(\mathbb{C})$ be the corresponding analytic isomorphism. Note that the composition $f'_{mn} \circ f'_m{}^{-1}$ is an automorphism of E^{σ}, say $f'_{mn} \circ f'_m{}^{-1} = [\xi] \in \mathrm{Aut}(E^{\sigma})$. Then for any $t \in m^{-1}\mathfrak{a}/\mathfrak{a}$ we have

$$
\begin{aligned}
[\xi] \circ f'_m(s^{-1}t) = f'_{mn}(s^{-1}t) \qquad & \text{by definition of } \xi \\
= f(t)^{\sigma} \qquad & \text{from construction of } f'_{mn} \\
= f'_m(s^{-1}t) \qquad & \text{from construction of } f'_m.
\end{aligned}
$$

This holds for all $t \in m^{-1}\mathfrak{a}$, so we conclude that

$$
[\xi]T = T \qquad \text{for all } T \in E^{\sigma}[m].
$$

Since $m \geq 3$, this can only happen if $[\xi] = [1]$, since for $[\xi] \neq [1]$ the kernel of $[1 - \xi]$ contains at most six points. (If $j(E) \neq 0, 1728$, then $\xi = \pm 1$, and in the two exceptional cases, $\xi^4 = 1$ or $\xi^6 = 1$. See [AEC III.10.1].) Therefore $f'_{mn} = f'_m$, which concludes the proof of (8.2). $\qquad \square$

§9. The Associated Grössencharacter

In this section we will use the main theorem of complex multiplication to define a, Grössencharacter associated to an elliptic curve with complex multiplication. Recall that a Grössencharacter on a number field L is a continuous homomorphism

$$
\psi : \mathbf{A}_L^* \longrightarrow \mathbb{C}^*
$$

with the property that $\psi(L^*) = 1$. Our first result describes a map $\mathbf{A}_L^* \to \mathbb{C}^*$ which, with some small modifications, will be the desired Grössencharacter.

Theorem 9.1. *Let E/L be an elliptic curve with complex multiplication by the ring of integers R_K of K, and assume that $L \supset K$. Let $x \in \mathbf{A}_L^*$ be an idele of L, and let $s = \mathrm{N}_K^L x \in \mathbf{A}_K^*$. Then there exists a unique $\alpha = \alpha_{E/L}(x) \in K^*$ with the following two properties:*
(i) $\alpha R_K = (s)$, where $(s) \subset K$ is the ideal of s.
(ii) For any fractional ideal $\mathfrak{a} \subset K$ and any analytic isomorphism

$$
f : \mathbb{C}/\mathfrak{a} \longrightarrow E(\mathbb{C}),
$$

the following diagram commutes:

$$
\begin{array}{ccc}
K/\mathfrak{a} & \xrightarrow{\alpha s^{-1}} & K/\mathfrak{a} \\
\downarrow{\scriptstyle f} & & \downarrow{\scriptstyle f} \\
E(L^{\mathrm{ab}}) & \xrightarrow{[x,L]} & E(L^{\mathrm{ab}}).
\end{array}
$$

Before beginning the proof of (9.1), we should make a few remarks. First, from (2.3) we know that $K(j(E), E_{\text{tors}})$ is an abelian extension of $K(j(E))$. Further, since E is defined over L, we know that $j(E) \in L$, and so we see that $L(E_{\text{tors}}) \subset L^{\text{ab}}$. This shows that the images of the vertical maps in (ii) do lie in $E(L^{\text{ab}})$. Second, since $K(j(E))$ is the Hilbert class field of K from (4.3), we know that $N_K^L \mathfrak{A}$ is principal for any ideal $\mathfrak{A}$ of L. Since $(s) = (N_K^L x) = N_K^L((x))$, we see that there always exists an $\alpha \in K^*$ satisfying (i), and (i) determines α up to a unit of K. It then remains for (ii) to pin α down precisely. Third, we note that (i) gives

$$\alpha s^{-1} \mathfrak{a} = \alpha(s)^{-1} \mathfrak{a} = \mathfrak{a},$$

so the top row of the diagram in (ii) is well-defined.

PROOF (of Theorem 9.1). Let

$$L' = L(E_{\text{tors}}).$$

Since $j(E) \in L$, it follows from (5.7) and (2.3) respectively that there are inclusions

$$K^{\text{ab}} \subset L' \subset L^{\text{ab}}.$$

Choose an automorphism $\sigma \in \text{Aut}(\mathbb{C})$ such that

$$\sigma|_{L^{\text{ab}}} = [x, L].$$

A standard property of the reciprocity map (3.5b) says that

$$\sigma|_{K^{\text{ab}}} = [x, L]\big|_{K^{\text{ab}}} = [s, K],$$

so applying the main theorem of complex multiplication (8.2), we find an analytic isomorphism $f' : \mathbb{C}/\mathfrak{a} \to E(\mathbb{C})$ and a commutative diagram

$$
\begin{array}{ccc}
K/\mathfrak{a} & \xrightarrow{s^{-1}} & K/s^{-1}\mathfrak{a} \\
\downarrow{f} & & \downarrow{f'} \\
E(\mathbb{C}) & \xrightarrow{\sigma} & E^{\sigma}(\mathbb{C}).
\end{array}
$$

Now $E^{\sigma} = E$, since σ fixes L. Hence $\mathfrak{a}$ and $s^{-1}\mathfrak{a}$ must be homothetic, so there is a $\beta \in K^*$ such that $\beta s^{-1}\mathfrak{a} = \mathfrak{a}$. Our commutative diagram then becomes

$$
\begin{array}{ccc}
K/\mathfrak{a} & \xrightarrow{\beta s^{-1}} & K/\mathfrak{a} \\
\downarrow{f} & & \downarrow{f''} \\
E(\mathbb{C}) & \xrightarrow{\sigma} & E(\mathbb{C}).
\end{array}
$$

Note that $f'' \circ f^{-1}$ is an automorphism of E, say $f'' = [\xi] \circ f$. If we set $\alpha = \xi\beta$ and use the facts that $\sigma|_{L^{\mathrm{ab}}} = [x, L]$ and $E_{\mathrm{tors}} \subset E(L^{\mathrm{ab}})$, we get

$$
\begin{array}{ccc}
K/\mathfrak{a} & \xrightarrow{\alpha s^{-1}} & K/\mathfrak{a} \\
\downarrow{\scriptstyle f} & & \downarrow{\scriptstyle f} \\
E(L^{\mathrm{ab}}) & \xrightarrow{[x, L]} & E(L^{\mathrm{ab}}),
\end{array}
$$

which is exactly (ii). Further, we have an equality of ideals

$$
\alpha s^{-1}\mathfrak{a} = \beta s^{-1}\mathfrak{a} = \mathfrak{a}, \qquad \text{so} \qquad \alpha R_K = (s).
$$

This proves that α satisfies both (i) and (ii), which completes the proof of the existence of part of (9.1).

Next we check that α is unique. Suppose that $\alpha' \in K^*$ also has properties (i) and (ii). From (ii) and the fact that f and $[x, L]$ are isomorphisms, we get a commutative triangle

$$
\begin{array}{ccc}
 & K/\mathfrak{a} & \\
{\scriptstyle \alpha s^{-1}} \swarrow & & \searrow {\scriptstyle \alpha' s^{-1}} \\
K/\mathfrak{a} & =\!=\!=\!=\!= & K/\mathfrak{a}
\end{array}
$$

Hence multiplication by $\alpha'\alpha^{-1}$ is the identity map on $K/\mathfrak{a}$, so $\alpha' = \alpha$.

Finally we must show that α is independent of the choice of f. Suppose that $f' : \mathbb{C}/\mathfrak{a}' \to E(\mathbb{C})$ is another analytic isomorphism. Then $\mathfrak{a} = \gamma\mathfrak{a}'$ for some $\gamma \in K^*$, and $f' \circ f^{-1}$ is an automorphism of E, so there is a unit $\xi \in R_K^*$ such that $f'(z) = f(\xi\gamma z)$. Then (ii) for f gives

$$
f'(t)^{[x, L]} = f(\xi\gamma t)^{[x, L]} = f(\alpha s^{-1}\xi\gamma t) = f'(\alpha s^{-1}t) \qquad \text{for all } t \in K/\mathfrak{a},
$$

so (ii) remains true if f is replaced by f'. Hence α is independent of f.

$\square$

Theorem 9.1 gives us a well-defined map

$$
\alpha_{E/L} : \mathbf{A}_L^* \longrightarrow K^* \subset \mathbb{C}^*,
$$

and it is clear that $\alpha_{E/L}$ is a homomorphism. However, it is easy to see that $\alpha_{E/L}(L^*) \neq 1$, so $\alpha_{E/L}$ is not a Grössencharacter. More precisely, if $\beta \in L^*$ and $x_\beta \in \mathbf{A}_L^*$ is the corresponding idele, then $[x_\beta, L] = 1$. So (9.1) says that $\alpha = \alpha_{E/L}(x_\beta)$ is the unique element of K^* such that $\alpha R_K = \mathrm{N}_K^L((x_\beta))R_K = \mathrm{N}_K^L(\beta)R_K$ and such that multiplication by $\alpha \mathrm{N}_K^L x_\beta^{-1}$ induces the identity map on $K/\mathfrak{a}$. Clearly, the required α is just $\mathrm{N}_K^L\beta$. In other words, we have proven that

$$
\alpha_{E/L}(x_\beta) = \mathrm{N}_K^L\beta \qquad \text{for all } \beta \in L^*,
$$

which is the first step in proving the following important result.

Theorem 9.2. *Let E/L be an elliptic curve with complex multiplication by the ring of integers R_K of K, assume that $L \supset K$, and let $\alpha_{E/L} : \mathbf{A}_L^* \to K^*$ be the map described in (9.1). For any idele $s \in \mathbf{A}_K^*$, let $s_\infty \in \mathbb{C}^*$ be the component of s corresponding to the unique archimedean absolute value on K. Define a map*

$$\psi_{E/L} : \mathbf{A}_L^* \longrightarrow \mathbb{C}^*, \qquad \psi_{E/L}(x) = \alpha_{E/L}(x) \mathrm{N}_K^L(x^{-1})_\infty.$$

(a) *$\psi_{E/L}$ is a Grössencharacter of L.*
(b) *Let $\mathfrak{P}$ be a prime of L. Then $\psi_{E/L}$ is unramified at $\mathfrak{P}$ if and only if E has good reduction at $\mathfrak{P}$. (Recall that a Grössencharacter $\psi : \mathbf{A}_L^* \to \mathbb{C}^*$ is said to be unramified at $\mathfrak{P}$ if $\psi(R_{\mathfrak{P}}^*) = 1$.)*

PROOF. (a) It is clear that $\psi_{E/L}$ is a homomorphism. We saw above that if $\beta \in L^*$, then $\alpha_{E/L}(x_\beta) = \mathrm{N}_K^L \beta$. On the other hand, untwisting the definitions we find

$$\mathrm{N}_K^L(x_\beta)_\infty = \prod_{\substack{\tau : L \hookrightarrow \mathbb{C} \\ \tau|_K = 1}} \beta^\tau = \mathrm{N}_K^L \beta.$$

Therefore $\psi_{E/L}(x_\beta) = 1$. This holds for all β, so $\psi_{E/L}(L^*) = 1$.

Next we are going to verify that $\alpha_{E/L} : \mathbf{A}_L^* \to \mathbb{C}$ is continuous. Fix an integer $m \geq 3$. We know from (2.3) that $L(E[m])$ is a finite abelian extension of L. Let $B_m \subset \mathbf{A}_L^*$ be the open subgroup corresponding to $L(E[m])$; that is, B_m is the subgroup so that the reciprocity map induces an isomorphism

$$\begin{aligned} \mathbf{A}_L^*/B_m & \xrightarrow{\ \sim\ } & \mathrm{Gal}\big(L(E[m])/L\big) \\ x & \longmapsto & [x, L]\big|_{L(E[m])}. \end{aligned}$$

Let

$$W_m = \big\{ s \in \mathbf{A}_K^* : s_\mathfrak{p} \in R_\mathfrak{p}^* \ \text{ and } \ s_\mathfrak{p} \equiv 1 \,(\mathrm{mod}\ m R \mathfrak{p}) \ \text{ for all } \mathfrak{p} \big\},$$

and let

$$U_m = B_m \cap \big\{ x \in \mathbf{A}_L^* : \mathrm{N}_K^L x \in W_m \big\}.$$

We note that U_m is an open subgroup $\mathbf{A}_L^*$. We are going to prove the

   Claim: $\alpha_{E/L}(x) = 1 \quad \text{for all } x \in U_m.$

Let $x \in U_m$, and to ease notation let $\alpha = \alpha_{E/L}(x)$. Also fix an analytic isomorphism

$$f : \mathbb{C}/\mathfrak{a} \xrightarrow{\ \sim\ } E(\mathbb{C})$$

as in (9.1). Then for any $t \in m^{-1}\mathfrak{a}/\mathfrak{a}$ we have $f(t) \in E[m]$, so

$$
\begin{aligned}
f(t) &= f(t)^{[x,L]} && \text{since } x \in B_m, \text{ so } [x,L] \text{ fixes } L\big(E[m]\big) \\
&= f\big(\alpha \mathrm{N}_K^L x^{-1} t\big) && \text{from (9.1ii)} \\
&= f(\alpha t) && \text{since } t \in m^{-1}\mathfrak{a}/\mathfrak{a} \text{ and} \\
& && \big(\mathrm{N}_K^L x\big)_\mathfrak{p} \in (1 + m R_\mathfrak{p}) \cap R_\mathfrak{p}^* \text{ for all } \mathfrak{p}.
\end{aligned}
$$

Hence multiplication by α fixes $m^{-1}\mathfrak{a}/\mathfrak{a}$ or, equivalently,

$$
(\alpha - 1)m^{-1}\mathfrak{a} \subset \mathfrak{a}.
$$

This inclusion of fractional ideals means that $(\alpha - 1)R_K \subset m R_K$, so

$$
\alpha \in R_K \quad \text{and} \quad \alpha \equiv 1 \,(\mathrm{mod}\, m R_K).
$$

On the other hand, for any prime $\mathfrak{p}$ of K we have

$$
\begin{aligned}
\mathrm{ord}_\mathfrak{p}\, \alpha &= \mathrm{ord}_\mathfrak{p}\big(\mathrm{N}_K^L x\big)_\mathfrak{p} && \text{from (9.1i)} \\
&= 1 && \text{since the } \mathfrak{p}\text{-component of } \mathrm{N}_K^L x \in W_m \text{ is a unit.}
\end{aligned}
$$

This holds for all $\mathfrak{p}$, so α must be a unit, $\alpha \in R_K^*$. But $\alpha \equiv 1 \,(\mathrm{mod}\, m R_K)$ from above, so the only possibility is $\alpha = 1$. This proves our claim.

It follows from the claim and the definition of $\psi_{E/L}$ that

$$
\psi_{E/L}(x) = \mathrm{N}_K^L(x^{-1})_\infty \quad \text{for all } x \in U_m.
$$

From this formula it is clear that $\psi_{E/L}$ is continuous on U_m. But U_m is an open subgroup of $\mathbf{A}_L^*$. Therefore $\psi_{E/L}$ is continuous on all of $\mathbf{A}_L^*$, which completes the proof that $\psi_{E/L}$ is a Grössencharacter.

(b) Let $I_\mathfrak{P}^{\mathrm{ab}} \subset \mathrm{Gal}(L^{\mathrm{ab}}/L)$ be the inertia group for $\mathfrak{P}$. The reciprocity map sends $R_\mathfrak{P}^*$ to $I_\mathfrak{P}^{\mathrm{ab}}$,

$$
[R_\mathfrak{P}^*, L] = I_\mathfrak{P}^{\mathrm{ab}},
$$

where we embed $R_\mathfrak{P}^*$ into $\mathbf{A}_L^*$ in the usual way,

$$
R_\mathfrak{P}^* \hookrightarrow \mathbf{A}_L^*, \qquad u \longmapsto [\dots, 1, 1, \underset{\underset{\mathfrak{P}\,-\,\text{component}}{\uparrow}}{u}, 1, 1, \dots].
$$

Let m be an integer with $\mathfrak{P} \nmid m$. We know from (2.3) that $E[m] \subset E(L^{\mathrm{ab}})$, so $I_\mathfrak{P}^{\mathrm{ab}}$ will act on $E[m]$. We want to characterize when this action is trivial in terms of values of the Grössencharacter $\psi_{E/L}$. Thus

$$
\begin{aligned}
\substack{I_\mathfrak{P}^{\mathrm{ab}} \text{ acts trivially} \\ \text{on } E[m]} \;\;&\Longleftrightarrow\;\; f(t)^\sigma = f(t) \quad \text{for all } \sigma \in I_\mathfrak{P}^{\mathrm{ab}} \text{ and all } t \in m^{-1}\mathfrak{a}/\mathfrak{a} \\
&\Longleftrightarrow\;\; f(t)^{[x,L]} = f(t) \text{ for all } x \in R_\mathfrak{P}^* \text{ and all } t \in m^{-1}\mathfrak{a}/\mathfrak{a} \\
&\Longleftrightarrow\;\; f\big(\alpha_{E/L}(x)\big(\mathrm{N}_K^L x^{-1}\big)t\big) = f(t) \\
& \qquad\qquad \text{for all } x \in R_\mathfrak{P}^* \text{ and all } t \in m^{-1}\mathfrak{a}/\mathfrak{a},
\end{aligned}
$$

where for the last equivalence we have used (9.1). We make two observations. First,

$$\psi_{E/L}(x) = \alpha_{E/L}(x) \qquad \text{for all } x \in R_{\mathfrak{P}}^*,$$

since the archimedean components of $x \in R_{\mathfrak{P}}^*$ are all 1. Second, multiplication by $N_K^L x^{-1}$ induces the identity map on $m^{-1}\mathfrak{a}/\mathfrak{a}$. This follows from Lemma 9.3 (see below) and the assumption that $\mathfrak{P} \nmid m$. Hence we find

$$\begin{aligned}
\begin{array}{c} I_{\mathfrak{P}}^{\text{ab}} \text{ acts trivially} \\ \text{on } E[m] \end{array} &\iff f(\psi_{E/L}(x)t) = f(t) \\
&\qquad \text{for all } x \in R_{\mathfrak{P}}^* \text{ and all } t \in m^{-1}\mathfrak{a}/\mathfrak{a} \\
&\iff \psi_{E/L}(x) \equiv 1 \,(\text{mod } mR_K) \\
&\qquad \text{for all } x \in R_{\mathfrak{P}}^*, \text{ since } f : m^{-1}\mathfrak{a}/\mathfrak{a} \xrightarrow{\sim} E[m].
\end{aligned}$$

Next we apply the criterion of Néron-Ogg-Shafarevich, which relates the action of $I_{\mathfrak{P}}^{\text{ab}}$ on $E[m]$ to the reduction of E modulo $\mathfrak{P}$. More precisely, [AEC VII.7.1] says that

$$\begin{array}{c} I_{\mathfrak{P}}^{\text{ab}} \text{ acts trivially on } E[m] \text{ for} \\ \text{infinitely many } m \text{ prime to } \mathfrak{P} \end{array} \iff E \text{ has good reduction at } \mathfrak{P}.$$

Combining this with the equivalence proved above, we obtain the desired result:

$$\begin{aligned}
\begin{array}{c} E \text{ has good} \\ \text{reduction at } \mathfrak{P} \end{array} &\iff \begin{array}{c} \text{there are infinitely many } m \text{ with } \mathfrak{P} \nmid m \text{ such that} \\ \psi_{E/L}(x) \equiv 1\,(\text{mod } mR_K) \quad \text{for all } x \in R_{\mathfrak{P}}^* \end{array} \\
&\iff \psi_{E/L}(x) = 1 \quad \text{for all } x \in R_{\mathfrak{P}}^* \\
&\iff \psi_{E/L} \text{ is unramified at } \mathfrak{P}.
\end{aligned}$$

$\qquad\qquad\qquad\qquad\qquad\qquad\qquad\qquad\qquad\qquad\qquad\qquad\qquad\qquad$ □

It remains to prove the elementary result used in the proof of (9.2b).

Lemma 9.3. *Let $\mathfrak{a}$ be a fractional ideal and $\mathfrak{b}$ an integral ideal of K. Let $s \in \mathbf{A}_K^*$ be an idele with the property that*

$$s_{\mathfrak{p}} = 1 \qquad \text{for all primes } \mathfrak{p} \text{ dividing } \mathfrak{b}.$$

Then the multiplication-by-s map $s : K/\mathfrak{a} \to K/\mathfrak{a}$ induces the identity map on $\mathfrak{b}^{-1}\mathfrak{a}/\mathfrak{a}$. In other words,

$$st = t \qquad \text{for all} \quad t \in \mathfrak{b}^{-1}\mathfrak{a}/\mathfrak{a}.$$

PROOF. From (8.1), the decomposition of $\mathfrak{b}^{-1}\mathfrak{a}/\mathfrak{a}$ into $\mathfrak{p}$-primary components is

$$\mathfrak{b}^{-1}\mathfrak{a}/\mathfrak{a} \xrightarrow{\sim} \bigoplus_{\mathfrak{p}} (\mathfrak{b}^{-1}\mathfrak{a}/\mathfrak{a})_{\mathfrak{p}} \cong \bigoplus_{\mathfrak{p}} \mathfrak{b}_{\mathfrak{p}}^{-1}\mathfrak{a}_{\mathfrak{p}}/\mathfrak{a}_{\mathfrak{p}} = \bigoplus_{\mathfrak{p}|\mathfrak{b}} \mathfrak{b}_{\mathfrak{p}}^{-1}\mathfrak{a}_{\mathfrak{p}}/\mathfrak{a}_{\mathfrak{p}}.$$

Here the last equality follows from the fact that $\mathfrak{b}_\mathfrak{p} = R_\mathfrak{p}$ for all $\mathfrak{p} \nmid \mathfrak{b}$, so the only non-zero terms in the direct sum are those with $\mathfrak{p} | \mathfrak{b}$. The multiplication-by-$s$ map on $\mathfrak{b}^{-1}\mathfrak{a}/\mathfrak{a}$ is now defined by the commutative diagram

$$
\begin{array}{ccc}
\mathfrak{b}^{-1}\mathfrak{a}/\mathfrak{a} & \xrightarrow{\ s\ } & \mathfrak{b}^{-1}\mathfrak{a}/\mathfrak{a} \\
\downarrow \wr & & \downarrow \wr \\
\displaystyle\bigoplus_{\mathfrak{p}|\mathfrak{b}} \mathfrak{b}_\mathfrak{p}^{-1}\mathfrak{a}_\mathfrak{p}/\mathfrak{a}_\mathfrak{p} & \longrightarrow & \displaystyle\bigoplus_{\mathfrak{p}|\mathfrak{b}} \mathfrak{b}_\mathfrak{p}^{-1}\mathfrak{a}_\mathfrak{p}/\mathfrak{a}_\mathfrak{p}. \\
(t_\mathfrak{p}) & \longmapsto & (s_\mathfrak{p} t_\mathfrak{p})
\end{array}
$$

But by assumption, $s_\mathfrak{p} = 1$ for all $\mathfrak{p} | \mathfrak{b}$, so multiplication-by-$s$ is just the identity map. $\qquad\square$

§10. The *L*-Series Attached to a CM Elliptic Curve

The L-series attached to an elliptic curve is an analytic function that is used to encode arithmetic information about the curve. One then hopes to deduce further arithmetic properties of the elliptic curve by studying the analytic properties of its L-series, much as one uses the Riemann zeta function to study the set of rational primes. In this section we will define the L-series of an elliptic curve E and show that if E has complex multiplication, then its L-series can be expressed in terms of Hecke L-series with Grössencharacter.

Let $L/\mathbb{Q}$ be a number field, and let E/L be an elliptic curve. For each prime $\mathfrak{P}$ of L, let

$$\mathbb{F}_\mathfrak{P} = \text{residue field of } L \text{ at } \mathfrak{P},$$

$$q_\mathfrak{P} = \mathrm{N}_\mathbb{Q}^L \mathfrak{P} = \#\mathbb{F}_\mathfrak{P}.$$

If E has good reduction at $\mathfrak{P}$, we define

$$a_\mathfrak{P} = q_\mathfrak{P} + 1 - \#\tilde{E}(\mathbb{F}_\mathfrak{P}),$$

$$L_\mathfrak{P}(E/L, T) = 1 - a_\mathfrak{P} T + q_\mathfrak{P} T^2.$$

The polynomial $L_\mathfrak{P}(E/L, T)$ is called the *local L-series of E at* $\mathfrak{P}$. If E has bad reduction at $\mathfrak{P}$, we define the local L-series according to the following rules:

$$
L_\mathfrak{P}(E/L, T) = \begin{cases}
1 - T & \text{if } E \text{ has split multiplicative reduction at } \mathfrak{P}, \\
1 + T & \text{if } E \text{ has non-split multiplicative} \\
& \qquad\qquad\qquad\qquad\quad \text{reduction at } \mathfrak{P}, \\
1 & \text{if } E \text{ has additive reduction at } \mathfrak{P}.
\end{cases}
$$

Remark 10.1. In the case that E has good reduction at $\mathfrak{P}$, we can give a more intrinsic definition of the local L-factor in terms of the action of Frobenius on the Tate module. Thus let $\phi_{\mathfrak{P}} : \tilde{E} \to \tilde{E}$ be the $q_{\mathfrak{P}}$-power Frobenius map, and let

$$\phi_{\mathfrak{P},\ell} : T_\ell(\tilde{E}) \longrightarrow T_\ell(\tilde{E})$$

be the associated map on the Tate module of $\tilde{E}$ (see [AEC III. §7]), where we take some ℓ relatively prime to the characteristic of $\mathbb{F}_{\mathfrak{P}}$. If we choose a basis for the Tate module, so $T_\ell(\tilde{E}) \cong \mathbb{Z}_\ell \times \mathbb{Z}_\ell$, then $\phi_{\mathfrak{P},\ell}$ is represented by a 2×2 matrix with coefficients in $\mathbb{Z}_\ell$. The characteristic polynomial of the linear transformation $\phi_{\mathfrak{P},\ell}$ is

$$\det(1 - \phi_{\mathfrak{P},\ell}T) = 1 - (\operatorname{tr}\phi_{\mathfrak{P},\ell})T + (\det\phi_{\mathfrak{P},\ell})T^2 \in \mathbb{Z}_\ell[T],$$

and this polynomial is independent of the chosen basis for $T_\ell(E)$.

In fact, [AEC V.2.3] says that the characteristic polynomial of $\phi_{\mathfrak{P},\ell}$ has coefficients in $\mathbb{Z}$ and is independent of ℓ. More precisely, we find that

$$\operatorname{tr}\phi_{\mathfrak{P},\ell} = 1 + \deg\phi - \deg(1 - \phi) \quad \text{from [AEC V.2.3]}$$
$$= 1 + q_v - \#\tilde{E}(\mathbb{F}_{\mathfrak{P}}) \quad \text{from [AEC II.2.11c] and [AEC V §1].}$$

Similarly,

$$\det\phi_{\mathfrak{P},\ell} = \deg\phi \quad \text{from [AEC V.2.3]}$$
$$= q_v \quad \text{from [AEC II.2.11c].}$$

Hence

$$L_{\mathfrak{P}}(E/L, T) = \det(1 - \phi_{\mathfrak{P},\ell}T).$$

For a general discussion of this material in terms of ℓ-adic cohomology, see Hartshorne [1, App. C].

We now piece together the local L-factors to form the global L-series of E.

Definition. The *(global) L-series of E/L* is defined by the Euler product

$$L(E/L, s) = \prod_{\mathfrak{P}} L_{\mathfrak{P}}(E/L, q_{\mathfrak{P}}^{-s})^{-1},$$

where the product is over all primes of L.

Using the estimate $|a_{\mathfrak{P}}| \le 2\sqrt{q_{\mathfrak{P}}}$ from [AEC V.2.4], it is not hard to show that the product converges and gives an analytic function for all s satisfying $\operatorname{Re}(s) > \frac{3}{2}$. Conjecturally, far more is true.

Conjecture 10.2. *Let E/L be an elliptic curve defined over a number field. The L-series of E/L has an analytic continuation to the entire complex plane and satisfies a functional equation relating its values at s and $2 - s$.*

We are going to verify this conjecture for elliptic curves with complex multiplication by showing that $L(E/L, s)$ is a product of Hecke *L*-series with Grössencharacter. In general, suppose that

$$\psi : \mathbf{A}_L^* \longrightarrow \mathbb{C}^*$$

is a Grössencharacter on L; that is, ψ is a continuous homomorphism which is trivial on L^*. Let $\mathfrak{P}$ be a prime of L at which ψ is unramified, so $\psi(R_{\mathfrak{P}}^*) = 1$. We then define $\psi(\mathfrak{P})$ to be

$$\psi(\mathfrak{P}) = \psi(\dots, 1, 1, \underset{\underset{\mathfrak{P} \ - \ \text{component}}{\uparrow}}{\pi}, 1, 1, \dots),$$

where π is a uniformizer at $\mathfrak{P}$. Note that since ψ is unramified at $\mathfrak{P}$, $\psi(\mathfrak{P})$ is well-defined independent of the choice of π. For convenience, we also set

$$\psi(\mathfrak{P}) = 0 \qquad \text{if } \psi \text{ is ramified at } \mathfrak{P}.$$

Definition. The *Hecke L-series attached to the Grössencharacter*

$$\psi : \mathbf{A}_L^* \longrightarrow \mathbb{C}^*$$

is defined by the Euler product

$$L(s, \psi) = \prod_{\mathfrak{P}} \bigl(1 - \psi(\mathfrak{P}) q_{\mathfrak{P}}^{-s}\bigr)^{-1},$$

where the product is over all primes of L.

Hecke *L*-series with Grössencharacter have the following important properties, whose proof we will omit.

Theorem 10.3. (Hecke) *Let $L(s, \psi)$ be the Hecke L-series attached to the Grössencharacter ψ. Then $L(s, \psi)$ has an analytic continuation to the entire complex plane. Further, there is a functional equation relating the values of $L(s, \psi)$ and $L(N - s, \bar{\psi})$ for some real number $N = N(\psi)$.*

PROOF. This was originally proven by Hecke. It was reformulated and reproven by Tate [8] using Fourier analysis on the adele ring $\mathbf{A}_L$. □

The key to expressing $L(E/L, s)$ in terms of Hecke *L*-series is to express the number of points in $\tilde{E}(\mathbb{F}_{\mathfrak{P}})$ in terms of the Grössencharacter attached to E/L.

Proposition 10.4. *Let E/L be an elliptic curve with complex multiplication by the ring of integers R_K of K, and assume that $L \supset K$. Let $\mathfrak{P}$ be a prime of L at which E has good reduction, let $\tilde{E}$ be the reduction of E modulo $\mathfrak{P}$, and let $\phi_\mathfrak{P} : \tilde{E} \to \tilde{E}$ be the associated $q_\mathfrak{P}$-power Frobenius map. Finally, let $\psi_{E/L} : \mathbf{A}_L^* \to \mathbb{C}^*$ be the Grössencharacter (9.2) attached to E/L. Then the following diagram commutes:*

$$
\begin{array}{ccc}
E & \xrightarrow{\;[\psi_{E/L}(\mathfrak{P})]\;} & E \\
\downarrow & & \downarrow \\
\tilde{E} & \xrightarrow{\quad \phi_\mathfrak{P} \quad} & \tilde{E},
\end{array}
$$

where the vertical maps are reduction modulo $\mathfrak{P}$.

PROOF. Before we begin the proof, two remarks are in order. First, $\psi_{E/L}$ is unramified at $\mathfrak{P}$ from (9.2b), so $\psi_{E/L}(\mathfrak{P})$ is well-defined. Second, since $\psi_{E/L}(\mathfrak{P})$ is the value of $\psi_{E/L}$ at an idele with 1's in its archimedean components, we have $\psi_{E/L}(\mathfrak{P}) = \alpha_{E/L}(\mathfrak{P}) \in R_K$, so it makes sense to talk about $[\psi_{E/L}(\mathfrak{P})]$ as an endomorphism of E.

Let $x \in \mathbf{A}_L^*$ be an idele with a uniformizer in its $\mathfrak{P}$-component and 1's elsewhere. Then as we just remarked,

$$
\psi_{E/L}(\mathfrak{P}) = \psi_{E/L}(x) = \alpha_{E/L}(x) \in R_K.
$$

The commutative diagram (9.1) used to define $\alpha_{E/L}$ tells us that

$$
f(t)^{[x,L]} = \left[\psi_{E/L}(x)\right] f\left(\mathrm{N}_K^L x^{-1} t\right) \qquad \text{for all } t \in K/\mathfrak{a}.
$$

Fix some integer m with $\mathfrak{P} \nmid m$. Then (9.3) says that $\mathrm{N}_K^L x^{-1} t = t$ for all $t \in m^{-1}\mathfrak{a}/\mathfrak{a}$, so we get

$$
f(t)^{[x,L]} = \left[\psi_{E/L}(x)\right] f(t) \qquad \text{for all } t \in m^{-1}\mathfrak{a}/\mathfrak{a}.
$$

Now consider what happens when we reduce modulo $\mathfrak{P}$. We have $[x, L] = (\mathfrak{P}, L^{\mathrm{ab}}/L)$ from (3.5), so $[x, L]$ reduces to the $q_\mathfrak{P}$-power Frobenius map. Hence

$$
\phi_\mathfrak{P}\left(\widetilde{f(t)}\right) = \widetilde{f(t)^{[x,L]}} = \left[\widetilde{\psi_{E/L}(x)}\right]\widetilde{f(t)} \qquad \text{for all } t \in m^{-1}\mathfrak{a}/\mathfrak{a}.
$$

Since this is true for all m prime to $\mathfrak{P}$, and since an endomorphism of $\tilde{E}$ is determined by its effect on torsion (or even on ℓ-primary torsion for a fixed prime ℓ [AEC III.7.4]), we conclude that

$$
\phi_\mathfrak{P} = \left[\widetilde{\psi_{E/L}(x)}\right].
$$

$\square$

Corollary 10.4.1. *With notation as in (10.4), we have*

(a) $$q_{\mathfrak{P}} = \mathrm{N}^L_{\mathbb{Q}}\mathfrak{P} = \mathrm{N}^K_{\mathbb{Q}}\big(\psi_{E/L}(\mathfrak{P})\big),$$

(b) $$\#\tilde{E}(\mathbb{F}_{\mathfrak{P}}) = \mathrm{N}^L_{\mathbb{Q}}\mathfrak{P} + 1 - \psi_{E/L}(\mathfrak{P}) - \overline{\psi_{E/L}(\mathfrak{P})},$$

(c) $$a_{\mathfrak{P}} = \psi_{E/L}(\mathfrak{P}) + \overline{\psi_{E/L}(\mathfrak{P})}.$$

(The bar indicates complex conjugation of elements of K.)

PROOF. (a) We compute

$$
\begin{aligned}
\mathrm{N}^L_{\mathbb{Q}}\mathfrak{P} &= \deg\phi_{\mathfrak{P}} && \text{from [AEC II.2.11c]}\\
&= \deg\big[\widetilde{\psi_{E/L}(\mathfrak{P})}\big] && \text{from (10.4)}\\
&= \deg\big[\psi_{E/L}(\mathfrak{P})\big] && \text{from (4.4)}\\
&= \mathrm{N}^K_{\mathbb{Q}}\big(\psi_{E/L}(\mathfrak{P})\big) && \text{from (1.5).}
\end{aligned}
$$

(b) Similarly, we compute

$$
\begin{aligned}
\#\tilde{E}(\mathbb{F}_{\mathfrak{P}}) &= \#\ker(1 - \phi_{\mathfrak{P}})\\
&= \deg(1 - \phi_{\mathfrak{P}}) && \text{from [AEC III.5.5]}\\
& && \text{and [AEC III.4.10c]}\\
&= \deg\big[1 - \widetilde{\psi_{E/L}(\mathfrak{P})}\big] && \text{from (10.4)}\\
&= \deg\big[1 - \psi_{E/L}(\mathfrak{P})\big] && \text{from (4.4)}\\
&= \mathrm{N}^K_{\mathbb{Q}}\big(1 - \psi_{E/L}(\mathfrak{P})\big) && \text{from (1.5)}\\
&= \big(1 - \psi_{E/L}(\mathfrak{P})\big)\big(1 - \overline{\psi_{E/L}(\mathfrak{P})}\big)\\
&= 1 - \psi_{E/L}(\mathfrak{P}) - \overline{\psi_{E/L}(\mathfrak{P})} + \mathrm{N}^L_{\mathbb{Q}}\mathfrak{P} && \text{from (a).}
\end{aligned}
$$

(c) Immediate from (a), (b), and the definition of $a_{\mathfrak{P}}$. $\qquad\square$

We now have all of the tools needed to relate the *L*-series of *E* to the *L*-series attached to its Grössencharacter, at least in the case that the field of definition of *E* contains the CM field. We will leave the proof in the other case to the reader.

Theorem 10.5. (Deuring) *Let E/L be an elliptic curve with complex multiplication by the ring of integers R_K of K.*
(a) *Assume that K is contained in L. Let $\psi_{E/L} : \mathbf{A}^*_L \to \mathbb{C}^*$ be the Grössencharacter (9.2) attached to E/L. Then*

$$L(E/L, s) = L(s, \psi_{E/L})L(s, \overline{\psi_{E/L}}).$$

(b) *Suppose that K is not contained in L, and let $L' = LK$. Further let $\psi_{E/L'} : \mathbf{A}_{L'}^* \to \mathbf{C}^*$ be the Grössencharacter attached to E/L'. Then*

$$L(E/L, s) = L(s, \psi_{E/L'}).$$

Using Hecke's theorem (10.3), we immediately deduce that the L-series of a CM elliptic curve has an analytic continuation and satisfies a functional equation. A more careful analysis yields the following result. We will leave the proof to the reader.

Corollary 10.5.1. *Let E/L be an elliptic curve with complex multiplication by the ring of integers R_K of K. The L-series of E admits an analytic continuation to the entire complex plane and satisfies a functional equation relating its values at s and $2 - s$.*

More precisely, define a function $\Lambda(E/L, s)$ as follows:
(i) *If $K \subset L$, let*

$$\Lambda(E/L, s) = \left(\mathrm{N}_{\mathbf{Q}}^{L}(\mathfrak{D}_{L/\mathbf{Q}}\mathfrak{c}_\psi)\right)^s \left((2\pi)^{-s}\Gamma(s)\right)^{[L:\mathbf{Q}]} L(E/L, s),$$

where $\mathfrak{c}_\psi$ is the conductor of the Grössencharacter $\psi_{E/L}$, $\mathfrak{D}_{L/\mathbf{Q}}$ is the different of $L/\mathbf{Q}$, and $\Gamma(s) = \int_0^\infty t^{s-1}e^{-t}\,dt$ is the usual Γ-function.
(ii) *If $K \not\subset L$, let $L' = LK$ and*

$$\Lambda(E/L, s) = \left(\mathrm{N}_{\mathbf{Q}}^{L'}(\mathfrak{D}_{L'/\mathbf{Q}}\mathfrak{c}'_\psi)\right)^{s/2} \left((2\pi)^{-s}\Gamma(s)\right)^{[L:\mathbf{Q}]} L(E/L, s),$$

where $\mathfrak{c}'_\psi$ is the conductor of the Grössencharacter $\psi_{E/L'}$.
Then Λ satisfies the functional equation

$$\Lambda(E/L, s) = w\Lambda(E/L, 2 - s),$$

where the quantity $w = w_{E/L} \in \{\pm 1\}$ is called the sign of the functional equation of E/L.

PROOF (of Theorem 10.5). We know from (6.1) and [AEC VII.5.5] that E has potential good reduction at every prime of L, so [AEC VII.5.4(b)] tells us that E has no multiplicative reduction. Hence

$$L_{\mathfrak{P}}(E/L, T) = \begin{cases} 1 - a_{\mathfrak{P}}T + q_{\mathfrak{P}}T^2 & \text{if } E \text{ has good reduction at } \mathfrak{P}, \\ 1 & \text{if } E \text{ has bad reduction at } \mathfrak{P}. \end{cases}$$

Now suppose E has good reduction at $\mathfrak{P}$. Then

$$
\begin{aligned}
L_{\mathfrak{P}}(E/L, T) &= 1 - a_{\mathfrak{P}}T + q_{\mathfrak{P}}T^2 \qquad \text{by definition of } L_{\mathfrak{P}}, \\
&= 1 - \left(\psi_{E/L}(\mathfrak{P}) + \overline{\psi_{E/L}(\mathfrak{P})}\right)T + \left(\mathrm{N}_{\mathbf{Q}}^{K}\psi_{E/L}(\mathfrak{P})\right)T^2 \\
&\hspace{7cm} \text{from (10.4.1)} \\
&= \left(1 - \psi_{E/L}(\mathfrak{P})T\right)\left(1 - \overline{\psi_{E/L}(\mathfrak{P})}T\right).
\end{aligned}
$$

On the other hand, (9.2b) says that $\psi_{E/L}$ is unramified at $\mathfrak{P}$ if and only if E has good reduction at $\mathfrak{P}$, and the same is true for $\overline{\psi_{E/L}}$. Thus

$$\psi_{E/L}(\mathfrak{P}) = \overline{\psi_{E/L}(\mathfrak{P})} = 0 \qquad \text{if } E \text{ has bad reduction at } \mathfrak{P},$$

so the formula given above for $L_{\mathfrak{P}}(E/L, T)$ is also true for primes of bad reduction, since it reduces to $L_{\mathfrak{P}}(E/L, T) = 1$. Therefore

$$L(E/L, s) = \prod_{\mathfrak{P}} L_{\mathfrak{P}}(E/L, q_{\mathfrak{P}}^{-s})^{-1}$$
$$= \prod_{\mathfrak{P}} \left(1 - \psi_{E/L}(\mathfrak{P}) q_{\mathfrak{P}}^{-s}\right)^{-1} \left(1 - \overline{\psi_{E/L}(\mathfrak{P})} q_{\mathfrak{P}}^{-s}\right)^{-1}$$
$$= L(s, \psi_{E/L}) L(s, \overline{\psi_{E/L}}).$$

(b) See exercises 2.30, 2.31, and 2.32. $\square$

Example 10.6. Let $D \in \mathbb{Z}$ be a non-zero integer, and let E be the elliptic curve

$$E : y^2 = x^3 + D$$

having complex multiplication by the ring of integers R_K of the field $K = \mathbb{Q}\left(\sqrt{-3}\right)$. Let $\mathfrak{p}$ be a prime of R_K with $\mathfrak{p} \nmid 6D$. Since R_K is a PID, we can write $\mathfrak{p} = (\pi)$, and one can check that there is a unique π generating $\mathfrak{p}$ which satisfies $\pi \equiv 2 \,(\text{mod } 3)$. It is then a moderately difficult exercise using Jacobi sums (see Ireland-Rosen [1, 18 §§5,7]) to show that

$$\#\tilde{E}(\mathbb{F}_{\mathfrak{p}}) = \mathrm{N}_{\mathbb{Q}}^K \mathfrak{p} + 1 + \overline{\left(\frac{4D}{\pi}\right)_6} \pi + \left(\frac{4D}{\pi}\right)_6 \overline{\pi},$$

where $\left(\frac{\alpha}{\pi}\right)_6$ is the 6th-power residue symbol; that is, $\left(\frac{\alpha}{\pi}\right)_6$ is the 6th-root of unity satisfying

$$\alpha^{(\mathrm{N}_{\mathbb{Q}}^K \mathfrak{p} - 1)/6} \equiv \left(\frac{\alpha}{\pi}\right)_6 \,(\text{mod } \pi).$$

Using (10.4.1), we see that the Grössencharacter attached to E is given either by

$$\psi_{E/K}(\mathfrak{p}) = -\overline{\left(\frac{4D}{\pi}\right)_6} \pi \qquad \text{or else by} \qquad \psi_{E/K}(\mathfrak{p}) = -\left(\frac{4D}{\pi}\right)_6 \overline{\pi}.$$

To determine which one it is, we use (5.4) to find a root of unity $\xi \in R_K^*$ such that the reduction of $[\xi\pi]$ modulo $\mathfrak{p}$ is $\mathrm{N}_{\mathbb{Q}}^K \mathfrak{p}$-power Frobenius. Note that (5.4) says this is possible for almost all degree 1 primes of K.

But (10.4) says that $\left[\psi_{E/K}(\mathfrak{p})\right]$ also reduces to Frobenius. We conclude that

$$\psi_{E/K}(\mathfrak{p}) = -\overline{\left(\frac{4D}{\pi}\right)}_6 \pi, \qquad \text{where } \mathfrak{p} = (\pi) \text{ and } \pi \equiv 2 \,(\mathrm{mod}\,3),$$

at least for almost all degree 1 primes $\mathfrak{p}$ of K. By the continuity of ψ and the reciprocity law for $\left(\frac{\cdot}{\pi}\right)_6$, we see that this formula holds for all $\mathfrak{p}$.

Using (10.5) and $\mathrm{N}_{\mathbb{Q}}^K \pi = \pi\bar{\pi}$, we find that the L-series of E over K and over $\mathbb{Q}$ can be written explicitly using residue symbols as

$$L(E/K, s) = \prod_{\substack{\pi \in R_K \text{ prime} \\ \pi \equiv 2 \,(\mathrm{mod}\,3)}} \left(1 + \overline{\left(\frac{4D}{\pi}\right)}_6 \pi^{1-s}\bar{\pi}^{-s} \right)^{-1}$$

$$\times \left(1 + \left(\frac{4D}{\pi}\right)_6 \pi^{-s}\bar{\pi}^{1-s} \right)^{-1},$$

$$L(E/\mathbb{Q}, s) = \prod_{\substack{\pi \in R_K \text{ prime} \\ \pi \equiv 2 \,(\mathrm{mod}\,3)}} \left(1 + \overline{\left(\frac{4D}{\pi}\right)}_6 \pi^{1-s}\bar{\pi}^{-s} \right)^{-1}.$$

EXERCISES

2.1. Let $K/\mathbb{Q}$ be a quadratic field with ring of integers R_K, and let $R \subset K$ be an order in K. Prove that there is a unique integer $f \geq 1$ such that

$$R = \mathbb{Z} + f \cdot R_K.$$

The integer f is called *conductor* of the order R.

2.2. *Let $\Lambda = \mathbb{Z}[i]$ be the lattice of Gaussian integers, and let λ be the elliptic integral

$$\lambda = \int_0^1 \frac{dt}{\sqrt{1 - t^4}}.$$

(a) Prove that

$$g_2(\Lambda) = 64\lambda^4.$$

(b) More generally, prove that for all integers $n \geq 1$, $G_{4n}(\Lambda)$ is a rational number multiplied by λ^{4n}.

2.3. Let $K/\mathbb{Q}$ be a quadratic imaginary field, and let $E/\mathbb{C}$ be an elliptic curve with $\mathrm{End}(E) \otimes \mathbb{Q} \cong K$. Let $E'/\mathbb{C}$ be another elliptic curve. Prove that E' is isogenous to E if and only if $\mathrm{End}(E') \otimes \mathbb{Q} \cong K$.

2.4. Let E be an elliptic curve defined over a number field L with complex multiplication by K, and let $\mathfrak{P}$ be a prime of L of characteristic p at which E has good ordinary reduction. Prove that $\mathbb{Q}_p$ contains a subfield isomorphic to K.

2.5. Let $E/\mathbb{Q}$ be an elliptic curve with complex multiplication by the ring of integers in $\mathbb{Q}\left(\sqrt{-7}\right)$. Without using an explicit Weierstrass equation for E, prove the following two facts:
(a) If $\phi : E \to E$ is an endomorphism of degree 2 and P is the non-zero point in the kernel of ϕ, then $P \notin E(\mathbb{Q})$ and $P \in E\left(\mathbb{Q}\left(\sqrt{-7}\right)\right)$.
(b) $E(\mathbb{Q})$ contains exactly one point of order 2.
(*Hint.* Use (2.2a).)

2.6. (a) Let F be a field, let G be a subgroup of $\mathrm{GL}_2(F)$, and let $C(G)$ be the centralizer of G; that is,

$$C(G) = \{\alpha \in \mathrm{GL}_2(F) : \alpha\gamma = \gamma\alpha \text{ for all } \gamma \in G\}.$$

Prove that one of the following two conditions is true.

(i) $C(G) = \left\{ \begin{pmatrix} c & 0 \\ 0 & c \end{pmatrix} : c \in F^* \right\}.$

(ii) G is abelian.

(*Hint.* If $C(G)$ contains a non-scalar matrix α, make a change of basis to put α into Jordan normal form and then calculate $C(\alpha)$.)
(b) Let L be a perfect field, let E/L be an elliptic curve, and let ℓ be a prime with $\ell \neq \mathrm{char}(L)$. Suppose that $\mathrm{End}_L(E)$ is strictly larger than $\mathbb{Z}$. Use (a) to prove that the action of $\mathrm{Gal}(\bar{L}/L)$ on the Tate module $T_\ell(E)$ is abelian.

2.7. The following fact (2.5.1) from commutative algebra was used in the proof of Proposition 2.4. Let R be a Dedekind domain, let $\mathfrak{a}$ be a fractional ideal of R, and let M be a torsion-free R-module. Prove that the natural map

$$\begin{array}{cccc} \phi : & \mathfrak{a}^{-1}M & \longrightarrow & \mathrm{Hom}_R(\mathfrak{a}, M) \\ & x & \longmapsto & (\phi_x : \alpha \mapsto \alpha x) \end{array}$$

is an isomorphism. (*Hint.* Prove that it is an isomorphism after you localize at any prime $\mathfrak{p}$ of R. Note that the localization $R_\mathfrak{p}$ is a principal ideal domain.)

2.8. Let L/K be a finite abelian extension of number fields.
(a) Let $\mathfrak{A}$ be a non-zero fractional ideal of L. Prove that

$$\left(\mathrm{N}_K^L\mathfrak{A}, L/K\right) = 1.$$

(b) Prove that an unramified prime $\mathfrak{p}$ of K splits completely in L if and only if $(\mathfrak{p}, L/K) = 1$.

2.9. Let $\mathfrak{a}$ be a fractional ideal of K, and let E be an elliptic curve corresponding to the lattice $\mathfrak{a}$.
(a) Prove that $j(E) \in \mathbb{R}$ if and only if $\bar{\mathfrak{a}}^2 = 1$ in $\mathcal{CL}(R_K)$.
(b) Prove that the following are equivalent:
   (i) $\mathbb{Q}(j(E))$ is Galois over $\mathbb{Q}$.
   (ii) $\mathbb{Q}(j(E))$ is totally real.
   (iii) Every element of $\mathcal{CL}(R_K)$ has order 2.

2.10. Let K be a number field, R_K its ring of integers, and c an integral ideal of R_K. Prove that there is an exact sequence

$$R_K^* \quad \rightarrow \quad (R_K/c)^* \quad \rightarrow \quad I(c)/P(c) \quad \rightarrow \quad \mathcal{CL}(R_K) \quad \rightarrow \quad 1.$$
$$\qquad \qquad \alpha(\mathrm{mod}\, c) \quad \mapsto \qquad (\alpha)$$

2.11. Let $K = \mathbb{Q}\left(\sqrt{-15}\right)$, and let R_K be the ring of integers of K.

(a) Prove that $\mathcal{CL}(R_K) = \mathbb{Z}/2\mathbb{Z}$.

(b) Let $L = \mathbb{Q}\left(\sqrt{-3}, \sqrt{5}\right)$. Prove that L/K is everywhere unramified, and deduce that L is the Hilbert class field H of K.

(c) Let $K = \mathbb{Q}\left(\sqrt{-23}\right)$. Prove that the Hilbert class field of K is given by $H = K(\alpha)$, where α satisfies $\alpha^3 - \alpha - 1 = 0$.

2.12. Let $E/\mathbb{C}$ be an elliptic curve such that $\mathrm{End}(E)$ is an order in the quadratic imaginary field K.

(a) Show that there exists an elliptic curve $E'/\mathbb{C}$ and an isogeny $\phi : E \to E'$ such that $\mathrm{End}(E') = R_K$.

(b) *If E is defined over the field L, prove that it is possible to choose E' and ϕ in (a) so that both are defined over L.

2.13. Let $K = \mathbb{Q}(i)$ and let K_N be the ray class field of K modulo N.

(a) Prove that

$$K_2 = K, \qquad K_3 = K(\sqrt{3}), \qquad K_4 = K(\sqrt{2}).$$

(b) Verify directly that the field

$$K\left(\sqrt[4]{\frac{8\sqrt{3} - 12}{9}}\right)$$

is an abelian extension of K, and compute its Galois group over K. (This exercise verifies some of the statements made in (4.9.1).)

2.14. Let E be the elliptic curve $y^2 = x^3 + 1$, and let $K = \mathbb{Q}\left(\sqrt{-3}\right)$. For each integer $N \geq 1$, let $K_N = K(h(E[N]))$ and $L_N = K(E[n])$.

(a) Calculate K_2, K_3, and K_4 explicitly and in each case verify that K_N is the ray class field of K modulo N.

(b) Calculate L_2, L_3, and L_4 explicitly and show that they are abelian extensions of K.

2.15. Let E be the elliptic curve $E : y^2 = x^3 + 4x^2 + 2x$, and let $K = \mathbb{Q}\left(\sqrt{-2}\right)$. From (2.3), this curve has complex multiplication by K. Redo the previous exercise for this curve E and field K.

2.16. For each of the quadratic imaginary fields in the following table, verify that the given α generates the Hilbert class field of K, and calculate the value of $j(R_K)$ explicitly as an element of $K(\alpha)$. (We have filled in the first row for you, see Example 6.2.2.)

	Disc $K/\mathbb{Q}$	h_K	α a root of	$j(R_K)$
(a)	-15	2	$x^2 - 5$	$-52515 - 85995\frac{1+\alpha}{2}$
(b)	-20	2	$x^2 + 1$	
(c)	-23	3	$x^3 - x - 1$	
(d)	-24	2	$x^2 + 3$	
(e)	-31	3	$x^3 + x - 1$	

2.17. Let $\mathcal{D}_n^*$ and $\mathcal{S}_n^*$ be the sets of matrices defined by

$$\mathcal{D}_n^* = \left\{ \begin{pmatrix} a & b \\ c & d \end{pmatrix} \in M_2(\mathbb{Z}) : ad - bc = n, \ \gcd(a,b,c,d) = 1 \right\},$$

$$\mathcal{S}_n^* = \left\{ \begin{pmatrix} a & b \\ 0 & d \end{pmatrix} \in \mathcal{D}_n^* : d > 0, \ 0 \le b < d \right\}.$$

(a) Prove that the natural map $\mathcal{S}_n^* \longrightarrow SL_2(\mathbb{Z})\backslash\mathcal{D}_n^*$ is a bijection.
(b) Prove that

$$\#\mathcal{S}_n^* = n \prod_{p|n} \left(1 + \frac{1}{p} \right).$$

(Notice that if n is squarefree, then $\mathcal{D}_n^*$ and $\mathcal{S}_n^*$ are just the sets $\mathcal{D}_n$ and $\mathcal{S}_n$ considered in §6.)

2.18. Let $\mathcal{S}_n^*$ be as in the previous exercise, and define

$$\Phi_n(X) = \prod_{\alpha \in \mathcal{S}_n^*} (X - j \circ \alpha).$$

Φ_n is called the *modular polynomial of order n*.
(a) Prove that $\Phi_n \in \mathbb{Z}[j][X]$. We will write $\Phi_n(j, X)$ to indicate that Φ_n is a polynomial in two variables.
(b) Prove that Φ_n is irreducible over $\mathbb{C}(j)$.
(c) Prove that $\Phi_n(Y, X) = \Phi_n(X, Y)$.
(d) Prove that if n is not a perfect square, then $\Phi_n(X, X)$ is a non-constant polynomial with leading coefficient ± 1.
(e) Let $F_n(j, X)$ be the polynomial from (6.3). Prove that

$$F_n(j, X) = \prod_{d^2|n, \, d \ge 1} \Phi_{n/d^2}(j, X).$$

In particular, if n is square-free, then $F_n = \Phi_n$.
(f) *Let $|\Phi_n|$ denote the magnitude of the largest coefficient of $\Phi_n(Y, X)$. Prove that

$$\lim_{n \to \infty} \frac{\log |\Phi_n|}{(\deg \Phi_n)(\log n)} = 6.$$

2.19. Let $F_n(Y, X)$ be the polynomial from (6.3), and let $\Phi_n(Y, X)$ be the polynomial from the previous exercise. Let $E_1/\mathbb{C}$ and $E_2/\mathbb{C}$ be elliptic curves.

(a) Prove that $F_n(j(E_1), j(E_2)) = 0$ if and only if there is an isogeny $E_1 \to E_2$ of degree n.

(b) Prove that $\Phi_n(j(E_1), j(E_2)) = 0$ if and only if there is an isogeny $E_1 \to E_2$ whose kernel is cyclic of degree n.

2.20. Let p be a prime. Prove *Kronecker's congruence relation*

$$F_p(Y, X) \equiv (X - Y^p)(X^p - Y) \pmod{p\mathbb{Z}[X, Y]},$$

where F_p is the polynomial defined in (6.3).

2.21. Let $f(\tau)$ be a modular function of weight 0 that is holomorphic on $\mathbf{H}$ and that has the q-expansion $f = \sum a_n q^n$. Let R be a ring containing all of the a_n's. Prove that $f \in R[j]$, where $j = j(\tau)$ is the modular j-function. This strengthens (I.4.2b), which says that $f \in \mathbb{C}[j]$.

2.22. Let $\ell \geq 3$ be a prime, let

$$\mathrm{GL}_2(\mathbb{Z}_\ell)_1 = \left\{ A \in \mathrm{GL}_2(\mathbb{Z}_\ell) \; : \; A \equiv \begin{pmatrix} 1 & 0 \\ 0 & 1 \end{pmatrix} \pmod{\ell} \right\},$$

and let $M_2(\ell\mathbb{Z}_\ell)$ be the additive group of 2×2 matrices with $\ell\mathbb{Z}_\ell$-coefficients. Prove that the map

$$\mathrm{GL}_2(\mathbb{Z}_\ell)_1 \longrightarrow M_2(\ell\mathbb{Z}_\ell), \qquad 1 + \ell A \longmapsto \log(1 + \ell A) = \sum_{n=1}^{\infty} \frac{(-1)^{n+1} \ell^n A^n}{n},$$

is a well-defined isomorphism.

2.23. This exercise generalizes the previous one. Let L be a finite extension of $\mathbb{Q}_\ell$, let R be the ring of integers of L, and let $\mathfrak{M}$ be the maximal ideal of R. For each integer $r \geq 1$ define a subgroup G_r of $\mathrm{GL}_n(R)$ by

$$G_r = \{A \in \mathrm{GL}_n(R) \; : \; A \equiv I_n \pmod{\mathfrak{M}^r}\},$$

where I_n is the $n \times n$ identity matrix.

(a) Prove that for every $r \geq 1$, the quotient G_r/G_{r+1} is a finite group whose order is a power of ℓ.

(b) Prove that

$$G_1 \cong \varprojlim G_1/G_r,$$

and deduce that G_1 is a pro-ℓ group.

(c) Prove that if r is sufficiently large, then G_r is isomorphic to the additive group of $n \times n$ matrices with coefficients in R. (*Hint.* For the case $n = 1$, see [AEC IV.6.4b].)

2.24. Let E and E' be elliptic curves given by Weierstrass equations

$$E : y^2 = x^3 + a_2 x^2 + a_4 x + a_6, \qquad E' : Y^2 = X^3 + a_2' X^2 + a_4' X + a_6',$$

and let $\phi : E \to E'$ be a non-zero isogeny.

(a) Prove that there is a rational function $R(x)$ and a constant c such that ϕ has the form

$$\phi(x, y) = \left(R(x), cy\frac{dR(x)}{dx} \right).$$

(*Hint.* Look at the invariant differentials.)

(b) Prove that there is a commutative diagram

$$
\begin{array}{ccc}
\mathbb{C}/\Lambda & \xrightarrow{\;z \mapsto c^{-1} z\;} & \mathbb{C}/\Lambda' \\
\downarrow & & \downarrow \\
E(\mathbb{C}) & \xrightarrow{\;\phi\;} & E'(\mathbb{C})
\end{array}
$$

where the vertical maps are complex analytic isomorphisms and c is the constant from part (a).

2.25. Let E/L be an elliptic curve defined over a number field L with complex multiplication by the ring of integers of K, and assume that $K \subset L$. Let $\chi \in H^1(G_{L/L}, \operatorname{Aut}(E))$, and let E^χ/L be the corresponding twist of E. (See [AEC X §5] for basic facts about twists.) If we identify $\operatorname{Aut}(E)$ with μ_n [AEC III.10.2], then χ gives a homomorphism

$$\chi : \operatorname{Gal}(L^{\mathrm{ab}}/L) \longrightarrow \mu_n \subset \mathbb{C}^*,$$

and we can extend χ to a homomorphism on the ideles by the rule

$$\chi : \mathbf{A}_L^* \longrightarrow \mathbb{C}^*, \qquad \chi(x) = \chi([x, L]).$$

Prove that

$$\psi_{E^\chi/L} = \chi^{-1} \psi_{E/L}.$$

2.26. Let E/L be an elliptic curve defined over a number field (not necessarily with complex multiplication). Prove that the infinite product defining the L-series $L(E/L, s)$ converges absolutely and uniformly in the half-plane $\operatorname{Re}(s) > \frac{3}{2}$.

2.27. Let E/L be an elliptic curve defined over a number field (not necessarily with complex multiplication), let $\mathfrak{P}$ be a prime of L at which E has good reduction, and let $\mathbb{F}_{\mathfrak{P}}$ be the residue field at $\mathfrak{P}$. For each integer $n \geq 1$, let $\mathbb{F}_{\mathfrak{P},n}$ be the extension of $\mathbb{F}_{\mathfrak{P}}$ of degree n. Recall [AEC V §2] that the zeta function of $\tilde{E}/\mathbb{F}_{\mathfrak{P}}$ is the formal power series

$$Z(\tilde{E}/\mathbb{F}_{\mathfrak{P}}, T) = \exp\left(\sum_{n=1}^{\infty} \#\tilde{E}(\mathbb{F}_{\mathfrak{P},n}) \frac{T^n}{n} \right).$$

(a) Prove that

$$Z(\tilde{E}/\mathbb{F}_\mathfrak{P}, T) = \frac{L_\mathfrak{P}(\tilde{E}/\mathbb{F}_\mathfrak{P}, T)}{(1 - T)(1 - q_\mathfrak{P}T)},$$

where $L_\mathfrak{P}$ is the local L-series described in §10 and $q_\mathfrak{P} = \mathrm{N}_\mathbb{Q}^L\mathfrak{P}$.

(b) The *zeta function of the field L* is given by the usual Euler product

$$\zeta_L(s) = \prod_\mathfrak{P}(1 - q_\mathfrak{P}^{-s})^{-1},$$

and the *(global) zeta function of E/L* is defined by the product

$$\zeta(E/L, s) = \prod_\mathfrak{P} Z(\tilde{E}/\mathbb{F}_\mathfrak{P}, q_\mathfrak{P}^{-s}).$$

Find the "correct" definition for the factor $Z(\tilde{E}/\mathbb{F}_\mathfrak{P}, T)$ in the case that E has bad reduction at $\mathfrak{P}$, and prove that

$$\zeta(E/L, s) = \zeta_L(s)\zeta_L(s - 1)L(E/L, s)^{-1}.$$

2.28. With notation as in §10, prove that

$$L_\mathfrak{P}(\tilde{E}/\mathbb{F}_\mathfrak{P}, q_\mathfrak{P}^{-1}) = \frac{\#\tilde{E}_{\mathrm{ns}}(\mathbb{F}_\mathfrak{P})}{q_\mathfrak{P}}.$$

Here $\tilde{E}_{\mathrm{ns}}$ is the non-singular part of $\tilde{E}$. Note that we do not assume E has good reduction at $\mathfrak{P}$. (See [AEC III §2] and [AEC exer. 3.5].)

2.29. Prove the functional equation (10.5.1) for the L-series of an elliptic curve with complex multiplication. (*Hint.* Use the functional equation for Hecke L-series with Grössencharacter as described, for example, in Tate [8].)

In the next three exercises we sketch the proof of Theorem 10.5(b). We set the following notation: Let E/L be an elliptic curve with complex multiplication by the ring of integers R_K of K, and assume that L does not contain K. Let $L' = LK$, so L' is a quadratic extension of L, and let $\mathfrak{P}$ be a prime of L. From (9.2) there is a Grössencharacter $\psi_{E/L'} : \mathbf{A}_{L'}^* \to \mathbb{C}^*$. Let $q_\mathfrak{P}$, $a_\mathfrak{P},\ldots$ be the quantities described in §10.

2.30. Assume that E has good reduction at $\mathfrak{P}$.

(a) Prove that $\mathfrak{P}$ is unramified in L'.

(b) Suppose $\mathfrak{P}$ splits in L' as $\mathfrak{P}R_{L'} = \mathfrak{P}'\mathfrak{P}''$. Prove that

$$q_\mathfrak{P} = q_{\mathfrak{P}'} = q_{\mathfrak{P}''} \qquad \text{and} \qquad a_\mathfrak{P} = \psi_{E/L'}(\mathfrak{P}') + \psi_{E/L'}(\mathfrak{P}'').$$

(c) Suppose $\mathfrak{P}$ remains inert in L', say $\mathfrak{P}R_{L'} = \mathfrak{P}'$. Prove that

$$q_\mathfrak{P}^2 = q_{\mathfrak{P}'}, \qquad a_\mathfrak{P} = 0, \qquad \psi_{E/L'}(\mathfrak{P}') = -q_\mathfrak{P}.$$

(d) Let $\tilde{E}$ be the reduction of E modulo $\mathfrak{P}$, and let p be the residue characteristic of $\mathfrak{P}$. Prove that

$$\tilde{E} \text{ is } \begin{cases} \text{ordinary} & \text{if } \mathfrak{P} \text{ splits in } L' \text{ and } p \text{ splits in } K, \\ \text{supersingular} & \text{if } \mathfrak{P} \text{ is inert in } L' \text{ and } p \text{ does not split in } K. \end{cases}$$

2.31. We continue with the notation and assumptions from above. Let $\mathfrak{P}'$ be a prime of L' lying over $\mathfrak{P}$.
 (a) If $\mathfrak{P}$ ramifies in L', prove that E has bad reduction at $\mathfrak{P}$.
 (b) If $\mathfrak{P}$ is unramified in L', prove that

$$E \text{ has good reduction at } \mathfrak{P} \iff E \text{ has good reduction at } \mathfrak{P}'.$$

2.32. We continue with the notation and assumptions from above.
 (a) Prove that the local L-series of E at $\mathfrak{P}$ is given by

$$L_{\mathfrak{P}}(E/L, T) = \begin{cases} (1 - \psi_{E/L'}(\mathfrak{P}')T)(1 - \psi_{E/L'}(\mathfrak{P}'')T) \\ \qquad \text{if } \mathfrak{P}R_{L'} = \mathfrak{P}'\mathfrak{P}'' \text{ splits in } L', \\ 1 - \psi_{E/L'}(\mathfrak{P}')T \quad \text{if } \mathfrak{P}R_{L'} = \mathfrak{P}' \text{ is inert in } L', \\ 1 \qquad\qquad\qquad \text{if } \mathfrak{P}R_{L'} = \mathfrak{P}'^2 \text{ ramifies in } L'. \end{cases}$$

 (b) Prove that the global L-series of E/L is given by

$$L(E/L, s) = L(s, \psi_{E/L'}).$$

2.33. Fix a non-zero integer $D \in \mathbb{Z}$ and let E be the elliptic curve

$$E : y^2 = x^3 - Dx.$$

Let $p \in \mathbb{Z}$ be a prime with $p \nmid 2D$.
 (a) If $p \equiv 3 \pmod 4$, prove that

$$\#\tilde{E}(\mathbb{F}_p) = p + 1 \qquad \text{and} \qquad \#\tilde{E}(\mathbb{F}_{p^2}) = (p+1)^2.$$

 (b) If $p \equiv 1 \pmod 4$, factor p in $\mathbb{Z}[i]$ as

$$p = \pi\bar{\pi} \qquad \text{with} \qquad \pi \equiv 1 \pmod{2 + 2i}.$$

Prove that

$$\#\tilde{E}(\mathbb{F}_p) = p + 1 - \overline{\left(\frac{D}{\pi}\right)_4} \pi - \left(\frac{D}{\pi}\right)_4 \bar{\pi},$$

where $\left(\frac{\alpha}{\pi}\right)_4$ is the 4th-power residue symbol.

2.34. Continuing with the notation from the previous exercise, let $\mathfrak{p} \subset \mathbb{Z}[i]$ be a prime ideal with $\mathfrak{p} \nmid 2D$. Write

$$\mathfrak{p} = (\pi) \text{ for an element } \pi \in \mathbb{Z}[i] \text{ satisfying } \pi \equiv 1 \pmod{2+2i}.$$

Prove that the Grössencharacter associated to $E/\mathbb{Q}(i)$ is given explicitly by the formula

$$\psi_{E/\mathbb{Q}(i)}(\mathfrak{p}) = \overline{\left(\frac{D}{\pi}\right)_4} \pi.$$

Here $\psi_{E/\mathbb{Q}(i)}(\mathfrak{p})$ equals the value of ψ at an idele with a uniformizer at the $\mathfrak{p}$-component and 1's elsewhere.

2.35. Let E/L be an elliptic curve with complex multiplication by the ring of integers R_K of K, and assume that $K \subset L$. Let $\mathfrak{p}$ be a degree 1 prime of K, let $p = \mathrm{N}_{\mathbb{Q}}^K \mathfrak{p}$, and let $R_\mathfrak{p}$ be the completion of R_K at $\mathfrak{p}$. Notice that $R_\mathfrak{p} \cong \mathbb{Z}_p$.

Fix an integer $h \geq 1$ so that $\mathfrak{p}^h$ is principal, say $\mathfrak{p}^h = \pi R_K$. We make the collection of groups $E[\mathfrak{p}^n]$, $n = 1, 2, \ldots$, into an inverse system using the maps

$$E[\mathfrak{p}^{n+h}] \xrightarrow{\;[\pi]\;} E[\mathfrak{p}^n], \qquad n = 1, 2, \ldots.$$

The $\mathfrak{p}$-*adic Tate module of* E is then defined to be the inverse limit

$$T_\mathfrak{p}(E) = \varprojlim E[\mathfrak{p}^n].$$

(a) Prove that $T_\mathfrak{p}(E)$ is a free $R_\mathfrak{p}$-module of rank 1. Deduce that $\mathrm{Aut}\, T_\mathfrak{p}(E)$ is isomorphic to $R_\mathfrak{p}^* \cong \mathbb{Z}_p^*$.

(b) Let $L_\mathfrak{p}$ be the compositum of the fields $L(E[\mathfrak{p}^n])$ for all $n \geq 1$ or, equivalently, the field defined by

$$\mathrm{Gal}(\bar{L}/L_\mathfrak{p}) = \ker\{\mathrm{Gal}(\bar{L}/L) \longrightarrow \mathrm{Aut}\, T_\mathfrak{p}(E)\}.$$

Prove that there is a finite extension L'/L contained in $L_\mathfrak{p}$ such that

$$\mathrm{Gal}(L_\mathfrak{p}/L') \cong \mathbb{Z}_p.$$

(c) Let $\Gamma = \mathbb{Z}_p$, and define a ring $\mathbb{Z}_p[\![\Gamma]\!]$ to be the inverse limit

$$\mathbb{Z}_p[\![\Gamma]\!] \stackrel{\mathrm{def}}{=} \varprojlim \mathbb{Z}_p[\Gamma/p^n\Gamma].$$

(Note that $\mathbb{Z}_p[\![\Gamma]\!]$ is not the same as the group ring $\mathbb{Z}_p[\Gamma]$.) Prove that

$$R_\mathfrak{p}[\![\mathrm{Gal}(L_\mathfrak{p}/L')]\!] \cong \mathbb{Z}_p[\![\Gamma]\!],$$

and hence that $T_\mathfrak{p}(E)$ is a $\mathbb{Z}_p[\![\Gamma]\!]$-module.

(d) Prove that $\mathbb{Z}_p[\![\Gamma]\!]$ is isomorphic to the power series ring $\mathbb{Z}_p[\![T]\!]$. (*Hint.* Let $\gamma \in \Gamma$ be a topological generator, send γ^n to $(1 + T)^n$, and show that this extends to an isomorphism.) $\mathbb{Z}_p[\![\Gamma]\!]$-modules are often called *Iwasawa modules*.

(e) Let L_p be the field of definition of $T_p(E) = \varprojlim E[p^n]$. Prove that there is a finite extension L''/L contained in L_p such that

$$\mathrm{Gal}(L_p/L'') \cong \mathbb{Z}_p \times \mathbb{Z}_p.$$

(*Hint.* Write $pR_K = \mathfrak{p}\mathfrak{p}'$. Show that $L_p = L_\mathfrak{p} L_{\mathfrak{p}'}$, and that $L_\mathfrak{p} \cap L_{\mathfrak{p}'}$ is a finite extension of L.)

CHAPTER III

Elliptic Surfaces

Elliptic surfaces appear in many guises. They are one-parameter algebraic families of elliptic curves, they are algebraic surfaces containing a pencil of elliptic curves, and they are elliptic curves over one-dimensional function fields. In this chapter we will see elliptic surfaces arising in all of these ways. Since our emphasis in this book is primarily on arithmetic questions, we will concentrate on those properties of elliptic surfaces which resemble the arithmetic properties of elliptic curves defined over number fields. This means we will neglect many of the fascinating geometric questions raised by the study of elliptic surfaces over algebraically closed fields, especially the classical theory of elliptic surfaces defined over $\mathbb{C}$. The interested reader will find a nice introduction to this material in Beauville [1], Griffiths-Harris [1, Ch. 4, §5] and Miranda [1].

We will also find it necessary to restrict attention to fields of characteristic zero. We do this in order to apply the results from [AEC], especially Chapters I, II, and III, to elliptic curves defined over the function field $k(C)$ of a projective curve C/k. All of the main theorems in [AEC] were proven for elliptic curves E/K under the assumption that the field K is perfect; if k has characteristic $p > 0$, then the field $k(C)$ will certainly not be perfect.

However, we must in all honesty point out that elliptic curves defined over non-perfect fields such as $\mathbb{F}_q(T)$ are also extensively studied. (Equivalently, one would study elliptic surfaces $\mathcal{E} \to \mathbb{P}^1$ defined over $\mathbb{F}_q$.) In fact, since the rings $\mathbb{F}_q[T]$ and $\mathbb{Z}$ share the property that all of their residue fields are finite, elliptic curves defined over $\mathbb{F}_q(T)$ behave arithmetically very much like elliptic curves defined over $\mathbb{Q}$. Thus conjectures about elliptic curves defined over $\mathbb{Q}$ are often first tested and proven in the easier setting of elliptic curves over $\mathbb{F}_q(T)$. Notice that $k(T)$ does not have this property if $\mathrm{char}(k) = 0$, so we will find that the theory of elliptic surfaces in characteristic zero differs in some respects from the theory of elliptic curves over number fields.

The main results proven in this chapter are the Mordell-Weil theorem for elliptic surfaces (6.1), two constructions of the canonical height pairing (4.3, 9.3), and specialization theorems for the canonical height

(11.1, 11.3.1) and for the homomorphism from sections to points on fibers (11.4). Unfortunately, it will not be possible to prove all of the background results we need from algebraic geometry. However, we will give a precise statement of the results we use and give at least some indication of the proofs. Briefly, we will use abelian varieties and Jacobian varieties in §2, rational maps between varieties in §3, intersection theory and minimal models of surfaces in §§7 and 8, and divisors on varieties in §10. Much of the material we need is contained in Hartshorne [1], but in any case, we will give references for all assertions that we do not prove.

§1. Elliptic Curves over Function Fields

One way to define an elliptic surface is as a one-parameter algebraic family of elliptic curves. In this guise we have already seen numerous examples of elliptic surfaces. For example, during the proof of [AEC III.1.4(c)] we wrote down the elliptic curve

$$E : y^2 + xy = x^3 - \frac{36}{j_0 - 1728} x - \frac{1}{j_0 - 1728}$$

with j-invariant j_0. In reality, E is a family of elliptic curves, one for each choice of the parameter j_0 (except that E is singular or non-existent for $j_0 = 0$ and $j_0 = 1728$). Similarly, in [AEC IX §7] and [AEC X §6] we looked at the elliptic curves

$$y^2 = x^3 + D \qquad \text{and} \qquad y^2 = x^3 + Dx$$

for varying values of D. Again these are families of elliptic curves, in this case parametrized by D, and each value of D other than $D = 0$ gives an elliptic curve.

More generally, if k is any field (of characteristic not equal to 2) and if $A(T), B(T) \in k(T)$ are rational functions of the parameter T, then we can look at the family of elliptic curves

$$E_T : y^2 = x^3 + A(T)x + B(T).$$

For most values of $t \in \bar{k}$ we can substitute $T = t$ to get an elliptic curve

$$E_t : y^2 = x^3 + A(t)x + B(t).$$

Precisely, E_t will be an elliptic curve provided

$$A(t) \neq \infty, \quad B(t) \neq \infty, \quad \text{and} \quad \Delta(t) = -16\big(4A(t)^3 + 27B(t)^2\big) \neq 0.$$

Later in this chapter we will pursue further this idea of an algebraic family of elliptic curves. But for now we want to alter our perspective a bit. Rather than considering the equation

$$E : y^2 = x^3 + A(T)x + B(T)$$

as defining a family of elliptic curves, one for each value of T, we will instead view E as a single elliptic curve defined over the field $k(T)$. As long as the discriminant

$$\Delta(T) = -16\big(4A(T)^3 + 27B(T)^2\big) \neq 0 \quad \text{in } k(T),$$

E will be an elliptic curve defined over the field $k(T)$. So we will be able to apply much of the general theory developed in [AEC] to the elliptic curve $E/k(T)$, at least provided that the field $k(T)$ is perfect. For this reason we will henceforth make the assumption:

$$\boxed{k \text{ is a field of characteristic zero.}}$$

(For further comments about this assumption, see the introduction to this chapter.)

Example 1.1.1. Consider the elliptic curve $E/\mathbb{Q}(T)$ given by the Weierstrass equation

$$E : y^2 = x^3 - T^2 x + T^2$$

with discriminant

$$\Delta = 16T^4(4T^2 - 27).$$

This curve has the rational point

$$P = (T, T) \in E\big(\mathbb{Q}(T)\big),$$

and one can easily use the addition law to compute

$$2P = (T^2 - 2T, -T^3 + 3T^2 - T),$$

$$3P = \left(\frac{T^3 - 2T^2 - 3T + 4}{(T-3)^2}, \frac{3T^4 - 15T^3 + 21T^2 - 9T + 8}{(T-3)^3} \right).$$

If we substitute $T = t$ for some $t \in \bar{\mathbb{Q}}$, then we will obtain an elliptic curve E_t unless $t = 0$ or $t = \pm\frac{3}{2}\sqrt{3}$.

Example 1.1.2. The elliptic curve

$$E : y^2 = x^3 + (T^2 - 1)x + T^2$$

has many rational points defined over $\mathbb{Q}(T)$, such as the point $(-1, 0)$ of order 2 and the point $(0, T)$ of infinite order. However, if we replace $\mathbb{Q}$ by $\mathbb{Q}\big(\sqrt{2}\big)$, we find a new point of infinite order:

$$(1, \sqrt{2}\,T) \in E\big(\mathbb{Q}(\sqrt{2}\,)(T)\big).$$

In general, if $E/\mathbb{Q}(T)$ is a non-constant elliptic curve (i.e., $j(E) \notin \mathbb{Q}$), then there exists a finite extension $k/\mathbb{Q}$ so that

$$E\big(k(T)\big) = E\big(\mathbb{C}(T)\big).$$

In practice it is often difficult to find k. See Kuwata [1,2] and exercise 3.17. Shioda [1] has constructed an interesting example for which the action of $\mathrm{Gal}(k/\mathbb{Q})$ on $E\big(k(T)\big)$ is a representation of type E_8, so $[k : \mathbb{Q}]$ may be quite large.

We have been taking the coefficients A and B of the elliptic curve

$$E : y^2 = x^3 + Ax + B$$

to lie in the field of rational functions $k(T)$. This is unnecessarily restrictive. We observe that $k(T)$ is the function field of the projective line $\mathbb{P}^1$, so we might consider choosing A and B from the function field of some other curve. Thus we can take a non-singular projective curve C/k and look at elliptic curves E defined over the field $k(C)$.

Example 1.1.3. Let $C/\mathbb{Q}$ be the (elliptic) curve

$$C : s^2 - s = t^3 - t^2;$$

that is, C is the projective curve corresponding to this affine equation. Then the equation

$$E : y^2 + (st + t - s^2)xy + s(s-1)(s-t)t^2y = x^3 + s(s-1)(s-t)tx^2$$

defines an elliptic curve E over the function field $\mathbb{Q}(C)$ of C. Notice that E contains the rational point

$$P = (0,0) \in E\big(\mathbb{Q}(C)\big).$$

It is not hard to verify (at least if you have access to a computer with a symbolic processor) that

$$[11]P = O,$$

so P is a point of order 11.

In fact, E is in some sense the universal family of elliptic curves containing a point of order 11. This means the following: Let A be any elliptic curve and $Q \in A$ a point of order 11. Then there is a unique point $(s_0, t_0) \in C$ such that if we substitute $(s, t) = (s_0, t_0)$ into the equations for E and P, we will obtain an elliptic curve E_0 and a point $P_0 \in E_0$ of order 11 such that there is an isomorphism $\phi : A \to E_0$ with $\phi(Q) = P_0$. In the literature, the curve C is called the modular curve $X_1(11)$, and E is usually denoted $E_1(11)$. For further details, see exercise 3.2.

§2. The Weak Mordell-Weil Theorem

Our task in this section is to prove the following weak Mordell-Weil theorem for elliptic curves defined over function fields. We emphasize again our assumption that the constant field k always has characteristic 0.

Theorem 2.1. (Weak Mordell-Weil Theorem) *Let k be an algebraically closed field, let $K = k(C)$ be the function field of a curve, and let E/K be an elliptic curve. Then the quotient group $E(K)/2E(K)$ is finite.*

The theory of function fields of curves is analogous to the theory of number fields, and the proof of the weak Mordell-Weil theorem is similar in both cases. In other words, we could save space here by merely quoting the proof given in [AEC, VIII §1] with the words "number field" replaced by the words "function field." However, there are enough differences that we feel it is worthwhile giving the proof. We will place our main emphasis on highlighting the differences between the two cases.

Recall that the proof for number fields has two main steps. The first step [AEC, VIII.1.5] depends on properties of the elliptic curve E/K. It says that the extension field $L = K\left([m]^{-1}E(K)\right)$ is an abelian extension of K, has exponent m, and is unramified outside a certain finite set of primes S. This step carries over word-for-word to the function field case once one has developed the theory of valuations. We will give a slight variant of the argument in [AEC, VIII §1], using divisors supported on a finite set of points, but the reader will have no trouble seeing that this is merely a matter of using geometric language to deal with the same ideas. We will also consider only the case $m = 2$, since this allows us to work more directly with the equation for E/K.

The second step [AEC, VIII.1.6] has nothing to do with elliptic curves. Instead, one uses Kummer theory to show that the maximal abelian extension of K of exponent m unramified outside of S is a finite extension. The proof of this proposition is not hard, but it ultimately relies upon the two fundamental finiteness theorems of algebraic number theory, namely the finiteness of the class group and the finite generation of the unit group. In general, neither of these last two results is true for function fields.

For example, if $K = k(C)$ is a function field, then the "unit group" in K^* is the constant field k^*. To see that this is the right analogy, note that for a number field K, the unit group can be described as the set of elements $\alpha \in K^*$ satisfying $v(\alpha) = 0$ for all discrete valuations on K^*. But the discrete valuations on a function field $K = k(C)$ correspond to the points of $C(k)$ (at least if k is algebraically closed). Thus if a function $f \in K$ has valuation 0 for all valuations, then it has no zeros or poles on C, so by [AEC, II.1.2] it is a constant. Hence the "unit group" of K will be k^*, and in general k^* will not be finitely generated.

Similarly, we will see during the proof of the weak Mordell-Weil theorem that the "ideal class group" of a function field $K = k(C)$ is the Picard group $\text{Pic}(C)$, that is, the group of divisors modulo linear equivalence. The Picard group need not be finitely generated; see (2.6) below.

However, all is not lost. A closer examination of [AEC, VIII.1.6] shows that it does not require the full strength of the finiteness theorems. Instead we used the facts that the ideal class group has only finitely many elements of order m and the unit group R^* has the property that the quotient R^*/R^{*m} is finite. These weaker results are true for function fields under appropriate assumptions on the constant field k of K. For example, if k is algebraically closed, then k^*/k^{*m} is certainly finite, since it is actually trivial.

Similarly, the Picard group $\text{Pic}(C)$ has only finitely many elements of order m. Unfortunately, the proof of this last statement requires results from the theory of Jacobians and abelian varieties which we will not be able to develop in full. So we will just state here the proposition that we need and postpone until the end of the section a sketch of the proof. For the proof of the weak Mordell-Weil theorem (2.1), we will only need to use the $m = 2$ case of the following proposition.

Proposition 2.2. *Let C be a non-singular projective curve defined over an algebraically closed field k. Then for any integer $m \geq 1$, the Picard group $\text{Pic}(C)$ has only finitely many elements of order m.*

PROOF. See (2.7) at the end of this section for a complete description of the torsion subgroup of $\text{Pic}(C)$. In particular, if C has genus g, then (2.7) implies that $\text{Pic}(C)[m]$ is isomorphic to $(\mathbb{Z}/m\mathbb{Z})^{2g}$.   □

PROOF (of Theorem 2.1). Our first observation is that if L/K is a finite Galois extension and if we can prove that $E(L)/2E(L)$ is finite, then it will follow that $E(K)/2E(K)$ is also finite. This is the content of [AEC, VIII.1.1.1], and we gave another proof using Galois cohomology in [AEC, VIII §2]. We leave it to the reader to verify that these proofs made no use of the assumption that the field K is a number field, so they are also valid in our case. Note that [AEC, II.2.5] ensures that any such L will be the function field of a curve over k.

Replacing K by a finite extension and C with the corresponding curve, it thus suffices to prove (2.1) under the assumption that $E(K)$ contains all of the points of order 2. Equivalently, we may assume that E has a Weierstrass equation of the form

$$E : y^2 = (x - e_1)(x - e_2)(x - e_3) \qquad \text{with } e_1, e_2, e_3 \in K.$$

Consider the map

$$\phi : E(K)/2E(K) \longrightarrow (K^*/K^{*2}) \times (K^*/K^{*2})$$

defined by

$$\phi : P = (x, y) \longrightarrow \begin{cases} (x - e_1, x - e_2) & \text{if } x \neq e_1, e_2, \\ ((e_1 - e_3)(e_1 - e_2), e_1 - e_2)) & \text{if } x = e_1, \\ (e_2 - e_1, (e_2 - e_3)(e_2 - e_1)) & \text{if } x = e_2, \\ (1, 1) & \text{if } x = \infty \; (P = O). \end{cases}$$

In the case that K is a number field, we proved in [AEC, X.1.4] that ϕ is an injective homomorphism, and the same proof works for an arbitrary field K.

The map ϕ can also be defined using group cohomology. We briefly sketch the proof. Taking $G_{\bar{K}/K}$-cohomology of the exact sequence

$$0 \longrightarrow E[2] \longrightarrow E(\bar{K}) \xrightarrow{[2]} E(\bar{K}) \longrightarrow 0$$

gives an exact sequence

$$E(K) \xrightarrow{[2]} E(K) \xrightarrow{\delta} H^1(G_{\bar{K}/K}, E[2]).$$

Now our assumption that $E[2] \subset E(K)$ implies that there are isomorphisms

$$H^1(G_{\bar{K}/K}, E[2]) \cong H^1(G_{\bar{K}/K}, (\mathbb{Z}/2\mathbb{Z})^2) \cong \mathrm{Hom}(G_{\bar{K}/K}, (\mathbb{Z}/2\mathbb{Z})^2)$$
$$\cong \mathrm{Hom}(G_{\bar{K}/K}, \mathbb{Z}/2\mathbb{Z})^2 \cong \mathrm{Hom}(G_{\bar{K}/K}, \mu_2)^2 \cong H^1(G_{\bar{K}/K}, \mu_2)^2.$$

Finally, the Kummer sequence for fields and Hilbert's theorem 90 give an isomorphism [AEC, VIII.2.2]

$$K^*/K^{*2} \xrightarrow{\delta} H^1(G_{\bar{K}/K}, \mu_2).$$

Combining the above maps, we obtain an injective homomorphism

$$E(K)/2E(K) \xrightarrow{\delta} H^1(G_{\bar{K}/K}, E[2]) \cong H^1(G_{\bar{K}/K}, \mu_2)^2$$
$$\xrightarrow{\delta^{-1} \times \delta^{-1}} (K^*/K^{*2}) \times (K^*/K^{*2}).$$

Of course, one still needs to unscramble the connecting homomorphisms and check that this map is the same as the map ϕ defined above. See [AEC, X §1] for the details.

We will prove that $E(K)/2E(K)$ is finite by proving that the image $\phi(E(K)/2E(K))$ is finite. The basic idea is to show that for any $P \in E(K)$, the two coordinates of $\phi(P)$ are almost squares in K. The following lemma quantifies this assertion.

Lemma 2.3.1. *Let k be an algebraically closed field, let $K = k(C)$ be the function field of a curve, let E/K be an elliptic curve, and suppose that E has a Weierstrass equation of the form*

$$E : y^2 = (x - e_1)(x - e_2)(x - e_3) \qquad \text{with } e_1, e_2, e_3 \in K.$$

Let $S \subset C$ be the set of points where any one of e_1, e_2, e_3 has a pole, together with the points where $\Delta = (e_1 - e_2)^2(e_1 - e_3)^2(e_2 - e_3)^2$ vanishes. Then for any point $P = (x, y) \in E(K)$ with $x \neq e_1$,

$$\mathrm{ord}_t(x - e_1) \equiv 0 \,(\mathrm{mod}\ 2) \qquad \text{for all } t \in C \text{ with } t \notin S.$$

Here $\mathrm{ord}_t : k(C)^* \to \mathbb{Z}$ is the normalized valuation on $k(C)$ which measures the order of vanishing of a function at t [AEC II §1].

PROOF. Let $t \in C$ with $t \notin S$, and let $n = \mathrm{ord}_t(x - e_1)$. Our choice of S implies that $\mathrm{ord}_t(e_i) \geq 0$. We consider three cases. First, if $n = 0$, then clearly $\mathrm{ord}_t(x - e_1) = n \equiv 0 \,(\mathrm{mod}\ 2)$.

Second, if $n < 0$, then t must be a pole of x and $n = \mathrm{ord}_t(x)$. It follows that

$$n = \mathrm{ord}_t(x - e_1) = \mathrm{ord}_t(x - e_2) = \mathrm{ord}_t(x - e_3).$$

Using the Weierstrass equation for E, we find that

$$2\,\mathrm{ord}_t(y) = \mathrm{ord}_t(y^2) = \mathrm{ord}_t((x - e_1)(x - e_2)(x - e_3)) = 3n,$$

which proves that $\mathrm{ord}_t(x - e_1) = n \equiv 0 \,(\mathrm{mod}\ 2)$.

Third, suppose that $n > 0$. This means that $x - e_1$ vanishes at t. We claim that $x - e_2$ and $x - e_3$ do not vanish at t. To see this, we let $i = 2$ or 3 and use the triangle inequality to compute

$$\min\{\mathrm{ord}_t(x-e_1), \mathrm{ord}_t(x-e_i)\} \leq \mathrm{ord}_t((x-e_1)-(x-e_i)) = \mathrm{ord}_t(e_i-e_1) = 0.$$

The last equality follows from the assumption that e_1, e_2, e_3 do not have poles at t and Δ does not vanish at t. But $\mathrm{ord}_t(x - e_1) = n \geq 1$, so we get $\mathrm{ord}_t(x - e_2) = \mathrm{ord}_t(x - e_3) = 0$. Therefore

$$2\,\mathrm{ord}_t(y) = \mathrm{ord}_t(y^2) = \mathrm{ord}_t((x - e_1)(x - e_2)(x - e_3)) = \mathrm{ord}_t(x - e_1).$$

Hence $\mathrm{ord}_t(x - e_1) \equiv 0 \,(\mathrm{mod}\ 2)$, which completes the proof of (2.3.1).

$$\square$$

We now resume the proof of (2.1). Let S be as in the statement of (2.3.1), and define a subgroup of K^*/K^{*2} by

$$K(S, 2) = \{f \in K^*/K^{*2} : \mathrm{ord}_t(f) \equiv 0 \,(\mathrm{mod}\ 2) \text{ for all } t \notin S\}.$$

(Compare with [AEC, X.1.4].) Then (2.3.1) tells us that the image of ϕ lies in $K(S, 2) \times K(S, 2)$, so we have an injective homomorphism

$$\phi : E(K)/2E(K) \longrightarrow K(S, 2) \times K(S, 2).$$

In order to complete the proof of (2.1), it suffices to prove that the group $K(S, 2)$ is finite. Note that the finiteness of $K(S, 2)$ is an assertion about the curve C; it has nothing to do with the elliptic curve E. We record this statement in the following lemma, whose proof also completes the proof of the weak Mordell-Weil theorem (2.1).

Lemma 2.3.2. *Let k be an algebraically closed field, let $K = k(C)$ be the function field of a curve, let $S \subset C$ be a finite set of points, and let $m \geq 1$ be an integer. Then the group*

$$K(S, m) = \left\{ f \in K^*/K^{*m} : \operatorname{ord}_t(f) \equiv 0 \,(\operatorname{mod} m) \text{ for all } t \notin S \right\}$$

is a finite subgroup of K^*/K^{*m}.

PROOF. Let $s = \#S$. Then there is an exact sequence

$$0 \longrightarrow K(\emptyset, m) \longrightarrow K(S, m) \longrightarrow (\mathbb{Z}/m\mathbb{Z})^s.$$
$$f \longmapsto \left(\operatorname{ord}_t(f) \right)_{t \in S}$$

It thus suffices to prove that

$$K(\emptyset, m) = \left\{ f \in K^*/K^{*m} : \operatorname{ord}_t(f) \equiv 0 \,(\operatorname{mod} m) \text{ for all } t \in C \right\}$$

is finite.

Let $f \,(\operatorname{mod} K^{*m}) \in K(\emptyset, m)$. Then $\operatorname{div}(f)$ has the form

$$\operatorname{div}(f) = mD_f \quad \text{for some } D_f \in \operatorname{Div}(C).$$

Notice that if we take some other representative fg^m for the coset of f in K^*/K^{*m}, then $D_{fg^m} = D_f + \operatorname{div}(g)$ changes by a principal divisor. Thus the divisor class of D_f in $\operatorname{Pic}(C)$ is independent of the choice of representative. Further, $mD_f = \operatorname{div}(f)$ is itself principal, so we get a well-defined homomorphism

$$K(\emptyset, m) \longrightarrow \operatorname{Pic}(C)[m], \qquad f \,(\operatorname{mod} K^{*m}) \longmapsto \operatorname{class}(D_f),$$

where $\operatorname{Pic}(C)[m]$ denotes the elements of $\operatorname{Pic}(C)$ of order dividing m.

Now suppose that $f \,(\operatorname{mod} K^{*m})$ is in the kernel of this homomorphism. Then D_f is principal, say $D_f = \operatorname{div}(F_f)$ for some function $F_f \in K^*$, which means that

$$\operatorname{div}(f F_f^{-m}) = \operatorname{div}(f) - m \operatorname{div}(F_f) = \operatorname{div}(f) - mD_f = 0.$$

Thus $f F_f^{-m}$ has no zeros or poles, so [AEC, II.3.1] tells us that it is constant. Using the assumption that k is algebraically closed, we can write this constant as c^m, so $f = (cF_f)^m$. In other words, $f \equiv 0 \,(\operatorname{mod} K^{*m})$. This proves that the homomorphism $K(\emptyset, m) \to \operatorname{Pic}(C)[m]$ is injective. Finally, we use the fact (2.2) that $\operatorname{Pic}(C)[m]$ is finite to conclude that $K(\emptyset, m)$ is finite. $\square$

In the remainder of this section we briefly discuss abelian varieties and Jacobians, including a sketch of the proof that the Picard group of a curve has only finitely many elements of a given order.

Definition. An *abelian variety* consists of a non-singular projective variety A, a point $O \in A$, and two morphisms

$$\mu : A \times A \longrightarrow A, \qquad i : A \longrightarrow A,$$

which make the points of A into an abelian group. In other words,

(i) $\mu(O, P) = \mu(P, O) = P$ for all $P \in A$,

(ii) $\mu(P, i(P)) = O$ for all $P \in A$,

(iii) $\mu(\mu(P, Q), R) = \mu(P, \mu(Q, R))$ for all $P, Q, R \in A$,

(iv) $\mu(P, Q) = \mu(Q, P)$ for all $P, Q \in A$.

If A, μ, and i are defined over a field k, and $O \in A(k)$, then we say that A *is defined over k*. Basic references for the theory of abelian varieties are Griffiths-Harris [1, Ch. 2, §6], Milne [2], and Mumford [1], but we will need very little of the general theory.

Example 2.4.1. An elliptic curve is an abelian variety of dimension one. The fact that the addition and negation operations on an elliptic curve satisfy (i)–(iv) is [AEC, III.2.2], and the fact that they are morphisms is [AEC, III.3.6]. Conversely, every abelian variety of dimension one is an elliptic curve, see exercise 3.5.

Example 2.4.2. Let A be an abelian variety of dimension d defined over $\mathbb{C}$. Then one can show that there is a lattice $\Lambda \subset \mathbb{C}^d$ and a complex analytic isomorphism $A(\mathbb{C}) \cong \mathbb{C}^d/\Lambda$. By lattice we mean a full sublattice, that is, a free subgroup of $\mathbb{C}^d$ of rank $2d$ which contains an $\mathbb{R}$-basis for $\mathbb{C}^d$. The isomorphism $A(\mathbb{C}) \cong \mathbb{C}^d/\Lambda$ is an isomorphism both as complex manifolds and as abelian groups. For elliptic curves, the existence of this isomorphism is the essential content of the uniformization theorem [AEC, VI.5.1.1]. Further, in the one dimensional case every lattice $\Lambda \subset \mathbb{C}$ corresponds to an elliptic curve [AEC, VI.3.6]. But in higher dimensions, a complex torus $\mathbb{C}^d/\Lambda$ will only be isomorphic to an abelian variety if the lattice Λ satisfies the Riemann conditions (see Griffiths-Harris [1, Ch. 2, §6]). In other words, there are certain restrictions on the lattice Λ in order for there to be a complex-analytic embedding of $\mathbb{C}^d/\Lambda$ into projective space $\mathbb{P}^n(\mathbb{C})$. Just as in the case of elliptic curves, this complex uniformization of abelian varieties is very useful in analyzing the group structure of $A(\mathbb{C})$.

Remark 2.5. A more succinct way to define abelian varieties is to say that an abelian variety is a group variety in the category of projective varieties. It turns out that the group law forces A to be non-singular. Further, the completeness of the variety A forces the group law to be commutative, so we actually did not need to include property (iv) in our definition. See Mumford [1, pp. 1, 41, 44] for three proofs of this fact.

The next proposition tells us that the Picard group of a curve is essentially an abelian variety, called the Jacobian variety of the curve. Its dimension is equal to the genus of the curve. Some general references for Jacobian varieties include Griffiths-Harris [1, Ch. 2, §§2, 3, 7], Milne [3], and Mumford [2].

Proposition 2.6. *Let C be a non-singular projective curve of genus g defined over an algebraically closed field k.*
(a) *The degree map* $\deg : \mathrm{Div}(C) \to \mathbb{Z}$ *induces an exact sequence*

$$0 \longrightarrow \mathrm{Pic}^0(C) \longrightarrow \mathrm{Pic}(C) \xrightarrow{\deg} \mathbb{Z} \longrightarrow 0,$$

where $\mathrm{Pic}^0(C)$ *is the group of degree 0 divisor classes on C.*
(b) *There exists an abelian variety* $\mathrm{Jac}(C)$ *of dimension g and a natural isomorphism of groups*

$$\mathrm{Pic}^0(C) \xrightarrow{\ \sim\ } \mathrm{Jac}(C).$$

(For the meaning of "natural", see (2.6.1) below.) $\mathrm{Jac}(C)$ *is called the Jacobian variety of C.*

PROOF. (a) The exact sequence merely defines the subgroup $\mathrm{Pic}^0(C)$, so all we need to do is verify that the degree map is well-defined on $\mathrm{Pic}(C)$. This follows immediately from the fact that every principal divisor has degree 0 [AEC, II.3.1].
(b) (Proof Sketch) We start with the two easy cases. First, if $g = 0$, then every divisor of degree 0 on $C \cong \mathbb{P}^1$ is principal [AEC, II.3.2]. It follows that $\mathrm{Pic}^0(C) = 0$, which is the desired result in this case.

Next suppose that $g = 1$. Fixing a point $O \in C(k)$, we turn C/k into an elliptic curve, so C is an abelian variety (2.4.1). Then there is a natural group isomorphism $\mathrm{Pic}^0(C) \xrightarrow{\sim} C$ as described in [AEC, III.3.4]. Hence C is its own Jacobian variety.

For curves of higher genus, one can construct the Jacobian variety analytically if $k = \mathbb{C}$ or algebraically in general. We briefly describe both approaches. For the algebraic method, we fix a basepoint $P_0 \in C$ and consider the map

$$\phi_g : C^g \longrightarrow \mathrm{Pic}^0(C), \qquad (P_1, \ldots, P_g) \longmapsto (P_1) + \cdots + (P_g) - g(P_0).$$

Let the symmetric group S_g act on C^g by permuting the coordinates. The map ϕ_g is clearly invariant under this action, so it induces a map on the quotient $C^{(g)} \stackrel{\text{def}}{=} C^g/S_g$. One checks that $C^{(g)}$ is a non-singular variety, that $\phi_g : C^{(g)} \to \mathrm{Pic}^0(C)$ is surjective, and that ϕ_g is injective off of a Zariski closed subset of $C^{(g)}$. Further, the group law on $\mathrm{Pic}^0(C)$ induces an algebraic (i.e., rational) map $C^{(g)} \times C^{(g)} \to C^{(g)}$. Unfortunately, this rational map is not defined everywhere, so one takes certain "group chunks" and glues together enough translates to form a group variety. This idea of gluing together group chunks is due to André Weil [3] and works in all characteristics. We refer the reader to the discussion in (IV, §6) for further details. See also exercise 4.29.

The analytic approach to constructing the Jacobian for curves over $\mathbb{C}$ is much older. The Riemann-Roch theorem [AEC, II.5.3, II.5.5a] says that on a curve of genus g, the space of holomorphic differential forms has dimension g. Let $\omega_1, \ldots, \omega_g$ be a basis for this space. Next, a curve C of genus g over $\mathbb{C}$ is a Riemann surface with g holes, so there are $2g$ independent cycles on C. Let $\Gamma_1, \ldots, \Gamma_{2g}$ be a basis for the space of cycles; that is, $\Gamma_1, \ldots, \Gamma_{2g}$ is a basis for the first homology $H_1(C, \mathbb{Z})$. We fix a basepoint $P_0 \in C$ and consider the map

$$\phi : C \longrightarrow \mathbb{C}^g, \qquad P \longmapsto \left(\int_{P_0}^{P} \omega_1, \ldots, \int_{P_0}^{P} \omega_g \right).$$

Here the integrals are to be computed along some path from P_0 to P. Unfortunately, the value of the integrals is not path-independent! (See [AEC, VI, §1] for a discussion when $g = 1$.)

In order to salvage this idea, we consider the subgroup $\Lambda \subset \mathbb{C}^g$ defined to be the image of the map

$$H_1(C, \mathbb{Z}) \longrightarrow \mathbb{C}^g, \qquad \Gamma \longmapsto \left(\int_{\Gamma} \omega_1, \ldots, \int_{\Gamma} \omega_g \right).$$

Then the integrals give a well-defined map $\phi : C \to \mathbb{C}^g/\Lambda$, since Λ eliminates the path dependence of the integrals. Next one proves that Λ is a lattice which satisfies Riemann's conditions (2.4.2), so $\mathbb{C}^g/\Lambda$ is complex-analytically isomorphic to an abelian variety. Denoting this abelian variety by $\mathrm{Jac}(C)$, one verifies that the map $\phi : C \to \mathrm{Jac}(C)$ is a morphism of algebraic varieties.

Extending ϕ by linearity gives a map

$$\phi : \mathrm{Div}^0(C) \longrightarrow \mathrm{Jac}(C), \qquad \sum n_i(P_i) \longrightarrow \sum [n_i]\phi(P_i).$$

In other words, use ϕ to map the points in a divisor to points of $\mathrm{Jac}(C)$, and then use the group law on $\mathrm{Jac}(C)$ to add them up. Finally, the theorems of Abel and Jacobi say that the map $\phi : \mathrm{Div}^0(C) \to \mathrm{Jac}(C)$ is a surjective homomorphism whose kernel consists of precisely the principal divisors. Hence ϕ induces the desired isomorphism $\mathrm{Pic}^0(C) \xrightarrow{\sim} \mathrm{Jac}(C)$. For further details, see Griffiths-Harris [1, Ch. 2, §§2,7].   □

Remark 2.6.1. What do we mean in (2.6b) when we say that the isomorphism $\mathrm{Pic}^0(C) \to \mathrm{Jac}(C)$ is "natural"? Recall [AEC, II.3.7] that a morphism $\phi : C_1 \to C_2$ of curves induces a homomorphism $\phi^* : \mathrm{Pic}^0(C_2) \to \mathrm{Pic}^0(C_1)$ of their Picard groups. Then one can prove that the corresponding map $\phi^* : \mathrm{Jac}(C_2) \to \mathrm{Jac}(C_1)$ is a morphism of varieties. In fancy language, the association $C \mapsto \mathrm{Jac}(C)$ is a functor from the category of (non-singular projective) curves to the category of abelian varieties.

Remark 2.6.2. If C is defined over an arbitrary field k, then its Jacobian variety $\mathrm{Jac}(C)$ will be an abelian variety defined over k. Further, the group isomorphism $\mathrm{Pic}^0(C) \to \mathrm{Jac}(C)$ will commute with the action of the Galois group $G_{\bar{k}/k}$. This is another way in which the Jacobian is a natural object. For example, if C is a curve of genus 1 with $C(k) = \emptyset$, then C cannot be its own Jacobian variety. However, we can always find an elliptic curve E/k so that C/k is a homogeneous space for E [AEC, exercise 10.3]. Then [AEC, X.3.8] says that there is a group isomorphism $\mathrm{Pic}^0(C) \xrightarrow{\sim} E$, so E is the Jacobian of C.

Remark 2.6.3. For hyperelliptic curves, it is possible to construct the Jacobian variety quite explicitly, see Mumford [3, Ch. IIIa, §§2,3]. In this case one can also precisely describe all of the elements of order 2 in $\mathrm{Pic}(C)$, see exercise 3.38.

The following corollary of (2.6) and (2.4.2) is a strengthened version of (2.2).

Corollary 2.7. *Let C/k be a non-singular projective curve of genus g defined over an algebraically closed field k. Then*

$$\mathrm{Pic}(C)_{\mathrm{tors}} \cong (\mathbb{Q}/\mathbb{Z})^{2g}.$$

In particular, for any integer $m \geq 1$, $\mathrm{Pic}(C)[m] \cong (\mathbb{Z}/m\mathbb{Z})^{2g}$, so $\mathrm{Pic}(C)$ has only finitely many elements of order dividing m.

PROOF. The field k has characteristic 0 by assumption, so the Lefschetz principle [AEC, VI §6] says that we may take k to be a subfield of $\mathbb{C}$. Let $J = \mathrm{Jac}(C)$ be the Jacobian variety of C. Then (2.6) tells us that $\mathrm{Pic}(C)_{\mathrm{tors}} \cong \mathrm{Pic}^0(C)_{\mathrm{tors}} \cong J_{\mathrm{tors}}$. On the other hand, (2.4.2) implies that there is a lattice $\Lambda \subset \mathbb{C}^g$ and a complex-analytic group isomorphism $J(\mathbb{C}) \cong \mathbb{C}^g/\Lambda$. (See also the proof of (2.6) for a concrete description of the isomorphism $\mathrm{Pic}^0(C) \xrightarrow{\sim} \mathbb{C}^g/\Lambda$.)

As abstract groups, $\mathbb{C}^g \cong \mathbb{R}^{2g}$ and $\Lambda \cong \mathbb{Z}^{2g}$, and if we use a $\mathbb{Z}$-basis for Λ as our $\mathbb{R}$-basis for $\mathbb{C}^g$, then we obtain a group isomorphism of the quotients $\mathbb{C}^g/\Lambda \cong (\mathbb{R}/\mathbb{Z})^{2g}$. Hence

$$\mathrm{Pic}(C)_{\mathrm{tors}} \cong J_{\mathrm{tors}} \cong J(\mathbb{C})_{\mathrm{tors}} \cong (\mathbb{C}^g/\Lambda)_{\mathrm{tors}} \cong (\mathbb{R}/\mathbb{Z})^{2g}_{\mathrm{tors}} \cong (\mathbb{Q}/\mathbb{Z})^{2g}.$$

This proves the first assertion of (2.7), and the other assertions are an immediate consequence.   □

§3. Elliptic Surfaces

We now return to the idea that an elliptic surface should be a one-parameter family of elliptic curves. For example, we might consider a family

$$E_T : y^2 = x^3 + A(T)x + B(T)$$

with rational functions $A(T), B(T) \in k(T)$. Or, more generally, we could fix a non-singular projective curve C/k and take

$$E : y^2 = x^3 + Ax + B$$

for some $A, B \in k(C)$ with $4A^3 + 27B^2 \neq 0$. Then for almost all points $t \in C(\bar{k})$ we can evaluate A and B at t to get an elliptic curve

$$E_t : y^2 = x^3 + A(t)x + B(t).$$

Suppose now that we do not evaluate A and B at particular points of C, but instead we treat t as a variable just like x and y. In other words, we look at the subset of $\mathbb{P}^2 \times C$ defined by

$$\mathcal{E} = \left\{ ([X, Y, Z], t) \in \mathbb{P}^2 \times C : Y^2 Z = X^3 + A(t)XZ^2 + B(t)Z^3 \right\}.$$

Note that $\mathcal{E}$ is a subvariety of $\mathbb{P}^2 \times C$ of dimension two; it is a surface formed from a family of elliptic curves.

Remark 3.1. Actually, in defining $\mathcal{E}$ we need to take a little more care with the points $t \in C$ where A or B has a pole. Here's one way to handle this problem. Consider the set

$$\left\{ ([X, Y, Z], t) \in \mathbb{P}^2 \times C : \begin{array}{l} A \text{ or } B \text{ has a pole at } t, \text{ or} \\ Y^2 Z = X^3 + A(t)XZ^2 + B(t)Z^3 \end{array} \right\}.$$

This set will consist of a number of irreducible components, all but one of which will look like

$$\mathbb{P}^2 \times \{t_0\}$$

for a pole t_0 of A or B. We take $\mathcal{E}$ to be the one component not of this form. Equivalently, $\mathcal{E}$ is the Zariski closure in $\mathbb{P}^2 \times C$ of the set

$$\left\{ ([X, Y, Z], t) \in \mathbb{P}^2 \times C : \begin{array}{l} t \text{ is not a pole of } A \text{ or } B, \text{ and} \\ Y^2 Z = X^3 + A(t)XZ^2 + B(t)Z^3 \end{array} \right\}.$$

Since $\mathcal{E}$ is a subvariety of $\mathbb{P}^2 \times C$, projection onto the second factor defines a morphism

$$\pi : \begin{array}{ccc} \mathcal{E} & \longrightarrow & C, \\ ([X, Y, Z], t) & \longmapsto & t. \end{array}$$

And for almost every[†] point $t \in C$, the *fiber*

$$\mathcal{E}_t = \pi^{-1}(t) = \{P \in \mathcal{E} \; : \; \pi(P) = t\}$$

is the curve E_t that we considered earlier. Further, since we have assumed that

$$\Delta = -16(4A^3 + 27B^2) \neq 0 \text{ in } k(C),$$

it will be true that almost every fiber $\mathcal{E}_t$ is an elliptic curve. We just need to choose points $t \in C$ such that $A(t) \neq \infty$, $B(t) \neq \infty$, and $\Delta(t) \neq 0$.

However, our family of elliptic curves $\mathcal{E}$ has one other important property. Recall [AEC III §3] that an elliptic curve is really a pair (E, O), where E is a curve of genus 1, and O is a point of E. The equation we used to define $\mathcal{E}$ gives a one-parameter family of elliptic curves. This means that for almost all values of t we get an elliptic curve $\mathcal{E}_t$, which we should really write as $(\mathcal{E}_t, O_t)$ to emphasize that each $\mathcal{E}_t$ comes equipped with a zero element $O_t \in \mathcal{E}_t$.

The family $\mathcal{E}$ is an *algebraic family*, which is a fancy way of saying that it is given by an equation whose coefficients A and B are algebraic functions, in our case functions on the curve C. The additional property that $\mathcal{E}$ possesses is that the collection of zero elements O_t is an *algebraic family of points*. In other words, we claim that the coordinates of O_t are algebraic functions of t; the coordinates of O_t are in the function field of C. Using the definition of $\mathcal{E}$ given above, we see that

$$O_t = \big([0, 1, 0], t\big) \in \mathcal{E}_t \subset \mathbb{P}^2 \times C.$$

So the coordinates $[0, 1, 0]$ of O_t are actually constant functions.

We can describe this property in another way. Since each fiber $\mathcal{E}_t$ is an elliptic curve with zero element O_t, we get a map

$$\sigma_0 : \quad C \quad \longrightarrow \quad \mathcal{E},$$
$$t \quad \longmapsto \quad O_t.$$

Clearly, this map has the property that

$$\pi\big(\sigma_0(t)\big) = t \qquad \text{for all } t \in C(\bar{k}).$$

Further, the fact that O_t is an algebraic family of points is equivalent to the fact that the map σ_0 is a rational map of varieties. (In fact, since C is a non-singular curve, σ_0 will be a morphism [AEC II.2.1].) This prompts the following definition.

[†] The phrases "for almost every" and "for almost all" are contractions of the expression "for all but finitely many."

Definition. Let $\pi : V \to W$ be a morphism of algebraic varieties. A *section to* π is a morphism

$$\sigma : W \longrightarrow V$$

such that the composition

$$\pi \circ \sigma : W \longrightarrow W$$

is the identity map on W.

Example 3.2. Consider the surface given by the equation

$$\mathcal{E} : Y^2 Z = X^3 - T^2 X Z^2 + T^2 Z^3.$$

(See (1.1) for our original discussion of this equation.) More precisely, let $\mathcal{E}$ be the projective surface in $\mathbb{P}^2 \times \mathbb{P}^1$ corresponding to this equation. The projection π is the map

$$\pi : \mathcal{E} \longrightarrow \mathbb{P}^1, \qquad ([X, Y, Z], T) \longmapsto T.$$

This map has a section

$$\sigma : \mathbb{P}^1 \longrightarrow \mathcal{E}, \qquad T \longmapsto ([T, T, 1], T).$$

To avoid excessive notation, one often says that $\pi : \mathcal{E} \to \mathbb{P}^1$ has the section $\sigma = [T, T, 1]$; one can even use the inhomogeneous equation

$$y^2 = x^3 - T^2 x + T^2$$

for $\mathcal{E}$ and say that $\sigma = (T, T)$ is a section to π.

We are now ready for the formal definition of an elliptic surface.

Definition. Let C be a non-singular projective curve. An *elliptic surface over* C consists of the following data:

    (i) a surface $\mathcal{E}$, by which we mean a two dimensional projective variety,

    (ii) a morphism

$$\pi : \mathcal{E} \longrightarrow C$$

such that for all but finitely many points $t \in C(\bar{k})$, the fiber

$$\mathcal{E}_t = \pi^{-1}(t)$$

is a non-singular curve of genus 1,

    (iii) a section to π,

$$\sigma_0 : C \longrightarrow \mathcal{E}.$$

Let $\mathcal{E} \to C$ be an elliptic surface. The *group of sections of* $\mathcal{E}$ *over* C is denoted by

$$\mathcal{E}(C) = \{\text{sections } \sigma : C \to \mathcal{E}\}.$$

Note that any rational map $C \to \mathcal{E}$ is automatically a morphism, since C is a non-singular curve and $\mathcal{E}$ is a projective variety [AEC, II.2.1], so every section is a morphism. We will see later (3.10) that $\mathcal{E}(C)$ is a group with zero element σ_0.

Remark 3.3.1. Since all but finitely many fibers of an elliptic surface have genus 1, one can show that any fiber which is a non-singular curve will automatically have genus 1. See Hartshorne [1, III.9.13]. These non-singular fibers are often called the *good fibers*. The fibers $\mathcal{E}_t$ which are not non-singular curves will be called the *singular fibers* or the *bad fibers*. Of course, when we refer to the non-singular fiber $\mathcal{E}_t$, we really mean the elliptic curve consisting of the pair $(\mathcal{E}_t, \sigma_0(t))$.

Remark 3.3.2. Our definition of elliptic surface is non-standard in two ways. First, most books require that $\mathcal{E}$ be a non-singular surface. In such a case we will call $\mathcal{E}$ a *non-singular elliptic surface*. Second, most algebraic geometers would define an elliptic surface to be a (non-singular) surface satisfying properties (i) and (ii) of our definition; they would not require that there be a section. This leads to many interesting geometric questions, such as the possible existence of multiple fibers. (See Griffiths-Harris [1, p. 564].) It is only our emphasis on questions with an arithmetic flavor which prompts us to require the existence of at least one section.

Remark 3.3.3. The classical theory of elliptic surfaces deals with surfaces defined over the field $k = \mathbb{C}$, or more generally over an algebraically closed field. We will also want to look at other fields, such as $k = \mathbb{Q}$. We will say that an elliptic surface $\mathcal{E}$ over C is *defined over k* if the curve C is defined over k, the surface $\mathcal{E}$ is defined over k, and both of the maps

$$\pi : E \longrightarrow C \quad \text{and} \quad \sigma_0 : C \longrightarrow E$$

are defined over k. In this case we write

$$\mathcal{E}(C/k) = \{\text{sections } \sigma : C \to \mathcal{E} \text{ such that } \sigma \text{ is defined over } k\}$$

for the group of sections defined over k. For example, the elliptic surface (3.2) is defined over $\mathbb{Q}$, and the section $\sigma = (T, T)$ is in $\mathcal{E}(C/\mathbb{Q})$.

Let $\mathcal{E}$ be an elliptic surface over C defined over k. We would like to associate to $\mathcal{E}$ an elliptic curve $E/k(C)$. Conversely, to each elliptic curve $E/k(C)$ we will assign a birational equivalence class of elliptic surfaces. In this way we will be able to apply our earlier results to study elliptic surfaces.

We begin by recalling some general definitions and basic facts about rational maps. For more details, see Hartshorne [1, I §4], Harris [1, Lecture 7], or Griffiths-Harris [1, 4 §2].

Definition. Let V and W be projective varieties. A *rational map* from V to W is an equivalence class of pairs (U, ϕ_U), where U is a non-empty Zariski open subset of V and $\phi_U : U \to W$ is a morphism. Two pairs (U, ϕ_U) and $(U', \phi_{U'})$ are deemed equivalent if $\phi_U = \phi_{U'}$ on $U \cap U'$. If ϕ is represented by a pair (U, ϕ_U), we say that ϕ *is defined on U*.

The *image* of a rational map ϕ, denoted $\overline{\phi(V)}$, is defined as follows: Let $U \subset V$ be an open subset on which ϕ is defined, take the Zariski closure in $V \times W$ of the graph

$$\{ (u, \phi(u)) \in V \times W : u \in U \},$$

and take the projection to W. (N.B. Since ϕ is not actually defined at all points of V, this "image" of ϕ may not agree with your intuition. For example, there may be points $w \in \overline{\phi(V)}$ which are not equal to $\phi(v)$ for any $v \in V$.)

A rational map $\phi : V \to W$ is *dominant* if $\overline{\phi(V)} = W$. Equivalently, ϕ is dominant if for one (hence every) open set U on which it is defined, the image $\phi(U)$ is Zariski dense in W.

The *domain of definition* of a rational map $\phi : V \to W$, which we denote by $\mathrm{Dom}(\phi)$, is the largest open subset of V on which ϕ is a morphism. (Such a largest set exists, see Hartshorne [1, I exercise 4.2].)

A rational map $\phi : V \to W$ is a *birational isomorphism* if it has a rational inverse $\psi : W \to V$; that is, ϕ and ψ are dominant and

$$\phi \circ \psi : W \longrightarrow W \qquad \text{and} \qquad \psi \circ \phi : V \longrightarrow V$$

are the identity maps at all points for which they are defined. If there is a birational isomorphism from V to W, then V and W are said to be *birationally equivalent*. If V, W, ϕ, and ψ are all defined over a field k, then we say that V and W are *birationally equivalent over k*.

Remark 3.4. In [AEC, I §3] we defined rational maps more naively using coordinates on $\mathbb{P}^n$. The reader will easily check that the two definitions are equivalent. For examples of rational maps that are not morphisms, see [AEC, I.3.6 and I.3.7]. Notice that [AEC, I.3.7] gives an example of non-isomorphic varieties that are birationally equivalent. Another important example of this phenomenon is provided by the process of "blowing-up"; see Hartshorne [1, I §4] or Harris [1, Lecture 7].

Proposition 3.5. *Let $\phi : V \to W$ be a rational map of projective varieties.*
(a) *The image $\overline{\phi(V)}$ is an algebraic subset of W. If V is irreducible, then so is $\overline{\phi(V)}$.*
(b) *Suppose that V is non-singular. Then ϕ is defined except on a set of codimension at least two. In other words, every component of the complement of $\mathrm{Dom}(\phi)$ in V has codimension at least two.*

PROOF. (a) See Harris [1, Lecture 7, p. 75] or Griffiths-Harris [1, 4 §2, p. 493] for the first part. The second part is immediate from the definition of irreducibility; see exercise 3.7.
(b) See Griffiths-Harris [1, 4 §2, p. 491]. $\square$

Proposition 3.6. *Let V/k and W/k be projective varieties. The following are equivalent.*

(i) *V and W are birationally equivalent over k.*

(ii) *The function fields $k(V)$ and $k(W)$ are isomorphic as k-algebras.*

(iii) *There are non-empty Zariski open sets $U_1 \subset V$ and $U_2 \subset W$ defined over k such that U_1 and U_2 are isomorphic over k.*

PROOF. This is a standard (and elementary) result in algebraic geometry. See, for example, Hartshorne [1, I.4.5] or Harris [1, exercise 7.10]. □

Let $\phi : V \to W$ be a dominant rational map, and let $f \in k(W)$ be a rational function on W. Then f is defined (i.e., regular) on a non-empty open subset of W, and $\phi(\mathrm{Dom}(\phi))$ is Zariski dense in W, so the composition $f \circ \phi$ is a regular function on a non-empty open subset of V. In other words, $f \circ \phi$ is a rational function on V, so we obtain a homomorphism of k-algebras

$$
\begin{array}{ccc}
k(W) & \longrightarrow & k(V), \\
f & \longmapsto & f \circ \phi.
\end{array}
$$

The next result says that the theory of varieties up to birational equivalence is essentially the same as the theory of their function fields. (See Hartshorne [1, I.4.4] for a more precise categorical statement.)

Proposition 3.7. *Let V/k and W/k be projective varieties. The association*

$$
\left\{
\begin{array}{c}
\text{dominant rational maps} \\
V \to W \text{ defined over } k
\end{array}
\right\}
\quad \longrightarrow \quad
\left\{
\begin{array}{c}
\text{k-algebra homomorphisms} \\
k(W) \to k(V)
\end{array}
\right\},
$$
$$
\phi \qquad\qquad \longmapsto \qquad (f \mapsto f \circ \phi)
$$

is a bijection.

PROOF. See Hartshorne [1, I.4.4]. □

We are now ready to apply the theory of rational maps to study elliptic surfaces. Recall that according to our definition, an elliptic surface consists of three pieces of data:

    (i) a projective surface $\mathcal{E}$,

    (ii) a projection map $\pi : \mathcal{E} \to C$,

    (iii) a zero section $\sigma_0 : C \to \mathcal{E}$.

Given two elliptic surfaces $\mathcal{E}$ and $\mathcal{E}'$ over the same base curve C, it thus makes sense to consider the rational maps $\mathcal{E} \to \mathcal{E}'$ which commute with projections and/or zero sections. This prompts us to make the following definitions.

Definition. Let $\pi : \mathcal{E} \to C$ and $\pi' : \mathcal{E}' \to C$ be elliptic surfaces over C. A *rational map from* $\mathcal{E}$ *to* $\mathcal{E}'$ *over* C is a rational map $\phi : \mathcal{E} \to \mathcal{E}'$ which commutes with the projection maps, $\pi' \circ \phi = \pi$. The elliptic surfaces $\mathcal{E}$ and $\mathcal{E}'$ are *birationally equivalent over* C if there is a birational isomorphism $\phi : \mathcal{E} \to \mathcal{E}'$ which commutes with the projection maps. If the elliptic surfaces and rational maps are defined over a field k, we will say that $\mathcal{E}$ and $\mathcal{E}'$ are *k-birationally equivalent over* C.

The next two propositions explain precisely how the theory of elliptic curves over $k(C)$ is the same as the birational theory of elliptic surfaces over C.

Proposition 3.8. (a) *Fix an elliptic curve* $E/k(C)$. *To each Weierstrass equation for* E,

$$E : y^2 = x^3 + Ax + B, \qquad A, B \in k(C),$$

we associate an elliptic surface

$$\mathcal{E}(A, B) = \left\{ ([X, Y, Z], t) \in \mathbb{P}^2 \times C : Y^2 Z = X^3 + A(t)XZ^2 + B(t)Z^3 \right\}$$

as described in (3.1). Then all of the $\mathcal{E}(A, B)$ *associated to* E *are k-birationally equivalent over* C.
(b) *Let* $\mathcal{E}$ *be an elliptic surface over* C *defined over* k. *Then* $\mathcal{E}$ *is k-birationally equivalent over* C *to* $\mathcal{E}(A, B)$ *for some* $A, B \in k(C)$. *Further, the elliptic curve*

$$E : y^2 = x^3 + Ax + B$$

over $k(C)$ *is uniquely determined (up to $k(C)$-isomorphism) by* $\mathcal{E}$.
(c) *Let* $E/k(C)$ *be an elliptic curve and* $\mathcal{E} \to C$ *an elliptic surface associated to* E *as in (a). Then*

$$k(\mathcal{E}) \cong k(C)(E) \quad \text{as } k(C)\text{-algebras.}$$

Here the projection map $\pi : \mathcal{E} \to C$ *induces an inclusion of fields* $k(C) \hookrightarrow k(\mathcal{E})$ *which makes* $k(\mathcal{E})$ *into a* $k(C)$-algebra.
We say that $E/k(C)$ is the *generic fiber* of $\mathcal{E} \to C$.

PROOF. (a) Suppose we take another Weierstrass equation for $E/k(C)$, say

$$E : y'^2 = x'^3 + A'x' + B', \qquad A', B' \in k(C).$$

Then there is a $u \in k(C)^*$ such that $u^4 A' = A$ and $u^6 B' = B$ [AEC III.1.3]. Now the map

$$\mathcal{E}(A', B') \longrightarrow \mathcal{E}(A, B), \qquad ([X', Y', Z'], t) \longmapsto ([u^2 X', u^3 Y', Z'], t)$$

shows that $\mathcal{E}(A, B)$ and $\mathcal{E}(A', B')$ are k-birationally equivalent over C.

(b,c) The projection map $\pi : \mathcal{E} \to C$ induces an inclusion of function fields $k(C) \hookrightarrow k(\mathcal{E})$ in the usual way, $f \mapsto f \circ \pi$. Further, $\mathcal{E}$ is a surface over k, and C is a curve over k, so $k(\mathcal{E})/k$ has transcendence degree 2 and $k(C)/k$ has transcendence degree 1. It follows that $k(\mathcal{E})/k(C)$ has transcendence degree 1, so there exists a curve $E/k(C)$, unique up to $k(C)$-isomorphism, whose function field $k(C)(E)$ is isomorphic to $k(\mathcal{E})$ as $k(C)$-algebras [AEC II.2.5].

We claim that E is a curve of genus 1. To see this, we write E as a subvariety of $\mathbb{P}^n$, so E is the set of zeros of a collection of homogeneous polynomials $\{f_i(\mathbf{x}) : 1 \le i \le r\}$ with coefficients in $k(C)$. Note that for almost all $t \in C$, we can evaluate the coefficients of the f_i's to get polynomials with coefficients in k. To indicate the dependence of the f_i's on t we will write $f_i(t, \mathbf{x})$. Then we can consider the algebraic variety in $\mathbb{P}^n \times C$ defined by

$$V \overset{\text{def}}{=} \big\{ (\mathbf{x}, t) \in \mathbb{P}^n \times C : f_i(t, \mathbf{x}) = 0 \text{ for } 1 \le i \le r \big\}.$$

Projection onto the second factor gives a map $V \to C$ which makes $k(V)$ into a $k(C)$-algebra, and by construction we see that $k(V)$ is isomorphic to $k(C)(E)$ as $k(C)$-algebras. Hence $k(V)$ is isomorphic to $k(\mathcal{E})$ as $k(C)$-algebras, so (3.6) tells us that V and $\mathcal{E}$ are birationally equivalent over C. In particular, for almost all $t \in C$ the fibers V_t and $\mathcal{E}_t$ are isomorphic, so almost all of the V_t's are curves of genus 1.

Now suppose that $\omega \in \Omega_{E/k(C)}$ is a differential form on E. (See [AEC, II §4] for general properties of differential forms on curves.) Any such differential can be written as a sum $\omega = \sum u_j \, dv_j$ with $u_j, v_j \in k(C)(E)$. For almost all $t \in C$ we can evaluate the u_j's and v_j's at t to get a differential form $\omega_t = \omega(t, \mathbf{x})$ on V_t. Further, if ω is a holomorphic differential form on E, then ω_t will be a holomorphic differential form on V_t for almost all $t \in C$.

Let $\omega_1, \omega_2 \in \Omega_{E/k(C)}$ be non-zero holomorphic differential forms. We claim that they are $k(C)$-linearly dependent. To prove this, we observe that for almost all $t \in C$, the forms $\omega_1(t, \mathbf{x})$ and $\omega_2(t, \mathbf{x})$ are holomorphic differentials on the curve V_t of genus 1, so they are k-linearly dependent from [AEC, II.5.3,II.5.5a]. In other words, there are non-zero constants $a_t, b_t \in k$ such that

$$a_t \omega_1(t, \mathbf{x}) + b_t \omega_2(t, \mathbf{x}) = 0.$$

But this means that the function $\omega_1(t, \mathbf{x})/\omega_2(t, \mathbf{x}) \in k(V_t)$ is constant; that is, it is in k. It follows that the function $\omega_1/\omega_2 \in k(V) = k(C)(E)$ is actually in $k(C)$, which proves that ω_1 and ω_2 are $k(C)$-linearly dependent.

To recapitulate, we have proven that the vector space of holomorphic differential forms in $\Omega_{E/k(C)}$ has $k(C)$-dimension at most one. It follows from [AEC, II.5.3,II.5.5a] that E has genus at most 1. Suppose that $E/k(C)$

has genus 0. We will see below that $E\bigl(k(C)\bigr)$ is non-empty, so we can assume that $E = \mathbb{P}^1$ and $V = \mathbb{P}^1 \times C$. But then the fibers $V_t \cong \mathbb{P}^1$ are all curves of genus 0, contradicting the fact that $V_t \cong \mathcal{E}_t$ has genus 1 for almost all t. This contradiction shows that $E/k(C)$ does not have genus 0, so the only possibility is that $E/k(C)$ has genus 1.

Further, the section $\sigma_0 : C \to \mathcal{E}$ corresponds to a point $P_0 \in E\bigl(k(C)\bigr)$. To see this we use the fact that V and $\mathcal{E}$ are birationally equivalent over C to get a section $\sigma_0' : C \to V$. This section is a map of the form $\sigma_0' = [h_0, \ldots, h_n]$ for certain functions $h_0, \ldots, h_n \in k(C)$, which is the same as saying that $P_0 = [h_0, \ldots, h_n] \in \mathbb{P}^n\bigl(k(C)\bigr)$ is a point in $E\bigl(k(C)\bigr)$. Taking P_0 to be the identity element, E becomes an elliptic curve defined over $k(C)$.

We have now proven that $E/k(C)$ is an elliptic curve, so we can take a Weierstrass equation for it, say

$$E : y^2 = x^3 + Ax + B \qquad \text{with } A, B \in k(C).$$

Then the corresponding surface V is precisely the elliptic surface $\mathcal{E}(A, B)$ described in (a), and we have already observed above that V is birationally equivalent to $\mathcal{E}$ over C. This completes the proof of the first part of (b) which asserts that every elliptic surface is birationally equivalent over C to some $\mathcal{E}(A, B)$. Further, we showed above that there are isomorphisms of $k(C)$-algebras

$$k(\mathcal{E}) \cong k(V) \cong k(C)(E),$$

which completes the proof of (c).

It remains to prove that E is determined up to $k(C)$-isomorphism by $\mathcal{E}$. Suppose that $\mathcal{E}$ is also birationally equivalent to $\mathcal{E}(A', B')$. Then $\mathcal{E}(A, B)$ and $\mathcal{E}(A', B')$ are birationally equivalent over C. This means that for almost all $t \in C$, there is an isomorphism on the fibers

$$\bigl\{y^2 = x^3 + A(t)x + B(t)\bigr\} \xrightarrow{\sim} \bigl\{y^2 = x^3 + A'(t)x + B'(t)\bigr\}.$$

From general principles [AEC, III.3.1b], we know that this isomorphism is given by a map of the form $(x, y) \mapsto (\alpha_t x, \beta_t y)$ for some $\alpha_t, \beta_t \in k$. But the birational equivalence $\mathcal{E}(A, B) \to \mathcal{E}(A', B')$ is an algebraic map, so we see that α and β are functions on $\mathcal{E}(A, B)$ which depend only on t. In other words, $\alpha, \beta \in k(C)$, which proves that the corresponding elliptic curves

$$y^2 = x^3 + Ax + B \qquad \text{and} \qquad y^2 = x^3 + A'x + B'$$

are isomorphic over $k(C)$. $\square$

Proposition 3.9. *There is a natural bijection*

$$\left\{ \begin{array}{c} \text{dominant rational maps} \\ \mathcal{E} \to \mathcal{E}' \text{ over } C \end{array} \right\} \longleftrightarrow \left\{ \begin{array}{c} \text{non-constant maps } E \to E' \\ \text{defined over } k(C) \end{array} \right\},$$

where we identify the set of elliptic curves defined over $k(C)$ with the set of (birational equivalence classes) of elliptic surfaces over C as described in Proposition 3.8.

PROOF. Let $\phi : \mathcal{E} \to \mathcal{E}'$ be a dominant rational map over C. Choose Weierstrass equations for $\mathcal{E}$ and $\mathcal{E}'$, or equivalently for E and E',

$$\mathcal{E} : y^2 = x^3 + Ax + B, \qquad \mathcal{E}' : y'^2 = x'^2 + A'x' + B'.$$

Then the map ϕ will have the form

$$\phi : ((x,y),t) \longmapsto ((f(t,x,y),g(t,x,y)),t),$$

where $f, g \in k(\mathcal{E}) \cong k(C)(E) = k(C)(x,y)$. Hence $F = (f,g)$ defines a map $F : E \to E'$.

Suppose that F is a constant map from E to E' over $k(C)$. This means that $f, g \in k(C)$, so $\phi((x,y),t) = ((f(t),g(t)),t)$ depends only on t, independent of x and y. It follows that the image of ϕ has dimension at most one, since it is the image of a map $C \to \mathcal{E}'$, so in particular ϕ is not dominant. This proves that if $\phi : \mathcal{E} \to \mathcal{E}'$ is dominant, then the associated map $F : E \to E'$ is non-constant.

The proof going the other direction is similar. Fix Weierstrass equations for E and E' as above, and let $F : E \to E'$ be a map defined over $k(C)$. Then F has the form $F = (f,g)$ for some $f, g \in k(C)(E) \cong k(\mathcal{E})$, and we can define a map $\phi : \mathcal{E} \to \mathcal{E}'$ over C by

$$\phi : ((x,y),t) \longmapsto ((f(t,x,y),g(t,x,y)),t).$$

If ϕ is not dominant, then its image must consist of a curve, since π maps the image $\overline{\phi(\mathcal{E})}$ onto C. But this means that if we fix (almost any point) $t \in C$ and vary x, y on the fiber $\mathcal{E}_t$, then $\phi((x,y),t)$ can assume only finitely many values. In other words, the map $\phi(\cdot, t) : \mathcal{E}_t \to \mathbb{P}^1$ takes on only finitely many values, so it is constant [AEC II.2.3]. Hence f and g do not depend on x, y, so $f, g \in k(C)$ and $F = (f,g) : E \to E'$ is a constant map. This proves that if $F : E \to E'$ is non-constant, then the associated map $\phi : \mathcal{E} \to \mathcal{E}'$ is dominant, which completes the proof of Proposition 3.9. $\square$

Our final task is to explain how the set of sections $\mathcal{E}(C)$ has a natural group structure. Recall that for almost all points $t \in C$, the fiber $\mathcal{E}_t$ is an elliptic curve (3.3.1), so given any two points on $\mathcal{E}_t$, we can add them or take their inverses. Let $\sigma_1, \sigma_2 \in \mathcal{E}(C)$ be two sections to $\mathcal{E}$. We define new sections $\sigma_1 + \sigma_2$ and $-\sigma_1$ by the rules

$$(\sigma_1 + \sigma_2)(t) = \sigma_1(t) + \sigma_2(t) \qquad \text{and} \qquad (-\sigma_1)(t) = -(\sigma_1(t)),$$

valid for all $t \in C$ such that the fiber $\mathcal{E}_t$ is non-singular. We will verify below (3.10) that $\sigma_1 + \sigma_2$ and $-\sigma_1$ define rational maps $C \to \mathcal{E}$, so in fact they define morphisms since C is a non-singular curve [AEC, II.2.2.1]. The next proposition says that this "fiber-by-fiber" addition makes $\mathcal{E}(C)$ into a group.

Proposition 3.10. *Let $\mathcal{E} \to C$ be an elliptic surface defined over k.*
(a) *Let $\sigma_1, \sigma_2 \in \mathcal{E}(C/k)$ be sections defined over k. Then the maps $\sigma_1 + \sigma_2$ and $-\sigma_2$ described above are in $\mathcal{E}(C/k)$.*
(b) *The operations $(\sigma_1, \sigma_2) \to \sigma_1 + \sigma_2$ and $\sigma \to -\sigma$ make $\mathcal{E}(C/k)$ into an abelian group.*
(c) *Let $E/k(C)$ be the elliptic curve associated to $\mathcal{E}$ as described in (3.8). Then there is a natural group isomorphism*

$$E\big(k(C)\big) \quad \xrightarrow{\sim} \quad \mathcal{E}(C/k),$$
$$P = (x_P, y_P) \quad \longmapsto \quad \big(\sigma_P : t \to ((x_P(t), y_P(t)), t)\big).$$

PROOF. (a) Take a Weierstrass equation for $\mathcal{E} \to C$ as described in (3.8),

$$\mathcal{E} : y^2 = x^3 + Ax + B, \qquad A, B \in k(C).$$

Then a section $\sigma_i : C \to \mathcal{E}$ is given by a pair of functions

$$\sigma_i : t \longmapsto \big(x_i(t), y_i(t)\big)$$

which satisfy the given Weierstrass equation for (almost all) $t \in C$. Equivalently $x_i, y_i \in k(C)$ are functions satisfying $y_i^2 = x_i^3 + Ax_i + B$ as elements of $k(C)$. By definition, $(\sigma_1 + \sigma_2)(t)$ is the sum of the two points $\sigma_1(t)$ and $\sigma_2(t)$ using the addition law on the elliptic curve

$$\mathcal{E}_t : y^2 = x^3 + A(t)x + B(t).$$

So the usual addition formula [AEC III.2.3] says that if $x_1(t) \neq x_2(t)$, then

$$(\sigma_1 + \sigma_2)(t) = \sigma_1(t) + \sigma_2(t) = (x_1(t), y_1(t)) + (x_2(t), y_2(t))$$
$$= \left(\left(\frac{y_2(t) - y_1(t)}{x_2(t) - x_1(t)} \right)^2 - x_1(t) - x_2(t), \cdots \right).$$

(We leave it to you to fill in the y-coordinate.) In other words, if $x_1 \neq x_2$ in $k(C)$, then the map $\sigma_1 + \sigma_2$ is given by the formula

$$\sigma_1 + \sigma_2 = \left(\left(\frac{y_2 - y_1}{x_2 - x_1} \right)^2 - x_1 - x_2, \cdots \right),$$

which shows that $\sigma_1 + \sigma_2$ is a rational map from C to $\mathcal{E}$ defined over k. Similarly, if $x_1 = x_2$ and $y_1 \neq y_2$, the duplication formula [AEC III.2.3(d)] yields the same conclusion. Finally, the map $-\sigma_1$ is given by $-\sigma_1 = (x_1, -y_1)$, so $-\sigma_1$ also gives a rational map $C \to \mathcal{E}$ defined over k. But C is a nonsingular curve, so all of these rational maps are morphisms [AEC II.2.2.1], which completes the proof that $\sigma_1 + \sigma_2$ and $-\sigma_1$ are in $\mathcal{E}(C/k)$.

(b) This is clear from the fact that the points on almost every fiber form a group. For example, for any three sections $\sigma_1, \sigma_2, \sigma_3 \in \mathcal{E}(C/k)$ and almost all points $t \in C$, we have

$$\big((\sigma_1 + \sigma_2) + \sigma_3\big)(t) = (\sigma_1 + \sigma_2)(t) + \sigma_3(t) = \big(\sigma_1(t) + \sigma_2(t)\big) + \sigma_3(t)$$
$$= \sigma_1(t) + \big(\sigma_2(t) + \sigma_3(t)\big) = \sigma_1(t) + (\sigma_2 + \sigma_3)(t) = \big(\sigma_1 + (\sigma_2 + \sigma_3)\big)(t).$$

Hence $\big((\sigma_1 + \sigma_2) + \sigma_3\big) = \big(\sigma_1 + (\sigma_2 + \sigma_3)\big)$ as sections, which verifies the associative law. The other group axioms can be verified in a similar fashion.

(c) Fix a Weierstrass equation for $\mathcal{E} \to C$ as in (a). If $P = (x_P, y_P) \in E\big(k(C)\big)$, then x_P and y_P satisfy the given Weierstrass equation as elements of the function field $k(C)$, so $\big(x_P(t), y_P(t)\big) \in \mathcal{E}_t$ for almost all $t \in C$. This shows that σ_P is a well-defined element of $\mathcal{E}(C/k)$. Similarly, any $\sigma \in \mathcal{E}(C/k)$ has the form $\sigma(t) = \big((x_\sigma(t), y_\sigma(t)), t\big)$ for some rational functions $x_\sigma, y_\sigma \in k(C)$ satisfying

$$y_\sigma(t)^2 = x_\sigma(t)^3 + A(t)x_\sigma(t) + B(t) \qquad \text{for almost all } t \in C.$$

It follows that $P_\sigma = (x_\sigma, y_\sigma)$ satisfies the given Weierstrass equation, so $P_\sigma \in E\big(k(C)\big)$. The identifications $P \mapsto \sigma_P$ and $\sigma \mapsto P_\sigma$ are clearly inverse to one another, so they define bijections $E\big(k(C)\big) \leftrightarrow \mathcal{E}(C/k)$. Finally, for any $P_1, P_2 \in E\big(k(C)\big)$ we have

$$(\sigma_{P_1} + \sigma_{P_2})(t) = \big(x_{P_1}(t), y_{P_1}(t)\big) + \big(x_{P_2}(t), y_{P_2}(t)\big) = \sigma_{P_1 + P_2}(t).$$

Similarly, $-\sigma_P = \sigma_{-P}$, which shows that the map $E\big(k(C)\big) \to \mathcal{E}(C/k)$ is a homomorphism, hence an isomorphism. $\qquad\qquad\qquad\qquad\qquad\qquad\qquad\qquad\square$

Remark 3.11. It is important to observe that we can only add points on $\mathcal{E}$ if they lie on the same (non-singular) fiber. This enables us to add two sections together, but it does not make the surface $\mathcal{E}$ itself into a group. Another way to say this is to use the notion of fiber product (Hartshorne [1, II §3]). The fiber product $\mathcal{E} \times_C \mathcal{E}$ of $\mathcal{E}$ with itself relative to the map $\pi : \mathcal{E} \to C$ is the set of pairs (z_1, z_2) in the ordinary product $\mathcal{E} \times \mathcal{E}$ with the property that $\pi(z_1) = \pi(z_2)$; that is, $\mathcal{E} \times_C \mathcal{E}$ consists of all pairs of points on $\mathcal{E}$ which lie on the same fiber over C. It is a variety, and the "group operation" is then the rational map

$$\mathcal{E} \times_C \mathcal{E} \longrightarrow \mathcal{E}$$

defined by addition on each (non-singular) fiber. One might say that $\mathcal{E}$ is a group relative to the projection map $\pi : \mathcal{E} \to C$. We will see this construction appearing in a much more general setting when we discuss group schemes in the next chapter.

§4. Heights on Elliptic Curves over Function Fields

Let E/K be an elliptic curve defined over a function field. We have proven (2.1) that the quotient $E(K)/2E(K)$ is finite. We would like to use the Descent Lemma [AEC VIII.3.1] to prove that $E(K)$ is a finitely generated group. This means we need a height function on $E(K)$ that satisfies certain properties. We begin by defining a height function on K and then use it to define a height on $E(K)$.

Definition. Let $K = k(C)$ be the function field of a non-singular algebraic curve C/k. The *height of an element $f \in K$* is defined to be the degree of the associated map from C to $\mathbb{P}^1$,

$$h(f) = \deg(f : C \to \mathbb{P}^1).$$

In particular, if $f \in k$, then the map is constant and we set $h(f) = 0$. Let E/K be an elliptic curve given by a Weierstrass equation

$$y^2 + a_1 xy + a_3 y = x^3 + a_2 x^2 + a_4 x + a_6.$$

The *height of a point $P \in E(K)$* is defined to be

$$h(P) = \begin{cases} 0 & \text{if } P = O, \\ h(x) & \text{if } P = (x, y). \end{cases}$$

(Note that $h(f)$ is really the height relative to the field K, and $h(P)$ depends on the choice of a Weierstrass equation for E, although our notation does not reflect this. See [AEC VIII §5].)

Remark 4.1. For each $t \in C$, let

$$\mathrm{ord}_t : k(C)^* \longrightarrow \mathbb{Z}$$

be the normalized valuation on $k(C)$ [AEC II §1]; that is, $\mathrm{ord}_t(f)$ is the order of vanishing of the function f at the point t. Then [AEC II.2.6a] implies that

$$h(f) = \deg f = \sum_{t \in C} \max\{\mathrm{ord}_t(f), 0\} = \sum_{t \in C} \max\{-\mathrm{ord}_t(f), 0\}.$$

This definition of the height as a sum of local values is analogous to the definition of the height for number fields described in [AEC VIII §5]. Notice that we can count either the total number of zeros of f or the total number of poles.

The properties that we want the height to possess are of two very different sorts. First, we want the height to satisfy certain transformation properties relative to the group law on E. We will be able to prove this below by a straightforward calculation that is very similar to the proof in the number field case. Indeed, there is a theory of height functions for a wide class of fields which includes number fields and function fields as special cases. (See Lang [4] for details.)

The second property we require of the height is a finiteness property, namely that a set of bounded height should contain only finitely many points. In the case of a number field K, we first showed that this was true of K itself; that is, a number field contains only finitely many elements of bounded height [AEC VIII.5.11]. This immediately implied the same result for $E(K)$. However, matters are more complicated for function fields, since a function field may have infinitely many elements of bounded height. For example, the elements of height 0 in $k(T)$ are precisely the elements of the field k. More generally, the elements of height at most d in $k(T)$ are the rational functions of the form

$$\frac{a_0 + a_1 T + a_2 T^2 + \cdots + a_d T^d}{b_0 + b_1 T + b_2 T^2 + \cdots + b_d T^d}.$$

Thus it is not clear whether $E(K)$ might possess infinite subsets of bounded height. We will postpone further discussion of this question until the next section.

The following proposition summarizes the principal geometric transformation properties of the height.

Theorem 4.2. Let E/K be an elliptic curve defined over a function field K.
(a) $h(2P) = 4h(P) + O(1)$ for all $P \in E(K)$.
(b) $h(P + Q) + h(P - Q) = 2h(P) + 2h(Q) + O(1)$ for all $P, Q \in E(K)$.
(The $O(1)$ bounds depend on the curve E and can be given explicitly; see exercise 3.11.)

Remark 4.2.1. We will prove Theorem 4.2 using the triangle inequality and elementary polynomial computations. Later we will give another proof using intersection theory on a non-singular model for the elliptic surface $\mathcal{E}$; see (9.3).

PROOF (of Theorem 4.2). Fix a Weierstrass equation for E of the form

$$E : y^2 = x^3 + Ax + B.$$

For any $t \in C$, we write $\mathrm{ord}_t(f)$ as usual for the order of vanishing of $f \in k(C)$ at t. To ease notation and avoid the inevitable confusion caused by repeated minus signs, we will also write

$$\pi_t(f) = (\text{order of the pole of } f \text{ at } t) = -\mathrm{ord}_t(f).$$

Note that the (non-archimedean) triangle inequality for π_t has the form

$$\pi_t(f + f') \leq \max\{\pi_t(f), \pi_t(f')\}, \quad \text{with equality unless } \pi_t(f) = \pi_t(f').$$

(a) Let $P = (x, y) \in E(K)$. The duplication formula [AEC III.2.3d] says that $x(2P) = \phi/\psi$, where

$$\phi = \phi(x) = x^4 - 2Ax^2 - 8Bx + A^2 \quad \text{and} \quad \psi = \psi(x) = 4x^3 + 4Ax + 4B.$$

We compute

$$\begin{aligned} h(2P) = h\big(x(2P)\big) \quad &\text{definition of height} \\ = \sum_{t \in C} \max\{\pi_t(\phi/\psi), 0\} \quad &\text{from (4.4)} \\ = \sum_{t \in C} \max\{\pi_t(\phi), \pi_t(\psi)\} \quad &\text{since } \sum_{t \in C} \pi_t(\psi) = 0. \end{aligned}$$

So we need to show that $\max\{\pi_t(\phi), \pi_t(\psi)\}$ and $4 \max\{\pi_t(x), 0\}$ are approximately equal. If x has a large order pole at t, this is fairly clear, since $\phi = x^4 + \cdots$ will then have a pole four times larger than x. On the other hand, if x does not have a large pole, we will be in good shape provided ϕ and ψ don't both have large order zeros at t. In order to make these vague comments precise, we define a quantity

$$\mu_t = \mu_t(E) \stackrel{\text{def}}{=} \max\left\{\frac{1}{2}\pi_t(A), \frac{1}{3}\pi_t(B), 0\right\}$$

and consider the following two cases.

$\boxed{\pi_t(x) > \mu_t \quad (x \text{ has a large pole at } t)}$

The definition of μ_t shows that in this case we have strict inequalities

$$\begin{aligned} \pi_t(x^4) &> \max\{\pi_t(2Ax^2), \pi_t(8Bx), \pi_t(A^2)\}, \\ \pi_t(x^3) &> \max\{\pi_t(Ax), \pi_t(B)\}. \end{aligned}$$

It follows from the triangle inequality that $\pi_t(\phi) = \pi_t(x^4)$ and $\pi_t(\psi) = \pi_t(x^3)$. We also have $\pi_t(x) > \mu_t \geq 0$, which proves that

$$\max\{\pi_t(\phi), \pi_t(\psi)\} = \pi_t(x^4) = 4 \max\{\pi_t(x), 0\}.$$

$\boxed{\pi_t(x) \leq \mu_t \quad (x \text{ has a small pole at } t)}$

The triangle inequality gives us trivial upper bounds

$$\begin{aligned} \pi_t(\phi) &\leq \max\{\pi_t(x^4), \pi_t(2Ax^2), \pi_t(8Bx), \pi_t(A^2)\} \\ &\leq \max\{4\mu_t, \pi_t(A) + 2\mu_t, \pi_t(B) + \mu_t, 2\pi_t(A)\} \\ &\leq 4\mu_t \quad \text{since } \pi_t(A) \leq 2\mu_t \text{ and } \pi_t(B) \leq 3\mu_t, \end{aligned}$$

and

$$\pi_t(\psi) \leq \max\{\pi_t(x^3), \pi_t(Ax), \pi_t(B)\}$$
$$\leq \max\{3\mu_t, \pi_t(A) + \mu_t, \pi_t(B)\}$$
$$\leq 3\mu_t.$$

Hence

$$\max\{\pi_t(\phi), \pi_t(\psi)\} \leq 4\mu_t \leq 4\mu_t + 4\max\{\pi_t(x), 0\}.$$

In order to get a lower bound, we need to know that ϕ and ψ cannot both vanish to high order at t. We define functions

$$\Phi = 12x^2 + 16A, \qquad \Psi = 3x^3 - 5Ax - 27B, \qquad \Delta = 4A^3 + 27B^2,$$

and observe that there is an identity

$$\Phi \cdot \phi - \Psi \cdot \psi = 4\Delta.$$

(We've used this identity many times before, for example [AEC VIII.4.3].) Note that the discriminant $\Delta = 4A^3 + 27B^2 \in k(C)$ is not identically zero, since $E/k(C)$ is assumed to be non-singular. Intuitively, our assumption that $\pi_t(x) \leq \mu_t$ implies that Φ and Ψ have bounded poles at t, and then the identity says that ϕ and ψ cannot both have high order zeros at t. More precisely, we start with the upper bounds

$$\pi_t(\Phi) \leq \max\{\pi_t(12x^2), \pi_t(16A)\} \leq 2\mu_t,$$
$$\pi_t(\Psi) \leq \max\{\pi_t(3x^3), \pi_t(5Ax), \pi_t(27B)\} \leq 3\mu_t.$$

Next, using the above identity, we find that

$$\mathrm{ord}_t(\Delta) = \mathrm{ord}_t(\Phi\phi - \Psi\psi) \geq \min\{\mathrm{ord}_t(\phi), \mathrm{ord}_t(\psi)\} - \max\{\pi_t(\Phi), \pi_t(\Psi)\}$$
$$\geq \min\{\mathrm{ord}_t(\phi), \mathrm{ord}_t(\psi)\} - 3\mu_t.$$

Now multiplying by -1 and using $-\mathrm{ord}_t = \pi_t$ yields

$$\max\{\pi_t(\phi), \pi_t(\psi)\} \geq -3\mu_t - \mathrm{ord}_t(\Delta)$$
$$\geq -3\mu_t - \mathrm{ord}_t(\Delta) - 4\left(\mu_t - \max\{\pi_t(x), 0\}\right)$$
$$\text{since } \mu_t \geq \max\{\pi_t(x), 0\}$$
$$= -7\mu_t - \mathrm{ord}_t(\Delta) + 4\max\{\pi_t(x), 0\}.$$

Combining the upper and lower bounds in this case, we obtain

$$-7\mu_t - \mathrm{ord}_t(\Delta) \leq \max\{\pi_t(\phi), \pi_t(\psi)\} - 4\max\{\pi_t(x), 0\} \leq 4\mu_t.$$

In both cases we have now proven bounds of the form

$$c_1(t) \leq \max\{\pi_t(\phi), \pi_t(\psi)\} - 4\max\{\pi_t(x), 0\} \leq c_2(t),$$

where the quantities $c_1(t)$ and $c_2(t)$ have the property that they are independent of the point $P = (x, y)$ and are equal to zero for all but finitely many $t \in C$. Summing this inequality over $t \in C$ gives the desired estimate

$$c_1 \leq h(2P) - 4h(P) \leq c_2$$

for certain constants $c_i = c_i(E)$ which do not depend on P.

(b) If $P = O$ or $Q = O$, the assertion is trivial, and if $P = \pm Q$, then (b) reduces to (a). So we will assume that $P, Q \neq O$ and $P \neq \pm Q$. We write

$$P = (x_1, y_1), \quad Q = (x_2, y_2), \quad P + Q = (x_3, y_3), \quad P - Q = (x_4, y_4).$$

The condition that $P \neq \pm Q$ ensures that the coordinates are all finite. The addition formula [AEC III.2.3d] on the elliptic curve gives

$$x_3 = \left(\frac{y_2 - y_1}{x_2 - x_1}\right)^2 - x_1 - x_2 = \frac{(A + x_1 x_2)(x_1 + x_2) + 2B - 2y_1 y_2}{(x_2 - x_1)^2},$$

$$x_4 = \left(\frac{y_2 + y_1}{x_2 - x_1}\right)^2 - x_1 - x_2 = \frac{(A + x_1 x_2)(x_1 + x_2) + 2B + 2y_1 y_2}{(x_2 - x_1)^2}.$$

Next we compute

$$h(P + Q) + h(P - Q) = h(x_3) + h(x_4)$$

$$= \sum_{t \in C} \max\{\pi_t(x_3), 0\} + \max\{\pi_t(x_4), 0\}$$

$$\leq \sum_{t \in C} \max\{\pi_t(x_3 x_4), \pi_t(x_3 + x_4), 0\},$$

where the last inequality needs some justification. In fact, for any functions $a, b, c, d \in k(C)^*$ we have

$$\max\{\pi_t(a), \pi_t(b)\} + \max\{\pi_t(c), \pi_t(d)\} = \max\{\pi_t(ac), \pi_t(ad + bc), \pi_t(bd)\}.$$

This is easily verified using the triangle inequality and checking the various cases. For reasons which will become apparent in a moment, we will add $0 = \sum \pi_t((x_1 - x_2)^2)$ to both sides of this inequality, which yields

$$h(P + Q) + h(P - Q) \leq \sum_{t \in C} \max\{\pi_t((x_1 - x_2)^2 x_3 x_4),$$

$$\pi_t((x_1 - x_2)^2(x_3 + x_4)), \pi_t((x_1 - x_2)^2)\}.$$

Next we use a little algebra and the fact that the points P_3 and P_4 lie on E to compute

$$(x_1 - x_2)^2 x_3 x_4 = (x_1 x_2 - A)^2 - 4B(x_1 + x_2),$$

$$(x_1 - x_2)^2(x_3 + x_4) = 2(x_1 + x_2)(A + x_1 x_2) + 4B.$$

Substituting in above and using the triangle inequality yields

$$
\begin{aligned}
h(P+Q) &+ h(P-Q)\\
&\leq \sum_{t\in C}\max\big\{\pi_t\big((x_1 x_2 - A)^2 - 4B(x_1+x_2)\big),\\
&\qquad\qquad \pi_t\big(2(x_1+x_2)(A+x_1 x_2)+4B\big), \pi_t\big((x_1-x_2)^2\big)\big\}\\
&\leq \sum_{t\in C}\max\big\{\pi_t(x_1^2 x_2^2), \pi_t(Ax_1 x_2), \pi_t(A^2), \pi_t(Bx_1), \pi_t(Bx_2),\\
&\qquad\qquad \pi_t(Ax_1), \pi_t(Ax_2), \pi_t(x_1^2 x_2), \pi_t(x_1 x_2^2),\\
&\qquad\qquad\qquad \pi_t(B), \pi_t(x_1^2), \pi_t(x_1 x_2), \pi_t(x_2^2)\big\}\\
&\leq \sum_{t\in C}\big(2\max\{\pi_t(x_1),0\} + 2\max\{\pi_t(x_2),0\}\\
&\qquad\qquad +2\max\{\pi_t(A),0\} + \max\{\pi_t(B),0\}\big)\\
&= 2h(x_1) + 2h(x_2) + 2h(A) + h(B)\\
&= 2h(P) + 2h(Q) + O(1).
\end{aligned}
$$

It remains to prove an inequality in the opposite direction. It is possible to do this directly, as is done, for example, in [AEC VIII.6.2]. But we will instead use the following clever trick which is due to Don Zagier. We have proven that the inequality

$$
2h(P) + 2h(Q) \geq h(P+Q) + h(P-Q) + O(1)
$$

holds for all $P, Q \in E(K)$. Given two points $P', Q' \in E(K)$, we apply this identity with $P = P' + Q'$ and $Q = P' - Q'$ and then use (a) to obtain

$$
\begin{aligned}
2h(P'+Q') + 2h(P'-Q') &\geq h(2P') + h(2Q') + O(1)\\
&= 4h(P') + 4h(Q') + O(1).
\end{aligned}
$$

Dividing by 2 gives the opposite inequality, which completes the proof of (b). $\qquad\square$

Just as in the number field case, we can construct a canonical height which is a quadratic form on the group $E(K)$. However, in order to prove that the canonical height is non-degenerate, we need to know that sets of bounded height are finite. The exact conditions under which this occurs will be described in the next section. We have included the construction of the canonical height here, since it seems to fit in better with the material in this section. We hope the reader will excuse this textual non-linearity.

Theorem 4.3. *Let E/K be an elliptic curve defined over a function field $K = k(C)$.*
(a) *For every point $P \in E(K)$, the limit*

$$
\hat{h}(P) = \frac{1}{2}\lim_{n\to\infty}\frac{1}{4^n}h(2^n P)
$$

exists. The quantity $\hat{h}(P)$ is called the *canonical (or Néron-Tate) height* of P.

(b) *The canonical height has the following properties:*

(i) $\hat{h}(P) = \frac{1}{2}h(P) + O(1)$ for all $P \in E(K)$.

(ii) $\hat{h}(mP) = m^2\hat{h}(P)$ for all $P \in E(K)$ and all $m \in \mathbb{Z}$.

(iii) $\hat{h}(P+Q) + \hat{h}(P-Q) = 2\hat{h}(P) + 2\hat{h}(Q)$ for all $P, Q \in E(K)$.

(c) *The canonical height is a quadratic form on $E(K)$. In other words,* $\hat{h}(-P) = \hat{h}(P)$, and the pairing

$$\langle\,\cdot\,,\,\cdot\,\rangle : E(K) \times E(K) \longrightarrow \mathbb{R}$$
$$\langle P, Q \rangle = \hat{h}(P+Q) - \hat{h}(P) - \hat{h}(Q)$$

is bilinear. (N.B. The pairing is normalized so that $\hat{h}(P) = \frac{1}{2}\langle P, P \rangle$.)

(d) *Assume that E does not split over k. (This means that E is not K-isomorphic to an elliptic curve defined over k. See §5 for more details.) Then $\hat{h}(P) \geq 0$, and*

$$\hat{h}(P) = 0 \quad \text{if and only if } P \text{ is a point of finite order.}$$

(e) *Any function $E(K) \to \mathbb{R}$ which satisfies (b)(i) and (b)(ii) for some integer $m \geq 2$ is equal to the canonical height.*

PROOF. We will just briefly sketch the proof, since it is exactly the same as in the number field case [AEC VIII.9.1, VIII.9.3]. For any integers $n \geq m \geq 0$ we have

$$\left|4^{-n}h(2^nP) - 4^{-m}h(2^mP)\right| = \left|\sum_{i=m}^{n-1} 4^{-i-1}h(2^{i+1}P) - 4^{-i}h(2^iP)\right|$$

$$\leq \sum_{i=m}^{n-1} 4^{-i-1}\left|h(2\cdot 2^iP) - 4h(2^iP)\right|$$

$$\leq \sum_{i=m}^{n-1} 4^{-i-1}O(1) \quad \text{from (4.2a)}$$

$$\leq \sum_{i=m}^{\infty} 4^{-i-1}O(1) \leq O(4^{-m}).$$

This shows that the sequence $4^{-n}h(2^nP)$ is Cauchy, hence converges, which proves (a). Further, taking $m = 0$ and letting $n \to \infty$ gives

$$\left|2\hat{h}(P) - h(P)\right| \leq O(1),$$

which is (b)(i).

Next we apply (4.2b) to the points $2^n P$ and $2^n Q$ to obtain

$$h\big(2^n(P+Q)\big) + h\big(2^n(P-Q)\big) = 2h(2^n P) + 2h(2^n Q) + O(1).$$

Dividing by 4^n and letting $n \to \infty$ gives (b)(iii). Evaluating (b)(iii) at $P = mQ$ yields the identity

$$\hat{h}((m+1)Q) + \hat{h}((m-1)Q) = 2\hat{h}(mQ) + 2\hat{h}(Q).$$

Taking $m = 0$ gives $\hat{h}(-Q) = \hat{h}(Q)$, and then an easy induction (up and down) on m gives (b)(ii). This completes the proof of (b).

It is a standard computation to show that a function satisfying the parallelogram law (b)(iii) is a quadratic form; see for example the proof of [AEC VIII.9.3c]. This gives (c).

If $P \in E(K)$ has finite order, then $2^n P$ takes on only finitely many values, so it is obvious from the definition that $\hat{h}(P) = 0$. Conversely, suppose that $\hat{h}(P) = 0$. Then for all $m \in \mathbb{Z}$ we use (b)(ii) and (b)(i) to compute

$$h(mP) = 2\hat{h}(mP) + O(1) = 2m^2 \hat{h}(P) + O(1) = O(1).$$

It follows that $\{mP : m \in \mathbb{Z}\}$ is a set of bounded height. We will prove in the next section (5.1) that if E does not split over k, then sets of bounded height are finite. Hence P is a point of finite order, which completes the proof of (d).

Finally, suppose that $\hat{g} : E(K) \to \mathbb{R}$ satisfies (b)(i) and (b)(ii) for some $m \geq 2$. Then we compute for every $P \in E(K)$ and every $i \geq 1$,

$$\begin{aligned}
2\hat{h}(P) - 2\hat{g}(P) &= m^{-2i}\big(2\hat{h}(m^i P) - 2\hat{g}(m^i P)\big) \\
&= m^{-2i}\big((h(m^i P) + O(1)) - (h(m^i P) + O(1))\big) \\
&= O(m^{-2i}).
\end{aligned}$$

Letting $i \to \infty$ shows that $\hat{h}(P) = \hat{g}(P)$. $\square$

Remark 4.3.1. It is also possible to construct the canonical height using intersection theory on the minimal elliptic surface associated to E. See §9 for details, especially (9.3). One consequence of the geometric construction, which is not at all evident from the definition, is that for function fields the canonical height $\hat{h}(P)$ is always a rational number. This is (probably) false for number fields, where it is conjectured that $\hat{h}(P)$ is transcendental for all non-torsion points.

§5. Split Elliptic Surfaces and Sets of Bounded Height

We proved in the last section that the height function $h : E(K) \to \mathbb{R}$ behaves nicely with respect to the group law on E. In order to prove that $E(K)$ is finitely generated, it remains to show that sets of bounded height in $E(K)$ are necessarily finite. The reader will recall that in the case of number fields, this was comparatively easy to do. Unfortunately, for function fields it is easy to construct a counterexample to this assertion!

For example, let E_0/k be an elliptic curve, let $\mathcal{E} = E_0 \times C$ be the elliptic surface with $\mathcal{E} \to C$ being projection onto the second factor, and let E/K be the corresponding elliptic curve over K. Then every point $\gamma \in E_0(k)$ gives a section

$$\sigma_\gamma : C \longrightarrow \mathcal{E} = E_0 \times C, \qquad \sigma_\gamma(t) = (\gamma, t),$$

and this section corresponds to a point $P_\gamma \in E(K)$. Clearly, distinct γ's give distinct P_γ's, and just as clearly the map

$$E_0(k) \longrightarrow E(K), \qquad \gamma \longmapsto P_\gamma,$$

is a homomorphism. It follows that $E(K)$ cannot possibly be finitely generated, since the fact that k is algebraically closed means that $E_0(k)$ is not finitely generated. (For example, if $k = \mathbb{C}$, then $E_0(k) \cong \mathbb{C}/\Lambda$ for some lattice $\Lambda \subset \mathbb{C}$.)

It will turn out that this is the only way in which $E(K)$ can fail to be finitely generated, which suggests that we make the following definition.

Definition. An elliptic surface $\mathcal{E} \to C$ *splits (over k)* if there is an elliptic curve E_0/k and a birational isomorphism

$$i : \mathcal{E} \xrightarrow{\sim} E_0 \times C$$

such that the following diagram commutes:

$$\begin{array}{ccc} \mathcal{E} & \xrightarrow{\ i\ } & E_0 \times C \\ {\scriptstyle \pi} \searrow & & \swarrow {\scriptstyle \mathrm{proj}_2} \\ & C & \end{array}$$

There are several other ways of characterizing split elliptic surfaces. The following one will be used later in this section. For others, see exercises 3.9 and 3.10.

Proposition 5.1. *Let $\mathcal{E} \to C$ be an elliptic surface over k, and let E/K be the corresponding elliptic curve over the function field $K = k(C)$. The following are equivalent:*

(i) *The elliptic surface* $\mathcal{E} \to C$ *splits over* k.
(ii) *There is an elliptic curve* E_0/k *and an isomorphism* $E \xrightarrow{\sim} E_0$ *defined over* K.

PROOF. Suppose first that $\pi : \mathcal{E} \to C$ splits. This means that there is a birational isomorphism $i : \mathcal{E} \to E_0 \times C$ so that $\text{proj}_2 \circ i = \pi$. A dominant rational map induces a corresponding map on function fields (3.7), so we obtain an isomorphism $k(\mathcal{E}) \cong k(E_0 \times C)$ which is compatible with the inclusions

$$k(C) \hookrightarrow k(\mathcal{E}) \qquad \text{and} \qquad k(C) \hookrightarrow k(E_0 \times C).$$

In other words, if we let $K = k(C)$ as usual, then the fields $k(\mathcal{E}) = K(E)$ and $k(E_0 \times C) = K(E_0)$ are isomorphic as K-algebras. Each of them is a field of transcendence 1 over K, so each corresponds to a unique non-singular curve defined over K (see [AEC, II.2.5] or Hartshorne [1, I.6.12]). In other words, there is an isomorphism $E \cong E_0$ defined over K. This completes the proof that (i) implies (ii).

Conversely, suppose that we are given an elliptic curve E_0/k and an isomorphism $E \xrightarrow{\sim} E_0$ defined over K. Then $K(E) \cong K(E_0)$ as K-algebras, which is the same as saying that $k(\mathcal{E}) \cong k(E_0 \times C)$ as $k(C)$-algebras. Again using (3.7), this isomorphism of fields induces a birational isomorphism of varieties $\mathcal{E} \to E_0 \times C$ commuting with the maps to C, which shows that $\mathcal{E} \to C$ is split over k. Hence (ii) implies (i), which completes the proof of (5.1). $\square$

Example 5.2. Note that the isomorphism in (5.1ii) is not required to be defined over the constant field k. In fact, since E is only defined over K, it really only makes sense to talk about maps being defined over K. For example, take $C = \mathbb{P}^1$ and $K = k(T)$, and consider the elliptic surfaces

$$\mathcal{E}_1 : y^2 = x^3 + 1, \qquad\qquad \mathcal{E}_2 : y^2 = x^3 + T^6,$$
$$\mathcal{E}_3 : y^2 = x^3 + T, \qquad\qquad \mathcal{E}_4 : y^2 = x^3 + x + T.$$

Also let E_0/k be the elliptic curve

$$E_0 : y^2 = x^3 + 1.$$

Then $\mathcal{E}_1$ is clearly split over k, since it is precisely $E_0 \times C$. The surface $\mathcal{E}_2$ also splits over k, as can be seen from the isomorphism

$$\mathcal{E}_2 \xrightarrow{\sim} E_0 \times C, \qquad ((x,y),t) \longmapsto ((t^{-2}x, t^{-3}y), t).$$

The elliptic surface $\mathcal{E}_3$ does not split over k, although it will split if we replace the base field $k(T)$ by the larger field $k(T^{1/6})$. Finally, $\mathcal{E}_4$ does not split over k; and since its j-invariant is non-constant, it will still not split even if we replace $k(T)$ by a finite extension. See exercises 3.9 and 3.10 for general statements.

Remark 5.3. For a number field K, it was not hard to show that there are only finitely many elements of K having bounded height [AEC VIII.5.11], which immediately gave the same result for $E(K)$. Unfortunately, this assertion is clearly false for function fields, since there may be infinitely many maps $C \to \mathbb{P}^1$ of any given degree. In other words, for number fields we proved that a set of bounded height

$$\{P \in E(K) : h(P) \le d\}$$

is finite by reducing to an assertion about elements of bounded height in K. But for function fields we will need a new sort of argument which makes use of the fact that the coordinates of $P = (x, y)$ satisfy the equation of an elliptic curve. Further, we need to rule out split elliptic curves, since they will have infinitely many points of bounded height. All of this will help to explain why the proof of the following result is far from trivial and requires techniques different from those used when studying number fields. The proof will take us the rest of this section.

Theorem 5.4. *Let $\mathcal{E} \to C$ be an elliptic surface over an algebraically closed field k, let E/K be the corresponding elliptic curve over the function field $K = k(C)$, and let d be a constant. If the set*

$$\{P \in E(K) : h(P) \le d\}$$

contains infinitely many points, then $\mathcal{E}$ splits over k.

PROOF (of Theorem 5.4). We will divide the proof of Theorem 5.4 into two steps. The first step says that if $\mathcal{E}$ has infinitely many sections of bounded degree (i.e., $E(K)$ has infinitely many points of bounded height), then there is a one-parameter family of such sections. The second step says that if there is a one-parameter family, then $\mathcal{E}$ splits. We begin with the existence of the family.

Proposition 5.5. *Under the assumptions of Theorem 5.4, there is a (nonsingular projective) curve Γ/k and a dominant rational map $\phi : \Gamma \times C \to \mathcal{E}$ such that the following diagram commutes:*

$$
\begin{array}{ccc}
\Gamma \times C & \xrightarrow{\;\;\phi\;\;} & \mathcal{E} \\
& {\scriptstyle \mathrm{proj}_2} \searrow \quad \swarrow {\scriptstyle \pi} & \\
& C &
\end{array}
$$

PROOF. We fix a Weierstrass equation for E/K of the form

$$E : y^2 = x^3 + Ax + B \qquad \text{with } A, B \in K = k(C),$$

and we define a set

$$E(K, d) \overset{\text{def}}{=} \{ P \in E(K) \, : \, h(P) \le d \}.$$

Our assumption is that $E(K, d)$ is infinite, and we wish to use this fact to find an appropriate curve Γ and map $\Gamma \times C \to \mathcal{E}$

The first step is to parametrize the set of maps from C to $\mathbb{P}^2$. Recall that if D is a divisor on C, then $L(D)$ is defined to be the space of rational functions

$$L(D) = \{ f \in k(C) \, : \, \text{div}(f) + D \ge 0 \}.$$

This is a finite dimensional vector space whose dimension is denoted by $\ell(D)$. (For basic facts about $L(D)$, see [AEC, II §5].) Taking three functions from $L(D)$ will define a map from C to $\mathbb{P}^2$, so we get a natural map

$$\begin{array}{ccc} L(D)^3 \smallsetminus \{0\} & \longrightarrow & \text{Map}(C, \mathbb{P}^2), \\ (F_0, F_1, F_2) & \longmapsto & (t \mapsto [F_0(t), F_1(t), F_2(t)]). \end{array}$$

Multiplying (F_0, F_1, F_2) by a scalar clearly gives the same map $C \to \mathbb{P}^2$, so we actually have an association

$$\mathbb{P}^{3\ell(D)-1} \cong \frac{L(D)^3 \smallsetminus \{0\}}{k^*} \longrightarrow \text{Map}(C, \mathbb{P}^2).$$

The key here is that we have taken a collection of maps in $\text{Map}(C, \mathbb{P}^2)$ and have parametrized this collection using the points of the algebraic variety $\mathbb{P}^{3\ell-1}$, where to ease notation we will write ℓ for $\ell(D)$. Some of these maps $C \to \mathbb{P}^2$ will actually correspond to elements of $E(K)$. The next step is to show that the maps corresponding to $E(K)$ form an algebraic subset of $\mathbb{P}^{3\ell-1}$.

For simplicity, we will assume henceforth that $D \ge 0$, and we fix a basis $f_1, \ldots, f_\ell$ for $L(D)$. Further, we choose a divisor $D' \ge 3D$ large enough so that $1, A, B \in L(D' - 3D)$, and we let $h_1, \ldots, h_r$ be a basis for $L(D')$.

Every element in $L(D)^3$ can be written uniquely in the form

$$F = (F_a, F_b, F_c) = \left(\sum_{i=1}^{\ell} a_i f_i, \sum_{i=1}^{\ell} b_i f_i, \sum_{i=1}^{\ell} c_i f_i \right).$$

Such an F will give an element of $E(K)$ if and only if F_a, F_b, F_c satisfy the homogeneous equation of E,

$$F_b^2 F_c = F_a^3 + A F_a F_c^2 + B F_c^3.$$

In other words, F will give an element of $E(K)$ if

$$\left(\sum b_i f_i \right)^2 \left(\sum c_i f_i \right) = \left(\sum a_i f_i \right)^3 + A \left(\sum a_i f_i \right) \left(\sum c_i f_i \right)^2$$
$$+ B \left(\sum c_i f_i \right)^3.$$

Multiplying this out gives a sum involving monomials of the form $f_i f_j f_k$ and $A f_i f_j f_k$ and $B f_i f_j f_k$. Our choice of D' ensures that each of these monomials is in $L(D')$, hence can be written uniquely as a linear combination of $h_1, \ldots, h_r$. So finally we end up with an equation that looks like

$$\sum_{i=1}^{r} \Phi_i(\mathbf{a}, \mathbf{b}, \mathbf{c}) h_i = 0,$$

where each Φ_i is a homogeneous polynomial in the coordinates

$$[\mathbf{a}, \mathbf{b}, \mathbf{c}] = [a_1, \ldots, a_\ell, b_1, \ldots, b_\ell, c_1, \ldots, c_\ell] \in \mathbb{P}^{3\ell - 1}.$$

Now the maps $C \to \mathbb{P}^2$ from above which correspond to elements of $E(K)$ are associated to the points of the variety

$$V_D \overset{\text{def}}{=} \{ [\mathbf{a}, \mathbf{b}, \mathbf{c}] \in \mathbb{P}^{3\ell(D) - 1} : \Phi_i(\mathbf{a}, \mathbf{b}, \mathbf{c}) = 0 \text{ for all } 1 \le i \le r \}.$$

Example 5.5.1. We briefly interrupt the proof of Proposition 5.5 to present an example. If we take $C = \mathbb{P}^1$ and $K = k(T)$, then E/K has a Weierstrass equation of the form

$$y^2 = x^3 + A(T)x + B(T) \qquad \text{with } A, B \in k[T].$$

We let $D = n(\infty)$, which means that $L(D)$ is the set of polynomials in $k[T]$ of degree at most n. (Here $\ell(D) = n + 1$.) The corresponding family of maps $\mathbb{P}^1 \to \mathbb{P}^2$ is parametrized by $\mathbb{P}^{3n+2}$ as described above,

$$\mathbb{P}^{3n+2} \qquad \longrightarrow \qquad \mathrm{Map}(\mathbb{P}^1, \mathbb{P}^2),$$

$$[a_0, \ldots, a_n, b_0, \ldots, b_n, c_0, \ldots, c_n] \longmapsto \left[\sum_{i=0}^{n} a_i T^i, \sum_{i=0}^{n} b_i T^i, \sum_{i=0}^{n} c_i T^i \right].$$

Writing $A(T) = \sum A_i T^i$ and $B(T) = \sum B_i T^i$, this map $\mathbb{P}^1 \to \mathbb{P}^2$ will give an element of $E(K)$ if it satisfies the equation

$$\left(\sum b_i T^i \right)^2 \left(\sum c_i T^i \right) = \left(\sum a_i T^i \right)^3$$
$$+ \left(\sum A_i T^i \right) \left(\sum a_i T^i \right) \left(\sum c_i T^i \right)^2 + \left(\sum B_i T^i \right) \left(\sum c_i T^i \right)^3.$$

Multiplying this out and writing it as a polynomial in T gives a formula that looks like

$$\sum_{i=0}^{r} \Phi_i(\mathbf{a}, \mathbf{b}, \mathbf{c}) T^i = 0,$$

and the system of homogeneous equations $\Phi_0 = \Phi_1 = \cdots = \Phi_r = 0$ defines the variety V_D.

To illustrate how this procedure works in practice, we will consider the elliptic curve

$$E : y^2 = x^3 + T^2 x - 1.$$

We will find all points of degree at most 1 by setting

$$P = [a_0 + a_1 T, b_0 + b_1 T, c_0 + c_1 T]$$

and substituting P into the (homogenized) equation for E. Multiplying everything out, we obtain the equation

$$a_1 c_1^2 T^5 + (2a_1 c_0 c_1 + a_0 c_1^2) T^4 + (a_1^3 + a_1 c_0^2 - b_1^2 c_1 + 2a_0 c_0 c_1 - c_1^3) T^3$$
$$+ (3a_0 a_1^2 - b_1^2 c_0 + a_0 c_0^2 - 2b_0 b_1 c_1 - 3c_0 c_1^2) T^2$$
$$+ (3a_0^2 a_1 - 2b_0 b_1 c_0 - b_0^2 c_1 - 3c_0^2 c_1) T + (a_0^3 - b_0^2 c_0 - c_0^3) = 0.$$

Setting the coefficients equal to 0 gives six homogeneous equations for the six variables $a_0, \ldots, c_1$. These six equations define the variety $V_D \subset \mathbb{P}^5$. After some work one finds that V_D consists of the 3 lines

$$\{[0, 0, u, v, 0, 0]\} \cup \{[0, 0, iu, iv, u, v]\} \cup \{[0, 0, -iu, -iv, u, v]\}$$

and the 22 isolated points

$$\{[0, \zeta_1, \zeta_2, 0, 1, 0] : \zeta_1^2 = \zeta_2^2 = -1\} \cup \{[\rho^2, 0, 0, \rho, 1, 0] : \rho^6 = 1\}$$
$$\cup \{[-2\rho^2, \zeta, 3\zeta\rho^3, 2\rho, 1, 0] : \zeta^2 = -1, \ \rho^6 = 1\}.$$

Note that although V_D itself is not zero-dimensional, its image in $E(K)$ consists of a finite set of points, namely

$$[0, 1, 0], \quad [0, \zeta, 1], \quad [\zeta_1 T, \zeta_2, 1], \quad [\rho^2, \rho T, 1], \quad [-2\rho^2 + \zeta T, 3\zeta\rho^3 + 2\rho T, 1],$$

where ζ, ζ_1, ζ_2, and ρ satisfy $\zeta^2 = \zeta_1^2 = \zeta_2^2 = -1$ and $\rho^6 = 1$. For example, all of the points on the line $[0, 0, iu, iv, u, v]$ in V_D are mapped to the single point $[0, iu + ivT, u + vT] = [0, i, 1] \in E(K)$. We also observe that of these 24 points in $E(K)$, only the 3 points $[0, 1, 0]$ and $[1, \pm T, 1]$ are in $E(\mathbb{Q}(T))$.

We now resume the (regularly scheduled) proof of Proposition 5.5. Recall that we have constructed a variety V_D and a map $V_D \to E(K)$. The following lemma shows that if $\deg(D)$ is sufficiently large, then the image of V_D in $E(K)$ will contain $E(K, d)$.

Lemma 5.5.2. *Let g be the genus of the curve C. If*

$$\deg D \geq g + \frac{5}{2}d + \frac{1}{2}(\deg A + \deg B),$$

then the image of V_D in $E(K)$ contains $E(K, d)$. (For a more accurate estimate, see exercise 3.13.)

PROOF (of Lemma 5.5.2). This is an exercise using the Riemann-Roch theorem. Let $P = (x_P, y_P) \in E(K, d)$, so, by definition, $\deg(x_P) = h(P) \le d$. Using the Weierstrass equation for E, we can check that $\deg(y_P)$ is also bounded,

$$2 \deg(y_P) = \deg(x_P^3 + Ax_P + B) \le 3 \deg(x_P) + \deg(A) + \deg(B),$$

$$\deg(y_P) \le \frac{3}{2} d + \frac{1}{2} \big(\deg(A) + \deg(B) \big).$$

In order to prove the lemma, we need to find functions $F_0, F_1, F_2 \in L(D)$ such that $(x_P, y_P) = (F_0/F_2, F_1/F_2)$.

Recall that any function $f \in K$ defines a map $f : C \to \mathbb{P}^1$, and the divisor of f has the form

$$\mathrm{div}(f) = \mathrm{div}_0(f) - \mathrm{div}_\infty(f) = f^*((0)) + f^*((\infty)),$$

where $\mathrm{div}_0(f)$ and $\mathrm{div}_\infty(f)$ are the divisors of zeros and poles of f respectively. (See [AEC II.3.5].) We also note from [AEC, II.3.6(a)] that

$$\deg(f) = \deg\big(\mathrm{div}_0(f)\big) = \deg\big(\mathrm{div}_\infty(f)\big).$$

We are going to apply the Riemann-Roch theorem to the divisor

$$D'' \overset{\mathrm{def}}{=} D - \mathrm{div}_\infty(x_P) - \mathrm{div}_\infty(y_P),$$

whose degree we estimate as

$$\deg(D'') = \deg(D) - \deg(x_P) - \deg(y_P)$$
$$\ge \deg(D) - d - \left(\frac{3}{2} d + \frac{1}{2} \big(\deg(A) + \deg(B) \big) \right) \ge g.$$

The Riemann-Roch theorem [AEC, II.5.4] then tells us that

$$\ell(D'') \ge \deg(D'') - g + 1 \ge 1,$$

so there exists a non-zero function $F \in L(D'')$. We claim that the three functions

$$F_0 = F x_P, \qquad F_1 = F y_P, \qquad F_2 = F,$$

are all in $L(D)$, which will complete the proof of the lemma.

To check this last assertion, we use the fact that $\operatorname{div}(F) + D'' \geq 0$ and compute

$$
\begin{aligned}
\operatorname{div}(Fx_P) + D &= \operatorname{div}(Fx_P) + D'' + \operatorname{div}_\infty(x_P) + \operatorname{div}_\infty(y_P) \\
&= \operatorname{div}(F) + D'' + \operatorname{div}_0(x_P) + \operatorname{div}_\infty(y_P) \geq 0, \\
\operatorname{div}(Fy_P) + D &= \operatorname{div}(Fy_P) + D'' + \operatorname{div}_\infty(x_P) + \operatorname{div}_\infty(y_P) \\
&= \operatorname{div}(F) + D'' + \operatorname{div}_\infty(x_P) + \operatorname{div}_0(y_P) \geq 0, \\
\operatorname{div}(F) + D &= \operatorname{div}(F) + D'' + \operatorname{div}_\infty(x_P) + \operatorname{div}_\infty(y_P) \geq 0.
\end{aligned}
$$

This completes the proof of Lemma 5.5.2. □

Continuing with the proof of Proposition 5.5, we fix a divisor $D \geq 0$ whose degree is large enough so that we can apply (5.5.2). Then (5.5.2) and our assumption that $E(K, d)$ is infinite tell us that the image of V_D in $E(K)$ is infinite.

We now change perspective a bit and consider the associated elliptic surface $\mathcal{E} \to C$. We have assigned to each point $\gamma \in V_D$ a point $P_\gamma \in E(K)$, and (3.10c) says that the point P_γ corresponds to a section $\sigma_\gamma : C \to \mathcal{E}$. In this way we get a natural rational map

$$
\phi : V_D \times C \longrightarrow \mathcal{E}, \qquad (\gamma, t) \longmapsto \sigma_\gamma(t).
$$

It is clear that ϕ is an algebraic map, since using notation from above, we see that ϕ can be written as

$$
\phi([\mathbf{a}, \mathbf{b}, \mathbf{c}], t) = \left[\sum a_i f_i(t), \sum b_i f_i(t), \sum c_i f_i(t) \right].
$$

If there exists an irreducible curve $\Gamma \subset V_D$ such that the map

$$
\phi : \Gamma \times C \longrightarrow \mathcal{E}, \qquad (\gamma, t) \longmapsto \sigma_\gamma(t)
$$

is dominant, then the proof of Proposition 5.5 will be complete. So we assume that $\phi : \Gamma \times C \to \mathcal{E}$ is not dominant for every irreducible curve $\Gamma \subset V_D$ and derive a contradiction.

Let $\Gamma \subset V_D$ be an irreducible curve. We are assuming that $\phi : \Gamma \times C \to \mathcal{E}$ is not dominant, so the image $\overline{\phi(\Gamma \times C)}$ has dimension at most one. However, we know that $\pi(\sigma_\gamma(t)) = t$, which shows that π maps $\overline{\phi(\Gamma \times C)}$ onto C. It follows that the image $\overline{\phi(\Gamma \times C)}$ must have dimension exactly one. Further, the product $\Gamma \times C$ is irreducible, so $\phi(\Gamma \times C)$ is also irreducible by (3.5a). Therefore $\overline{\phi(\Gamma \times C)}$ must consist of a single irreducible curve. On the other hand, for any given $\gamma \in \Gamma$ the map σ_γ is a section to $\pi : \mathcal{E} \to C$, so we have

$$
\phi(\{\gamma\} \times C) = \sigma_\gamma(C) \cong C
$$

is already an irreducible curve contained in $\overline{\phi(\Gamma \times C)}$. Hence $\phi(\{\gamma\} \times C)$ must be equal to $\overline{\phi(\Gamma \times C)}$ for every $\gamma \in \Gamma$. Equivalently, every $\gamma \in \Gamma$ gives the same section σ_γ, which means that the image of Γ in $E(K)$ consists of a single point.

We have now shown that every irreducible curve $\Gamma \subset V_D$ maps to a single point in $E(K)$. But any two points in any connected component of V_D can be linked by a connected chain of irreducible curves. (In fact, on any irreducible component of V_D, any two points can be connected by a single irreducible curve; see exercise 3.14.) Since V_D has only finitely many connected components, it follows that the image of V_D in $E(K)$ is finite. This contradicts the fact shown above that the image of V_D contains $E(K, d)$. Hence there exists an irreducible curve $\Gamma \subset V_D$ with the property that $\phi : \Gamma \times C \to \mathcal{E}$ is dominant. Replacing Γ with a non-singular model for Γ (see Hartshorne [1, I.6.11]) completes the proof of Proposition 5.5. $\square$

The following proposition, taken together with (5.5), completes the proof of Theorem 5.4.

Proposition 5.6.   *Let* $\pi : \mathcal{E} \to C$ *be an elliptic surface over* k, *let* Γ/k *be a non-singular projective curve, and suppose that there exists a dominant rational map* $\phi : \Gamma \times C \to \mathcal{E}$ *so that the following diagram commutes:*

$$\begin{array}{ccc} \Gamma \times C & \xrightarrow{\quad \phi \quad} & \mathcal{E} \\ & \text{proj}_2 \searrow \qquad \swarrow \pi & \\ & C & \end{array}$$

Then $\mathcal{E}$ *splits.*

PROOF.   The fact that ϕ is a dominant map of varieties of the same dimension means that there is a non-empty Zariski open subset $\mathcal{E}^0$ of $\mathcal{E}$ over which ϕ is a finite map, say of degree m. Let $t_0 \in C$ be a point such that the fiber $\mathcal{E}_{t_0}$ is non-singular and such that ϕ is well-defined at every point of $\Gamma \times \{t_0\}$. Note that the set of such t_0's is a non-empty Zariski open subset of C, since $\mathcal{E}_t$ is non-singular for all but finitely many $t \in C$, and ϕ is well-defined except at finitely many points of $\Gamma \times C$ by (3.5b). To ease notation, we let $E_0 = \mathcal{E}_{t_0}$. Note that E_0/k is an elliptic curve.

We define a map

$$\begin{array}{ccc} \psi : & \mathcal{E}^0 & \longrightarrow & E_0 \times C, \\ & ((x, y), t) & \longmapsto & \left(\sum_{i=1}^{m} \phi(\gamma_i, t_0), t \right) \end{array}$$

where the points γ_i are determined by the formula

$$\phi^*\big((x, y), t\big) = \sum_{i=1}^{m} (\gamma_i, t).$$

In other words, ψ takes a point on $\mathcal{E}^0$, pulls it back by ϕ to get a collection of m points on $\Gamma \times C$ (counted with multiplicity), changes the t-coordinate to t_0 to get a collection of points on $\Gamma \times \{t_0\}$, uses ϕ to push them forward to a collection of points on E_0, and finally uses the group law on E_0 to add them up.

Note that the map ψ is a well-defined rational map on $\mathcal{E}^0$. This is true despite the fact that the definition of ψ involves applying ϕ^{-1}, since ultimately we take a symmetric expression of the points in $\phi^{-1}((x,y),t)$, so the resulting point can be expressed as a rational combination of x, y, t. This is clearest for those points $((x,y),t) \in \mathcal{E}^0$ for which $\phi^{-1}((x,y),t)$ consists of m distinct points, which suffices for our purposes since we only need to define ψ on an open subset of $\mathcal{E}^0$.

(Aside: An alternative description of ψ is as follows. Let $\Gamma^{(m)}$ be the m-fold symmetric product of Γ. Then the map ϕ defines in a natural way a morphism from $\mathcal{E}^0$ to $\Gamma^{(m)} \times C$ which sends a point in $\mathcal{E}^0$ to the collection of points in its inverse image. Next we use the map $\phi(\,\cdot\,, t_0) : \Gamma \to E_0$ to map $\Gamma^{(m)} \times C \to E_0^{(m)} \times C$. Finally, the summation map $E_0^{(m)} \to E_0$ using the group law on E_0 gets us to $E_0 \times C$, and the composition of all these maps is $\psi : \mathcal{E}^0 \to E_0 \times C$. For information about the symmetric product, see Harris [1, Lecture 10, especially 10.23].)

Note that if ψ were a (birational) isomorphism, we would be done. Unfortunately, there is no reason that this should be true. We begin our analysis of the map ψ by computing it on the fiber over t_0.

$$\psi((x,y),t_0) = \Big(\sum_{(\gamma,t')\in\phi^*((x,y),t_0)} \phi(\gamma,t_0), t_0 \Big) = (m(x,y), t_0).$$

Thus $\psi : \mathcal{E}_{t_0} \to E_0 \times \{t_0\}$ is just the multiplication-by-m map on E_0. In particular, since the multiplication-by-m map is surjective, we see that $\psi(\mathcal{E}^0)$ contains $E_0 \times \{t_0\}$. This implies that the rational map $\mathcal{E} \to E_0 \times C$ is dominant, since otherwise the irreducibility of $\psi(\mathcal{E}^0)$ would imply that $\psi(\mathcal{E}^0) = E_0 \times \{t_0\}$, contradicting the fact that $\psi(\mathcal{E}^0)$ maps onto C (i.e., $\psi(\mathcal{E}^0)$ must contain at least one point on each fiber on $\Gamma \times C \to C$).

We now consider the elliptic curve E/K associated to the elliptic surface $\mathcal{E}$. We also take the elliptic curve E_0/k and think of it as the elliptic curve E_0/K associated to the split elliptic surface $E_0 \times C$. Then the dominant rational map $\psi : \mathcal{E} \to E_0 \times C$ defined above corresponds to a non-constant map $E \to E_0$ of elliptic curves over K (3.9). Any such map can be written as the composition of a translation followed by an isogeny [AEC, III.4.7], so we obtain a non-zero isogeny

$$\lambda : E \longrightarrow E_0$$

defined over K. Taking the dual isogeny [AEC, III §4] gives a map $\hat{\lambda} : E_0 \to E$ defined over K, and this induces an isomorphism [AEC, III.4.11,III.4.12]

$$\hat{\lambda} : E_0/\ker(\hat{\lambda}) \xrightarrow{\;\cong/K\;} E.$$

Now $\ker(\hat{\lambda})$ is a finite subgroup of $E_0(\bar{K})$, so the fact that E_0 is defined over k implies that $\ker(\hat{\lambda}) \subset E_0(k)$. To see why this is true, take a Weierstrass equation for E_0 with coefficients in k. Then the n-torsion points of E have coordinates which are roots of certain polynomials having coefficients in k, so $E[n] \subset E(k)$. (For explicit formulas, see [AEC, exercise 3.7]. Note we are assuming that k is algebraically closed.) It follows that the elliptic curve $E_1 \stackrel{\text{def}}{=} E_0/\ker(\hat{\lambda})$ is defined over k.

We have now produced an elliptic curve E_1/k and an isomorphism of elliptic curves $E_1 \to E$ defined over K. It follows from (5.1) that E/K splits over k. This completes the proof of Proposition 5.6 and, in conjunction with Proposition 5.5, also completes the proof of Theorem 5.4.   □

§6. The Mordell-Weil Theorem for Function Fields

We have now assembled all of the tools needed to prove the following important result.

Theorem 6.1. (Mordell-Weil Theorem for Function Fields) *Let $\mathcal{E} \to C$ be an elliptic surface defined over a field k, and let E/K be the corresponding elliptic curve over the function field $K = k(C)$. If $\mathcal{E} \to C$ does not split, then $E(K)$ is a finitely generated group.*

PROOF. Suppose first that k is algebraically closed. The weak Mordell-Weil theorem (2.1) tells us that the quotient group $E(K)/2E(K)$ is finite. Next let $h : E(K) \to \mathbb{Z}$ be the height function defined in §4. This height function satisfies

(i) $h(P + Q) = 2h(P) + 2h(Q) - h(P - Q) + O(1)$
                    $\leq 2h(P) + O_Q(1)$ for all $P, Q \in E(K)$,

(ii) $h(2P) = 4h(P) + O(1)$ for all $P \in E(K)$,

(iii) $\{P \in E(K) : h(P) \leq C\}$ is finite.

The first two statements are (4.2a,b), whereas the third statement is (5.4) and uses the assumption that $\mathcal{E} \to C$ does not split. We now have all of the hypotheses needed to apply the Descent theorem [AEC, VIII.3.1], which completes the proof that $E(K)$ is finitely generated under the assumption that k is algebraically closed. Finally, for arbitrary constant fields k, it suffices to observe that $E(K) = E(k(C))$ is a subgroup of $E(\bar{k}(C))$, so $E(K)$ is finitely generated.   □

Remark 6.2.1. If $\mathcal{E} \to C$ splits over k, then the group $E(K)$ need not be finitely generated. More precisely, if $\mathcal{E} \cong E_0 \times C$, then each point $z \in E_0(k)$

is also a point in $E_0(K) = E(K)$. Thus there is an inclusion $E_0(k) \hookrightarrow E(K)$. So for example, if $k = \mathbb{C}$, then $E_0(k)$ will certainly not be finitely generated, and the same is then true of $E(K)$. However, the quotient group $E(K)/E_0(k)$ will be finitely generated. This relative version of the Mordell-Weil theorem is due to Lang and Néron; see exercise 3.15.

Remark 6.2.2. If k is a number field, or more generally if k is a finitely generated extension of $\mathbb{Q}$, then the Mordell-Weil theorem (6.1) is true regardless of whether or not $\mathcal{E} \to C$ splits. This generalization of the original Mordell-Weil theorem for number fields is due to Néron; see exercise 3.4.

§7. The Geometry of Algebraic Surfaces

All of our previous work in this chapter has dealt with the birational geometry of elliptic surfaces. In order to investigate the finer structure of elliptic surfaces, we will need to study them up to isomorphism. This section reviews the basic theory of non-singular algebraic surfaces, including especially intersection theory and minimal models. Our main reference will be Chapter 5 of Hartshorne [1], specifically §1 for intersection theory and section 5 for the theory of minimal models, although we will also need a few additional facts from other sources concerning minimal models. For more information about surfaces, the reader might consult Beauville [1] and Griffiths-Harris [1, Ch. 4].

Let S/k be a non-singular surface defined over an algebraically closed field k of characteristic 0. A *divisor* on S is a formal sum

$$D = \sum_{i=1}^{n} a_i \Gamma_i,$$

where $a_i \in \mathbb{Z}$ and the $\Gamma_i \subset S$ are irreducible curves lying on the surface S. The Γ_i's that appear in the sum are called the *components of the divisor* D. The group of divisors on S is denoted $\mathrm{Div}(S)$.

Recall that the *local ring of S at a point* $P \in S$ is defined to be

$$\mathcal{O}_{S,P} = \{f \in k(S) : f \text{ is defined at } P\}.$$

Similarly, for any irreducible curve $\Gamma \subset S$, the *local ring of S at Γ* is

$$\mathcal{O}_{S,\Gamma} = \{f \in k(S) : f \text{ is defined at some point } P \in \Gamma\} = \bigcup_{P \in \Gamma} \mathcal{O}_{S,P}.$$

Our assumption that S is non-singular implies that each $\mathcal{O}_{S,\Gamma}$ is a discrete valuation ring. We denote its valuation by ord_Γ, since intuitively $\mathrm{ord}_\Gamma(f)$

is the order of vanishing of f along Γ. We extend this as usual to a homomorphism

$$\operatorname{ord}_\Gamma : k(S)^* \longrightarrow \mathbb{Z},$$

and then use this to define a homomorphism

$$
\begin{aligned}
\operatorname{div} : \quad k(S)^* &\longrightarrow \quad \operatorname{Div}(S), \\
f &\longmapsto \sum_{\Gamma \subset S} \operatorname{ord}_\Gamma(f)\Gamma.
\end{aligned}
$$

A divisor is *principal* if it is the divisor of a function $\operatorname{div}(f)$. Two divisors $D_1, D_2 \in \operatorname{Div}(S)$ are *linearly equivalent* if their difference $D_1 - D_2$ is principal, in which case we write $D_1 \sim D_2$. Linear equivalence is an equivalence relation on the divisor group $\operatorname{Div}(S)$, and the *Picard group of S* is the corresponding quotient group,

$$\operatorname{Pic}(S) = \operatorname{Div}(S)/\sim .$$

Example 7.1. Consider a divisor $D = \sum a_i \Gamma_i \in \operatorname{Div}(\mathbb{P}^2)$ in the projective plane. To each irreducible curve $\Gamma_i \subset \mathbb{P}^2$ we can associate its degree, and by extending linearly we obtain a homomorphism

$$
\begin{aligned}
\deg : \quad \operatorname{Div}(\mathbb{P}^2) &\longrightarrow \quad \mathbb{Z}, \\
\sum a_i \Gamma_i &\longmapsto \quad \sum a_i \deg(\Gamma_i).
\end{aligned}
$$

If $D = \operatorname{div}(f)$ is principal, then it is easy to check that $\deg(D) = 0$, so the degree map induces a homomorphism

$$\deg : \operatorname{Pic}(\mathbb{P}^2) \longrightarrow \mathbb{Z}.$$

We will leave it to you (exercise 3.18) to verify that this last map is an isomorphism, so $\operatorname{Pic}(\mathbb{P}^2) \cong \mathbb{Z}$. This is the analogue for surfaces of [AEC, II.3.1b]. It says that a rational function on $\mathbb{P}^2$ has the "same number" of zeros and poles.

In order to study the geometry of a surface, we will look at the curves it contains and how those curves intersect. For example, a curve of degree m and a curve of degree n in $\mathbb{P}^2$ will "usually" intersect in mn distinct points, and they will always intersect in mn points if we count tangencies and singularities with the correct multiplicities. This famous result is known as Bezout's theorem; see (7.3) below. Our next step is to define an intersection index for curves and divisors on arbitrary non-singular surfaces.

Let Γ_1 and Γ_2 be irreducible curves on S, and let $P \in \Gamma_1 \cap \Gamma_2$. Fix local equations $f_1, f_2 \in k(S)^*$ for Γ_1, Γ_2 around P; that is, choose $f_i \in \mathcal{O}_{S,P}$ so that $\operatorname{ord}_{\Gamma_i}(f_i) = 1$ and $\operatorname{ord}_\Gamma(f_i) = 0$ for every other irreducible curve Γ containing P. We say that Γ_1 and Γ_2 *intersect transversally* at P if f_1 and f_2 generate the maximal ideal of the local ring $\mathcal{O}_{S,P}$. (See exercise 3.19 for the intuition behind this definition.)

If Γ_1 and Γ_2 are irreducible curves that meet everywhere transversally, then it is natural to define the intersection $\Gamma_1 \cdot \Gamma_2$ to be the number of intersection points. The next theorem says that this definition can be extended in a natural way to all divisors.

Theorem 7.2. *There is a unique symmetric bilinear pairing*

$$\mathrm{Div}(X) \times \mathrm{Div}(X) \longrightarrow \mathbb{Z}, \qquad (D_1, D_2) \longmapsto D_1 \cdot D_2,$$

with the following two properties:
(i) *If Γ_1 and Γ_2 are irreducible curves on S that meet everywhere transversally, then $\Gamma_1 \cdot \Gamma_2 = \#(\Gamma_1 \cap \Gamma_2)$.*
(ii) *If $D, D_1, D_2 \in \mathrm{Div}(S)$ are divisors with $D_1 \sim D_2$, then $D \cdot D_1 = D \cdot D_2$.*

PROOF. (Sketch) Given divisors $D_1, D_2 \in \mathrm{Div}(S)$, one uses ampleness and a Bertini theorem to find divisors $D_1', D_2' \in \mathrm{Div}(S)$ with $D_1' \sim D_1$, $D_2' \sim D_2$, and such that D_1' and D_2' are sums of irreducible curves that meet each other transversally. Then $D_1' \cdot D_2'$ is defined using linearity and (i). One then checks that the answer is independent of the choice of D_1' and D_2'. For details, see Hartshorne [1, V.1.1].   □

Example 7.3. We have seen ((7.1) and exercise 3.18) that the degree map defines an isomorphism $\mathrm{Pic}(\mathbb{P}^2) \cong \mathbb{Z}$. Let $\Gamma_1, \Gamma_2 \subset \mathbb{P}^2$ be curves of degrees n_1, n_2 respectively, and let $H_1, H_2 \subset \mathbb{P}^2$ be (distinct) lines. Then

$$\deg(\Gamma_i) = n_i = \deg(n_i H_i), \quad \text{which implies that} \quad \Gamma_i \sim n_i H_i.$$

Further, $H_1 \cdot H_2 = 1$, since distinct lines in $\mathbb{P}^2$ intersect transversally in a single point, so we can compute

$$\Gamma_1 \cdot \Gamma_2 = (n_1 H_1) \cdot (n_2 H_2) = n_1 n_2 (H_1 \cdot H_2) = n_1 n_2 = \deg(\Gamma_1) \deg(\Gamma_2).$$

The equality $\Gamma_1 \cdot \Gamma_2 = \deg(\Gamma_1) \deg(\Gamma_2)$ is called Bezout's theorem for the plane.

Theorem 7.2 is a powerful existence theorem, but it does not give a very practical method for computing the intersection $D_1 \cdot D_2$. In principle, one can find divisors $D_1' \sim D_1$ and $D_2' \sim D_2$ which intersect transversally and then count the number of points in $D_1' \cap D_2'$, but in practice it is better to assign multiplicities to the points in $D_1 \cap D_2$. This is done in the following way.

Let $D \in \mathrm{Div}(S)$ be a divisor, and let $P \in S$. A *local equation for D at P* is a function $f \in k(S)^*$ with the property that

$$P \notin D - \mathrm{div}(f).$$

Notice that if $D = \Gamma$ is an irreducible curve, then this is equivalent to the condition that $\mathrm{ord}_\Gamma(f) = 1$ and that $\mathrm{ord}_{\Gamma'}(f) = 0$ for all other irreducible curves Γ' containing P.

Now let $D_1, D_2 \in \mathrm{Div}(S)$ be divisors, and let $P \in S$ be a point which does not lie on a common component of D_1 and D_2. Choose local equations $f_1, f_2 \in k(S)^*$ for D_1, D_2 respectively. The *(local) intersection index of D_1 and D_2 at P* is defined to be the quantity

$$(D_1 \cdot D_2)_P = \dim_k \mathcal{O}_{S,P}/(f_1, f_2).$$

Notice that $(D_1 \cdot D_2)_P = 0$ if $P \notin D_1 \cap D_2$, since if $P \notin D_i$, then $f_i = 1$ will be a local equation for D_i at P. The next result explains how the local intersection indices can be used to calculate the global intersection number $D_1 \cdot D_2$.

Proposition 7.4. Let $D_1, D_2 \in \mathrm{Div}(S)$ be divisors with no common components. Then the local intersection index $(D_1 \cdot D_2)_P$ is finite for all $P \in S$, and

$$D_1 \cdot D_2 = \sum_{P \in D_1 \cap D_2} (D_1 \cdot D_2)_P.$$

PROOF. See Hartshorne [1, V.1.4]. □

Example 7.5. The local intersection indices are comparatively easy to calculate. As illustration, we will compute the intersection index of

$$\Gamma_1 : Y^2 Z = X^3 \qquad \text{and} \qquad \Gamma_2 : YZ = X^2$$

in $\mathbb{P}^2$ at the point $P = [0, 0, 1]$. (See Figure 3.1.) We dehomogenize $x = X/Z$, $y = Y/Z$, so the local ring at P is

$$\mathcal{O}_{\mathbb{P}^2, P} = k[x, y]_{(0,0)} = \left\{ \frac{f}{g} \in k(x, y) : g(0, 0) \neq 0 \right\}.$$

Then

$$\frac{k[x, y]_{(0,0)}}{(y^2 - x^3, y - x^2)} \cong \frac{k[x]_0}{(x^4 - x^3)} \cong \frac{k[x]_0}{(x^3)} \cong k + kx + kx^2 + kx^3,$$

where the middle equality follows from the fact that $x - 1$ is a unit in $k[x]_0$. Hence

$$(\Gamma_1 \cdot \Gamma_2)_P = \dim_k \frac{k[x, y]_{(0,0)}}{(y^2 - x^3, y)} = \dim_k (k + kx + kx^2 + kx^3) = 4.$$

Notice that Γ_1 is singular at P, but it has a unique tangent line there which is the same as the tangent line to Γ_2 at P. This explains why $(\Gamma_1 \cdot \Gamma_2)_P$ is so large.

Remark 7.6. Proposition 7.4 gives a method for computing the intersection number $D_1 \cdot D_2$ when the divisors D_1 and D_2 have no common components. However, one frequently wants to compute the intersection of divisors with components in common. An important example is the *self-intersection* $D^2 = D \cdot D$ of a divisor D. This cannot be calculated directly using (7.4). One approach to calculating D^2 is to find a $D' \sim D$ such that D and D' have no common components, and then compute $D \cdot D'$. For example, if $\Gamma \subset \mathbb{P}^2$ is a curve of degree n, then the computation in (7.3) shows that $\Gamma^2 = n^2$. The argument in (7.3) works for self-intersections because $\Gamma \sim nH$ for any line $H \subset \mathbb{P}^2$. In general, it may be difficult to find an appropriate D'. Another approach to computing self-intersections is to use the adjunction formula; see Hartshorne [1, V.1.5].

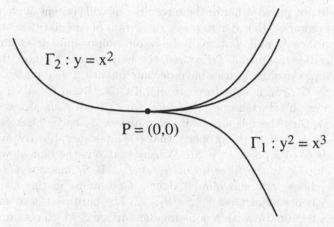

The Curves $\Gamma_1 : y^2 = x^3$ and $\Gamma_2 : y = x^2$ Near $P = (0,0)$

Figure 3.1

If C is a (possibly singular) curve, then there is a unique non-singular projective curve C' such that C is birationally equivalent to C'. (See [AEC, II.2.5] or Hartshorne [1, I.6.11,I.6.12].) The situation for surfaces is more complicated. It is true that every surface is birationally equivalent to a non-singular projective surface. However, if S is a non-singular projective surface, we can always blow up a point $P \in S$ to produce a new non-singular surface S' that is birationally equivalent to S but not isomorphic to S. The blown-up surface has the property that there is a *birational morphism* $S' \to S$; that is, the map $S' \to S$ is a morphism, and it has an inverse $S \to S'$ that is a rational map. This leads us to make the following definition.

Definition. A surface S is *relatively minimal* if it has the following two properties:

(i) S is a non-singular projective surface.

(ii) If S' is another non-singular projective surface, and if $\phi : S \to S'$ is a birational morphism, then ϕ is an isomorphism.

Theorem 7.7. *Every surface S_0 is birationally equivalent to a relatively minimal surface S. If the original surface S_0 is non-singular, then there is a birational morphism $S_0 \to S$.*

PROOF. This theorem is really an amalgamation of two important results in the theory of algebraic surfaces. First, resolution of singularities tells us that every surface is birationally equivalent to a non-singular projective surface. See Hartshorne [1, V.3.8.1] for a discussion of resolution and a

history of its proof. Second, every non-singular projective surface is birationally equivalent to a relatively minimal surface (Hartshorne [1, V.5.8]).

We will not prove either of these results but will content ourselves with
a few brief remarks. In order to prove resolution of singularities, one starts
with the surface S_0 and continually blows up points and curves until all of
the singularities disappear. Of course, the hard part is to show that after
each blow-up, the singularities have become quantitatively better.

A curve $C \subset S_0$ is called *exceptional* if $C \cong \mathbb{P}^1$ and $C^2 = -1$. Castelnuovo's criterion (Hartshorne [1, V.5.7]) says that if C is an exceptional
curve on S_0, then there is a non-singular surface S_1 and a birational morphism $\phi : S_0 \to S_1$ with the property that ϕ is an isomorphism away from C
and ϕ sends C to a point $P_1 \in S_1$. We say that ϕ is the blow-down of the
curve C, since ϕ is the blow-up of S_1 at P_1. If S_1 has any exceptional
curves, we choose one and blow it down. Continuing in this fashion, we
obtain a sequence of surfaces $S_0, S_1, S_2, \ldots$. The hard part is to show that
this process terminates with a non-singular surface S which contains no exceptional curves. Then one shows that such a surface is relatively minimal.
See Hartshorne [1, V.5.8] for details. This also proves the last part of (7.7),
since the blow-down maps $S_0 \to S_1 \to S_2 \to \cdots$ are all morphisms. $\quad\square$

§8. The Geometry of Fibered Surfaces

An elliptic surface $\pi : \mathcal{E} \to C$ is an example of a *fibered surface*; that is, a
surface that is described as a collection of fibers $\mathcal{E}_t = \pi^{-1}(t)$ parametrized
by the points t of a curve C. Other examples of fibered surfaces include
ruled surfaces, which are surfaces whose fibers are all isomorphic to $\mathbb{P}^1$
(Hartshorne [1, V §2]), and products $C_1 \times C_2$, which can be made into
fibered surfaces in two ways by using the projections onto either the first
or second factor. In this section, we will prove some geometric properties
that are true for all fibered surfaces. In subsequent sections, we will apply
these results to our study of elliptic surfaces.

Definition. A *fibered surface* is a non-singular projective surface S, a non-
singular curve C, and a surjective morphism $\pi : S \to C$. For any $t \in C$,
the *fiber of S lying over t* is the curve $S_t = \pi^{-1}(t)$. Note that S_t will be a
non-singular curve for all but finitely many $t \in S$.

The irreducible divisors on a fibered surface naturally divide into two
different sorts, those that lie in a single fiber and those that cover C. More
precisely, let $\Gamma \subset S$ be an irreducible curve lying on a fibered surface $\pi :
S \to C$. Then π induces a map of curves $\pi : \Gamma \to C$ that is either constant
or surjective [AEC, II.2.3]. If it is constant, say $\pi(\Gamma) = \{t\}$, then Γ lies

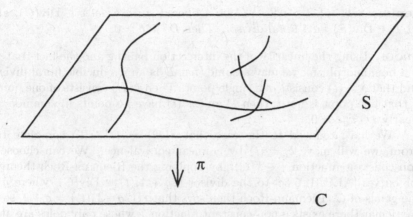

A Fibered Surface with Horizontal and Fibral Curves

Figure 3.2

entirely in the fiber S_t, and we call Γ *fibral*. If not, then $\pi : \Gamma \to C$ is a finite map of positive degree, and we call Γ *horizontal*. See Figure 3.2.

Definition. A divisor $D \in \mathrm{Div}(S)$ on a fibered surface S is called *fibral* if all of its components are fibral. D is called *horizontal* if all of its components are horizontal. Note that every divisor can be uniquely written as the sum of a horizontal divisor and a fibral divisor, since every irreducible curve is either horizontal or fibral.

Let $\pi : S \to C$ be a fibered surface. If $t \in C$, then the components of $\pi^{-1}(t)$ are irreducible fibral divisors. We assign multiplicities to these components in the following way. Let $u_t \in k(C)$ be a uniformizer at t, that is, $\mathrm{ord}_t(u) = 1$. Then $u_t \circ \pi$ is a function on S, so we can take its divisor, or more precisely that part of its divisor lying in the fiber S_t. Extending linearly, this gives us a homomorphism from $\mathrm{Div}(C)$ to $\mathrm{Div}(S)$.

Definition. Let $\pi : S \to C$ be a fibered surface, and for each $t \in C$, fix a uniformizer $u_t \in k(C)$ at t. We define a homomorphism

$$\pi^* : \mathrm{Div}(C) \longrightarrow \mathrm{Div}(S),$$

$$\sum_{t \in C} n_t(t) \longmapsto \sum_{t \in C} n_t \sum_{\Gamma \subset S_t} \mathrm{ord}_\Gamma(u_t \circ \pi)\Gamma,$$

where the inner sum on the right is over all irreducible curves Γ contained in the fiber $S_t = \pi^{-1}(t)$. It is easy to see that π^* is independent of the choice of uniformizers u_t, since if u_t' is another uniformizer, then $(u_t/u_t')(t) \neq 0, \infty$ at t. Hence $(u_t/u_t') \circ \pi$ is not identically 0 or ∞ on any component of S_t, so $\mathrm{ord}_\Gamma(u_t \circ \pi) = \mathrm{ord}_\Gamma(u_t' \circ \pi)$.

It is clear from the definition of π^* that the divisors in $\pi^*\big(\mathrm{Div}(C)\big)$ are fibral. We begin by showing that they have trivial intersection with all other fibral divisors.

Lemma 8.1. *Let* $\pi : S \to C$ *be a fibered surface, let* $\delta \in \mathrm{Div}(C)$, *and let* $D \in \mathrm{Div}(S)$ *be a fibral divisor. Then* $D \cdot \pi^*\delta = 0$.

PROOF. Using the linearity of the intersection pairing and the fact that π^* is a homomorphism, we may assume that D is an irreducible fibral divisor and that $\delta = (t)$ consists of a single point. Then $\pi(D)$ consists of one point. If that one point is not t, then D and $\pi^*(t)$ have no points in common, so clearly $D \cdot \pi^*\delta = 0$.

We have reduced to the case that $\pi(D) = \{t\}$. To complete the proof, we will move $\delta = (t)$ by a linear equivalence. We can choose a non-constant function $f \in k(C)^*$ by applying the Riemann-Roch theorem for curves [AEC, II.5.5c] to the divisor $(2g + 1)(t) \in \mathrm{Div}(C)$, where g is the genus of C. Riemann-Roch then says that $\ell\big((2g+1)(t)\big) = g+2$, so in particular there exists a non-constant function f whose only poles are at t.

Let $\mathrm{ord}_t(f) = -n$, and consider the divisor

$$\pi^*\big(n\delta + \mathrm{div}(f)\big) = n\pi^*\delta + \mathrm{div}(f \circ \pi).$$

The point t does not appear in the divisor $n\delta + \mathrm{div}(f)$, so the left-hand side has no points in common with S_t, and hence it has trivial intersection with D. On the other hand, $\mathrm{div}(f \circ \pi)$ is linearly equivalent to 0, so it has trivial intersection with every divisor on S. Intersecting both sides with D, we find that

$$0 = D \cdot \pi^*\big(n\delta + \mathrm{div}(f)\big) = D \cdot \big(n\pi^*\delta + \mathrm{div}(f \circ \pi)\big) = nD \cdot \pi^*\delta.$$

Hence $D \cdot \pi^*\delta = 0$.   □

We now show that the intersection pairing is negative semi-definite when it is restricted to fibral divisors, and we calculate its null space.

Proposition 8.2. *Let* $\pi : S \to C$ *be a fibered surface, and let* $D \in \mathrm{Div}(S)$ *be a fibral divisor on* S.
(a) $D^2 \leq 0$.
(b) $D^2 = 0$ *if and only if* $D \in \pi^*\big(\mathrm{Div}(C) \otimes \mathbb{Q}\big)$. *In other words,* $D^2 = 0$ *if and only if there is a divisor* $\delta \in \mathrm{Div}(C)$ *such that* $aD = b\pi^*\delta$ *for some non-zero integers* $a, b \in \mathbb{Z}$.

PROOF. (a) Write $D = D_1 + \cdots + D_n$, where each D_i is contained in a different fiber. Then $D_i \cdot D_j = 0$ for $i \neq j$, since they have no points in common, which implies that $D^2 = D_1^2 + \cdots + D_n^2$. It thus suffices to prove the proposition for each D_i, so we may assume that D is contained in a single fiber, say $D \subset S_t$.

Write

$$F \stackrel{\text{def}}{=} \pi^*(t) = \sum_{i=0}^{r} n_i \Gamma_i$$

as a sum of irreducible divisors. It is clear from the definition of π^* that the n_i's are all positive. Our assumption that $D \subset S_t$ means that D has the form

$$D = \sum_{i=0}^{r} a_i \Gamma_i \quad \text{for some integers } a_i.$$

We rewrite D and define another divisor D' by the formulas

$$D = \sum_{i=0}^{r} \frac{a_i}{n_i}(n_i \Gamma_i), \quad \text{and} \quad D' = \sum_{i=0}^{r} \frac{a_i^2}{n_i^2}(n_i \Gamma_i).$$

Proposition 8.1 tells us that $D' \cdot F = F \cdot D' = 0$. We use this to compute

$$-2D^2 = D' \cdot F - 2D^2 + F \cdot D'$$

$$= \sum_{i,j=0}^{r} \frac{a_i^2}{n_i^2}(n_i \Gamma_i) \cdot (n_j \Gamma_j) - 2 \sum_{i,j=0}^{r} \frac{a_i a_j}{n_i n_j}(n_i \Gamma_i) \cdot (n_j \Gamma_j)$$

$$+ \sum_{i,j=0}^{r} \frac{a_j^2}{n_j^2}(n_i \Gamma_i) \cdot (n_j \Gamma_j)$$

$$= \sum_{i,j=0}^{r} \left(\frac{a_i}{n_i} - \frac{a_j}{n_j}\right)^2 (n_i \Gamma_i) \cdot (n_j \Gamma_j).$$

The terms with $i = j$ in this last sum are zero, so we find that

$$D^2 = -\frac{1}{2} \sum_{\substack{i,j=0 \\ i \neq j}}^{r} \left(\frac{a_i}{n_i} - \frac{a_j}{n_j}\right)^2 (n_i \Gamma_i) \cdot (n_j \Gamma_j).$$

For $i \neq j$, the divisors Γ_i and Γ_j are distinct irreducible divisors, so $\Gamma_i \cdot \Gamma_j \geq 0$. Further, as noted above, the multiplicities $n_0, \ldots, n_r$ are all positive, so

$$(n_i \Gamma_i) \cdot (n_j \Gamma_j) \geq 0 \quad \text{for all } i \neq j.$$

This immediately implies that $D^2 \leq 0$, which completes the proof of (a).
(b) Suppose now that $D^2 = 0$. Then the formula for D^2 shows that

$$\frac{a_i}{n_i} = \frac{a_j}{n_j} \quad \text{for all } i, j \text{ such that } \Gamma_i \cdot \Gamma_j > 0.$$

In other words, the ratios a_i/n_i and a_j/n_j will be the same if the divisors Γ_i and Γ_j have a point in common. On the other hand, it is a general fact that the fibers of a fibered surface are connected. This is a special case of Hartshorne [1, III.11.3], or see exercise 3.21. So given any two components Γ_i and Γ_j, we can find a sequence of components

$$\Gamma_i = \Gamma_{i_0}, \Gamma_{i_1}, \Gamma_{i_2}, \ldots, \Gamma_{i_m} = \Gamma_j$$

with $\Gamma_{i_k} \cdot \Gamma_{i_{k+1}} > 0$ for all $k = 0, 1, \ldots, m-1$. Hence $a_i/n_i = a_j/n_j$ for all i and j. Let $a = a_0/n_0 \in \mathbb{Q}$ be this common ratio. Then

$$D = \sum_{i=0}^{r} a_i \Gamma_i = \sum_{i=0}^{r} \frac{a_i}{n_i} n_i \Gamma_i = a \sum_{i=0}^{r} n_i \Gamma_i = aF \in \pi^*\big(\mathrm{Div}(C) \otimes \mathbb{Q}\big),$$

which completes the proof of Proposition 8.2.   □

Remark 8.2.3. With notation as in the proof of (8.2), consider the incidence matrix

$$I = (\Gamma_i \cdot \Gamma_j)_{0 \le i, j \le r}$$

which describes how the components of the fiber S_t intersect one another. Then (8.2) may be restated as follows: The quadratic form

$$\mathbb{Q}^r \longrightarrow \mathbb{Q}, \qquad \mathbf{a} \longmapsto {}^t\mathbf{a}I\mathbf{a},$$

is negative semi-definite, with one dimensional null space spanned by the vector $(n_0, \ldots, n_r)$. In particular, $\det(I) = 0$, but every $\det(I_{ii}) \ne 0$, where I_{ii} is the minor obtained by deleting the ith row and i^{th} column of I.

Next we show that for a large class of divisors it is possible to add on a fibral divisor so that the sum will have trivial intersection with all fibral divisors. We will use this construction in the next section to describe the canonical height in terms of intersection theory.

Proposition 8.3. *Let $\pi : S \to C$ be a fibered surface, and let $D \in \mathrm{Div}(S)$ be a divisor on S with the property that*

$$D \cdot \pi^*(t) = 0 \quad \text{for some (every) } t \in C.$$

(The quantity $D \cdot \pi^*(t)$ is independent of t; see exercises 3.22 and 3.23 or Hartshorne [1, exercise V.1.7].) Then there exists a fibral divisor $\Phi_D \in \mathrm{Div}(S) \otimes \mathbb{Q}$ such that

$$(D + \Phi_D) \cdot F = 0 \quad \text{for all fibral divisors } F \in \mathrm{Div}(S).$$

If Φ'_D is another divisor with this property, then

$$\Phi_D - \Phi'_D \in \pi^*\big(\mathrm{Div}(C) \otimes \mathbb{Q}\big).$$

In other words, Φ_D is uniquely determined by D up to divisors that come from $\tilde{C}$.

PROOF. We are going to try to write Φ_D in the form $\sum a_\Gamma \Gamma$ and solve for the coefficients a_Γ. More precisely, for every point $t \in C$, write

$$\pi^*(t) = \sum_{i=0}^{r_t} n_{ti} \Gamma_{ti}$$

as a sum of irreducible components. We set $a_{t0} = 0$ for all t. Further, when $r_t \geq 1$ we consider the following system of linear equations:

$$\sum_{i=1}^{r_t} a_{ti} \Gamma_{ti} \cdot \Gamma_{tj} = -D \cdot \Gamma_{tj}, \qquad 1 \leq j \leq r_t.$$

Note we are discarding the 0th-component Γ_{t0}, so this is a system of r_t equations in the r_t variables a_{ti}. Proposition 8.2 says that the incidence matrix

$$(\Gamma_{ti} \cdot \Gamma_{tj})_{1 \leq i,j \leq r_t}$$

has non-zero determinant (see also (8.2.3)), so this system of equations has a unique solution in rational numbers $a_{ti} \in \mathbb{Q}$.

We claim that the divisor

$$\Phi_D = \sum_{t \in C} \sum_{i=0}^{r_t} a_{ti} \Gamma_{ti}$$

has the desired property. Note that this is a finite sum, since $r_t = 0$ for all but finitely many t, and $a_{t0} = 0$. To check that Φ_D works, it suffices by linearity to show that $(D + \Phi_D) \cdot F = 0$ for every irreducible fibral divisor F. Each irreducible fibral divisor has the form $F = \Gamma_{tj}$ for some $t \in C$ and some $0 \leq j \leq r_t$. We consider three cases.

First, if $r_t = 0$, then $F = \Gamma_{t0} = \pi^*(t)$. Using (8.1) and the assumption that $D \cdot \pi^*(t) = 0$, we find that

$$(D + \Phi_D) \cdot F = D \cdot \pi^*(t) + \Phi_D \cdot \pi^*(t) = 0.$$

Second, suppose that $r_t \geq 1$ and $F = \Gamma_{tj}$ with $j \geq 1$. Then the fact that the a_{ti}'s give a solution to the system of linear equations allows us to compute

$$(D + \Phi_D) \cdot F = D \cdot \Gamma_{tj} + \sum_{t \in C} \sum_{i=1}^{r_t} a_{ti} \Gamma_{ti} \cdot \Gamma_{tj} = 0.$$

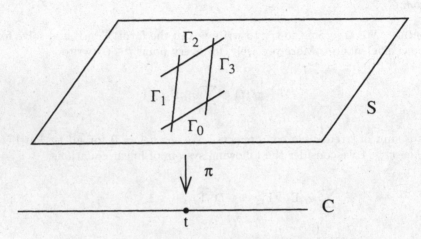

A Reducible Fiber on a Fibered Surface

Figure 3.3

Finally, we consider that case that $r_t \geq 1$ and $F = \Gamma_{t0}$. Then

$$0 = (D + \Phi_D) \cdot \pi^*(t) \quad \text{from (8.1)}$$

$$= \sum_{i=0}^{r_t} n_{ti}(D + \Phi_D) \cdot \Gamma_{ti} \quad \text{since } \pi^*(t) = \sum n_{ti}\Gamma_{ti}$$

$$= n_{t0}(D + \Phi_D) \cdot \Gamma_{t0} \quad \text{from the previous case.}$$

This completes the proof that $(D + \Phi_D) \cdot F = 0$ for all fibral divisors F.

It remains to show that Φ_D is unique up to addition of a divisor from C. Suppose Φ'_D is another divisor with the same property. Then for every fibral divisor F we have

$$(\Phi_D - \Phi'_D) \cdot F = (D + \Phi_D) \cdot F - (D + \Phi'_D) \cdot F = 0.$$

But $\Phi_D - \Phi'_D$ is itself fibral, so $(\Phi_D - \Phi'_D)^2 = 0$. It follows from (8.2b) that $\Phi_D - \Phi'_D$ is in $\pi^*(\text{Div}(C) \otimes \mathbb{Q})$. $\qquad \square$

Example 8.3.1. Let $\pi : S \to C$ be a fibered surface, and suppose that the fiber S_t consists of four components arranged in the shape of a square with transversal intersections, as illustrated in Figure 3.3. In other words,

$$\pi^*(t) = \Gamma_0 + \Gamma_1 + \Gamma_2 + \Gamma_3 \quad \text{with} \quad \Gamma_i \cdot \Gamma_j = \begin{cases} 1 & \text{if } i - j \equiv \pm 1 \,(\text{mod } 4), \\ 0 & \text{if } i - j \equiv 2 \,(\text{mod } 4). \end{cases}$$

We can use (8.1) to compute the self-intersections of the components. For example,

$$0 = \Gamma_0 \cdot \pi^*(t) = \Gamma_0^2 + \Gamma_0 \cdot \Gamma_1 + \Gamma_0 \cdot \Gamma_2 + \Gamma_0 \cdot \Gamma_3 = \Gamma_0^2 + 1 + 0 + 1,$$

so $\Gamma_0^2 = -2$, and similarly $\Gamma_i^2 = -2$ for the other i's. Thus the incidence matrix for this fiber is

$$I = (\Gamma_i \cdot \Gamma_j)_{0 \leq i,j \leq 3} = \begin{pmatrix} -2 & 1 & 0 & 1 \\ 1 & -2 & 1 & 0 \\ 0 & 1 & -2 & 1 \\ 1 & 0 & 1 & -2 \end{pmatrix}.$$

Suppose now that $D \in \text{Div}(S)$ is a (horizontal) divisor with

$$D \cdot \Gamma_0 = -1, \quad D \cdot \Gamma_1 = 1, \quad D \cdot \Gamma_2 = 0, \quad D \cdot \Gamma_3 = 0.$$

For example, D might consist of two curves $D = D_1 - D_0$ each of which maps isomorphically $\pi : D_i \to C$, with D_1 going through Γ_1 and D_0 going through Γ_0. To find the part of Φ_D lying over t, call it $\Phi_{D,t}$, we take $\Phi_{D,t} = a_1\Gamma_1 + a_2\Gamma_2 + a_3\Gamma_3$, set $(D + \Phi_{D,t}) \cdot \Gamma_i = 0$ for $i = 1, 2, 3$, and solve for the a_i's. Doing this gives

$$\Phi_{D,t} = \frac{3}{4}\Gamma_1 + \frac{1}{2}\Gamma_2 + \frac{1}{4}\Gamma_3.$$

The reader can check that $(D + \Phi_{D,t}) \cdot \Gamma_i = 0$ for $0 \leq i \leq 4$. Similarly, if D were to satisfy

$$D \cdot \Gamma_0 = -1, \quad D \cdot \Gamma_1 = 0, \quad D \cdot \Gamma_2 = 1, \quad D \cdot \Gamma_3 = 0,$$

then

$$\Phi_{D,t} = \frac{1}{2}\Gamma_1 + \Gamma_2 + \frac{1}{2}\Gamma_3.$$

See exercise 3.24 for a generalization to the case that the fiber is an n-gon with transversal intersections.

The final topic for this section is minimal models of fibered surfaces. These will be minimal models which respect the fact that the surface is fibered. More precisely, we might say that a fibered surface $S \to C$ is *relatively minimal* if for every fibered surface $S' \to C$, every birational map $S \to S'$ commuting with the maps to C is a morphism. In the case that the non-singular fibers of $S \to C$ have genus at least 1, then it turns out that there is a unique relatively minimal model. Further, this model will have the stronger minimality property described in the next theorem.

Theorem 8.4. *Let $S \to C$ be a fibered surface with the property that its non-singular fibers are curves of genus at least 1. Then there exists a fibered surface $S^{\min} \to C$ and a birational morphism $\phi : S \to S^{\min}$ commuting with the maps to C with the following property:*

Let $S' \to C$ be a fibered surface, and let $\phi' : S' \to S$ be a birational map commuting with the maps to C. Then the rational map $\phi \circ \phi'$ extends to a morphism. In other words, the top line of the commutative diagram

$$S' \quad \overset{\phi'}{\dashrightarrow} \quad S \quad \overset{\phi}{\longrightarrow} \quad S^{\min}$$

$$\searrow \quad \downarrow \quad \swarrow$$

$$C$$

extends to a morphism.

PROOF. The basic idea is as follows. For any given S, let $S^{\min}$ be obtained from S by blowing down all of the exceptional curves on the reducible fibers. Next, given an S' birational to S, take the resulting birational map $S'^{\min} \to S^{\min}$ and factor it into the smallest number of quadratic transformations

$$S'^{\min} = S_0 \dashrightarrow S_1 \dashrightarrow S_2 \dashrightarrow \cdots \dashrightarrow S_n = S^{\min}.$$

Then by studying the behavior of the exceptional curves on these quadratic transformations, one shows that it is possible eliminate one of the "blow-up–blow-down pairs." In other words, if $n \geq 1$, then one shows that $S'^{\min}$ and $S^{\min}$ are connected by a smaller chain of quadratic transformations. Hence $S'^{\min}$ and $S^{\min}$ are isomorphic, which gives the desired result.

Unfortunately, we do not have at our disposal the tools needed to turn this brief sketch into a rigorous proof. We refer the reader to Lichtenbaum [1, Thm. 4.4] or Shafarevich [2, p. 131] for the complete proof of Theorem 8.4. □

Definition. It is clear that the surface $S^{\min}$ described in (8.4) is uniquely determined up to a unique isomorphism commuting with the maps to C. A fibered surface $S \to C$ is called a *minimal fibered surface (over C)* if it is equal to $S^{\min}$.

Corollary 8.4.1. *Let $\pi : S \to C$ be a minimal fibered surface over C, and let $\tau : S \to S$ be a birational map commuting with the map to C (i.e., $\pi \circ \tau = \pi$). Then τ is a morphism.*

PROOF. By assumption, S is minimal, so the map $\phi : S \to S^{\min}$ in (8.4) is an isomorphism. Now applying (8.4) with $S = S'$ and $\phi' = \tau$, we deduce that the composition

$$S \overset{\tau}{\longrightarrow} S \overset{\phi}{\longrightarrow} S^{\min}$$

is a morphism. Hence the same is true of $\tau = \phi^{-1} \circ (\phi \circ \tau)$. □

§9. The Geometry of Elliptic Surfaces

Let $\pi : \mathcal{E} \to C$ be a minimal elliptic surface, and let E/K be the associated elliptic curve over the function field $K = k(C)$ of C (3.8). Recall (3.10c) that each point $P \in E(K)$ corresponds to a section $\sigma_P : C \to \mathcal{E}$. We define a *translation-by-P map*

$$\tau_P : \mathcal{E} \longrightarrow \mathcal{E}$$

on $\mathcal{E}$ by using the translation-by-$\sigma_P(t)$ map on each non-singular fiber $\mathcal{E}_t$. It is clear that τ_P is a birational map, since it is certainly given by rational functions and it has the rational inverse τ_{-P}. The minimality of $\mathcal{E}$ then implies that τ_P extends to a morphism. We record this important fact in the following proposition.

Proposition 9.1. *Let $\pi : \mathcal{E} \to C$ be a minimal elliptic surface with associated elliptic curve E/K.*
(a) *For any point $P \in E(K)$, the translation-by-P map*

$$\tau_P : \mathcal{E} \longrightarrow \mathcal{E}$$

extends to an automorphism of $\mathcal{E}$.
(b) *Let*

$$\operatorname{Aut}(\mathcal{E}/C) = \{\text{automorphisms } \tau : \mathcal{E} \to \mathcal{E} \text{ satisfying } \pi \circ \tau = \pi\}.$$

Then the map
$$E(K) \longrightarrow \operatorname{Aut}(\mathcal{E}/C), \qquad P \longmapsto \tau_P,$$

is a homomorphism.

PROOF. (a) This a special case of Corollary 8.4.1, which says that any birational map of a minimal fibered surface to itself extends to a morphism.
(b) If the fiber $\mathcal{E}_t$ is non-singular, then τ_P maps $\mathcal{E}_t$ to itself by definition. It follows that $\pi \circ \tau_P = \pi$ on all non-singular fibers. But the non-singular fibers are Zariski dense in $\mathcal{E}$, and a morphism is determined by its values on any Zariski dense set (Hartshorne [1, I.4.1]), so $\pi \circ \tau_P = \pi$ on all of $\mathcal{E}$. This proves that $\tau_P \in \operatorname{Aut}(\mathcal{E}/C)$. Similarly, the identity $\tau_{P+Q} = \tau_P \circ \tau_Q$ is clearly true on all non-singular fibers, so it is true everywhere. Finally, τ_O is the identity map, which completes the proof that $E(K) \to \operatorname{Aut}(\mathcal{E}/C)$ is a homomorphism. $\square$

Let $P \in E(K)$ with corresponding section $\sigma_P : C \to \mathcal{E}$. The image $\sigma_P(C)$ of σ_P is a curve on the surface $\mathcal{E}$, which we can think of as a divisor on $\mathcal{E}$. We will write

$$(P) \in \operatorname{Div}(\mathcal{E})$$

for this divisor. It is important to note that the divisors

$$(P) + (Q) \qquad \text{and} \qquad (P + Q)$$

are very different. The former is the sum of the two divisors (P) and (Q) in $\text{Div}(\mathcal{E})$, whereas the latter is the image of the section $\sigma_{P+Q} = \sigma_P + \sigma_Q$ which is defined using the group law on E. The following proposition shows how they are related.

Proposition 9.2. *With notation as above, let $P_1, \ldots, P_r \in E(K)$ be points, and let $n_1, \cdots, n_r \in \mathbb{Z}$ be integers such that*

$$[n_1]P_1 + \cdots + [n_r]P_r = O.$$

Let $n = n_1 + \cdots + n_r$. Then the divisor

$$n_1(P_1) + \cdots + n_r(P_r) - n(O) \in \text{Div}(\mathcal{E})$$

is linearly equivalent to a fibral divisor.
In particular, for all $P, Q \in E(K)$, the divisor

$$(P + Q) - (P) - (Q) + (O)$$

is linearly equivalent to a fibral divisor.

PROOF. If $P \in E(K)$, our notation (P) is potentially ambiguous, since we could mean either the divisor on the curve E/K consisting of the point P, or the divisor on the surface $\mathcal{E}$ consisting of the curve $\sigma_P(C)$. To resolve this difficulty, we will denote the former by $(P)_E$ and the latter by $(P)_{\mathcal{E}}$.

Fix a Weierstrass equation for E/K, say

$$E : y^2 = x^3 + Ax + B, \qquad A, B \in K,$$

and consider the divisor

$$D = n_1(P_1)_E + \cdots + n_r(P_r)_E - n(O)_E \in \text{Div}(E).$$

By assumption, D has degree 0 and sums to the zero element of $E(K)$. Applying [AEC, III.3.5] to the divisor D on the elliptic curve E/K, we find that D is linearly equivalent to 0. Thus there is a function $f \in K(E)$ such that

$$D = (\text{div } f)_E \in \text{Div}(E).$$

The relationship (3.8) between E and $\mathcal{E}$ says that $K(E) \cong k(\mathcal{E})$, so we can consider f as an algebraic function on the surface $\mathcal{E}$. When we compute its divisor on $\mathcal{E}$, we find that

$$(\text{div } f)_{\mathcal{E}} \qquad \text{and} \qquad n_1(P_1)_{\mathcal{E}} + \cdots + n_r(P_r)_{\mathcal{E}} - n(O)_{\mathcal{E}}$$

are almost the same. To see this, note that $f \in K(E) = k(C)(x, y)$ is a rational function in x and y with coefficients in $k(C)$. Hence for all but finitely many $t \in C$, we can evaluate those coefficients to get a function $f_t \in k(\mathcal{E}_t)$ whose divisor will be precisely

$$n_1\big(\sigma_{P_1}(t)\big) + \cdots + n_r\big(\sigma_{P_r}(t)\big) - n\big(\sigma_0(t)\big) \in \mathrm{Div}(\mathcal{E}_t).$$

This proves that the difference

$$(\mathrm{div}\, f)_\mathcal{E} - \big(n_1(P_1)_\mathcal{E} + \cdots + n_r(P_r)_\mathcal{E} - n(O)_\mathcal{E}\big)$$

is contained in finitely many fibers, hence it is fibral. $\qquad\square$

For any point $P \in E(K)$, the divisor $(P) - (O) \in \mathrm{Div}(\mathcal{E})$ satisfies

$$\big((P) - (O)\big) \cdot \pi^*(t) = 0 \qquad \text{for all } t \in C.$$

This is true because the image of a section will intersect a fiber $\pi^*(t)$ exactly once. (See exercise 3.22.) This shows that we can apply (8.3) to the divisor $(P) - (O)$, as in the following definition.

Definition. For each point $P \in E(K)$, let $\Phi_P \in \mathrm{Div}(\mathcal{E}) \otimes \mathbb{Q}$ be a fibral divisor so that the divisor

$$D_P \stackrel{\mathrm{def}}{=} (P) - (O) + \Phi_P$$

satisfies

$$D_P \cdot F = 0 \qquad \text{for all fibral divisors } F \in \mathrm{Div}(\mathcal{E}).$$

Such a divisor exists by (8.3) and the remarks made above. Then we define a pairing on $E(K)$ by the formula

$$\langle \cdot, \cdot \rangle : E(K) \times E(K) \longrightarrow \mathbb{Q},$$
$$\langle P, Q \rangle = -D_P \cdot D_Q.$$

The next result shows that this geometrically defined pairing is equal to the canonical height pairing (4.3), which justifies our use of the same notation for the two pairings! This geometric construction of the canonical height is due to Manin [1]. See also Shioda [2] for a more detailed analysis of the induced Euclidean structure on the lattice $E(K)/E(K)_{\mathrm{tors}}$.

Theorem 9.3. (Manin [1]) *Let* $\pi : \mathcal{E} \to C$ *be a minimal elliptic surface with associated elliptic curve* E/K. *The pairing*

$$\langle \cdot, \cdot \rangle : E(K) \times E(K) \longrightarrow \mathbb{Q}, \qquad \langle P, Q \rangle = -D_P \cdot D_Q,$$

defined above has the following two properties:
(a) $\langle \cdot, \cdot \rangle$ *is bilinear.*

(b) $\langle P, P \rangle = h(P) + O(1)$ for all $P \in E(K)$, where we recall from §4 that $h(P) = h(x_P)$ is the degree of the map $x_P : C \to \mathbb{P}^1$.

Hence this pairing agrees with the canonical height pairing defined in (4.3). In particular, $\hat{h}(P) = \frac{1}{2}\langle P, P \rangle \in \mathbb{Q}$ for all $P \in E(K)$.

PROOF. (a) Let $P, Q, R \in E(K)$ be any three points. Applying (9.2), we choose a fibral divisor F such that

$$(Q + R) - (Q) - (R) + (O) \sim F.$$

Then using standard properties of the intersection pairing (7.2), we compute

$$
\begin{aligned}
\langle P, Q + R \rangle &- \langle P, Q \rangle - \langle P, R \rangle \\
&= -D_P \cdot D_{Q+R} + D_P \cdot D_Q + D_P \cdot D_R \\
&= -D_P \cdot \big((Q+R) - (O) + \Phi_{Q+R}\big) + D_P \cdot \big((Q) - (O) + \Phi_Q\big) \\
&\quad + D_P \cdot \big((R) - (O) + \Phi_R\big) \\
&= -D_P \cdot \big((Q+R) - (Q) - (R) + (O) + \Phi_{Q+R} - \Phi_Q - \Phi_R\big) \\
&= -D_P \cdot (F + \Phi_{Q+R} - \Phi_Q - \Phi_R) \\
&= 0.
\end{aligned}
$$

The last line follows from the fact that D_P has trivial intersection with all fibral divisors. Hence

$$\langle P, Q + R \rangle - \langle P, Q \rangle - \langle P, R \rangle = 0.$$

It is also easy to check that the pairing is symmetric,

$$\langle P, Q \rangle = -D_P \cdot D_Q = -D_Q \cdot D_P = \langle Q, P \rangle.$$

This completes the proof that the pairing is bilinear.

(b) Directly from the definition we find that

$$
\begin{aligned}
\langle P, P \rangle &= -D_P \cdot D_P \\
&= -\big((P) - (O) + \Phi_P\big) \cdot D_P \\
&= -\big((P) - (O)\big) \cdot D_P \quad \text{since } D_P \cdot (\text{fibral}) = 0 \\
&= 2(P) \cdot (O) - (P)^2 - (O)^2 + \big((P) - (O)\big) \cdot \Phi_P.
\end{aligned}
$$

Our first claim is that $(P)^2$ does not depend on P. To see this, consider the translation-by-P map

$$\tau_P : \mathcal{E} \longrightarrow \mathcal{E}.$$

We know from (9.1a) that τ_P extends to an automorphism of $\mathcal{E}$. It follows that $\tau_P^* D_1 \cdot \tau_P^* D_2 = D_1 \cdot D_2$ for any two divisors $D_1, D_2 \in \mathrm{Div}(\mathcal{E})$. Hence

$$(P) \cdot (P) = \tau_P^*(P) \cdot \tau_P^*(P) = (O) \cdot (O)$$

is independent of τ_P. Alternative approach: first show that the canonical divisor on $\mathcal{E}$ has the form $\pi^*\delta$, and then use the adjunction formula (Hartshorne [1, V.1.5]) to compute

$$(P)^2 = 2g(C) - 2 - (P) \cdot \pi^*\delta = 2g(C) - 2 - \deg \delta.$$

Our second observation is that although Φ_P depends on P, there are essentially only finitely many choices for Φ_P. More precisely, for each $t \in C$, write

$$\pi^*(t) = \sum_{i=0}^{r_t} n_{ti} \Gamma_{ti}$$

as a sum of irreducible components. Note that $r_t = 0$ for all but finitely many $t \in C$. Looking back at the proof of (8.3), we see that Φ_P can be written in the form

$$\Phi_P = \sum_{t \in C} \sum_{i=1}^{r_t} a_{ti} \Gamma_{ti} + \pi^*(\delta)$$

for some $\delta \in \mathrm{Div}(C)$, where the integers a_{ti} are uniquely determined by the finitely many intersection indices

$$((P) - (O)) \cdot \Gamma_{tj}, \qquad t \in C, 1 \le j \le r_t.$$

But every $(P) \cdot \Gamma_{tj}$ is either 0 or 1, so as we take different points $P \in E(K)$, there will be only finitely many possibilities for the a_{ti}'s. Further,

$$((P) - (O)) \cdot \pi^*(\delta) = (P) \cdot \pi^*(\delta) - (O) \cdot \pi^*(\delta) = \deg(\delta) - \deg(\delta) = 0$$

from exercise 3.22(b), so we find that

$$((P) - (O)) \cdot \Phi_P = ((P) - (O)) \cdot \left(\sum_{t \in C} \sum_{i=1}^{r_t} a_{ti} \Gamma_{ti} \right)$$

can take on only finitely many values as we vary $P \in E(K)$.

Combining these two observations with the calculation from above yields

$$\langle P, P \rangle = 2(P) \cdot (O) + O(1) \qquad \text{for } P \in E(K).$$

It remains to calculate the intersection index $(P) \cdot (O)$. Adjusting the $O(1)$ if necessary, we may assume that $[2]P \ne O$. Fix a Weierstrass equation for E,

$$E : y^2 = x^3 + Ax + B,$$

and write $P = (x_P, y_P)$. Changing coordinates if necessary, we may assume x_P and y_P have no poles in common with the poles of A and B.

Let $t \in C$. We will compute the local intersection index of (P) and (O) at the point $\sigma_O(t)$. We denote this local intersection by $(P \cdot O)_t$. If $\mathrm{ord}_t(x_P) \geq 0$, that is, if $x_P(t) \neq \infty$, then (P) and (O) do not intersect on the fiber $\mathcal{E}_t$, so $(P \cdot O)_t = 0$. Suppose now that $\mathrm{ord}_t(x_P) < 0$, so the equation for E tells us that

$$3 \,\mathrm{ord}_t(x_P) = 2 \,\mathrm{ord}_t(y_P).$$

We make a change of coordinates $w = x/y$, $z = 1/y$, so E now has the equation

$$E : z = w^3 + Awz^2 + Bz^3,$$

and $P = (w_P, z_P) = (x_P/y_P, 1/y_P)$. Also let $u \in k(C) \subset k(\mathcal{E})$ be a uniformizer at t, so we are looking at the intersection of (P) and (O) at the point $(w, z, u) = (0, 0, 0)$. The local ring of $\mathcal{E}$ at this point is

$$\frac{k[w, z, u]_{(0,0,0)}}{(z - w^3 - Awz^2 - Bz^3)}.$$

Further, in this ring, (P) has the local equation $w - w_P = 0$ and (O) has the local equation $w = 0$, so by definition the intersection index $(P, O)_t$ is equal to the dimension over k of the vector space

$$\frac{k[w, z, u]_{(0,0,0)}}{(z - w^3 - Awz^2 - Bz^3, w - w_P, w)} \cong \frac{k[z, u]_{(0,0)}}{(z - Bz^3, w_P)} \cong \frac{k[u]_0}{(w_P)}.$$

Note the last equality follows from the fact that $z - Bz^3 = z(1 - Bz^2)$ and $1 - Bz^2$ is a unit in $k[z, u]_{(0,0)}$. If we write $w_P = u^e w_P'$ for some function w_P' that is neither 0 nor ∞ at t, then we have

$$\dim_k \frac{k[u]_0}{(w_P)} = \dim_k \frac{k[u]_0}{(u^e)} = e,$$

and also

$$e = \mathrm{ord}_t \, w_P = \mathrm{ord}_t(x_P/y_P) = -\frac{1}{2}\,\mathrm{ord}_t(x_P).$$

This proves that

$$(P \cdot O)_t = \begin{cases} 0 & \text{if } \mathrm{ord}_t(x_P) \geq 0, \\ -\frac{1}{2}\,\mathrm{ord}_t(x_P) & \text{if } \mathrm{ord}_t(x_P) < 0. \end{cases}$$

Adding over $t \in C$ gives

$$(P){\cdot}(O) = \sum_{t \in C}(P{\cdot}O)_t = \sum_{t \in C,\, \mathrm{ord}_t(x_P) < 0} -\frac{1}{2}\,\mathrm{ord}_t(x_P) = \frac{1}{2}\deg(x_P) = \frac{1}{2}h(P).$$

Hence

$$\langle P, P \rangle = 2(P) \cdot (O) + O(1) = h(P) + O(1),$$

which completes the proof of (b).

Let $g(P) = \frac{1}{2}\langle P, P \rangle$ be the quadratic form associated to our pairing. Then $g(P) = \frac{1}{2}h(P) + O(1)$ from (b), whereas the bilinearity in (a) tells us that $g(2P) = 4g(P)$. This shows that g satisfies properties (i) and (ii) of (4.3b), so by the uniqueness (4.3e) of the canonical height, we have $g = \hat{h}$.

$\square$

Theorem 9.3 shows that the canonical height on $E(K)$ can be computed using intersection theory. Our next goal is to define a natural pairing on a certain subgroup $E(K)_0$ of $E(K)$. This pairing takes its values in $\operatorname{Pic}(C)$, and the composition

$$E(K)_0 \times E(K)_0 \longrightarrow \operatorname{Pic}(C) \xrightarrow{\deg} \mathbb{Z}$$

will be the canonical height pairing. We begin by describing $E(K)_0$.

Let $P \in E(K)$ be a point and $\tau_P : \mathcal{E} \to \mathcal{E}$ the translation-by-P automorphism. We know from (9.1b) that τ_P gives an automorphism of each fiber $\mathcal{E}_t$. In particular, it must permute each of the components of $\mathcal{E}_t$.

Definition. Define a subset $E(K)_0$ of $E(K)$ by

$$E(K)_0 = \big\{ P \in E(K) : \tau_P(\Gamma) = \Gamma \text{ for all fibral curves } \Gamma \subset \mathcal{E} \big\}.$$

Lemma 9.4. $E(K)_0$ is a subgroup of finite index in $E(K)$.

PROOF. Let $P, Q \in E(K)_0$. From (9.1b) we know that $\tau_{P+Q} = \tau_P \circ \tau_Q$, so for any fibral curve Γ we have $\tau_{P+Q}(\Gamma) = \tau_P(\tau_Q(\Gamma)) = \tau_P(\Gamma) = \Gamma$. Therefore $P + Q \in E(K)_0$. Similarly, $\Gamma = \tau_O(\Gamma) = \tau_{-P+P}(\Gamma) = \tau_{-P}(\Gamma)$, so $-P \in E(K)_0$. This proves that $E(K)_0$ is a subgroup of $E(K)$.

For the second part, we observe that if $\mathcal{E}_t$ is an irreducible fiber, then clearly $\tau_P(\mathcal{E}_t) = \mathcal{E}_t$. Let $\{\Gamma_1, \ldots, \Gamma_r\}$ be the set of all components of the reducible fibers of $\mathcal{E}$. It is a finite set, since $\mathcal{E}$ has finitely many reducible fibers, and each reducible fiber has finitely many components. Then $E(K)$ acts on this set by

$$\begin{array}{ccc} E(K) & \longrightarrow & \operatorname{Aut}\{\Gamma_1, \ldots, \Gamma_r\} \cong \mathcal{S}_r, \\ P & \longmapsto & (\Gamma_i \mapsto \tau_P(\Gamma_i)). \end{array}$$

In other words, there is a homomorphism from $E(K)$ into the symmetric group $\mathcal{S}_r$ on r letters. From the definition of $E(K)_0$, the quotient group $E(K)/E(K)_0$ injects into $\mathcal{S}_r$, which proves that $E(K)_0$ is a subgroup of finite index.

$\square$

Remark 9.4.1. There is another way to characterize $E(K)_0$ in terms of the sections $\sigma_P : C \to \mathcal{E}$ associated to points $P \in E(K)$. Let $\sigma_0 : C \to \mathcal{E}$ be the zero-section. Then a point $P \in E(K)$ is in $E(K)_0$ if and only if the curves $\sigma_P(C)$ and $\sigma_0(C)$ hit the same component of every fiber of $\mathcal{E}$. We will study the group $E(K)_0$ and the quotient $E(K)/E(K)_0$ in greater generality and detail in the next chapter; see (IV.6.12), (IV.9.1), (IV.9.2) and exercise 4.25.

For any two points $P, Q \in E(K)$, Proposition 9.2 tells us that there is a fibral divisor $\Phi_{P,Q} \in \mathrm{Div}(\mathcal{E})$ satisfying

$$(P+Q) - (P) - (Q) + (O) \sim \Phi_{P,Q}.$$

Clearly, $\Phi_{P,Q}$ is determined by P and Q up to principal divisors, so its class in $\mathrm{Pic}(\mathcal{E})$ is well-defined. This gives a pairing on $E(K)$ with values in $\mathrm{Pic}(\mathcal{E})$. The next result shows that this pairing is quite nice when restricted to the subgroup $E(K)_0$.

Theorem 9.5. *Let $\pi : \mathcal{E} \to C$ be a minimal elliptic surface with associated elliptic curve E/K.*
(a) *Let $P, Q \in E(K)_0$. Then there exists a divisor $[P, Q] \in \mathrm{Div}(C)$ such that*

$$(P+Q) - (P) - (Q) + (O) \sim \pi^*([P, Q]).$$

The divisor $[P, Q]$ is determined by P and Q up to linear equivalence.
(b) *The pairing*

$$E(K)_0 \times E(K)_0 \longrightarrow \mathrm{Pic}(C), \qquad (P, Q) \longmapsto \mathrm{class}[P, Q],$$

is a well-defined symmetric bilinear pairing. (See also exercise 3.26.)
(c)

$$\langle P, Q \rangle = \deg[P, Q] \quad \text{for all } P, Q \in E(K)_0.$$

In particular, $\hat{h}(P) = \frac{1}{2} \deg[P, P]$ for all $P \in E(K)_0$.

PROOF. (a) For any two points $P, Q \in E(K)$, let $\Phi_{P,Q} \in \mathrm{Div}(\mathcal{E})$ be a fibral divisor satisfying

$$(P+Q) - (P) - (Q) + (O) \sim \Phi_{P,Q}$$

as described in (9.2). Then for any fibral divisor $F \in \mathrm{Div}(\mathcal{E})$,

$$
\begin{aligned}
\Phi_{P,Q} \cdot F &= (P+Q) \cdot F - (P) \cdot F - (Q) \cdot F + (O) \cdot F \\
&= \tau_Q(P) \cdot F - (P) \cdot F - \tau_Q(O) \cdot F + (O) \cdot F \\
&= \tau_Q((P) - (O)) \cdot F - ((P) - (O)) \cdot F \\
&= ((P) - (O)) \cdot \tau_{-Q}(F) - ((P) - (O)) \cdot F \\
&= 0 \quad \text{since } Q \in E(K)_0 \text{ implies } \tau_{-Q}(F) = F.
\end{aligned}
$$

But $\Phi_{P,Q}$ itself is fibral, so we deduce that $\Phi_{P,Q}^2 = 0$. It follows from (8.2b) that there is a divisor $[P,Q] \in \mathrm{Div}(C) \otimes \mathbb{Q}$ such that $\Phi_{P,Q} = \pi^*([P,Q])$. This will suffice for our purposes in this chapter, so we will leave it for the reader (exercise 3.28c) to show that $[P,Q]$ is actually in $\mathrm{Div}(C)$.

The divisor $\Phi_{P,Q}$ is clearly determined by P and Q up to linear equivalence on $\mathcal{E}$. In order to show that $[P,Q]$ is determined up to linear equivalence on C, we will prove that if $\delta \in \mathrm{Div}(C)$ satisfies $\pi^*\delta \sim 0$, then $\delta \sim 0$. Write $\pi^*\delta = \mathrm{div}(f)$ for some $f \in k(\mathcal{E})$. For all but finitely many $t \in C$ we can restrict f to the fiber $\mathcal{E}_t$ to get a rational function $f_t \in k(\mathcal{E}_t)$. By assumption, the poles and zero of f lie on finitely many fibers, so for almost all $t \in C$ we see that $f_t \in k(\mathcal{E}_t)$ has no zeros or poles. It follows that f_t is constant. Let $\sigma : C \to \mathcal{E}$ be any section, for example the zero section. Then the fact that f is constant on almost all fibers means that the function $f - f \circ \sigma \circ \pi$ is identically 0 on those fibers. But a rational function is determined by its values on any non-empty open set, so $f = f \circ \sigma \circ \pi$. Therefore

$$\pi^*\delta = \mathrm{div}(f) = \mathrm{div}(f \circ \sigma \circ \pi) = \pi^*(\mathrm{div}(f \circ \sigma)),$$

so $\delta = \mathrm{div}(f \circ \sigma)$ is a principal divisor on C. Notice that what this result really says is that the natural map $\pi^* : \mathrm{Pic}(C) \to \mathrm{Pic}(\mathcal{E})$ is injective.

(b) The pairing is well-defined from (a), and it is clearly symmetric. To see that it is bilinear, we let $P, Q, R \in E(K)_0$ and compute

$$\begin{aligned}
\Phi_{P,Q+R} &\sim (P+Q+R) - (P) - (Q+R) + (O)\\
&= (P+Q+R) - (P+R) - (Q+R) + (R)\\
&\qquad\qquad\qquad\qquad + (P+R) - (P) - (R) + (O)\\
&= \tau_R\big((P+Q) - (P) - (Q) + (O)\big) + \big((P+R) - (P) - (R) + (O)\big)\\
&\sim \tau_R(\Phi_{P,Q}) + \Phi_{P,R}\\
&= \Phi_{P,Q} + \Phi_{P,R}.
\end{aligned}$$

Note that the last equality is true because $R \in E(K)_0$, so τ_R fixes the fibral divisor $\Phi_{P,Q}$. Now write each $\Phi_{X,Y}$ as $\pi^*([X,Y])$ and use the fact proven above that $\pi^* : \mathrm{Pic}(C) \to \mathrm{Pic}(\mathcal{E})$ is injective. This yields the desired result,

$$[P,Q+R] \sim [P,Q] + [P,R].$$

(c) Let $P \in E(K)_0$. Then for every fibral divisor $F \in \mathrm{Div}(\mathcal{E})$ we have

$$(P) \cdot F = \tau_P(O) \cdot F = (O) \cdot \tau_{-P}(F) = (O) \cdot F.$$

In other words,

$$((P) - (O)) \cdot F = 0 \quad \text{for all fibral divisors } F \in \mathrm{Div}(\mathcal{E}),$$

so in the notation of (9.2), $D_P = (P) - (O)$. Note that this is only valid for points in $E(K)_0$. Let $P, Q \in E(K)_0$. We compute

$$
\begin{aligned}
-\langle P, Q \rangle &= D_P \cdot D_Q \qquad \text{by definition of } \langle \cdot , \cdot \rangle \\
&= ((P) - (O)) \cdot ((Q) - (O)) \qquad \text{from above} \\
&= (P) \cdot (Q) - (P) \cdot (O) - (Q) \cdot (O) + (O) \cdot (O) \\
&= \tau_{-P}(P) \cdot \tau_{-P}(Q) - \tau_{-P}(P) \cdot \tau_{-P}(O) - (Q) \cdot (O) + (O) \cdot (O) \\
&= (O) \cdot (-P + Q) - (O) \cdot (-P) - (Q) \cdot (O) + (O) \cdot (O) \\
&= ((-P + Q) - (-P) - (Q) + (O)) \cdot (O) \\
&= \pi^*([-P, Q]) \cdot (O) \qquad \text{by definition of } [\cdot , \cdot] \text{ in (a)} \\
&= \deg[-P, Q] \qquad \text{from exercise 3.22(b)} \\
&= -\deg[P, Q] \qquad \text{by linearity of } [\cdot , \cdot] \text{ from (b).}
\end{aligned}
$$

This proves that $\langle P, Q \rangle = \deg[P, Q]$ for all $P, Q \in E(K)_0$. Putting $P = Q$ gives $\hat{h}(P) = \frac{1}{2}\langle P, P \rangle = \frac{1}{2}\deg[P, P]$, which completes the proof of Theorem 9.5.

$\square$

Remark 9.6. Theorem 9.5 says that the canonical height pairing $\langle \cdot , \cdot \rangle$ endows $E(K)_0$ with the structure of a Euclidean lattice whose inner product takes integer values. Similarly, the height pairing gives $E(K)$ a Euclidean structure with an inner product taking rational values having severely limited denominators. It is an interesting problem to classify the possible lattice structures on $E(K)_0$ and $E(K)$. In a series of papers, T. Shioda [1,2,4–7] has investigated these Mordell-Weil lattices and proven many interesting results, including the construction of examples for which $E(K)_0$ is isomorphic to a root lattice of type E_6, E_7, and E_8.

Remark 9.7. Let $\mathcal{E} \to C$ be a non-split minimal elliptic surface, and let E/K be the associated elliptic curve. The Néron-Severi group of $\mathcal{E}$, denoted by $\mathrm{NS}(\mathcal{E})$, is the group of divisors modulo algebraic equivalence. (For the definition of algebraic equivalence, see Hartshorne [1, exercise V.1.7].) One can prove that $\mathrm{NS}(\mathcal{E})$ is a finitely generated group and that the intersection pairing on $\mathrm{Div}(\mathcal{E})$ gives a well-defined pairing on $\mathrm{NS}(\mathcal{E})$. It is thus an interesting question to relate $\mathrm{NS}(\mathcal{E})$ and its intersection pairing to $E(K)$ and its height pairing. Shioda [3, Thm. 1.1] has shown how to find generators for $\mathrm{NS}(\mathcal{E})$ by using generators for $E(K)$ and fibral components of $\mathcal{E}$. In particular, he proves the fundamental rank relation (Shioda [3, Cor. 1.5])

$$
\mathrm{rank}\,\mathrm{NS}(\mathcal{E}) = \mathrm{rank}\,E(K) + 2 + \sum_{t \in C}(r_t - 1),
$$

where r_t is the number of irreducible components in the fiber $\mathcal{E}_t$. He also gives a formula relating the intersection regulator of $\mathrm{NS}(\mathcal{E})$ to the canonical height regulator of $E(K)$.

§10. Heights and Divisors on Varieties

Let $\mathcal{E} \to C$ be an elliptic surface defined over a number field k. For each point $t \in C(\bar{k})$ such that $\mathcal{E}_t$ is non-singular, there is a canonical height function $\hat{h} : \mathcal{E}_t(\bar{k}) \to \mathbb{R}$ on the elliptic curve $\mathcal{E}_t$. In the next section we will investigate how the canonical height varies from fiber to fiber, especially for points lying on the image of a section $\sigma : C \to \mathcal{E}$. To carry out this investigation, we will need to develop more fully the theory of height functions on varieties.

For our purposes in this chapter, it would suffice to consider only curves and surfaces, but the theory is hardly more difficult for general varieties. We will, however, need to assume that the reader is familiar with standard properties of divisors on varieties, as covered for example in Hartshorne [1, II §§6,7]. Some of the proofs in this section are fairly technical, so some readers may want to read the definitions and statements of the main results (10.1, 10.2, 10.3) and then proceed directly to the next section.

For this section, we set the following notation:

k a number field, with algebraic closure $\bar{k}$,

V/k a non-singular projective variety defined over k,

$\mathrm{Div}(V)$ the group of divisors on V,

$h_{\mathbb{P}}$ the (absolute logarithmic) height function $h_{\mathbb{P}} : \mathbb{P}^r(\bar{k}) \to \mathbb{R}$ on projective space as defined in [AEC, VIII §5].

A morphism $\phi : V \to W$ between non-singular varieties induces a homomorphism of their divisor groups $\phi^* : \mathrm{Div}(W) \to \mathrm{Div}(V)$ in the following way. Let $\Gamma \in \mathrm{Div}(W)$ be an irreducible divisor and fix a function u_Γ which vanishes to order 1 along Γ. Equivalently, u_Γ is a generator for the maximal ideal in the discrete valuation ring $\mathcal{O}_{W,\Gamma}$. Then

$$\phi^* \Gamma = \sum_{\Delta \in \mathrm{Div}(V)} \mathrm{ord}_\Delta(u_\Gamma \circ \phi) \Delta,$$

where the sum is over all irreducible divisors $\Delta \in \mathrm{Div}(V)$ and we are writing $\mathrm{ord}_\Delta : k(V)^* \to \mathbb{Z}$ for the normalized valuation on the local ring $\mathcal{O}_{V,\Delta}$. Of course, we have cheated a little bit. The divisor $\phi^* \Gamma$ will only be defined if the image $\phi(V)$ is not contained in Γ, since otherwise $u_\Gamma \circ \phi$ is identically zero. However, ϕ^* sends principal divisors to principal divisors, so it induces a map $\phi^* : \mathrm{Pic}(W) \to \mathrm{Pic}(V)$ which is well-defined on all of $\mathrm{Pic}(W)$, since we can always move Γ by a linear equivalence so that it intersects $\phi(V)$ properly.

The following theorem is often called the "Height Machine." It is the main result of this section. The Height Machine associates a height function to each divisor on V, or, more precisely, it associates an equivalence class of height functions to each divisor class on V. The power of the height machine is that it takes geometric relations involving divisor classes on V and translates them into height relations between points on V. It is thus a tool for transforming geometric information into arithmetic information. We have already seen this machine in action in [AEC, VIII §6], where the geometric group law on an elliptic curve was transformed into the arithmetic statement $h(P + Q) + h(P - Q) = 2h(P) + 2h(Q) + O(1)$. The general formulation of the Height Machine is due to André Weil. For further details and additional properties of heights, see Lang [4], Hindry-Silverman [1], and exercises 3.31 and 3.32.

Theorem 10.1. (Weil's Height Machine, Weil [2]) *Let V be a non-singular projective variety defined over a number field k. There is a map*

$$h : \mathrm{Div}(V) \longrightarrow \{\text{functions } V(\bar{k}) \to \mathbb{R}\},$$

uniquely determined up to bounded functions on $V(\bar{k})$, with the following two properties:

(a) *(Normalization) Let $\phi : V \to \mathbb{P}^r$ be a morphism, and let $H \in \mathrm{Div}(\mathbb{P}^r)$ be a hyperplane with the property that $\phi(V) \not\subset H$. Then*

$$h_{\phi^* H}(P) = h_{\mathbb{P}}\big(\phi(P)\big) + O(1) \qquad \text{for all } P \in V(\bar{k}).$$

(b) *(Additivity) Let $D, D' \in \mathrm{Div}(V)$. Then*

$$h_{D+D'}(P) = h_D(P) + h_{D'}(P) + O(1) \qquad \text{for all } P \in V(\bar{k}).$$

The height mapping has the following additional properties:

(c) *(Equivalence) Let $D, D' \in \mathrm{Div}(V)$ be linearly equivalent divisors. Then*

$$h_D(P) = h_{D'}(P) + O(1) \qquad \text{for all } P \in V(\bar{k}).$$

(d) *(Functoriality) Let $\psi : V \to W$ be a morphism of non-singular projective varieties over k, and let $D \in \mathrm{Div}(W)$. Then*

$$h_{V, \psi^* D}(P) = h_{W, D}\big(\psi(P)\big) + O(1) \qquad \text{for all } P \in V(\bar{k}).$$

Remark 10.1.1. Another way to formulate Theorem 10.1 is as follows. For every $V/\bar{\mathbb{Q}}$ there is a unique homomorphism

$$h_V : \mathrm{Pic}(V) \longrightarrow \frac{\{\text{functions } V(\bar{\mathbb{Q}}) \to \mathbb{R}\}}{\{\text{bounded functions } V(\bar{\mathbb{Q}}) \to \mathbb{R}\}}$$

such that $h_{\mathbb{P}^r}$ is the usual height on projective space [AEC, VIII §5] and such that $h_{V, \psi^* D} = h_{W, D} \circ \psi$ for every morphism $\psi : V \to W$.

Example 10.1.2. Let E/k be an elliptic curve. Consider the divisor relation $[2]^*(O) \sim 4(O)$, which follows from [AEC, III.3.5] and the fact that the four points in $E[2]$ sum to zero. Then (10.1) gives the height relation

$$h_{(O)}([2]P) = h_{[2]^*(O)}(P) + O(1) = h_{4(O)} + O(1) = 4h_{(O)}(P) + O(1).$$

This is one of the properties of the height that was used in the proof of the Mordell-Weil theorem [AEC, VIII.6.7].

Example 10.1.3. Two divisors $D, D' \in \mathrm{Div}(\mathbb{P}^1)$ on the projective line are linearly equivalent if and only if they have the same degree. It follows from the additivity and equivalence properties (10.1b,c) that

$$\deg(D')h_D(P) = \deg(D)h_{D'}(P) + O(1) \qquad \text{for all } P \in \mathbb{P}^1(\bar{k}).$$

In particular, (10.1a) implies that $h_D(P) = \deg(D)h_{\mathbb{P}}(P) + O(1)$.

For curves of higher genus, the identity (10.1.3) will not be true, since divisors of the same degree need not be linearly equivalent. However, a slightly weaker result is valid, as described in the following result. For a generalization to varieties of arbitrary dimension, see exercise 3.32.

Theorem 10.2. Let C be a curve, let $D, D' \in \mathrm{Div}(C)$ be divisors with $\deg(D) \neq 0$, and let $h_D, h_{D'}$ be associated height functions. Then

$$\lim_{P \in C(\bar{k}),\, h_D(P) \to \infty} \frac{h_{D'}(P)}{h_D(P)} = \frac{\deg(D')}{\deg(D)}.$$

The last theorem that we will be proving in this section is a finiteness result. Recall [AEC, VIII.5.11] that in projective space $\mathbb{P}^r(k)$, there are only finitely many points of bounded height. Of course, here height means $h_{\mathbb{P}}$, the standard height on projective space. It is clear that this result cannot be true for every height h_D on every variety. For example, if $V(k)$ is infinite, then it cannot be true for both h_D and h_{-D}, since $h_{-D} = -h_D + O(1)$. In order to give the correct statement, we need one definition.

Definition. A divisor $D \in \mathrm{Div}(V)$ is called *very ample* if there is an embedding $\phi : V \to \mathbb{P}^r$ and a hyperplane $H \in \mathrm{Div}(\mathbb{P}^r)$ not containing $\phi(V)$ so that $D = \phi^*H$. (To say that $\phi : V \to \mathbb{P}^r$ is an embedding means that ϕ maps V isomorphically onto its image.) The divisor D is called *ample* if there is an integer $n > 0$ so that nD is very ample.

Theorem 10.3. Let $D \in \mathrm{Div}(V)$ be an ample divisor on V, and let $h_D : V(\bar{k}) \to \mathbb{R}$ be an associated height function. Then for all $a, b > 0$, the set

$$\{P \in V(\bar{k}) : h_D(P) \leq a \quad \text{and} \quad [k(P) : k] \leq b\}$$

is finite. In particular, the set $\{P \in V(k') : h(P) \leq a\}$ is finite for any finite extension k'/k.

Example 10.3.1. Let C be a non-singular curve of genus g, and let $D \in \mathrm{Div}(C)$ be a divisor. Then D is ample if $\deg(D) > 0$, and D is very ample if $\deg(D) \geq 2g + 1$. See Hartshorne [1, IV.3.2, IV.3.3] or [AEC, exercise III.3.6].

Example 10.3.2. Let S be a non-singular surface. The criterion of Nakai-Moishezon (Hartshorne [1, V.1.10]) says that a divisor $D \in \mathrm{Div}(S)$ is ample if and only if $D^2 > 0$ and $D \cdot \Gamma > 0$ for all irreducible curves $\Gamma \subset S$.

As mentioned above, the reader may at this point wish to proceed directly to the next section, where we will apply the Height Machine to study the specialization map on elliptic surfaces. The remainder of this section is devoted to proving Theorems 10.1–10.3.

Definition. Let $\phi : V \to \mathbb{P}^r$ be a morphism of V into projective space. A divisor $D \in \mathrm{Div}(V)$ is said to be *associated to* ϕ if there is a hyperplane $H \in \mathrm{Div}(\mathbb{P}^r)$, not containing $\phi(V)$, such that $D = \phi^* H$. Note that the divisor class of D is uniquely determined by ϕ, since any two hyperplanes in $\mathbb{P}^r$ are linearly equivalent.

The *height on V associated to* ϕ is the height function

$$h_\phi : V(\bar{k}) \longrightarrow \mathbb{R}, \qquad h_\phi(P) = h_{\mathbb{P}}\big(\phi(P)\big).$$

Our ultimate goal is to associate to every divisor D on V a height function h_D with the properties described in (10.1). In particular, we will want the heights attached to linearly equivalent divisors to be essentially the same. The following important proposition will be crucial for this construction.

Lemma 10.4. *Let $\phi : V \to \mathbb{P}^r$ and $\psi : V \to \mathbb{P}^s$ be morphisms which are associated to the same divisor class. Then*

$$h_\phi(P) = h_\psi(P) + O(1) \qquad \textit{for all } P \in V(\bar{k}).$$

Here the $O(1)$ depends on ϕ and ψ but is independent of P.

PROOF. Let D be any positive divisor in the divisor class associated to ϕ and ψ. This means that on the complement of D we can write ϕ and ψ in the form $\phi = [f_0, \ldots, f_r]$ and $\psi = [g_0, \ldots, g_s]$ with rational functions $f_i, g_j \in \bar{k}(V)$ satisfying

$$\mathrm{div}(f_i) = D_i - D \quad \text{and} \quad \mathrm{div}(g_j) = D'_j - D \quad \text{for divisors } D_i, D'_j \geq 0.$$

Further, the fact that ϕ is a morphism means that the D_i's have no points in common, and similarly for the D'_j's. (For general facts about the relationship between morphisms $\phi : V \to \mathbb{P}^r$ and divisors, see Hartshorne [1, II §7], especially the section on linear systems.)

Now fix some j, let $V_j = V \smallsetminus D'_j$ be the complement of D'_j, and let $F_i = f_i/g_j$ for $0 \leq i \leq r$. Notice that

$$\mathrm{div}(F_i) = \mathrm{div}(f_i/g_j) = (D_i - D) - (D'_j - D) = D_i - D'_j,$$

so F_i is a regular function on V_j. Taken together, the F_i's define a morphism

$$F = (F_0, \ldots, F_r) : V_j \longrightarrow \mathbf{A}^{r+1}.$$

We also observe that the F_i's have no common zeros on V_j, since any common zero would lie on all of the D_i's.

We need to recall how the maximal ideals in the ring $\mathfrak{R} = \bar{k}[F_0, \ldots, F_r]$ correspond to the points of V_j. If $\mathfrak{M} \subset \mathfrak{R}$ is a maximal ideal, then $\mathfrak{R}/\mathfrak{M}$ is a finitely generated $\bar{k}$-algebra which is also a field. It follows from the weak Nullstellensatz (Atiyah-MacDonald [1, 5.24, 7.10], Lang [7, X §2 Cor. 2.2]) that $\mathfrak{R}/\mathfrak{M}$ is isomorphic to $\bar{k}$. More precisely, the natural inclusion $\bar{k} \to \mathfrak{R}/\mathfrak{M}$ is an isomorphism. This means that there are unique elements $\alpha_0, \ldots, \alpha_r \in \bar{k}$ so that $F_i \equiv \alpha_i \,(\mathrm{mod}\ \mathfrak{M})$, and then there is a unique point $P_{\mathfrak{M}} \in V_j(\bar{k})$ with $F(P_{\mathfrak{M}}) = (\alpha_0, \ldots, \alpha_r)$. Equivalently, the point $P_{\mathfrak{M}}$ is determined by the congruences

$$F_i \equiv F_i(P_{\mathfrak{M}}) \quad (\mathrm{mod}\ \mathfrak{M}) \qquad \text{for } 0 \le i \le r.$$

Now consider the ideal $\mathfrak{I} = (F_0, \ldots, F_r) \subset \mathfrak{R}$ generated by the F_i's. We claim that $\mathfrak{I}$ must be the unit ideal. To prove this, we assume that $\mathfrak{I}$ is not the unit ideal and derive a contradiction. Every non-unit ideal is contained in at least one maximal ideal, so we take a maximal ideal $\mathfrak{M}$ with $\mathfrak{I} \subset \mathfrak{M}$. Then $\mathfrak{M}$ corresponds to a point $P_{\mathfrak{M}} \in V_j$ as described above. On the other hand, we have $F_i \in \mathfrak{I} \subset \mathfrak{M}$ from the definition of $\mathfrak{I}$, so

$$F_i(P_{\mathfrak{M}}) \equiv F_i \equiv 0 \quad (\mathrm{mod}\ \mathfrak{M}).$$

Hence $F_i(P_{\mathfrak{M}}) \in \bar{k} \cap \mathfrak{M}$, so $F_i(P_{\mathfrak{M}}) = 0$. In other words, $P_{\mathfrak{M}}$ is a common zero of $F_0, \ldots, F_r$, which is a contradiction. This completes the proof that $\mathfrak{I}$ is the unit ideal.

We can rephrase this last argument in slightly fancier language. The scheme $\mathrm{Spec}(\mathfrak{R})$ is isomorphic to V_j, and by the weak Nullstellentsatz, maximal ideals in $\mathrm{Spec}(\mathfrak{R})$ correspond to $\bar{k}$-valued points in V_j. But then any maximal ideal $\mathfrak{M}$ containing $\mathfrak{I}$ would correspond to a point P in the zero set of $\mathfrak{I}$, contradicting the fact that the F_i's have no common zero.

The fact that $\mathfrak{I} = (F_0, \ldots, F_r)$ is the unit ideal in $\mathfrak{R}$ means that we can find a polynomial $A_j(T_0, \ldots, T_r) \in \bar{k}[T_0, \ldots, T_r]$ with no constant term such that

$$1 = A_j(F_0, \ldots, F_r).$$

For any finite extension k'/k, any point $P \in V_j(k')$, and any absolute value v on k', we evaluate this identity at P, take the v-adic absolute value, and use the triangle inequality to get an estimate of the form

$$1 \le c_1 \max\{|F_0(P)|_v, \ldots, |F_r(P)|_v\}.$$

Here $c_1 = c_1(v, \phi, \psi, D, A_j) > 0$ is a constant that does not depend on P. Further, for all but finitely many absolute values on k', we can take $c_1 = 1$. (See the proof of [AEC, VIII.5.6] for a similar calculation.)

Recall that $F_i = f_i/g_j$, so if we multiply both sides by $|g_j(P)|_v$ we obtain the estimate

$$|g_j(P)|_v \leq c_1 \max\{|f_0(P)|_v, \ldots, |f_r(P)|_v\}.$$

Notice that this bound is still valid if $g_j(P) = 0$, so it holds for all points at which the f_i's and g_j's are defined, that is, at all points on the complement of D. Taking the maximum for $0 \leq j \leq s$ gives

$$\max\{|g_0(P)|_v, \ldots, |g_s(P)|_v\} \leq c_2 \max\{|f_0(P)|_v, \ldots, |f_r(P)|_v\},$$

where $c_2 = c_2(v, \phi, \psi, D)$ is again positive and is equal to 1 for all but finitely many v. Now we take the logarithm of both sides, multiply by the local degrees $[k'_v : \mathbb{Q}_v]/[k' : \mathbb{Q}]$, and sum over all absolute values on k' to obtain

$$h([g_0(P), \ldots, g_s(P)]) \leq h([f_0(P), \ldots, f_r(P)]) + c_3,$$

where $c_3 = c_3(\phi, \psi, D)$ is independent of P. In other words, we have shown that

$$h(\psi(P)) \leq h((\phi(P)) + c_3 \qquad \text{for all } P \in (V \smallsetminus D)(\bar{k}).$$

The divisor D was chosen to be any positive divisor in the divisor class associated to ϕ and ψ. In other words, we can take D to be $\phi^* H$ for any hyperplane $H \subset \mathbb{P}^r$ not containing $\phi(V)$. Let $H_1, \ldots, H_m \subset \mathbb{P}^r$ be hyperplanes not containing $\phi(V)$ with the property that $H_1 \cap \cdots \cap H_m = \emptyset$. Then the corresponding divisors $\phi^* H_1, \ldots, \phi^* H_m$ have no points in common, so their complements cover V. Applying the above estimate to each of these D's and letting c_4 be the maximum of the c_3's gives

$$h(\psi(P)) \leq h((\phi(P)) + c_4 \qquad \text{for all } P \in V(\bar{k}).$$

This is one of the inequalities we are trying to prove, and the opposite inequality follows if we interchange the role of ϕ and ψ. $\qquad \square$

If $D \in \mathrm{Div}(V)$ is a very ample divisor, then we can choose an embedding $\phi : V \to \mathbb{P}^r$ associated to D and attach to D the height function $h_D = h_\phi$. For arbitrary divisors D, we will write $D = D_1 - D_2$ as a difference of very ample divisors and define h_D linearly by $h_D = h_{D_1} - h_{D_2}$. The following lemma shows that every divisor can be decomposed in this way.

Lemma 10.5. *Every divisor on V can be written as a difference of two very ample divisors.*

PROOF. This is a basic result from algebraic geometry. Before giving the general proof, we consider a special case. Suppose V is a curve of genus g. Then a divisor on V is very ample if it has degree at least $2g + 1$ (10.3.1). So for an arbitrary divisor $D \in \mathrm{Div}(V)$, say of degree d, we can write D as a difference of very ample divisors

$$D = (D + n(P)) - n(P)$$

by choosing $n = 2g + 1 + |d|$. Similarly, let V be a surface, $D \in \mathrm{Div}(V)$ an arbitrary divisor, and $H \in \mathrm{Div}(V)$ an ample divisor. Then one can use the Nakai-Moishezon criterion (10.2.2) to show that $nH + D$ is ample for all sufficiently large n, after which it is easy to write D as a difference of very ample divisors. We will leave the details to the reader (exercise 3.30) and go on to the general case.

Let $D \in \mathrm{Div}(V)$ be an arbitrary divisor, and fix a very ample divisor $H \in \mathrm{Div}(V)$. Serre's theorem (Hartshorne [1, II.5.17, II.7.4.3]) says that there is an integer $n \geq 1$ so that $D + nH$ is ample. (Note we have translated from the language of invertible sheaves into the language of divisors, as explained in the last part of Hartshorne [1, II §6].) It follows from Hartshorne [1, II.7.6] that $m(D+nH)$ is very ample for all sufficiently large integers m. Further, nH is very ample, so $nH + m(D+nH)$ is very ample, since it is the sum of two very amples. Hence

$$D = (m + 1)(D + nH) - \big(nH + m(D + nH)\big)$$

is a difference of very ample divisors.   □

Lemma 10.5 lets us decompose a divisor D into a difference $D_1 - D_2$ of very ample divisors. In particular, D_1 and D_2 are associated to morphisms from V into projective space. The next result gives some basic properties of height functions associated to such morphisms.

Lemma 10.6. *Let $\phi_1 : V \to \mathbb{P}^r$ and $\phi_2 : V \to \mathbb{P}^s$ be morphisms, and let D_1 and D_2 be divisors associated to ϕ_1 and ϕ_2 respectively.*
(a) There exists a morphism $\phi_3 : V \to \mathbb{P}^{rs+r+s}$ associated to $D_1 + D_2$.
(b) If $\phi : V \to \mathbb{P}^n$ is any morphism associated to $D_1 + D_2$, then

$$h_\phi(P) = h_{\phi_1}(P) + h_{\phi_2}(P) + O(1) \qquad \text{for all } P \in V(\bar{k}).$$

PROOF. (a) The Segre embedding (Hartshorne [1, exercise I.2.14] or Harris [1, 2.11–2.29]) is the map

$$
\begin{array}{ccc}
\mathbb{P}^r \times \mathbb{P}^s & \longrightarrow & \mathbb{P}^{rs+r+s} \\
([x_0, \ldots, x_r], [y_0, \ldots, y_s]) & \longmapsto & [x_0 y_0, \ldots, x_i y_j, \ldots, x_r y_s].
\end{array}
$$

It is clear from this definition that if we pull a hyperplane back by the Segre embedding, we will get $H \times \mathbb{P}^s + \mathbb{P}^r \times H$, where the H's are hyperplanes in the appropriate projective spaces. Hence $D_1 + D_2$ is associated to the morphism

$$V \overset{\text{diagonal}}{\longrightarrow} V \times V \overset{\phi_{D_1} \times \phi_{D_2}}{\longrightarrow} \mathbb{P}^r \times \mathbb{P}^s \overset{\text{Segre}}{\longrightarrow} \mathbb{P}^{rs+r+s}.$$

(b) Write

$$\phi_1 = [f_0, \ldots, f_r] : V \to \mathbb{P}^r \qquad \text{and} \qquad \phi_2 = [g_0, \ldots, g_s] : V \to \mathbb{P}^s$$

with rational functions $f_0, \ldots, g_s \in \bar{k}(V)$. Lemma 10.4 says that it suffices to prove (b) for any one morphism ϕ associated to $D_1 + D_2$, so we will take ϕ to be the map using the Segre embedding described in (a). In other words,

$$\phi = [f_0 g_0, \ldots, f_i g_j, \ldots, f_r g_s] : V \to \mathbb{P}^{rs+r+s}.$$

Let $P \in V(\bar{k})$ be any point. Replacing k be a finite extension, we may assume that $P \in V(k)$. Then directly from the definition of the height on projective space we have

$$
\begin{aligned}
h_\phi(P) &= h\big(\phi(P)\big) \\
&= h\big([f_0 g_0(P), \ldots, f_i g_j(P), \ldots, f_r g_s(P)]\big) \\
&= \sum_{v \in M_k} \frac{[k_v : \mathbb{Q}_v]}{[k : \mathbb{Q}]} \log \left(\max_{0 \le i \le r, 0 \le j \le s} |(f_i g_j)(P)| \right) \\
&= \sum_{v \in M_k} \frac{[k_v : \mathbb{Q}_v]}{[k : \mathbb{Q}]} \log \left(\max_{0 \le i \le r} |f_i(P)| \cdot \max_{0 \le j \le s} |g_j(P)| \right) \\
&= \sum_{v \in M_k} \frac{[k_v : \mathbb{Q}_v]}{[k : \mathbb{Q}]} \left(\log \left(\max_{0 \le i \le r} |f_i(P)| \right) + \log \left(\max_{0 \le j \le s} |g_j(P)| \right) \right) \\
&= h\big([f_0(P), \ldots, f_r(P)]\big) + h\big([g_0(P), \ldots, g_s(P)]\big).
\end{aligned}
$$

$\square$

After these lengthy preliminaries, we are finally ready to tackle the proof of the Height Machine.

PROOF (of the Height Machine (10.1)). Take each divisor $D \in \text{Div}(V)$ and write it as a difference $D = D_1 - D_2$ of divisors with the property that there are morphisms $\phi_1 : V \to \mathbb{P}^r$ and $\phi_2 : V \to \mathbb{P}^s$ associated to D_1 and D_2 respectively. Note that (10.5) assures us that this is possible; in fact, (10.5) says that we can even choose D_1 and D_2 so that ϕ_1 and ϕ_2 are embeddings. In any case, having fixed D_1, D_2, ϕ_1, ϕ_2, we define h_D to be

$$h_D(P) = h_{\phi_1}(P) - h_{\phi_2}(P) \qquad \text{for all } P \in V(\bar{k}).$$

Our first observation is that if the Height Machine exists, then up to a bounded function, this is the only choice for h_D. This follows from (10.1a) and (10.1b), which let us compute

$$h_D = h_{D_1 - D_2} = h_{D_1} - h_{D_2} + O(1) = h_{\phi_1}(P) - h_{\phi_2}(P) + O(1).$$

This gives the uniqueness assertion in Theorem 10.1.

Next we show that up to bounded functions, h_D is independent of the choice of D_1, D_2, ϕ_1, ϕ_2. Once we have proven this independence, the rest of (10.1) will follow very easily. So suppose that $D = D_1' - D_2'$ is another decomposition, and let $\phi_1' : V \to \mathbb{P}^{r'}$ and $\phi_2' : V \to \mathbb{P}^{s'}$ be morphisms associated to D_1' and D_2' respectively. Then $D_1' + D_2 = D_1 + D_2'$. Lemma 10.6 says that there exists a morphism $\phi : V \to \mathbb{P}^n$ associated to this divisor, and then two applications of (10.6b) yields

$$h_{\phi_1'} + h_{\phi_2} = h_\phi + O(1) = h_{\phi_1} + h_{\phi_2'} + O(1).$$

Therefore

$$h_{\phi_1}(P) - h_{\phi_2}(P) = h_{\phi_1'}(P) - h_{\phi_2'}(P) + O(1),$$

which proves that up to bounded functions, the definition of h_D is independent of the choice of D_1, D_2, ϕ_1, ϕ_2.

(a) The divisor $D = \phi^* H$ is already associated to a morphism, so we can write D as $D = D - 0$. Note that the divisor 0 is associated to the trivial morphism $\psi : V \to \mathbb{P}^0$ which maps V to a point. Then

$$h_D = h_\phi - h_\psi + O(1) = h_\phi + O(1),$$

since $h_\psi(P) = h_{\mathbb{P}}(\psi(P))$ is a constant.

(b) We decompose each of the given divisors into a difference of divisors that are associated to morphisms, say $D = D_1 - D_2$ and $D' = D_1' - D_2'$. Then $D_1 + D_1'$ and $D_2 + D_2'$ are also associated to morphisms (10.6a), and their difference is $D + D'$, so we can use (a) and (10.6b) to compute

$$h_{D+D'} = h_{D_1 + D_1'} - h_{D_2 + D_2'} + O(1) = h_{D_1} + h_{D_1'} - h_{D_2} - h_{D_2'} + O(1)$$
$$= h_D + h_{D'} + O(1).$$

(c) Write $D - D'$ as a difference of divisors associated to morphisms, say $D - D' = D_1 - D_2$, with D_1 associated to ϕ_1 and D_2 associated to ϕ_2. Note that D_1 and D_2 are linearly equivalent by assumption, so (10.4) tells us that $h_{\phi_1} = h_{\phi_2} + O(1)$. Now using (b) we obtain the desired result,

$$h_D - h_{D'} = h_{D-D'} + O(1) = h_{\phi_1} - h_{\phi_2} + O(1) = O(1).$$

(d) By the linearity proven in (b), it suffices to prove (d) for a divisor D associated to a morphism $\phi : W \to \mathbb{P}^r$. Then the divisor $\psi^* D$ is associated to the morphism $\phi \circ \psi : V \to W \to \mathbb{P}^r$. Using (a) twice, we find

$$h_{V, \psi^* D} = h_{V, \phi \circ \psi} + O(1) = h_{W, \phi} \circ \psi + O(1) = h_{W, D} \circ \psi + O(1).$$

Note that the middle equality is trivial, since $h_{\phi \circ \psi} = h_{\mathbb{P}} \circ \phi \circ \psi$ by definition. $\square$

PROOF (of Theorem 10.2). Let $d = \deg(D)$ and $d' = \deg(D')$. Replacing D by $-D$ if necessary, we may assume that $d \geq 1$. For any integer n we consider the divisor

$$H_n = (2g + 1)D + n(d'D - dD').$$

Notice that $\deg(H_n) = d(2g + 1)$, so (10.3.1) says that H_n is a very ample divisor on C. In particular, there is an embedding $\phi_n : C \to \mathbb{P}^r$ associated to H_n, so (10.1a) gives

$$h_{H_n}(P) = h_{\mathbb{P}}(\phi_n(P)) + O(1) \qquad \text{for all } P \in C(\bar{k}).$$

Now using the definition of H_n, the linearity property of height functions (10.1b), and the fact that the height on projective space is nonnegative [AEC, VIII.5.4b], we obtain the estimate

$$(2g + 1)h_D(P) + nd'h_D(P) - ndh_{D'}(P) = h_{H_n}(P) + O(1) \geq -c_n.$$

Note that the constant c_n depends on n, but it is independent of P. Assuming $h_D(P) > 0$, a little algebra then gives the inequality

$$n\left(\frac{d'}{d} - \frac{h_{D'}(P)}{h_D(P)}\right) \geq -\frac{2g + 1}{d} - \frac{c_n}{dh_D(P)}.$$

Taking the lim inf as $h_D(P) \to \infty$, we obtain

$$\liminf_{P \in C(\bar{k}),\, h_D(P) \to \infty} n\left(\frac{d'}{d} - \frac{h_{D'}(P)}{h_D(P)}\right) \geq -\frac{2g + 1}{d}.$$

This is true for every value of n (positive and negative), which gives the desired result,

$$\lim_{P \in C(\bar{k}),\, h_D(P) \to \infty} \left(\frac{d'}{d} - \frac{h_{D'}(P)}{h_D(P)}\right) = 0.$$

$\square$

PROOF (of Theorem 10.3). Replacing D by nD and using the fact (10.1b) that $h_{nD} = nh_D + O(1)$, we may assume that D is very ample. Let ϕ :

$V \to \mathbb{P}^r$ be an embedding and $H \in \text{Div}(\mathbb{P}^r)$ a hyperplane with $\phi^* H = D$. Taking a finite extension of k if necessary, we may assume that ϕ is defined over k. Then (10.1a) implies that there is a constant c so that

$$h_D(P) = h_{\phi^* H}(P) \geq h_{\mathbb{P}}(\phi(P)) - c \qquad \text{for all } P \in V(\bar{k}).$$

It follows that ϕ maps the set

$$\{P \in V(\bar{k}) : h_D(P) \leq a \quad \text{and} \quad [k(P) : k] \leq b\}$$

injectively into the set

$$\{Q \in \mathbb{P}^r(\bar{k}) : h_{\mathbb{P}}(Q) \leq a + c \quad \text{and} \quad [k(Q) : k] \leq b\}.$$

This last set is finite from [AEC, VIII.5.11], which proves the first part of (10.3). The second part follows by setting $b = 1$. $\qquad\square$

§11. Specialization Theorems for Elliptic Surfaces

In this section we will prove a theorem of Tate which describes how the canonical height $\hat{h}(\sigma_P(t))$ varies as one moves along a section of an elliptic surface. As a corollary we obtain a theorem of Silverman, strengthening earlier results of Néron, Dem'janenko, and Manin, which says that the specialization homomorphism $E(K) \to \mathcal{E}_t(k)$ is injective for all but finitely many $t \in C(k)$.

Let $\pi : \mathcal{E} \to C$ be a minimal elliptic surface with corresponding elliptic curve E/K, and let $P \in E(K)$. To ease notation, we will write

$$P_t = \sigma_P(t)$$

for the image of a point $t \in C$ by the section $\sigma_P : C \to \mathcal{E}$ associated to P.

Theorem 11.1. (Tate [4]) *Assume that the elliptic surface $\mathcal{E} \to C$ is defined over a number field k. For each $t \in C(\bar{k})$ such that the fiber $\mathcal{E}_t$ is non-singular, let*

$$\langle \cdot, \cdot \rangle_t : \mathcal{E}_t(\bar{k}) \times \mathcal{E}_t(\bar{k}) \longrightarrow \mathbb{R}$$

be the canonical height pairing on the elliptic curve $\mathcal{E}_t$ [AEC, VIII §9].

Fix two points $P, Q \in E(K)_0$, let $[P, Q] \in \text{Div}(C)$ be the divisor described in (9.5), and let $h_{[P,Q]} : C(\bar{k}) \to \mathbb{R}$ be an associated height function on C (10.1). Then

$$\langle P_t, Q_t \rangle_t = h_{[P,Q]}(t) + O(1) \quad \text{for all } t \in C(\bar{k}) \text{ such that } \mathcal{E}_t \text{ is non-singular.}$$

Note that the $O(1)$ bound depends on P and Q but is independent of t.

Remark 11.1.1. Putting $P = Q$ in (11.1) gives $\hat{h}_{\mathcal{E}_t}(P_t) = h_{[P,P]}(t) + O(1)$, where

$$\hat{h}_{\mathcal{E}_t} : \mathcal{E}_t(\bar{k}) \to \mathbb{R}$$

is the canonical height on $\mathcal{E}$ [AEC, III §9]. In other words, for any point $P \in E(K)_0$, the map

$$C(\bar{k}) \longrightarrow \mathbb{R}, \qquad t \longmapsto \hat{h}_{\mathcal{E}_t}(P_t),$$

is a height function on $C(\bar{k})$ corresponding to the divisor $[P, P]$. Silverman [6] shows that it is possible to choose the height $h_{[P,P]}$ in such a way that the difference $\hat{h}_{\mathcal{E}_t}(P_t) - h_{[P,P]}(t)$ varies quite regularly as a function of t. For example, consider the elliptic surface and section

$$E : y^2 = x^3 + t^2(1 - t^2)x, \qquad P = (t^2, t^2).$$

Then there is a power series $f(z) \in \mathbb{R}[\![z]\!]$ with $f(0) = 0$ so that for all sufficiently large integers $t \in \mathbb{Z}$,

$$\hat{h}_{\mathcal{E}_t}(P_t) = \frac{1}{2}h(t) + \frac{1}{4}\log 2 + f\left(\frac{1}{t^2}\right).$$

For details, see Silverman [5,6].

Before beginning the proof of Tate's theorem (11.1), we need to describe how to use a height function on the surface $\mathcal{E}$ to compute canonical heights on the individual fibers. For any integer n and any non-singular fiber $\mathcal{E}_t$, we will write

$$[n]_t : \mathcal{E}_t \to \mathcal{E}_t$$

for the multiplication-by-n map on $\mathcal{E}_t$. These maps clearly fit together to give a rational map on the surface $\mathcal{E}$,

$$[n] : \mathcal{E} \to \mathcal{E}, \qquad [n](x, y, t) = ([n]_t(x, y), t).$$

(N.B. Even if the surface $\mathcal{E}$ is minimal, the rational map $[n] : \mathcal{E} \to \mathcal{E}$ will generally not extend to a morphism.) With these preliminaries completed, we are ready for the following lemma.

Lemma 11.2. *Let $\pi : \mathcal{E} \to C$ be an elliptic surface defined over a number field k, and let $h_{\mathcal{E},(O)} : \mathcal{E}(\bar{k}) \to \mathbb{R}$ be a height function on $\mathcal{E}$ associated to the divisor $(O) \in \mathrm{Div}(\mathcal{E})$. Then for all $t \in C(\bar{k})$ such that $\mathcal{E}_t$ is non-singular, and all points $(x, y) \in \mathcal{E}_t(\bar{k})$,*

$$\hat{h}_{\mathcal{E}_t}(x, y) = \lim_{n \to \infty} \frac{1}{n^2} h_{\mathcal{E},(O)}([n](x, y, t)).$$

PROOF. Fix a Weierstrass equation for $\mathcal{E}$,

$$\mathcal{E} : y^2 = x^3 + Ax + B, \qquad A, B \in k(C).$$

For the moment we will restrict attention to points $t \in C(\bar{k})$ such that A and B are defined at t and $\Delta(t) \neq 0$. Then $\mathcal{E}_t$ is obtained by evaluating A and B at t.

For each such t we let p_t be the map

$$p_t : \mathcal{E}_t \longrightarrow \mathbb{P}^1, \qquad p_t(x, y) = x.$$

This x-coordinate function on $\mathcal{E}_t$ has a double pole at O_t, which means that

$$p_t^*(\infty) = 2(O_t).$$

We also write $\phi_t : \mathcal{E}_t \to \mathcal{E}$ for the inclusion of the fiber $\mathcal{E}_t$ into the surface $\mathcal{E}$.

Now let $(x, y) \in \mathcal{E}_t(\bar{k})$ be any point on a non-singular fiber. We use standard properties of the Height Machine (10.1) to compute

$$
\begin{aligned}
& h_{\mathcal{E},(O)}(x, y, t) \\
&= h_{\mathcal{E},(O)}\big(\phi_t(x, y)\big) && \text{definition of } \phi_t \\
&= h_{\mathcal{E}_t, \phi_t^*(O)}(x, y) + O(1) && \text{functoriality of height (10.1d)} \\
&= h_{\mathcal{E}_t,(O_t)}(x, y) + O(1) && \text{since } \phi_t^*(O) = (O_t) \\
&= \frac{1}{2} h_{\mathcal{E}_t, 2(O_t)}(x, y) + O(1) && \text{additivity of height (10.1b)} \\
&= \frac{1}{2} h_{\mathcal{E}_t, p_t^*(\infty)}(x, y) + O(1) && \text{since } p_t^*(\infty) = 2(O_t) \\
&= \frac{1}{2} h_{\mathbb{P}^1,(\infty)}\big(p_t(x, y)\big) + O(1) && \text{functoriality of height (10.1d)} \\
&= \frac{1}{2} h(x) + O(1). && \text{definition of } p_t.
\end{aligned}
$$

It is important to note that in this computation the $O(1)$ bounds will depend on t. This dependence arises because we have used the morphism $\phi_t : \mathcal{E}_t \to \mathcal{E}$, and the $O(1)$ in the functoriality property (10.1d) depends on the morphism. However, for a given t, the $O(1)$'s are independent of the point $(x, y) \in \mathcal{E}_t(\bar{k})$. To make the dependence visible, we will write

$$h_{\mathcal{E},(O)}(x, y, t) = \frac{1}{2} h(x) + O_t(1) \qquad \text{for all } (x, y) \in \mathcal{E}_t(\bar{k}).$$

For any point $(x, y) \in \mathcal{E}_t(\bar{k})$ and integer n, write $[n]_t(x, y) = (x_n, y_n)$. Then standard properties of the canonical height [AEC, VIII.9.3b,e] allow us to compute

$$
\begin{aligned}
\frac{1}{2} \lim_{n \to \infty} \frac{1}{n^2} h(x_n) &= \lim_{n \to \infty} \frac{1}{n^2} \big\{ \hat{h}_{\mathcal{E}_t}\big([n]_t(x, y)\big) + O_t(1) \big\} \\
&= \lim_{n \to \infty} \frac{1}{n^2} \big\{ n^2 \hat{h}_{\mathcal{E}_t}(x, y) + O_t(1) \big\} \\
&= \hat{h}_{\mathcal{E}_t}(x, y).
\end{aligned}
$$

Now combining these two formulas gives the desired result,

$$\lim_{n\to\infty} \frac{1}{n^2} h_{\mathcal{E},(O)}([n](x,y,t)) = \lim_{n\to\infty} \frac{1}{n^2} h_{\mathcal{E},(O)}(x_n, y_n, t)$$

$$= \lim_{n\to\infty} \frac{1}{n^2}\left(\frac{1}{2}h(x_n) + O_t(1)\right) = \frac{1}{2}\lim_{n\to\infty} \frac{1}{n^2}h(x_n) = \hat{h}_{\mathcal{E}_t}(x,y).$$

It is instructive to note that the $O_t(1)$ disappears in the limit because the "t-coordinate" of $[n](x,y,t)$ is independent of n.

This completes the proof of (11.2) for all points $t \in C(\bar{k})$ such that A and B are defined at t and $\Delta(t) \neq 0$. But by choosing different Weierstrass equations for $\mathcal{E}$, we can cover $\{t \in C(\bar{k}) : \mathcal{E}_t \text{ is non-singular}\}$ by finitely many such sets. $\qquad\square$

PROOF (of Theorem 11.1). The divisor $[P,Q]$ is determined up to linear equivalence by the relation (9.5a),

$$(P+Q) - (P) - (Q) + (O) \sim \pi^*([P,Q]).$$

For any point $z \in \mathcal{E}(\bar{k})$ lying on a non-singular fiber $\mathcal{E}_t$, we use standard properties of the Height Machine (10.1) to compute

$$\begin{aligned}
h_{[P,Q]}(t) &= h_{[P,Q]}(\pi(z)) &&\text{since } z \in \mathcal{E}_t \\
&= h_{\pi^*[P,Q]}(z) + O(1) &&\text{functoriality of height (10.1d)} \\
&= h_{(P+Q)-(P)-(Q)+(O)}(z) + O(1) &&\text{equivalence of heights (10.1c)} \\
&= h_{(P+Q)}(z) - h_{(P)}(z) - h_{(Q)}(z) + h_{(O)}(z) + O(1) \\
&&&\text{additivity of height (10.1b)} \\
&= h_{\tau^*_{-P-Q}(O)}(z) - h_{\tau^*_{-P}(O)}(z) - h_{\tau^*_{-Q}(O)}(z) + h_{(O)}(z) + O(1) \\
&&&\text{where } \tau_R : \mathcal{E} \to \mathcal{E} \text{ is translation-by-}R \\
&= h_{(O)}\big(\tau_{-P-Q}(z)\big) - h_{(O)}\big(\tau_{-P}(z)\big) - h_{(O)}\big(\tau_{-Q}(z)\big) \\
&\quad + h_{(O)}(z) + O(1) &&\text{functoriality of height (10.1d)} \\
&= h_{(O)}(-P_t - Q_t + z) - h_{(O)}(-P_t + z) - h_{(O)}(-Q_t + z) \\
&\quad + h_{(O)}(z) + O(1) &&\text{since } z \in \mathcal{E}_t(\bar{k}), \text{ so } \tau_R(z) = R_t + z.
\end{aligned}$$

Note that the $O(1)$ constants appearing in this calculation depend on P and Q, but they are independent of z and $t = \pi(z)$. To indicate this dependence, we will write $O_{P,Q}(1)$. For each pair of integers $1 \leq i,j \leq n$, we evaluate at the point $z = iP_t + jQ_t \in \mathcal{E}(\bar{k})$ to obtain

$$\begin{aligned}
h_{[P,Q]}(t) = h_{(O)}\big((i-1)P_t + (j-1)Q_t\big) - h_{(O)}\big((i-1)P_t + jQ_t\big) \\
- h_{(O)}\big(iP_t + (j-1)Q_t\big) + h_{(O)}\big(iP_t + jQ_t\big) + O_{P,Q}(1).
\end{aligned}$$

Summing this identity over $1 \le i, j \le n$, we find that most of the terms telescope, leaving

$$n^2 h_{[P,Q]}(t) = h_{(O)}(nP_t + nQ_t) - h_{(O)}(nP_t) - h_{(O)}(nQ_t)$$
$$+ h_{(O)}(O_t) + O_{P,Q}(n^2).$$

Now dividing by n^2, letting $n \to \infty$, and using (11.2) yields

$$h_{[P,Q]}(t) = \lim_{n \to \infty} \frac{1}{n^2} h_{(O)}(nP_t + nQ_t) - \lim_{n \to \infty} \frac{1}{n^2} h_{(O)}(nP_t)$$
$$- \lim_{n \to \infty} \frac{1}{n^2} h_{(O)}(nQ_t) + O_{P,Q}(1)$$
$$= \hat{h}_{\mathcal{E}_t}(P_t + Q_t) - \hat{h}_{\mathcal{E}_t}(P_t) - \hat{h}_{\mathcal{E}_t}(Q_t) + O_{P,Q}(1)$$
$$= \langle P_t, Q_t \rangle_t + O_{P,Q}(1).$$

This completes the proof of Theorem 11.1. $\qquad\qquad\qquad\qquad$ □

Taking the limit of (11.1) as the height of t goes to infinity, we can recover the following result of Silverman which will be used below to prove the injectivity of the specialization map. In the special case that the elliptic surface $\mathcal{E} \to C$ is split, this result had earlier been proven by Dem'janenko [1] and Manin [2].

Corollary 11.3.1. (Silverman [1], [7]) *Let $\mathcal{E} \to C$ be an elliptic surface defined over a number field k, fix two points $P, Q \in E(K)$, and let $\langle P, Q \rangle$ denote the canonical height pairing (4.3, 9.3) of P and Q on $E(K)$. Further, let $h_\delta : C(\bar{k}) \to \mathbb{R}$ be a height function on C corresponding to a divisor $\delta \in \text{Div}(C)$ of degree 1, and for each $t \in C(\bar{k})$ such that $\mathcal{E}_t$ is non-singular let $\langle \cdot, \cdot \rangle_t$ be the canonical height pairing on $\mathcal{E}_t(\bar{k})$. Then*

$$\lim_{t \in C(\bar{k}), \, h_\delta(t) \to \infty} \frac{\langle P_t, Q_t \rangle_t}{h_\delta(t)} = \langle P, Q \rangle.$$

Notice that (11.3.1) applies to all points in $E(K)$, not just those in the subgroup $E(K)_0$. It is possible to improve (11.3.1) as described in our next result, but we will only give the proof in the case that the base curve C is $\mathbb{P}^1$.

Corollary 11.3.2. (Tate [4]) *Let $\mathcal{E} \to C$, $P, Q \in E(K)$, h_δ, and $\langle \cdot, \cdot \rangle_t$ be as in (11.3.1).*
(a) *Suppose that the base curve C is isomorphic to $\mathbb{P}^1$. Then*

$$\langle P_t, Q_t \rangle_t = \langle P, Q \rangle h_\delta(t) + O(1) \quad \text{for } t \in \mathbb{P}^1(\bar{k}) \text{ with } \mathcal{E}_t \text{ non-singular.}$$

Notice that in this result we can take h_δ to be the usual height function on $\mathbb{P}^1$.

(b) *For an arbitrary base curve C, we have*

$$\langle P_t, Q_t \rangle_t = \langle P, Q \rangle h_\delta(t) + O\left(\sqrt{h_\delta(t)}\right) \quad \text{for } t \in C(\bar{k}) \text{ with } \mathcal{E}_t \text{ non-singular.}$$

(See exercise 3.34 for the case that C has genus 1.)

We will prove (11.3.1) and (11.3.2) simultaneously.

PROOF (of Corollaries 11.3.1 and 11.3.2). The subgroup $E(K)_0$ has finite index in $E(K)$ from (9.4), so given any two points $P, Q \in E(K)$, we can find an integer N such that $NP, NQ \in E(K)_0$. Note that N depends only on P and Q. Further, the canonical height pairings on $E(K)$ and $\mathcal{E}_t(\bar{k})$ are bilinear, so

$$\langle NP_t, NQ_t \rangle_t = N^2 \langle P_t, Q_t \rangle_t \quad \text{and} \quad \langle NP, NQ \rangle = N^2 \langle P, Q \rangle.$$

Replacing P, Q by NP, NQ and dividing each of the formulas in (11.3.1) and (11.3.2) by N^2, we see that it suffices to prove the two corollaries for points $P, Q \in E(K)_0$.

Assuming now that P and Q are in $E(K)_0$, (9.5c) tells us that

$$\langle P, Q \rangle = \deg[P, Q].$$

Hence the divisor

$$\beta = [P, Q] - \langle P, Q \rangle \delta \in \text{Div}(C)$$

is a divisor of degree 0. Using (11.1) and the additivity of heights (10.1b) yields the estimate

$$\langle P_t, Q_t \rangle_t = h_{[P,Q]}(t) + O(1) = \langle P, Q \rangle h_\delta(t) + h_\beta(t) + O(1).$$

We consider three cases.

First, to prove (11.3.1), we divide by $h_\delta(t)$ and take the limit

$$\lim_{h_\delta(t) \to \infty} \frac{\langle P_t, Q_t \rangle_t}{h_\delta(t)} = \langle P, Q \rangle + \lim_{h_\delta(t) \to \infty} \frac{h_\beta(t)}{h_\delta(t)}.$$

Now (10.2) and the fact that $\deg(\beta) = 0$ imply that

$$\lim_{h_\delta(t) \to \infty} \frac{h_\beta(t)}{h_\delta(t)} = \frac{\deg \beta}{\deg \delta} = 0,$$

which completes the proof of (11.3.1).

Next, suppose that $C = \mathbb{P}^1$. Then $\beta \sim 0$, since two divisors in $\text{Div}(\mathbb{P}^1)$ are linearly equivalent if and only if they have the same degree [AEC, II.3.2].

The equivalence property of heights (10.1c) implies that $h_\beta(t)$ is bounded, so the above estimate becomes

$$\langle P_t, Q_t\rangle_t = \langle P, Q\rangle h_\delta(t) + O(1).$$

This finishes the proof of (11.3.2a).

We will not give the proof of (11.3.2b), other than to say that one uses an estimate

$$h_\beta(t) = O\big(\sqrt{h_\delta(t)}\,\big) + O(1),$$

valid for divisors β of degree 0. This in turn follows from properties of the canonical height on the Jacobian variety of the curve C. For the complete proof, see Lang [4, 12 corollary 5.4] or Tate [4]. The special case that C has genus 1 is discussed in exercises 3.33 and 3.34.   □

Let $\mathcal{E} \to C$ be an elliptic surface. Each point $P \in E(K)$ defines a morphism $\sigma_P : C \to \mathcal{E}$ which we have been denoting by $t \mapsto P_t$. Turning this around, we can also say that each point $t \in C(\bar{k})$ determines a map

$$\sigma_t : E(K) \longrightarrow \mathcal{E}_t(\bar{k}), \qquad P \longmapsto P_t,$$

called the *specialization map of* $\mathcal{E}$ *at* t. If the fiber $\mathcal{E}_t$ is non-singular, then it is clear that the specialization map is a homomorphism,

$$\sigma_t(P + Q) = (P + Q)_t = P_t + Q_t = \sigma_t(P) + \sigma_t(Q).$$

This follows from the fact that on a non-singular fiber $\mathcal{E}_t$, the section σ_{P+Q} is defined by the relation $\sigma_{P+Q}(t) = \sigma_P(t) + \sigma_Q(t)$. We now show that for "most" values of t the specialization homomorphism is injective.

Theorem 11.4. (Silverman [1], [7]) *Let* $\mathcal{E} \to C$ *be a non-split elliptic surface defined over a number field* k, *and let* $\delta \in \mathrm{Div}(C)$ *be a divisor of positive degree. Then there is a constant* $c > 0$ *so that*

$$\sigma_t : E(K) \to \mathcal{E}_t(\bar{k}) \text{ is injective for all } t \in C(\bar{k}) \text{ satisfying } h_\delta(t) \geq c.$$

(One says that the set of points where σ_t *fails to be injective is a set of bounded height.) In particular, the specialization map* $\sigma_t : E(K) \to \mathcal{E}_t(\bar{k})$ *is injective for all but finitely many points* $t \in C(k)$.

Remark 11.4.1. In the case that the elliptic surface is split, there is a version of (11.4) due to Dem'janenko [1] and Manin [2]. Both Dem'janenko and Manin used their results to prove the Mordell conjecture (now Faltings' theorem) for certain curves. See exercise 3.16.

Remark 11.4.2. There is an earlier result, due to Néron [3] using a Hilbert irreducibility argument, which says that if $C(k)$ is infinite, then there are infinitely many $t \in C(k)$ for which the specialization map $\sigma_t : E(K) \rightarrow \mathcal{E}_t(\bar{k})$ is injective. This is sufficient for one of the main applications of (11.4), namely the construction of elliptic curves of elevated rank over $\mathbb{Q}$ or over number fields k. The idea is to find an elliptic surface $\mathcal{E} \rightarrow C$ over k for which $C(k)$ is infinite and such that $E(K)$ has high rank. Then specializing $t \in C(k)$ gives elliptic curves over k of high rank. Néron [3] used this procedure to construct infinitely many elliptic curves over $\mathbb{Q}$ with rank at least 10. More recently, Mestre [2] constructed an elliptic surface $\mathcal{E} \rightarrow \mathbb{P}^1$ defined over $\mathbb{Q}$ so that $E(\mathbb{Q}(t))$ has rank at least 12, and Nagao [2] extended this result to get rank at least 13. By taking particular values for $t \in \mathbb{Q}$, it is possible to find specific elliptic curves over $\mathbb{Q}$ with even higher ranks. See Fermigier [1], Nagao [1], and Nagao-Kouya [1] for examples with ranks at least 19, 20, and 21. And the quest continues!

PROOF (of Theorem 11.4). Our assumption that $\mathcal{E} \rightarrow C$ is non-split means that the Mordell-Weil theorem (6.1) is valid, so the group $E(K)$ is finitely generated. In particular, the torsion subgroup $E(K)_{\text{tors}}$ is finite.

Let $P \in E(K)$ be any non-zero point. Then there are only finitely many $t \in C(\bar{k})$ for which $P_t = O_t$, since the two divisors (P) and (O) intersect in only finitely many points. This holds for each of the finitely many points in $E(K)_{\text{tors}}$, so we see that on torsion points, the specialization map

$$\sigma_t : E(K)_{\text{tors}} \rightarrow \mathcal{E}_t(\bar{k})$$

is injective for all but finitely many $t \in C(\bar{k})$. (In fact, the specialization map is injective on torsion whenever $\mathcal{E}_t$ is non-singular, since the residue field k has characteristic 0. This follows from the identification of the kernel of the specialization map with the formal group of the elliptic curve; see [AEC, IV.3.2b, VII.2.2].)

Next let $P^1, \ldots, P^r \in E(K)$ be generators for the free part of $E(K)$; that is, $P^1, \ldots, P^r$ give a basis for the free group $E(K)/E(K)_{\text{tors}}$. Then the non-degeneracy of the canonical height pairing on $E(K)$ described in (4.3cd) implies that

$$\det\left(\langle P^i, P^j \rangle\right)_{1 \leq i, j \leq r} \neq 0.$$

(This is an elementary property of non-degenerate bilinear forms; see Lemma 11.5 below.)

Next we specialize the P^i's to the fiber $\mathcal{E}_t$, take the height regulator of the resulting P_t^i's, and use (11.3.1) to compute

$$\lim_{h_\delta(t)\to\infty} \frac{\det\big((\langle P_t^i, P_t^j\rangle_t)_{1\le i,j\le r}\big)}{h_\delta(t)^r} = \det\left(\lim_{h_\delta(t)\to\infty} \frac{\langle P_t^i, P_t^j\rangle_t}{h_\delta(t)}\right)_{1\le i,j\le r}$$

$$= \det\big((\langle P^i, P^j\rangle)\big)_{1\le i,j\le r}$$

$$\ne 0.$$

Hence there is a constant c so that

$$\det\big((\langle P_t^i, P_t^j\rangle_t)_{1\le i,j\le r}\big) \ne 0 \qquad \text{for all } t \in C(\bar{k}) \text{ with } h_\delta(t) > c.$$

It follows from (11.5) and the non-degeneracy of the height pairing on $\mathcal{E}_t(\bar{k})$ [AEC, VIII.9.3 or VIII.9.6] that the points $P_t^1, \ldots, P_t^r$ are linearly independent provided that $h_\delta(t) > c$.

Adjusting c if necessary to account for the finitely many points in the torsion subgroup $E(K)_{\text{tors}}$, we have now proven that for all $t \in C(\bar{k})$ with $h_\delta(t) > c$, both of the specialization maps

$$E(K)_{\text{tors}} \longrightarrow \mathcal{E}_t(\bar{k})_{\text{tors}} \qquad \text{and} \qquad E(K)/E(K)_{\text{tors}} \longrightarrow \mathcal{E}_t(\bar{k})/\mathcal{E}_t(\bar{k})_{\text{tors}}$$

are injective. Now a simple diagram chase using the commutative diagram

$$
\begin{array}{ccccccccc}
0 & \longrightarrow & E(K)_{\text{tors}} & \longrightarrow & E(K) & \longrightarrow & E(K)/E(K)_{\text{tors}} & \longrightarrow & 0 \\
& & \downarrow & & \downarrow & & \downarrow & & \\
0 & \longrightarrow & \mathcal{E}_t(\bar{k})_{\text{tors}} & \longrightarrow & \mathcal{E}_t(\bar{k}) & \longrightarrow & \mathcal{E}_t(\bar{k})/\mathcal{E}_t(\bar{k})_{\text{tors}} & \longrightarrow & 0
\end{array}
$$

shows that $E(K) \to \mathcal{E}_t(\bar{k})$ is injective, which completes the proof of the first part of (11.4).

The second part is then an immediate consequence of the first part and of (10.3) once we observe (10.3.1) that on a curve, any divisor of positive degree is ample. $\qquad \square$

It remains to prove the elementary property of non-degenerate bilinear forms used in the proof of (11.4).

Lemma 11.5. *Let Γ be a free abelian group, let $\langle \cdot, \cdot \rangle$ be a positive definite bilinear form on Γ with values in $\mathbb{Q}$, and let $x_1, \ldots, x_r \in \Gamma$. Then*

$$x_1, \ldots, x_r \text{ are linearly independent} \iff \det\big((\langle x_i, x_j\rangle)\big)_{1\le i,j\le r} \ne 0.$$

PROOF. Suppose first that the determinant is 0. This means that there are integers $a_1, \ldots, a_r$, not all zero, so that

$$\sum_{i=1}^r a_i \langle x_i, x_j\rangle = 0 \qquad \text{for all } 1 \le j \le r.$$

Multiplying by a_j, summing over j, and using the bilinearity gives

$$0 = \sum_{j=1}^{r} a_j \sum_{i=1}^{r} a_i \langle x_i, x_j \rangle = \Big\langle \sum_{i=1}^{r} a_i x_i, \sum_{j=1}^{r} a_j x_j \Big\rangle.$$

In other words, $y = \sum a_i x_i \in \Gamma$ satisfies $\langle y, y \rangle = 0$, so the positivity of the bilinear form implies that $y = 0$. Hence the x_i's are linearly dependent.

Conversely, if the x_i's are linearly dependent, say $\sum a_i x_i = 0$, then the linearity of the pairing implies that the rows of the matrix $(\langle x_i, x_j \rangle)$ are linearly dependent, so the determinant is zero. $\square$

§12. Integral Points on Elliptic Curves over Function Fields

There is a theory of S-integral points on elliptic curves over function fields which is completely analogous to the theory over number fields as described in [AEC, Ch. IX]. However, for function fields it is possible to prove much stronger results using relatively elementary techniques. In this section we will give a short and elegant proof of the analogue of Siegel's theorem [AEC, IX.3.2.1] which asserts that an elliptic curve has only finitely many S-integral points. We will also state and briefly sketch the proof of an effective version of this result.

The simplest function field analogue of integral points are "polynomial points." Thus let $E/k(T)$ be an elliptic curve over a rational function field, say given by a Weierstrass equation

$$E : y^2 = x^3 + A(T)x + B(T) \qquad \text{with } A(T), B(T) \in k[T].$$

The the set of polynomial points of E is the set

$$\{ P = (x, y) \in E(k(T)) \,:\, x, y \in k[T] \}.$$

For example, if

$$E : y^2 = x^3 - T^2 x + T^2 \qquad \text{and} \qquad P = (T, T),$$

then (1.1.1) says that P and $2P$ are polynomial points, but $3P$ is not.

One way to characterize the polynomial ring $k[T]$ is to observe that it is the subring of $k(T)$ consisting of functions with no (finite) poles. More generally, we define the ring of S-integers of an arbitrary function field in the following way.

Definition. Let $K = k(C)$ be the function field of a curve, and let $S \subset C$ be a non-empty finite set of points of C. The *ring of S-integers of K* is the ring

$$R_S = \{ f \in K \ : \ \mathrm{ord}_t(f) \geq 0 \text{ for all } t \notin S \}.$$

Here $\mathrm{ord}_t(f)$ is the order of vanishing of f at t; see [AEC, II §1].

The following function field analogue of Siegel's theorem is a special case of a result of Lang [8], who proved a general finiteness theorem for integral points on curves of arbitrary genus over function fields.

Theorem 12.1. *Let $K = k(C)$ be the function field of a curve, let $S \subset C$ be a non-empty finite set of points of C, let E/K be an elliptic curve that does not split over $\bar{k}$, and let $F \in K(E)$ be a non-constant function on E. Then*

$$\{ P \in E(K) \ : \ F(P) \in R_S \}$$

is a finite set.

PROOF. Our first observation is that it suffices to prove (12.1) for the special case that F is taken to be the x-coordinate on some Weierstrass equation for E/K. The reduction from the general case to this special case is given for number fields in [AEC, IX.3.2.2], but the proof is the same for function fields. So we are reduced to showing that

$$\{ P \in E(K) \ : \ x(P) \in R_S \}$$

is a finite set.

If P is a point in this set, then the height of P is a sum of local contributions coming from the points in S. More precisely, we have

$$
\begin{aligned}
h(P) &= h\big(x(P)\big) && \text{definition of } h(P) \\
&= \sum_{t \in C} \max\{ -\mathrm{ord}_t\big(x(P)\big), 0 \} && \text{from (4.1)} \\
&= \sum_{t \in S} \max\{ -\mathrm{ord}_t\big(x(P)\big), 0 \} && \text{since } \mathrm{ord}_t\big(x(P)\big) \geq 0 \text{ for } t \notin S \\
&\leq \#S \cdot \max_{t \in S}\{ -\mathrm{ord}_t\big(x(P)\big) \}.
\end{aligned}
$$

Further, our assumption that E does not split combined with (5.4) tells us that $E(K)$ has only finitely many points of bounded height. Hence the following result (12.2) completes the proof of (12.1). $\square$

Theorem 12.2. (Manin [3]) *Let $K = k(C)$ be the function field of a curve, let $t \in C$ be any point of C, and let E/K be an elliptic curve that does not split over $\bar{k}$, say given by a Weierstrass equation*

$$E : y^2 = x^3 + Ax + B.$$

Then the function $\operatorname{ord}_t\big(x(P)\big)$ is bounded below as P ranges over $E(K)$.

Before giving the proof of (12.2), we want to observe just how strong a statement it is. For example, the number field analogue of (12.2) is certainly false. Thus, if we let $p \in \mathbb{Z}$ be a prime and $E/\mathbb{Q}$ be an elliptic curve with a rational point

$$P \in E(\mathbb{Q}) \text{ satisfying } \operatorname{ord}_p\big(x(P)\big) < 0,$$

then consideration of the formal group of $E(\mathbb{Q}_p)$ shows immediately that

$$\operatorname{ord}_p\big(x(p^n P)\big) \to -\infty \text{ as } n \to \infty.$$

The following short proof of (12.2), which is due to Voloch [1], uses the formal group and depends crucially on the fact that the base field k has characteristic 0.

PROOF. (of Theorem 12.2, Voloch [1]) We may replace the constant field k by its algebraic closure, since this will only have the effect of making $E(K)$ larger. Further, replacing x by $u^2 x$ for some $u \in K^*$, we may assume that the given Weierstrass equation is a minimal equation for the valuation ord_t.

For each integer $n \geq 1$, let

$$E_n(K) = \big\{P \in E(K) : \operatorname{ord}_t\big(x(P)\big) \leq -2n\big\}.$$

This is the standard filtration on the formal group of E, see [AEC, Ch. IV]. The crucial facts to note here are that each $E_n(K)$ is a subgroup of $E(K)$, and each quotient group $E_n(K)/E_{n+1}(K)$ is isomorphic to a subgroup of k. To see this, let K_t be the completion of the field K for the valuation ord_t, let R_t be the ring of integers of K_t, let $\mathfrak{M}_t$ be the maximal ideal of R_t, and let $\hat{E}$ be the formal group of E/K_t. It follows from [AEC, IV.3.1.3] that $E_n(K_t) \cong \hat{E}(\mathfrak{M}_t^n)$, and then [AEC, IV.3.2(a)] tells us that

$$E_n(K_t)/E_{n+1}(K_t) \cong \hat{E}(\mathfrak{M}_t^n)/\hat{E}(\mathfrak{M}_t^{n+1}) \cong \mathfrak{M}_t^n/\mathfrak{M}_t^{n+1} \cong k.$$

(The last isomorphism uses $R_t/\mathfrak{M}_t \cong k$.) Now the fact that $E_n(K) = E(K) \cap E_n(K_t)$ immediately implies our two assertions that $E_n(K)$ is a subgroup of $E(K)$ and $E_n(K)/E_{n+1}(K)$ is isomorphic to a subgroup of k. In particular, our assumption that k has characteristic zero means that the quotient groups $E_n(K)/E_{n+1}(K)$ have no elements of finite order.

Suppose now that $\operatorname{ord}_t\big(x(P)\big)$ is not bounded below on $E(K)$. Then we can choose a sequence of points $P_1, P_2, \ldots$ with

$$\operatorname{ord}_t\big(x(P_i)\big) = -2n_i \qquad \text{and} \qquad 0 < n_1 < n_2 < n_3 < \cdots.$$

We claim that the points $P_1, P_2, \ldots$ are linearly independent, which will contradict the Mordell-Weil theorem (6.1) and thus complete the proof of (12.2).

Suppose to the contrary that $P_1, P_2, \ldots$ are linearly dependent. Discarding the first few P_i's if necessary and relabeling, we may assume that there is a relation

$$m_1 P_1 + m_2 P_2 + \cdots + m_r P_r = O \qquad \text{with } m_1 \neq 0.$$

Using the fact that $n_1 < n_2 < \cdots < n_r$ and that the $E_n(K)$'s are nested subgroups of $E(K)$, we see that

$$m_1 P_1 = -m_2 P_2 - \cdots - m_r P_r \in E_{n_2}(K), \qquad \text{so } m_1 P_1 \in E_{n_1+1}(K).$$

But as noted above, the quotient $E_{n_1}(K)/E_{n_1+1}(K)$ is isomorphic to a subgroup of k, and k is a field of characteristic zero. So the fact that $m_1 P_1$ is zero in this quotient implies that P_1 itself is zero in the quotient. In other words, $P_1 \in E_{n_1+1}(K)$, which contradicts the fact that $\mathrm{ord}_t\big(x(P_1)\big) = -2n_1$. Hence the P_i's are independent. $\qquad \square$

The proofs of (12.1) and (12.2) are ineffective because they depend on the Mordell-Weil theorem (6.1). Notice that (12.2) is analogous to Siegel's theorem [AEC, IX.3.1], although (12.2) is both stronger and considerably easier to prove.

Similarly, one can prove effective bounds for S-integral points in the function field case which are analogous to the bounds provided by linear-forms-in-logarithms methods for number fields [AEC, §5]. Again the function field estimates are stronger and much easier to prove. A number of people have given such bounds, including for example Schmidt [1], Mason [1], and Hindry-Silverman [2]. We will briefly sketch the proof of the following version. The argument is the same for elliptic or hyperelliptic curves, so we give the more general case.

Theorem 12.3. Let $K = k(C)$ be the function field of a curve C of genus g, let $S \subset C$ be a non-empty finite set of points of C, let R_S be the ring of S-integers of K, and let $f(x) \in R_S[x]$ be a monic polynomial with discriminant Δ satisfying $\Delta \in R_S^*$. Suppose that $x, y \in R_S$ satisfy

$$y^2 = f(x).$$

Then

$$h\big(y^4/\Delta\big) \leq 4n(n-1)\max\{2g - 2 + \#S, 0\}.$$

(Recall from §4 that the height of an element $f \in K$ is defined to be the degree of the map $f : C \to \mathbb{P}^1$. Also note that the set S has to be chosen large enough to contain all of the zeros and poles of Δ.)

Remark 12.3.1. The bound in (12.3) is stated for y^4/Δ because this quantity is invariant under linear change of variables. However, it is easy to use (12.3), the relation $y^2 = f(x)$, and elementary properties of height functions to give a bound for $h(x)$ in terms of the coefficients of f. See exercise 3.39 for the particular case of an elliptic curve.

PROOF. (Sketch of Theorem 12.3) The first step is to reduce the problem of S-integral points on elliptic curves to the problem of solving the S-unit equation

$$u + v = 1, \qquad u, v \in R_S^*.$$

This reduction procedure is due to Siegel and is described in [AEC, IX.4.3]. Next, one proves that if $u, v \in R_S^*$ satisfy $u + v = 1$, then

$$h(u) \leq \max\{2g - 2 + \#S, 0\}.$$

This bound is the function field analogue of the abc-conjecture of Masser and Osterlé. There are several elementary proofs available; see for example Mason [1], Silverman [8], or Vojta [1]. Tracing back through Siegel's argument gives the estimate described in (12.3). We leave the details to the reader, or see Hindry-Silverman [2, Prop. 8.2]. $\square$

EXERCISES

3.1.   For any pair (a, b), let $E_{a,b}$ be the curve

$$E_{a,b} : y^2 + axy + by = x^3 + bx^2.$$

Notice that the point $(0, 0)$ is on each of these curves.

(a) Let E/k be an elliptic curve, let $P \in E(k)$ be a point, and assume that $P, 2P, 3P \neq 0$. Prove that E has a Weierstrass equation of the form $E_{a,b}$ with P corresponding to $(0, 0)$. (*Hint.* Move P to $(0, 0)$, rotate so that the tangent line at $(0, 0)$ is the x-axis, and make a dilation to get $a_2 = a_3$.)

(b) Prove that $5P = O$ if and only if $a = b + 1$. Conclude that every elliptic curve E/k with a point P of exact order 5 is isomorphic to some fiber of the elliptic surface

$$\mathcal{E} : y^2 + (t + 1)xy + ty = x^3 + tx^2$$

by an isomorphism taking P to the point $(0, 0)$ on that fiber.

(c) More precisely, if E is defined over k and if $P \in E(k)$ is a point of exact order 5, prove that there is a unique point $t_0 \in \mathbb{P}^1(k)$ and an isomorphism $\phi : E \to \mathcal{E}_{t_0}$ defined over k such that $\phi(P) = (0, 0)$.

(d) Using a similar construction, find an elliptic surface $\mathcal{E} \to \mathbb{P}^1$ which classifies elliptic curves with a given point of order 7.

3.2. Let $C/\mathbb{Q}$ be the (elliptic) curve

$$C : s^2 - s = t^3 - t^2,$$

let $E/\mathbb{Q}(C)$ be the elliptic curve

$$E : y^2 + (st + t - s^2)xy + s(s-1)(s-t)t^2 y = x^3 + s(s-1)(s-t)tx^2,$$

and let $P = (0,0) \in E(\mathbb{Q}(C))[11]$ be the point of order 11 as described in (1.1.3).
(a) Let A be an elliptic curve and $Q \in A$ a point of order 11. Prove that there is a unique point $(s_0, t_0) \in C$ such that if we substitute $(s,t) = (s_0, t_0)$ into the equations for E and P, then we obtain an elliptic curve E_0 and a point $P_0 \in E_0$ of order 11 such that there is an isomorphism $\phi : A \to E_0$ satisfying $\phi(Q) = P_0$.
(b) If A is defined over k and $Q \in E(k)$, prove that the point (s_0, t_0) obtained in (a) will lie in $C(k)$.

3.3. This exercise gives a function field analogue of [AEC, exercise 8.1]. Let C/k be a curve of genus g with function field $K = k(C)$, and let E/K be an elliptic curve which does not split over K.
(a) Suppose that E/K has a Weierstrass equation of the form

$$E : y^2 = (x - e_1)(x - e_2)(x - e_3) \qquad \text{with } e_1, e_2, e_3 \in K.$$

Let $S \subset C$ be the set of points where any one of e_1, e_2, e_3 has a pole together with the points where the product $(e_1 - e_2)(e_1 - e_3)(e_2 - e_3)$ vanishes. Prove that

$$\operatorname{rank} E(K) \le 4g + 2\#S - 2.$$

(b) Suppose that E/K has a Weierstrass equation of the form

$$E : y^2 = x^3 + Ax + B \qquad \text{with } A, B \in K.$$

Let $S \subset C$ be the set of points where A or B has a pole together with the points where $\Delta = 4A^3 + 27B^2$ vanishes. Find an explicit bound for the rank of $E(K)$ in terms of g and $\#S$.

3.4. Let $k/\mathbb{Q}$ be a finitely generated field extension, that is, $k = \mathbb{Q}(\alpha_1, \ldots, \alpha_r)$ for some $\alpha_1, \ldots, \alpha_r \in \mathbb{C}$. Let C/k be a curve with function field $K = k(C)$, and let E/K be an elliptic curve. Prove that $E(K)$ is a finitely generated group. In particular, this is true if k is a number field. (*Hint.* You may find exercise 3.15 below useful for doing this problem.)

3.5. Let A be an abelian variety of dimension one, so in particular A is a non-singular projective curve. (See §2 for the definition of abelian variety.) Prove that A has genus 1, so A is an elliptic curve.

3.6. Let $\phi : V \to W$ be a rational map of projective varieties.
(a) Prove that the image $\overline{\phi(V)}$ is an algebraic subset of W.
(b) Suppose that V is non-singular. Prove that the set of points where ϕ is not defined has codimension at least two in V.

3.7. Recall that a topological space X is *irreducible* if it cannot be written as a union $X = X_1 \cup X_2$ of non-empty closed subsets of X.
(a) Let X be a topological space, and let $Z \subset X$ be a subset taken with the induced topology. If Z is irreducible, prove that the closure of Z in X is also irreducible.
(b) Let $\phi : X \to Y$ be a continuous map of topological spaces. If X is irreducible, prove that its image $\phi(X)$ is irreducible.
(c) Use (a) and (b) to deduce the second part of Proposition 3.5(a).

3.8. Let $\mathcal{E} \to C$ be an elliptic surface over k. Define a map

$$j_{\mathcal{E}} : C \longrightarrow \mathbb{P}^1, \qquad j_{\mathcal{E}}(t) = j(\mathcal{E}_t).$$

More precisely, define $j_{\mathcal{E}}(t)$ to be the j-invariant of the elliptic curve $\mathcal{E}_t$ provided that the fiber $\mathcal{E}_t$ is non-singular, and for the moment leave it undefined for the remaining points of C. Prove that $j_{\mathcal{E}}$ is an algebraic map, and conclude that it extends to a morphism from C to $\mathbb{P}^1$.

3.9. Let $\mathcal{E} \to C$ be an elliptic surface over k, and let $j_{\mathcal{E}} : C \to \mathbb{P}^1$ be the morphism defined in exercise 3.8.
(a) If $\mathcal{E} \to C$ splits over k, prove that $j_{\mathcal{E}}$ is a constant map.
(b) Give an example (with proof) of an elliptic surface $\mathcal{E} \to C$ that does not split over k for which $j_{\mathcal{E}}$ is a constant map.

3.10. Let $\mathcal{E} \to C$ be an elliptic surface over k, and let $j_{\mathcal{E}} : C \to \mathbb{P}^1$ be the morphism defined in exercise 3.8. Choose a Weierstrass equation

$$\mathcal{E} : y^2 + a_1 xy + a_3 y = x^3 + a_2 x^2 + a_4 x + a_6$$

for $\mathcal{E}$, where $a_1, \ldots, a_6 \in k(C)$, and let c_4, c_6 be the usual associated quantities [AEC III §1]. Prove that $\mathcal{E} \to C$ splits over k if and only if one of the following three conditions is true.
(i) $j_{\mathcal{E}}(C) = \{0\}$ and c_6 is a sixth power in $k(C)$.
(ii) $j_{\mathcal{E}}(C) = \{1728\}$ and c_4 is a fourth power in $k(C)$.
(iii) $j_{\mathcal{E}}(C) = \{\alpha\}$ with $\alpha \neq 0, 1728$, and c_6/c_4 is a square in $k(C)$.
(Keep in mind we are assuming that $\operatorname{char}(k) = 0$, although this exercise remains true if $\operatorname{char}(k) \geq 5$.)

3.11. Let E/K be an elliptic curve defined over a function field $K = k(C)$ by an equation of the form

$$E : y^2 = x^3 + Ax + B.$$

Further define the "height of E" to be $h(E) = h(A^3) + h(B^2)$.
(a) Prove that for all $P \in E(K)$,

$$4h(P) - 3h(E) \leq h(2P) \leq 4h(P) + h(E).$$

(b) Prove that for all $P, Q \in E(K)$,

$$4h(E) \leq h(P + Q) + h(P - Q) - 2h(P) - 2h(Q) \leq h(E).$$

(c) Prove that for all $P \in E(K)$,

$$\left| \hat{h}(P) - \frac{1}{2} h(P) \right| \leq \frac{1}{2} h(E).$$

(In all three parts of this problem, the constants in front of the $h(E)$'s are far from best possible. See how much you can improve them.)

3.12. Let $E/\mathbb{C}(T)$ be the elliptic curve

$$E : y^2 = x^3 - (T^2 + T)x + T^2.$$

Find all points in $E(\mathbb{C}(T))$ of the form

$$P = [a_0 + a_1 T, b_0 + b_1 T, c_0 + c_1 T]$$

by substituting P into the equation for E and solving for $a_0, \ldots, c_1$. How many of these points are defined over $\mathbb{Q}(T)$?

3.13. With notation as in the statement of Lemma 5.5.2, prove that if

$$\deg D \geq g + \frac{3}{2}d + \frac{1}{2}\deg A + \frac{1}{2}\deg B,$$

then the image of V_D in $E(K)$ contains $E(K, d)$. This provides a strengthened version of (5.5.2).

3.14. Let V be an irreducible projective variety defined over an algebraically closed field, and let $\gamma_1, \gamma_2 \in V$ be distinct points. Prove that there exists an irreducible curve $\Gamma \subset V$ with $\gamma_1, \gamma_2 \in \Gamma$.

3.15. This exercise describes the Mordell-Weil theorem (6.1) for split elliptic surfaces. Let C/k be a curve, let E_0/k be an elliptic curve, and let $\mathcal{E} = E_0 \times C$ be the corresponding split elliptic surface. Further, let $\mathrm{Map}_k(C, E_0)$ be the set of morphisms from C to E_0 defined over k. We use the group structure on E_0 to make $\mathrm{Map}_k(C, E_0)$ into a group in the usual way, $(\phi + \psi)(t) = \phi(t) + \psi(t)$. Notice that the constant maps in $\mathrm{Map}_k(C, E_0)$ form a subgroup isomorphic to $E_0(k)$.
(a) Prove that there is a natural isomorphism $\mathcal{E}(C/k) \cong \mathrm{Map}_k(C, E_0)$. In particular, the group of sections $\mathcal{E}(C/k)$ contains a subgroup isomorphic to $E_0(k)$.
(b) Prove that the quotient group $\mathcal{E}(C/k)/E_0(k)$ is finitely generated.
(c) If k is a number field, prove that $\mathcal{E}(C/k)$ is finitely generated.

3.16. Let k be a number field, and let C/k, E_0/k, $\mathcal{E} = E_0 \times C$, and $\mathrm{Map}_k(C, E_0)$ be as in the previous exercise.
(a) Let $\delta \in \mathrm{Div}(C)$ be a divisor of degree 1, and fix a height function $h_\delta : C(\bar{k}) \to \mathbb{R}$ associated to δ. Prove that for any map $\phi \in \mathrm{Map}_k(C, E_0)$,

$$\lim_{t \in C(\bar{k}),\, h_\delta(t) \to \infty} \frac{\hat{h}_{E_0}(\phi(t))}{h_\delta(t)} = \deg(\phi).$$

(b) Fix a basepoint $t_0 \in C(k)$. Prove that there is a constant c so that if $t \in C(\bar{k})$ satisfies $h_\delta(t) \geq c$, then the map

$$\begin{aligned}
\{\phi \in \mathrm{Map}_k(C, E_0) : \phi(t_0) = O\} &\longrightarrow E_0(\bar{k}) \\
\phi &\longrightarrow \phi(t)
\end{aligned}$$

is an injective homomorphism.

(c) Suppose that the quotient group $\mathrm{Map}_k(C, E_0)/E_0(k)$ has free rank r, say generated by the maps $\phi_1, \ldots, \phi_r : C \to E_0$. Suppose further that the group $E_0(k)$ has rank strictly less than r. Prove that $C(k)$ is finite.

(d) Fix an element $b \in k^*$, and let C/k and E_0/k be the curves

$$C : X^6 + Y^6 = bZ^6, \qquad E_0 : y^2 z = x^3 + bz^3.$$

Prove that the group $\mathrm{Map}_k(C, E_0)/E_0(k)$ has rank at least two by showing that the maps

$$\phi_1([X, Y, Z]) = [-X^2 Z, Y^3, Z^3], \qquad \phi_2([X, Y, Z]) = [-Y^2 Z, X^3, Z^3]$$

give independent elements. Use this to prove that if rank $E_0(k) \le 1$, then $C(k)$ is finite.

3.17. Let E be an elliptic curve defined over $\mathbb{Q}(T)$ that does not split over $\mathbb{C}(T)$.

(a) Suppose that $k_1/\mathbb{Q}$ and $k_2/\mathbb{Q}$ are fields with the property that

$$E(k_1(T)) = E(\mathbb{C}(T)) \qquad \text{and} \qquad E(k_2(T)) = E(\mathbb{C}(T)).$$

Let $k = k_1 \cap k_2$. Prove that $E(k(T)) = E(\mathbb{C}(T))$. Deduce that there is a smallest field with this property. We will call this field the *field of definition for* $E(\mathbb{C}(T))$ and will denote it by k_E.

(b) Prove that k_E is a finite extension of $\mathbb{Q}$.

(c) More precisely, find an explicit bound for the degree $[k_E : \mathbb{Q}]$ that depends only on the rank of $E(\mathbb{C}(T))$.

(d) *Fix a Weierstrass equation for E of the form

$$E : y^2 = x^3 + A(T)x + B(T)$$

with $A(T), B(T) \in \mathbb{Z}[T]$, and let $\Delta(T) = 4A(T)^3 + 27B(T)^2$. Prove that the extension $k_E/\mathbb{Q}$ is unramified except possibly at 2, 3, and the primes dividing the discriminant of the polynomial $\Delta(T) \in \mathbb{Z}[T]$.

3.18. (a) Let $f \in k(\mathbb{P}^2)$. Prove that $\deg(\mathrm{div}(f)) = 0$, and deduce that the degree map $\deg : \mathrm{Pic}(\mathbb{P}^2) \to \mathbb{Z}$ described in (7.1) is a well-defined homomorphism.

(b) Prove that the degree map $\deg : \mathrm{Pic}(\mathbb{P}^2) \to \mathbb{Z}$ is an isomorphism.

(c) Generalize (a) and (b) to $\mathbb{P}^n$.

3.19. Let $P = (0, 0) \in \mathbb{A}^2$, and let $f_1, f_2 \in k(\mathbb{P}^2)$ be rational functions satisfying $f_1(P) = f_2(P) = 0$. Let Γ_1 and Γ_2 be the curves $f_1 = 0$ and $f_2 = 0$ respectively. Prove that Γ_1 and Γ_2 intersect transversally at P if and only if the following three conditions are true:

(i) Γ_1 is non-singular at P;

(ii) Γ_2 is non-singular at P;

(iii) the tangent line to Γ_1 at P is distinct from the tangent line to Γ_2 at P.

3.20. For each of the following curves $\Gamma_1, \Gamma_2 \subset \mathbb{P}^2$, calculate the local intersection index $(\Gamma_1 \cdot \Gamma_2)_P$ at the point $P = [0, 0, 1]$. (*Hint.* First dehomogenize by setting $Z = 1$.)

(a) $\Gamma_1 : Y^2 Z = X^3 + X^2 Z$, $\qquad \Gamma_2 : Y = 0$.

(b) $\Gamma_1 : Y^2 Z = X^3 + X^2 Z$, $\qquad \Gamma_2 : Y = X$.

(c) $\Gamma_1 : Y^2 Z = X^3$, $\qquad\qquad\quad \Gamma_2 : Y = X$.

(d) $\Gamma_1 : Y^2 Z = X^3 + X^2 Z$, $\qquad \Gamma_2 : Y^2 Z = X^3$.

3.21. *Let $\pi : S \to C$ be a fibered surface. Prove that every fiber S_t is connected. (You may want to take $k \Rightarrow \mathbb{C}$.)

3.22. Let $\pi : S \to C$ be a fibered surface, let $\Gamma \subset S$ be a non-singular irreducible horizontal curve, and let $\phi : \Gamma \to C$ be the restriction of π to Γ.
(a) Let $P \in \Gamma$ and $t = \phi(P) \in C$. Prove that

$$(\Gamma \cdot \pi^*(t))_P = e_P(\phi),$$

where $e_P(\phi)$ is the ramification index of $\phi : \Gamma \to C$ at P (see [AEC, II §2]).
(b) Prove that $\Gamma \cdot \pi^*(\delta) = \deg(\phi) \deg(\delta)$ for all divisors $\delta \in \mathrm{Div}(C)$.
(c) *Prove that $\Gamma \cdot \pi^*(\delta) = \deg(\phi) \deg(\delta)$ remains true even if the irreducible horizontal curve Γ is allowed to be singular.

3.23. Let $\pi : S \to C$ be a fibered surface, and let $\delta \in \mathrm{Div}(C)$ be a divisor of degree 0. Prove that $D \cdot \pi^*\delta = 0$ for every divisor $D \in \mathrm{Div}(S)$. (*Hint.* Use (8.1) and the previous exercise.)

3.24. Generalize (8.3.1) as follows. Let $\pi : S \to C$ be a fibered surface, and suppose that the fiber S_t consists of n components arranged in the shape of an n-gon with transversal intersections. In other words,

$$\pi^*(t) = \Gamma_0 + \Gamma_1 + \cdots + \Gamma_{n-1} \quad \text{with} \quad \Gamma_i \cdot \Gamma_j = \begin{cases} 1 & \text{if } i \equiv j \pm 1 \,(\mathrm{mod}\ n), \\ -2 & \text{if } i = j, \\ 0 & \text{otherwise.} \end{cases}$$

(a) Draw a picture illustrating this fiber, and show that the self-intersection values $\Gamma_i^2 = -2$ follow from the values of $\Gamma_i \cdot \Gamma_j$ for $i \neq j$.
(b) Let $I = (\Gamma_i \cdot \Gamma_j)_{0 \leq i,j \leq n-1}$ be the incidence matrix of the fiber, and let I_{00} be the minor obtained by deleting the first row and column from I. Find the value of $\det(I_{00})$ in terms of n.
(c) Let k be an integer between 0 and $n - 1$, and let $D \in \mathrm{Div}(S)$ be a divisor satisfying

$$D \cdot \Gamma_0 = -1, \quad D \cdot \Gamma_k = 1, \quad D \cdot \Gamma_i = 0 \text{ for } i \neq 0, k.$$

Find a fibral divisor

$$\Phi_D = \sum_{i=1}^{n-1} a_i \Gamma_i \quad \text{such that } (D + \Phi_D) \cdot \Gamma_i = 0 \text{ for all } 0 \leq i \leq n - 1;$$

that is, find an explicit formula for a_i in terms of i, k, and n.

3.25. Let $\pi : S \to C$ be a fibered surface.
(a) Let $\Gamma \in \mathrm{Div}(S)$ be an irreducible fibral divisor. Prove that there exists an irreducible horizontal divisor $D \in \mathrm{Div}(S)$ satisfying $D \cdot \Gamma > 0$.
(b) Prove that there exists a horizontal divisor $D \in \mathrm{Div}(S)$ with the property that $D \cdot \Gamma > 0$ for every irreducible fibral divisor $\Gamma \in \mathrm{Div}(S)$.
(c) Let $D \in \mathrm{Div}(S)$ be a divisor as in (b), and let $t \in C$. Use the Nakai-Moishezon criterion (10.3.2) to prove that the divisor $D + n\pi^*(t)$ is ample for all sufficiently large integers n.

3.26. Let $\pi : \mathcal{E} \to C$ be a minimal elliptic surface with associated elliptic curve E/K. Define an action of $E(K)$ on $\mathrm{Div}(\mathcal{E})$ by having $P \in E(K)$ send a divisor D to the divisor $\tau_P(D)$.

(a) Prove that the action of $E(K)$ on $\mathrm{Div}(\mathcal{E})$ descends to give a well-defined action of $E(K)$ on $\mathrm{Pic}(\mathcal{E})$.

(b) For $P, Q \in E(K)$, let $\Phi_{P,Q} \in \mathrm{Div}(\mathcal{E})$ be a fibral divisor with the property $(P + Q) - (P) - (Q) + (O) \sim \Phi_{P,Q}$. (See (9.5).) Prove that for a fixed $Q \in E(K)$, the map

$$E(K) \longrightarrow \mathrm{Pic}(\mathcal{E}), \qquad P \longmapsto \mathrm{class}\,\Phi_{P,Q},$$

is a one-cocycle from $E(K)$ to $\mathrm{Pic}(\mathcal{E})$, where $\mathrm{Pic}(\mathcal{E})$ is an $E(K)$-module as described in (a). (See (9.5) for a stronger result for $E(K)_0$.)

3.27. Let $P \in E(K)$. Prove that $P \in E(K)_0$ if and only if $(P) \cdot F = (O) \cdot F$ for every fibral divisor $F \in \mathrm{Div}(\mathcal{E})$.

3.28. Let $\pi : S \to C$ be a fibered surface.

(a) Let $\Gamma \subset S$ be a curve with the property that $\pi : \Gamma \to C$ is an isomorphism. Prove that $\Gamma \cdot \pi^*(t) = 1$ for all $t \in C$. Use this to deduce that

$$\Gamma \cdot \pi^*(\delta) = \deg(\delta) \quad \text{for all divisors } \delta \in \mathrm{Div}(C).$$

(b) Fix a point $t \in C$. Prove that the image of a section $\sigma : C \to S$ intersects exactly one component of the fiber S_t.

(c) Let $F \subset \mathrm{Div}(S)$ be a fibral divisor, and suppose that $F = \pi^*\delta$ for some $\delta \in \mathrm{Div}(C) \otimes \mathbb{Q}$. Suppose further that there exists a section $\sigma : C \to S$. Prove that $\delta \in \mathrm{Div}(C)$.

(d) Let $\delta \in \mathrm{Div}(C)$ be a divisor such that $\pi^*\delta$ is a principal divisor on $\mathcal{E}$. Prove that δ is a principal divisor on C. Deduce that π induces an injective homomorphism $\pi^* : \mathrm{Pic}(C) \to \mathrm{Pic}(S)$. (This is easier if you assume that there exists a section $\sigma : C \to S$.)

3.29. Let $\pi : \mathcal{E} \to C$ be an elliptic surface, let E/K be the associated elliptic curve over the function field $K = k(C)$ of C, and let $P_1, \ldots, P_r \in E(K)$. Prove that if

$$n_1(P_1) + \cdots + n_r(P_r) - n(O) \in \mathrm{Div}(\mathcal{E})$$

is linearly equivalent to a fibral divisor, then

$$[n_1]P_1 + \cdots + [n_r]P_r = O.$$

This gives the converse to Proposition 9.2.

3.30. Let S be a non-singular surface, let $D \in \mathrm{Div}(S)$ be a divisor, and let $H \in \mathrm{Div}(S)$ be an ample divisor.

(a) Use the Nakai-Moishezon criterion (10.3.2) to prove that the divisor $nH + D$ is ample for all sufficiently large integers n.

(b) Use (a) to prove (10.5) for surfaces; that is, prove that D can be written as the difference of two very ample divisors on the surface S.

3.31. Let V/k be a non-singular projective variety defined over a number field, let $D \in \mathrm{Div}(V)$ be a positive divisor, and let $h_D : V(\bar{k}) \to \mathbb{R}$ be an associated height function. (A divisor is positive if it can be written as a sum $\sum n_i D_i$, where the D_i's are irreducible subvarieties of V and the n_i's are positive.)

(a) Prove that there is a constant $c = c(V, D, h_D)$ such that

$$h_D(P) \geq c \quad \text{for all } P \in V(\bar{k}), \ P \notin D.$$

(Note that c need not be positive.)

(b) Give an example to show that (a) need not true for all $P \in V(\bar{k})$. (*Hint.* Take D to be the exceptional curve on $\mathbb{P}^2$ blown up at one point. A harder example is to let D be the diagonal in $C \times C$, where C is a curve of genus $g \geq 2$.)

3.32. Let V/k be a non-singular projective variety defined over a number field, and let $D, H \in \mathrm{Div}(V)$ be divisors with H ample and D algebraically equivalent to zero (see Hartshorne [1, exercise 1.7]). Prove that

$$\lim_{P \in V(\bar{k}), h_H(P) \to \infty} \frac{h_D(P)}{h_H(P)} = 0.$$

(This generalization of (10.3) is due to Lang.)

3.33. Let E/k be an elliptic curve defined over a number field k. For any divisor $\beta = \sum b_i(P_i) \in \mathrm{Div}(E)$, we define the canonical height $\hat{h}_\beta$ associated to β by the formula

$$\hat{h}_\beta(P) = \sum b_i \hat{h}(P - P_i).$$

Thus the usual canonical height $\hat{h}$ is the height associated to the divisor (O).

(a) If the divisor β is symmetric, that is, $[-1]^* \beta = \beta$, prove that

$$\hat{h}_\beta(P) = \lim_{n \to \infty} \frac{1}{n^2} h_\beta([n]P).$$

(b) If the divisor β is anti-symmetric, that is, $[-1]^* \beta = -\beta$, prove that

$$\hat{h}_\beta(P) = \lim_{n \to \infty} \frac{1}{n} h_\beta([n]P).$$

Also prove in this case that the map $\hat{h}_\beta : E(K) \to \mathbb{R}$ is a homomorphism.

(c) Let $\beta \in \mathrm{Div}(E)$ be a divisor of degree 0. Prove that there is a constant $c = c(E, \beta)$ so that

$$|\hat{h}_\beta(P)| \leq c\sqrt{\hat{h}(P)} \quad \text{for all } P \in E(\bar{k}).$$

3.34. Let $\mathcal{E} \to C$ be an elliptic surface defined over a number field k, and fix two points $P, Q \in E(K)$. Suppose that the base curve C is an elliptic curve, and let $\hat{h}_C$ be the canonical height on C. With notation as in (11.1) and (11.3), prove that

$$\langle P_t, Q_t \rangle_t = \langle P, Q \rangle \hat{h}_C(t) + O\left(\sqrt{\hat{h}_C(t)}\right)$$

$$\text{for all } t \in C(\bar{k}) \text{ with } \mathcal{E}_t \text{ non-singular.}$$

This is a special case of (11.3.2b).

3.35. Let $K = k(C)$ be the function field of a curve over an algebraically closed field k, let E/K be an elliptic curve, and choose a Weierstrass equation for E/K of the form

$$y^2 = x^3 + Ax + B \quad \text{for some } A, B \in K.$$

Let $\Delta = 4A^3 + 27B^2$, and for each point $t \in C$ let

$$n_t = \mathrm{ord}_t(\Delta) - 12\left[\frac{\min\{\mathrm{ord}_t(A^3), \mathrm{ord}_t(B^2)\}}{12}\right].$$

(Here $[r]$ is the greatest integer in r.) Define the *minimal discriminant divisor of E/K* to be the divisor

$$\mathcal{D}_{E/K} = \sum_{t \in C} n_t(t) \in \mathrm{Div}(C).$$

(This is the analogue of the minimal discriminant ideal for an elliptic curve defined over a number field [AEC, VIII §8].)

(a) Prove that $\mathcal{D}_{E/K}$ is a positive divisor and that it is independent of the choice of the Weierstrass equation for E/K.

(b) Let $\mathcal{E} \to C$ be a minimal elliptic surface associated to E. Prove that the fiber $\mathcal{E}_t$ is non-singular if and only if $\mathrm{ord}_t(\mathcal{D}_{E/K}) = 0$.

(c) Prove that if $\mathcal{D}_{E/K} = 0$, then the j-invariant $j(\mathcal{E}_t)$ is constant.

(d) Prove that if $\mathcal{D}_{E/K} = 0$ and $C = \mathbb{P}^1$, then $\mathcal{E}$ splits over C.

(e) Let C/k and E/K be given by

$$C : v^2 = u^4 - 7u^2 + 6u, \qquad E : y^2 = x^3 - 7u^2x + 6u^3.$$

Prove that $\mathcal{D}_{E/K} = 0$ and that the associated elliptic surface $\mathcal{E} \to C$ does not split.

(f) For the example in (e), show that $P = (u^2, uv) \in E(K)$ is a point of infinite order.

3.36. We continue with the notation from the previous exercise, with the additional assumption that E does not split over K. Let g be the genus of C. For each point $t \in C$, we take a Weierstrass equation for E that is minimal for the valuation ord_t and we let E_t be curve obtained by evaluating the coefficients at t. We define integers f_t by

$$f_t = \begin{cases} 0 & \text{if } E_t \text{ is non-singular (good reduction),} \\ 1 & \text{if } E_t \text{ has a node (multiplicative reduction),} \\ 2 & \text{if } E_t \text{ has a cusp (additive reduction).} \end{cases}$$

Then the *conductor divisor of* E/K is defined to be the quantity

$$\mathfrak{f}_{E/K} = \sum_{t \in C} f_t(t) \in \mathrm{Div}(C).$$

(For another description of the conductor, see (IV §10).)

(a) Prove that $\mathrm{ord}_t(\mathfrak{f}_{E/K}) \leq \mathrm{ord}_t(\mathcal{D}_{E/K})$ for all $t \in C$, and deduce that

$$\deg(\mathfrak{f}_{E/K}) \leq \deg(\mathcal{D}_{E/K}).$$

(b) Prove that

$$\deg(\mathcal{D}_{E/K}) \leq 6 \deg(\mathfrak{f}_{E/K}) + 12(g - 1).$$

This inequality is a precise function field analogue of Szpiro's conjecture (IV.10.6). It was originally discovered by Kodaira (see Shioda [3, Proposition 2.8]) long before Szpiro formulated his conjecture.

3.37. *This exercise is the function field analogue of Lang's conjectural lower bound for the canonical height [AEC, VIII.9.9]. We continue with the notation from the previous two exercises. Let g be the genus of C, and let $\hat{h} : E(K) \to \mathbb{R}$ be the canonical height on E (4.3, 9.3).

(a) Prove that there is a constant $c_1(g) > 0$, depending only on g, such that if $P \in E(K)$ is a point of infinite order, then $\hat{h}(P) \geq c_1(g) \deg \mathcal{D}_{E/K}$.

(b) Prove that there is an absolute constant $c_2 > 0$ so that if $\deg \mathcal{D}_{E/K} \geq 2g - 2$ and if $P \in E(K)$ is a point of infinite order, then $\hat{h}(P) \geq c_2 \deg \mathcal{D}_{E/K}$.

3.38. Let C be a non-singular hyperelliptic curve of genus two given by an equation

$$C : y^2 = ax^5 + bx^4 + cx^3 + dx^3 + ex + f,$$

let $i : C \to C$ be the involution $i(x, y) = (x, -y)$, and let $P_0 \in C$ be the point at infinity on C. (See [AEC, II.2.5.1 and exercise 2.14] for basic facts about hyperelliptic curves.)

(a) Prove that $i(P_0) = P_0$. Find all other points satisfying $i(P) = P$.

(b) Let $P, Q \in C$ be any two points. Prove that the divisors $(P) + (i(P))$ and $(Q) + (i(Q))$ are linearly equivalent.

(c) Let $D \in \mathrm{Div}^0(C)$. Prove that there exist points $P, Q \in C$ such that $D \sim (P) + (Q) - 2(P_0)$.

(d) Prove that the points P and Q in (c) are uniquely determined by D unless $P = i(Q)$.

(e) Prove that $\mathrm{Pic}(C)[2]$ is finite. More precisely, prove that it is isomorphic to $(\mathbb{Z}/2\mathbb{Z})^4$.

(f) Generalize (a)–(e) to the case of a hyperelliptic curve $C : y^2 = f(x)$ of genus g, where $f(x)$ is a polynomial of degree $2g + 1$ with distinct roots.

3.39. Let $K = k(C)$ be the function field of a curve C of genus g, let $S \subset C$ be a non-empty finite set of points of C, let R_S be the ring of S-integers of K, and let E/K be an elliptic curve given by a Weierstrass equation

$$E : y^2 = x^3 + Ax + B \qquad \text{with } A, B \in R_S \text{ and } \Delta = 4A^3 + 27B^2 \in R_S^*.$$

Let $P \in E(K)$ be a point satisfying $x(P), y(P) \in R_S$. Prove that

$$h(P) = h(x(P)) \leq 4 \max\{2g - 2 + \#S, 0\} + \frac{1}{6}(h(\Delta) + h(A^3) + h(B^2)).$$

(*Hint.* Use (12.3) and elementary properties of height functions.)

3.40. Let $\mathcal{E} \to C$ be a minimal elliptic surface, let E/K be the associated elliptic curve, and let $E(K)_0$ be the subgroup of $E(K)$ described in §9.

(a) Suppose that $\mathcal{E}$ has non-constant j-invariant; that is, the function $j_{\mathcal{E}} : C \to \mathbb{P}^1$ defined in exercise 3.8 is non-constant. Prove that $E(K)_0$ has no non-trivial torsion.

(b) For each integer $m \in \{2, 3, 6\}$, give an example of a non-split elliptic surface (necessarily with constant j-invariant) such that $E(K)_0$ contains a point of exact order m. Prove that these are the only orders possible.

CHAPTER IV

The Néron Model

Let R be a discrete valuation ring with maximal ideal $\mathfrak{p}$ and fraction field K, and let E/K be an elliptic curve given by a Weierstrass equation

$$E : y^2 + a_1 xy + a_3 y = x^3 + a_2 x^2 + a_4 x + a_6,$$

say with coefficients $a_1, a_2, a_3, a_4, a_6 \in R$. This equation can be used to define a closed subscheme $\mathcal{W} \subset \mathbb{P}_R^2$. An elementary property of closed subschemes of projective space says that every point of $E(K)$ extends to give a point of $\mathcal{W}(R)$, that is, a section $\mathrm{Spec}(R) \to \mathcal{W}$.

An important property of the elliptic curve E is that it has the structure of a group variety, which means that there is a group law given by a morphism $E \times E \to E$. This group law will extend to a rational map $\mathcal{W} \times_R \mathcal{W} \to \mathcal{W}$, but in general it will not be a morphism, so $\mathcal{W}$ will not be a group scheme over R. However, if we discard all of the singular points on the special fiber of $\mathcal{W}$ (i.e., the singular points on the reduction of E modulo $\mathfrak{p}$) and call the resulting scheme $\mathcal{W}^0$, then we will prove that the group law on E does extend to a morphism $\mathcal{W}^0 \times_R \mathcal{W}^0 \to \mathcal{W}^0$. This makes $\mathcal{W}^0$ into a group scheme over R, but, unfortunately, we may have lost the point extension property. In other words, not every point of $E(K)$ will extend to give a point in $\mathcal{W}^0(R)$.

A Néron model for E/K is a scheme $\mathcal{E}/R$ which has both of these desirable properties. Thus every point in $E(K)$ extends to a point in $\mathcal{E}(R)$, and further the group law on E extends to a morphism $\mathcal{E} \times_R \mathcal{E} \to \mathcal{E}$ which makes $\mathcal{E}$ into a group scheme over R. It is by no means clear that such a scheme exists. Our main goals in this chapter are to construct Néron models, describe what they look like, and give some applications.

The material in this chapter is of a more technical nature than most of the rest of this book. We will assume that the reader has some familiarity with basic scheme theory, as described for example in Hartshorne [1, Ch. II] or Eisenbud-Harris [1]. When we need more advanced material, we will give at least a brief explanation together with a reference for further reading.

We begin in §1 with a brief discussion of group varieties. This material is not strictly necessary for the remainder of the chapter but may prove helpful for those readers who have not studied group schemes previously. Section 2 contains some basic material on abstract schemes and schemes over a base S, including material on fiber products, special fibers, regularity, properness, and smoothness. In §3 we define group schemes and describe some of their elementary properties. Section 4 is devoted to the theory of arithmetic surfaces. An arithmetic surface $\mathcal{C}$ over a discrete valuation ring R is a "nice" scheme whose generic fiber is a non-singular curve C/K. We give several examples and prove that the smooth part $\mathcal{C}^0$ of a regular proper arithmetic surface $\mathcal{C}$ has the point extension property $C(K) = \mathcal{C}^0(R)$. We also state the fundamental existence theorems concerning minimal regular models of arithmetic surfaces.

In §5 we define Néron models and show that the smooth part $\mathcal{W}^0$ of a Weierstrass equation is a group scheme. In some cases, for example when E/K has good reduction, this will imply that $\mathcal{W}^0$ is a Néron model for E/K. The general construction of Néron models is given in §6, where we prove that the smooth part $\mathcal{E}/R$ of a minimal proper regular model $\mathcal{C}/R$ for E/K is a Néron model. The proof is quite technical and may be omitted on first reading. This is especially true for those readers who are mainly interested in applications of the theory of Néron models. Frequently, it is enough to know that a Néron model for E/K is a group scheme $\mathcal{E}$ over R whose generic fiber is K and which has the property that $\mathcal{E}(R) = E(K)$.

We next take up the question of what the special fibers $\tilde{\mathcal{E}}\,(\text{mod }\mathfrak{p})$ and $\tilde{\mathcal{C}}\,(\text{mod }\mathfrak{p})$ look like. Section 7 contains a discussion of intersection theory on general arithmetic surfaces, and then in §8 we apply this theory to give the Kodaira-Néron classification of the special fibers of an elliptic fibration. Section 9 contains a description and verification of an algorithm of Tate which gives an efficient method of computing the special fiber $\tilde{\mathcal{C}}\,(\text{mod }\mathfrak{p})$ from a given Weierstrass equation. In §10 we define the conductor of an elliptic curve and give some of its properties. Finally, in §11, we state and mostly verify an important formula of Ogg which gives a relation between the conductor, the minimal discriminant, and the special fiber of the minimal proper regular model of an elliptic curve.

In order to simplify the discussion in this chapter, we will make the following convention:

> All Dedekind domains and all discrete
> valuation rings have perfect residue fields.

Notice this includes Dedekind domains and discrete valuation rings whose residue fields are finite, which is the case that will mainly interest us.

§1. Group Varieties

A group variety is an algebraic variety that is also a group. In slightly fancier language, a group variety is a group in the category of algebraic varieties. This means that the group law is given by algebraic functions.

Definition. A *group variety* (or *algebraic group*) is an algebraic variety G and two morphisms

$$\mu : G \times G \longrightarrow G \quad \text{and} \quad i : G \longrightarrow G$$

satisfying the following group axioms:

(i) There is a point $O \in G$ such that $\mu(P, O) = \mu(O, P) = P$ for all $P \in G$.

(ii) $\mu\big(P, i(P)\big) = \mu\big(i(P), P\big) = O$ for all $P \in G$.

(iii) $\mu\big(P, \mu(Q, R)\big) = \mu\big(\mu(P, Q), R\big)$ for all $P, Q, R \in G$.

G is called a *commutative group variety* if it further satisfies

(iv) $\mu(P, Q) = \mu(Q, P)$ for all $P, Q \in G$.

The group variety G is *defined over K* if G is defined over K, the morphisms μ and i are defined over K, and $O \in G(K)$.

Example 1.1.1. An elliptic curve E/K is a commutative group variety defined over K. This follows from [AEC, III.2.2] and [AEC, III.3.6].

Example 1.1.2. The *additive group* $\mathbb{G}_a$ and the *multiplicative group* $\mathbb{G}_m$ are the commutative group varieties

$$\mathbb{G}_a \cong \mathbb{A}^1 \quad \text{and} \quad \mathbb{G}_m \cong \{x \in \mathbb{A}^1 : x \neq 0\}.$$

The group laws on $\mathbb{G}_a$ and $\mathbb{G}_m$ are defined by the formulas

$$\mu : \mathbb{G}_a \times \mathbb{G}_a \longrightarrow \mathbb{G}_a \qquad \text{and} \qquad \mu : \mathbb{G}_m \times \mathbb{G}_m \longrightarrow \mathbb{G}_m$$
$$(x, y) \longmapsto x + y \qquad\qquad\qquad (x, y) \longmapsto xy.$$

The additive group is clearly an affine variety. The multiplicative group is also an affine variety, since there is an isomorphism

$$\mathbb{G}_m \longrightarrow \{(x, y) \in \mathbb{A}^2 : xy = 1\}, \qquad x \longmapsto (x, 1/x).$$

Example 1.1.3. The general linear group GL_n is defined by

$$\mathrm{GL}_n = \left\{ M = \begin{pmatrix} x_{11} & \cdots & x_{1n} \\ \vdots & \ddots & \vdots \\ x_{n1} & \cdots & x_{nn} \end{pmatrix} \in \mathbb{A}^{n^2} : \det(M) \neq 0 \right\}.$$

It is a group variety with group law given by matrix multiplication. Note that the inverse map $i(M) = M^{-1}$ is a morphism on GL_n, since the function $1/\det(M)$ is a regular function on GL_n. Just as with the multiplicative group, we observe that GL_n is an affine variety, since it is the complement of a hypersurface in $\mathbb{A}^{n^2}$. (In general, the complement of $f = 0$ in $\mathbb{A}^m$ is affine, since it is isomorphic to $\{(\mathbf{x}, y) \in \mathbb{A}^{m+1} : yf(\mathbf{x}) = 1\}$.) We have $\mathrm{GL}_1 = \mathbb{G}_m$, and GL_n is non-commutative for $n \geq 2$.

Definition. A *homomorphism of group varieties* $\phi : G \to H$ is a morphism of varieties that is also a homomorphism of groups; that is, ϕ is a morphism, $\phi(O_G) = O_H$, and

$$\phi\big(\mu_G(P,Q)\big) = \mu_H\big(\phi(P), \phi(Q)\big) \quad \text{for all } P, Q \in G.$$

Example 1.2.1. An isogeny $E_1 \to E_2$ of elliptic curves is a homomorphism of group varieties. This follows from [AEC, III.4.8], which says that any morphism $\phi : E_1 \to E_2$ satisfying $\phi(O_{E_1}) = O_{E_2}$ is automatically a homomorphism.

Example 1.2.2. The determinant map defines a homomorphism of group varieties

$$\det : \mathrm{GL}_n \longrightarrow \mathbb{G}_m.$$

The kernel of the determinant map is another affine group variety called the *special linear group*,

$$\mathrm{SL}_n = \{M \in \mathrm{GL}_n \ : \ \det(M) = 1\}.$$

In general, an algebraic subgroup of GL_n is called a *linear group*. It turns out that every connected affine group variety is a linear group (Waterhouse [1, §3.4]). For some other examples of linear groups, see exercises 4.1 and 4.2.

Proposition 1.3. *Let G be a group variety defined over a field K. Then the set of K-rational points $G(K)$ is a subgroup of G.*

PROOF. The identity element O of G is in $G(K)$ by definition. Further, the morphisms $\mu : G \times G \to G$ and $i : G \to G$ are defined over K, so $G(K)$ is closed under the group operations. Hence $G(K)$ is a subgroup of G. $\square$

Example 1.4.1. Let E/K be an elliptic curve. Then $E(K)$ is the group of K-rational points of E. If K is a number field, then $E(K)$ is a finitely generated group [AEC, VIII.6.7].

Example 1.4.2. For any field K we have $\mathbb{G}_a(K) = K$ and $\mathbb{G}_m(K) = K^*$. Similarly, $\mathrm{GL}_n(K)$ is the group of $n \times n$ invertible matrices with coefficients in K.

Proposition 1.5. *Let G be a group variety.*
(a) *G is a non-singular variety.*
(b) *Every connected component of G is irreducible.*
(c) *The connected component of G which contains the identity element is a normal subgroup of G of finite index.*

Definition. Let G be a group variety. The connnected component of G containing the identity element is denoted by G^0 and is called the *identity component of G*. The quotient group G/G^0 called the *group of components of G*.

PROOF (of Proposition 1.5). (a) There is a Zariski open subset $U \subset G$ which is non-singular (Hartshorne [1, I.5.3]). For any $P \in G$, let $\tau_P : G \to G$ be the translation-by-P map, $\tau_P(Q) = \mu(P, Q)$. Note that τ_P is an isomorphism from G to itself. Now G is covered by the non-singular open sets $\tau_P(U)$, $P \in G$, so every point of G is non-singular.

(b) Suppose that G has a connected component that consists of more than one irreducible component. Then that connected component would contain distinct irreducible components that have a point in common, and the common point would be singular. This contradicts (a). Hence every connected component of G is irreducible.

(c) A variety has only finitely many connected components, since it actually has only finitely many irreducible components (Hartshorne [1, I.1.6]). We label the connected components of G as $G^0, G^1, \ldots, G^n$, where G^0 is the connected component of G containing the identity element. Let $P \in G^0$. The translation-by-P map τ_P permutes the connected components of G, so $\tau_P(G^0) = G^j$ for some j. But

$$P = \tau_P(O) \in \tau_P(G^0) = G^j,$$

so the connected components G^0 and G^j have the common point P. Hence $G^0 = G^j$. This means that $\mu(P, Q) = \tau_P(Q) \in G^0$ for all $P, Q \in G^0$. Similarly, $O \in G^0 \cap i(G^0)$, so $i(G^0) = G^0$. This proves that G^0 is a subgroup of G.

Next, fix a point $Q \in G$ and consider the conjugation-by-Q map

$$\phi : G \longrightarrow G, \qquad \phi(P) = \mu\big(i(Q), \mu(P, Q)\big).$$

ϕ is an automorphism of G, so it permutes the components of G. Further, $\phi(O) = O$, so as above we conclude that $\phi(G^0) = G^0$. Therefore G^0 is a normal subgroup of G.

Finally, for each $0 \leq j \leq n$ we fix a point $P_j \in G^j$. Then the maps

$$\phi_j : G \longrightarrow G, \qquad \phi_j(P) = \mu\big(P, i(P_j)\big),$$

permute the components of G and satisfy $\phi_j(P_j) = O$, from which we conclude that $\phi_j(G^j) = G^0$. Hence $P_0, \ldots, P_n$ includes a complete set of coset representatives for G/G^0, so G^0 has finite index in G. $\qquad \square$

We have seen above (1.1.1, 1.1.2) that the additive group, the multiplicative group, and elliptic curves are group varieties. We will now prove that these are the only connected group varieties of dimension one.

Theorem 1.6. *Let G be a connected group variety of dimension one defined over an algebraically closed field. Then either $G \cong \mathbb{G}_a$, $G \cong \mathbb{G}_m$, or G is an elliptic curve. (For non-algebraically closed fields, see exercise 4.13.)*

Before beginning the proof, we prove a lemma that gives conditions under which a curve has only finitely many automorphisms.

Lemma 1.7. *Let C be a non-singular projective curve of genus g, let $S \subset C$ be a finite set of points, and suppose that S satisfies one of the following conditions:*

(i) $\#S \geq 3$ *if* $g = 0$.   (ii) $\#S \geq 1$ *if* $g = 1$.   (iii) S *is arbitrary if* $g \geq 2$.

Then

$$\operatorname{Aut}(C; S) \stackrel{\text{def}}{=} \{\phi \in \operatorname{Aut}(C) \; : \; \phi(S) \subset S\}$$

is a finite set.

PROOF. Suppose first that $g = 0$, so we can take $C = \mathbb{P}^1$. Fix three distinct points $P_1, P_2, P_3 \in S$. Every automorphism of $\mathbb{P}^1$ is given by a linear fractional transformation (Hartshorne [1, II.7.1.1])

$$\phi([x, y]) = [ax + by, cx + dy].$$

An automorphism ϕ will thus be determined by the images of P_1, P_2, P_3, which proves that the map (of sets)

$$\operatorname{Aut}(C; S) \longrightarrow S^3, \qquad \phi \longmapsto (\phi(P_1), \phi(P_2), \phi(P_3))$$

is injective. But S is finite by assumption, so $\operatorname{Aut}(C; S)$ is finite.

Next suppose that $g = 1$. We make C into an elliptic curve by taking the origin O to be a point in S. Then every isomorphism $C \to C$ is a translation followed by an isomorphism fixing O [AEC, III.4.7]. But there are only finitely many isomorphisms $C \to C$ that fix O [AEC, III.10.1], so the map (of sets)

$$\operatorname{Aut}(C; S) \longrightarrow C, \qquad \phi \longmapsto \phi(O),$$

is finite-to-one. Since $\phi(O) \in S$ and S is finite, this proves that $\operatorname{Aut}(C; S)$ is finite.

Finally, we recall a theorem of Hurwitz which says that if a curve has genus $g \geq 2$, then it has at most $84(g - 1)$ automorphisms. (See Hartshorne [1, exercises IV.2.5, IV.5.2, V.1.11].) This completes the proof of Lemma 1.7. $\square$

PROOF (of Theorem 1.6). We know that the variety G is non-singular, irreducible (1.5), and has dimension one, so we can embed it as a Zariski open subset of a non-singular projective curve, say $G \subset C$ (Hartshorne [1, I §6]). Let $S = C \smallsetminus G$ be the complement of G in C.

For every point $P \in G$, the translation-by-P map $\tau_P : G \to G$ is an automorphism of G as a variety. Then τ_P induces a rational map $\tau_P : C \to C$ which extends to an isomorphism since C is non-singular [AEC, II.2.1]. Clearly, we have $\tau_P(S) \subset S$, since $\tau_P(G) = G$. In this way, we obtain an inclusion $G \hookrightarrow \text{Aut}(C;S)$ given by $P \mapsto \tau_P$, where $\text{Aut}(C;S)$ is as in (1.7). But G is a variety of dimension one, so it has infinitely many points. It follows from (1.7) that C has genus less than 2, that $S = \emptyset$ if C has genus 1, and $\#S \le 2$ if C has genus 0. There are thus four cases to consider.

Suppose first that C has genus 0 and $S = \emptyset$. Then $G = C = \mathbb{P}^1$, and the group law on G is a morphism

$$\mu : \mathbb{P}^1 \times \mathbb{P}^1 \longrightarrow \mathbb{P}^1.$$

Such a map has the form $\mu(\mathbf{x}, \mathbf{y}) = [f(\mathbf{x}, \mathbf{y}), g(\mathbf{x}, \mathbf{y})]$, where f and g are bihomogeneous polynomials. The fact that μ is a morphism means that f and g can have no common roots in $\mathbb{P}^1 \times \mathbb{P}^1$, which implies that μ must look like either

$$\mu(\mathbf{x}, \mathbf{y}) = [f(\mathbf{x}), g(\mathbf{x})] \quad \text{or} \quad \mu(\mathbf{x}, \mathbf{y}) = [f(\mathbf{y}), g(\mathbf{y})].$$

But then either $\mu(O, P)$ or $\mu(P, O)$ is constant, so μ cannot define a group law. This proves that it is not possible to have $G = \mathbb{P}^1$.

Next suppose that C has genus 0 and that $\#S = 1$. Then we can identify C with $\mathbb{P}^1$ in such a way that the point in S is the point at infinity and the identity element of G is the point 0. In other words, we have $C = \mathbb{P}^1$, $G = \mathbb{A}^1$, and $0 \in \mathbb{A}^1$ is the identity element of G. The group law on G is a morphism

$$\mu : \mathbb{A}^1 \times \mathbb{A}^1 \longrightarrow \mathbb{A}^1,$$

so μ is a polynomial map, $\mu(x, y) \in K[x, y]$. For every fixed value of x, we know that the map $y \mapsto \mu(x, y)$ is an automorphism of $\mathbb{A}^1$, so $\mu(x, y)$ must be linear in y. Similarly for $x \mapsto \mu(x, y)$, so $\mu(x, y)$ is also linear in x. Further, $\mu(x, 0) = x$ and $\mu(0, y) = y$, so we conclude that μ has the form

$$\mu(x, y) = x + y + cxy \quad \text{for some } c \in K.$$

Finally, we observe that if $c \ne 0$, then c^{-1} would not have an inverse, since $\mu(-c^{-1}, y) = -c^{-1}$ is constant. Hence $c = 0$ and $\mu(x, y) = x + y$, which proves that $G \cong \mathbb{G}_a$.

The next case to consider is a curve C of genus 0 and $\#S = 2$. This time we identify C with $\mathbb{P}^1$ so that the two points in S are 0 and ∞ and so that the identity element of G is the point 1. Then the group law on $G = \mathbb{A}^1 \smallsetminus \{0\}$ is a morphism

$$\mu : (\mathbb{A}^1 \smallsetminus \{0\}) \times (\mathbb{A}^1 \smallsetminus \{0\}) \longrightarrow (\mathbb{A}^1 \smallsetminus \{0\}),$$

so μ is a Laurent polynomial, $\mu(x, y) \in K[x, x^{-1}, y, y^{-1}]$. As above, the map $y \mapsto \mu(x, y)$ is an automorphism for every fixed x, which means it must have the form

$$\mu(x, y) = a(x) + b(x)y \quad \text{or} \quad \mu(x, y) = a(x) + b(x)y^{-1}.$$

We further know that $\mu(1, y) = y$, which rules out the second possibility and tells us that $a(1) = 0$ and $b(1) = 1$. In particular, $b(x) \neq 0$.

If $a(x) \neq 0$, then we can find an $\alpha \in K^*$ so that $a(\alpha) \neq 0$ and $\beta(\alpha) \neq 0$. Then

$$\mu\left(\alpha, -\frac{a(\alpha)}{b(\alpha)}\right) = 0,$$

which contradicts the fact that $\mu(G \times G) \subset G$. Therefore $a(x) = 0$. We have now shown that $\mu(x, y) = b(x)y$. Reversing the roles of x and y and using the fact that $\mu(1, 1) = 1$, we conclude that $\mu(x, y) = xy$, which completes the proof that $G \cong \mathbb{G}_m$.

It remains to consider the case that $C = G$ is a curve of genus 1. The group variety G has an identity element O, and we use this point to give (C, O) the structure of an elliptic curve. It remains to show that the identification G and C as curves is also an isomorphism of groups. In other words, we need to prove that

$$\mu_G(P, Q) = P + Q \qquad \text{and} \qquad i_G(P) = -P,$$

where $\mu_G : G \times G \to G$ is the given group law on G and $+$ is the group law on the elliptic curve (C, O). Note that we do not assume, a priori, that G is commutative.

Consider the map

$$\phi : C \times C \longrightarrow C, \qquad \phi(P, Q) = \mu_G(P, Q) - P - Q.$$

The point $O \in C$ is the identity element for both group laws, so we find that $\phi(P, O) = O$ and $\phi(O, Q) = O$ for all $P, Q \in C$. It follows from an elementary rigidity lemma (1.8) which we will prove below that ϕ is constant. Hence

$$\phi(P, Q) = \phi(O, O) = O, \quad \text{and so} \quad \mu_G(P, Q) = P + Q \quad \text{for all } P, Q \in C.$$

Finally, we observe that

$$P + i_G(P) = \mu_G(P, i_G(P)) = O,$$

which proves that $i_G(P) = -P$. This completes the proof of (1.6), subject to our proving the following lemma. $\qquad\square$

Lemma 1.8. (Rigidity Lemma) *Let C_1, C_2, C_3 be irreducible projective curves, and let*

$$\phi : C_1 \times C_2 \longrightarrow C_3$$

be a morphism. Suppose that there are points $P_1 \in C_1$ and $P_2 \in C_2$ with the property that each of $\phi(P_1 \times C_2)$ and $\phi(C_1 \times P_2)$ consists of a single point. Then ϕ is a constant map.

PROOF. We are given that $\phi(P_1 \times C_2)$ consists of a single point, say

$$\phi(P_1 \times C_2) = R.$$

Choose a point $R' \in C_3$, $R' \neq R$, and consider the set
$$U_1 = \{Q \in C_1 \ : \ R' \notin \phi(Q \times C_2)\}.$$
Notice that the complement of U_1 in C_1 is the set
$$C_1 \smallsetminus U_1 = \text{proj}_1\big(\phi^{-1}(R')\big),$$
where $\text{proj}_1 : C_1 \times C_2 \to C_1$ is projection onto the first factor. The projection map sends closed sets to closed sets. This follows from the fact that C_2 is projective, hence proper (Hartshorne [1, II.4.9]), and the definition of properness implies that any projection $V \times C_2 \to V$ is a closed morphism.

Now the set $\phi^{-1}(R')$ is closed, so the same is true of $\text{proj}_1\big(\phi^{-1}(R')\big)$, which shows that U_1 is an open subset of C_1. Further, it is clear that that $P_1 \in U_1$, so U_1 is non-empty. For any $Q \in U_1$, we consider the morphism
$$C_2 \longrightarrow C_3, \qquad S \longmapsto \phi(Q, S).$$
The fact that $Q \in U_1$ tells us that R' is not in the image of this map. In other words, this map is not surjective, so it follows from [AEC, II.2.3] that it is constant. In other words, if $Q \in U_1$, then $\phi(Q, S)$ is independent of $S \in C_2$. Equivalently, $\phi(Q \times C_2)$ consists of a single point.

We now repeat the above argument using the fact that $\phi(C_1 \times P_2)$ consists of one point. Doing so yields a non-empty open set $U_2 \subset C_2$ with the property that for all $S \in U_2$, the set $\psi(C_1 \times S)$ consists of a single point.

Combining these two facts, we find that $\phi(U_1 \times U_2)$ consists of one point. But $U_1 \times U_2$ is Zariski dense in $C_1 \times C_2$, and a morphism is determined by its values on a Zariski dense set. Therefore ϕ is constant.

$\square$

§2. Schemes and S-Schemes

In this section we are going to review some basic notions about schemes, especially schemes over a fixed base scheme. We assume that the reader has some familiarity with this material, as covered for example in Hartshorne [1, II §§2,3] or Eisenbud-Harris [1].

Definition. Let S be a fixed scheme. An S-*scheme* is a scheme X equipped with a morphism $X \to S$. A *morphism of S-schemes* (or S-*morphism*) is a morphism $X \to Y$ so that the diagram

$$
\begin{array}{ccc}
X & \longrightarrow & Y \\
& \searrow \quad \swarrow & \\
& S &
\end{array}
$$

is commutative. If $S = \text{Spec}(R)$, we will often refer to R-*schemes* and R-*morphisms* instead of $\text{Spec}(R)$-schemes and $\text{Spec}(R)$-morphisms.

Intuitively, an S-scheme $X \to S$ can be viewed as an algebraic family of schemes, namely the family of fibers X_s parametrized by the points $s \in S$. We have seen an example of this in Chapter III, where the elliptic surface $\mathcal{E} \to C$ is a C-scheme whose fibers $\mathcal{E}_t$ form a family of elliptic curves.

Two other important examples are provided by affine and projective space over a ring R. These are defined to be

$$\mathbb{A}_R^n = \operatorname{Spec} R[x_1, \ldots, x_n] \quad \text{and} \quad \mathbb{P}_R^n = \operatorname{Proj} R[x_0, \ldots, x_n].$$

See Hartshorne [1, II.2.5.1] for details.

In Chapter III we studied the group of sections $C \to \mathcal{E}$ to the elliptic surface. Similarly, we can look at the set of sections $S \to X$ of an S-scheme. These are precisely the S-morphisms from S to X. More generally, for any S-scheme T, we can consider the set of S-morphisms from T to X.

Definition. Let X and T be S-schemes. The *set of T-valued points of X* is the set

$$X(T) = \operatorname{Hom}_S(T, X) = \{S\text{-morphisms } T \to X\}.$$

If $T = S$, we will sometimes call $X(S)$ the *set of sections of the S-scheme X*. Similarly, if $S = T = \operatorname{Spec}(R)$, we will refer to the *$R$-valued points of X* and write $X(R)$.

Example 2.1.1. Let K be a field, let $S = \operatorname{Spec}(K)$, and let X/K be an affine scheme, say given by equations

$$f_1 = f_2 = \cdots = f_r = 0 \quad \text{with } f_1, \ldots, f_r \in K[x_1, \ldots, x_n].$$

Then

$$
\begin{aligned}
X(S) &= \{K\text{-morphisms } \operatorname{Spec}(K) \to X\} \\
&\cong \left\{ K\text{-algebra homomorphisms } \frac{K[x_1, \ldots, x_n]}{(f_1, \ldots, f_r)} \to K \right\} \\
&= \{P \in K^n : f_1(P) = \cdots = f_r(P) = 0\}.
\end{aligned}
$$

Thus $X(S)$ agrees with our intuition of what $X(K)$ should be. More generally, if R is any ring and $X \subset \mathbb{A}_R^n$ is an affine scheme given by equations $f_1 = \cdots = f_r = 0$ with $f_i \in R[x_1, \ldots, x_n]$, then $X(R)$ is naturally identified with the set of n-tuples $(x_1, \ldots, x_n) \in R^n$ satisfying the equations.

Remark 2.1.2. Each S-scheme X defines a functor F_X on the category of S-schemes by assigning to an S-scheme T the set of T-valued points of X. Thus

$$F_X : (S\text{-schemes}) \longrightarrow (\text{Sets}), \quad T \longmapsto X(T).$$

If $\phi : T \to T'$ is an S-morphism of S-schemes, the associated map $F_X(\phi)$ is given by composition,

$$F_X(\phi) : F_X(T') \longrightarrow F_X(T), \qquad \sigma \longmapsto \sigma \circ \phi.$$

Notice that F_X is a contravariant (i.e., arrow reversing) functor. It is a basic categorical fact (called Yoneda's lemma, see Eisenbud-Harris [1, Lemma IV.1]) that the functor F_X determines the scheme X. Similarly, morphisms of functors $F_X \to F_Y$ correspond bijectively with S-morphisms $X \to Y$. We will not make use of this functorial approach, but the reader should be aware that it is a convenient language which is in common usage.

Next we describe one of the most important constructions of algebraic geometry.

Definition. Let X and Y be S-schemes. The *fiber product of X and Y over S* is an S-scheme, denoted $X \times_S Y$, together with projection morphisms

$$p_1 : X \times_S Y \longrightarrow X \qquad \text{and} \qquad p_2 : X \times_S Y \longrightarrow Y$$

over S with the following universal property:

> Let Z be an S-scheme, and let $f : Z \to X$ and $g : Z \to Y$ be S-morphisms. Then there exists a unique S-morphism $Z \to X \times_S Y$ so that the following diagram commutes:

$$
\begin{array}{ccccc}
& & Z & & \\
& {}^{f}\swarrow & \downarrow & \searrow^{g} & \\
X & \xleftarrow{p_1} & X \times_S Y & \xrightarrow{p_2} & Y
\end{array}
$$

The fiber product exists and is unique up to unique isomorphism; see Hartshorne [1, II.3.3] or Eisenbud-Harris [1, I.C.i, IV.B]). If $S = \operatorname{Spec} R$, we will often write $X \times_R Y$.

The fiber product is the smallest scheme that fits into the commutative diagram

$$
\begin{array}{ccc}
X \times_S Y & \xrightarrow{p_1} & X \\
\downarrow^{p_2} & & \downarrow \\
Y & \longrightarrow & S.
\end{array}
$$

In some sense, $X \times_S Y$ should "look like" the set of ordered pairs (x, y) having the property that x and y have the same image in S. This is literally true in the category of sets, but care must be taken when applying this intuition in the category of schemes. In fact, $X \times_S Y$ will generally be quite large, even when S consists of a single point; see for example Hartshorne [1, II, exercise 3.1].

Example 2.2.1. Let $s \in S$, and let $Y = \{s\} \hookrightarrow S$ be the subscheme of S consisting of the point s. More precisely, if we write $k(s) = \mathcal{O}_{S,s}/\mathcal{M}_{S,s}$ for the residue field of the local ring at s, then Y is the scheme $\operatorname{Spec} k(s)$. The *fiber of X over s* is defined to be the scheme

$$X_s \overset{\text{def}}{=} X \times_S \{s\}.$$

It is a scheme over $k(s)$. In this case the underlying topological space of $X \times_S \{s\}$ actually equals the set of points $x \in X$ such that the image of x in S is s, see Hartshorne [1, II exercise 3.10]. Thus at the level of points, this definition of the fiber X_s agrees with our intuition of what the fiber should be.

Example 2.2.2. Let R be a ring, let $\mathfrak{p}$ be a maximal ideal of R, and let X be an R-scheme. Then the fiber

$$X_{\mathfrak{p}} = X \times_R \mathfrak{p}$$

is the *reduction of X modulo $\mathfrak{p}$*. It is a scheme over the residue field $R/\mathfrak{p}$. This agrees with our intuition, since an (affine) scheme X over R is defined by a system of polynomial equations with coefficients in R, and $X_{\mathfrak{p}}$ is the scheme defined by reducing the coefficients of the polynomials modulo $\mathfrak{p}$.

Example 2.2.3. Let R be an integral domain, and let $\eta = (0) \in \operatorname{Spec} R$ be the generic point of $\operatorname{Spec} R$. If X is an R-scheme, then the fiber

$$X_{\eta} = X \times_R \eta$$

is called the *generic fiber of X*. It is a scheme over the fraction field K of R. In particular, if R is a discrete valuation ring with maximal ideal $\mathfrak{p}$, then X has two fibers, its *generic fiber* X_{η}/K and its *special* (or *closed*) *fiber* $X_{\mathfrak{p}}/k$, where $k = R/\mathfrak{p}$ is the residue field of R.

For example, suppose that $X \subset \mathbb{P}^2_R$ is given by a single homogeneous equation $f(x, y, z) = 0$ with coefficients in R. Then the generic fiber $X_{\eta} \subset \mathbb{P}^2_K$ is the variety defined by the same equation $f(x, y, z) = 0$, and the special fiber $X_{\mathfrak{p}} \subset \mathbb{P}^2_k$ is the variety defined by the equation $\tilde{f}(x, y, z) = 0$, where $\tilde{f}$ is obtained by reducing the coefficients of f modulo $\mathfrak{p}$.

Example 2.3. Let $\pi : X \to S$ be an S-scheme. In the definition of the fiber product, if we take $Z = X$ and f and g to be the identity map $X \to X$, then we obtain the *diagonal morphism*

$$\delta_X : X \longrightarrow X \times_S X;$$

that is, δ_X is the unique map to the fiber product with the property that $p_1 \circ \delta_X$ and $p_2 \circ \delta_X$ are each the identity map on X.

More generally, let $\phi : X \to Y$ be an S-morphism. Then the *graph of ϕ* is the unique morphism

$$\delta_{\phi} : X \longrightarrow X \times_S Y$$

such that $p_1 \circ \delta_{\phi}$ is the identity map on X and $p_2 \circ \delta_{\phi} = \phi$. Notice the diagonal morphism is the graph of the identity map $X \to X$.

Example 2.4. Let $\pi : \mathcal{E} \to C$ be an elliptic surface, say defined over an algebraically closed field k. Then the fiber product $\mathcal{E} \times_C \mathcal{E}$, or more precisely, the set of k-valued points on the fiber product, is the set

$$(\mathcal{E} \times_C \mathcal{E})(k) = \{(P,Q) : P,Q \in \mathcal{E}(k) \text{ and } \pi(P) = \pi(Q)\}.$$

Thus $(\mathcal{E} \times_C \mathcal{E})(k)$ consists of pairs of points which lie on the same fiber. In particular, if P and Q lie on a non-singular fiber $\mathcal{E}_t$, then we can add them using the group law on $\mathcal{E}_t$. In this way (most of) the fiber product $(\mathcal{E} \times_C \mathcal{E})(k)$ becomes a group.

Example 2.5. Recall that every scheme S admits a unique morphism $S \to \operatorname{Spec} \mathbb{Z}$ (Hartshorne [1, exercise II.2.4]). Affine and projective space over S are defined to be

$$\mathbb{A}_S^n = \mathbb{A}_{\mathbb{Z}}^n \times_{\mathbb{Z}} S \qquad \text{and} \qquad \mathbb{P}_S^n = \mathbb{P}_{\mathbb{Z}}^n \times_{\mathbb{Z}} S.$$

Projection onto the second factor makes $\mathbb{A}_S^n$ and $\mathbb{P}_S^n$ into S-schemes. Notice that if $S = \operatorname{Spec} R$ is an affine scheme, then $\mathbb{A}_S^n \cong \mathbb{A}_R^n$ and $\mathbb{P}_S^n \cong \mathbb{P}_R^n$, so these definitions are compatible with the definitions of affine and projective space over a ring.

We are now faced with the task of discussing three important properties of schemes and morphisms, namely regularity, properness, and smoothness. The definition of each of these properties is somewhat technical, and in truth we will make very little use of the formal definitions of properness and smoothness in subsequent sections. On the other hand, the intuitions underlying all of these properties are quite easy to understand, especially if one works in a "nice" setting. So we are going to begin with an informal discussion, including examples and basic material which we will give without proof. This discussion should suffice for reading the remainder of this chapter, except possibly for parts of §6. Then, at the end of this section, we will give precise definitions and provide references for further reading.

Intuitive "Definitions." A scheme X is *regular* if it is non-singular, by which we mean that every point of X has a tangent space of the correct dimension.

A morphism of schemes $X \to S$ is *proper* if all of its fibers are complete and separated. (These are algebraic analogues of compact and Hausdorff.) Essentially, this means that the fibers of $X \to S$ are not missing any points and do not have too many points. We also say that X is a *proper S-scheme*.

A morphism of schemes $X \to S$ is *smooth* if all of its fibers are non-singular, or, to put it another way, if X is a family of regular schemes. We also say that X is a *smooth S-scheme*.

In order to define regularity, we recall that the *Krull dimension* of a ring A is the largest integer d such that there is a chain of distinct prime ideals of A,

$$\mathfrak{M}_0 \subset \mathfrak{M}_1 \subset \cdots \subset \mathfrak{M}_d.$$

A local ring A with maximal ideal $\mathfrak{M}$ is called *regular* if the dimension of $\mathfrak{M}/\mathfrak{M}^2$ as an $A/\mathfrak{M}$-vector space is equal to the Krull dimension of A. (See Matsumura [1, Ch. 7] or Atiyah-MacDonald [1, Ch. 11] for basic material on regular local rings.) Intuitively, $\mathfrak{M}/\mathfrak{M}^2$ is the cotangent space of $\mathrm{Spec}(A)$ at the point $\mathfrak{M}$, and the regularity of A is an assertion that $\mathfrak{M}$ is a non-singular point of $\mathrm{Spec}(A)$. For arbitrary schemes one defines dimension and regularity in terms of the local rings as follows:

Definition. The *dimension* of a point P of a scheme X is the Krull dimension of the local ring $\mathcal{O}_P$ at P. If every closed point of X has the same dimension, we call this the *dimension of X*.

Definition. A point P of a scheme X is said to be *regular* (or *non-singular*) if the local ring $\mathcal{O}_P$ is a regular local ring. The scheme X is *regular* (or *non-singular*) if every point of X is regular. In fact, it suffices to check that every closed point is regular; see exercise 4.5.

Example 2.6.1. Let R be a Dedekind domain. Then $\mathrm{Spec}(R)$ is a regular scheme of dimension one. To see this, note that in a Dedekind domain, every non-zero prime ideal is maximal by definition. Hence the longest chain of prime ideals is $(0) \subset \mathfrak{p}$, so R has dimension one. Further, each localization $R_\mathfrak{p}$ is a discrete valuation ring, so its maximal ideal $M_\mathfrak{p}$ is principal. It follows that $M_\mathfrak{p}/M_\mathfrak{p}^2$ has dimension one as an $R_\mathfrak{p}/M_\mathfrak{p}$-vector space, so $R_\mathfrak{p}$ is regular.

Example 2.6.2. If R is a regular local ring, or more generally if $\mathrm{Spec}(R)$ is regular, then both $\mathbb{A}_R^n$ and $\mathbb{P}_R^n$ are regular schemes.

Example 2.6.3. Let R be a discrete valuation ring, let π be a uniformizer for R, and assume that $2, 3 \in R^*$. Let $a \in R$, and define a scheme $X \subset \mathbb{P}_R^2$ by the equation

$$X : y^2 z = x^3 + az^3.$$

Then X is a regular scheme if and only if $a \not\equiv 0 \pmod{\pi^2}$. To see this, one first checks that the only possible singular point is the point $\gamma = [0, 0, 1]$ on the special fiber $X_\mathfrak{p}$ and that this can only occur if $a \equiv 0 \pmod{\pi}$. Dehomogenizing the equation by setting $z = 1$, we find that the maximal ideal M_γ of the local ring $\mathcal{O}_\gamma$ is generated by x, y, and π, and that these quantities are related by the equation

$$y^2 = x^3 + a.$$

If $a \not\equiv 0 \pmod{\pi^2}$, then a is itself a uniformizer for R. Hence

$$\pi \in aR = (y^2 - x^3)R \subset M_\gamma^2,$$

so x and y generate M_γ/M_γ^2, which shows that $\mathcal{O}_\gamma$ is regular. Conversely, if $a \equiv 0 \pmod{\pi^2}$, then M_γ/M_γ^2 cannot be generated by fewer than three elements, so $\mathcal{O}_\gamma$ is not regular.

Next we look at proper morphisms, which, recall, are supposed to be morphisms $X \to S$ whose fibers are separated and complete. For example, suppose that we are given a "curve" C, a point $\gamma \in C$, and a commutative diagram of morphisms

$$
\begin{array}{ccc}
C \smallsetminus \gamma & \xrightarrow{\ F\ } & X \\
\downarrow & & \downarrow \\
C & \xrightarrow{\ f\ } & S.
\end{array}
$$

If $X \to S$ is a proper morphism, then the fiber of X over $f(\gamma)$ is separated and complete, so there should be a unique way to extend F to all of C. This statement is essentially the following valuative criterion for properness.

Theorem 2.7.   (Valuative Criterion of Properness) *Let* $\phi : X \to S$ *be a morphism of finite type of Noetherian schemes. The map* ϕ *is proper if and only if for every (discrete) valuation ring* R *with fraction field* K *and every commutative square of morphisms*

$$
\begin{array}{ccc}
\mathrm{Spec}(K) & \longrightarrow & X \\
\downarrow & & \downarrow{\scriptstyle \phi} \\
\mathrm{Spec}(R) & \longrightarrow & S,
\end{array}
$$

there is a unique morphism $\mathrm{Spec}(R) \to X$ *fitting into the diagram. (In other words, there is a unique morphism* $\mathrm{Spec}(R) \to X$ *so that the composition* $\mathrm{Spec}(R) \to X \to S$ *agrees with the bottom line of the square.)*

PROOF. Hartshorne [1, II.4.7]. See also Hartshorne [1, exercise II.4.11] for the assertion that it suffices to consider only discrete valuation rings.   □

In order to better understand what the valuative criterion is saying, we note that if R is a discrete valuation ring with fraction field K, then $\mathrm{Spec}(R)$ is a regular one-dimensional scheme (2.6.1), and $\mathrm{Spec}(K)$ is $\mathrm{Spec}(R)$ with its closed point removed. Thus $\mathrm{Spec}(K)$ looks like a curve with one point removed.

An important collection of proper S-schemes is the set of projective schemes over S, as described in the following result.

Theorem 2.8.   *Let* S *be a Noetherian scheme, and let* $X \subset \mathbb{P}^n_S$ *be a closed subscheme of projective space over* S. *Then* X *is proper over* S. *In particular,* $\mathbb{P}^n_S$ *itself is proper over* S.

PROOF. See Hartshorne [1, II.4.9].   □

We continue our informal discussion by looking at smooth morphisms. For our purposes, the most important examples of smooth morphisms will be schemes which are smooth over a discrete valuation ring or Dedekind domain R. In this situation, the condition that X be smooth over R is essentially equivalent to the assertion that all of its fibers are non-singular and have the same dimension.

Proposition 2.9. *Let R be a discrete valuation ring with fraction field K, residue field k, and maximal ideal $\mathfrak{p}$. Let X be an integral (i.e., reduced and irreducible) R-scheme of finite type over R whose generic fiber X_η/K is non-empty. Then X is a smooth R-scheme if and only if $X_\eta(\bar{K})$ and $X_\mathfrak{p}(\bar{k})$ contain no singular points.*

PROOF. The scheme X is irreducible, so it has a unique generic point. Our assumption that the generic fiber X_η is non-empty shows that the generic point of X maps to the generic point of $\operatorname{Spec} R$, so X is flat over $\operatorname{Spec} R$ from Hartshorne [1, III.9.7]. If R contains its residue field k (the so-called function field case), the desired result then follows from Hartshorne [1, III.10.2]. The general case is Milne [4, I Prop. 3.24], see also Bosch-Lütkebohmert-Raynaud [1, §2.4, Prop. 8]. $\qquad\square$

There are many theorems in algebraic geometry which say that some property of morphisms, such as properness, smoothness, separability, finiteness, etc., is preserved under composition, base extension, and products. The only result of this sort that we will need is the following assertion that the composition of smooth morphisms is again smooth.

Proposition 2.10. *If $\phi : X \to Y$ and $\psi : Y \to Z$ are smooth morphisms, then the composition $\psi \circ \phi : X \to Z$ is a smooth morphism.*

PROOF. See Hartshorne [1, III.10.1c] or Altman-Kleiman [1, VII.1.7ii]. $\qquad\square$

We now look at some examples of regular schemes and proper and smooth morphisms.

Example 2.11.1. Let R be a discrete valuation ring with uniformizer π. We assume that $2 \in R^*$. Let $X \subset \mathbb{P}^2_R$ be the scheme given by the equation

$$X : x^2 + \pi y^2 = z^2.$$

Then X is proper over R from (2.8), since it is a closed subscheme of $\mathbb{P}^2_R$. It is also easy to check that the scheme X is irreducible and regular. However, the special fiber of X is given by the equation $x^2 = z^2$, so the special fiber is reducible and singular. Hence X is not smooth over R.

Example 2.11.2. We continue analyzing example (2.6.3), so R is a discrete valuation ring with uniformizer π and $2, 3 \in R^*$, and $X \subset \mathbb{P}^2_R$ is the scheme defined by the equation

$$X : y^2 z = x^3 + az^3 \qquad \text{for some } a \in R.$$

We also let K be the fraction field of R, k the residue field of R, and $\mathfrak{p}$ the maximal ideal of R. Notice that X is proper over R by (2.8), since it is a closed subscheme of $\mathbb{P}^2_R$.

Suppose first that $a \in R^*$. Then the special fiber $X_{\mathfrak{p}}/k$ is a nonsingular curve, so X is smooth over R from (2.9).

Next, suppose that $a \equiv 0 \,(\mathrm{mod}\,\mathfrak{p})$. Then the special fiber $X_{\mathfrak{p}}/k$ is given by the equation $y^2 z = x^3$, so the special fiber is singular and X is not smooth over R. Let $\gamma \in X_{\mathfrak{p}} \subset X$ be the singular point on the special fiber, and let $X^0 = X \smallsetminus \gamma$ be the scheme obtained by removing γ from X. This makes the special fiber $X_{\mathfrak{p}}^0/k$ non-singular, so X^0 is smooth over R. However, removing the point γ has destroyed the completeness of the special fiber, so X^0 is not proper over R.

Finally, we observe that if a is a uniformizer for R, then (2.6.3) says that X is regular. This is true despite the fact that its special fiber $X_{\mathfrak{p}}$ is singular and so X is not smooth over R. We will prove later (4.4) that since X is regular, every R-valued point of X lies in the smooth part X^0. In other words, the natural inclusion $X^0(R) \subset X(R)$ is an equality, so in this situation X^0 retains a sort of properness property over R.

This last example (2.11.2) illustrates an important general phenomenon. Let X be a (nice) scheme which is proper over a discrete valuation ring R and which has a smooth generic fiber. Then X need not be smooth over R, since its special fiber $X_{\mathfrak{p}}$ may have singularities. We can create a smooth R-scheme X^0 by removing from X the singular points on its special fiber, but then X^0 will not be proper over R. Thus the attributes of properness and smoothness are somewhat antithetical to one another.

However, if the original scheme X is regular, then it turns out that every R-valued point in $X(R)$ actually lies in $X^0(R)$. So for regular schemes, X^0 still behaves to some extent as if it were proper over R. We will prove this later (4.4) when X has relative dimension one over R. The general case we leave as an exercise.

We are now ready to define properness and smoothness, but we want to stress that the most important thing is for the reader to understand the underlying intuitions and the examples described above. For further reading on this material, see Hartshorne [1, III §§9,10], Altman-Kleiman [1, V,VI], and Bosch-Lütkebohmert-Raynaud [1, 2.1–2.4].

Definition. Let $\phi : X \to S$ be a morphism of finite type. The map ϕ is *separated* if the diagonal morphism $\delta_X : X \to X \times_S X$ (2.3) is a closed immersion. The map ϕ is *universally closed* if for any base extension $S' \to S$, the map $X \times_S S' \to S'$ sends closed set to closed sets. The map ϕ is *proper* if it is separated and universally closed. We also say that X *is proper over S*, or that X *is a proper S-scheme*.

Definition. Let $\phi : X \to S$ be a morphism of finite type, let $x \in X$, and let $s = \phi(x) \in S$. The map ϕ is *smooth (of relative dimension r) at a point $x \in X$* if there are affine open neighborhoods

$$s \in \operatorname{Spec} R \subset S \quad \text{and} \quad x \in \operatorname{Spec} A \subset X$$

with

$$A = R[t_1, \ldots, t_{n+r}]/(f_1, \ldots, f_n) \quad \text{for some } f_1, \ldots, f_n \in R[t_1, \ldots, t_{n+r}]$$

so that the $n \times n$ minors of the Jacobian matrix $(\partial f_i/\partial t_j)$ generate the unit ideal in A.

We say that ϕ is *smooth* (or that X *is smooth over* S) if ϕ is smooth at all points of X. A morphism that is smooth of relative dimension zero is called an *étale morphism*.

Remark 2.12. The Jacobian condition in the definition of smoothness is similar to the criterion we used to define non-singular points on varieties in [AEC, I §1]. In particular, it is clear that if $X \to \operatorname{Spec} R$ is a smooth morphism and $\mathfrak{p} \in \operatorname{Spec} R$ is a maximal ideal, then the fiber $X_{\mathfrak{p}}$ is a non-singular variety over the residue field $R/\mathfrak{p}$, which is a special case of (2.9). There are many other ways to define smoothness. One of the most useful is in terms of the sheaf of relative differentials of X/S, see for example Hartshorne [1, II §8, III §10], Altman-Kleiman [1, VI,VII], Milne [4, 1§3] or Bosch-Lütkebohmert-Raynaud [1, §2.1,2.2]. For a fancy functorial definition, see Milne [4, I.3.22].

§3. Group Schemes

A group scheme over S is an S-scheme G whose fibers form an algebraic family of groups, similar to the example described in (2.4). This means we should be able to multiply two points provided they lie on the same fiber, and the multiplications should fit together to give a group law on the fiber product $G \times_S G$. More formally, a group scheme over S is a group in the category of S-schemes. This leads to the following definition. Note that we must be careful to define everything in terms of maps, rather than in terms of points. (An alternative approach is to define a group scheme in terms of its associated functor of points; see (2.1.2) or Eisenbud-Harris [1, IV.A.v].)

Definition. Let S be a scheme. A *group scheme over* S is an S-scheme $\pi : G \to S$ and S-morphisms

$$\sigma_0 : S \longrightarrow G, \qquad i : G \longrightarrow G, \qquad \mu : G \times_S G \longrightarrow G,$$

such that the following diagrams commute:
(i) *(identity element)*

$$
\begin{array}{ccc}
 & G \times_S G & \\
{\scriptstyle \sigma_0 \times 1} \nearrow & \Big\downarrow {\scriptstyle \mu} & \\
S \times_S G \xrightarrow{\ p_2\ } & G &
\end{array}
\qquad
\begin{array}{ccc}
 & G \times_S G & \\
{\scriptstyle 1 \times \sigma_0} \nearrow & \Big\downarrow {\scriptstyle \mu} & \\
G \times_S S \xrightarrow{\ p_1\ } & G &
\end{array}
$$

(ii) *(inverse)*

$$
\begin{array}{ccc}
G \times_S G & \xrightarrow{\;1 \times i\;} & G \times_S G \\
\uparrow{\scriptstyle \delta_G} & & \downarrow{\scriptstyle \mu} \\
G & \xrightarrow{\;\pi\;} S \xrightarrow{\;\sigma_0\;} & G
\end{array}
\qquad
\begin{array}{ccc}
G \times_S G & \xrightarrow{\;i \times 1\;} & G \times_S G \\
\uparrow{\scriptstyle \delta_G} & & \downarrow{\scriptstyle \mu} \\
G & \xrightarrow{\;\pi\;} S \xrightarrow{\;\sigma_0\;} & G
\end{array}
$$

(Here $G \xrightarrow{\;\delta_G\;} G \times_S G$ is the diagonal map (2.3).)

(iii) *(associativity)*

$$
\begin{array}{ccc}
G \times_S G \times_S G & \xrightarrow{\;\mu \times 1\;} & G \times_S G \\
\downarrow{\scriptstyle 1 \times \mu} & & \downarrow{\scriptstyle \mu} \\
G \times_S G & \xrightarrow{\;\mu\;} & G
\end{array}
$$

Example 3.1.1. Let G be a group variety defined over a field K as discussed in §1. Then G is a group scheme over the one point scheme $S = \mathrm{Spec}(K)$. This is clear from the definitions. Note that the identity morphism $\sigma_0 : S \to G$ sends the one point in S to the identity element of G.

Example 3.1.2. The *additive group scheme* $\mathbb{G}_a$ *over* $\mathbb{Z}$ is the scheme $\mathbb{G}_a = \mathrm{Spec}\,\mathbb{Z}[T]$. The group law on $\mathbb{G}_a$ is given by

$$
\begin{array}{ccc}
\mathbb{G}_a \times_\mathbb{Z} \mathbb{G}_a & \longrightarrow & \mathbb{G}_a \\
\| & & \| \\
(\mathrm{Spec}\,\mathbb{Z}[T_1]) \times_\mathbb{Z} (\mathrm{Spec}\,\mathbb{Z}[T_2]) & & \\
\| & & \| \\
\mathrm{Spec}(\mathbb{Z}[T_1] \otimes_\mathbb{Z} \mathbb{Z}[T_2]) & & \\
\| & & \| \\
\mathrm{Spec}\,\mathbb{Z}[T_1, T_2] & \longrightarrow & \mathrm{Spec}\,\mathbb{Z}[T],
\end{array}
$$

where the morphism $\mathrm{Spec}\,\mathbb{Z}[T_1, T_2] \to \mathrm{Spec}\,\mathbb{Z}[T]$ is induced by the ring homomorphism

$$
\mathbb{Z}[T] \longrightarrow \mathbb{Z}[T_1, T_2], \qquad T \longmapsto T_1 + T_2.
$$

For any ring R, we have $\mathbb{G}_a(R) = R$ with group law given by addition on R. The *additive group scheme* $\mathbb{G}_{a/S}$ over an arbitrary scheme S is the group scheme $\mathbb{G}_a \times_\mathbb{Z} S$ obtained by base extension. In particular, $\mathbb{G}_{a/R} = \mathrm{Spec}\,R[T]$.

Example 3.1.3. The *multiplicative group scheme* $\mathbb{G}_m$ *over* $\mathbb{Z}$ is the scheme $\mathbb{G}_m = \operatorname{Spec}\mathbb{Z}[T, T^{-1}]$. The group law on $\mathbb{G}_m$ is given by

$$
\begin{array}{ccc}
\mathbb{G}_m \times_{\mathbb{Z}} \mathbb{G}_m & \longrightarrow & \mathbb{G}_m \\
\| & & \| \\
(\operatorname{Spec}\mathbb{Z}[T_1, T_1^{-1}]) \times_{\mathbb{Z}} (\operatorname{Spec}\mathbb{Z}[T_2, T_2^{-1}]) & & \\
\| & & \\
\operatorname{Spec}\mathbb{Z}[T_1, T_1^{-1}, T_2, T_2^{-1}] & \longrightarrow & \operatorname{Spec}\mathbb{Z}[T, T^{-1}],
\end{array}
$$

where the morphism $\operatorname{Spec}\mathbb{Z}[T_1, T_1^{-1}, T_2, T_2^{-1}] \to \operatorname{Spec}\mathbb{Z}[T, T^{-1}]$ is induced by the ring homomorphism

$$
\operatorname{Spec}\mathbb{Z}[T, T^{-1}] \longrightarrow \mathbb{Z}[T_1, T_1^{-1}, T_2, T_2^{-1}], \qquad T \longmapsto T_1 T_2.
$$

For any ring R, we have $\mathbb{G}_m(R) = R^*$ with group law given by multiplication on R^*. The *multiplicative group scheme* $\mathbb{G}_{m/S}$ over an arbitrary scheme S is the group scheme $\mathbb{G}_m \times_{\mathbb{Z}} S$ obtained by base extension. In particular, $\mathbb{G}_{m/R} = \operatorname{Spec}R[T, T^{-1}]$.

Example 3.1.4. Let R be a discrete valuation ring with maximal ideal $\mathfrak{p}$ and fraction field K, and let E/K be an elliptic curve with good reduction at $\mathfrak{p}$. Fix a minimal Weierstrass equation for E,

$$
E : y^2 z + a_1 xy + a_3 y = x^3 + a_2 x^2 + a_4 x + a_6.
$$

The coefficients of this equation are in R, so we can use the equation to define an R-scheme $\mathcal{E} \subset \mathbb{P}_R^2$. (Of course, we need to homogenize the equation first.) The fact that E has good reduction implies that the scheme $\mathcal{E}$ is smooth over R, since good reduction is equivalent to the fact that the special fiber $\mathcal{E}_{\mathfrak{p}}$ of $\mathcal{E} \to \operatorname{Spec}R$ is a smooth elliptic curve over the residue field $R/\mathfrak{p}$.

The addition law on E is given by rational functions with coefficients in R, so it induces a rational map

$$
\mu : \mathcal{E} \times_R \mathcal{E} \longrightarrow \mathcal{E}.
$$

We know from [AEC, III.3.6] that the addition law $E \times E \to E$ on the generic fiber is a morphism. We will later give two proofs that μ itself is an R-morphism. The first proof (5.3) uses explicit equations and is similar to the argument in [AEC, III.3.6, III.3.6.1]. The second proof (6.1) uses fancier machinery to prove a much stronger result.

The next proposition shows that the set of T-valued points of a group scheme form a group.

Proposition 3.2. *Let G be a group scheme over S, let T be an arbitrary S-scheme, and let $G(T)$ be the set of T-valued points of G, which recall is the set of S-morphisms $T \to G$. For any two elements $\phi, \psi \in G(T)$, define a new element $\phi * \psi \in G(T)$ by the commutativity of the diagram*

$$
\begin{array}{ccc}
T \times_S T & \xrightarrow{\phi \times \psi} & G \times_S G \\
\uparrow{\scriptstyle \delta_T} & & \downarrow{\scriptstyle \mu} \\
T & \xrightarrow{\phi * \psi} & G,
\end{array}
$$

where δ_T is the diagonal map (2.3). In other words,

$$
\phi * \psi = \mu \circ (\phi \times \psi) \circ \delta_T \in G(T).
$$

This operation gives $G(T)$ the structure of a group. The identity element is $\sigma_0 \circ \pi_T$, where $\pi_T : T \to S$ is the map making T into an S-scheme. The inverse of ϕ is $i \circ \phi$.

   More precisely, the association $T \to G(T)$ is a contravariant functor from the category of S-schemes to the category of groups.

PROOF. All of this follows from the definitions and elementary diagram chases. For example, to verify that $\sigma_0 \circ \pi_T$ is the identity element of $G(T)$, we observe that the following diagram is commutative:

$$
\begin{array}{ccc}
G \times_S T & \xrightarrow{1 \times \pi_T} & G \times_S S \\
\uparrow{\scriptstyle \phi \times 1} & & \downarrow{\scriptstyle 1 \times \sigma_0} \\
T \times_S T & \xrightarrow{\phi \times (\sigma_0 \circ \pi_T)} & G \times_S G \\
\uparrow{\scriptstyle \delta_T} & & \downarrow{\scriptstyle \mu} \\
T & \xrightarrow{\phi * (\sigma_0 \circ \pi_T)} & G
\end{array}
$$

But the definition of σ_0 tells us that the map $\mu \circ (1 \times \sigma_0) : G \times_S S \to G$ down the right-hand side of this diagram is projection onto the first factor. Hence tracing around the boundary of the diagram yields

$$
\phi * (\sigma_0 \circ \pi_T) = p_1 \circ (1 \times \pi_T) \circ (\phi \times 1) \circ \delta_T = p_1 \circ (\phi \times 1) \circ \delta_T = \phi.
$$

   We will leave it to the reader to perform the similar computations needed to check that $i \circ \phi$ is the inverse of ϕ and that the associative law holds, which completes the proof that $G(T)$ is a group.

   Finally, the functoriality statement means that if $f : T' \to T$ is an S-morphism of S-schemes, then the map

$$
G(T) \longrightarrow G(T'), \qquad \phi \longmapsto \phi \circ f,
$$

is a homomorphism of groups. It is clear that the identity element is mapped to the identity element, so we must verify that

$$(\phi * \psi) \circ f = (\phi \circ f) * (\psi \circ f) \qquad \text{for all } \phi, \psi \in G(T).$$

The definition of $\phi * \psi$ says that the right-hand square of the following diagram is commutative, and the left-hand square is clearly commutative:

$$
\begin{array}{ccccc}
T' \times_S T' & \xrightarrow{f \times f} & T \times_S T & \xrightarrow{\phi \times \psi} & G \times G \\
\Big\uparrow{\scriptstyle \delta_{T'}} & & \Big\uparrow{\scriptstyle \delta_T} & & \Big\downarrow{\scriptstyle \mu} \\
T' & \xrightarrow{f} & T & \xrightarrow{\phi * \psi} & G
\end{array}
$$

The map $(\phi \times \psi) \circ (f \times f)$ along the top row of this diagram is equal to $(\phi \circ f) \times (\psi \circ f)$, so by definition the map along the bottom row equals $(\phi \circ f) * (\psi \circ f)$. This is the desired result, which completes the proof of Proposition 3.2. $\qquad\qquad\qquad\qquad\qquad\qquad\qquad\qquad\qquad\qquad\qquad\qquad$ □

Remark 3.3. In our study of elliptic curves and group varieties, the translation-by-P maps provided an important tool. A group scheme G/S is not a group, so we cannot translate G by a point of the scheme G. Instead, G is a family of groups parametrized by the points of S. So in order to translate G, we need to start with a family of points on G parametrized by S. Then we can translate each group in the family by the appropriate point. We formalize this idea in the following manner.

Let G/S be a group scheme, and let $\sigma \in G(S)$ be an S-valued point of G. Then the (*right*) *translation-by-σ morphism* is the S-morphism $\tau_\sigma : G \to G$ defined by the composition

$$\tau_\sigma : G \;\cong\; G \times_S S \;\xrightarrow{1 \times \sigma}\; G \times_S G \;\xrightarrow{\mu}\; G.$$

To understand τ_σ further, we note that the S-valued point σ is a map $\sigma : S \to G$. In particular, for each point $s \in S$, we get a point $\sigma(s)$ on the fiber G_s, where G_s is a group variety over the residue field at s. The restriction of τ_σ to the fiber G_s is precisely translation by the point $\sigma(s) \in G_s$. Thus τ_σ can be viewed as a family of translations of the fibers of G.

Remark 3.4. Another important tool in our study of elliptic curves was the collection of multiplication-by-m maps. These maps can be defined inductively on every group scheme in the following way. Let $\pi : G \to S$ be a group group scheme over S with identity element $\sigma_0 : S \to G$, inverse map $i : G \to G$, and group multiplication $\mu : G \times_S G \to G$. Also let $\mathrm{id}_G : G \to G$ be the identity map on G. For each integer m, the *multiplication-by-m* map on G is the morphism

$$[m] : G \longrightarrow G$$

defined inductively by the rules

$$[1] = \mathrm{id}_G, \qquad [m+1] = \mu \circ ([m] \times [1]), \qquad [m-1] = \mu \circ ([m] \times i).$$

§4. Arithmetic Surfaces

Let R be a Dedekind domain. An arithmetic surface over $\mathrm{Spec}(R)$ is the arithmetic analogue of the fibered surfaces studied in III §8. Here $\mathrm{Spec}(R)$ plays the role of the base curve, and an arithmetic surface is an R-scheme $\mathcal{C} \to \mathrm{Spec}(R)$ whose fibers are curves. For example, if R is a discrete valuation ring, then there will be two fibers. The generic fiber will be a curve over the fraction field of R and the special fiber will be a curve over the residue field of R. Just as in the case of fibered surfaces, an arithmetic surface $\mathcal{C}$ may be regular (non-singular) even if it has singular fibers.

Definition. Let R be a Dedekind domain with fraction field K. For example, R could be a discrete valuation ring. Intuitively, an *arithmetic surface (over R)* is a "nice" R-scheme $\mathcal{C}$ whose generic fiber is a non-singular connected projective curve C/K and whose special fibers are unions of curves over the appropriate residue fields. Note that the special fibers may be reducible or singular or even non-reduced. This intuitive definition will suffice for our purposes, but for the technically inclined, we indicate that the word "nice" is an abbreviation which means that $\mathcal{C}$ is an integral, normal, excellent scheme which is flat and of finite type over R.[†]

Remark 4.1.1. An arithmetic surface $\mathcal{C}$ is a one-dimensional family of one-dimensional varieties, so it is a scheme of dimension two. One might instead call $\mathcal{C}$ a curve over R, since it has relative dimension one over R (i.e., the fibers are one-dimensional). We will frequently be interested in arithmetic surfaces which are regular, or proper over R, or smooth over R. We recall the intuitions from section 2. An arithmetic surface $\mathcal{C}$ is regular if it is non-singular as a surface, $\mathcal{C}$ is proper over R if its fibers are complete, and $\mathcal{C}$ is smooth over R if its fibers are non-singular. If $\mathcal{C}$ is smooth over R, then it is automatically regular, but in general the converse is not true.

Remark 4.1.2. The definition of an arithmetic surface $\mathcal{C}$ ensures that even if $\mathcal{C}$ is not regular, its set of singular points is a finite set of closed points. In other words, an arithmetic surface is regular in codimension one. This means that there is a theory of Weil divisors on $\mathcal{C}$. In particular, for any irreducible curve $F \subset \mathcal{C}$ (equivalently, any point $F \in \mathcal{C}$ of codimension one), the local ring $\mathcal{O}_F$ of $\mathcal{C}$ at F is a discrete valuation ring. We denote the corresponding normalized valuation by

$$\mathrm{ord}_F : K(\mathcal{C})^* \longrightarrow \mathbb{Z},$$

[†]   Integral is equivalent to reduced and irreducible (Hartshorne [1, II.3.1]), normal means that all local rings are integrally closed (Hartshorne [1, II exercise 3.8]), flat means that the fibers vary "nicely" (Hartshorne [1, III §9]), finite type means the extensions of local rings are finitely generated algebras (Hartshorne [1, II §3]), and excellent is a somewhat technical condition which won't concern us, but see for example Matsumura [1, Ch. 13].

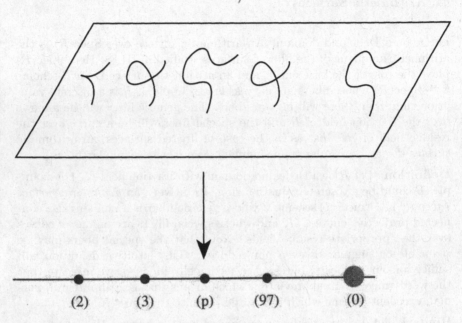

The Arithmetic Surface $\mathcal{C} : y^2 = x^3 + 2x^2 + 6$ over $\mathrm{Spec}(\mathbb{Z})$

Figure 4.1

where $K(\mathcal{C})$ is the function field of $\mathcal{C}$. For the basic thoery of Weil divisors, principal divisors, and the divisor class group, see Hartshorne [1, II §6]. We will continue our discussion of divisors on arithmetic surfaces in section 7.

Example 4.2.1. The projective line $\mathbb{P}^1_R$ over R is an arithmetic surface over R. For any maximal ideal $\mathfrak{p}$ of R, the fiber over $\mathfrak{p}$ is $\mathbb{P}^1_k$, the projective line over the residue field $k = R/\mathfrak{p}$. Notice that $\mathbb{P}^1_R$ is both proper and smooth over R.

Example 4.2.2. Let $\mathcal{C} \subset \mathbb{P}^2_{\mathbb{Z}}$ be the closed subscheme of $\mathbb{P}^2_{\mathbb{Z}}$ given by the equation

$$\mathcal{C} : y^2 = x^3 + 2x^2 + 6.$$

The generic fiber of $\mathcal{C}$ is an elliptic curve $E/\mathbb{Q}$ with discriminant $\Delta = -2^6 \cdot 3 \cdot 97$, so for all primes $p \neq 2, 3, 97$, the fiber $\mathcal{C}_p$ is a (non-singular) elliptic curve over $\mathbb{F}_p$. The fibers over the "bad" primes are

$$\mathcal{C}_2 : y^2 = x^3, \qquad \mathcal{C}_3 : y^2 = x^2(x+2), \qquad \mathcal{C}_{97} : y^2 = (x+66)^2(x+64).$$

The arithmetic surface $\mathcal{C}/\mathbb{Z}$ is illustrated in Figure 4.1.

The arithmetic surface $\mathcal{C}$ is proper over $\mathbb{Z}$, since it is a closed subscheme of $\mathbb{P}^2_{\mathbb{Z}}$ (2.8). It is clear that $\mathcal{C}$ is not smooth over $\mathbb{Z}$, since it has singular fibers. We claim that $\mathcal{C}$ is a regular scheme. To see this, it suffices to check that $\mathcal{C}$ is regular at the singular points on the fibers. We will check that the point $P \in \mathcal{C}$ corresponding to the cusp $x = y = 2 = 0$ on the fiber $\mathcal{C}_2$ is non-singular. The maximal ideal $\mathcal{M}_P$ of the local ring $\mathcal{O}_P$ at P is generated by x, y, and 2, and the residue field at P is $\mathcal{O}_P/\mathcal{M}_P \cong \mathbb{F}_2$. By definition, $\mathcal{C}$ is regular at P if

$$\dim_{\mathbb{F}_2} \mathcal{M}_P/\mathcal{M}_P^2 = 2.$$

This dimension cannot be less than two, so we must show that $\mathcal{M}_P/\mathcal{M}_P^2$ can be generated by two of $x, y, 2$. Using the equation for $\mathcal{C}$, we see that

$$2 = 3^{-1}(y^2 - x^3 - 2x^2) \in \mathcal{M}_P^2,$$

so x and y are generators. This proves that $\mathcal{C}$ is regular at P. We will leave for the reader the analogous calculations at $x = y = 3 = 0$ and $x + 66 = y = 97 = 0$ (exercise 4.14).

The scheme $\mathcal{C}$ is thus regular and proper over $\mathbb{Z}$. If we discard the three singular points on the three singular fibers, we obtain an open sub-scheme $\mathcal{C}^0 \subset \mathcal{C}$ with the property that $\mathcal{C}^0$ is smooth over $\mathbb{Z}$. Of course, $\mathcal{C}^0$ will not be proper over $\mathbb{Z}$, since some of its fibers are missing points.

Example 4.2.3. Let $\mathcal{C} \subset \mathbb{P}_{\mathbb{Z}}^2$ be the closed subscheme of $\mathbb{P}_{\mathbb{Z}}^2$ given by the equation

$$\mathcal{C} : y^2 = x^3 + 2x^2 + 4.$$

The singular fibers of $\mathcal{C}$ are $\mathcal{C}_2$, $\mathcal{C}_5$, and $\mathcal{C}_7$. The scheme $\mathcal{C}$ is not regular, since one easily checks that the point $x = y = 2 = 0$ is a singular point of $\mathcal{C}$.

Let $\pi : \mathcal{C} \to \mathrm{Spec}(R)$ be an arithmetic surface and let $\mathfrak{p} \in \mathrm{Spec}(R)$ be a point with residue field $k_{\mathfrak{p}} = R/\mathfrak{p}$. The fiber

$$\mathcal{C}_{\mathfrak{p}} = \mathcal{C} \times_R \mathfrak{p} = \mathcal{C} \times_{\mathrm{Spec}(R)} \mathrm{Spec}(k_{\mathfrak{p}})$$

is a curve, but it may be reducible or singular or even non-reduced. More precisely, we can write the fiber as a union

$$\mathcal{C}_{\mathfrak{p}} = \sum_{i=1}^{r} n_i F_i$$

for certain irreducible curves $F_1, \ldots, F_r/k_{\mathfrak{p}}$ and multiplicities $n_1, \ldots, n_r \geq 1$ in the following manner. Fix a uniformizer $u \in R$ for $\mathfrak{p}$, that is, $\mathrm{ord}_{\mathfrak{p}}(u) = 1$. Then $\pi^*(u) = u \circ \pi$ is a rational function on $\mathcal{C}$, and the fiber of $\mathcal{C}$ over $\mathfrak{p}$ is given by

$$\mathcal{C}_{\mathfrak{p}} = \sum_{F \subset \pi^{-1}(\mathfrak{p})} \mathrm{ord}_F(\pi^* u) F.$$

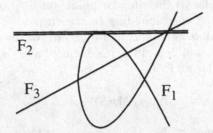

The Special Fiber $\mathcal{C}_5 : (y^2 - x^3 - 3x^2)(y-2)^2(2y - x - 3) = 0$

Figure 4.2

Here the sum is over the irreducible components of the fiber over $\mathfrak{p}$, and ord_F is the normalized valuation on $K(\mathcal{C})$ corresponding to F (4.1.2).

There are several ways in which a point $x \in \mathcal{C}_\mathfrak{p}$ can be a singular point of the fiber $\mathcal{C}_\mathfrak{p}$. It may lie on a component F with multiplicity $n \geq 2$, it may be a point where two or more components intersect, or it may be a singular point of a particular component. The following example illustrates these ideas.

Example 4.2.4. Consider the arithmetic surface $\mathcal{C} \subset \mathbb{A}_{\mathbb{Z}}^2$ defined by the equation

$$\mathcal{C} : 2y^5 - (x+1)y^4 - (2x^3 + x^2 + x)y^3 \\ + (x^4 - x^3 + 3x^2 + x - 2)y^2 + (x^4 + 3x^3)y - x^4 - x^3 + x^2 = 5.$$

We are going to look at the special fiber $\mathcal{C}_5$ of $\mathcal{C}$ over the point $(5) \in \operatorname{Spec} \mathbb{Z}$. This special fiber is the curve in $\mathbb{A}_{\mathbb{F}_5}^2$ defined by reducing the equation of $\mathcal{C}$ modulo 5, so after some algebra we find

$$\mathcal{C}_5 : (y^2 - x^3 - 3x^2)(y-2)^2(2y - x - 3) = 0.$$

Thus $\mathcal{C}_5$ consists of three irreducible components, which we label as

$$F_1 : y^2 = x^3 + 3x^2, \quad F_2 : y = 2, \quad F_3 : 2y = x + 3.$$

We have illustrated the special fiber $\mathcal{C}_5$ in Figure 4.2. Such illustrations can be very useful for visualizing components, multiplicities, and intersections, as long as one keeps in mind that one is looking at a drawing in $\mathbb{R}^2$ which purports to represent a curve in characteristic p! In particular, there may be intersection points which are "hidden" because $\mathbb{R}$ is not algebraically closed.

The component F_2 of $\mathcal{C}_5$ appears with multiplicity 2, and each of the other components has multiplicity 1, so as a scheme the special fiber has the form

$$\mathcal{C}_3 = F_1 + 2F_2 + F_3.$$

In particular, the scheme $\mathcal{C}_5$ is neither irreducible nor reduced. Every point on F_2 is a singular point of $\mathcal{C}_5$, since F_2 itself appears with multiplicity greater than 1. The other singular points on the special fiber are the node $(0,0)$ on F_1 and the points where the various F_i's intersect, such as the point $(3,2)$ on $F_1 \cap F_2$ and the points $(2,0)$ and $(3,3)$ on $F_1 \cap F_3$. (Remember that the fiber $\mathcal{C}_5$ lives in characteristic 5.)

The next proposition says that if an arithmetic surface $\mathcal{C}$ is regular and if a point $x \in \mathcal{C}_\mathfrak{p}$ on its special fiber lies in the image of an R-valued point $P \in \mathcal{C}(R)$ (i.e., if $x = P(\mathfrak{p})$), then x is automatically a non-singular point of $\mathcal{C}_\mathfrak{p}$.

Proposition 4.3. *Let* $\pi : \mathcal{C} \to \mathrm{Spec}(R)$ *be a regular arithmetic surface over a Dedekind domain R, and let* $\mathfrak{p} \in \mathrm{Spec}(R)$.
(a) *Let* $x \in \mathcal{C}_\mathfrak{p} \subset \mathcal{C}$ *be a closed point on the fiber of $\mathcal{C}$ over $\mathfrak{p}$. Then*

$$\mathcal{C}_\mathfrak{p} \text{ is non-singular at } x \iff \pi^*(\mathfrak{p}) \not\subset \mathcal{M}_{\mathcal{C},x}^2.$$

Here π^* is the natural map $\pi^* : R \to \mathcal{O}_{\mathcal{C},x}$ induced by π, and $\mathcal{M}_{\mathcal{C},x}$ is the maximal ideal of the local ring $\mathcal{O}_{\mathcal{C},x}$ of $\mathcal{C}$ at x.
(b) *Let* $P \in \mathcal{C}(R)$. *Then $\mathcal{C}_\mathfrak{p}$ is non-singular at $P(\mathfrak{p})$.*

PROOF. To ease notation, we will write

$$\tilde{\mathcal{C}} = \mathcal{C}_\mathfrak{p} = \mathcal{C} \times_R (R/\mathfrak{p})$$

for the fiber of $\mathcal{C}$ over $\mathfrak{p}$, and we will let $\mathfrak{P} = \pi^*(\mathfrak{p})\mathcal{O}_{\mathcal{C},x}$. Notice that $\mathfrak{P} \subset \mathcal{M}_{\mathcal{C},x}$, since x lies on the special fiber over $\mathfrak{p}$.
(a) We first assume that $\mathfrak{P} \not\subset \mathcal{M}_{\mathcal{C},x}^2$ and prove that x is a non-singular point of $\tilde{\mathcal{C}}$. We are given that $\mathcal{C}$ is regular, so, by definition, $\mathcal{O}_{\mathcal{C},x}$ is a regular local ring of dimension two. This means that we can find elements $f_1, f_2 \in \mathcal{M}_{\mathcal{C},x}$ so that

$$\mathcal{M}_{\mathcal{C},x} = f_1 \mathcal{O}_{\mathcal{C},x} + f_2 \mathcal{O}_{\mathcal{C},x} + \mathcal{M}_{\mathcal{C},x}^2.$$

If we write $\mathfrak{p} = tR$, then $\pi^*(t) \in \mathfrak{P} \subset \mathcal{M}_{\mathcal{C},x}$, so

$$\pi^*(t) \equiv a_1 f_1 + a_2 f_2 \pmod{\mathcal{M}_{\mathcal{C},x}^2} \qquad \text{for some } a_1, a_2 \in \mathcal{O}_{\mathcal{C},x}.$$

Our assumption is that $\pi^*(t)\mathcal{O}_{\mathcal{C},x} = \mathfrak{P} \not\subset \mathcal{M}_{\mathcal{C},x}^2$, which means that at least one of a_1 and a_2 is not in $\mathcal{M}_{\mathcal{C},x}$, and hence at least one of them is a unit in $\mathcal{O}_{\mathcal{C},x}$. Switching f_1 and f_2 if necessary, this means that

$$\mathcal{M}_{\mathcal{C},x} = \pi^*(t)\mathcal{O}_{\mathcal{C},x} + f_2 \mathcal{O}_{\mathcal{C},x} + \mathcal{M}_{\mathcal{C},x}^2 = \mathfrak{P} + f_2 \mathcal{O}_{\mathcal{C},x} + \mathcal{M}_{\mathcal{C},x}^2.$$

The fiber $\tilde{\mathcal{C}}$ (as a scheme, which includes multiplicities associated to non-reduced components) is $\tilde{\mathcal{C}} = \mathcal{C} \times_R (R/\mathfrak{p})$, so its local ring at x is obtained from the local ring of $\mathcal{C}$ by reduction modulo $\mathfrak{p}$. In other words,

$$\mathcal{O}_{\tilde{\mathcal{C}},x} = \mathcal{O}_{\mathcal{C},x}/\mathfrak{P} \qquad \text{and} \qquad \mathcal{M}_{\tilde{\mathcal{C}},x} = \mathcal{M}_{\mathcal{C},x}/\mathfrak{P}.$$

Therefore

$$\mathcal{M}_{\tilde{\mathcal{C}},x} = (\mathfrak{P} + f_2 \mathcal{O}_{\mathcal{C},x} + \mathcal{M}_{\mathcal{C},x}^2)/\mathfrak{P} = f_2 \mathcal{O}_{\tilde{\mathcal{C}},x} + \mathcal{M}_{\tilde{\mathcal{C}},x}^2.$$

Hence $\mathcal{M}_{\tilde{\mathcal{C}},x}/\mathcal{M}_{\tilde{\mathcal{C}},x}^2$ is generated by the single element f_2, which shows that $\mathcal{O}_{\tilde{\mathcal{C}},x}$ is a regular local ring of dimension one, and so x is a non-singular point of $\tilde{\mathcal{C}}$.

This proves the implication that we will need for part (b). We will leave the proof of the opposite implication as an exercise for the reader (exercise 4.17).

(b) We assume that $\pi^*(\mathfrak{p}) \subset \mathcal{M}_{\mathcal{C},x}^2$ and derive a contradiction. Using the fact that $\pi \circ P$ is the identity map on $\mathrm{Spec}(R)$, we compute

$$\mathfrak{p} = (\pi \circ P)^*(\mathfrak{p}) = P^* \circ \pi^*(\mathfrak{p}) \subset P^*(\mathcal{M}_{\mathcal{C},x}^2) = (P^*\mathcal{M}_{\mathcal{C},x})^2 = \mathfrak{p}^2.$$

The last equality follows from the fact that $P : \mathrm{Spec}(R) \to \mathcal{C}$ is a morphism of schemes, so by definition (Hartshorne [1, II §2]) the induced map $P^* : \mathcal{O}_{\mathcal{C},x} \to R_{\mathfrak{p}}$ is a local homomorphism of local rings. This means in particular that $P^*\mathcal{M}_{\mathcal{C},x} = \mathfrak{p}$.

But $\mathfrak{p}$ is a maximal ideal of the Dedekind domain R, so the inclusion $\mathfrak{p} \subset \mathfrak{p}^2$ is impossible. Therefore $\pi^*(\mathfrak{p}) \not\subset \mathcal{M}_{\mathcal{C},x}^2$. Applying (a), we conclude that x is a non-singular point of the fiber $\mathcal{C}_{\mathfrak{p}}$, which concludes the proof of (b). $\square$

The following important corollary says that the smooth part of a proper regular arithmetic surface is large enough to contain all of the rational points on the generic fiber. For an example which shows that the regularity condition is necessary, see (5.4.4) in the next section.

Corollary 4.4.   *Let R be a Dedekind domain with fraction field K, let $\mathcal{C}/R$ be an arithmetic surface, and let C/K be the generic fiber of $\mathcal{C}$.*
(a) *If $\mathcal{C}$ is proper over R, then*

$$C(K) = \mathcal{C}(R).$$

(b) *Suppose that the scheme $\mathcal{C}$ is regular, and let $\mathcal{C}^0 \subset \mathcal{C}$ be the largest subscheme of $\mathcal{C}$ such that the map $\mathcal{C}^0 \to \mathrm{Spec}(R)$ is a smooth morphism. Then*

$$\mathcal{C}(R) = \mathcal{C}^0(R).$$

(c) *In particular, if $\mathcal{C}$ is regular and proper over R, then*

$$C(K) = \mathcal{C}(R) = \mathcal{C}^0(R).$$

PROOF. (a) This is really just a special case of the valuative criterion of properness. Any point in $\mathcal{C}(R)$ can be specialized to the generic fiber to give a point in $C(K)$, so there is a natural map $\mathcal{C}(R) \to C(K)$. This map is clearly one-to-one, since two morphisms $\mathrm{Spec}(R) \to \mathcal{C}$ which agree generically (i.e., on a dense open set) are the same. Thus $\mathcal{C}(R) \hookrightarrow C(K)$.

Let $P \in C(K)$ be a point. We are given that $\mathcal{C}$ is proper over R, so the valuative criterion (2.7) says that there is a morphism $\sigma_P : \mathrm{Spec}(R) \to \mathcal{C}$ making the following diagram commute:

$$
\begin{array}{ccc}
C = & \mathcal{C} \times_R K & \longrightarrow & \mathcal{C} \\
& \Big\uparrow{\scriptstyle P} & & \Big\uparrow{\scriptstyle \sigma_P} \\
& \mathrm{Spec}(K) & \longrightarrow & \mathrm{Spec}(R).
\end{array}
$$

This proves that every point in $C(K)$ comes from a point in $\mathcal{C}(R)$, so $\mathcal{C}(R) = C(K)$.

(b) Proposition 4.3 says that every point in $\mathcal{C}(R)$ intersects each fiber at a non-singular point of the fiber. But, by definition, $\mathcal{C}^0$ is the complement in $\mathcal{C}$ of the singular points on the fibers. Therefore the natural inclusion $\mathcal{C}^0(R) \hookrightarrow \mathcal{C}(R)$ is a bijection.

(c) This is immediate from (a) and (b). $\qquad\qquad\qquad\qquad\qquad\qquad\square$

The previous corollary (4.4) says that if $\mathcal{C}$ is a regular arithmetic surface that is proper over R, then the smooth part $\mathcal{C}^0$ of $\mathcal{C}$ is large enough so that all of the K-valued points on the generic fiber extend to R-valued points of $\mathcal{C}^0$. This raises two questions. First, given a (non-singular projective) curve C defined over K, does there exist a regular arithmetic surface $\mathcal{C}$ proper over R whose generic fiber is C/K? Second, assuming such proper regular models exist, to what extent is there a minimal such model? The following theorem gives the answer to these questions. It is the arithmetic analogue of the geometric results described in (III.7.7) and (III.8.4). We will discuss the construction of these minimal models further in §7.

Theorem 4.5. *Let R be a Dedekind domain with fraction field K, and let C/K be a non-singular projective curve of genus g.*

(a) (Resolution of Singularities for Arithmetic Surfaces, Abhyankar [1,2], Lipman [1,2]) *There exists a regular arithmetic surface $\mathcal{C}/R$, proper over R, whose generic fiber is isomorphic to C/K. We call $\mathcal{C}/R$ a proper regular model for C/K.*

(b) (Minimal Models Theorem, Lichtenbaum [1], Shafarevich [2]) *Assume that $g \geq 1$. Then there exists a proper regular model $\mathcal{C}^{\min}/R$ for C/K with the following minimality property:*

Let $\mathcal{C}/R$ be any other proper regular model for C/K. Fix an isomorphism from the generic fiber of $\mathcal{C}$ to the generic fiber of $\mathcal{C}^{\min}$. Then the induced R-birational map

$$\mathcal{C} \dashrightarrow \mathcal{C}^{\min}$$

is an R-isomorphism. We call $\mathcal{C}^{\min}/R$ the *minimal proper regular model for C/K*. It is unique up to unique R-isomorphism.

PROOF. (a) See Abhyankar [1,2] and Lipman [1,2]. There is also a nice exposition of Lipman's proof in Artin [2]. In the case that C has genus 1, we will explicitly construct a proper regular model for C/K in §9.

(b) See Lichtenbaum [1, Thm. 4.4] and Shafarevich [2, lectures 6,7,8]. There is a nice summary of the main results with sketches of the proofs in Chinburg [1]. See also §7 for a further discussion.   □

Just as in (III.8.4.1), the importance of the minimal regular model lies in the fact that every automorphism of its generic fiber extends to a morphism of the entire scheme.

Proposition 4.6. *Let R be a Dedekind domain with fraction field K, and let C/K be a non-singular projective curve of genus $g \geq 1$. Let $\mathcal{C}/R$ be a minimal proper regular model for C/K, and let $\mathcal{C}^0 \subset \mathcal{C}$ be the largest subscheme of $\mathcal{C}$ which is smooth over R. Then every K-automorphism $\tau : C/K \to C/K$ of the generic fiber of $\mathcal{C}$ extends to give R-automorphisms*

$$\tau : \mathcal{C} \longrightarrow \mathcal{C} \quad \text{and} \quad \tau : \mathcal{C}^0 \longrightarrow \mathcal{C}^0.$$

PROOF. The fact that τ extends to an R-automorphism $\mathcal{C} \to \mathcal{C}$ is exactly the definition of minimality given in (4.5b). Next take any point $x \in \mathcal{C}^0$ and choose some neighborhood $U \subset \mathcal{C}^0$ of x. Then U is smooth over R. Further, U is an open subset of $\mathcal{C}$, since $\mathcal{C}^0$ is open in $\mathcal{C}$. We know that $\tau : \mathcal{C} \to \mathcal{C}$ is an R-isomorphism, so $\tau(U)$ is an open neighborhood of $\tau(x)$ and is smooth over R. Therefore $\tau(x) \in \mathcal{C}^0$, which proves that $\tau(\mathcal{C}^0) \subset \mathcal{C}^0$. Applying the same argument to τ^{-1} gives $\tau^{-1}(\mathcal{C}^0) \subset \mathcal{C}^0$, which completes the proof that τ gives an R-automorphism of $\mathcal{C}^0$.   □

§5. Néron Models

Let K be the fraction field of a discrete valuation ring R. The Néron model of an elliptic curve E/K is an arithmetic surface $\mathcal{E}/R$ whose generic fiber is the given elliptic curve. The scheme $\mathcal{E}/R$ is characterized by the fact that it is large enough so that every point of E gives a point of $\mathcal{E}$, but small enough so that the group law on E extends to make $\mathcal{E}$ into a group scheme over R. Of course, when we talk of "points of $\mathcal{E}$," we mean more than just the points of the underlying scheme. This leads to the following definition.

Definition. Let R be a Dedekind domain with fraction field K, and let E/K be an elliptic curve. A *Néron model for* E/K is a (smooth) group scheme $\mathcal{E}/R$ whose generic fiber is E/K and which satisfies the following universal property:

> Let $\mathcal{X}/R$ be a smooth R-scheme (i.e., $\mathcal{X}$ is smooth over R) with generic fiber X/K, and let $\phi_K : X_{/K} \to E_{/K}$ be a rational map defined over K. Then there exists a unique R-morphism $\phi_R : \mathcal{X}_{/R} \to \mathcal{E}_{/R}$ extending ϕ_K.
> $\left(\begin{array}{c}\text{Néron}\\\text{Mapping}\\\text{Property}\end{array}\right)$

Remark 5.1.1. A Néron model $\mathcal{E}/R$ is a smooth R-scheme. This means that for every point $\mathfrak{p} \in \mathrm{Spec}(R)$, the fiber $\mathcal{E}_\mathfrak{p}$ of $\mathcal{E} \to \mathrm{Spec}(R)$ is a nonsingular variety defined over the residue field $k(\mathfrak{p})$; see (2.9). However, as we will soon see, the fiber $\mathcal{E}_\mathfrak{p}$ over a closed point $\mathfrak{p}$ may have several components and may not be complete, so in general $\mathcal{E}$ will not be proper over R.

Remark 5.1.2. In the Néron mapping property we have only required that the map $\phi_K : X_{/K} \to E_{/K}$ on the generic fiber be a rational map. It turns out that any rational map from a non-singular variety to an elliptic curve is a morphism. See (6.2b) below for an even more general statement.

Remark 5.1.3. The most important instance of the Néron mapping property is the case that $\mathcal{X} = \mathrm{Spec}(R)$ and $X = \mathrm{Spec}(K)$. Then the set of K-maps $X_{/K} \to E_{/K}$ is precisely the group of K-rational points $E(K)$, and the set of R-morphisms $\mathcal{X}_{/R} \to \mathcal{E}_{/R}$ is the group of sections $\mathcal{E}(R)$. So in this situation the Néron mapping property says that the natural inclusion

$$\mathcal{E}(R) \lhook\joinrel\longrightarrow E(K)$$

is a bijection. If R is a complete discrete valuation ring with algebraically closed residue field, then one can show that the equality $\mathcal{E}(R) = E(K)$ suffices to ensure that the group scheme $\mathcal{E}/R$ is a Néron model for E/K. See exercise 4.30.

We begin our study of Néron models by proving that they are unique and behave well under unramified base extension.

Proposition 5.2. *Let R be a Dedekind domain with fraction field K, and let E/K be an elliptic curve.*
(a) *Suppose that $\mathcal{E}_1/R$ and $\mathcal{E}_2/R$ are Néron models for E/K. Then there exists a unique R-isomorphism $\psi : \mathcal{E}_1/R \to \mathcal{E}_2/R$ whose restriction to the generic fiber is the identity map on E/K. In other words, the Néron model of E/K is unique up to unique isomorphism.*
(b) *Let K'/K be a finite unramified extension, and let R' be the integral closure of R in K'. Let $\mathcal{E}/R$ be a Néron model for E/K. Then $\mathcal{E} \times_R R'$ is a*

Néron model for E/K'. (N.B. If K'/K is ramified, this result will generally not be true.)

PROOF. (a) The identity map $E/K \to E/K$ is a rational map from the generic fiber of $\mathcal{E}_1$ to the generic fiber of $\mathcal{E}_2$, and $\mathcal{E}_1$ is smooth over R, so the Néron mapping property for $\mathcal{E}_2$ says that the identity map extends uniquely to an R-morphism $\psi : \mathcal{E}_1/R \to \mathcal{E}_2/R$. In a similar fashion we obtain a unique R-morphism $\phi : \mathcal{E}_2/R \to \mathcal{E}_1/R$ which is the identity map on the generic fiber. But then $\phi \circ \psi : \mathcal{E}_1/R \to \mathcal{E}_1/R$ and the identity map $\mathcal{E}_1/R \to \mathcal{E}_1/R$ are R-morphisms which are the same on the generic fiber, so the uniqueness part of the Néron mapping property says that $\phi \circ \psi$ equals the identity map. This proves that ϕ and ψ are isomorphisms.

(b) Let $\mathcal{X}'/R'$ be a smooth R'-scheme with generic fiber X'/K', and let $\phi_{K'} : X'_{/K'} \to E_{/K'}$ be a rational map. The composition

$$\mathcal{X}' \longrightarrow \mathrm{Spec}(R') \longrightarrow \mathrm{Spec}(R)$$

makes $\mathcal{X}'$ into an R-scheme. Further, our assumptions on K' imply that the map

$$\mathrm{Spec}(R') \longrightarrow \mathrm{Spec}(R)$$

is a smooth morphism. (See exercise 4.19.) Hence the composition is a smooth morphism (2.10), so $\mathcal{X}'$ is a smooth R-scheme.

Now the Néron mapping property for $\mathcal{E}/R$ tells us that there is an R-morphism

$$\mathcal{X}' \xrightarrow{\phi_R} \mathcal{E}$$

whose restriction to the generic fiber is the composition

$$X' \xrightarrow{\phi_{K'}} E \times_K K' \xrightarrow{p_1} E.$$

The two R-morphisms $\phi_R : \mathcal{X}' \to \mathcal{E}$ and $\mathcal{X}' \to \mathrm{Spec}(R')$ determine an R-morphism (and thus an R'-morphism) to the fiber product,

$$\phi_{R'} : \mathcal{X}' \longrightarrow \mathcal{E} \times_R R'.$$

Further, the restriction of $\phi_{R'}$ to the generic fiber is $\phi_{K'}$. This gives the existence part of the Néron mapping property. We will leave it to the reader to prove the uniqueness part, which completes the proof that $\mathcal{E} \times_R R'$ is a Néron model for E/K'. $\qquad\square$

Let R be a discrete valuation ring with fraction field K. We are going to use the Weierstrass equation of an elliptic curve E/K directly to construct an R-group scheme $\mathcal{W}^0 \subset \mathbb{P}^2_R$ whose generic fiber is E/K. If the closure $\mathcal{W}$ of $\mathcal{W}^0$ in $\mathbb{P}^2_R$ is regular, then we will also prove that $\mathcal{W}^0(R) = E(K)$, so $\mathcal{W}^0$ satisfies the most important instance of the Néron mapping property (5.1.3). In particular, if E has good reduction, then we will see (6.3) that a minimal Weierstrass equation for E/K already defines a Néron model.

Theorem 5.3. *Let R be a discrete valuation ring with fraction field K, let E/K be an elliptic curve, and choose a Weierstrass equation for E/K with coefficients in R,*

$$E : y^2 + a_1 xy + a_3 y = x^3 + a_2 x^2 + a_4 x + a_6.$$

This Weierstrass equation defines a scheme $\mathcal{W} \subset \mathbb{P}_R^2$. Let $\mathcal{W}^0 \subset \mathcal{W}$ be the largest subscheme of $\mathcal{W}$ which is smooth over R.
(a) Both $\mathcal{W}/R$ and $\mathcal{W}^0/R$ have generic fiber E/K.
(b) The natural map $\mathcal{W}(R) \to E(K)$ is a bijection. If $\mathcal{W}$ is regular, then the natural map $\mathcal{W}^0(R) \to \mathcal{W}(R)$ is also a bijection, so in this case there is a natural identification $\mathcal{W}^0(R) = E(K)$.
(c) The addition and negation maps on E extend to R-morphisms

$$\mathcal{W}^0 \times_R \mathcal{W}^0 \longrightarrow \mathcal{W}^0 \qquad and \qquad \mathcal{W}^0 \longrightarrow \mathcal{W}^0$$

which make $\mathcal{W}^0$ into a group scheme over R. The addition map further extends to an R-morphism

$$\mathcal{W}^0 \times_R \mathcal{W} \longrightarrow \mathcal{W}$$

giving a group scheme action of $\mathcal{W}^0$ on $\mathcal{W}$.

Remark 5.4.1. If E/K has good reduction and if we take a minimal Weierstrass equation for E/K, then $\mathcal{W}$ itself is smooth over R. So in this case (5.3) says that $\mathcal{W} = \mathcal{W}^0$ is a group scheme over R.

Remark 5.4.2. If E/K has bad reduction, then there is exactly one singular point on the reduction $\tilde{E} \pmod{\mathfrak{p}}$. In other words, the special fiber $\tilde{\mathcal{W}}$ of $\mathcal{W}$ contains exactly one singular point, say $\gamma \in \tilde{\mathcal{W}} \subset \mathcal{W}$, and then $\mathcal{W}^0$ is obtained by discarding that point,

$$\mathcal{W}^0 = \mathcal{W} \smallsetminus \{\gamma\}.$$

Remark 5.4.3. The scheme $\mathcal{W}/R$ in (5.3) is proper over R, since it is a closed subscheme of $\mathbb{P}_R^2$ (2.8). It follows from the valuative criterion (4.4a) that $\mathcal{W}(R) = E(K)$. However, in general, the scheme $\mathcal{W}$ will not be regular, since a singular point on the special fiber will often be a singular point of $\mathcal{W}$. So, in general, we cannot use (4.4b) to deduce that $\mathcal{W}^0(R) = E(K)$. Intuitively, if $\mathcal{W}$ is singular, then there will be points $P \in E(K) = \mathcal{W}(R)$ which go through the singular point of $\mathcal{W}$. Thus, in general, $\mathcal{W}^0$ will not be large enough to be a Néron model, because $\mathcal{W}^0(R) \neq E(K)$, whereas $\mathcal{W}$ itself will be too large to be a Néron model, because the group law on E will not extend to all of $\mathcal{W}$.

Example 5.4.4. We illustrate the previous remark by looking at the curve

$$E : y^2 + xy = x^3 + a_6.$$

If $a_6 \in R^*$, then $\mathcal{W}$ is smooth over R (exercise 4.20(b)), so $\mathcal{W}$ is a Néron model for E. If $v(a_6) \geq 1$, then the special fiber

$$\widetilde{\mathcal{W}} : y^2 + xy = x^3$$

has the singular point $(\tilde{0}, \tilde{0})$, so

$$\mathcal{W}^0 = \mathcal{W} \smallsetminus \{(\tilde{0}, \tilde{0})\}.$$

We consider two cases.

First, if $v(a_6) = 1$ (i.e., a_6 is a uniformizer in R), then $\mathcal{W}$ is a regular scheme from exercise 4.20(a). It follows from (4.4c) that $\mathcal{W}^0(R) = \mathcal{W}(R) = E(K)$. We can also see this directly as follows. If $P \in \mathcal{W}(R)$ were to go through the singular point on the special fiber, then we would have $P = (x, y) \equiv (\tilde{0}, \tilde{0}) \pmod{\mathfrak{p}}$, which means that $x, y \in \mathfrak{p}$. But then the equation for E would give

$$a_6 = y^2 + xy - x^3 \in \mathfrak{p}^2,$$

contradicting the assumption that $v(a_6) = 1$. Hence $\mathcal{W}^0(R) = E(K)$.

Second, if $v(a_6) \geq 2$, then $\mathcal{W}$ is not a regular scheme and $\mathcal{W}^0$ will not be a Néron model for E, despite the fact that $\mathcal{W}^0$ is a group scheme with generic fiber E. For example, suppose that $a_6 = \alpha^2$ with $v(\alpha) \geq 1$. Then the point $P = (0, \alpha) \in E(K) = \mathcal{W}(R)$ is not in $\mathcal{W}^0(R)$, since $P \equiv (\tilde{0}, \tilde{0}) \pmod{\mathfrak{p}}$.

PROOF (of Theorem 5.3). (a) $\mathcal{W}$ is projective over R, since it is the closed subscheme of $\mathbb{P}_R^2 = \operatorname{Proj} R[X, Y, Z]$ defined by the single homogeneous equation

$$\mathcal{W} : Y^2 Z + a_1 XYZ + a_3 YZ^2 = X^3 + a_2 X^2 Z + a_4 XZ^2 + a_6 Z^3.$$

Its generic fiber is the variety in $\mathbb{P}_K^2$ defined by this same equation. Thus the generic fiber of $\mathcal{W}$ is precisely E/K.

If $\mathcal{W}^0$ is not equal to $\mathcal{W}$, then as described above in (5.4.2), $\mathcal{W}^0$ consists of $\mathcal{W}$ with one point on the special fiber removed. In particular, $\mathcal{W}^0$ and $\mathcal{W}$ have the same generic fiber, so the generic fiber of $\mathcal{W}^0$ is also E/K.

(b) The scheme $\mathcal{W}$ is a closed subscheme of $\mathbb{P}_R^2$, so it is proper over R (2.8). Further, its generic fiber is E/K from (a), so (4.4a) tells us that $E(K) = \mathcal{W}(R)$. This proves the first part of (b). If in addition $\mathcal{W}$ is regular, then (4.4b) says that $\mathcal{W}(R) = \mathcal{W}^0(R)$, which gives the second equality $E(K) = \mathcal{W}^0(R)$.

(c) Let
$$\mu : W \times_R W \dashrightarrow W \qquad \text{and} \qquad i : W \dashrightarrow W$$
be the rational maps on W induced by the addition and negation laws on
the generic fiber E/K of W. The fact that the generic fiber is a group
variety means that μ and i satisfy all of the group axioms on a non-empty
open subscheme of W, so they will satisfy the group axioms on the largest
open subscheme on which they are defined. In other words, if we can show
that μ is a morphism on $W^0 \times_R W$ and that i is a morphism on W, then
the group axioms are automatically true.

If W is smooth over R, we are going to prove that μ is a morphism on
all of $W \times_R W$. If W is not smooth over R, then as explained above (5.4.2),
the special fiber $\widetilde{W}$ contains a unique singular point γ and $W^0 = W \smallsetminus \{\gamma\}$.
In this situation we will show that μ is a morphism except at the single
point
$$(\gamma, \gamma) \in \widetilde{W} \times_k \widetilde{W} \subset W \times_R W.$$

In particular, it is a morphism on $W^0 \times_R W$.

In order to simplify our calculations, we will assume that the residue
field k does not have characteristic 2 or 3 (equivalently, 2 and 3 are units
in R). The general case is similar, but the formulas are considerably longer.
This assumption allows us to make a change of variables in $\mathbb{P}^2_R$ and put our
Weierstrass equation in the form

$$W : Y^2 Z = X^3 + AXZ^2 + BZ^3.$$

In other words, W is the closed subscheme of $\mathbb{P}^2_R$ defined by this homogeneous equation.

Let W_{aff} be the affine open subscheme of W defined by

$$W_{\mathrm{aff}} = \{Z \neq 0\} \subset W,$$

and let $x = X/Z$ and $y = Y/Z$ be affine coordinates on W_{aff}. The addition
map μ is then given by the usual formula [AEC, III.2.3]

$$\mu = \big[(x_2 - x_1)\big((y_2 - y_1)^2 - (x_2 - x_1)^2(x_2 + x_1)\big),$$
$$(y_2 - y_1)^3 + (x_2 - x_1)^2(x_1 y_1 - x_2 y_2 + 2x_2 y_1 - 2x_1 y_2), (x_2 - x_1)^3\big].$$

More precisely, this formula gives the restriction of μ to $W_{\mathrm{aff}} \times_R W_{\mathrm{aff}}$,
and on this affine scheme μ will be a morphism except possibly along the
closed subscheme where its three coordinate functions vanish. Looking
at the third coordinate and then at the second coordinate, we see that μ
is a morphism on $W_{\mathrm{aff}} \times_R W_{\mathrm{aff}}$ except possibly on the closed subscheme
defined by the equations

$$x_2 - x_1 = y_2 - y_1 = 0.$$

In other words, μ is a morphism off of the diagonal.

To deal with points on the diagonal, we use the relations

$$y_1^2 = x_1^3 + Ax_1 + B \qquad \text{and} \qquad y_2^2 = x_2^3 + Ax_2 + B$$

which hold identically on $\mathcal{W}_{\text{aff}} \times_R \mathcal{W}_{\text{aff}}$ to rewrite the addition map μ as

$$
\begin{aligned}
\mu = \big[&(y_1 + y_2)\big((x_1 + x_2)(y_1 + y_2)^2 + (x_1^2 + x_1 x_2 + x_2^2 + A)^2\big), \\
&(x_1^2 + x_1 x_2 + x_2^2 + A)^3 \\
&- (y_1 + y_2)^2\big((x_1 + x_2)^3 + A(x_1 + x_2) + B - y_1 y_2\big), (y_1 + y_2)^3 \big].
\end{aligned}
$$

(See [AEC, III.3.6.1] for a similar calculation.) Just as above, we see that μ is a morphism on $\mathcal{W}_{\text{aff}} \times_R \mathcal{W}_{\text{aff}}$ except possibly on the closed subscheme defined by the equations

$$y_1 + y_2 = x_1^2 + x_1 x_2 + x_2^2 + A = 0.$$

We have now proven that μ is a morphism on $\mathcal{W}_{\text{aff}} \times_R \mathcal{W}_{\text{aff}}$ except on the subscheme defined by the four equations

$$x_2 - x_1 = y_2 - y_1 = y_1 + y_2 = x_1^2 + x_1 x_2 + x_2^2 + A = 0.$$

A little algebra and the fact that $2 \in R^*$ shows that this subscheme is defined by the equations

$$x_2 = x_1, \qquad y_1 = y_2 = 0, \qquad 3x_1^2 + A = 0.$$

In particular, it is contained in the diagonal of $\mathcal{W}_{\text{aff}} \times_R \mathcal{W}_{\text{aff}}$, so if we identify $\mathcal{W}_{\text{aff}}$ with this diagonal, then μ is a morphism except on the subscheme

$$y = 3x^2 + A = 0.$$

Using the relation $y^2 = x^3 + Ax + B$ and the fact that $3 \in R^*$, we see that the discriminant $4A^3 + 27B^2$ is contained in the ideal generated by y and $3x^2 + A$. Hence if W is smooth over R, which implies that its discriminant is a unit in R, then μ is a morphism on all of $\mathcal{W}_{\text{aff}} \times_R \mathcal{W}_{\text{aff}}$. Similarly, if W is not smooth over R, then μ will be a morphism on $\mathcal{W}_{\text{aff}} \times_R \mathcal{W}_{\text{aff}}$ away from the subscheme

$$x_2 = x_1, \qquad y_1 = y_2 = 0, \qquad 3x_1^2 + A = 0, \qquad 4A^3 + 27B^2 = 0,$$

which is precisely the singular point on the special fiber of the diagonal.

Next let $\mathcal{W}'_{\text{aff}}$ be the affine open subscheme of W defined by

$$\mathcal{W}'_{\text{aff}} = \{Y \neq 0\} \subset \mathcal{W}.$$

Notice that $\mathcal{W} \times_R \mathcal{W}$ is covered by the four affine subschemes

$$\mathcal{W}_{\text{aff}} \times_R \mathcal{W}_{\text{aff}}, \qquad \mathcal{W}_{\text{aff}} \times_R \mathcal{W}'_{\text{aff}}, \qquad \mathcal{W}'_{\text{aff}} \times_R \mathcal{W}_{\text{aff}}, \qquad \mathcal{W}'_{\text{aff}} \times_R \mathcal{W}'_{\text{aff}},$$

since $\mathcal{W}$ does not intersect the scheme $Y = Z = 0$. We have already dealt with the restriction of μ to the first set, so it remains to show that μ is a morphism on each of the last three. We will leave this task for the reader (exercise 4.22), since the proof is similar to the argument given above.

Finally, we observe that the negation map

$$i : \mathcal{W} \times_R \mathcal{W} \longrightarrow \mathcal{W}, \qquad [X, Y, Z] \longmapsto [X, -Y, Z],$$

is a morphism on $\mathcal{W}$, since it is actually the restriction of a morphism on $\mathbb{P}^2_R \times_R \mathbb{P}^2_R$. This completes the proof of Theorem 5.3. $\square$

§6. Existence of Néron Models

In this section we are going to prove the existence of Néron models for elliptic curves. The proof, which closely follows the exposition of Artin [1, §1], is largely scheme-theoretic and is at a more advanced level than the other material in this chapter. The reader who is willing to accept the statement of Theorem 6.1 should read Remarks 6.1.1–6.1.3 and can then proceed to the next section with no loss of continuity.

Theorem 6.1.   *Let R be a Dedekind domain with fraction field K, let E/K be an elliptic curve, let $\mathcal{C}/R$ be a minimal proper regular model for E/K (4.5), and let $\mathcal{E}/R$ be the largest subscheme of $\mathcal{C}/R$ which is smooth over R. Then $\mathcal{E}/R$ is a Néron model for E/K.*

Remark 6.1.1.   The generic fiber of $\mathcal{C}$ is the non-singular curve E, so $\mathcal{C}$ has only finitely many singular fibers. Each fiber of $\mathcal{C}$ consists of one or more irreducible components, possibly with multiplicities (see §4), say

$$\mathcal{C}_{\mathfrak{p}} = \sum_{i=1}^{r_{\mathfrak{p}}} n_{\mathfrak{p}i} F_{\mathfrak{p}i}.$$

Then $\mathcal{E}$ is formed by discarding from $\mathcal{C}$ all $F_{\mathfrak{p}i}$'s with $n_{\mathfrak{p}i} \geq 2$, all singular points on each $F_{\mathfrak{p}i}$, and all points where the $F_{\mathfrak{p}i}$'s intersect one another.

Remark 6.1.2. Continuing with the notation from (6.1.1), the group scheme $\mathcal{E}/R$ comes equipped with an identity element, which is an R-valued point $\sigma_0 \in \mathcal{E}(R)$. The image $\sigma_0(R)$ of the identity element is a curve on $\mathcal{C}$ which will intersect the fiber $\mathcal{C}_\mathfrak{p}$ at the point $\sigma_0(\mathfrak{p})$. Proposition 4.3 tells us that $\sigma_0(\mathfrak{p})$ will be a non-singular point of $\mathcal{C}_\mathfrak{p}$, so it will lie on an $F_{\mathfrak{p}i}$ having multiplicity $n_{\mathfrak{p}i} = 1$. The component of $\mathcal{C}_\mathfrak{p}$ containing $\sigma_0(\mathfrak{p})$ is called the *identity component of* $\mathcal{C}_\mathfrak{p}$. The scheme obtained by removing all non-identity components from the fibers of $\mathcal{C}$ is called the *connected component (of the identity) of* $\mathcal{C}$. The image $\sigma_0(R)$ of the identity element lies in $\mathcal{E}$, so we can define the *identity component of* $\mathcal{E}_\mathfrak{p}$ and the *connected component (of the identity) of* $\mathcal{E}$ in an analogous manner.. The connected component of $\mathcal{E}$ is a subgroup scheme of $\mathcal{E}$; see exercise 4.25. We will see later (9.1) that it is isomorphic to the smooth part of the scheme defined by a minimal Weierstrass equation for E.

Remark 6.1.3. With notation as in (6.1), one can prove that the group law $\mathcal{E} \times_R \mathcal{E} \to \mathcal{E}$ extends to give a group scheme action

$$\mathcal{E} \times_R \mathcal{C} \longrightarrow \mathcal{C}.$$

See exercise 4.23. We proved a special case of this in (5.3c).

Before beginning the proof of (6.1), we want to say a few words explaining why the smooth part of a minimal proper regular model for E/K turns out to be a Néron model. In other words, how will we use the four properties "smooth," "minimal," "proper," and "regular"? First, the properness of $\mathcal{C}$ over R ensures that $E(K) = \mathcal{C}(R)$ (4.4a). Next, the regularity of $\mathcal{C}$ tells us that every R-point lies in the smooth part of $\mathcal{C}$ (4.4b), so $\mathcal{C}(R) = \mathcal{E}(R)$. This gives $E(K) = \mathcal{E}(R)$, which is an important case of the Néron mapping property (5.1.3). Thus the properness and regularity of $\mathcal{C}$ are mainly used to obtain the Néron mapping property. On the other hand, the smoothness and minimality of $\mathcal{E}$ are used to prove that $\mathcal{E}$ is a group scheme over R. In particular, the minimality implies that any K-automorphism of E, such as a translation-by-P map for some point $P \in E(K)$, will extend to give an R-automorphism of $\mathcal{E}$. These translation maps on $\mathcal{E}$ will be essential for showing that the group law on E/K extends to give $\mathcal{E}$ the structure of a group scheme over R.

With these preliminary comments completed, we begin the proof of Theorem 6.1. The first step in the proof is the following generalization of a theorem of Weil [3]. Weil's theorem asserts that a rational map from a smooth variety to a complete group variety is automatically a morphism, and Artin [1, Prop. 1.3] has extended Weil's theorem to a scheme-theoretic setting.

Proposition 6.2. (Weil [3], Artin [1]) *Let R be a Dedekind domain, let G/R be a group scheme over R, let $\mathcal{X}/R$ be a smooth R-scheme, and* *let $\phi : \mathcal{X} \dashrightarrow G$ be a rational map over R. Write $\mathrm{Dom}(\phi)$ for the domain of ϕ, and suppose that $\mathrm{Dom}(\phi)$ is dense in every fiber of $\mathcal{X}/R$.*

(a) The complement $\mathcal{X} \setminus \mathrm{Dom}(\phi)$ is a subscheme of $\mathcal{X}$ of pure codimension one.

(b) If G is proper over R, then $\mathrm{Dom}(\phi) = \mathcal{X}$. In other words, ϕ is a morphism.

PROOF. We will write $\mu(g, h) = gh$ and $i(g) = g^{-1}$. Further, to simplify our exposition and help reveal the underlying ideas, we will phrase our argument in terms of points. But the reader should be aware that in order to be completely rigorous, our "points" should be T-valued points for arbitrary R-schemes T. As an alternative, the proof can be given purely scheme-theoretically, a task which we will leave for the reader.

Having made this disclaimer, we begin by considering the rational map

$$F : \mathcal{X} \times_R \mathcal{X} \dashrightarrow G, \qquad F(x, y) = \phi(x)\phi(y)^{-1}.$$

We claim that there is a natural identification

$$\mathrm{Dom}(\phi) \xleftarrow{\sim} \Delta \cap \mathrm{Dom}(F), \qquad x \longleftrightarrow (x, x),$$

where Δ is the diagonal in $\mathcal{X} \times_R \mathcal{X}$. To see this, take a point $x \in \mathrm{Dom}(\phi)$. Then $F(x, x) = \phi(x)\phi(x)^{-1}$ is defined, so $(x, x) \in \mathrm{Dom}(F)$. Conversely, if $(x, x) \in \mathrm{Dom}(F)$, we can use the fact that $\mathrm{Dom}(F)$ is open to find a non-empty open set $U \subset \mathcal{X}$ so that $x \times_R U \subset \mathrm{Dom}(F)$. Next, since $\mathrm{Dom}(\phi)$ is open, we can find a point $y \in U \cap \mathrm{Dom}(\phi)$. It then follows from

$$\phi(x) = F(x, y)\phi(y)$$

that ϕ is defined at x, so $x \in \mathrm{Dom}(\phi)$. This completes the proof of the claim.

Let $K(\mathcal{X} \times \mathcal{X})$ be the function field of the scheme $\mathcal{X} \times_R \mathcal{X}$, and let $\mathcal{O}_{G,0}$ be the local ring of G along the identity section; that is, $\mathcal{O}_{G,0}$ is the ring of rational functions on G which are well-defined at some point on the image of the map $\sigma_0 : \mathrm{Spec}(R) \to G$, where σ_0 is the identity element of the group scheme G.

The rational map F defines a ring homomorphism

$$F^* : \mathcal{O}_{G,0} \longrightarrow K(\mathcal{X} \times \mathcal{X}), \qquad f \longmapsto f \circ F.$$

Let $f \in \mathcal{O}_{G,0}$. If $x \in \mathrm{Dom}(\phi)$, then $(x, x) \in \mathrm{Dom}(F)$ from the claim proven above, and further $F(x, x) = \phi(x)\phi(x)^{-1}$ is the identity element of G, so $F^*(f)$ is defined at (x, x). Conversely, if $F^*(f) = f \circ F$ is defined at (x, x) for all functions $f \in \mathcal{O}_{G,0}$, then F must be defined at (x, x). This proves that

$$x \in \mathrm{Dom}(\phi) \iff (x, x) \in \mathrm{Dom}(F^*f) \text{ for all } f \in \mathcal{O}_{G,0}$$
$$\iff F^*(\mathcal{O}_{G,0}) \subset \mathcal{O}_{\mathcal{X} \times \mathcal{X}, (x,x)},$$

where $\mathcal{O}_{\mathfrak{X}\times\mathfrak{X},(x,x)} \subset K(\mathfrak{X}\times\mathfrak{X})$ is the local ring of $\mathfrak{X}\times_R\mathfrak{X}$ at (x,x).

The scheme $\mathfrak{X}\times_R\mathfrak{X}$ is smooth over R, so in particular it is normal. This implies that a function $f \in K(\mathfrak{X}\times\mathfrak{X})$ will be defined at (x,x) unless its polar divisor $\mathrm{div}_\infty(f)$ goes through (x,x). (This is a standard property of normal schemes. It follows, for example, from Hartshorne [1, II.6.3A].) In other words, the local ring $\mathcal{O}_{\mathfrak{X}\times\mathfrak{X},(x,x)}$ can be characterized as

$$\mathcal{O}_{\mathfrak{X}\times\mathfrak{X},(x,x)} = \big\{ g \in K(\mathfrak{X}\times\mathfrak{X})^* : (x,x) \notin \mathrm{div}_\infty(g) \big\} \cup \{0\}.$$

Combining this with our description of the domain of ϕ from above yields

$$\begin{aligned}
\mathfrak{X}\smallsetminus\mathrm{Dom}(\phi) &= \big\{ x \in \mathfrak{X} : F^*(\mathcal{O}_{G,0}) \not\subset \mathcal{O}_{\mathfrak{X}\times\mathfrak{X},(x,x)} \big\} \\
&= \big\{ x \in \mathfrak{X} : (x,x) \in \mathrm{div}_\infty(F^*f) \text{ for some } f \in \mathcal{O}_{G,0} \big\} \\
&\cong \Delta \cap \bigcup_{f\in\mathcal{O}_{G,0}} \mathrm{div}_\infty(F^*f) \\
&= \bigcup_{f\in\mathcal{O}_{G,0}} \big(\Delta \cap \mathrm{div}_\infty(F^*f) \big).
\end{aligned}$$

The diagonal Δ is a complete intersection in $\mathfrak{X}\times_R\mathfrak{X}$, and each divisor $\mathrm{div}_\infty(F^*f)$ has pure codimension one in $\mathfrak{X}\times_R\mathfrak{X}$, so each of the intersections $\Delta \cap \mathrm{div}_\infty(F^*f)$ has pure codimension one in Δ. It follows that the union over $f \in \mathcal{O}_{G,0}$ also has pure codimension one in Δ, since we know a priori that it is a proper closed subset of Δ. This completes the proof of (a).

(b) The following lemma (6.2.1) says that a rational map from a smooth scheme to a proper scheme is defined off of a subset of codimension at least two. Then (6.2.1) and (a) imply (b).

Lemma 6.2.1. *Let R be a Dedekind domain, let $\mathfrak{X}/R$ be a smooth R-scheme, let $\mathcal{Y}/R$ be a proper R-scheme, and let $\phi : \mathfrak{X} \dashrightarrow \mathcal{Y}$ be a dominant rational map defined over R. Then every component of $\mathfrak{X}\smallsetminus\mathrm{Dom}(\phi)$ has codimension at least two in $\mathfrak{X}$.*

PROOF. Let $\mathcal{Z} \subset \mathfrak{X}$ be an irreducible subscheme of codimension one in $\mathfrak{X}$. We need to show that ϕ is defined at the generic point of $\mathcal{Z}$ (i.e., ϕ is defined on a non-empty open subset of $\mathcal{Z}$). Consider the local ring $\mathcal{O}_{\mathfrak{X},\mathcal{Z}}$ of $\mathfrak{X}$ at $\mathcal{Z}$. It is a local ring of dimension one, and it is regular since $\mathfrak{X}/R$ is smooth, so it is a discrete valuation ring.

The dominant rational map $\phi : \mathfrak{X} \dashrightarrow \mathcal{Y}$ induces a morphism

$$\mathrm{Spec}\, K(\mathfrak{X}) \to \mathrm{Spec}\, K(\mathcal{Y})$$

from the generic point of $\mathfrak{X}$ to the generic point of $\mathcal{Y}$. In other words, composition with ϕ induces an inclusion of function fields $K(\mathcal{Y}) \hookrightarrow K(\mathfrak{X})$.

This gives us the commutative diagram

$$
\begin{array}{ccc}
\mathcal{X} & \overset{\phi}{\dashrightarrow} & \mathcal{Y} \\
\uparrow & & \\
\operatorname{Spec} \mathcal{O}_{\mathcal{X},\mathcal{Z}} & & \uparrow \\
\uparrow & & \\
\operatorname{Spec} K(\mathcal{X}) & \longrightarrow & \operatorname{Spec} K(\mathcal{Y}).
\end{array}
$$

The discrete valuation ring $\mathcal{O}_{\mathcal{X},\mathcal{Z}}$ has fraction field $K(\mathcal{X})$, and we are given that $\mathcal{Y}$ is proper over R, so the valuative criterion of properness (2.7) implies that the rational map

$$\operatorname{Spec} \mathcal{O}_{\mathcal{X},\mathcal{Z}} \longrightarrow \mathcal{X} \overset{\phi}{\dashrightarrow} \mathcal{Y}$$

extends to a morphism $\operatorname{Spec} \mathcal{O}_{\mathcal{X},\mathcal{Z}} \to \mathcal{Y}$. This says precisely that ϕ is defined at the generic point of $\mathcal{Z}$, which completes the proof of (6.2.1), and with it also the proof of (6.2b).   □

We can use (6.2) to find the Néron model of an elliptic curve with good reduction.

Corollary 6.3.   *Let R be a Dedekind domain with fraction field K, let E/K be an elliptic curve given by a Weierstrass equation*

$$y^2 + a_1 xy + a_3 y = x^3 + a_2 x^2 + a_4 x + a_6$$

having coefficients in R, and let $\mathcal{W} \subset \mathbb{P}^2_R$ be the closed subscheme of $\mathbb{P}^2_R$ defined by this Weierstrass equation. Suppose that $\mathcal{W}$ is smooth over R or, equivalently, that the Weierstrass equation has good reduction at every prime of R. Then $\mathcal{W}/R$ is a Néron model for E/K.

PROOF. Theorem 5.3 says that the addition law on E/K extends to make $\mathcal{W}$ into a group scheme over the localization of R at each of its prime ideals. These group laws are given by the same equations, so they fit together to make $\mathcal{W}$ into a group scheme over R. It remains to verify that $\mathcal{W}$ has the Néron mapping property.

Let $\mathcal{X}/R$ be a smooth R-scheme with generic fiber X/K, take any rational map $\phi_K : X_{/K} \dashrightarrow E_{/K}$ defined over K, and let $\phi : \mathcal{X} \dashrightarrow \mathcal{W}$ be the associated rational map over R. The fact that $\mathcal{W}$ is a closed subscheme of $\mathbb{P}^2_R$ implies that it is proper over R (2.8), so we can use (6.2b) to deduce that the rational map ϕ extends to a morphism. This proves that $\mathcal{W}/R$ has the Néron mapping property, so it is a Néron model for E/K.   □

Our next step is to prove that if the scheme $\mathcal{E}$ in (6.1) is a group scheme, then it will be a Néron model for E, at least provided that the ring R is large enough. The precise property we will require R to have is described in the following definition.

Definition. A discrete valuation ring R is called *Henselian* if it satisfies Hensel's lemma; that is, R is Henselian if for any monic polynomial $f(x) \in R[x]$ and any element $a \in R$ satisfying

$$f(a) \equiv 0 \, (\mathrm{mod} \, \mathfrak{p}) \qquad \text{and} \qquad f'(a) \not\equiv 0 \, (\mathrm{mod} \, \mathfrak{p}),$$

there exists a unique element $\alpha \in R$ satisfying

$$\alpha \equiv a \, (\mathrm{mod} \, \mathfrak{p}) \qquad \text{and} \qquad f(\alpha) = 0.$$

The ring R is called *strictly Henselian* if it is Henselian and if its residue field $k = R/\mathfrak{p}$ is algebraically closed. (Remember our residue fields are perfect. The usual definition of strictly Henselian requires k to be separably closed.)

For example, if R is a discrete valuation ring, then the completion of R with respect to its maximal ideal $\mathfrak{p}$ is Henselian. We have seen many instances in which it is helpful to work with complete discrete valuation rings, for example in our study of formal groups [AEC, IV §6] and the reduction theory of elliptic curves [AEC, VII §2]. In particular, we used Hensel's lemma for complete discrete valuation rings to prove the surjectivity of the reduction map $E_0(K) \to \tilde{E}_{\mathrm{ns}}(k)$ in [AEC, VII.2.1].

However, for many purposes the completion is too large, since the completion of R will generally not be flat over R. The following generalization of [AEC, VII.2.1] says that the reduction map is surjective for any Henselian ring.

Proposition 6.4. *Let R be a discrete valuation ring with maximal ideal $\mathfrak{p}$ and residue field k, let $\mathcal{X}/R$ be a smooth R-scheme, and let $\tilde{\mathcal{X}}/k$ be its special fiber. Consider the reduction map*

$$\mathcal{X}(R) \longrightarrow \tilde{\mathcal{X}}(k).$$

(a) *If R is Henselian, then the reduction map is surjective.*
(b) *If R is strictly Henselian, then the image of the reduction map is dense in $\tilde{\mathcal{X}}$.*

PROOF. (a) Replacing $\mathcal{X}$ by an affine neighborhood, we can assume that

$$\mathcal{X} = \operatorname{Spec} A \quad \text{with} \quad A = R[t_1, \ldots, t_{n+r}]/(f_1, \ldots, f_n)$$

for certain polynomials $f_1, \ldots, f_n \in R[t_1, \ldots, t_{n+r}]$. Further, the assumption that $\mathcal{X}$ is smooth over R means that the $n \times n$ minors of the Jacobian matrix

$$J = \left(\frac{\partial f_i}{\partial t_j} \right)_{\substack{1 \le i \le n \\ 1 \le j \le n+r}}$$

generate the unit ideal in A.

Choose any point $\tilde{b} = (\tilde{b}_1, \ldots, \tilde{b}_n) \in \tilde{\mathfrak{X}}(k)$; that is, let $b \in \mathbb{A}^n(R)$ satisfy

$$f_1(b) \equiv \cdots \equiv f_n(b) \equiv 0 \pmod{\mathfrak{p}}.$$

We need to construct a point $\beta \in \mathbb{A}^n(R)$ with the property that

$$\beta \equiv b \,(\mathrm{mod}\,\mathfrak{p}) \qquad \text{and} \qquad f_1(\beta) = \cdots = f_n(\beta) = 0.$$

For example, suppose that $\mathfrak{X}$ is locally a hypersurface (i.e., $m = 1$), given by the single equation $f(t_1, \ldots, t_{n+r}) = 0$. Then the Jacobian condition says that the partial derivatives $\partial f / \partial t_i$ generate A, so in particular one of the values $(\partial f / \partial t_i)(b)$ must be a unit in A, say $(\partial f / \partial t_1)(b) \in A^*$. Thus the polynomial $F(t) = f(t, b_2, \ldots, b_{n+r})$ satisfies the hypotheses of Hensel's lemma, so it has a root $\beta_1 \in R$ with $\beta_1 \equiv b_1 \,(\mathrm{mod}\,\mathfrak{p})$. The point $\beta = (\beta_1, b_2, \ldots, b_n) \in \mathfrak{X}(R)$ reduces modulo $\mathfrak{p}$ to the original point $\tilde{b}$ in $\tilde{\mathfrak{X}}(k)$, which completes the proof in this case. We will leave the general case, which is somewhat more difficult, for the reader to complete, or see the references for exercise 4.27.

(b) The residue field k is algebraically closed, so it is clear that $\mathfrak{X}_{\mathfrak{p}}(k)$ is dense in $\mathfrak{X}_{\mathfrak{p}}$. Now (b) follows from (a). $\qquad\square$

As the next proposition explains, every discrete valuation ring R can be embedded in a minimal (strictly) Henselian ring.

Proposition 6.5. *Let R be a discrete valuation ring with maximal ideal $\mathfrak{p}$, residue field k, and fraction field K. Let K^s be a separable closure for K, let R^s be the integral closure of R in K^s, and choose an ideal $\mathfrak{p}^s$ of K^s lying above $\mathfrak{p}$. Let*

$$D = \left\{ \sigma \in G_{K^s/K} : \sigma(\mathfrak{p}^s) = \mathfrak{p}^s \right\},$$
$$I = \left\{ \sigma \in D : \sigma(x) - x \in \mathfrak{p}^s \text{ for all } x \in R^s \right\}$$

be the associated decomposition and inertia groups.

(a) Let $R^s(D)$ denote the subring of R^s fixed by D, and define R^{h} to be the localization of $R^s(D)$ at the maximal ideal $\mathfrak{p}^s \cap R^s(D)$. Then R^{h} is Henselian. It is called the Henselization of R.

(b) Let $R^s(I)$ denote the subring of R^s fixed by I, and define R^{sh} to be the localization of $R^s(I)$ at the maximal ideal $\mathfrak{p}^s \cap R^s(I)$. Then R^{sh} is strictly Henselian. It is called the strict Henselization of R.

(c) With the obvious notation, we have

$$\mathfrak{p}^{\mathrm{h}} = \mathfrak{p}R^{\mathrm{h}}, \quad k^{\mathrm{h}} = k, \quad \mathfrak{p}^{\mathrm{sh}} = \mathfrak{p}R^{\mathrm{sh}}, \quad k^{\mathrm{sh}} = \bar{k}.$$

Further the natural map

$$G_{K^{\mathrm{sh}}/K^{\mathrm{h}}} \longrightarrow G_{\bar{k}/k}$$

of Galois groups is an isomorphism.

PROOF. (a) We are going to verify that R^h has the Henselian property. Let $f(x) \in R^h[x]$ be a monic polynomial, which we may assume to be irreducible and separable over K^h. (If f is inseparable, then $f'(x)$ is identically 0, so the Henselian property is vacuously true!) Factor f over K^s as

$$f(x) = (x - \alpha_1)(x - \alpha_2) \cdots (x - \alpha_d).$$

Suppose that $a \in R^h$ satisfies $f(a) \in \mathfrak{p}^h$ and $f'(a) \notin \mathfrak{p}^h$. This implies that there is exactly one root, say $\alpha = \alpha_i$, with the property that $a - \alpha \in \mathfrak{p}^s$. Further, if $\sigma \in D$ is any element of the decomposition group of $\mathfrak{p}^s$, then we have

$$a - \sigma(\alpha) = \sigma(a - \alpha) \in \sigma(\mathfrak{p}^s) = \mathfrak{p}^s.$$

But there is exactly one root of f which is congruent to a, so $\sigma(\alpha)$ must equal α. This shows that α is fixed by D, and hence $\alpha \in R^h$.

(b) The proof that R^{sh} is Henselian is similar to the proof of (a), and it is a standard fact that the residue field is an algebraic closure of k; see for example Serre [4, I §7]. (Remember we are assuming that k is perfect.)

(c) Again these are standard properties of Galois extensions of local fields.

$\square$

Remark 6.6.1. It is clear from the construction (6.5) that the fraction fields K^h and K^{sh} are separable algebraic extensions of K. Note that in general the fraction field of the completion of R will be transcendental over K, in fact, of infinite transcendence degree. This is one reason why it is often better to work with the Henselization. The moral is that one should work with Henselizations if one only needs to solve polynomial equations, but one has to go to the completion in order to use convergent power series. We also mention that strictly Henselian rings play the same role for the étale topology that local rings play for the Zariski topology.

Remark 6.6.2. The Henselization can also be described in terms of a universal mapping property, which essentially says that it is the smallest Henselian extension of R, and similarly for the strict Henselization. See exercise 6.28.

Proposition 6.7. *Let R be a strictly Henselian discrete valuation ring with fraction field K, let E/K be an elliptic curve, let $\mathcal{C}/R$ be a minimal proper regular model for E/K (4.5), and let $\mathcal{E}/R$ be the largest subscheme of $\mathcal{C}/R$ which is smooth over R (6.1.1). If the group law on E/K extends to make $\mathcal{E}$ into a group scheme over R, then $\mathcal{E}/R$ is a Néron model for E/K.*

PROOF. Let $\mathcal{X}/R$ be a smooth R-scheme with generic fiber X/K, and let $\phi_K : X \dashrightarrow E$ be a rational map. In order to verify the Néron mapping property, we must show that ϕ_K extends to a morphism $\mathcal{X} \to \mathcal{E}$.

The elliptic curve E is a proper group scheme over K, and X is smooth over K, so applying (6.2b) with $R = K$ to the map $\phi_K : X \dashrightarrow E$, we see that ϕ_K extends to a morphism. This means that the rational map

$$\phi : X \dashrightarrow \mathcal{E}$$

induced by ϕ_K is a morphism on the generic fiber.

We suppose that ϕ is not a morphism and derive a contradiction. By assumption, $\mathcal{E}$ is a group scheme, so (6.2a) tells us that the set of points where ϕ is not defined is a set of pure codimension one in X. Hence there is an irreducible closed subscheme $Z \subset X$ such that ϕ is not defined at the generic point η_Z of Z. Note that the generic point of Z is given by $\eta_Z = \operatorname{Spec} \mathcal{O}_{X,Z}$, and that the local ring $\mathcal{O}_{X,Z}$ is a discrete valuation ring because X is regular and Z has codimension one. We now have the following picture:

$$
\begin{array}{ccccc}
X & \overset{\phi}{\dashrightarrow} & \mathcal{E} & \subset & \mathcal{C} \\
\uparrow & & & & \\
\eta_Z = \operatorname{Spec} \mathcal{O}_{X,Z} & & \uparrow & & \uparrow \\
\uparrow & & & & \\
\operatorname{Spec} K(X) & \overset{\phi_K}{\longrightarrow} & \operatorname{Spec} K(\mathcal{E}) & = & \operatorname{Spec} K(\mathcal{C}).
\end{array}
$$

The scheme $\mathcal{C}$ is proper over R, and $\mathcal{O}_{X,Z}$ is a discrete valuation ring, so the valuative criterion of properness (2.7) says that ϕ extends to a morphism $\phi : \eta_Z \to \mathcal{C}$. In other words, if we map to $\mathcal{C}$ rather than to the smaller scheme $\mathcal{E}$, then ϕ is defined generically on Z.

We are assuming that $\phi : X \dashrightarrow \mathcal{E}$ does not extend generically to Z, or equivalently that $\phi(\eta_Z) \in \mathcal{C}$ is not contained in $\mathcal{E}$. In particular, if we let k be the residue field of R and take any point $x_0 \in Z(k)$ so that $\phi : X \dashrightarrow \mathcal{C}$ is defined at x_0, then $\phi(x_0) \notin \mathcal{E}$. (This is another way of saying that $\phi : X \dashrightarrow \mathcal{E}$ is not defined generically on Z.)

The set of R-valued points $X(R)$ maps to a dense set of points in the special fiber of X by (6.4b). Note that this is where we use our assumption that R is strictly Henselian. In particular we can find a point $x \in X(R)$ which intersects Z at a point, call it $x_0 \in Z(k)$, at which the map $\phi : X \dashrightarrow \mathcal{C}$ is defined. Composing x with ϕ gives a rational map

$$\operatorname{Spec}(R) \overset{x}{\longrightarrow} X \overset{\phi}{\dashrightarrow} \mathcal{C}$$

which by the valuative criterion of properness (2.7) extends to a morphism $\operatorname{Spec}(R) \to \mathcal{C}$. In other words $\phi \circ x \in \mathcal{C}(R)$, and by our construction it is clear that $\phi \circ x \notin \mathcal{E}(R)$. However, (4.4b) says that $\mathcal{C}(R) = \mathcal{E}(R)$. This contradiction completes the proof that ϕ extends to a morphism $X \to \mathcal{E}$, and hence that $\mathcal{E}$ has the Néron mapping property. $\qquad\square$

In order to complete the proof of Theorem 6.1 for strictly Henselian rings, it remains to show that the scheme $\mathcal{E}$ is a group scheme over R. This is done in two steps. First, we prove that there exists some group scheme $\mathcal{A}$ over R such that $\mathcal{E}$ and $\mathcal{A}$ have isomorphic dense open subsets. Second, using the group operation on $\mathcal{A}$, we show that $\mathcal{E}$ and $\mathcal{A}$ must be isomorphic. The proof of the first part uses an argument of Weil to construct a group variety (or scheme) by pasting together group chunks. We will not give the full proof of Weil's result but will be content to give a brief sketch and refer the reader to Artin [1, §2] for the details.

Definition. Let R be a Dedekind domain, let $\mathcal{V}/R$ be a smooth R-scheme with non-empty fibers, and let

$$\mu : \mathcal{V} \times_R \mathcal{V} \dashrightarrow \mathcal{V}$$

be a rational map defined over R. The map μ is called a *normal law on* $\mathcal{V}$ if it satisfies the following two conditions:

(i) The map μ is associative; that is,

$$\mu\big(\mu(x,y),z\big) = \mu\big(x,\mu(y,z)\big) \quad \text{whenver both sides are defined.}$$

(ii) Define rational maps

$$\phi : \mathcal{V} \times_R \mathcal{V} \dashrightarrow \mathcal{V} \times_R \mathcal{V} \qquad \psi : \mathcal{V} \times_R \mathcal{V} \dashrightarrow \mathcal{V} \times_R \mathcal{V}$$
$$(x,y) \longmapsto (x, \mu(x,y)), \qquad (x,y) \longmapsto (y, \mu(x,y)).$$

Then the domains of definition of ϕ and ψ contain a dense subset of each fiber of $\mathcal{V} \times_R \mathcal{V}$, and the restriction of ϕ and ψ to each fiber is a birational isomorphism.

Remark 6.8. If G is a group scheme over a Dedekind domain R, then its group law is a normal law. Condition (i) is true because the group law on G is associative by definition, and condition (ii) is immediate since ϕ and ψ are isomorphisms. For example, the inverse of ψ is the map

$$G \times_R G \longrightarrow G \times_R G, \qquad (x,y) \longmapsto \big((\mu(y,i(x)),x\big).$$

For a general normal law μ, the requirement that the rational maps ϕ and ψ satisfy condition (ii) provides a sort of inverse for the hidden group law that μ is trying to emulate.

The following theorem says that a normal group law on $\mathcal{V}$ makes a large chunk of $\mathcal{V}$ into a large chunk of a group scheme.

Theorem 6.9. (Weil [3]) *Let R be a Dedekind domain, and let $\mathcal{V}/R$ be a smooth R-scheme of finite type over R. Suppose that every fiber of $\mathcal{V}$ is non-empty and that $\mathcal{V}(R)$ is dense in each fiber.*

Let μ be a normal law on $\mathcal{V}$. Then there exists a group scheme G/R of finite type over R, an open subscheme $U/R \subset \mathcal{\dot{V}}/R$, and an open subscheme $U'/R \subset G/R$ with the following two properties:

(i) U and U' are dense in every fiber.

(ii) There is an R-isomorphism $U \cong U'$ so that the normal law μ restricted to U coincides with the group law of G restricted to U'.

PROOF. (Sketch) The underlying idea is to start with a good open subscheme U of $\mathcal{V}$ and construct G as a union of translates of U. More precisely, for each $x \in U(R)$, let U_x be a copy of U. Then one treats U_x as if it were "U translated by x" and uses μ to provide gluing data to attach U_x to U. For further details, see Artin [1, Thm. 1.12]. $\square$

The next result, combined with (6.7), will complete the proof that $\mathcal{E}/R$ is a Néron model for E/K, at least over strictly Henselian rings.

Proposition 6.10. *Let R be a strictly Henselian discrete valuation ring with fraction field K, let E/K be an elliptic curve, let $\mathcal{C}/R$ be a minimal proper regular model for E/K (4.5), and let $\mathcal{E}/R$ be the largest subscheme of $\mathcal{C}/R$ which is smooth over R (6.1.1). Then the group law on E extends to make $\mathcal{E}$ into a group scheme over R.*

PROOF. The proof consists of two steps. First, we verify that the group law on E defines a normal law on $\mathcal{E}$. This allows us to apply (6.9), which yields a group scheme G/R that is birational to $\mathcal{E}$. The second step is to show that the resulting birational map $\mathcal{E} \to G$ is actually an isomorphism. Notice that the proof is somewhat indirect. Rather than proving that the group law on E extends to $\mathcal{E}$, we instead construct an auxiliary group scheme G which extends the group law on E, and then we show that G must equal $\mathcal{E}$.

We begin with the assertion that the group law on E defines a normal law

$$\mu : \mathcal{E} \times_R \mathcal{E} \dashrightarrow \mathcal{E}.$$

The associativity of μ is clear, since μ is associative on the generic fiber of $\mathcal{E}$, and a rational map is determined by its restriction to any dense open subset. So it remains to verify that the maps

$$\phi : \mathcal{E} \times_R \mathcal{E} \dashrightarrow \mathcal{E} \times_R \mathcal{E} \qquad \psi : \mathcal{E} \times_R \mathcal{E} \dashrightarrow \mathcal{E} \times_R \mathcal{E}$$
$$(x, y) \longmapsto (x, \mu(x, y)), \qquad (x, y) \longmapsto (y, \mu(x, y)).$$

are defined on a dense subset of the special fiber and that their restrictions to the special fiber are birational isomorphisms.

Suppose that R' is any discrete valuation ring which is the localization of a smooth R-scheme, and let K' be the fraction field of R'. Then the minimal proper regular model of E over the field K' is $\mathcal{C} \times_R \operatorname{Spec}(R')$. We apply this fact, taking R' to be the ring

$$R' = \mathcal{O}_{\mathcal{E},\xi} = \begin{pmatrix} \text{the local ring of } \mathcal{E} \text{ at a generic} \\ \text{point } \xi \text{ of its special fiber } \tilde{\mathcal{E}} \end{pmatrix}.$$

Note that the special fiber $\tilde{\mathcal{E}}$ need not be irreducible, so it may have several generic points, one for each component. We can take R' to be the local ring at any one of these generic points. Note also that here is where we use the fact that $\mathcal{E}$ is smooth over R, since this fact implies that R' is the localization of a smooth R-scheme.

There is a natural map $\operatorname{Spec}(R') = \xi \to \mathcal{E}$, in other words an R'-valued point of $\mathcal{E}$. We let τ be the corresponding translation map,

$$\tau : \operatorname{Spec}(R') \times_R \mathcal{E} \longrightarrow \operatorname{Spec}(R') \times_R \mathcal{E}.$$

This translation map is an automorphism on the generic fiber, so it follows from (4.6) and the minimality of $\mathcal{C} \times_R \operatorname{Spec}(R')$ that τ is actually a morphism.

The map τ on $\operatorname{Spec}(R') \times_R \mathcal{E}$ is translation on the second factor by the generic point of the first factor, so we obtain a commutative diagram

$$
\begin{array}{ccc}
\operatorname{Spec}(R') \times_R \mathcal{E} & \overset{\tau}{\longrightarrow} & \operatorname{Spec}(R') \times_R \mathcal{E} \\
\downarrow & & \downarrow \\
\mathcal{E} \times_R \mathcal{E} & \overset{\phi}{\longrightarrow} & \mathcal{E} \times_R \mathcal{E}.
\end{array}
$$

This proves that ϕ is defined at every generic point of the special fiber of $\mathcal{E} \times_R \mathcal{E}$ lying over ξ. But ξ is an arbitrary generic point of the special fiber $\tilde{\mathcal{E}}$, which implies that ϕ is defined at every generic point of the special fiber of $\mathcal{E} \times_R \mathcal{E}$. A similar argument can be applied to ψ, which proves that the domains of definition of ϕ and ψ contain a dense subset of the special fiber of $\mathcal{E} \times_R \mathcal{E}$. This verifies the first part of property (ii) in the definition of a normal law.

Now take any point $P \in \mathcal{E}(R)$, and let $\tau_P : \mathcal{E} \to \mathcal{E}$ be the automorphism of $\mathcal{E}$ extending the translation-by-P morphism on the generic fiber of E (4.6). Then the map

$$\mathcal{E} = \operatorname{Spec}(R) \times_R \mathcal{E} \overset{P \times 1}{\longrightarrow} \mathcal{E} \times_R \mathcal{E} \overset{\phi}{\longrightarrow} \mathcal{E} \times_R \mathcal{E}$$

is precisely the map $P \times \tau_P$, so it is one-to-one. It follows that the fibers of ϕ are not all positive dimensional. Therefore ϕ must be generically surjective, and hence a birational isomorphism on the closed fiber of $\mathcal{E} \times_R \mathcal{E}$. Similarly for ψ, which completes the verification that μ is a normal law on $\mathcal{E}$.

We can now apply (6.9) to deduce the existence of a group scheme G/R, open subschemes $U/R \subset \mathcal{E}$ and $U'/R \subset G/R$ which are dense in every fiber, and an R-isomorphism $U \cong U'$ so that the restriction of μ to U coincides with the group law of G restricted to U'. The isomorphism $U \cong U'$ extends to give an R-birational map $\lambda : G \to \mathcal{E}$, and the proof of (6.10) will be complete if we can show that λ is an isomorphism.

As described in (3.3), any point $P \in G(R)$ defines a translation morphism $\tau_P : G \to G$. The generic fiber of G is E/K, so P gives a point in $E(K)$. Translation-by-P on E/K is an automorphism of the generic fiber of $\mathcal{E}$. It follows from (4.6) and the minimality of $\mathcal{E}$ that this induces a translation map on $\mathcal{E}$, which we will also denote by $\tau_P : \mathcal{E} \to \mathcal{E}$. Now if $g \in G$ is any point, then we can find a $P \in G(R)$ so that $\tau_P(g) \in U'$. (Note R is strictly Henselian, so $G(R)$ maps to a dense subset of the special fiber of G from (6.4b).) Then we can extend the definition of λ to a neighborhood of g by using the fact that $\lambda = \tau_{-P} \circ \lambda \circ \tau_P$ at every point where it is defined, since the right-hand side is clearly defined at g. This proves that λ is a morphism.

On the other hand, the map $\lambda^{-1} : \mathcal{E} \to G$ in the opposite direction is a rational map from a scheme smooth over R to a group scheme over R. Suppose that λ^{-1} is not a morphism. Then Weil's theorem (6.2) tells us that there is an irreducible curve $Z \subset \mathcal{E}$ such that λ^{-1} is undefined at the generic point η_Z of Z. But we know that λ^{-1} is defined on U, and U is dense in the special fiber, so Z cannot be a component of the special fiber. It follows that η_Z is contained in the generic fiber E of $\mathcal{E}$. In other words, there is a point of E at which λ^{-1} is not defined. But on the generic fiber, λ is the identity map $E \to E$. This contradiction shows that λ^{-1} is defined everywhere on $\mathcal{E}$. Therefore λ is an isomorphism, and hence $\mathcal{E}$ is a group scheme over R. $\square$

We now have all of the tools needed to prove the existence of Néron models for elliptic curves over Dedekind domains.

PROOF (of Theorem 6.1). If the ring R is strictly Henselian, then combining (6.7) and (6.10) shows that $\mathcal{E}$ is a Néron model for E/K. There are two more steps needed to complete the proof of (6.1). First, we have to descend from the strict Henselization of a discrete valuation ring down to the ring itself. Second, we have to glue together Néron models over discrete valuation rings to create a Néron model over a Dedekind domain.

So suppose first that R is a discrete valuation ring, and let R^{sh} be the strict Henselization of R (6.5). Then $\mathcal{C}^{\mathrm{sh}} = \mathcal{C} \times_R R^{\mathrm{sh}}$ will be a minimal proper regular model for E/R^{sh}. To see this, we note first that proper morphisms are stable under base extension (Hartshorne [1, II.4.8c]), so $\mathcal{C}^{\mathrm{sh}}$ is proper over R^{sh}. Next, the regularity of $\mathcal{C}^{\mathrm{sh}}$ follows, since $\mathcal{C}$ is regular and R^{sh} is flat and unramified over R. Finally, the minimality of $\mathcal{C}^{\mathrm{sh}}$ is a consequence of the construction of the minimal proper regular model in terms of a regular model with all exceptional curves blown down. See Lichtenbaum [1], Shafarevich [1], Chinburg [1], and the discussion in §7 of this chapter.

Letting $\mathcal{E}^{\mathrm{sh}} = \mathcal{E} \times_R R^{\mathrm{sh}}$, it is clear that $\mathcal{E}^{\mathrm{sh}}$ is the largest open sub-scheme of $\mathcal{C}^{\mathrm{sh}}$ which is smooth over R^{sh}, so our previous work implies that $\mathcal{E}^{\mathrm{sh}}/R^{\mathrm{sh}}$ is a Néron model for E/K^{sh}. We are going to use this fact to verify that $\mathcal{E}$ has the Néron mapping property over R.

Let $\mathcal{X}/R$ be a smooth R-scheme with generic fiber X/K, and let $\phi_K : X_{/K} \to E_{/K}$ be a rational map defined over K. Consider the extension of $\mathcal{X}$ and ϕ_K to R^{sh}, say $\mathcal{X}^{\mathrm{sh}} = \mathcal{X} \times_R R^{\mathrm{sh}}$ and $\phi_K^{\mathrm{sh}} : X_{/K^{\mathrm{sh}}}^{\mathrm{sh}} \to E_{/K^{\mathrm{sh}}}$. The scheme $\mathcal{X}^{\mathrm{sh}}$ is smooth over R^{sh} from Hartshorne [1, III.10.1b] or Altman-Kleiman [1, VII.1.7], so the Néron mapping property for $\mathcal{E}^{\mathrm{sh}}$ tells us that ϕ_K^{sh} extends to a unique morphism $\phi_R^{\mathrm{sh}} : \mathcal{X}_{/R^{\mathrm{sh}}}^{\mathrm{sh}} \to \mathcal{E}_{/R^{\mathrm{sh}}}^{\mathrm{sh}}$. This gives us the commutative diagram

$$
\begin{array}{ccc}
\mathcal{X}^{\mathrm{sh}} & \xrightarrow{\phi_R^{\mathrm{sh}}} & \mathcal{E}^{\mathrm{sh}} \\
\downarrow & & \downarrow \\
\mathcal{X} & \xdashrightarrow{\phi_K} & \mathcal{E},
\end{array}
$$

where the top row is obtained from the bottom row using the base extension $\operatorname{Spec} R^{\mathrm{sh}} \to \operatorname{Spec} R$. The strict Henselization R^{sh} is faithfully flat over R (Bosch-Lütkebohmert-Raynaud [1, 2.4, corollary 9]), so this is exactly the situation in which we can apply faithfully flat descent (see, e.g., Bosch-Lütkebohmert-Raynaud [1, Chap. 6] or Milne [4, I §2]) to conclude that the rational map on the bottom row is a morphism. Therefore $\mathcal{E}/R$ has the Néron mapping property, which concludes the proof that $\mathcal{E}/R$ is a Néron model for E/K in the case that R is a discrete valuation ring.

Finally, suppose that R is a Dedekind domain. From what we have already done, we know that for each prime $\mathfrak{p} \in \operatorname{Spec} R$, the localization $\mathcal{E} \times_R R_{\mathfrak{p}}$ is a Néron model for E over $K_{\mathfrak{p}}$. Further, (6.3) tells us that if we fix a Weierstrass equation $\mathcal{W}/R$ for E/K, and if we let $S \subset \operatorname{Spec} R$ be the set of primes for which $\mathcal{W}$ has bad reduction, then $\mathcal{W} \times_R R_S$ (i.e., the part of $\mathcal{W}$ lying over R_S) is a Néron model for E over R_S. This gives the Néron model over a dense open subset of $\operatorname{Spec} R$, and gluing this large piece to the finitely many bad fibers produces a Néron model over all of $\operatorname{Spec} R$. This completes the proof of (6.1). $\square$

§7. Intersection Theory, Minimal Models, and Blowing-Up

In Chapter III we saw amply demonstrated the power of intersection theory as a tool for studying the geometry of surfaces. In this section we will describe, without proof, the analogous theory on arithmetic surfaces due to Lichtenbaum [1] and Shafarevich [1]. Unfortunately, the fact that an arithmetic surface is not complete means that it is not possible to define an intersection theory on the full divisor group, but we will be able to compute intersections with divisors which lie on the special fiber. In the next section we will use intersection theory to completely describe all of the possible special fibers for a minimal proper regular model of an elliptic curve. We will work over a discrete valuation ring, rather than a Dedekind domain, since everything we do in this section can be done fiber-by-fiber.

Let R be a discrete valuation ring with maximal ideal $\mathfrak{p}$ and residue field $k = R/\mathfrak{p}$, and let $\mathcal{C}/R$ be an arithmetic surface over R. The scheme $\mathcal{C}$ is normal by definition, so there is a good theory of Weil divisors on $\mathcal{C}$ as described in Hartshorne [1, II §6]. An irreducible divisor Γ on $\mathcal{C}$ is a closed integral subscheme of dimension one, in other words a curve, and the divisor group $\mathrm{Div}(\mathcal{C})$ of $\mathcal{C}$ is the free abelian group generated by the irreducible divisors. Further, each non-zero function $f \in K(\mathcal{C})$ defines a *principal divisor*

$$\mathrm{div}(f) = \sum_{\Gamma} \mathrm{ord}_\Gamma(f)\Gamma \in \mathrm{Div}(\mathcal{C}),$$

and as usual we say that two divisors are *linearly equivalent* if the difference is principal. Here ord_Γ is the normalized valuation associated to the irreducible divisor Γ; see (4.1.2).

Let $\Gamma \in \mathrm{Div}(\mathcal{C})$ be an irreducible divisor and let $x \in \mathcal{C}_\mathfrak{p}$ be a point on the special fiber of $\mathcal{C}$. Informally, a *uniformizer for* Γ *at* x is a function which vanishes to order 1 along Γ and has no other zeros or poles in a neighborhood of x. More precisely, a uniformizer for Γ at x is a function $f \in \mathcal{O}_{\mathcal{C},x}$ in the local ring of $\mathcal{C}$ at x with the property that

$$\mathrm{ord}_\Gamma(f) = 1, \quad \text{and } \mathrm{ord}_{\Gamma'}(f) = 0 \text{ for all irreducible } \Gamma' \neq \Gamma \text{ with } x \in \Gamma'.$$

To see that such a function exists, we need merely note that if $x \in \Gamma$, then $\mathcal{O}_{\mathcal{C},\Gamma}$ is a discrete valuation ring containing the integrally closed local ring $\mathcal{O}_{\mathcal{C},x}$.

Definition. Let $\Gamma_1, \Gamma_2 \in \mathrm{Div}(\mathcal{C})$ be distinct irreducible divisors and let $x \in \mathcal{C}$ be a closed point on the special fiber $\mathcal{C}_\mathfrak{p}$ of $\mathcal{C}$. Choose uniformizers $f_1, f_2 \in \mathcal{O}_{\mathcal{C},x}$ for Γ_1, Γ_2 respectively. The *(local) intersection index of* Γ_1 *and* Γ_2 *at* x is the quantity

$$(\Gamma_1 \cdot \Gamma_2)_x = \dim_k \mathcal{O}_{\mathcal{C},x}/(f_1, f_2).$$

Notice that this is the same definition that we gave in (III §7) for the local intersection index on geometric surfaces. Just as in Chapter III, we would like to add up these local intersection indices to get a global theory. Further, we would like our intersection theory to have the important functorial properties described in (III.7.2). In particular, linearly equivalent divisors should give the same intersection index. Unfortunately, it is not possible to define an intersection theory on $\mathrm{Div}(\mathcal{C})$ with this property. The problem is that $\mathcal{C}$ is not complete. This is true even if $\mathcal{C}$ is proper over R, which will ensure that the fibers are complete, because the base $\mathrm{Spec}(R)$ itself is not complete. This non-completeness means that it is possible to use a linear equivalence to move an intersection point "out to infinity," where it then disappears. The following simple example illustrates this difficulty.

Example 7.1. Let $\mathcal{C} = \mathbb{P}^1_R = \mathrm{Proj}\,R[X,Y]$, and consider the two divisors

$$\Gamma_1 = \{X = 0\} \qquad \text{and} \qquad \Gamma_2 = \{X + \pi^n Y = 0\}.$$

Here $\pi \in R$ is a uniformizer for the maximal ideal $\mathfrak{p}$ of R, and $n \geq 1$ is an integer. These two divisors intersect at the point

$$x = \{X = \pi = 0\} \in \mathcal{C}_\mathfrak{p} = \mathbb{P}^1_k$$

on the special fiber. To compute the local intersection index we dehomogenize by setting $Y = 1$ and then compute

$$(\Gamma_1 \cdot \Gamma_2)_x = \dim_k R[X]_{(X)}/(X, X + \pi^n) = \dim_k R/(\pi^n) = n.$$

The divisor Γ_2 is linearly equivalent to the divisor Γ_3 defined by

$$\Gamma_3 = \Gamma_2 + \mathrm{div}\left(\frac{X + \pi^n Y}{Y}\right) = \{Y = 0\}.$$

Notice that Γ_1 and Γ_3 have no points in common. So the linear equivalence $\Gamma_2 \sim \Gamma_3$ has caused the intersection point of Γ_1 and Γ_2 to disappear.

The preceding example gives us two options. Either we can drop the requirement that intersections be invariant under linear equivalence, or we can restrict the allowable divisors. As we will see in the next section, it is extremely important to be able to compute the intersection of a divisor with itself, and to do this we need to be able to move the divisor in some way while not changing total the intersection index. So we will adopt the second alternative and restrict the set of divisors.

The irreducible divisors on an arithmetic surface $\mathcal{C}/R$ come in two flavors. First, there are the components of the special fiber, as described in §4. Second, if $\Gamma \subset \mathcal{C}$ is an irreducible divisor which does not lie in the special fiber, then the map $\Gamma \to \mathrm{Spec}(R)$ will be surjective. An irreducible divisor which is a component of the special fiber is called a *fibral divisor*, and an irreducible divisor which maps onto $\mathrm{Spec}(R)$ is called a *horizontal divisor*. For example, the image $\sigma(\mathrm{Spec}\,R)$ of a section $\sigma \in \mathcal{C}(R)$ is a horizontal divisor.

Definition. A divisor $D = \sum n_i \Gamma_i$ is called *fibral* if every component Γ_i of D is a fibral divisor. The *group of fibral divisors on* $\mathcal{C}$ is denoted

$$\mathrm{Div}_\mathfrak{p}(\mathcal{C}) = \{D \in \mathrm{Div}(\mathcal{C}) : D \text{ is fibral}\}.$$

Notice that $\mathrm{Div}_\mathfrak{p}(\mathcal{C})$ is a subgroup of $\mathrm{Div}(\mathcal{C})$.

Definition. A divisor $D = \sum n_i \Gamma_i$ is called *positive* if every $n_i \geq 1$. The set of positive divisors clearly does not form a group.

The fibral divisors are fairly rigid, at least in the sense that their intersections generally cannot be moved off of the special fiber by a linear equivalence. This makes them suitable for intersection theory as described in the following result.

Theorem 7.2. *Let R be a discrete valuation ring, and let $\mathcal{C}/R$ be a regular arithmetic surface which is proper over R. There is a unique bilinear pairing*

$$\mathrm{Div}(\mathcal{C}) \times \mathrm{Div}_\mathfrak{p}(\mathcal{C}) \longrightarrow \mathbb{Z}, \qquad (D, F) \longmapsto D \cdot F,$$

with the following properties.

(i) *If $\Gamma \in \mathrm{Div}(\mathcal{C})$ and $F \in \mathrm{Div}_\mathfrak{p}(\mathcal{C})$ are distinct irreducible divisors, then*

$$\Gamma \cdot F = \sum_{x \in \Gamma \cap F} (\Gamma \cdot F)_x.$$

(ii) *If $D_1, D_2 \in \mathrm{Div}(\mathcal{C})$ and $F \in \mathrm{Div}_\mathfrak{p}(\mathcal{C})$ are divisors with D_1 linearly equivalent to D_2, then $D_1 \cdot F = D_2 \cdot F$. In particular,*

$$\mathrm{div}(f) \cdot F = 0 \qquad \text{for all } f \in K(\mathcal{C})^* \text{ and all } F \in \mathrm{Div}_\mathfrak{p}(\mathcal{C}).$$

The intersection pairing also has the following symmetry property.

(iii) *If $F_1, F_2 \in \mathrm{Div}_\mathfrak{p}(\mathcal{C})$ are fibral divisors, then $F_1 \cdot F_2 = F_2 \cdot F_1$.*

PROOF. Just as in the geometric case (III.7.2), the main idea is to use linearity and the linear equivalence property (ii) to reduce the computation of $D \cdot F$ to the case of distinct irreducible divisors, and then apply (i). The principal difficulty, as always, is to show that the result is independent of the various choices made. For details, see Lichtenbaum [1], Shafarevich [1], Lang [6, III §§2,3], or Chinburg [2, §4]. ☐

Remark 7.2.1. Let $\mathcal{C} = \mathbb{P}^1_R = \mathrm{Proj}\, R[X, Y]$, let $a \in R$ be an element which is not a square in R, and consider the two irreducible divisors

$$\Gamma = \{X^2 = aY^2\} \qquad \text{and} \qquad F = \mathcal{C}_\mathfrak{p} = \mathbb{P}^1_k.$$

Then $\Gamma \cap F$ consists of either one or two points, depending on whether or not a is a square in the residue field k. If $\sqrt{a} \notin k$, then Γ and F intersect at the one point

$$x = \{\pi = X^2 - aY^2 = 0\} \in \mathcal{C}_\mathfrak{p},$$

where π is a uniformizer for R. We compute the intersection index at x by dehomogenizing $Y = 1$,

$$(\Gamma \cdot F)_x = \dim_k R[X]_{(X^2-a)}/(\pi, X^2 - a) = \dim_k k[X]/(X^2 - a) = 2.$$

Similarly, if $\sqrt{a} \in k$, then Γ and F intersect at the two points

$$y = \{\pi = X - \sqrt{a}\, Y = 0\} \qquad \text{and} \qquad z = \{\pi = X + \sqrt{a}\, Y = 0\}.$$

We will leave it to the reader to check that $(\Gamma \cdot F)_y = (\Gamma \cdot F)_z = 1$.

We proved (III.8.2) that the intersection pairing on a fibered surface is negative semi-definite on fibral divisors, with kernel the entire fiber. The argument given in Chapter III carries over almost verbatim to give the same result for arithmetic surfaces.

Proposition 7.3. *Let R be a discrete valuation ring with maximal ideal $\mathfrak{p}$, and let $\mathcal{C}/R$ be a regular arithmetic surface proper over R.*
(a) *The special fiber $\mathcal{C}_\mathfrak{p}$ is connected.*
(b) *Let $F \in \mathrm{Div}_\mathfrak{p}(\mathcal{C})$ be a fibral divisor. Then $F^2 \leq 0$, and the following three conditions are equivalent:*

(i) *$F^2 = 0$.*
(ii) *$F \cdot F' = 0$ for every $F' \in \mathrm{Div}_\mathfrak{p}(\mathcal{C})$.*
(iii) *$F = a\mathcal{C}_\mathfrak{p}$ for some $a \in \mathbb{Q}$, where $\mathcal{C}_\mathfrak{p} = \mathcal{C} \times_R \mathfrak{p}$ is the special fiber of $\mathcal{C}$ with appropriate multiplicities; see §4.*

PROOF. (a) This is a special case of Hartshorne [1, III.11.3].
(b) Clearly, (ii) implies (i). Further, the divisor $\mathcal{C}_\mathfrak{p}$ is principal, since it is equal to $\mathrm{div}(\pi)$ for a uniformizer $\pi \in R$, so (7.2ii) shows that (iii) implies (ii). The fact that $F^2 \leq 0$ and the remaining implication (i) $\Rightarrow$ (iii) are proven in exactly the same way as the geometric case (III.8.2). For further details, see Lichtenbaum [1], Shafarevich [1], or Lang [6, III Prop. 3.5].   □

To simplify our discussion for the remainder of this section, we are going to assume that our discrete valuation ring R has an algebraically closed residue field k. In practice, most of what we say will remain true with some slight modifications.

With this assumption, an irreducible fibral divisor F on $\mathcal{C}$ is an irreducible curve defined over k. If we further assume that $\mathcal{C}$ is proper over R, then F will be proper (hence projective) over k. Recall that the *arithmetic genus* of such a curve F/k is defined to be

$$p_a(F) = \dim_k H^1(F, \mathcal{O}_F).$$

More generally, if we write the special fiber as $\mathcal{C}_\mathfrak{p} = \sum n_i F_i$, then any positive fibral divisor $F = \sum a_i F_i$ is a one-dimensional scheme over k. If F is connected, the arithmetic genus of F is defined by exactly the same formula. See Hartshorne [1, exercise III.5.3] for a discussion of the arithmetic genus. The next proposition describes the few facts that we will need to know about the arithmetic genus.

Proposition 7.4. *Let R be a discrete valuation ring with maximal ideal $\mathfrak{p}$, fraction field K, and algebraically closed residue field k. Let $\mathcal{C}/R$ be a regular arithmetic surface proper over R.*
(a) *(Adjunction Formula) There is a divisor $K_\mathcal{C} \in \mathrm{Div}(\mathcal{C})$ with the property that*

$$F^2 + K_\mathcal{C} \cdot F = 2p_a(F) - 2 \quad \text{for every } F \in \mathrm{Div}_\mathfrak{p}(\mathcal{C}).$$

The divisor $K_{\mathcal{C}}$ is called a canonical divisor on $\mathcal{C}$. (N.B. The adjunction formula is only valid for fibral divisors.)

(b) *Let C/K be the generic fiber of $\mathcal{C}$. Then*

$$p_a(\mathcal{C}_{\mathfrak{p}}) = p_a(C) = g(C),$$

where $g(C)$ is the usual genus of C/K [AEC, II.5.4].

(c) *Let $F \in \mathrm{Div}_{\mathfrak{p}}(\mathcal{C})$ be an irreducible fibral divisor. Then $p_a(F) \geq {}^\cdot 0$, and $p_a(F) = 0$ if and only if F is isomorphic to $\mathbb{P}^1_k$.*

PROOF. (a) The classical adjunction formula for a non-singular curve on a non-singular surface is proven in Hartshorne [1, V.1.5], and the case of singular curves is described in Hartshorne [1, exercise V.1.3]. The adjunction formula for arithmetic surfaces is due to Lichtenbaum [1, Thm. 3.2]. For further information, see also the discussion in Lang [6, remark 1 on p. 117].

(b) The first equality follows from the general fact that in a flat family, the arithmetic genus of the fibers remains constant (Hartshorne [1, III.9.10]). The second equality is Hartshorne [1, IV.1.1], since C is non-singular.

(c) The inequality $p_a(F) \geq 0$ is clear, since by definition the arithmetic genus is the dimension of a certain cohomology group. For the second assertion, see Hartshorne [1, IV exercise 1.8(b)]. $\qquad\square$

Remark 7.4.1. The description of the canonical divisor $K_{\mathcal{C}}$ in (7.4a) is not, of course, the usual definition. A canonical divisor is normally defined to be the divisor of a differential form of top dimension, so in the case of an arithmetic surface $\mathcal{C}$, the divisor of a differential 2-form on $\mathcal{C}$. But for our main application (8.1) in the next section, we will only need to know that there exists some divisor satisfying the adjunction formula (7.4a). In fact, it would suffice to know that the map

$$\mathrm{Div}_{\mathfrak{p}}(\mathcal{C}) \longrightarrow \mathbb{Z}, \qquad F \longmapsto 2p_a(F) - 2 - F^2,$$

is a homomorphism; see Lichtenbaum [1, Thm. 3.2].

Earlier (4.5) we stated the existence of a minimal proper regular model for a curve C/K. We now want to briefly describe how such models are constructed and give Castelnuovo's criterion for minimality. If $\mathcal{C}/R$ is any regular model for C/K, and if $x \in \mathcal{C}_{\mathfrak{p}}$ is a point on the special fiber of $\mathcal{C}$, then we can blow up x to get another regular model $\mathcal{C}'/R$ and a birational morphism

$$\phi : \mathcal{C}' \longrightarrow \mathcal{C}.$$

(See Hartshorne [1, I §4, II §7, V §3] and the discussion (7.7, 7.7.1) at the end of this section.) The map ϕ is an isomorphism away from x, and the inverse image of x is a divisor $D = \phi^{-1}(x)'$ with the property that

$$D \cong \mathbb{P}^1_k \qquad \text{and} \qquad D^2 = -1.$$

Thus blowing up x has the effect of replacing x by a projective line whose self-intersection is -1.

In general, an irreducible fibral divisor $D \in \mathrm{Div}_\mathfrak{p}(\mathcal{C})$ which satisfies $D \cong \mathbb{P}^1_k$ and $D^2 = -1$ is called an *exceptional divisor* or an *exceptional curve*. Castelnuovo showed (in the geometric setting) that such curves can always be blown back down.

Proposition 7.5. (Castelnuovo's criterion) *Let R be a discrete valuation ring with fraction field K and algebraically closed residue field k, and let C/K be a non-singular projective curve of genus $g \geq 1$.*
(a) *Let $\mathcal{C}'/R$ be a proper regular model for C/K (4.5a), and let $D \in \mathrm{Div}_\mathfrak{p}(\mathcal{C}')$ be an exceptional divisor on $\mathcal{C}'$. Then there exists a proper regular model $\mathcal{C}/R$ for C/K and a birational morphism $\phi : \mathcal{C}' \to \mathcal{C}$ so that $x = \phi(D)$ is a single point and ϕ is the blow-up of $\mathcal{C}$ at x.*
(b) *Let $\mathcal{C}$ be a minimal proper regular model for C/K. Then $\mathcal{C}$ contains no exceptional divisors.*

PROOF. (a) See Hartshorne [1, V.5.7] for a proof in the geometric situation. The arithmetic version is due to Lichtenbaum [1, Thm. 3.9] and Shafarevich [1, p. 102], see also Chinburg [2, Thm. 3.1].
(b) If $\mathcal{C}$ contains an exceptional curve, then (a) says that we can blow it down to get a smaller regular model for C/K. But a blow-down map is clearly not an isomorphism, which contradicts the assumed minimality of $\mathcal{C}$. $\square$

Remark 7.5.1. Continuing with the notation from (7.5b), it can be shown that a proper regular model $\mathcal{C}/R$ for C/K is minimal if and only if it contains no exceptional divisors. Further, starting with any proper regular model for $\mathcal{C}/R$, one can produce a minimal model by blowing down exceptional curves until none are left. The geometric case is described in Hartshorne [1, V §§3,5]. The arithmetic case is due to Lichtenbaum [1, Thm. 4.4] and Shafarevich [1, p. 126]; see also Chinburg [2, Thm. 1.2].

Remark 7.6. Let R be a Dedekind domain, say the ring of integers of a number field K, and let $\mathcal{C}/R$ be an arithmetic surface. Then we can define an intersection pairing

$$\mathrm{Div}(\mathcal{C}) \times \mathrm{Div}_{\mathrm{fib}}(\mathcal{C}) \longrightarrow \mathbb{Z},$$

where $\mathrm{Div}_{\mathrm{fib}}(\mathcal{C})$ denotes the group of divisors generated by the components of the special fibers and the pairing is defined linearly using (7.2). Unfortunately, if we want to retain the linear equivalence property (7.2ii), then it still is not possible to extend this pairing to all of $\mathrm{Div}(\mathcal{C})$. As before, the underlying problem is that $\mathrm{Spec}(R)$ is not complete, so intersection points can move out to infinity and disappear. Arakelov [1] had the brilliant idea of adding in some extra fibers "at infinity." More precisely, he adds one fiber for each archimedean absolute value of K, and then uses tools from differential geometry to define real-valued local intersection indices on these archimedean fibers. Arakelov's intersection theory extends to the full divisor group while retaining the linear equivalence property, and many of the most important theorems from the classical geometry of surfaces, such as the Riemann-Roch and adjunction formulas, extend to the Arakelov setting. For more information about Arakelov intersection theory on arithmetic surfaces, see for example Chinburg [1], Faltings [1], or Lang [6].

The final topic we want to discuss in this section is the blowing-up process. This is described with varying degrees of generality in Hartshorne [1, I §4, II §7, V §3]. With an eye towards the explicit computations we will be doing in §9, we offer the following brief primer on blowing-up surface singularities.

Remark 7.7. (Blowing-Up Singularities on Arithmetic Surfaces) Let R be a discrete valuation ring with uniformizing element π and residue field k. Let $\mathcal{C} \subset \mathbb{A}_R^2$ be an arithmetic surface defined by a single equation

$$f(x, y) = 0 \qquad \text{for some polynomial } f(x, y) \in R[x, y].$$

In other words, $\mathcal{C} = \operatorname{Spec} R[x, y]/(f)$. In order to ensure that $\mathcal{C}$ is a two-dimensional scheme whose special fiber has dimension one, we will assume that f is not a constant polynomial and that at least one coefficient of f is a unit in R. In fancier terminology, this means that $\mathcal{C}$ is flat over R.

Keep in mind that $\mathcal{C}$ is a "surface" (i.e., a two-dimensional scheme) sitting inside the three-dimensional scheme $\mathbb{A}_R^2$. Intuitively, the three "coordinate functions" on $\mathbb{A}_R^2$ are π, x, and y, and in order to calculate the special fiber we always set $\pi = 0$.

We are going to assume that $\mathcal{C}$ has a singularity at the point $\pi = x = y = 0$ on the special fiber. In other words, we assume that

$$f(0, 0) \equiv \frac{\partial f}{\partial x}(0, 0) \equiv \frac{\partial f}{\partial y}(0, 0) \equiv 0 \pmod{\pi}.$$

Let $\mathbf{m} = (\pi, x, y) \in \mathcal{C}$ be the singular point on the special fiber of $\mathcal{C}$. Then the *blow-up of $\mathcal{C}$ at* $\mathbf{m}$ is formed by taking the following three schemes and gluing them together as explained below.

Chart 1. Define new variables

$$x = \pi x_1 \qquad \text{and} \qquad y = \pi y_1,$$

and let ν be the largest integer so that

$$f(\pi x_1, \pi y_1) = \pi^\nu f_1(x_1, y_1) \qquad \text{with } f_1(x_1, y_1) \in R[x_1, y_1].$$

In other words, factor out a power of π so that the coefficients of f_1 are in R and at least one coefficient is a unit. Then the first coordinate chart for the blow-up of $\mathcal{C}$ at $\mathfrak{m}$ is the scheme $\mathcal{C}_1 \subset \mathbb{A}_R^2 = \operatorname{Spec} R[x_1, y_1]$ defined by

$$\mathcal{C}_1 : \operatorname{Spec} R[x_1, y_1]/(f_1(x_1, y_1)).$$

Chart 2. The second chart is formed using new variables π', x', y' defined by

$$\pi = \pi' y', \qquad x = x' y', \qquad y = y'.$$

We substitute these into the polynomial $f(x, y)$. This means we do two things. First, we replace x and y by $x'y'$ and y'. Second, we take each coefficient a of $f(x, y)$ and replace the largest power of π dividing a by that power of $\pi'y'$. For example, if $\pi^2 | a$ and $\pi^3 \nmid a$, then we would replace a by $(\pi'y')^2 \pi^{-2} a$. We factor out the largest possible power of y' to get

$$f(x'y', y') = (y')^{\nu'} f'(x', y') \qquad \text{with } f'(x', y') \in R[\pi', x', y'],$$

and then the second coordinate chart of the blow-up is the scheme

$$\mathcal{C}' : \operatorname{Spec} R[\pi', x', y']/(\pi - \pi'y', f'(x', y')).$$

Note that π' is a new variable, just like the variables x' and y'. The scheme $\mathcal{C}'$ is the closed subscheme of $\mathbb{A}_R^3 = \operatorname{Spec} R[\pi', x', y']$ defined by the two equations $\pi = \pi'y'$ and $f'(x', y') = 0$.

Chart 3. The third chart is formed similarly to the second chart using the variables π'', x'', y'' defined by

$$\pi = \pi'' x'', \qquad x = x'', \qquad y = y'' x''.$$

Substituting these into $f(x, y)$ as explained above and pulling out the largest power of x'' gives

$$f(x'', y''x'') = (x'')^{\nu''} f''(x'', y'') \qquad \text{with } f''(x'', y'') \in R[x'', y''].$$

Then the third coordinate chart of the blow-up is the scheme

$$\mathcal{C}'' : \operatorname{Spec} R[\pi'', x'', y'']/(\pi - \pi''x'', f''(x'', y'')).$$

It is easy to see how to glue the three coordinate charts together. For example, in order to map $\mathcal{C}_1$ to $\mathcal{C}'$, we just need to solve for (π', x', y') in terms of (π, x_1, y_1). Thus

$$\pi' = \frac{\pi}{y'} = \frac{\pi}{y} = \frac{1}{y_1}, \quad x' = \frac{x}{y'} = \frac{x}{y} = \frac{x_1}{y_1}, \quad y' = y = \pi y_1.$$

These equations define a birational map $\mathcal{C}_1 \to \mathcal{C}'$ which is defined everywhere except at the points of $\mathcal{C}_1$ with $y_1 = 0$. Similarly, we get a birational map $\mathcal{C}_1 \to \mathcal{C}''$ by using the equations

$$\pi'' = \frac{\pi}{x''} = \frac{\pi}{x} = \frac{1}{x_1}, \quad x'' = x = \pi x_1, \quad y'' = \frac{y}{x''} = \frac{y}{x} = \frac{y_1}{x_1},$$

and a birational map $\mathcal{C}' \to \mathcal{C}''$ using

$$\pi'' = \frac{\pi}{x''} = \frac{\pi}{x} = \frac{\pi'}{x'}, \quad x'' = x = x'y', \quad y'' = \frac{y}{x''} = \frac{y}{x} = \frac{1}{x'}.$$

These maps are used to glue the three coordinate charts together, and the resulting scheme is the blow-up of $\mathcal{C}$ at $\mathfrak{m}$.

In order to find the special fiber of the blow-up, we take the special fibers of each of the coordinate charts and then glue them together. The special fiber of a coordinate chart is calculated by setting $\pi = 0$ and looking at the resulting curve defined over k. The first coordinate chart is easiest, and we find that

$$\tilde{\mathcal{C}}_1 = \operatorname{Spec} k[x_1, y_1]/(\tilde{f}_1(x_1, y_1)).$$

In other words, $\tilde{\mathcal{C}}_1$ is the curve in $\mathbb{A}_k^2$ defined by the single equation $\tilde{f}_1 = 0$.

Similarly, the special fiber of $\mathcal{C}'$ is obtained by setting $\pi = 0$, which means that

$$\tilde{\mathcal{C}}' = \operatorname{Spec} k[\pi', x', y']/(\pi'y', \tilde{f}'(x', y')).$$

Here π' is to be treated as a variable, so $\tilde{\mathcal{C}}'$ consists of two pieces, one obtained by setting $\pi' = 0$ and the other obtained by setting $y' = 0$. Of course, each piece may consist of several components, or a piece could be empty. Finally, $\tilde{\mathcal{C}}''$ is given by

$$\tilde{\mathcal{C}}'' = \operatorname{Spec} k[\pi'', x'', y'']/(\pi''x'', \tilde{f}''(x'', y'')),$$

so $\tilde{\mathcal{C}}''$ also consists of two pieces, one with $\pi'' = 0$ and the other with $x'' = 0$.

Example 7.7.1. We are going to illustrate (7.7) by blowing-up the arithmetic surface

$$\mathcal{C} : x^2 + y^5 = \pi^4$$

at its singular point $\pi = x = y = 0$. For simplicity, we will assume that the residue field does not have characteristic 2, 3, or 5.

To find the first coordinate chart of the blow-up, we substitute $x = \pi x_1$ and $y = \pi y_1$ into the equation for $\mathcal{C}$ and cancel π^2, which yields

$$\mathcal{C}_1 : x_1^2 + \pi^3 y_1^5 = \pi^2.$$

The second coordinate chart is obtained by substituting $\pi = \pi'y'$, $x = x'y'$, and $y = y'$, and then canceling y'^2 to obtain

$$\mathcal{C}' : x'^2 + y'^3 = \pi'^4 y'^2, \quad \pi = \pi'y'.$$

Finally, to get the third chart we substitute $\pi = \pi''x''$, $x = x''$, and $y = y''x''$ and cancel x''^2 to get

$$\mathcal{C}'' : 1 + x''^3 y''^5 = \pi''^4 x''^2, \quad \pi = \pi''x''.$$

Next we compute the special fibers by setting $\pi = 0$. The special fiber of $\mathcal{C}_1$ is

$$\tilde{\mathcal{C}}_1 : x_1^2 = 0,$$

so $\tilde{\mathcal{C}}_1$ is a non-singular rational curve appearing with multiplicity 2.

To find the special fiber of $\mathcal{C}'$, we must set $\pi = \pi'y' = 0$. This gives us two pieces, one with $\pi' = 0$ and one with $y' = 0$. In this way we find two fibral components, which we will denote by F_1' and F_2':

$$\tilde{\mathcal{C}}' = \begin{cases} F_1' : \pi' = 0, \ x'^2 + y'^3 = 0, \\ 2F_2' : y' = 0, \ x'^2 = 0. \end{cases}$$

Keep in mind that F_1' and F_2' are curves in $\mathbb{A}_k^3 = \operatorname{Spec} k[\pi', x', y']$. Thus F_1' is a rational curve with a cusp, whereas F_2' is a non-singular rational curve which appears with multiplicity 2 in the fiber. In other words, as a divisor we have $\tilde{\mathcal{C}}' = F_1' + 2F_2'$.

Similarly, the special fiber of $\mathcal{C}''$ is obtained by setting $\pi = \pi''x'' = 0$. However, when we set $x'' = 0$ we obtain the equation $1 = 0$, so $x'' = 0$ does not give any components of $\mathcal{C}''$. Hence $\mathcal{C}''$ consists of the single rational curve

$$\mathcal{C}'' : \pi'' = 0, \ 1 = x''^3 y''^5.$$

We claim that when we glue $\mathcal{C}_1$, $\mathcal{C}'$, and $\mathcal{C}''$ together, the special fiber $\tilde{\mathcal{C}}''$ is identified with F_1', and the special fiber $\tilde{\mathcal{C}}_1$ is identified with $2F_2'$. To verify the first statement, we observe that the special fiber $\tilde{\mathcal{C}}''$ is defined by the equations $\pi'' = 0$ and $1 = x''^3 y''^5$. According to (7.7), the gluing map $\mathcal{C}' \to \mathcal{C}''$ is given by the substitutions

$$\pi'' = \pi'/x', \quad x'' = x'y', \quad y'' = 1/x'.$$

Substituting these into the equations for $\tilde{\mathcal{C}}''$ yields

$$\pi'/x' = 0 \qquad \text{and} \qquad 1 = (x'y')^3 (1/x')^5 = y'^3/x'^2,$$

which are exactly the equations of F_1'. We leave it to the reader to verify the assertion that $\tilde{\mathcal{C}}_1$ is glued to $2F_2'$.

Thus the special fiber of $\mathcal{C}'$ contains all of the components of the special fiber of the blow-up. If $\mathcal{C}'$ were regular, we would have completed our construction of a regular model, but unfortunately it is not regular. In fact, every point on the component F_2' of the special fiber is a singular point, so next we want to blow up $\mathcal{C}'$ along the entire curve $y' = 0$. We didn't discuss blowing up a surface along a curve in (7.7), but the procedure is very similar.

Recall that $\mathcal{C}' \subset \mathbb{A}_R^3 = \operatorname{Spec} R[\pi', x', y']$ is given by the equations

$$\mathcal{C}' : x'^2 + y'^3 = \pi'^4 y'^2, \quad \pi = \pi'y'.$$

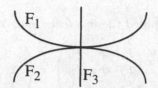

The Special Fiber $\tilde{\mathbb{S}}$ of the Scheme $\mathbb{S} : X^2 + Y = T^4$, $\pi = TY$

Figure 4.3

To blow up $\mathbb{C}'$ along the curve $y' = 0$, we make the substitutions

$$x' = XY, \quad y' = Y, \quad \pi' = T,$$

and cancel Y^2 from the first equation. This yields the arithmetic surface $\mathbb{S} \subset \mathbb{A}_R^3 = \operatorname{Spec} R[T, X, Y]$ given by the equations

$$\mathbb{S} : X^2 + Y = T^4, \quad \pi = TY.$$

Note that $\mathbb{S}$ is regular at $X = Y = T = \pi = 0$, since the equations for $\mathbb{S}$ show that the maximal ideal at that point is generated by the two variables X and T. One can similarly verify that $\mathbb{S}$ is regular at all other points, so it is a regular model for $\mathbb{C}$.

To compute the special fiber $\tilde{\mathbb{S}}$ of $\mathbb{S}$, we set $\pi = TY = 0$. The part with $T = 0$ is the non-singular rational curve $X^2 + Y = 0$, whereas the part with $Y = 0$ factors as

$$X^2 - T^4 = (X - T^2)(X + T^2) = 0,$$

so it consists of two non-singular rational curves which intersect tangentially. All three of the components of $\tilde{\mathbb{S}}$ intersect at the point $X = Y = T = 0$, so the special fiber $\tilde{\mathbb{S}}$ looks as illustrated in Figure 4.3.

Label the three components of $\tilde{\mathbb{S}}$ as indicated in Figure 4.3,

$$F_1 : Y = X - T^2 = 0, \quad F_2 : Y = X + T^2 = 0, \quad F_3 : T = X^2 + Y = 0.$$

Looking at Figure 4.3 or directly from the equations for the components, we can compute the pairwise intersections

$$F_1 \cdot F_2 = 2, \qquad F_1 \cdot F_3 = 1, \qquad F_2 \cdot F_3 = 1.$$

Next, using the fact (7.3) that the intersection of a component with the entire fiber $\tilde{\mathbb{S}} = F_1 + F_2 + F_3$ is zero, we compute the self-intersections,

$$F_1^2 = -(F_1 \cdot F_2 + F_1 \cdot F_3) = -3,$$
$$F_2^2 = -(F_2 \cdot F_1 + F_2 \cdot F_3) = -3,$$
$$F_3^2 = -(F_3 \cdot F_1 + F_3 \cdot F_2) = -2.$$

This shows that F_1, F_2, and F_3 are not exceptional curves, so $\mathcal{S}$ is a minimal regular model.

We have now computed the *incidence matrix* of the special fiber,

$$
\begin{pmatrix} F_1 \cdot F_1 & F_1 \cdot F_2 & F_1 \cdot F_3 \\ F_2 \cdot F_1 & F_2 \cdot F_2 & F_2 \cdot F_3 \\ F_3 \cdot F_1 & F_3 \cdot F_2 & F_3 \cdot F_3 \end{pmatrix} = \begin{pmatrix} -3 & 2 & 1 \\ 2 & -3 & 1 \\ 1 & 1 & -2 \end{pmatrix}.
$$

The incidence matrix of the special fiber of a minimal proper regular model of a curve can be used to compute the group of components of the Néron model of its Jacobian variety, see Raynaud [1] and exercises 4.32 and 4.33. In this example, the 2×2 minors of the incidence matrix have determinant 5. Raynaud's theorem then implies that the group of components of the Néron model of the Jacobian variety is a cyclic group of order 5.

§8. The Special Fiber of a Néron Model

In this section we are going to describe the Kodaira-Néron classification of special fibers on minimal proper regular models of elliptic curves. Our main tool will be the intersection theory described in the previous section. We will work over a discrete valuation ring with algebraically closed residue field. In the next section we will give an algorithm of Tate which computes the special fiber and also provides some additional information, including a description of what happens when the residue field is not algebraically closed.

We begin with a proposition which describes the intersection properties of the components of the special fiber. The most important part of this proposition is the last formula in (d), since it is this formula which puts severe constraints on the possible configurations of the components.

Proposition 8.1. *Let R be a discrete valuation ring with maximal ideal $\mathfrak{p}$, fraction field K, and algebraically closed residue field k. Let E/K be an elliptic curve, and let $\mathcal{C}/R$ be a minimal proper regular model for E/K. Suppose that the special fiber of $\mathcal{C}$ contains r irreducible components, say $F_1, \ldots, F_r$, and write the special fiber as*

$$
\mathcal{C}_\mathfrak{p} = \sum_{i=1}^{r} n_i F_i.
$$

(a) *At least one of the n_i's is equal to 1.*
(b) *Let $K_\mathcal{C}$ be a canonical divisor on $\mathcal{C}$ (7.4a). Then*

$$
K_\mathcal{C} \cdot F = 0 \quad \text{for all fibral divisors } F \in \mathrm{Div}_\mathfrak{p}(\mathcal{C}).
$$

(c) *If $r = 1$, then $F_1^2 = 0$ and $p_a(F_1) = 1$.*

(d) *Suppose that $r \geq 2$. Then for each $1 \leq i \leq r$,*

$$F_i^2 = -2, \quad F_i \cong \mathbb{P}_k^1, \quad \text{and} \quad \sum_{1 \leq j \leq r,\, j \neq i} n_j F_j \cdot F_i = 2n_i.$$

PROOF. The scheme $\mathcal{C}$ is proper over R, so $\mathcal{C}(R) \cong E(K)$ from (4.4a). By definition, an elliptic curve E/K has at least one K-rational point, namely its identity element, so we can find a point $P \in \mathcal{C}(R)$. Let $P(\mathfrak{p}) \in \mathcal{C}_\mathfrak{p}$ be the image of P on the special fiber of $\mathcal{C}$, and let F_i be a component of $\mathcal{C}_\mathfrak{p}$ containing $P(\mathfrak{p})$. The scheme $\mathcal{C}$ is regular, so (4.3b) says that $P(\mathfrak{p})$ is a non-singular point of $\mathcal{C}_\mathfrak{p}$. It follows that $n_i = 1$, since if $n_i \geq 2$, then every point of F_i would be a singular point of $\mathcal{C}_\mathfrak{p}$. This completes the proof of (a).

Next we consider the special fiber $\mathcal{C}_\mathfrak{p}$ as a divisor on $\mathcal{C}$. It has the following three properties:

$$\mathcal{C}_\mathfrak{p}^2 = 0 \qquad \text{from (7.3b)}$$
$$p_a(\mathcal{C}_\mathfrak{p}) = g(E) = 1 \qquad \text{from (7.4b)}$$
$$\mathcal{C}_\mathfrak{p}^2 + K_\mathcal{C} \cdot \mathcal{C}_\mathfrak{p} = 2p_a(\mathcal{C}_\mathfrak{p}) - 2 \qquad \text{adjunction formula (7.4a).}$$

Substituting the first two equations into the third gives

$$K_\mathcal{C} \cdot \mathcal{C}_\mathfrak{p} = 0.$$

We next apply the adjunction formula (7.4a) to an irreducible fibral component F_i to get

$$F_i^2 + K_\mathcal{C} \cdot F_i = 2p_a(F_i) - 2.$$

The arithmetic genus of an irreducible divisor is non-negative (7.4c), since by definition it is the dimension of a certain cohomology group. Thus the right-hand side of the adjunction formula is at least -2. On the other hand, we know that $F_i^2 \leq 0$ from (7.3b). This leads to the following possibilities:

(i) $F_i^2 = 0$ for some $1 \leq i \leq r$.

(ii) $F_i^2 < 0$ and $K_\mathcal{C} \cdot F_i < 0$ for some $1 \leq i \leq r$.

(iii) $F_i^2 < 0$ and $K_\mathcal{C} \cdot F_i \geq 0$ for all $1 \leq i \leq r$.

In case (i) we know from (7.3b) that F_i must be a multiple of the entire special fiber. More precisely, we must have $r = 1$ and $\mathcal{C}_\mathfrak{p} = n_1 F_1$. Then (a) tells us that $n_1 = 1$, so $F_1 = \mathcal{C}_\mathfrak{p}$. Now the equality $K_\mathcal{C} \cdot \mathcal{C}_\mathfrak{p} = 0$ proven above gives us (b), and the facts $\mathcal{C}_\mathfrak{p}^2 = 0$ and $p_a(\mathcal{C}_\mathfrak{p}) = 1$ noted above give us (c). This completes the proof of Proposition 8.1 in case (i).

Next consider case (ii). Each term in the left-hand side of the adjunction formula is negative, whereas the right-hand side is at least -2. The only way for this to happen is if

$$F_i^2 = K_{\mathcal{C}} \cdot F_i = -1 \qquad \text{and} \qquad p_a(F_i) = 0.$$

Now (7.4c) says that $F_i \cong \mathbb{P}^1_k$, and hence F_i is an exceptional divisor. But Castelnuovo's criterion (7.5b) says that a minimal proper regular model contains no exceptional divisors, so case (ii) cannot occur.

It remains to consider case (iii). The strict inequality $F_i^2 < 0$ implies in particular that $r \geq 2$. We take the equality $K_{\mathcal{C}} \cdot \mathcal{C}_{\mathfrak{p}} = 0$ proven above and write it out in terms of the fibral components as

$$\sum_{i=1}^{r} n_i K_{\mathcal{C}} \cdot F_i = 0.$$

Each $n_i \geq 1$, and since we are in case (iii), each $K_{\mathcal{C}} \cdot F_i \geq 0$, so the only way that this can be true is if we have

$$K_{\mathcal{C}} \cdot F_i = 0 \qquad \text{for all } 1 \leq i \leq r.$$

This proves (b).

Next we substitute $K_{\mathcal{C}} \cdot F_i = 0$ into the adjunction formula for F_i, which yields
$$F_i^2 = 2p_a(F_i) - 2.$$

We are in case (iii), so $F_i^2 < 0$, whereas $p_a(F_i) \geq 0$ from (7.4c). It follows that

$$F_i^2 = -2 \qquad \text{and} \qquad p_a(F_i) = 0,$$

and then (7.4c) tells us that $F_i \cong \mathbb{P}^1_k$. Finally, we note that $\mathcal{C}_{\mathfrak{p}} \cdot F_i = F_i \cdot \mathcal{C}_{\mathfrak{p}} = 0$ from (7.3b), which allows us to compute

$$0 = \mathcal{C}_{\mathfrak{p}} \cdot F_i = \sum_{j=1}^{r} n_j F_j \cdot F_i = -2n_i + \sum_{1 \leq j \leq r, \, j \neq i} n_j F_j \cdot F_i.$$

This completes the proof of (d).   □

We are now going to use (8.1) and a combinatorial argument to give the Kodaira-Néron classification of the fibers of minimal proper regular model of elliptic curves.

Theorem 8.2. (Kodaira [1], Néron [1]) *Let R be a discrete valuation ring with maximal ideal $\mathfrak{p}$, fraction field K, and algebraically closed residue field k. Let E/K be an elliptic curve, and let $\mathcal{C}/R$ be a minimal proper regular model for E/K. Then the special fiber $\mathcal{C}_{\mathfrak{p}}$ of $\mathcal{C}$ has one of the* *following forms. (See Figure 4.4. Note that the small numbers in Figure 4.4 indicate the multiplicities of the components.)*

Type I_0. $\mathcal{C}_p$ *is a non-singular curve of genus 1.*

Type I_1. $\mathcal{C}_p$ *is a rational curve with a node.*

Type I_n. $\mathcal{C}_p$ *consists of n non-singular rational curves arranged in the shape of an n-gon, where $n \geq 2$.*

Type II. $\mathcal{C}_p$ *is a rational curve with a cusp.*

Type III. $\mathcal{C}_p$ *consists of two non-singular rational curves which intersect tangentially at a single point.*

Type IV. $\mathcal{C}_p$ *consists of three non-singular rational curves intersecting at a single point.*

Type I_0^*. $\mathcal{C}_p$ *is a non-singular rational curve of multiplicity 2 with four non-singular rational curves of multiplicity 1 attached.*

Type I_n^*. $\mathcal{C}_p$ *consists of a chain of $n + 1$ non-singular rational curves of multiplicity 2, with two non-singular rational curves of multiplicity 1 attached at either end.*

Type IV^*. $\mathcal{C}_p$ *consists of seven non-singular rational curves arranged as pictured in Figure 4.4.*

Type III^*. $\mathcal{C}_p$ *consists of eight non-singular rational curves arranged as pictured in Figure 4.4.*

Type II^*. $\mathcal{C}_p$ *consists of nine non-singular rational curves arranged as pictured in Figure 4.4.*

Remark 8.2.1. If the residue field k of R is not algebraically closed, then $\mathcal{C}_p$ may have some components which are irreducible over k but become reducible over a finite extension of k. In other words, the Galois group $G_{\bar{k}/k}$ may act non-trivially on the $\bar{k}$-irreducible components of $\mathcal{C}_p$, and then the k-irreducible components of $\mathcal{C}_p$ are the orbits. We will discuss this situation further in the next section when we describe Tate's algorithm.

Remark 8.2.2. The dual graphs of the pictures in Figure 4.4 turn out to be extended Dynkin diagrams. There is a discussion of this in Miranda [1, I §6], as well as a proof of (8.2) based on the negative semi-definite quadratic forms attached to the extended Dynkin diagrams. The proof relies only on the facts proven in (7.3), namely that $\mathcal{C}_p$ is connected and that the intersection pairing on $\mathcal{C}_p$ is negative semi-definite with kernel equal to the entire fiber.

Remark 8.2.3. We will use Kodaira's [1] notation I_n, II, ..., III^*, II^* to describe the various types of special fibers (8.1). There is a second notational system, due to Néron [1], which is also in common use. For the convenience of the reader, we briefly list the equivalences.

Kodaira	I_0	I_n	II	III	IV	I_0^*	I_n^*	IV^*	III^*	II^*
Néron	a	b_n	$c1$	$c2$	$c3$	$c4$	$c5_n$	$c6$	$c7$	$c8$

Reduction Type	Number of Components	Configuration (with multiplicity)
I_0	1	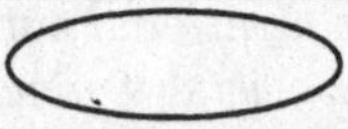
I_1	1	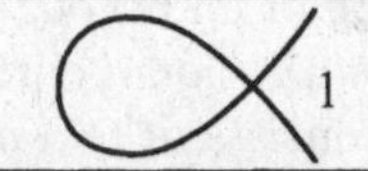
I_n	n	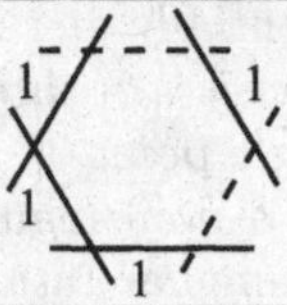
II	1	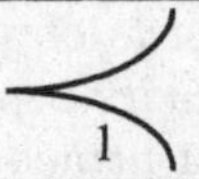
III	2	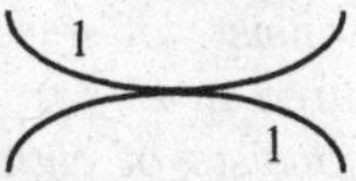
IV	3	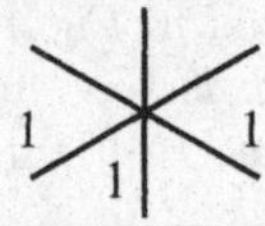
I_0^*	5	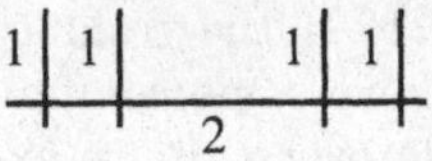
I_n^*	n + 5	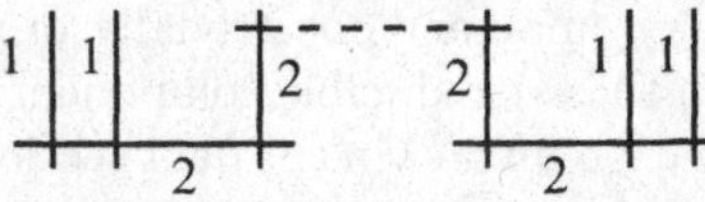
IV^*	7	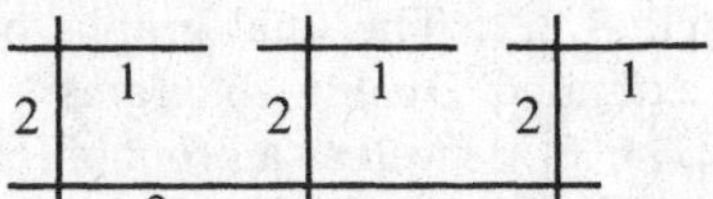
III^*	8	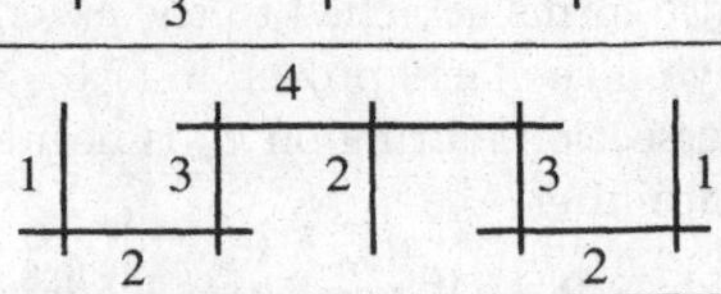
II^*	9	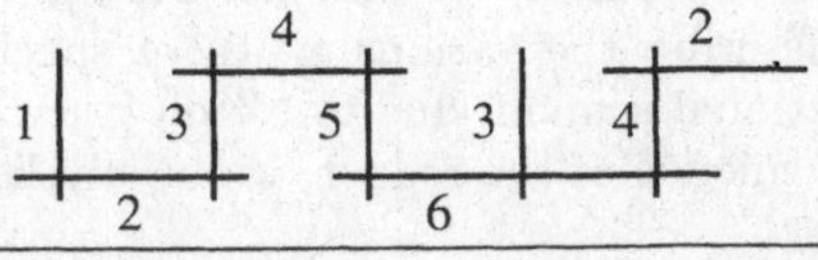

The Kodaira-Néron Classification of Special Fibers

Figure 4.4

Remark 8.2.4. Ogg [3] and Namikawa and Ueno [1] have given a classification, similar to (8.2), for the special fibers of proper regular minimal models of curves of genus 2. It turns out that there are more than 100 configurations!

PROOF (of Theorem 8.2). We will write the special fiber of $\mathcal{C}_{\mathfrak{p}}$ as usual as

$$\mathcal{C}_{\mathfrak{p}} = \sum_{i=1}^{r} n_i F_i.$$

We are going to consider a number of cases, which we will box for clarity.

$\boxed{\mathcal{C}_{\mathfrak{p}} \text{ has } r = 1 \text{ component}}$

Proposition 8.1(a,c) tells us that $n_1 = 1$ and $p_a(\mathcal{C}_{\mathfrak{p}}) = 1$, so $\mathcal{C}_{\mathfrak{p}} = F_1$ is an irreducible curve of arithmetic genus 1. If $\mathcal{C}_{\mathfrak{p}}$ is non-singular, then it is a non-singular curve of genus 1 (Hartshorne [1, IV.1.1]), so we have Type I_0. If $\mathcal{C}_{\mathfrak{p}}$ is singular, then Hartshorne [1, V.3.7] and the fact that $p_a(\mathcal{C}_{\mathfrak{p}}) = 1$ means that a single blow-up of a singular point on $\mathcal{C}_{\mathfrak{p}}$ will produce a non-singular rational curve. Hence $\mathcal{C}_{\mathfrak{p}}$ is a rational curve with exactly one singular point of multiplicity 2, from which it is not hard to show that the singular point is either an ordinary node or an ordinary cusp. This gives Types I_1 and II.

(Alternative proof for $r = 1$. We will see in the next section that if $r = 1$, then the scheme $\mathcal{W}/R$ defined by a minimal Weierstrass equation for E is already a regular scheme. It will follow that $\mathcal{C} = \mathcal{W}$, so the special fiber $\mathcal{C}_{\mathfrak{p}}$ is obtained by reducing the minimal Weierstrass equation modulo $\mathfrak{p}$. But we already know from [AEC, VII §5] that the reduction $\mathcal{W}_{\mathfrak{p}}$ is either a non-singular curve of genus 1, a rational curve with a node, or a rational curve with a cusp.)

We assume henceforth that $r \geq 2$, which means that we can apply the formula given in (8.1d),

$$\sum_{1 \leq j \leq r, \, j \neq i} n_j F_j \cdot F_i = 2n_i. \qquad (*)$$

Note that each $n_j \geq 1$, so every term in the sum is non-negative. We will be making frequent use of this important formula $(*)$. We also note from (8.1d) that every component is a non-singular rational curve; that is, $F_i \cong \mathbb{P}^1_k$ for every $1 \leq i \leq r$.

Proposition 8.1(a) says that one of the n_i's equals 1, so relabeling the F_i's if necessary, we may assume that $n_1 = 1$. We further know that $\mathcal{C}_{\mathfrak{p}}$ is connected (7.3a), and $r \geq 2$ by assumption, so after further relabeling we may also assume that $F_1 \cdot F_2 \geq 1$.

$\boxed{\mathcal{C}_{\mathfrak{p}} \text{ has } r = 2 \text{ components}}$

This means that $\mathcal{C}_{\mathfrak{p}} = F_1 + n_2 F_2$, so applying $(*)$ for $i = 1$ and $i = 2$ gives

$$n_2 F_2 \cdot F_1 = 2 \qquad \text{and} \qquad F_1 \cdot F_2 = 2n_2.$$

Further, $F_1 \cdot F_2 = F_2 \cdot F_1$ from (7.2iii), so we deduce that

$$F_1 \cdot F_2 = 2 \quad \text{and} \quad n_2 = 1.$$

This means either that F_1 and F_2 intersect tangentially in a single point, which gives Type III, or else they intersect transversally at two distinct points, which gives Type I_2. This completes the analysis of the special fiber in the case that $\mathcal{C}_\mathfrak{p}$ has exactly two components.

$\boxed{\mathcal{C}_\mathfrak{p} \text{ has } r \geq 3 \text{ components}}$

We claim in this case that intersecting components always intersect transversally; that is, we claim that

$$F_i \cdot F_{i'} \leq 1 \quad \text{for all } 1 \leq i, i' \leq r, \ i \neq i'.$$

To see this, we use the fact that $\mathcal{C}_\mathfrak{p}$ is connected and contains at least three components to find a third component, say F_k, so that F_k intersects at least one of F_i or $F_{i'}$, say $F_k \cdot F_i \geq 1$. Applying (∗) to F_i and to $F_{i'}$ gives the inequalities

$$n_{i'} F_{i'} \cdot F_i < n_{i'} F_{i'} \cdot F_i + n_k F_k \cdot F_i \leq 2n_i \quad \text{and} \quad n_i F_i \cdot F_{i'} \leq 2n_{i'}.$$

We now multiply these two estimates to obtain the strict inequality

$$n_i n_{i'} \left(F_i \cdot F_{i'} \right)^2 < 4 n_i n_{i'}.$$

Therefore $F_i \cdot F_{i'} < 2$, and since the intersection index is an integer, we find that $F_i \cdot F_{i'} \leq 1$ as desired.

In particular, we have $F_1 \cdot F_2 = 1$. Using this and the fact that $n_1 = 1$, we can apply (∗) to F_1 to obtain the bound

$$n_2 = n_2 F_1 \cdot F_2 \leq 2n_1 = 2.$$

Thus n_2 equals either 1 or 2, which leads to two further subcases.

$\boxed{r \geq 3, \ n_2 = 1}$

Applying (∗) to F_2 gives

$$n_1 F_1 \cdot F_2 + \sum_{j=3}^{r} n_j F_j \cdot F_2 = 2n_2, \quad \text{and hence} \quad \sum_{j=3}^{r} n_j F_j \cdot F_2 = 1,$$

since $F_1 \cdot F_2 = 1$ and $n_1 = n_2 = 1$. This means that there is exactly one more component intersecting F_2, call it F_3, and $n_3 = 1$. If F_3 also intersects F_1, then we get two possible configurations depending on whether or not $F_1 \cap F_3$ is the same point as $F_2 \cap F_3$. If they are the same point, then we get a fiber of Type IV, and if they are not, then we get three rational curves arranged in a triangle, which is Type I_3.

Suppose now that F_3 does not intersect F_1. Then applying $(*)$ to F_3 gives

$$n_2 F_2 \cdot F_3 + \sum_{j=4}^{r} n_j F_j \cdot F_3 = 2n_3, \quad \text{and hence} \quad \sum_{j=4}^{r} n_j F_j \cdot F_3 = 1,$$

since $F_2 \cdot F_3 = 1$ and $n_2 = n_3 = 1$. Therefore there is exactly one more component intersecting F_3, call it F_4, and $n_4 = 1$. If F_4 also intersects F_1, then we have four rational curves arranged in a square, which is Type I_4. If F_4 does not intersect F_1, then applying $(*)$ to F_4 gives in the same way one more component F_5 intersecting F_4, and $n_5 = 1$. The fiber $\mathcal{C}_\mathfrak{p}$ has only finitely many components, so this process must eventually terminate. More precisely, since $\mathcal{C}_\mathfrak{p}$ has r components, the process will terminate with F_r intersecting F_1. At this point we will have r rational curves, each of multiplicity 1, arranged in the shape of a polygon, which means that the fiber $\mathcal{C}_\mathfrak{p}$ is of Type I_r.

$\boxed{r \geq 3,\ n_2 = 2}$

Applying $(*)$ to F_1 gives

$$n_2 F_2 \cdot F_1 + \sum_{j=3}^{r} n_j F_j \cdot F_1 = 2n_1, \quad \text{and hence} \quad \sum_{j=3}^{r} n_j F_j \cdot F_1 = 0,$$

since $F_1 \cdot F_2 = 1$, $n_1 = 1$, and $n_2 = 2$. This means that there are no more components intersecting F_1. Next applying $(*)$ to F_2 gives

$$n_1 F_1 \cdot F_2 + \sum_{j=3}^{r} n_j F_j \cdot F_2 = 2n_2, \quad \text{and hence} \quad \sum_{j=3}^{r} n_j F_j \cdot F_2 = 3.$$

Thus F_2 intersects either one, two, or three additional components.

Suppose first that F_2 intersects three additional components, which we label F_3, F_4, and F_5. Then $n_3 = n_4 = n_5 = 1$, which gives a fiber of Type I_0^*, and applying $(*)$ to F_3, F_4, and F_5 shows that $\mathcal{C}_\mathfrak{p}$ contains no other components.

Next suppose that F_2 intersects exactly two other components, say F_3 and F_4. Switching these two components if necessary, we have $n_3 = 1$ and $n_4 = 2$. Thus $\mathcal{C}_\mathfrak{p}$ contains the configuration illustrated in Figure 4.5(a), where the small 2's next to F_2 and F_4 indicate that they are components of multiplicity 2. Of course, $\mathcal{C}_\mathfrak{p}$ may contain some additional components. Let $s \geq 4$ be the largest integer so that $\mathcal{C}_\mathfrak{p}$ contains the configuration illustrated in Figure 4.5(b). Applying $(*)$ to F_i for any $4 \leq i < s$ gives

$$n_{i-1} F_{i-1} \cdot F_i + n_{i+1} F_{i+1} \cdot F_i + \sum_{j=s+1}^{r} n_j F_j \cdot F_i = 2n_i, \quad \text{so} \quad \sum_{j=s+1}^{r} n_j F_j \cdot F_i = 0,$$

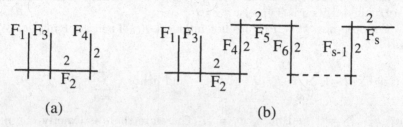

(a)   (b)

Building a Fiber of Type I_n^*

Figure 4.5

since $F_{i-1} \cdot F_i = F_{i+1} \cdot F_i = 1$ and $n_{i-1} = n_i = n_{i+1} = 2$. Thus there are no more components intersecting any of $F_4, \ldots, F_{s-1}$. On the other hand, applying $(*)$ to F_s gives

$$n_{s-1}F_{s-1} \cdot F_s + \sum_{j=s+1}^{r} n_j F_j \cdot F_s = 2n_s, \quad \text{and hence} \quad \sum_{j=s+1}^{r} n_j F_j \cdot F_s = 2.$$

If F_s intersects exactly one additional component, then that component will have multiplicity 2, which means that $\mathcal{C}_p$ contains the configuration in Figure 4.5(b) with one more multiplicity-2 component. This contradicts our choice of s, so F_2 must intersect two additional components, each of which has multiplicity 1. This gives us a fiber of Type I_n^* (with $n = s - 3$), and it is then easy to check using $(*)$ that there are no more components. This completes the proof in the case that F_2 intersects exactly two components in addition to F_1.

Finally we suppose that F_2 intersects exactly one additional component, say F_3, with multiplicity $n_3 = 3$. This means that $\mathcal{C}_p$ contains the configuration illustrated in Figure 4.6(a). Let $t \geq 3$ be the largest integer so that $\mathcal{C}_p$ contains the configuration illustrated in Figure 4.6(b). Applying $(*)$ to F_i for any $3 \leq i < t$ gives

$$n_{i-1}F_{i-1} \cdot F_i + n_{i+1}F_{i+1} \cdot F_i + \sum_{j=t+1}^{r} n_j F_j \cdot F_i = 2n_i, \quad \text{so} \quad \sum_{j=t+1}^{r} n_j F_j \cdot F_i = 0,$$

since

$$F_{i-1} \cdot F_i = F_{i+1} \cdot F_i = 1, \quad n_{i-1} = i-1, \quad n_i = i, \quad \text{and} \quad n_{i+1} = i+1.$$

Thus there are no more components intersecting any of $F_3, \ldots, F_{t-1}$. On the other hand, applying $(*)$ to F_t gives

$$n_{t-1}F_{t-1} \cdot F_t + \sum_{j=t+1}^{r} n_j F_j \cdot F_t = 2n_t, \quad \text{and hence} \quad \sum_{j=t+1}^{r} n_j F_j \cdot F_t = t+1.$$

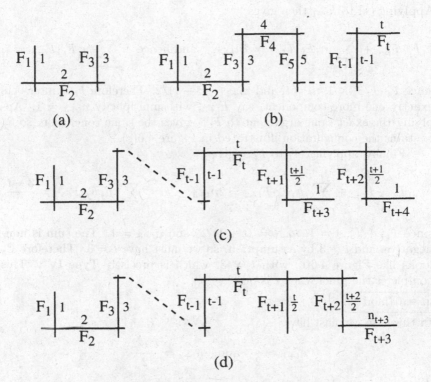

Building Fibers of Type IV*, III*, II*

Figure 4.6

If F_t were to intersect exactly one more component, then that component would have multiplicity $t + 1$, contradicting the fact that we chose the largest t so that $\mathcal{C}_p$ contains the configuration in Figure 4.6(b). Thus F_t intersects at least two additional components, say F_{t+1} and F_{t+2}.

Let F_i be any component intersecting F_t. Applying $(*)$ to F_i gives

$$n_t F_t \cdot F_i \le 2n_i, \quad \text{and hence} \quad n_i \ge t/2.$$

It follows that F_t intersects only the two additional components F_{t+1} and F_{t+2}, and we have the estimates

$$n_{t+1} + n_{t+2} = t + 1, \qquad n_{t+1} \ge t/2, \qquad n_{t+2} \ge t/2.$$

Switching F_{t+1} and F_{t+2} if necessary, we may assume that $n_{t+1} \le n_{t+2}$. Then there are only two possibilities, depending on the parity of t.

$\boxed{t \equiv 1 \, (\mathrm{mod} \, 2)}$

In this case we must have

$$n_{t+1} = n_{t+2} = \frac{t+1}{2}.$$

Applying (∗) to F_{t+1} then gives

$$n_t F_t \cdot F_{t+1} + \sum_{j=t+3}^{r} n_j F_j \cdot F_{t+1} = 2n_{t+1}, \quad \text{and hence} \quad \sum_{j=t+3}^{r} n_j F_j \cdot F_{t+1} = 1,$$

since $F_t \cdot F_{t+1} = 1$, $n_t = t$, and $n_{t+1} = (t+1)/2$. Therefore F_{t+1} intersects exactly one more component, say F_{t+3}, with multiplicity $n_{t+3} = 1$. Applying the exact same argument to F_{t+2} gives the same conclusion, so $\mathcal{C}_p$ contains the configuration illustrated in Figure 4.6(c).

Finally, applying (∗) to F_{t+3} gives

$$n_{t+1} F_{t+1} \cdot F_{t+3} + \sum_{j=t+5}^{r} n_j F_j \cdot F_{t+3} = 2n_{t+3}, \quad \text{so} \quad \sum_{j=t+5}^{r} n_j F_j \cdot F_{t+3} = \frac{3-t}{2},$$

since $F_{t+1} \cdot F_{t+3} = 1$, $n_{t+1} = (t+1)/2$, and $n_{t+3} = 1$. The sum is non-negative, and $t \geq 3$ by assumption, so we must have $t = 3$. Therefore $\mathcal{C}_p$ looks like Figure 4.6(c) with $t = 3$, which is precisely Type IV∗. This completes the proof when t is odd.

$\boxed{t \equiv 0 \,(\text{mod } 2)}$

In this case we must have

$$n_{t+1} = \frac{t}{2} \quad \text{and} \quad n_{t+2} = \frac{t+2}{2}.$$

Applying (∗) to F_{t+1} gives

$$n_t F_t \cdot F_{t+1} + \sum_{j=t+3}^{r} n_j F_j \cdot F_{t+1} = 2n_{t+1}, \quad \text{and hence} \quad \sum_{j=t+3}^{r} n_j F_j \cdot F_{t+1} = 0,$$

since $F_t \cdot F_{t+1} = 1$, $n_t = t$, and $n_{t+1} = t/2$. Thus there are no additional components intersecting F_{t+1}.

Next we apply (∗) to F_{t+2}. This gives

$$n_t F_t \cdot F_{t+2} + \sum_{j=t+3}^{r} n_j F_j \cdot F_{t+2} = 2n_{t+2}, \quad \text{and hence} \quad \sum_{j=t+3}^{r} n_j F_j \cdot F_{t+2} = 2,$$

since $F_t \cdot F_{t+2} = 1$, $n_t = t$, and $n_{t+2} = (t+2)/2$. Hence F_{t+2} intersects at least one additional component, say F_{t+3}, whose multiplicity n_{t+3} is either 1 or 2. So we now know that $\mathcal{C}_p$ contains the configuration illustrated in Figure 4.6(d).

Don't despair, we're almost done! Applying (∗) to F_{t+3} gives

$$n_{t+2} F_{t+2} \cdot F_{t+3} + \sum_{j=t+4}^{r} n_j F_j \cdot F_{t+3} = 2n_{t+3},$$

and hence

$$\sum_{j=t+4}^{r} n_j F_j \cdot F_{t+3} = 2n_{t+3} - \frac{t+2}{2} = \begin{cases} (2-t)/2 & \text{if } n_{t+3} = 1, \\ (6-t)/2 & \text{if } n_{t+3} = 2, \end{cases}$$

since $F_{t+2} \cdot F_{t+3} = 1$ and $n_{t+2} = (t+2)/2$. But the sum is non-negative, and t is even and ≥ 3 by assumption, so we must have

$$t = 4 \text{ or } 6 \quad \text{and} \quad n_{t+3} = 2.$$

If $t = 6$, then there are no additional components, and Figure 4.6(d) is exactly Type II*. Finally, if $t = 4$, then there is one more component F_{t+4} hitting F_{t+3}, and its multiplicity is $n_{t+4} = 1$. This gives Type III*, which completes the proof of Theorem 8.2.   □

§9. Tate's Algorithm to Compute the Special Fiber

In this section we are going to describe an algorithm of Tate which computes, among other things, the reduction type of an elliptic curve given by a Weierstrass equation. We set the following notation, which will be used throughout this section.

R \qquad a discrete valuation ring with maximal ideal $\mathfrak{p}$, uniformizing element π, fraction field K, perfect residue field k of characteristic p, and normalized valuation v.

E/K \qquad an elliptic curve given by a Weierstrass equation

$$E : y^2 + a_1 xy + a_3 y = x^3 + a_2 x^2 + a_4 x + a_6.$$

$\mathcal{C}/R$ \qquad a minimal proper regular model of E over R (4.5b).

$\mathcal{E}/R$ \qquad the largest subscheme of $\mathcal{C}/R$ which is smooth over R (6.1.1). Note that $\mathcal{E}/R$ is a Néron model for E/K (6.1).

$\tilde{\mathcal{C}}/k$ \qquad $= \mathcal{C} \times_R k$, the special fiber of $\mathcal{C}$.

$\tilde{\mathcal{E}}/k$ \qquad $= \mathcal{E} \times_R k$, the special fiber of $\mathcal{E}$. It is a group variety over k.

$\mathcal{E}^0/R$ \qquad the identity component of $\mathcal{E}$ (6.1.2); that is, $\mathcal{E}^0$ is the open subset of $\mathcal{E}$ obtained by discarding the non-identity components of the special fiber. It is a subgroup scheme of $\mathcal{E}$ (exercise 4.25).

$\tilde{\mathcal{E}}^0/k$ \qquad the identity component of the group variety $\tilde{\mathcal{E}}/k$ (1.5c).

Tate's algorithm is essentially a set of instructions for computing $\mathcal{C}$ and $\mathcal{E}$ from a given Weierstrass equation. For this reason its statement as a formal theorem has the unsatisfying form: "The following 11 step procedure leads to the stated results." So before we describe the algorithm itself, we want to give two corollaries. This will serve to explain (if not to excuse) why the following results are called "corollaries," when in fact they are really conclusions which can be deduced from the description and validity of Tate's algorithm.

Corollary 9.1. *Take a minimal Weierstrass equation for E/K, and let $W \subset \mathbb{P}_R^2$ be the closed subscheme defined by this equation. Further, let W^0/R be the largest subscheme of W which is smooth over R. Then $W^0 \cong \mathcal{E}^0$; that is, W^0 is the identity component of a Néron model for E/K.*

Corollary 9.2. *Take a minimal Weierstrass equation for E/K. We recall from* [AEC, VII §2] *the following notation:*

$\tilde{E}/k$ *the reduction of the given Weierstrass equation modulo* $\mathfrak{p}$.

$\tilde{E}_{ns}(k)$ *the set of non-singular points of* $\tilde{E}(k)$.

$E_0(K)$ $= \{P \in E(K) : \tilde{P} \in \tilde{E}(k)_{ns}\}$, *the set of points of $E(K)$ with non-singular reduction.*

$E_1(K)$ $= \{P \in E(K) : \tilde{P} = \tilde{O}\}$, *the set of points of $E(K)$ which reduce to the identity element.*

Further, let

$\mathcal{E}^1(R)$ $= \{\sigma \in \mathcal{E}(R) : \sigma(\mathfrak{p}) = \tilde{O} \in \tilde{E}(k)\}$.

The isomorphism $E(K) \cong \mathcal{E}(R)$ described in (5.1.3) induces the following identifications:

(a)
$$E(K) \quad \supset \quad E_0(K) \quad \supset \quad E_1(K)$$
$$\| \qquad\qquad\quad \| \qquad\qquad\quad \|$$
$$\mathcal{E}(R) \quad \supset \quad \mathcal{E}^0(R) \quad \supset \quad \mathcal{E}^1(R).$$

(b)
$$E(K)/E_0(K) \quad \cong \quad \mathcal{E}(R)/\mathcal{E}^0(R) \quad \longhookrightarrow \quad \tilde{E}(k)/\tilde{\mathcal{E}}^0(k)$$
$$\uparrow \qquad\qquad\qquad \uparrow \qquad\qquad\qquad\qquad \uparrow$$
$$E(K)/E_1(K) \quad \cong \quad \mathcal{E}(R)/\mathcal{E}^1(R) \quad \longhookrightarrow \quad \tilde{E}(k).$$

If K is complete, or even merely Henselian, then both inclusions in (b) are isomorphisms.

(c) $\tilde{E}_{ns}(k) = \tilde{\mathcal{E}}^0(k)$.

(d) *The group $E(K)/E_0(K)$ is finite. More precisely, if E has split multiplicative reduction, then $E(K)/E_0(K)$ is a cyclic group of order $-v(j(E))$; otherwise, $E(K)/E_0(K)$ has order 1, 2, 3, or 4.*

Remark 9.2.1. Note that Corollary 9.2(d) is exactly [AEC, VII.6.1], a result which was left unproven in [AEC]. The fact that $E(K)/E_0(K)$ is finite, even if the residue field k is infinite, played an important role in the proof of the criterion of Néron-Ogg-Shafarevich [AEC, VII.7.1], and we will use it again in the next section (10.2) when we prove a generalization of this criterion.

Remark 9.2.2. Corollary 9.2 can be used to bound the torsion subgroup of an elliptic curve defined over a local field or a number field. Continuing with the notation set at the beginning of this section, we recall from [AEC, VII.3.1(a)] that the subgroup $E_1(K)$ contains no prime-to-p torsion, since it is isomorphic to the formal group of E. Suppose now that E has additive reduction. Then $\tilde{\mathcal{E}}^0(k) = \tilde{E}_{\mathrm{ns}}(k) = k^+$ is a p-group from (9.2c), whereas the quotient $E(K)/E_0(K) \cong \tilde{\mathcal{E}}(k)/\tilde{\mathcal{E}}^0(k)$ is a group of order 1, 2, 3, or 4 from (9.2d). Using these facts and the exact sequences

$$0 \longrightarrow E_1(K) \longrightarrow E_0(K) \longrightarrow \tilde{\mathcal{E}}^0(k),$$

$$0 \longrightarrow E_0(K) \longrightarrow E(K) \longrightarrow E(K)/E_0(K) \longrightarrow 0,$$

we conclude that if E/K has additive reduction, then $E(K)_{\mathrm{tors}}$ has order ap^e for some $a \in \{1, 2, 3, 4\}$ and some $e \geq 0$.

Now consider an elliptic curve E defined over a number field K. The torsion subgroup of $E(K)$ injects into the torsion subgroup of $E(K_{\mathfrak{p}})$ for each completion of K, so the local estimate we just proved can often be used to obtain strong global estimates. For example, suppose that E has additive reduction at primes $\mathfrak{p}_1, \mathfrak{p}_2$ of K with distinct residue characteristics p_1, p_2. Then $E(K)_{\mathrm{tors}}$ has order dividing 12, and if $p_1 \geq 5$, then $E(K)_{\mathrm{tors}}$ has order at most 4.

Tate's algorithm, which we are now going to describe, computes the following quantities associated to the elliptic curve E/K:

Type   the reduction type of the special fiber $\tilde{\mathcal{C}}$ over the algebraic closure $\bar{k}$ of k. We will use the Kodaira symbols (8.2) to describe the reduction type.

$m(E/K)$ the number of components, defined over $\bar{k}$ and counted without multiplicity, on the special fiber $\tilde{\mathcal{C}}$.

$v(\mathcal{D}_{E/K})$ the valuation of the minimal discriminant of E/K.

$f(E/K)$ the exponent of the conductor of E/K. This quantity will be defined in §10, but for now we note that it can be computed using Ogg's formula (11.1),

$$f(E/K) = v(\mathcal{D}_{E/K}) - m(E/K) + 1.$$

$c(E/K)$ the order of the group of components $\tilde{\mathcal{E}}(k)/\tilde{\mathcal{E}}^0(k)$. Equivalently, $c(E/K)$ is the number of components of the special fiber $\tilde{\mathcal{C}}$ which have multiplicity 1 and are defined over k.

To ease notation, we will sometimes write

$$m = m(E/K), \quad f = f(E/K), \quad c = c(E/K),$$

when the curve E and field K are clear from the context.

Remark 9.3. If K is a complete local field, then (9.2b) says that $c(E/K)$ equals the order of $E(K)/E_0(K)$. This quantity is the so-called Birch-Swinnerton-Dyer "fudge factor." It is a sort of p-adic period and appears in the conjectural formula for the leading coefficient of $L(E,s)$ around $s = 1$ [AEC, C.16.5]. See Tate [2, §5] for details.

Tate's Algorithm 9.4.   (Tate [2]) *The following algorithm computes the reduction type, the values of $m(E/K)$, $v(\mathcal{D}_{E/K})$, $f(E/K)$, and $c(E/K)$, and the various other quantities described during the course of the algorithm.*

Remark 9.4.1. At the conclusion of Tate's algorithm, one obtains a minimal Weierstrass equation for the given elliptic curve. In practice, however, it is considerably easier to implement Tate's algorithm if one knows, a priori, that the initial Weierstrass equation is minimal. Further, if one only wants a minimal equation and is not interested in computing other quantities, such as the reduction type, then there are easier methods available. For example, if the characteristic p of k satisfies $p \geq 5$, then a given Weierstrass equation is minimal if and only if either $v(c_4) < 4$ or $v(c_6) < 6$. In general, one can use a short algorithm of Laska [1] to find a minimal Weierstrass equation. See also exercise 4.36 for another method.

Remark 9.4.2. In the case that the residue field k is algebraically closed, we have assembled information about the various reduction types in Table 4.1. This table is taken, with minor modifications, from Tate [2, §6]. Notice that if $\operatorname{char}(k) \neq 2, 3$, then everything about E (reduction type, exponent of conductor, group of components $E(K)/E_0(K)$) can be read off from Table 4.1 once one has a minimal Weierstrass equation for E/K.

Our description of Tate's algorithm follows very closely Tate's exposition [2]. The idea is to begin with an arbitrary Weierstrass equation

$$E : y^2 + a_1 xy + a_3 y = x^3 + a_2 x^2 + a_4 x + a_6$$

for E/K and manipulate it to produce a minimal proper regular model $\mathcal{C}$. Once we have this model, we will be able to read off all of the information we want. As we go along we will be making various assumptions. These assumptions are **cumulative**, and will be $\boxed{\text{boxed}}$ for clarity. We will delay the proofs of the various steps until after describing the complete algorithm.

Making a change of variables, we may assume that the Weierstrass equation has coefficients $\boxed{a_1, a_2, a_3, a_4, a_6 \in R}$. We let

$$b_2 = a_1^2 + 4a_2, \quad b_4 = a_1 a_3 + 2a_4, \quad b_6 = a_3^2 + 4a_6,$$
$$b_8 = a_1^2 a_6 + 4a_2 a_6 - a_1 a_3 a_4 + a_2 a_3^2 - a_4^2 = (b_2 b_6 - b_4^2)/4,$$
$$\Delta = -b_2^2 b_8 - 8b_4^3 - 27b_6^2 + 9b_2 b_4 b_6$$

	I_0	I_n $(n\geq 1)$	II	III	IV	I_0^*	I_n^* $(n\geq 1)$	IV^*	III^*	II^*
Kodaira symbol										
Special fiber $\tilde{C}$ (The numbers indicate multiplicities)	(diagram)	(diagram)	(diagram)	(diagram)	(diagram)	(diagram)	(diagram)	(diagram)	(diagram)	(diagram)
m = number of irred. components	1	n	1	2	3	5	$5+n$	7	8	9
$E(K)/E_0(K)$ $\cong \tilde{\mathcal{E}}(k)/\tilde{\mathcal{E}}^0(k)$	(0)	$\dfrac{\mathbb{Z}}{n\mathbb{Z}}$	(0)	$\dfrac{\mathbb{Z}}{2\mathbb{Z}}$	$\dfrac{\mathbb{Z}}{3\mathbb{Z}}$	$\dfrac{\mathbb{Z}}{2\mathbb{Z}}\times\dfrac{\mathbb{Z}}{2\mathbb{Z}}$	$\dfrac{\mathbb{Z}}{2\mathbb{Z}}\times\dfrac{\mathbb{Z}}{2\mathbb{Z}}$ n even $\dfrac{\mathbb{Z}}{4\mathbb{Z}}$ n odd	$\dfrac{\mathbb{Z}}{3\mathbb{Z}}$	$\dfrac{\mathbb{Z}}{2\mathbb{Z}}$	(0)
$\tilde{\mathcal{E}}^0(k)$	$\bar{E}(k)$	k^*	k^+	k^+	k^+	k^+	k^+	k^+	k^+	k^+

Entries below this line only valid for char$(k) = p$ as indicated

	I_0	I_n $(n\geq 1)$	II	III	IV	I_0^*	I_n^* $(n\geq 1)$	IV^*	III^*	II^*
char$(k)=p$			$p\neq 2,3$	$p\neq 2$	$p\neq 3$	$p\neq 2$	$p\neq 2$	$p\neq 3$	$p\neq 2$	$p\neq 2,3$
$v(\mathcal{D}_{E/K})$ (discriminant)	0	n	2	3	4	6	$6+n$	8	9	10
$f(E/K)$ (conductor)	0	1	2	2	2	2	2	2	2	2
behavior of j	$v(j)\geq 0$	$v(j)=-n$	$\bar{j}=0$	$\bar{j}=1728$	$\bar{j}=0$	$v(j)\geq 0$	$v(j)=-n$	$\bar{j}=0$	$\bar{j}=1728$	$\bar{j}=0$

Table 4.1: A Table of Reduction Types

be the usual quantities [AEC, III §1] associated to the given Weierstrass equation.

It is not necessary to assume that the original Weierstrass equation is minimal. When the algorithm terminates, the resulting Weierstrass equation will be minimal, so its discriminant will equal $v(\mathcal{D}_{E/K})$. Further, we will see that the smooth part of the final Weierstrass equation, considered as a scheme over R, is the identity component of $\mathcal{C}$, which will prove (9.1).

Step 1. If $\pi \nmid \Delta$, then the special fiber $\tilde{\mathcal{E}}/k$ is an elliptic curve, and we have

$$\text{Type I}_0, \quad v(\Delta) = 0, \quad m = 1, \quad f = 0, \quad c = 1.$$

Step 2. Assume $\boxed{\pi | \Delta}$. This means that $\tilde{E}$ has a singular point. Make a change of variables to move the singular point to $(0,0)$. Then $\boxed{\pi | a_3, \, a_4 \text{ and } a_6}$. If $\pi \nmid b_2$, then we have Type I$_n$ with $n = v(\Delta)$. More precisely, let k' be the splitting field over k of the polynomial $T^2 + a_1 T - a_2$. Then we have

$$\text{Type I}_n, \quad v(\Delta) = n \geq 1, \quad m = n, \quad f = 1.$$

Further, if $k' = k$, then E has split multiplicative reduction,

$$\tilde{\mathcal{E}}^0(k) \cong k^* \quad \text{and} \quad c = n;$$

whereas if $k' \neq k$, then E has non-split multiplicative reduction,

$$\tilde{\mathcal{E}}^0(k) \cong \{\alpha \in k' : \mathrm{N}_k^{k'}(\alpha) = 1\} \quad \text{and} \quad c = \begin{cases} 1 & \text{if } n \text{ is odd,} \\ 2 & \text{if } n \text{ is even.} \end{cases}$$

From now on, $\tilde{E}$ has a cusp and $\boxed{\tilde{\mathcal{E}}^0(k) \cong k^+}$. We are going transform the Weierstrass equation so as to make the a_i's more and more divisible by π. To keep track, we introduce the convenient notation

$$a_{i,r} = \pi^{-r} a_i.$$

Step 3. Assume now that $\boxed{\pi | b_2}$. If $\pi^2 \nmid a_6$, then

$$\text{Type II}, \quad m = 1, \quad f = v(\Delta), \quad c = 1.$$

Step 4. Assume that $\boxed{\pi^2 | a_6}$ (which implies that $\pi^2 | b_6$ and $\pi^2 | b_8$). If $\pi^3 \nmid b_8$, then

$$\text{Type III}, \quad m = 2, \quad f = v(\Delta) - 1, \quad c = 2.$$

Step 5. Assume that $\boxed{\pi^3 | b_8}$ (which implies that $\pi^2 | b_4$). If $\pi^3 \nmid b_6$, then Type IV. Let k' be the splitting field over k of $T^2 + a_{3,1} T - a_{6,2} = 0$. Then

$$\text{Type IV}, \quad m = 3, \quad f = v(\Delta) - 2, \quad c = \begin{cases} 3 & \text{if } k' = k, \\ 1 & \text{if } k' \neq k. \end{cases}$$

Step 6. Assume that $\boxed{\pi^3 | b_6}$. Then we can change coordinates to get

$$\boxed{\pi | a_1 \text{ and } a_2, \quad \pi^2 | a_3 \text{ and } a_4, \quad \text{and } \pi^3 | a_6}.$$

More precisely, the boxed assumptions up to this point show that we can factor

$$Y^2 + a_1 Y - a_2 \equiv (Y - \alpha)^2 \pmod{\pi},$$
$$Y^2 + a_{3,1} Y - a_{6,2} \equiv (Y - \beta)^2 \pmod{\pi},$$

and then the substitution $y' = y + \alpha x + \beta \pi$ will have the desired effect. Having done this, we consider the factorization over $\bar{k}$ of the polynomial

$$P(T) = T^3 + a_{2,1} T^2 + a_{4,2} T + a_{6,3}.$$

To assist in explicit computations, we note that P has discriminant

$$\mathrm{Disc}(P) = \pi^{-6}(-4a_2^3 a_6 + a_2^2 a_4^2 - 4a_4^3 - 27a_6^2 + 18a_2 a_4 a_6).$$

If $P(T)$ has distinct roots in $\bar{k}$ (i.e., if $\pi \nmid \mathrm{Disc}(P)$), then

$$\text{Type } I_0^*, \quad m = 5, \quad f = v(\Delta) - 4, \quad c = 1 + \#\{\alpha \in k : P(\alpha) = 0\}.$$

Step 7. If $P(T)$ has one simple root and one double root in $\bar{k}$, then

$$\text{Type } I_n^*, \quad m = n + 5, \quad f = v(\Delta) - 4 - n, \quad c = 2 \text{ or } 4.$$

If $p \neq 2$, then $n = v(\Delta) - 6$, so $m = v(\Delta) - 1$ and $f = 2$. For arbitrary p, one can calculate the values of n and c using the following subprocedure to Step 7.

Step 7. *(Subprocedure)* Translate x so that the double root of $P(T)$ is $T = 0$. Then $\pi^2 \nmid a_2$, $\pi^3 | a_4$, and $\pi^4 | a_6$. If the polynomial $Y^2 + a_{3,2} Y - a_{6,4}$ has distinct roots in $\bar{k}$, let k' be its splitting field. Then

$$\text{Type } I_1^*, \quad m = 6, \quad f = v(\Delta) - 5, \quad c = \begin{cases} 4 & \text{if } k' = k, \\ 2 & \text{if } k' \neq k. \end{cases}$$

If $Y^2 + a_{3,2} Y - a_{6,4}$ has a double root in $\bar{k}$, translate y so that the root is $Y = 0$. Then $\pi^3 | a_3$ and $\pi^5 | a_6$. If the polynomial $a_{2,1} X^2 + a_{4,3} X + a_{6,5}$ has distinct roots in $\bar{k}$, let k' be its splitting field. Then

$$\text{Type } I_2^*, \quad m = 7, \quad f = v(\Delta) - 6, \quad c = \begin{cases} 4 & \text{if } k' = k, \\ 2 & \text{if } k' \neq k. \end{cases}$$

If $a_{2,1}X^2 + a_{4,3}X + a_{6,5}$ has a double root in $\bar{k}$, translate x so that the root is $X = 0$. Then $\pi^4 | a_4$ and $\pi^6 | a_6$. If the polynomial $Y^2 + a_{3,3}Y - a_{6,6}$ has distinct roots in $\bar{k}$, let k' be its splitting field. Then

$$\text{Type } I_3^*, \quad m = 8, \quad f = v(\Delta) - 7, \quad c = \begin{cases} 4 & \text{if } k' = k, \\ 2 & \text{if } k' \neq k. \end{cases}$$

If $Y^2 + a_{3,3}Y - a_{6,6}$ has a double root in $\bar{k}$, etc. Continue this procedure until the quadratic polynomial which appears has distinct roots in $\bar{k}$. The process will terminate because after each two steps we have will have forced a_3, a_4, and a_6 to each be divisible by at least one additional power of π. This means that b_4, b_6, and b_8 are also divisible by at least one additional power of π, and hence the same is true of Δ. But the discriminant Δ is invariant under all of the translations involved, so the process will stop.

Step 8. Suppose now that $P(T)$ has a triple root in $\bar{k}$. Making a translation on x, we may assume that the root is $T = 0$, which means that $\boxed{\pi^2 | a_2, \ \pi^3 | a_4, \text{ and } \pi^4 | a_6}$. If the polynomial $Y^2 + a_{3,2}Y - a_{6,4}$ has distinct roots in $\bar{k}$, let k' be its splitting field. Then

$$\text{Type IV}^*, \quad m = 7, \quad f = v(\Delta) - 6, \quad c = \begin{cases} 3 & \text{if } k' = k, \\ 1 & \text{if } k' \neq k. \end{cases}$$

Step 9. Suppose now that $Y^2 + a_{3,2}Y - a_{6,4}$ has a double root in $\bar{k}$. Making a translation on y, we may assume that the root is $Y = 0$, which means that $\boxed{\pi^3 | a_3 \text{ and } \pi^5 | a_6}$. If $\pi^4 \nmid a_4$, then

$$\text{Type III}^*, \quad m = 8, \quad f = v(\Delta) - 7, \quad c = 2.$$

Step 10. Suppose that $\boxed{\pi^4 | a_4}$. If $\pi^6 \nmid a_6$, then

$$\text{Type II}^*, \quad m = 9, \quad f = v(\Delta) - 8, \quad c = 1.$$

Step 11. Finally, suppose that $\boxed{\pi^6 | a_6}$. Then the original Weierstrass equation was not minimal. The substitution $(x, y) = (\pi^2 x', \pi^3 y')$ leads to the equation

$$y'^2 + a_{1,1}x'y' + a_{3,3}y' = x'^3 + a_{2,2}x'^2 + a_{4,4}x' + a_{6,6}$$

with coefficients in R and discriminant $\Delta' = \pi^{-12}\Delta$. Go back to Step 1 and begin the algorithm again with this new equation. Note that we can only get to Step 11 a finite number of times, since each time we get here, the discriminant of the original Weierstrass equation must be divisible by an additional factor of π^{12}. Therefore the algorithm will terminate.

This concludes our description of Tate's algorithm. We are now going to give some indication of why the various steps in Tate's algorithm yield the stated conclusions. The idea is to start with the given Weierstrass equation and perform a sequence of blow-ups to produce a minimal regular model for E. In practice, we will really only need to carry out the blowing-up process until we are able to recognize which type of fiber is emerging.

Let $\mathcal{W} \subset \mathbb{P}^2_R$ be the scheme defined by the given Weierstrass equation, and let $\mathcal{W}^0/R$ be the largest subscheme of $\mathcal{W}$ that is smooth over R. In other words, $\mathcal{W}^0$ is formed by removing from $\mathcal{W}$ all singular points (if any) on its special fiber $\widetilde{\mathcal{W}}$. Just as in the description of Tate's algorithm, we will put a box around cumulative assumptions as we make them.

Proof of Step 1. The condition $\pi \nmid \Delta$ means that the special fiber $\widetilde{\mathcal{W}}$ is non-singular, so $\mathcal{W}$ itself is smooth over R. Hence $\mathcal{E} = \mathcal{C} = \mathcal{W}$, which shows that the special fiber is of Type I_0.

We assume now that $\boxed{\pi \mid \Delta}$, which means that the reduction $\tilde{E}$ has a singular point. (Equivalently, the special fiber $\widetilde{\mathcal{W}}$ is singular.) Making a linear change of variables, we may assume that the singular point is $(0,0) \in \tilde{E}$. This means that if we write

$$f(x, y) = y^2 + a_1 xy + a_3 y - x^3 - a_2 x^2 - a_4 x - a_6,$$

then $f(0,0) \equiv 0 \pmod{\pi}$, and further both partial derivatives $(\partial f/\partial x)(0,0)$ and $(\partial f/\partial y)(0,0)$ vanish modulo π. Hence $\boxed{\pi \mid a_3, a_4, a_6}$.

Proof of Step 2. This is the case that E has multiplicative reduction. We are going to leave it to the reader (exercise 4.37) to perform the blow-ups necessary to resolve the singularity in this case. At the end of this section (9.6) we will briefly explain another approach to analyzing multiplicative reduction using Tate's analytic models for elliptic curves over complete local fields. We will also prove the following lemma which covers Types I_1 and II.

Lemma 9.5. *Let R be a discrete valuation ring with fraction field K, let E/K be an elliptic curve given by a Weierstrass equation*

$$y^2 + a_1 xy + a_3 y = x^3 + a_2 x^2 + a_4 x + a_6$$

with coefficients in R, let $\mathcal{W} \subset \mathbb{P}^2_R$ be the R-scheme defined by this equation, and let $\mathcal{W}^0/R$ be the largest subscheme of $\mathcal{W}$ that is smooth over R.
(a) If $v(\Delta) = 1$, then $\mathcal{W}$ is regular, $\mathcal{C} = \mathcal{W}$, and $\mathcal{E} = \mathcal{W}^0$. The curve E has Type I_1 reduction.
(b) If $\pi \mid a_3, a_4, a_6$ and $\pi^2 \nmid a_6$, then $\mathcal{W}$ is regular, $\mathcal{C} = \mathcal{W}$, and $\mathcal{E} = \mathcal{W}^0$. The curve E has Type I_1 reduction if $\pi \nmid b_2$, and Type II reduction if $\pi \mid b_2$.

PROOF. (a) As described above, the fact that $\pi \mid \Delta$ means that we can make a linear change of variables to get $\pi \mid a_3, a_4, a_6$. This implies that $\pi \mid b_4, b_6, b_8$.

. If we make the assumption that $v(\Delta) = 1$, then we must have $v(b_8) = 1$, since all of the other terms in the formula for Δ are divisible by at least π^2. Writing b_8 in the form

$$b_8 = b_2 a_6 - a_1 a_3 a_4 + a_2 a_3^2 - a_4^2 \equiv b_2 a_6 \pmod{\pi^2},$$

we find that if $v(\Delta) = 1$, then $v(a_6) = 1$ and $\pi \nmid b_2$.

We now drop the assumption that $v(\Delta) = 1$ and prove that $\mathcal{W}$ is regular assuming only that $\pi | a_3, a_4, a_6$ and $\pi^2 \nmid a_6$. This will verify the first statements in both (a) and (b). We need to prove that $\mathcal{W}$ is regular at the singular point $(0,0) \in \widetilde{\mathcal{W}}$ on its special fiber. In other words, if we let $\mathfrak{m} = (\pi, x, y)$ be the maximal ideal corresponding to the singular point on the special fiber, then we must show that the local ring of $\mathcal{W}$ at $\mathfrak{m}$,

$$\mathcal{O}_{\mathcal{W},\mathfrak{m}} = \frac{R[x,y]_{\mathfrak{m}}}{(y^2 + a_1 xy + a_3 y - x^3 - a_2 x^2 - a_4 x - a_6)},$$

is a regular local ring. By assumption, $v(a_6) = 1$, so a_6 is a uniformizer for R. On the other hand, a_6 is in the ideal of $\mathcal{O}_{\mathcal{W},\mathfrak{m}}$ generated by x and y, since

$$a_6 = y^2 + a_1 xy + a_3 y - x^3 - a_2 x^2 - a_4 x \in x\mathcal{O}_{\mathcal{W},\mathfrak{m}} + y\mathcal{O}_{\mathcal{W},\mathfrak{m}}.$$

Therefore the maximal ideal (π, x, y) of $\mathcal{O}_{\mathcal{W},\mathfrak{m}}$ is generated by the two elements x and y, so $\mathcal{O}_{\mathcal{W},\mathfrak{m}}$ is a regular local ring. This proves that $\mathcal{W}$ is regular, and since it is clearly also proper over R, we find that $\mathcal{C} = \mathcal{W}$ and $\mathcal{E} = \mathcal{W}^0$. This proves the first part of (a) and (b).

The special fiber $\tilde{\mathcal{C}} = \widetilde{\mathcal{W}}$ is the curve

$$y^2 + \tilde{a}_1 xy = x^3 + \tilde{a}_2 x^2$$

in $\mathbb{A}_{\bar{k}}^2$. It will have a node (respectively cusp) at $(0,0)$ if the quadratic form $y^2 + \tilde{a}_1 xy - \tilde{a}_2 x^2$ has distinct roots (respectively a double root) in $\bar{k}$. The discriminant of this quadratic form is $\tilde{a}_1^2 + 4\tilde{a}_2 = \tilde{b}_2$, so $\widetilde{\mathcal{W}}$ has a node if $\pi \nmid b_2$ and a cusp if $\pi | b_2$. By definition, the special fiber is of Type I_1 if it has a node, and of Type II if it has a cusp, and we saw above that if $v(\Delta) = 1$, then $\pi \nmid b_2$. This completes the proof of (9.5). $\square$

Continuing on past Step 2, we now assume that $\boxed{\pi | b_2}$. Notice that b_2 is the discriminant of the quadratic form $y^2 + a_1 xy - a_2 x^2$, so this form has a double root in $\bar{k}$, say

$$y^2 + a_1 xy - a_2 x^2 \equiv (y - \alpha x)^2 \pmod{\pi}.$$

The substitution $y \to y + \alpha x$ allows us to assume that $\boxed{\pi | a_1, a_2}$. Notice that this substitution leaves the other a_i's and all of the b_i's unchanged.

Proof of Step 3. We are given that $\pi | a_3, a_4, a_6, b_2$ and that $\pi^2 \nmid a_6$. This is exactly the situation in (9.5b), which shows that the special fiber is of Type II.

Proof of Steps 4 and 5. We now add the assumption that $\boxed{\pi^2 | a_6}$, and recall that our model satisfies $\pi | a_1, a_2, a_3, a_4$. We are going to blow up the singular point $\pi = x = y = 0$ of $\mathcal{W}$ using the procedure described in (7.7). Thus the blow-up consists of the following three coordinate charts glued together in an appropriate fashion:

$$\mathcal{W}_1 : y_1^2 + a_1 x_1 y_1 + a_{3,1} y_1 = \pi x_1^3 + a_2 x_1^2 + a_{4,1} x_1 + a_{6,2},$$

$$\mathcal{W}' : 1 + a_1 x' + a_{3,1} x' = x'^3 y' + a_2 x'^2 + a_{4,1}\pi' x' + a_{6,2}\pi'^2, \quad \pi' y' = \pi,$$

$$\mathcal{W}'' : y''^2 + a_1 y'' + a_{3,1}\pi'' y'' = x'' + a_2 + a_{4,1}\pi'' + a_{6,2}\pi''^2, \quad \pi'' x'' = \pi.$$

Note that we are using the notation $a_{i,r} = \pi^{-r} a_i$ introduced earlier.

Looking at the special fibers of each of $\mathcal{W}_1$, $\mathcal{W}'$, and $\mathcal{W}''$, it is easy to verify that the special fiber of $\mathcal{W}''$ contains all of the components and all of the singular points of the special fiber of the blow-up. Thus all of the action will be happening on $\mathcal{W}''$. Further, it is not hard to see that the projection $\mathbb{A}_R^3 \to \mathbb{A}_R^2$ induced by the natural inclusion $R[y'', \pi''] \hookrightarrow R[x'', y'', \pi'']$ maps $\mathcal{W}''$ isomorphically to the subscheme of $\mathbb{A}_R^2 = \operatorname{Spec} R[y'', \pi'']$ given by the single equation

$$y''^2 \pi'' + \pi a_{1,1} y'' \pi'' + a_{3,1} y'' \pi''^2 = \pi + \pi a_{2,1}\pi'' + a_{4,1}\pi''^2 + a_{6,2}\pi''^3.$$

(The map in the other direction is $x'' \to y''^2 + a_1 y'' + a_{3,1}\pi'' y'' - a_2 - a_{4,1}\pi'' - a_{6,2}\pi''^2$.)

We next take the closure of this scheme in $\mathbb{P}_R^2$. This means we homogenize

$$y'' = Y/X \qquad \text{and} \qquad \pi'' = Z/X,$$

which yields the scheme $\mathcal{V} \subset \mathbb{P}_R^2 = \operatorname{Proj} R[X, Y, Z]$ given by the equation

$$\mathcal{V} : Y^2 Z + \pi a_{1,1} XYZ + a_{3,1} YZ^2 = \pi X^3 + \pi a_{2,1} X^2 Z + a_{4,1} XZ^2 + a_{6,2} Z^3.$$

Notice that $\mathcal{V}$ is a model for E/K, since its generic fiber is isomorphic over K to the original Weierstrass equation defining E.

To find the special fiber of $\mathcal{V}$, we set $\pi = 0$, so $\tilde{\mathcal{V}}$ is the curve in $\mathbb{A}_k^2$ given by the equation

$$\tilde{\mathcal{V}} : (Y^2 + \tilde{a}_{3,1} YZ - \tilde{a}_{4,1} XZ - \tilde{a}_{6,2} Z^2) Z = 0.$$

Thus the special fiber consists of the line $Z = 0$ and the (possibly degenerate) conic

$$Y^2 + \tilde{a}_{3,1} YZ - \tilde{a}_{6,2} Z^2 = \tilde{a}_{4,1} XZ. \tag{$*$}$$

This line and conic intersect at the point $Y = Z = 0$ with multiplicity 2.

Suppose first that $\pi^3 \nmid b_8$, which is the condition for Step 4. The formula defining b_8 is

$$b_8 = a_1^2 a_6 + 4a_2 a_6 - a_1 a_3 a_4 + a_2 a_3^2 - a_4^2,$$

and our cumulative assumptions imply that every term except the last one is divisible by π^3. Hence $\pi^3 \nmid b_8$ if and only if $\pi^2 \nmid a_4$, and this in turn is equivalent to the assertion that $(*)$ is a non-singular conic (as opposed to being two lines). So $\pi^3 \nmid b_8$ implies that $\tilde{V}$ consists of two non-singular rational curves intersecting at a single point with multiplicity 2, which is exactly Type III.

Next assume that $\boxed{\pi^3 | b_8}$, or equivalently that $\pi^2 | a_4$. This means that the special fiber $\tilde{V}$ is given by the equation

$$(Y^2 + \tilde{a}_{3,1} YZ - \tilde{a}_{6,2} Z^2)Z = 0, \qquad (**)$$

so over $\bar{k}$ it consists of three lines. These lines will be distinct if and only if the quadratic form $Y^2 + \tilde{a}_{3,1} YZ - \tilde{a}_{6,2} Z^2$ has distinct roots, which is equivalent to the condition that its discriminant $\tilde{a}_{3,1}^2 + 4\tilde{a}_{6,2} = \tilde{b}_{6,2}$ does not vanish. So if we assume that $\pi^3 \nmid b_6$, which is exactly the condition for Step 5, then $\tilde{V}$ consists of three non-singular lines intersecting transversally at a single point, which is a fiber of Type IV. Further, the number c of components defined over k will be 3 if $Y^2 + \tilde{a}_{3,1} YZ - \tilde{a}_{6,2} Z^2$ splits into linear factors over k and will be 1 otherwise. This completes our consideration of Steps 4 and 5 of Tate's algorithm.

We now assume that $\boxed{\pi^3 | b_6}$. This means that the quadratic form in $(**)$ has a double root in $\bar{k}$. Making a translation $Y \to Y + \beta Z$ moves the double root to $Y = 0$. We now have $\boxed{\pi | a_1, a_2, \ \pi^2 | a_3, a_4, \text{ and } \pi^3 | a_6}$. The equation for V can be written as

$$V : Y^2 Z + \pi a_{1,1} XYZ + \pi a_{3,2} YZ^2 = \pi X^3 + \pi a_{2,1} X^2 Z + \pi a_{4,2} XZ^2 + \pi a_{6,3} Z^3,$$

and its special fiber $Y^2 Z = 0$ consists of the line $Y = 0$ with multiplicity 2 and the line $Z = 0$ with multiplicity 1.

The next step is to blow-up the double line $\pi = Y = 0$. To ease notation, we are going to dehomogenize at the same time, so we set

$$X = x_1, \quad Y = \pi y_2, \quad Z = 1$$

and divide the equation for V by π. (The reason for the subscripts on x_1 and y_2 is that they are related to our original Weierstrass coordinates by the formulas $x = \pi x_1$ and $y = \pi^2 y_2$. For the rest of this proof we will use the notation $x = \pi^r x_r$ and $y = \pi^r y_r$.)

We now have the scheme

$$\mathcal{V}_0 : \pi y_2^2 + \pi a_{1,1} x_1 y_2 + \pi a_{3,2} y_2 = x_1^3 + a_{2,1} x_1^2 + a_{4,2} x_1 + a_{6,3},$$

and the total blow-up consists of $\mathcal{V}$ and $\mathcal{V}_0$ glued together in the natural way. The special fiber of the total blow-up is thus formed by gluing together the two pieces

$$\tilde{\mathcal{V}} : Y^2 Z = 0,$$

$$\tilde{\mathcal{V}}_0 : 0 = x_1^3 + \tilde{a}_{2,1} x_1^2 + \tilde{a}_{4,2} x_1 + \tilde{a}_{6,3}.$$

There are now three cases to consider, depending on the number of distinct roots (in $\bar{k}$) of the polynomial

$$P(T) = T^3 + \tilde{a}_{2,1} T^2 + \tilde{a}_{4,2} T + \tilde{a}_{6,3}.$$

Proof of Step 6. For Step 6 we assume that $P(T)$ has distinct roots in $\bar{k}$. Then $\tilde{\mathcal{V}}_0$ consists of three distinct lines, so the blow-up is composed of the double line $Y^2 = 0$ together with four lines of multiplicity 1 intersecting it. This means we have a fiber of Type I_0^*. Further, there is always one component $Z = 0$ of multiplicity 1 defined over k, and the other multiplicity-1 components correspond to the roots of $P(T)$. Hence the number c of multiplicity-1 components is one more than the number of roots of $P(T)$ in k.

Proof of Step 7. For Step 7 we assume that $P(T)$ has one simple root and one double root. Making a translation of the form $x_1 \to x_1 + \gamma$, we may assume that the double root is $T = 0$, which implies that $\pi^2 \nmid a_2$, $\pi^3 | a_4$, and $\pi^4 | a_6$. The special fiber of $\mathcal{V}_0$ is now

$$\tilde{\mathcal{V}}_0 : (x_1 + \tilde{a}_{2,1}) x_1^2 = 0,$$

so we need to blow-up $\mathcal{V}_0$ along the double line $\pi = x_1 = 0$. To do this, we make the substitution $x_1 = \pi x_2$ and divide by π to obtain the scheme

$$\mathcal{V}_1 : y_2^2 + \pi a_{1,1} x_2 y_2 + a_{3,2} y_2 = \pi^2 x_2^3 + \pi a_{2,1} x_2^2 + \pi a_{4,3} x_2 + a_{6,4}.$$

Our total special fiber is now composed of the following components: the simple lines $Z = 0$ and $x_1 + \tilde{a}_{2,1} = 0$, the double lines $Y = 0$ and $x_1 = 0$, and the special fiber of $\mathcal{V}_1$. Notice how a fiber of Type I_n^* is emerging.

The special fiber of $\mathcal{V}_1$ is

$$\tilde{\mathcal{V}}_1 : y_2^2 + \tilde{a}_{3,2} y_2 - \tilde{a}_{6,4} = 0.$$

If this quadratic equation has distinct roots in $\bar{k}$, then $\tilde{\mathcal{V}}_1$ consists of two distinct lines, and we have a fiber of Type I_1^*. Further, there are already two multiplicity-1 components defined over k, namely the lines $Z = 0$

and $x_1 + \tilde{a}_{2,1} = 0$, so $c = 4$ if the polynomial $y_2^2 + \tilde{a}_{3,2}y_2 - \tilde{a}_{6,4}$ has its roots in k, and $c = 2$ otherwise.

If the polynomial $y_2^2 + \tilde{a}_{3,2}y_2 - \tilde{a}_{6,4}$ has a double root, then making a translation on y_2 allows us to take the double root to be $y_2 = 0$. This means that $\pi^3 | a_3$ and $\pi^5 | a_6$, and the special fiber of $\mathcal{V}_1$ is $y_2^2 = 0$. We blow-up $\mathcal{V}_1$ along this double line by making the substitution $y_2 = \pi y_3$ and dividing by π, which gives the scheme

$$\mathcal{V}_2 : \pi y_3^2 + \pi a_{1,1} x_2 y_3 + \pi a_{3,3} y_3 = \pi x_2^3 + a_{2,1} x_2^2 + a_{4,3} x_2 + a_{6,5}.$$

The special fiber of $\mathcal{V}_2$ is

$$\tilde{\mathcal{V}}_2 : \tilde{a}_{2,1} x_2^2 + \tilde{a}_{4,3} x_2 + \tilde{a}_{6,5} = 0.$$

If this quadratic equation has distinct roots in $\bar{k}$, then $\tilde{\mathcal{V}}_2$ consists of two distinct lines, we have a fiber of Type I_2^*, and we're done. Otherwise the quadratic equation has a double root and $\tilde{\mathcal{V}}_2$ is a double line, so we translate to make the double line $x_2^2 = 0$, blow it up using $x_2 = \pi x_3$, and continue on our merry way.

As explained during the description of the Step 7 subprocedure, this process will eventually terminate. The point is that the special fiber at each stage looks like

$$\tilde{\mathcal{V}}_n : \begin{cases} y_\nu + \tilde{a}_{3,\nu}y_\nu - \tilde{a}_{6,2\nu} = 0 & \text{if } n = 2\nu - 3 \text{ is odd,} \\ \tilde{a}_{2,1} x_\nu^2 + \tilde{a}_{4,\nu+1} x_\nu + \tilde{a}_{6,2\nu+1} = 0 & \text{if } n = 2\nu - 2 \text{ is even.} \end{cases}$$

So each two steps of the algorithm force a_3, a_4, and a_6 to be divisible by an additional power of π. This implies the same for b_4, b_6, and b_8, and hence also for Δ. But Δ is invariant under the various translations we are using, which shows that eventually we must get a quadratic polynomial with distinct roots. We will leave for the reader the easy verification that if the residue characteristic $p \neq 2$, then the fiber $\tilde{\mathcal{V}}_n$ consists of two distinct lines precisely when $n = v(\Delta) - 6$. This concludes our discussion of Step 7 of Tate's algorithm.

Proof of Step 8. We now assume that the polynomial $P(T)$ has a triple root in $\bar{k}$, which after a translation we can take to be $T = 0$. This means that $\boxed{\pi^2 | a_2, \ \pi^3 | a_4, \text{ and } \pi^4 | a_6}$, so the special fiber $\tilde{\mathcal{V}}_0$ is the triple line $x_1^3 = 0$. Our total special fiber now consists of the simple line $Z = 0$, the double line $Y^2 = 0$, and the triple line $x_1^3 = 0$. The scheme $\mathcal{V}_0$ is regular except at the points on the special fiber satisfying

$$\pi = x_1 = y_2^2 + a_{3,2}y_2 - a_{6,4} = 0.$$

Making a translation $y_2 \to y_2 + \gamma$ allows us to assume that the polynomial $y_2^2 + a_{3,2}y_2 - a_{6,4}$ has $y_2 = 0$ as a root in $\bar{k}$. This may require making a quadratic extension of k, in which case the components of the special fiber corresponding to the two roots of the quadratic polynomial will not be defined over k.

We now have $\boxed{\pi^5 | a_6}$, and we blow-up $\mathcal{V}_0$ at the point $\pi = x_1 = y_2 = 0$ by making the change of variables

$$\pi = \pi' y', \qquad x_1 = x' y', \qquad y = y'.$$

(This is chart 2 of the blow-up as described in (7.7).) This yields the scheme $\mathcal{U}' \subset \mathbb{A}_R^3 = \operatorname{Spec} R[\pi', x', y']$ given by the equations

$$\mathcal{U}' : \pi + \pi a_{1,1} x' + a_{3,2} \pi' = x'^3 y' + \pi a_{2,2} x'^2 + a_{4,3} x' \pi' + a_{6,5} \pi'^2, \quad \pi = \pi' y'.$$

The special fiber of $\mathcal{U}'$ consists of three components, which we label as

$$F_1' : \pi' = x' = 0, \quad F_2' : \pi' = y' = 0, \quad F_3' : y' = \tilde{a}_{3,2} - \tilde{a}_{4,3} x' - \tilde{a}_{6,5} \pi' = 0.$$

Notice that when we glue $\mathcal{U}'$ to $\mathcal{V}_0$, F_1' is identified with the multiplicity-3 component $x_1^3 = 0$ of $\tilde{\mathcal{V}}_0$. Our next step is to compute the multiplicities of the new components F_2' and F_3'. To do this, we rewrite the equation for $\mathcal{U}'$ as

$$\pi' \{ y' + a_{1,1} x' y' + a_{3,2} - a_{2,2} x'^2 y' - a_{4,3} x' - a_{6,5} \pi' \} = x'^3 y'.$$

The function x' does not vanish identically on F_2', so it is a unit in the local ring $\mathcal{O}_{F_2'}$. Similarly, since we are making the Step 8 assumption that $\pi^3 \nmid a_3$, the quantity in braces is also a unit in $\mathcal{O}_{F_2'}$. It follows that both π' and y' are uniformizers for $\mathcal{O}_{F_2'}$; that is, they each vanish to order 1 on F_2', so

$$\operatorname{ord}_{F_2'}(\tilde{\mathcal{U}}') = \operatorname{ord}_{F_2'}(\pi) = \operatorname{ord}_{F_2'}(\pi' y') = 2.$$

We leave for the reader the analogous verification that $\operatorname{ord}_{F_3'}(\tilde{\mathcal{U}}') = 1$.

But we're not done with Step 8, because we have to perform an identical blow-up of $\mathcal{V}_0$ at the singular point $\pi = x_1 = y_2 + a_{3,2} = 0$. This gives another pair of components, one of multiplicity 2 and one of multiplicity 1. The resulting configuration is of Type IV^*, which completes the verification of Step 8 of Tate's algorithm.

Proof of Step 9. For this step we have $\boxed{\pi^3 | a_3}$, so the scheme $\mathcal{U}'$ given above is singular at the point $\pi' = x' = y' = 0$. We blow it up at that point by making the substitution

$$\pi' = \pi'' x'', \qquad x' = x'', \qquad y' = y'' x''.$$

This gives the scheme $\mathcal{U}'' \subset \mathbb{A}_R^3 = \operatorname{Spec} R[\pi'', x'', y'']$ defined by the equations

$$\mathcal{U}'' : y'' \pi'' + a_{1,1} x'' y'' \pi'' + a_{3,3} x'' y'' \pi''^2$$
$$= x''^2 y'' + a_{2,2} x''^2 y'' \pi'' + a_{4,3} \pi'' + a_{6,5} \pi''^2,$$
$$\pi = x''^2 y'' \pi''.$$

Under our Step 9 assumption that $\pi^4 \nmid a_4$, we find that $\tilde{\mathcal{U}}''$ consists of the following four components:

$$F_1'' : x'' = \pi'' = 0, \qquad F_2'' : y'' = \pi'' = 0,$$
$$F_3'' : x'' = y'' - \tilde{a}_{4,3} - \tilde{a}_{6,5}\pi'' = 0,$$
$$F_4'' : y'' = \tilde{a}_{4,3} + \tilde{a}_{6,5}\pi'' = 0.$$

We compute the multiplicity of F_1'' in the fiber by writing

$$\pi'' \left\{ y'' + a_{1,1}x''y'' + a_{3,3}x''y''\pi'' - a_{2,2}x''^2 y'' - a_{4,3} - a_{6,5}\pi'' \right\} = x''^2 y''.$$

The function in braces does not vanish identically on F_1'', so it is a unit in $\mathcal{O}_{F_1''}$. This means that x'' is a uniformizer for F_1'' and $\text{ord}_{F_1''}(\pi'') = 2\,\text{ord}_{F_1''}(x'') = 2$. Hence

$$\text{ord}_{F_1''}(\tilde{\mathcal{U}}'') = \text{ord}_{F_1''}(\pi) = \text{ord}_{F_1''}(x''^2 y'' \pi'') = 4.$$

So we now have a chain of components of multiplicities 1, 2, 3, and 4. Notice how the fibers of Type III* and II* are emerging.

A similar calculation shows that $\text{ord}_{F_2''}(\tilde{\mathcal{U}}'') = 2$. Further, our Step 9 assumption that $\pi^4 \nmid a_4$ implies that $\mathcal{U}''$ is regular at the point $\pi'' = x'' = y'' = 0$ where F_1'' and F_2'' intersect. Hence there is a multiplicity-2 component attached to the multiplicity-4 component of the regular minimal model. This means that the fiber is of Type III*, which completes our analysis of Step 9. For those who wish to recover the full Type III* fiber, we mention that $\mathcal{U}''$ is singular at the intersection of F_1 and F_3, that is, at the point $\pi'' = x'' = y'' - a_{4,3} = 0$.

Proof of Step 10. We now assume that $\boxed{\pi^4 | a_4}$. Then $\mathcal{U}''$ is singular at the point $\pi'' = x'' = y'' = 0$, so we blow it up using the substitution

$$\pi'' = \pi'''y''', \qquad x'' = x'''y''', \qquad y'' = y'''.$$

This gives the scheme

$$\mathcal{U}''' : \pi''' + a_{1,1}x'''y'''\pi''' + a_{3,3}x'''y'''^2\pi'''^2$$
$$= x'''^2 y''' + a_{2,2}x'''^2 y'''^2\pi''' + a_{4,4}x'''^2 y'''^3\pi'''^2 + a_{6,5}\pi'''^2,$$
$$\pi = x'''^2 y'''^4 \pi'''.$$

The special fiber $\tilde{\mathcal{U}}'''$ consists of four components,

$$F_1''' : x''' = \pi''' = 0, \qquad\qquad F_2''' : y''' = \pi''' = 0,$$
$$F_3''' : x''' = 1 - \tilde{a}_{6,5}\pi''' = 0, \qquad F_4''' : y''' = 1 - \tilde{a}_{6,5}\pi''' = 0.$$

We compute the multiplicity of F_2''' in the special fiber by writing

$$\pi'''\{1 + a_{1,1}x'''y''' + a_{3,3}x'''y'''^2\pi''' - a_{2,2}x'''^2y'''^2$$
$$- a_{4,4}x'''^2y'''^3\pi''' - a_{6,5}\pi'''\} = x'''^2y'''.$$

Both x''' and the quantity in braces are units in $\mathcal{O}_{F_2'''}$, which shows that π''' and y''' each vanish to order 1 on F_2'''. This allows us to compute

$$\mathrm{ord}_{F_2'''}(\tilde{\mathcal{U}}''') = \mathrm{ord}_{F_2'''}(\pi) = \mathrm{ord}_{F_2'''}(x'''^2y'''^4\pi''') = 5.$$

Further, the Step 10 assumption that $\pi^6 \nmid a_6$ implies that $\mathcal{U}'''$ is regular at the intersection point $\pi''' = x''' = y''' = 0$ of F_1''' and F_2'''. Hence the appearance of the multiplicity-5 component F_2''' tells us that the fiber is of Type II*. As usual, we leave for the reader the enthralling task of performing the additional blow-ups necessary to find the other II* components.

Proof of Step 11. Finally, suppose that $\boxed{\pi^6 | a_6}$. Our cumulative assumptions to this point are that $\pi | a_1$, $\pi^2 | a_2$, $\pi^3 | a_3$, $\pi^4 | a_4$, and $\pi^6 | a_6$. In $\mathcal{U}'''$ we make the substitutions

$$\pi''' = 1/\pi y_2^2, \qquad x''' = x_3^2/\pi y_2^3 \qquad y''' = \pi y_2^2/x_3,$$

which leads to the R-scheme

$$y_3^2 + a_{1,1}x_2y_3 + a_{3,3}y_3 = x_2^3 + a_{2,2}x_2^2 + a_{4,4}x_2 + a_{6,6}$$

defined by a Weierstrass equation whose discriminant is $\pi^{-12}\Delta$. We can now begin again at Step 1 using this "smaller" Weierstrass equation. Note that each time we pass through Step 11, we will have shown that the original discriminant is divisible by an additional π^{12}. Therefore the algorithm will terminate. This concludes the proof of Tate's algorithm (9.4). $\qquad\square$

Remark 9.6. During our verification of Tate's algorithm (9.4), we left the case of multiplicative reduction (Type I_n) for the reader to analyze. There is another approach to multiplicative reduction using Tate's p-adic analytic uniformization. We will describe Tate's uniformization in the next chapter (V.3.1, V.5.3), but briefly, if E has split multiplicative reduction and K is a complete local field, then there is a $q \in K^*$ with $v(q) = v(\Delta) > 0$ and an isomorphism of groups

$$K^*/q^{\mathbb{Z}} \xrightarrow{\sim} E(K).$$

This isomorphism is given by v-adically convergent power series. Further, the isomorphism identifies the subgroups $R^* \cong E_0(K)$, so we get isomorphisms

$$E(K)/E_0(K) \xrightarrow{\sim} K^*/q^{\mathbb{Z}}R^* \xrightarrow{\sim} \mathbb{Z}/n\mathbb{Z}.$$

Here the second map is induced by the valuation $v : K^* \to \mathbb{Z}$, and

$$n = v(q) = v(\Delta).$$

We know that $E(K) = \mathcal{E}(R)$, and it is clear from the definitions that $E_0(K) = \mathcal{W}^0(R)$. Further, one can show that $\mathcal{W}^0(R) = \mathcal{E}^0(R)$, either by a direct calculation or using the argument in (V §4). It follows that $\mathcal{E}(R)/\mathcal{E}^0(R) \cong \mathbb{Z}/n\mathbb{Z}$, and then (9.2b) gives $c = n$.

Similarly, if E has non-split reduction, then (V.5.4) says that we can find an unramified quadratic extension K'/K with residue field k' such that E has split reduction over K'. Then

$$E(K') \cong K'^*/q^{\mathbb{Z}},$$
$$E(K) \cong \{u \in K'^*/q^{\mathbb{Z}} : \mathrm{N}_K^{K'}(u) \in q^{\mathbb{Z}}/q^{2\mathbb{Z}}\},$$
$$E_0(K) \cong \{u \in R'^* : \mathrm{N}_K^{K'}(u) = 1\} \cong \{u \in k'^* : \mathrm{N}_k^{k'}(u) = 1\}.$$

Note the last isomorphism depends on the fact that K'/K is unramified. Finally, we have

$$\tilde{\mathcal{E}}(k)/\tilde{\mathcal{E}}^0(k) \cong \mathcal{E}(R)/\mathcal{E}^0(R) \cong E(K)/E_0(K) \cong (\mathrm{N}_K^{K'})^{-1}(q^{\mathbb{Z}})/q^{\mathbb{Z}}(\mathrm{N}_K^{K'})^{-1}(1).$$

The fact that K'/K is unramified means that the norm map is surjective on units, $\mathrm{N}_K^L : R'^* \to R^*$, from which one easily deduces that this last group is trivial if n is odd, and has order 2 if n is even.

PROOF (of Corollary 9.1). If we start with a minimal Weierstrass equation for E/K, then we never get to Step 11 of Tate's algorithm, so the original equation defining $\mathcal{W}$ never changes. Tracing through the various stages of Tate's algorithm, we see that the non-singular part $\mathcal{W}^0$ of the Weierstrass equation ends up as an open subset of the minimal regular model $\mathcal{C}$. Since $\mathcal{W}^0$ clearly contains the image of the zero section, and since the special fiber $\tilde{\mathcal{W}}$ is irreducible, we see that $\tilde{\mathcal{W}}$ is the identity component of the special fiber of $\mathcal{C}$. Equivalently, $\mathcal{W}^0 = \mathcal{E}^0$. (For an alternative proof of (9.1) which uses a bit more algebro-geometric machinery and does not rely on a case-by-case analysis, see Liu [1].)   □

PROOF (of Corollary 9.2). (a) First, the equality $E(K) = \mathcal{E}(R)$ follows from the definition of the Néron model (5.1.3). Next we observe that the definitions of E_0 and $\mathcal{W}^0$ are both given in terms of the reduction of the given Weierstrass equation, so $E_0(K) = \mathcal{W}^0(R)$ is automatic. Now (9.1) says that $\mathcal{W}^0 = \mathcal{E}^0$, so we get the middle equality $E_0(K) = \mathcal{E}^0(R)$. Finally, $E_1(K)$ and $\mathcal{E}^1(R)$ each consists of the points which reduce to the identity on the special fiber, and these reductions are compatible since we already know that $E_0(K) = \mathcal{E}^0(R)$. This proves the third equality $E_1(K) = \mathcal{E}^1(R)$.

(b) The isomorphisms $E(K)/E_0(K) \cong \mathcal{E}(R)/\mathcal{E}^0(R)$ and $E(K)/E_1(K) \cong \mathcal{E}(R)/\mathcal{E}^1(R)$ are immediate from the identifications proven in (a). Further, the reduction map $\mathcal{E}(R) \to \tilde{\mathcal{E}}(k)$ has kernel $\mathcal{E}^1(R)$, and the inverse image of $\tilde{\mathcal{E}}^0(k)$ is $\mathcal{E}^0(R)$ by definition, which gives the injectivity of the right-hand maps. Finally, if R is complete (or merely Henselian), then the reduction map $\mathcal{E}(R) \to \tilde{\mathcal{E}}(k)$ is surjective (6.4a), so in this case the right-hand maps are isomorphisms.

(c) We have $\tilde{E}_{\mathrm{ns}}(k) = \widetilde{W}^0(k)$ directly from the definitions. Now (9.1) implies that $\widetilde{W}^0(k) = \tilde{\mathcal{E}}^0(k)$, which gives the desired result.

(d) From (b) we have an injection $E(K)/E_0(K) \hookrightarrow \tilde{\mathcal{E}}(k)/\tilde{\mathcal{E}}^0(k)$. The group $\tilde{\mathcal{E}}(k)/\tilde{\mathcal{E}}^0(k)$ is formed by looking at the special fiber $\tilde{\mathcal{C}}$ of the minimal proper regular model and taking the components that have multiplicity 1 and are defined over k. A quick perusal of the list of reduction types shows that only Type I_n has more than four multiplicity-1 components. Further, Tate's algorithm (Step 2, see also (9.6)) says that if a fiber of Type I_n has non-split reduction, then it has at most two components defined over k. This proves that $E(K)/E_0(K)$ has order at most 4 unless E/K has split multiplicative reduction. Finally, if E/K has split multiplicative reduction, say with a Type I_n fiber, then Step 2 of Tate's algorithm says that n equals the valuation of the minimal discriminant, which is also equal to $-v(j(E))$. $\qquad\square$

§10. The Conductor of an Elliptic Curve

The conductor of an elliptic curve E/K is a quantity which measures the arithmetic complexity of E/K, similar in some ways to the minimal discriminant. Just like the discriminant, the conductor is a product over the primes $\mathfrak{p}$ at which E has bad reduction, but the exponent of $\mathfrak{p}$ is defined in terms of the representation of the inertia group on the torsion subgroup of E. The conductor is an important quantity which appears in the functional equation of the L-series of E, in the modular parametrization of elliptic curves over $\mathbb{Q}$, and in various questions concerning the cohomology of E.

Before defining the conductor, we briefly recall some standard facts about local fields. For more details, see Serre [4]. Let K be a local field of residue characteristic p, let L/K be a finite Galois extension with normalized valuation v_L and ring of integers R_L, and let $G(L/K)$ be the Galois group of L/K. Then for each integer $i \geq -1$, the ith-*higher ramification group of L/K* is the subgroup of $G(L/K)$ defined by

$$G_i(L/K) = \{\sigma \in G(L/K) : v_L(\alpha^\sigma - \alpha) \geq i + 1 \text{ for all } \alpha \in R_L\}.$$

We write
$$g_i(L/K) = \#G_i(L/K)$$
for the order of the ith-ramification group. One should think of the higher ramification groups as measuring the extent to which the extension L/K is wildly ramified. The following lemma records some basic facts about the higher ramification groups.

Lemma 10.1. *Let L/K be a finite Galois extension of local fields.*
(a) *The higher ramification groups $G_i(L/K)$ are normal subgroups of $G(L/K)$.*
(b) *$G_{-1}(L/K) = G(L/K)$.*
(c) *$G_0(L/K)$ is the inertia group of L/K.*
(c) *$\big[G_0(L/K) : G_1(L/K)\big]$ is relatively prime to p.*
(d) *$G_1(L/K)$ is a p-group. Thus L/K is wildly ramified if and only if $G_1(L/K) \neq 1$.*

PROOF. See Serre [4], especially Chapter IV, Proposition 1 and Corollaries 1 and 3 to Proposition 7. $\square$

The conductor of an elliptic curve consists of two pieces, a tame part and a wild part. It turns out that if the residue characteristic p is at least 5, then the wild part will be zero. So if one is willing to ignore residue characteristics 2 and 3, then $\delta(E/K)$ can just be set equal to 0 in the following definition.

Definition. Let E/K be an elliptic curve defined over a local field of residue characteristic p, and let $I(\bar{K}/K)$ be the absolute inertia group of K. Fix a prime ℓ different from p, let $V_\ell(E) = T_\ell(E) \otimes_{\mathbb{Z}_\ell} \mathbb{Q}_\ell$ be the ℓ-adic Tate module of E, and write $V_\ell(E)^{I(\bar{K}/K)}$ for the subspace of $V_\ell(E)$ that is fixed by $I(\bar{K}/K)$. The *tame part of the conductor of E/K* is the quantity
$$\varepsilon(E/K) = \dim_{\mathbb{Q}_\ell}\left(V_\ell(E)/V_\ell(E)^{I(\bar{K}/K)}\right) = 2 - \dim_{\mathbb{Q}_\ell}\left(V_\ell(E)^{I(\bar{K}/K)}\right).$$

Next let $L = K\big(E[\ell]\big)$. Then the *wild part of the conductor of E/K* is the quantity
$$\delta(E/K) = \sum_{i=1}^{\infty} \frac{g_i(L/K)}{g_0(L/K)} \dim_{\mathbb{F}_\ell}\left(E[\ell]/E[\ell]^{G_i(L/K)}\right).$$

The *exponent of the conductor of E/K* is the sum of the tame and wild parts,
$$f(E/K) = \varepsilon(E/K) + \delta(E/K).$$

The conductor is a representation-theoretic quantity, since it is defined in terms of the action of the Galois group $G(\bar{K}/K)$ on the torsion subgroup of E. The following generalization of the criterion of Néron-Ogg-Shafarevich [AEC, VII.7.1] provides a geometric interpretation for the tame part of the conductor. We note that in many books (including [AEC, C §16]) one finds this geometric description (10.2b) used as the "definition" of the conductor.

Theorem 10.2. *Let K be a local field of residue characteristic p, and let E/K be an elliptic curve.*
(a) *The tame part of the conductor of E/K is given by*

$$\varepsilon(E/K) = \begin{cases} 0 & \text{if } E \text{ has good reduction,} \\ 1 & \text{if } E \text{ has multiplicative reduction,} \\ 2 & \text{if } E \text{ has additive reduction.} \end{cases}$$

(b) *If E/K has good or multiplicative reduction, or if $p \geq 5$, then*

$$\delta(E/K) = 0 \quad \text{and} \quad f(E/K) = \begin{cases} 0 & \text{if } E \text{ has good reduction,} \\ 1 & \text{if } E \text{ has multiplicative reduction,} \\ 2 & \text{if } E \text{ has additive reduction.} \end{cases}$$

(c) *In all cases, the exponent of the conductor $f(E/K)$ is an integer which is independent of the choice of ℓ.*

PROOF. (a) Notice that

$$\begin{aligned} \varepsilon(E/K) = 0 &\Longleftrightarrow V_\ell(E)^{I(\bar{K}/K)} = V_\ell(E) \\ &\Longleftrightarrow I(\bar{K}/K) \text{ acts trivially on } T_\ell(E) \\ &\Longleftrightarrow T_\ell(E) \text{ is unramified.} \end{aligned}$$

So the assertion that

$$\varepsilon(E/K) = 0 \Longleftrightarrow E/K \text{ has good reduction}$$

is precisely the criterion of Néron-Ogg-Shafarevich [AEC, VII.7.1]. We are going to mimic the proof of [AEC, VII.7.1] to obtain a somewhat stronger result. This proof is taken from Serre-Tate [1].

Let K^{nr} be the maximal unramified extension of K, and consider the two exact sequences

$$0 \longrightarrow E_0(K^{\mathrm{nr}}) \longrightarrow E(K^{\mathrm{nr}}) \longrightarrow E(K^{\mathrm{nr}})/E_0(K^{\mathrm{nr}}) \longrightarrow 0,$$
$$0 \longrightarrow E_1(K^{\mathrm{nr}}) \longrightarrow E_0(K^{\mathrm{nr}}) \longrightarrow \tilde{E}_{\mathrm{ns}}(\bar{k}) \longrightarrow 0.$$

Here $\bar{k}$, the residue field of K^{nr}, is the algebraic closure of the residue field of K. We note that $E(K^{\mathrm{nr}})/E_0(K^{\mathrm{nr}})$ is a finite group from (9.2d), and that $E_1(K^{\mathrm{nr}})$ has no ℓ-torsion from [AEC, VII.3.1], since it is isomorphic to the formal group of E.

For any abelian group A, we let $T_\ell(A)$ denote the Tate module of A,

$$T_\ell(A) = \varprojlim A[\ell^n],$$

and we set

$$V_\ell(A) = T_\ell(A) \otimes_{\mathbb{Z}_\ell} \mathbb{Q}_\ell.$$

We observe that $V_\ell(A)$ will be 0 if A has no ℓ-torsion, or if A is a finite group. In particular,

$$V_\ell\big(E(K^{\mathrm{nr}})/E_0(K^{\mathrm{nr}})\big) = 0 \quad\text{and}\quad V_\ell\big(E_1(K^{\mathrm{nr}})\big) = 0.$$

Hence the two exact sequences given above yield isomorphisms

$$V_\ell\big(E(K^{\mathrm{nr}})\big) \xleftarrow{\;\sim\;} V_\ell\big(E_0(K^{\mathrm{nr}})\big) \xrightarrow{\;\sim\;} V_\ell\big(\tilde{E}_{\mathrm{ns}}(\bar{k})\big).$$

On the other hand, we clearly have

$$V_\ell\big(E(K^{\mathrm{nr}})\big) = V_\ell\big(E(\bar{K})\big)^{G(\bar{K}/K^{\mathrm{nr}})} = V_\ell\big(E(\bar{K})\big)^{I(\bar{K}/K)},$$

which proves the fundamental isomorphism

$$V_\ell\big(E(\bar{K})\big)^{I(\bar{K}/K)} \cong V_\ell\big(\tilde{E}_{\mathrm{ns}}(\bar{k})\big).$$

Now we compute

$$\varepsilon(E/K) = 2 - \dim_{\mathbb{Q}_\ell}\left(V_\ell(E)^{I(\bar{K}/K)}\right)$$

$$= 2 - \dim_{\mathbb{Q}_\ell}\left(V_\ell\big(\tilde{E}_{\mathrm{ns}}(\bar{k})\big)\right)$$

$$= 2 - \begin{cases} \dim_{\mathbb{Q}_\ell}\big(V_\ell(\tilde{E})\big) & \text{if } E \text{ has good reduction,} \\ \dim_{\mathbb{Q}_\ell}\big(V_\ell(\bar{k}^*)\big) & \text{if } E \text{ has multiplicative reduction,} \\ \dim_{\mathbb{Q}_\ell}\big(V_\ell(\bar{k}^+)\big) & \text{if } E \text{ has additive reduction,} \end{cases}$$

where the last line follows from the standard description of the various reduction types [AEC, VII.5.1]. Using the fact that $\ell \neq p$, we find that

$$V_\ell(\tilde{E}) \cong \mathbb{Q}_\ell^2, \qquad V_\ell(\bar{k}^*) \cong \mathbb{Q}_\ell, \qquad V_\ell(\bar{k}^+) = 0,$$

which completes the proof of (10.2a).

(b) Fix a prime $\ell \neq p$, and let $L = K(E[\ell])$. If E/K has good reduction, then L/K is unramified from [AEC, VII.4.1], so the inertia group $G_0(L/K)$ is trivial. It is then clear from the definition that $\delta(E/K) = 0$.

Next suppose that E/K has non-integral j-invariant, that is, $v_K(j_E) < 0$. Let $\ell \neq p$ be a prime. We will prove later (see (V.5.3) and exercise 5.11) that there is an extension K'/K with the following properties:

(1) $[K' : K] = 1$ or 2.
(2) K'/K is unramified (respectively ramified) if E/K has multiplicative (respectively additive) reduction.
(3) There is a $q \in K'$ such that $K'(E[\ell]) = K'(\mu_\ell, q^{1/\ell})$.

Here μ_ℓ denotes the group of ℓth-roots of unity.

Thus the extension $K'(E[\ell])/K'$ is composed of an unramified cyclotomic extension $K'(\mu_\ell)/K'$ and a Kummer extension $K'(\mu_\ell, q^{1/\ell})/K'(\mu_\ell)$ whose order divides ℓ. This shows that the extension $K'(E[\ell])/K'$ is at worst tamely ramified. It follows from properties (1) and (2) that $K(E[\ell])/K$ is tamely ramified if either E/K has multiplicative reduction or if $p \geq 3$, which completes the proof of (10.2b) in the case that j_E is non-integral.

Finally we consider the case that E/K has integral j-invariant, or equivalently from [AEC, VII.5.5], E has potential good reduction. A key tool in proving (10.2) in this case is the following strengthening of the criterion of Néron-Ogg-Shafarevich.

Proposition 10.3. *Let K be a local field of residue characteristic p, and let E/K be an elliptic curve with integral j-invariant.*
(a) *The following are equivalent.*
  (i) *E has good reduction over K.*
  (ii) *$E[m]$ is unramified for every integer $m \geq 1$ relatively prime to p.*
  (iii) *$E[m]$ is unramified for at least one integer $m \geq 3$ relatively prime to p.*
(b) *Let $m \geq 3$ be an integer relatively prime to p. Then E has good reduction over $K(E[m])$.*

PROOF. (a) The equivalence of (i) and (ii) is [AEC, VII.7.1], and the implication (ii) $\Rightarrow$ (iii) is trivial. So it suffices to prove that (iii) implies (i).

We are given that E/K has potential good reduction, so we can find a finite Galois extension L/K such that E has good reduction over L. We are also given an integer $m \geq 3$ such that $E[m]$ is unramified over K. Let ℓ be the largest prime dividing m, and let

$$\ell' = \ell \quad \text{if } \ell \neq 2 \quad \text{and} \quad \ell' = 4 \quad \text{if } \ell = 2.$$

Notice that $\ell'|m$ since $m \geq 3$, so $E[\ell'] \subset E[m]$. Thus $E[\ell']$ is unramified over K.

The fact that E has good reduction over L means that the inertia group $I_{\bar{L}/L}$ acts trivially on the Tate module $T_\ell(E)$ [AEC, VII.4.1b], so the inertia group $I_{L/K}$ of L/K acts on $T_\ell(E)$. This action gives us a homomorphism

$$\rho : I_{L/K} \longrightarrow \text{Aut}(T_\ell(E)).$$

Further, we are given that $E[\ell']$ is unramified over K, so $I_{L/K}$ acts trivially on $E[\ell']$. In other words, the image $\rho(I_{L/K})$ is contained in the kernel of the natural map

$$\operatorname{Aut}(T_\ell(E)) \longrightarrow \operatorname{Aut}(E[\ell']).$$

If we choose bases $T_\ell(E) \cong \mathbb{Z}_\ell^2$ and $E[\ell'] \cong (\mathbb{Z}/\ell'\mathbb{Z})^2$, then this last map becomes

$$\operatorname{GL}_2(\mathbb{Z}_\ell) \longrightarrow \operatorname{GL}_2(\mathbb{Z}/\ell'\mathbb{Z}).$$

It is an elementary exercise to verify that the kernel of this map, namely

$$\{M \in \operatorname{GL}_2(\mathbb{Z}_\ell) : M \equiv 1 \,(\mathrm{mod}\ \ell')\},$$

has no elements of finite order. (See exercise 4.38. This is the point at which we need $\ell' = 4$ if $\ell = 2$, since if we took $\ell' = 2$, then the matrix $\begin{pmatrix} -1 & 0 \\ 0 & -1 \end{pmatrix}$ would be in the kernel.)

We saw above that the image $\rho(I_{L/K})$ is contained in this kernel. But the group $I_{L/K}$ is finite, so it follows that its image $\rho(I_{L/K})$ is trivial. In other words, the inertia group $I_{L/K}$ acts trivially on $T_\ell(E)$, which proves that $T_\ell(E)$ is unramified over K. Now [AEC, VII.7.1] tells us that E has good reduction over K, which completes the proof that (iii) implies (i).
(b) This follows immediately from (a), since $E[m]$ is clearly unramified over the field $K(E[m])$. $\qquad\qquad\square$

We now resume the proof of Theorem 10.2(b), where, recall, we are assuming that E has integral j-invariant and that $p \geq 5$, and we are trying to verify that $\delta(E/K) = 0$. Without loss of generality, we may replace K by its maximal unramified extension.

For each integer $m \geq 3$ relatively prime to p, let $L_m = K(E[m])$. Now Proposition 10.3(a) tells us that E has good reduction over L_m, and then another application of (10.3a) says that for any other m', the set $E[m']$ is unramified over L_m, which means that the compositum $L_{m'}L_m$ is an unramified extension of L_m. But we took $K = K^{\mathrm{nr}}$, so L_m has no unramified extensions, and hence $L_{m'} \subset L_m$. Reversing the role of m and m' gives the opposite inclusion, which proves that all of the L_m's are the same. We write L for this common field.

Now let $\ell \geq 3$ be a prime with $\ell \neq p$. (We will deal with the case $\ell = 2$ later.) The action of the Galois group $G(L_\ell/K)$ on $E[\ell]$ gives an injection

$$G(L_\ell/K) \hookrightarrow \operatorname{Aut}(E[\ell]) \cong \operatorname{GL}_2(\mathbb{Z}/\ell\mathbb{Z}).$$

It follows that

$$\#G(L_\ell/K) \mid \#\operatorname{GL}_2(\mathbb{Z}/\ell\mathbb{Z}).$$

But we showed above that the field $L_\ell = L$ is independent of ℓ, so we find that $\#G(L/K)$ divides $\#\operatorname{GL}_2(\mathbb{Z}/\ell\mathbb{Z})$ for all $\ell \neq 2, p$. The group $\operatorname{GL}_2(\mathbb{Z}/\ell\mathbb{Z})$ has order $\ell(\ell - 1)^2(\ell + 1)$, and it is easy to see using Dirichlet's theorem on primes in arithmetic progressions that

$$\gcd_{\ell \neq 2, p} \{\ell(\ell - 1)^2(\ell + 1)\} = 48.$$

Hence $G(L/K)$ has order dividing 48.

In particular, $[L : K]$ is not divisible by p, since $p \geq 5$ by assumption. This proves that the extension L/K is at worst tamely ramified, so the higher ramification groups $G_i(L/K)$ are trivial for $i \geq 1$ (10.1d). It follows directly from the definition of $\delta(E/K)$ that $\delta(E/K) = 0$. This completes the proof of (10.2b) for $\ell \geq 3$.

Finally, if $\ell = 2$, we use the fact that $L = L_4$ and that E has good reduction over L to conclude that

$$G(L/K) \longhookrightarrow \mathrm{Aut}\big(E[4]\big) \cong \mathrm{GL}_2(\mathbb{Z}/4\mathbb{Z}).$$

This last group has order $\#\,\mathrm{GL}_2(\mathbb{Z}/4\mathbb{Z}) = 96 = 2^5 \cdot 3$, so L/K is not wildly ramified at p, since $p \geq 5$. Hence $\delta(E/K) = 0$.

(c) If $p \geq 5$, then (b) says that $f(E/K)$ is an integer which depends only on the reduction type of E/K, hence is independent of ℓ. The general case, which is due to Ogg [2], uses more machinery than we want to develop here. We have sketched the proof in exercise 4.46. For further details, see Ogg [2], Serre [7, chapter 19], and Serre-Tate [1, §3]. $\square$

If the residue characteristic of K is not equal to 2 or 3, then we have seen that $\delta(E/K) = 0$, and so the exponent of the conductor satisfies $f(E/K) \leq 2$ from (10.2b). When the residue characteristic is 2 or 3, the exponent of the conductor is still bounded as described in the following result.

Theorem 10.4. (Lockhart-Rosen-Silverman [1], Brumer-Kramer [1]) *Let $K/\mathbb{Q}_p$ be a local field with normalized valuation v_K, and let E/K be an elliptic curve. Then the exponent of the conductor of E/K is bounded by*

$$f(E/K) \leq 2 + 3v_K(3) + 6v_K(2).$$

(Here $v_K(p)$ is the ramification index of $K/\mathbb{Q}_p$.) Further, this bound is best possible in the sense that for every finite extension $K/\mathbb{Q}_p$ there is an elliptic curve E/K whose conductor attains this bound.

PROOF. We are going to prove the slightly weaker bound

$$f(E/K) \leq 2 + 3v_K(3) + 8v_K(2),$$

since the proof is easier and the weaker estimate suffices for most applications. For an elementary, but involved, proof of the stronger inequality in certain cases, see Lockhart-Rosen-Silverman [1]. The proof for general K requires heavier machinery from representation theory; see Brumer-Kramer [1].

We begin with the observation that if $p \geq 5$, then $\delta(E/K) = 0$ from (10.2b) and $\varepsilon(E/K) \leq 2$ directly from its definition, so $f(E/K) \leq 2$. It remains to deal with the cases $p = 2$ and $p = 3$.

Let ℓ be any prime other than p, let $L = K(E[\ell])$, let $\mathfrak{D}_{L/K}$ be the different of L/K, and let r be the smallest integer such that $G_r(L/K) = 1$. We will need the following elementary properties of local fields:

(i) $$v_L(\mathfrak{D}_{L/K}) = \sum_{i=0}^{\infty} \big(g_i(L/K) - 1\big),$$

(ii) $$v_L(\mathfrak{D}_{L/K}) \le g_0(L/K) - 1 + v_L\big(g_0(L/K)\big),$$

(iii) $$r \le \frac{v_L(p)}{p-1} + 1.$$

See Serre [4, IV, §1, Prop. 4] for (i), Serre [4, III, § 7, remark following Prop. 13] for (ii), and Serre [4, IV, §2, exercise 3(c)] or Lockhart-Rosen-Silverman [1, Lemma 1.2(b)] for (iii).

We are now ready to compute.

$$
\begin{aligned}
f(E/K) &= \varepsilon(E/K) + \delta(E/K) \\
&\le 2 + \delta(E/K) && \text{since clearly } \varepsilon(E/K) \le 2 \\
&= 2 + \sum_{i=1}^{r-1} \frac{g_i(L/K)}{g_0(L/K)} \dim_{\mathbb{F}_\ell}\left(E[\ell]/E[\ell]^{G_i(L/K)}\right) \\
&&& \text{since } G_i(L/K) = 1 \text{ for } i \ge r \\
&\le 2 + \frac{2}{g_0(L/K)} \sum_{i=1}^{r-1} g_i(L/K) && \text{since } \dim_{\mathbb{F}_\ell}\big(E[\ell]\big) = 2 \\
&= \frac{2}{g_0(L/K)} \sum_{i=0}^{r-1} g_i(L/K) \\
&= \frac{2}{g_0(L/K)} \left(r + \sum_{i=0}^{\infty}\big(g_i(L/K) - 1\big)\right) \\
&= \frac{2}{g_0(L/K)} \big(r + v_L(\mathfrak{D}_{L/K})\big) && \text{from property (i)} \\
&\le \frac{2}{g_0(L/K)} \left(\frac{v_L(p)}{p-1} + 1 + g_0(L/K) - 1 + v_L\big(g_0(L/K)\big)\right) \\
&&& \text{from properties (ii) and (iii)} \\
&= \frac{2v_K(p)}{p-1} + 2 + 2v_K\big(g_0(L/K)\big) && \text{since } g_0(L/K) \text{ is the} \\
&&& \text{ramification index of } L/K.
\end{aligned}
$$

Now suppose that $p = 3$, and take (say) $\ell = 5$. Then

$$G(L/K) \hookrightarrow \mathrm{Aut}\big(E[5]\big) \cong \mathrm{GL}_2(\mathbb{Z}/5\mathbb{Z}),$$

so in particular $g_0(L/K)$ divides $\#\,\mathrm{GL}_2(\mathbb{Z}/5\mathbb{Z}) = 480 = 2^5 \cdot 3 \cdot 5$. Hence

$$v_K\big(g_0(L/K)\big) \le v_K(480) = v_K(3),$$

and substituting this in above with $p = 3$ gives the desired estimate,

$$f(E/K) \le 2 + 3v_K(3).$$

Similarly, if $p = 2$, then we can take (say) $\ell = 3$ and use the injection

$$G(L/K) \longhookrightarrow \mathrm{Aut}(E[3]) \cong \mathrm{GL}_2(\mathbb{Z}/3\mathbb{Z})$$

to conclude that $g_0(L/K)$ divides $\# \mathrm{GL}_2(\mathbb{Z}/3\mathbb{Z}) = 48 = 2^4 \cdot 3$. However, we can easily save a little bit. We are allowed to make an unramified extension of K, so we may adjoin to K a primitive cube root of unity. Then basic properties of the Weil pairing [AEC, III.8.1] imply that the image of $G(L/K)$ lies in $\mathrm{SL}_2(\mathbb{Z}/3\mathbb{Z})$. (We sketch the proof below.) Hence the ramification index $g_0(L/K)$ divides $\# \mathrm{SL}_2(\mathbb{Z}/3\mathbb{Z}) = 24 = 2^3 \cdot 3$. It follows that

$$v_K(g_0(L/K)) \le v_K(24) = v_K(2^3),$$

and then substituting this in above with $p = 2$ yields

$$f(E/K) \le 2v_K(2) + 2 + 2v_K(2^3) = 2 + 8v_K(2).$$

Let π be a uniformizer for K. If $p \ge 5$, then (10.2b) says that any elliptic curve E/K with additive reduction will hit the maximum conductor exponent $f(E/K) = 2$. If $p = 3$, then we claim that the elliptic curve

$$E : y^2 = x^3 + \pi$$

satisfies $f(E/K) = 2 + 3v_K(3)$. Similarly, if $p = 2$, we claim that the elliptic curve

$$E : y^2 + 2xy = x^3 - x^2 + \pi x$$

satisfies $f(E/K) = 2 + 6v_K(2)$. We could verify these claims by a lengthy direct calculation, but instead we will leave them for the reader to check (exercises 4.52 and 4.53) using Tate's algorithm (9.4) and a formula of Ogg (11.1) to be proven in the next section.

It remains to prove the assertion from above that the image of $G(L/K)$ lies in $\mathrm{SL}_2(\mathbb{Z}/3\mathbb{Z})$. Fix a basis $S, T \in E[3]$. Then $e_3(S, T)$ is a primitive cube root of unity, so it is in K. Let $\sigma \in G(L/K)$, and let $\rho(\sigma) = \left(\begin{smallmatrix} a & b \\ c & d \end{smallmatrix} \right) \in$ $\mathrm{GL}_2(\mathbb{Z}/3\mathbb{Z})$ be the matrix giving the action of σ on $E[3]$ relative to the chosen basis. Using [AEC, III.8.1(a,b,c,d)], we compute

$$e_3(S, T) = e_3(S, T)^\sigma = e_3(S^\sigma, T^\sigma) = e_3(aS + cT, bS + dT)$$
$$= e_3(S, T)^{ad-bc} = e_3(S, T)^{\det \rho(\sigma)}.$$

Therefore $\det \rho(\sigma) = 1$, so the image of $G(L/K)$ lies in $\mathrm{SL}_2(\mathbb{Z}/3\mathbb{Z})$. $\qquad\square$

The conductor of an elliptic curve over a number field is defined by combining all of the local conductor exponents, just as the minimal discriminant was defined as a product of the local discriminants.

Definition. Let E/K be an elliptic curve defined over a number field K, and for each prime $\mathfrak{p}$ of K, let $f(E/K_\mathfrak{p})$ be the exponent of the conductor of E consider as an elliptic curve over the local field $K_\mathfrak{p}$. The *conductor of E/K* is the ideal

$$\mathfrak{f}(E/K) = \prod_\mathfrak{p} \mathfrak{p}^{f(E/K_\mathfrak{p})}.$$

Example 10.5. Let E/K be an elliptic curve defined over a number field, and suppose that E has everywhere semi-stable reduction, by which we mean that E has either good or multiplicative reduction at every prime. Then the conductor of E/K is the product of the primes of bad reduction,

$$\mathfrak{f}(E/K) = \prod_{\mathfrak{p}\mid\mathcal{D}_{E/K}} \mathfrak{p},$$

where $\mathcal{D}_{E/K}$ is the minimal discriminant of E/K [AEC, VII §8]. Conversely, if the conductor $\mathfrak{f}(E/K)$ is square-free, then E/K has everywhere semi-stable reduction. This follows from (10.2), which says that the conductor exponent satisfies $f(E/K_\mathfrak{p}) \geq 2$ if and only if $E/K_\mathfrak{p}$ has additive reduction.

Both the minimal discriminant and the conductor measure the extent to which an elliptic curve has bad reduction. We will see in the next section (11.2) that the exponent of the minimal discriminant is always greater than the exponent of the conductor, so we always have an inequality of the form

$$\mathrm{N}_\mathbb{Q}^K(\mathfrak{f}_{E/K}) \leq \mathrm{N}_\mathbb{Q}^K(\mathcal{D}_{E/K}).$$

Szpiro has conjectured that there should be an inequality in the other direction.

Szpiro's Conjecture 10.6. *Fix a number field K and an $\varepsilon > 0$. There is a constant $c(K, \varepsilon)$ so that for every elliptic curve E/K,*

$$\mathrm{N}_\mathbb{Q}^K(\mathcal{D}_{E/K}) \leq c(K, \varepsilon)\mathrm{N}_\mathbb{Q}^K(\mathfrak{f}_{E/K})^{6+\varepsilon}.$$

This conjecture, if true, lies very deep. Its validity would imply the solution to many other Diophantine problems, including for example the assertion that if $a, b \in \mathbb{Q}^*$ are fixed, if $n \geq 2$, and if m is sufficiently large, then the equation $ax^n + by^m = 1$ has no non-trivial solutions $x, y \in \mathbb{Q}$.

Surprisingly, it is quite easy to prove a function field analogue of Szpiro's conjecture; see exercise 3.36. One can even prove such a result with $\varepsilon = 0$ and with an explicit constant c. The function field version of Szpiro's conjecture was originally discovered by Kodaira (see Shioda [3, Prop. 2.8]) long before Szpiro formulated his conjecture, and it has been frequently rediscovered since that time; see for example Hindry-Silverman [2, Thm. 5.1] and Szpiro [1].

§11. Ogg's Formula

Let E/K be an elliptic curve defined over a local field. Ogg's formula relates the minimal discriminant of E/K, the exponent of the conductor of E/K, and the number of components on the special fiber of the minimal model of E over the ring of integers of K. This formula was originally proven by Ogg [2] in all cases except when K is a field of characteristic 0 with residue field of characteristic 2. Ogg's proof relies on a lengthy case-by-case analysis. A more conceptual proof using scheme-theoretic techniques and working in all residue characteristics has been given by Saito [1], who proves Ogg's formula as a special case of a general result for curves of arbitrary positive genus. An expanded exposition of Saito's proof just in the case of elliptic curves can be found in Liu [1].

Ogg's Formula 11.1. (Ogg [2], Saito [1]) *Let $K/\mathbb{Q}_p$ be a local field, let E/K be an elliptic curve, and let*

$v_K(\mathcal{D}_{E/K}) = $ *the valuation of the minimal discriminant of E/K,*

$f(E/K) = $ *the exponent of the conductor of E/K,*

$m(E/K) = $ *the number of components on the special fiber of E/K.*

Then

$$v_K(\mathcal{D}_{E/K}) = f(E/K) + m(E/K) - 1.$$

Remark 11.1.1. The number $m(E/K)$ in (10.1) is the number of irreducible components defined over $\bar{k}$ on the special fiber of the minimal proper regular model of E/K. This includes all of the components, not just the multiplicity-1 components which make up the special fiber of the Néron model. Further, each component is counted once, regardless of its multiplicity. For example, if E/K has Type I_n reduction, then $m(E/K) = n$, and if E/K has Type III^* reduction, then $m(E/K) = 8$. The value of $m(E/K)$ for these and the other reduction types can be found in Table 4.1.

Remark 11.1.2. The minimal discriminant and special fiber of E/K can be computed in a straightforward manner using Tate's algorithm (9.4). For this reason, Ogg's formula (11.1) is frequently used to compute the exponent of the conductor of E/K for residue characteristics 2 and 3. See for example exercises 4.52 and 4.53, as well as the conductor tables contained in Birch-Kuyk [1] and Cremona [1].

PROOF (of Ogg's Formula 11.1). If E/K has good reduction, then

$$v_K(\mathcal{D}_{E/K}) = 0, \quad f(E/K) = 0, \quad \text{and} \quad m(E/K) = 1.$$

These three equalities follow from [AEC, VII.5.1a], (10.2b), and (6.3) respectively. (Notice that (6.3) says that the minimal Weierstrass equation for E/K is already the Néron model, so its special fiber is irreducible.) This verifies Ogg's formula when E/K has good reduction.

Next suppose that E/K has multiplicative reduction. Then $f(E/K) = 1$ from (10.2b). Further, Step 2 of Tate's algorithm (9.4) tells us that the special fiber of the minimal model of E/K is an $m(E/K)$-sided polygon, with $m(E/K) = v_K(\mathcal{D}_{E/K})$. This verifies Ogg's formula when E/K has multiplicative reduction.

Finally, suppose that E/K has additive reduction. Consider first the case that $p \geq 5$. Then (10.2b) tells us that $f(E/K) = 2$, so Ogg's formula becomes

$$v_K(\mathcal{D}_{E/K}) = m(E/K) + 1.$$

It is now a simple matter using Table 4.1 to verify Ogg's formula case-by-case, checking each of the reduction types II, III,..., II*.

It remains to consider $p = 3$ and $p = 2$ when E/K has additive reduction. We will give a direct case-by-case verification for $p = 3$, since it only takes a few pages. A similar proof for $p = 2$ would be extremely lengthy, so for this last case we refer the reader to Saito's proof [1] which works in all residue characteristics and does not rely on a case-by-case analysis. (See also Liu [1].) Unfortunately, the papers of Saito [1] and Liu [1] use techniques which are beyond the scope of this book.

So we now assume that $p = 3$ and that E/K has additive reduction. In particular, (9.2a) tells us that the tame part of the conductor is $\varepsilon(E/K) = 2$. Further, the fact that $p = 3$ means that we can find a minimal Weierstrass equation for E/K of the form

$$E : y^2 = x^3 + a_2 x^2 + a_4 x + a_6.$$

The discriminant of this equation is

$$\Delta = -16(4a_2^3 a_6 - a_2^2 a_4^2 + 4a_4^3 + 27a_6^2 - 18a_2 a_4 a_6).$$

Using this simplified form for E will make all of our calculations easier.

Let $L = K\big(E[2]\big)$ be the field generated by the 2-torsion points of E, so L is the splitting field over K of the cubic polynomial

$$f(x) = x^3 + a_2 x^2 + a_4 x + a_6.$$

Further, let $M = K(\sqrt{\Delta})$. Notice that the discriminant of the polynomial $f(x)$ satisfies $\Delta = -16 \operatorname{Disc}(f)$. Thus we see that $K \subset M \subset L$ and that

$$[M : K] = \begin{cases} 2 & \text{if } [L : K] = 2 \text{ or } 6, \\ 1 & \text{if } [L : K] = 1 \text{ or } 3. \end{cases}$$

Suppose first that E/K has Type III reduction, so $m(E/K) = 2$. A quick perusal of Step 4 of Tate's algorithm (9.4) shows that E has a Weierstrass equation satisfying

$$v_K(a_2) \geq 1, \quad v_K(a_4) = 1, \quad v_K(a_6) \geq 2, \quad \text{and} \quad v_K(\Delta) = 3.$$

We claim that L/K is at worst tamely ramified. To see this, we let π_K be a uniformizer for K and use the fact that $v_K(\Delta) = 3$ to observe that $\pi_M = \pi_K^{-1}\sqrt{\Delta}$ is a uniformizer for M. Notice in particular that M/K is a ramified extension of degree 2. Now consider the polynomial

$$g(x) = \pi_M^{-3} f(\pi_M x) = x^3 + \pi_M^{-1} a_2 x^2 + \pi_M^{-2} a_4 x + \pi_M^{-3} a_6 \in M[x].$$

The coefficients of g are in the ring of integers of M, and the discriminant of g has valuation

$$v(\operatorname{Disc} g) = v(\pi_M^{-6} \operatorname{Disc} f) = v(\pi_K^6/\Delta^2) = 0.$$

Hence the splitting field of g over M, which is L, is unramified over M. This proves that L/K is not wildly ramified, so $\delta(E/K) = 0$. We have now computed all of the pieces in

$$v_K(\mathcal{D}_{E/K}) - f(E/K) - m(E/K) + 1 = 3 - (2 + 0) - 2 + 1 = 0,$$

which completes the proof of Ogg's formula for $p = 3$ and Type III reduction.

If E/K has Type III* reduction, then a similar calculation shows that $v_K(\Delta) = 9$ and that $K(E[2])$ is tamely ramified over K. So again we find that

$$v_K(\mathcal{D}_{E/K}) - f(E/K) - m(E/K) + 1 = 9 - (2 + 0) - 8 + 1 = 0.$$

We leave the details to the reader (exercise 4.54a).

For the remaining reduction types (II, III, IV, I_n^*, IV*, III*, II*), we are first going to show that if Ogg's formula is true for E/M, then it is also true for E/K. More precisely, if we write $\operatorname{Ogg}(E/K)$ for the quantity

$$\operatorname{Ogg}(E/K) = v_K(\mathcal{D}_{E/K}) - f(E/K) - m(E/K) + 1,$$

then we will show that

$$\operatorname{Ogg}(E/M) = e(M/K) \operatorname{Ogg}(E/K),$$

where $e(M/K)$ is the ramification index of M/K. It is clear that this equality holds if M/K is unramified, since none of the quantities in $\operatorname{Ogg}(E/K)$ will change, so we only need to consider the case that M/K is ramified.

Assuming now that M/K is ramified, we have $e(M/K) = 2$, so the ramification is tame. It follows that the higher ramification groups for L/K and L/M are the same,

$$G_i(L/M) = G_i(L/K) \qquad \text{for all } i \geq 1.$$

Further, M/K is a ramified extension of degree 2, so $g_0(L/K) = 2g_0(L/M)$. Using these two facts and the definition of wild part of the conductor, we compute

$$\delta(E/K) = \sum_{i=1}^{\infty} \frac{g_i(L/K)}{g_0(L/K)} \dim_{\mathbb{F}_2} \left(E[2]/E[2]^{G_i(L/K)} \right)$$

$$= \sum_{i=1}^{\infty} \frac{g_i(L/M)}{2g_0(L/M)} \dim_{\mathbb{F}_2} \left(E[2]/E[2]^{G_i(L/M)} \right)$$

$$= \frac{1}{2}\delta(E/M).$$

Notice that the full conductor does not satisfy such a simple relation, since $\varepsilon(E/K) = 2$, and we will see that $\varepsilon(E/M)$ may be any of 0, 1, or 2.

The next step in our proof that $\mathrm{Ogg}(L/M) = 2\,\mathrm{Ogg}(L/K)$ is to verify the following table describing how various quantities change when we make the ramified quadratic extension M/K. (For more extensive tables, see exercises 4.48, 4.49, and 4.50.)

Type(E/K)	II	IV	I_0^*	I_n^* $(n \geq 1)$	IV*	II*
Type(E/M)	IV	IV*	I_0	I_{2n}	IV	IV*
$2v_K(\mathcal{D}_{E/K}) - v_M(\mathcal{D}_{E/M})$	0	0	12	12	12	12
$2f(E/K) - f(E/M)$	2	2	4	3	2	2
$2m(E/K) - m(E/M)$	-1	-1	9	10	11	11

Notice that the value of $2m(E/K) - m(E/M)$ in the last line is easy to compute by using the first two lines and reading off the number of components for each reduction type from Table 4.1. Similarly, the identity $\delta(E/K) = \delta(E/M)/2$ that we proved above implies that

$$2f(E/K) - f(E/M) = 2\varepsilon(E/K) - \varepsilon(E/M),$$

so the penultimate line of the table follows immediately from the first two lines and the fact (9.2a) that $\varepsilon = 0$ for good reduction, $\varepsilon = 1$ for multiplicative reduction, and $\varepsilon = 2$ for additive reduction. The verification of the remainder of the table is now simply a matter of tracing the various reduction types through Tate's algorithm. We will do Type IV* reduction to illustrate the idea, and leave the other cases for the reader.

So suppose that E/K has Type IV* reduction. Then Step 8 of Tate's algorithm gives a minimal Weierstrass for E/K satisfying

$$v_K(a_2) \geq 2, \qquad v_K(a_4) \geq 3, \qquad v_K(a_6) = 4.$$

(Remember we can assume that $a_1 = a_3 = 0$.) Fix a uniformizer π_M for M and make the change of variables $x = \pi_M^2 x'$, $y = \pi_M^3 y'$. This gives a Weierstrass equation for E/M of the form

$$E : y'^2 = x'^3 + a_2' x'^2 + a_4' x' + a_6',$$

where $a_2' = \pi_M^{-2} a_2$, $a_4' = \pi_M^{-4} a_4$, and $a_6' = \pi_M^{-6} a_6$. In particular, we have

$$v_M(a_2') = 2v_K(a_2) - 2 \geq 2, \quad v_M(a_4') = 2v_K(a_4) - 4 \geq 2,$$
$$v_M(a_6') = 2v_K(a_6) - 6 = 2, \quad v_M(\Delta') = 2v_K(\Delta) - 12.$$

Now a quick check of Step 5 of Tate's algorithm shows that this is a minimal equation for E/M and that E/M has Type IV reduction. Further, $\Delta' = \pi_M^{-12}\Delta$, so

$$v_M(\mathcal{D}_{E/M}) = v_M(\Delta') = 2v_K(\Delta) - 12 = 2v_K(\mathcal{D}_{E/K}) - 12.$$

Finally, a fiber of Type IV* has seven components, and a fiber of Type IV has three components, so $2m(E/K) - m(E/M) = 11$. This completes the verification of the Type IV* column in the above table. The other columns may be verified similarly.

We now use this table to compute

$$2\,\mathrm{Ogg}(E/K) - \mathrm{Ogg}(E/M)$$
$$= \{2v_K(\mathcal{D}_{E/K}) - v_M(\mathcal{D}_{E/M})\} - \{2f(E/K) - f(E/M)\}$$
$$\qquad - \{2m(E/K) - m(E/M)\} + \{2 - 1\}$$
$$= \begin{cases} 0 - 2 - (-1) + 1 = 0 & \text{if Type}(E/K)=\text{II or IV}, \\ 12 - 4 - 9 + 1 = 0 & \text{if Type}(E/K)=\text{I}_0^*, \\ 12 - 3 - 10 + 1 = 0 & \text{if Type}(E/K)=\text{I}_n^*, \, n \geq 1, \\ 12 - 2 - 11 + 1 = 0 & \text{if Type}(E/K)=\text{IV* or II*}. \end{cases}$$

This completes the proof that $\mathrm{Ogg}(E/M) = 2\,\mathrm{Ogg}(E/K)$, so it now suffices to prove Ogg's formula for E/M. We note from the table that E/M is of Type IV, IV*, I_0, or I_n, so it suffices to consider these four cases.

If E/M has Type I_0 reduction, which is to say E/M has good reduction, or if E/M has Type I_n reduction with $n \geq 1$, which means E/M has multiplicative reduction, then we are done, since we have already verified Ogg's formula for good and multiplicative reduction.

Suppose now that E/M has Type IV reduction. Using Step 5 of Tate's algorithm, we see that our Weierstrass equation satisfies

$$v_M(a_2) \geq 1, \qquad v_M(a_4) \geq 2, \qquad v_M(a_6) = 2.$$

Let $\alpha \in L$ be a root of the polynomial $f(x) = x^3 + a_2 x^2 + a_4 x + a_6$. Note that the degree of L/M is either 1 or 3, and L is the splitting field of f, so we have $L = M(\alpha)$. We want to use the fact that α satisfies the equation

$$f(\alpha) = \alpha^3 + a_2 \alpha^2 + a_4 \alpha + a_6 = 0$$

to compute its valuation. First we observe that

$$3v_M(\alpha) = v_M(a_2\alpha^2 + a_4\alpha + a_6) \geq \min\{v_M(a_2\alpha^2), v_M(a_4\alpha), v_M(a_6)\}$$
$$\geq \min\{1 + 2v_M(\alpha), 2 + v_M(\alpha), 2\},$$

so $v_M(\alpha) \geq 2/3$. Similarly,

$$2 = v_M(a_6) = v_M(\alpha^3 + a_2\alpha^2 + a_4\alpha)$$
$$\geq \min\{v_M(\alpha^3), v_M(a_2\alpha^2), v_M(a_4\alpha)\}$$
$$\geq \min\{3v_M(\alpha), 1 + 2v_M(\alpha), 2 + v_M(\alpha)\},$$

which gives the opposite inequality $2/3 \geq v_M(\alpha)$. Hence $v_M(\alpha) = 2/3$. In particular, this proves that L/M is totally ramified of degree 3, so L/M is wildly ramified. Further, if we choose a uniformizer π_M for M, then

$$\pi_L \stackrel{\text{def}}{=} \pi_M/\alpha \in L$$

will be a uniformizer for L, since $v_M(\pi_L) = 1/3$.

We can use π_L to determine the higher ramification groups for L/M. The Galois group $G(L/M)$ is a cyclic group of order 3, so there is an integer $r \geq 1$ such that

$$G(L/M) = G_1(L/M) = \cdots = G_{r-1}(L/M),$$

and

$$G_r(L/M) = G_{r+1}(L/M) = \cdots = 1.$$

We want to compute this integer r. Writing $G(L/M) = \{1, \sigma, \sigma^2\}$, the definition of the higher ramification groups says that

$$r = v_L(\pi_L^\sigma - \pi_L).$$

We substitute $\pi_L = \pi_M/\alpha$ and use the fact that $\pi_M \in M$ to get

$$r = v_L\left(\frac{\pi_M}{\alpha^\sigma} - \frac{\pi_M}{\alpha}\right) = v_L(\pi_M) + v_L(\alpha - \alpha^\sigma) - v_L(\alpha\alpha^\sigma).$$

The extension L/M is totally ramified, so $v_L = 3v_M$. This means that $v_L(\pi_M) = 3$ and $v_L(\alpha) = v_L(\alpha^\sigma) = 2$, and hence

$$r = v_L(\alpha - \alpha^\sigma) - 1.$$

Further, the exact same calculation gives $r = v_L(\alpha - \alpha^{\sigma^2}) - 1$, and since the valuation is Galois invariant, we also find that $r = v_L(\alpha^\sigma - \alpha^{\sigma^2}) - 1$. Adding these three expressions for r yields

$$3r = v_L(\alpha - \alpha^\sigma) + v_L(\alpha - \alpha^{\sigma^2}) + v_L(\alpha^\sigma - \alpha^{\sigma^2}) - 3$$
$$= v_L\left((\alpha - \alpha^\sigma)(\alpha - \alpha^{\sigma^2})(\alpha^\sigma - \alpha^{\sigma^2})\right) - 3.$$

Notice that α, α^σ, and α^{σ^2} are the three roots of $f(x)$, so the discriminant of $f(x)$ is

$$\mathrm{Disc}(f) = -\Big((\alpha - \alpha^\sigma)(\alpha - \alpha^{\sigma^2})(\alpha^\sigma - \alpha^{\sigma^2})\Big)^2.$$

We observed above that $\Delta = -16\,\mathrm{Disc}(f)$, which gives us the formula

$$r = \frac{1}{6}v_L\big(\mathrm{Disc}(f)\big) - 1 = \frac{1}{6}v_L(\Delta) - 1 = \frac{1}{2}v_M(\Delta) - 1.$$

This relation is the key to proving Ogg's formula, since it relates the conductor, via the higher ramification groups, to the discriminant of the minimal Weierstrass equation.

The non-trivial elements of $G(L/M)$ act on the non-zero elements of $E[2]$ via a permutation of order 3, so the only element of $E[2]$ fixed by $G(L/M)$ is O. This means that $E[2]^{G_i(L/M)}$ equals $E[2]$ for $i \geq r$ and is trivial for $i < r$, and hence

$$\dim_{\mathbb{F}_2}\Big(E[2]/E[2]^{G_i(L/M)}\Big) = \begin{cases} 2 & \text{if } i < r, \\ 0 & \text{if } i \geq r. \end{cases}$$

Using this and the value for r computed above, we can determine the wild part of the conductor directly from the definition:

$$\begin{aligned}
\delta(E/M) &= \sum_{i=1}^{\infty} \frac{g_i(L/M)}{g_0(L/M)} \dim_{\mathbb{F}_2}\Big(E[2]/E[2]^{G_i(L/M)}\Big) \\
&= 2(r - 1) \\
&= v_M(\Delta) - 4 \\
&= v_M(\mathcal{D}_{E/M}) - 4.
\end{aligned}$$

Adding this to the tame part $\varepsilon(E/M) = 2$ of the conductor gives the relation

$$f(E/M) = \varepsilon(E/M) + \delta(E/M) = v_M(\mathcal{D}_{E/M}) - 2.$$

It only remains to recall that we are working with a curve E/M having Type IV reduction. This means that $m(E/M) = 3$, so the last relation can be rewritten as

$$v_M(\mathcal{D}_{E/M}) = f(E/M) + 2 = f(E/M) + 3 - 1 = f(E/M) + m(E/M) - 1.$$

This completes the proof of Ogg's formula when E/M has Type IV reduction. The proof for Type IV* reduction is similar, so we leave it for the reader.   □

Corollary 11.2. *Let $K/\mathbb{Q}_p$ be a local field, let E/K be an elliptic curve, let $v_K(\mathcal{D}_{E/K})$ be the valuation of the minimal discriminant of E/K, and let $f(E/K)$ be the exponent of the conductor of E/K. Then*

$$f(E/K) \leq v_K(\mathcal{D}_{E/K}),$$

with equality if and only if E/K has reduction type I_0, I_1, or II.

PROOF. When $p \geq 5$, the stated inequality is quite easy to verify by a direct calculation, since it essentially comes down to showing that if the discriminant satisfies $v_K(\mathcal{D}_{E/K}) = 1$, then E/K has multiplicative reduction. We proved this fact earlier (9.5a). In general, Ogg's formula (11.1) tells us that

$$v_K(\mathcal{D}_{E/K}) - f(E/K) = m(E/K) - 1 \geq 0,$$

since the number of components certainly satisfies $m(E/K) \geq 1$. Further, there is equality if and only if $m(E/K) = 1$, which occurs exactly for reduction types I_0, I_1, and II. $\square$

EXERCISES

4.1. (a) The *special orthogonal group* SO_n is defined to be

$$\mathrm{SO}_n = \{M \in \mathrm{SL}_n : {}^t MM = I\},$$

where ${}^t M$ denotes the transpose of the matrix M. Prove that SO_n is an affine group variety.
  (b) The *orthogonal group* O_n is defined to be

$$\mathrm{O}_n = \{M \in \mathrm{GL}_n : {}^t MM = I\}.$$

Prove that there is an isomorphism $\mathrm{O}_n \cong \mathrm{SO}_n \times \mathbb{Z}/2\mathbb{Z}$ of group varieties.

4.2. Let A be the matrix $A = \begin{pmatrix} 0 & I_n \\ -I_n & 0 \end{pmatrix}$, where I_n is the $n \times n$ identity matrix. The *symplectic group* Sp_{2n} is defined to be

$$\mathrm{Sp}_{2n} = \{M \in \mathrm{SL}_{2n} : {}^t MAM = A\}.$$

Prove that Sp_{2n} is an affine group variety.

4.3. Let E be an elliptic curve, and let $\phi : E \to E$ be a morphism satisfying $\phi(O) = O$. Use the Rigidity Lemma 1.8 to prove that ϕ is a group homomorphism. (This provides an alternative proof of [AEC, III.4.8] not requiring the theory of Picard groups. This proof readily generalizes to abelian varieties of arbitrary dimension.)

4.4.   Let E be an elliptic curve, and let $\mu : E \times E \to E$ be a morphism satisfying
$$\mu(P, O) = \mu(O, P) = P \qquad \text{for all } P \in E.$$
Prove that $\mu(P, Q) = P + Q$ for all $P, Q \in E$.

4.5.   (a) Let A be a regular local ring, and let $\mathfrak{P}$ be a prime ideal of A. Prove that the localization of A at $\mathfrak{P}$ is a regular local ring.
(b) Let X be a scheme, let $x, y \in X$ be points, and suppose that x is in the closure of y. (N.B. Points of a scheme need not be closed.) If x is a regular point of X, prove that y is also a regular point. Hence X is regular if and only if all of its closed points are regular.
(c) With notation as in (b), give an example to show that it is possible to have x singular and y non-singular.

4.6.   Let R be a discrete valuation ring with fraction field K, residue field k, and maximal ideal $\mathfrak{p}$, and let π be a uniformizer for R. Let X/R be the R-scheme defined by $X = \operatorname{Spec} R[t]/(\pi t)$.
(a) Prove that the generic and special fibers of X are given by $X_\eta \cong \operatorname{Spec} K$ and $X_\mathfrak{p} \cong \mathbb{A}^1_k$. Note that X_η is smooth over K and that $X_\mathfrak{p}$ is smooth over k.
(b) Prove that X is not smooth over R. This shows that something like the irreducibility condition in (2.9) is necessary.

4.7.   *Let $\phi : X \to S$ be a morphism of finite type of Noetherian schemes, let $x \in X$, and let $s = \phi(x)$. Prove that ϕ is smooth at x if and only if ϕ is flat at x and the fiber X_s is smooth over the residue field of S at s. (This shows that what is really going wrong in the previous exercise is the fact that X is not flat over R, since its fibers have different dimensions.)

4.8.   Complete the proof that $G(T)$ is a group by verifying the following two facts, where we use the notation from (3.2).
(a) $(i \circ \phi) * \phi = \sigma_0 \circ \pi_T$ for all $\phi \in G(T)$.
(b) $\phi * (\psi * \lambda) = (\phi * \psi) * \lambda$ for all $\phi, \psi, \lambda \in G(T)$.

4.9.   Let S be a scheme, and let $\mathbb{G}_a$ and $\mathbb{G}_m$ be the additive and multiplicative group schemes respectively (3.1.2, 3.1.3).
(a) Prove that $\mathbb{G}_a(S) = \Gamma(S, \mathcal{O}_S)$.
(b) Prove that $\mathbb{G}_m(S) = \Gamma(S, \mathcal{O}_S^*)$.
Here $\mathcal{O}_S$ is the structure sheaf on S, and $\Gamma(S, \mathcal{F})$ denotes the global sections of the sheaf $\mathcal{F}$.

4.10. (a) For each integer $r \geq 1$, let μ_r be the scheme
$$\mu_r = \operatorname{Spec} \mathbb{Z}[T]/(T^r - 1).$$
Prove that there is a natural inclusion $\mu_r \hookrightarrow \mathbb{G}_m$ so that μ_r is a (closed) subgroup scheme of $\mathbb{G}_m$. The group scheme μ_r is called the *scheme of rth-roots of unity*.
(b) Let R be a ring of characteristic $p > 0$, and for each integer $r \geq 1$ let α_{p^r} be the R-scheme
$$\alpha_{p^r} = \operatorname{Spec} R[T]/(T^{p^r}).$$
Prove that there is a natural inclusion $\alpha_{p^r} \hookrightarrow \mathbb{G}_{a/R}$ so that α_{p^r} is a (closed) subgroup scheme of $\mathbb{G}_{a/R}$.
(c) Let R be as in (b). Prove that α_{p^r} and $\mu_{p^r/R}$ are isomorphic as schemes over R, but that they are not isomorphic as group schemes over R.

4.11. Let G be a group scheme over S.

(a) Prove that for each integer m, the multiplication-by-m map $[m] : G \to G$ described in (3.4) is a morphism.

(b) Prove that $[-m] = [-1] \circ [m]$.

(c) More generally, prove that $[mn] = [m] \circ [n]$.

(d) Prove that $[m+n] = \mu \circ ([m] \times [n])$.

(e) With notation as in (3.4), prove that $[0] = \sigma_0 \circ \pi$ and that $[-1] = i$.

4.12. Let R be a ring, let $d \in R$, and let $G_d \subset \mathbb{A}_R^2$ be the affine scheme given by the equation

$$G_d : x^2 - dy^2 = 1.$$

(a) Prove that the composition law

$$G_d \times_R G_d \longrightarrow G_d,$$
$$((x_1, y_1), (x_2, y_2)) \longmapsto (x_1 x_2 + d y_1 y_2, x_1 y_2 + x_2 y_1)$$

gives G_d the structure of a group scheme over R.

(b) Prove that G_0 fits into the following exact sequence of group schemes over R:

$$0 \longrightarrow \mathbb{G}_{a/R} \longrightarrow G_0 \longrightarrow \mu_{2/R} \longrightarrow 1.$$

(Here $\mu_{2/R} = \operatorname{Spec} R[T]/(T^2 - 1)$ is the scheme of square roots of unity; see exercise 4.10(a).)

(c) Prove that G_0 is isomorphic to $\mathbb{G}_{a/R} \times_R \mu_{2/R}$ as group schemes over R.

(d) Let $d_1, d_2 \in R$. Prove that G_{d_1} is isomorphic to G_{d_2} as group schemes over R if and only if there is a unit $u \in R^*$ such that $d_1 = u^2 d_2$.

(e) If $2 \in R^*$, prove that G_1 and $\mathbb{G}_{m/R}$ are isomorphic as group schemes over R.

4.13. Let K be a field of characteristic 0, and for each $d \in K$, let $G_d/K \subset \mathbb{A}_K^2$ be the group variety

$$G_d : x^2 - dy^2 = 1$$

described in the previous exercise. Prove that every connected group variety of dimension one defined over K is isomorphic over K to one of the other following group varieties:

(i) The additive group $\mathbb{G}_{a/K}$.

(ii) The group G_d for some $d \in K^*$. (Note that G_1 is isomorphic to the multiplicative group $\mathbb{G}_{m/K}$.)

(iii) An elliptic curve defined over K.

4.14. Let $\mathcal{C} \subset \mathbb{P}_{\mathbb{Z}}^2$ be the arithmetic surface given by the equation

$$\mathcal{C} : y^2 = x^3 + 2x^2 + 6.$$

Complete the proof (4.2.2) that $\mathcal{C}$ is a regular scheme by verifying that $\mathcal{C}$ is regular at the points $x = y = 3 = 0$ and $x + 66 = y = 97 = 0$.

4.15. Let R be a discrete valuation ring with normalized valuation v, and let $\mathcal{W} \subset \mathbb{P}^2_R$ be the arithmetic surface given by the Weierstrass equation
$$\mathcal{W} : y^2 + a_1 xy + a_3 y = x^3 + a_2 x^2 + a_4 x + a_6.$$
Let Δ and j be the associated discriminant and j-invariant.
(a) If $v(\Delta) = 1$, prove that $\mathcal{W}$ is a regular scheme.
(b) If $v(\Delta) = 2$ and $v(j) \geq 0$, prove that $\mathcal{W}$ is a regular scheme.
(The computations are simpler if you assume that 2 and 3 are units in R, but the results are true in general.)

4.16. Let $\mathcal{C}/\mathbb{Z}$ be the arithmetic surface in $\mathbb{A}^2_\mathbb{Z}$ defined by the equation
$$\mathcal{C} : (x^3 + 4x^2 + 3x - 1)y^3 - (2x^4 + x^3 - x^2 - 2x)y^2$$
$$- (x^6 - 3x^5 + 3x^4 - x^3)y + 2x^7 + x^6 - x^5 - 2x^4 = 7.$$
Describe the special fiber $\mathcal{C}_7$ of $\mathcal{C}$ over the point $(7) \in \operatorname{Spec} \mathbb{Z}$; that is, describe the components of $\mathcal{C}_7$, their multiplicities, and their intersection points. Draw a sketch (in $\mathbb{R}^2$) illustrating $\mathcal{C}_7$. (See (4.2.4) and Figure 4.2 for a similar calculation.)

4.17. Let $\pi : \mathcal{C} \to \operatorname{Spec}(R)$ be a regular arithmetic surface over a Dedekind domain R, let $\mathfrak{p} \in \operatorname{Spec}(R)$, and let $x \in \mathcal{C}_\mathfrak{p} \subset \mathcal{C}$ be a non-singular closed point on the fiber of $\mathcal{C}$ over $\mathfrak{p}$. Complete the proof of Proposition 4.3 by proving that $\pi^*(\mathfrak{p}) \not\subset \mathcal{M}^2_{\mathcal{C},x}$.

4.18. This exercise generalizes (4.4). Let R be a Dedekind domain with fraction field K, and let $\mathcal{X}$ be a "nice" scheme over R whose generic fiber X/K is a smooth, projective variety. (Here "nice" has the same meaning as in the definition of arithmetic surface; see §4.)
(a) If $\mathcal{X}$ is proper over R, prove that $X(K) = \mathcal{X}(R)$.
(b) Suppose that $\mathcal{X}$ is a regular scheme, and let $\mathcal{X}^0 \subset \mathcal{X}$ be the largest subscheme of $\mathcal{X}$ with the property that $\mathcal{X}^0$ is smooth over R. Prove that $\mathcal{X}(R) = \mathcal{X}^0(R)$.

4.19. Let R'/R be an extension of discrete valuation rings with maximal ideals $\mathfrak{p}$, $\mathfrak{p}'$, fraction fields K, K', and residue fields k, k' respectively. Suppose that K'/K is a finite extension. Prove that R' is the localization of a smooth R-scheme if and only if $\mathfrak{p}' = \mathfrak{p}R'$ (i.e., R'/R is unramified).

4.20. Let R be a discrete valuation ring with fraction field K, and let E/K be an elliptic curve given by a Weierstrass equation
$$E : y^2 + a_1 xy = x^3 + a_6 \qquad \text{with } a_1 \in R^* \text{ and } a_6 \in R.$$
Let $\mathcal{W} \subset \mathbb{P}^2_R$ be the R-scheme defined by this Weierstrass equation.
(a) Prove that $\mathcal{W}$ is a regular scheme if and only if $v(a_6) \leq 1$.
(b) Prove that $\mathcal{W}$ is smooth over R if and only if $v(a_6) = 0$.

4.21. Let E/K, $\mathcal{W}/R$ and $\mathcal{W}^0/R$ be as in the statement of Theorem 5.3, where we assume that we start with a minimal Weierstrass equation for E/K.
(a) If E/K has split multiplicative reduction, prove that the special fibers of $\mathcal{W}^0$ and $\mathbb{G}_{m/R}$ are isomorphic as group schemes over the residue field of R.
(b) If E/K has additive reduction, prove that the special fibers of $\mathcal{W}^0$ and $\mathbb{G}_{a/R}$ are isomorphic as group schemes over the residue field of R.
(c) Give a similar description of the special fiber of $\mathcal{W}^0$ in the case that the curve E/K has non-split multiplicative reduction.

4.22. Let R be a discrete valuation ring with fraction field K, let E/K be an elliptic curve, and choose a Weierstrass equation for E/K with coefficients in R,

$$E : Y^2 Z + a_1 XYZ + a_3 YZ^2 = X^3 + a_2 X^2 Z + a_4 XZ^2 + a_6 Z^3.$$

Let $\mathcal{W} \subset \mathbb{P}_R^2$ be the R-scheme defined by this Weierstrass equation, and let

$$\mu : \mathcal{W} \times \mathcal{W} \dashrightarrow \mathcal{W}$$

be the rational map induced by the addition law on the generic fiber. Define affine subsets of $\mathcal{W}$ by

$$\mathcal{W}_{\text{aff}} = \{Z \neq 0\} \qquad \text{and} \qquad \mathcal{W}'_{\text{aff}} = \{Y \neq 0\}.$$

Prove that μ is a morphism when restricted to each of the following sets:
(a) $\mathcal{W}_{\text{aff}} \times_R \mathcal{W}'_{\text{aff}}$.
(b) $\mathcal{W}'_{\text{aff}} \times_R \mathcal{W}_{\text{aff}}$.
(c) $\mathcal{W}'_{\text{aff}} \times_R \mathcal{W}'_{\text{aff}}$.
(The formulas will be easier if you assume that 2 and 3 are units in R and take a Weierstrass equation of the form $Y^2 Z = X^3 + AXZ^2 + BZ^3$. We described the behavior of μ on $\mathcal{W}_{\text{aff}} \times_R \mathcal{W}_{\text{aff}}$ during the proof of (5.3). This exercise is asking you to complete the proof of (5.3).)

4.23. Let R be a Dedekind domain with fraction field K, let E/K be an elliptic curve, let $\mathcal{C}/R$ be a minimal proper regular model for E/K, and let $\mathcal{E}/R$ be the largest subscheme of $\mathcal{C}/R$ which is smooth over R. Note that $\mathcal{E}/R$ is a Néron model for E/K, so in particular $\mathcal{E}$ is a group scheme over R (6.1).
(a) Let $P \in E(K) \cong \mathcal{E}(R)$, and let $\tau_P : \mathcal{E} \to \mathcal{E}$ be translation-by-P (3.3). Prove that τ_P extends to an R-morphism $\mathcal{C} \to \mathcal{C}$.
(b) Prove that every automorphism $\alpha : E/K \to E/K$ extends to an R-morphism $\mathcal{C} \to \mathcal{C}$.
(c) *Prove that the group law $\mathcal{E} \times_R \mathcal{E} \to \mathcal{E}$ extends to give a group scheme action $\mathcal{E} \times_R \mathcal{C} \to \mathcal{C}$.
(d) Prove that in general the group law $\mathcal{E} \times_R \mathcal{E} \to \mathcal{E}$ does not extend to give an R-morphism $\mathcal{C} \times_R \mathcal{C} \to \mathcal{C}$.

4.24. Let R be a discrete valuation ring with fraction field K, residue field k, and residue characteristic p. Let E/K and E'/K be elliptic curves, and let $\mathcal{E}/R$ and $\mathcal{E}'/R$ be Néron models for E/K and E'/K respectively. Let $\phi_K : E \to E'$ be an isogeny of degree $m \geq 1$ defined over K. Assume that either p does not divide m, or else that E/K does not have additive reduction.
(a) *Prove that ϕ_K extends to an R-morphism $\phi_R : \mathcal{E} \to \mathcal{E}'$.
(b) Prove that ϕ_R is a homomorphism of R-group schemes.
(c) Prove that the restriction of ϕ_R to the special fiber is a finite morphism $\phi_k : \tilde{\mathcal{E}}/k \to \tilde{\mathcal{E}}'/k$ which maps the identity component of $\tilde{\mathcal{E}}$ to the identity component of $\tilde{\mathcal{E}}'$.
(d) Prove that there is an R-morphism $\hat{\phi}_R : \mathcal{E}' \to \mathcal{E}$ with the property that the composition $\hat{\phi}_R \circ \phi_R : \mathcal{E} \to \mathcal{E}$ is the multiplication-by-m map on $\mathcal{E}$ (3.4). This generalizes the construction of the dual isogeny for elliptic curves over fields [AEC, III §6].

4.25. Let R be a Dedekind domain with fraction field K, let E/K be an elliptic curve, and let $\mathcal{E}/R$ be a Néron model for E/K. Prove that the connected component of $\mathcal{E}/R$ as described in (6.1.2) is a subgroup scheme of $\mathcal{E}/R$.

4.26. Let R be a Henselian discrete valuation ring with valuation v. Let $f(x) \in R[x]$ be a monic polynomial, and let $a \in R$ be an element with the property that

$$v(f(a)) > 2v(f'(a)).$$

Prove that there is a unique element $\alpha \in R$ satisfying

$$v(\alpha - a) > v(f'(a)) \qquad \text{and} \qquad f(\alpha) = 0.$$

(Note the strict inequalities.)

4.27. Let R be a discrete valuation ring with maximal ideal $\mathfrak{p}$ and residue field k, let $f_1, \ldots, f_m \in R[x_1, \ldots, x_n]$, let

$$\mathcal{X} = \operatorname{Spec} R[x_1, \ldots, x_n]/(f_1, \ldots, f_m)$$

be the scheme defined by the f_i's, and let

$$J = J(x_1, \ldots, x_n) = (\partial f_i/\partial x_j)_{1 \le i \le m, 1 \le j \le n}$$

be the associated Jacobian matrix.

(a) Prove that the generic fiber of $\mathcal{X}$ is empty if and only if some power of $\mathfrak{p}$ is contained in the ideal $(f_1, \ldots, f_m)$.

(b) Let $\tilde{a} = (\tilde{a}_1, \ldots, \tilde{a}_n) \in \tilde{\mathcal{X}}(k)$ be a point on the special fiber of $\mathcal{X}$. Prove that $\mathcal{X}_\mathfrak{p}$ is smooth over k at $\tilde{a}$ if and only if the matrix $J(\tilde{a})$ satisfies

$$\operatorname{rank} J(\tilde{a}) = n - \dim \mathcal{X}_\mathfrak{p}.$$

(c) Assume that $\mathcal{X}$ is irreducible, reduced, and has non-empty generic fiber. Suppose further that the ring R is Henselian, and let $a = (a_1, \ldots, a_n) \in \mathbb{A}^n(R)$ be a point satisfying

$$f_1(a) \equiv \cdots \equiv f_m(a) \equiv 0 \,(\operatorname{mod} \mathfrak{p}) \qquad \text{and} \qquad \operatorname{rank} \widetilde{J(a)} = n - \dim \mathcal{X}_\mathfrak{p}.$$

Prove that there exists a (unique) point $\alpha \in \mathbb{A}^n(R)$ such that

$$f_1(\alpha) = \cdots = f_m(\alpha) = 0 \qquad \text{and} \qquad \alpha \equiv a \,(\operatorname{mod} \mathfrak{p}).$$

This result is a multi-variable version of Hensel's lemma. It implies the surjectivity of the reduction map for smooth schemes over Henselian rings. (*Hint.* First prove (c) under the assumption that R is complete.)

4.28. Let R be a discrete valuation ring with residue field k, and let R^{h} and R^{sh} be the Henselization and strict Henselization of R respectively (6.5).

(a) Let R' be a Henselian discrete valuation ring, and let $i : R \to R'$ be a local homomorphism. Prove that there is a unique local homomorphism $R^{\mathrm{h}} \to R'$ extending i.

(b) Let R'' be a strictly Henselian discrete valuation ring with residue field k'', let $i : R \to R''$ be a local homomorphism, and let $u : k'' \to \bar{k}$ be a k-homomorphism. Prove that there exists a unique local homomorphism $R^{\mathrm{sh}} \to R''$ which extends i and which induces the map u on the residue fields.

4.29. *Let K be an algebraically closed field, let C/K be a curve of genus $g \geq 1$, and let V be the g-fold symmetric product of C. Note that the points of V can be naturally identified with the positive divisors of degree g on C.
(a) Fix a basepoint $P_0 \in C$. Prove that there is a rational map $\mu : V \times V \to V$ determined by the property

$$\mu(x, y) \sim x + y - g(P_0).$$

(Here $\sim$ denotes linear equivalence of divisors.)
(b) Prove that μ is a normal law, and hence from (6.9) that there is a group variety J/K associated to μ.
(c) Prove that the map $V \to J$ is a morphism, and deduce that J is proper over K.
(d) Prove that the map of sets

$$V \longrightarrow \mathrm{Pic}^0(C), \qquad x \longmapsto \mathrm{class}[x - g(P_0)],$$

induces an isomorphism of groups $J(K) \to \mathrm{Pic}^0(C)$.

The group variety J is an abelian variety called the *Jacobian variety of* C. This construction of the Jacobian variety is due to Weil [3]. For a further discussion, see the proof sketch of Proposition III.2.6(b).

4.30. Let R be a strictly Henselian discrete valuation ring with fraction field K. Let $\mathcal{E}/R$ be a group scheme over R whose generic fiber E/K is an elliptic curve. Prove that $\mathcal{E}/R$ is a Néron model for E/K if and only if the inclusion $\mathcal{E}(R) \to E(K)$ is a bijection.

4.31. Let R be a discrete valuation ring with uniformizing element π and algebraically closed residue field k. Assume that $\mathrm{char}(k) \neq 2, 3, 5$. Let $\mathcal{C} \subset \mathbb{A}_R^2$ be the affine scheme defined by

$$\mathcal{C} : y^2 = x^5 + \pi^2;$$

that is, $\mathcal{C} = \mathrm{Spec}\, R[x, y]/(y^2 - x^5 - \pi^2)$.
(a) Show that $\mathcal{C}$ is regular except at the one point $\pi = x = y = 0$ on the special fiber.
(b) Compute the blow-up of $\mathcal{C}$ at the singular point $\pi = x = y = 0$ as explained in (7.7). Show that the resulting scheme is still not regular.
(c) Continue blowing up until you get a regular scheme. Draw a picture of the special fiber similar to the diagrams in Figures 4.3 and 4.4.
(d) Repeat (a), (b), and (c) for the arithmetic surface $y^2 = x^5 + \pi^7$.

4.32. (a) For each of the Kodaira-Néron reduction types (8.2), compute the intersection incidence matrix of the special fiber.
(b) Show that each of the incidence matrices in (a) has determinant 0.
(c) Let M be the matrix obtained by taking any one of the incidence matrices in (a) and deleting a row and column corresponding to a multiplicity-1 component. Show that $\det(M)$ is equal to plus or minus the number of multiplicity-1 components on the special fiber. Equivalently, $|\det(M)|$ is the order of the group of components of the Néron model.

4.33. Let R be a discrete valuation ring, and let $\mathcal{C}/R$ be a proper regular model of a curve of positive genus. Generalize part (b) of the previous exercise by showing that the incidence matrix of its special fiber has determinant 0. (For a generalization of part (c), see Raynaud [1].)

4.34. Let R be a discrete valuation ring with maximal ideal $\mathfrak{p}$, fraction field K, and algebraically closed residue field k. Let C/K be a non-singular projective curve of genus $g \geq 1$, and let $\mathcal{C}/R$ be a minimal proper regular model for C/K. Suppose that the special fiber $\mathcal{C}_\mathfrak{p}$ of $\mathcal{C}$ contains a configuration with t components of the form shown in Figure 4.6(b), where each of the illustrated components satisfies $p_a(F_i) = 0$ and $F_i^2 = -2$.
(a) Prove that $t \leq 4g + 2$.
(b) If $t = 4g + 2$, complete the picture of $\mathcal{C}_\mathfrak{p}$. In particular, show that $\mathcal{C}_\mathfrak{p}$ has exactly $4g + 5$ components. (For $g = 1$, you'll get a fiber of Type II*.)

4.35. Let R be a complete discrete valuation ring with fraction field K and algebraically closed residue field k of characteristic $p \neq 2, 3$. Let E/K be an elliptic curve with additive reduction, and let $E_0(K)$ be the subgroup of $E(K)$ consisting of points with non-singular reduction (9.2).
(a) Prove that $E_0(K)$ is uniquely divisible by 2 and 3.
(b) Prove that $E(K)/E_0(K)$ is killed by 12.
(c) Prove that the natural map $E(K)[12] \to E(K)/E_0(K)$ is an isomorphism.

4.36. (a) Let $C_4, C_6 \in \mathbb{Z}$ be integers with $C_4^3 - C_6^2 \neq 0$. Prove that there exists a Weierstrass equation

$$y^2 + a_1 xy + a_3 y = x^3 + a_2 x^2 + a_4 x + a_6$$

with coefficients $a_1, a_2, a_3, a_4, a_6 \in \mathbb{Z}$ satisfying $c_4 = C_4$ and $c_6 = C_6$ if and only if one of the following two conditions is true:
   (i) $\mathrm{ord}_3(C_6) \neq 2$ and $C_6 \equiv -1 \pmod 4$.
   (ii) $\mathrm{ord}_3(C_6) \neq 2$, $\mathrm{ord}_2(C_4) \geq 4$, and $C_6 \equiv 0$ or $8 \pmod{32}$.

(b) Use the criteria in (a) to devise a quick algorithm to check whether a given Weierstrass equation with integer coefficients is a minimal Weierstrass equation.
(c) Generalize the criteria in (a) to an arbitrary field $K/\mathbb{Q}$.

4.37. Let R be a discrete valuation ring with fraction field K, let E/K be an elliptic curve given by a Weierstrass equation

$$y^2 + a_1 xy + a_3 y = x^3 + a_2 x^2 + a_4 x + a_6$$

with coefficients in R, and assume that $\pi | a_3, a_4$, $\pi^2 | a_6$, and $\pi \nmid b_2$. Resolve the singularity on the special fiber by a sequence of blow-ups and show that the special fiber is of Type I_n with $n = v(\Delta)$.

4.38. Let ℓ be a prime, let $\ell' = \ell$ if $\ell \neq 2$, and let $\ell' = 4$ if $\ell = 2$. Prove that the group

$$\{ M \in \mathrm{GL}_2(\mathbb{Z}_\ell) : M \equiv 1 \pmod{\ell'} \}$$

contains no elements of finite order.

4.39. This exercise generalizes the previous exercise. Let $K/\mathbb{Q}_p$ be a p-adic field with ring of integers R, maximal ideal $\mathfrak{p}$, and normalized valuation v_K.
(a) Suppose that there is a matrix $M \in \mathrm{GL}_n(R)$ of exact order $m \geq 2$ satisfying

$$M \equiv 1 \pmod{\mathfrak{p}^r}.$$

Prove that

$$m = p^n \quad \text{and} \quad r \leq \frac{v_K(p)}{p^n - p^{n-1}}.$$

(b) If $r > v_K(p)/(p-1)$, prove that the group

$$\{M \in \mathrm{GL}_n(R) \; : \; M \equiv 1 \pmod{\mathfrak{p}^r}\}$$

contains no elements of finite order other than the identity element.

4.40. Let E/K and E'/K be elliptic curves defined over a local field, and let $\phi : E \to E'$ be a non-constant isogeny defined over K. Prove that

$$\varepsilon(E/K) = \varepsilon(E'/K), \qquad \delta(E/K) = \delta(E'/K), \qquad f(E/K) = f(E'/K).$$

Notice that this generalizes the assertion [AEC, VII.7.2] that E and E' either both have good reduction or both have bad reduction, since (10.1) says that good reduction is equivalent to $\varepsilon = 0$.

4.41. Let $K/\mathbb{Q}_3$ be a 3-adic field, let E/K be an elliptic curve, let $\ell \neq 2$ be a prime, and let $L = K(E[\ell])$.
(a) Prove that the first higher ramification group $G_1(L/K)$ is either trivial or a cyclic group of order 3. (*Hint.* Show that $G_1(L/K)$ is independent of ℓ, and then take $\ell = 2$.)
(b) Prove that $G_1(L/K)$ is a cyclic group of order 3 if and only if E/K has reduction type II, IV, IV*, or II*.

4.42. Let $K/\mathbb{Q}_2$ be a 2-adic field, let E/K be an elliptic curve with potential good reduction, let $\ell \geq 3$ be a prime, let $L = K(E[\ell])$, and let $G_1(L/K)$ be the first higher ramification group of L/K. Prove that

$$G_1(L/K) \cong \{1\} \quad \text{or} \quad \mathbb{Z}/2\mathbb{Z} \quad \text{or} \quad \mathbb{Z}/4\mathbb{Z} \quad \text{or} \quad H_8,$$

where H_8 is the quaternion group of order 8. (*Hint.* Show that $G_1(L/K)$ is independent of ℓ, and then take $\ell = 3$.)

4.43. For each of the following elliptic curves $E/\mathbb{Q}_2$, let $L = \mathbb{Q}_2(E[3])$ and compute the first higher ramification group $G_1(L/\mathbb{Q}_2)$.
(a) $E : y^2 + 2y = x^3$.
(b) $E : y^2 + 2xy + 8y = x^3$.
(c) $E : y^2 + 2y = x^3 + 2x$.

4.44. Let K be a number field. Prove that for any constant B there are only finitely many elliptic curves E/K whose conductor $\mathfrak{f}_{E/K}$ satisfies

$$|\mathrm{N}^K_{\mathbb{Q}}(\mathfrak{f}_{E/K})| \leq B.$$

4.45. Let L/K be a finite extension of local fields as described at the beginning of §10, let π_L be a uniformizer for L, and define an index function

$$i_{L/K} : G(L/K) \longrightarrow \mathbb{Z} \cup \{\infty\}, \qquad i_{L/K}(\sigma) = v_L(\pi_L^\sigma - \pi_L).$$

(a). Prove that $i_{L/K}(\sigma) \geq i+1$ if and only if $\sigma \in G_i(L/K)$.
(b) Prove that $i_{L/K}(\tau\sigma\tau^{-1}) = i_{L/K}(\sigma)$ for all $\sigma, \tau \in G(L/K)$.
(c) Prove that $i_{L/K}(\sigma\tau) \geq \min\{i_{L/K}(\sigma), i_{L/K}(\tau)\}$ for all $\sigma, \tau \in G(L/K)$.

4.46. We continue with the notation from the previous exercise, with the additional assumption that the extension L/K is totally ramified. For basic material on the representation theory used in this exercise, see for example Serre [7]. The *Artin character* $\mathrm{Ar}_{L/K}$ and the *Swan character* $\mathrm{Sw}_{L/K}$ are defined to be the functions

$$\mathrm{Ar}_{L/K} : G_{L/K} \longrightarrow \mathbb{Z}, \qquad\qquad \mathrm{Sw}_{L/K} : G_{L/K} \longrightarrow \mathbb{Z},$$
$$\mathrm{Ar}_{L/K}(\sigma) = -i_{L/K}(\sigma) \quad \text{if } \sigma \neq 1, \qquad \mathrm{Sw}_{L/K}(\sigma) = 1 - i_{L/K}(\sigma) \quad \text{if } \sigma \neq 1,$$
$$\mathrm{Ar}_{L/K}(1) = \sum_{\sigma \neq 1} i_{L/K}(\sigma), \qquad\qquad \mathrm{Sw}_{L/K}(1) = \sum_{\sigma \neq 1}(i_{L/K}(\sigma) - 1).$$

(a) A function ψ on $G(L/K)$ is called a *class function* if $\psi(\tau\sigma\tau^{-1}) = \psi(\sigma)$ for all $\sigma, \tau \in G(L/K)$. Prove that $\mathrm{Ar}_{L/K}$ and $\mathrm{Sw}_{L/K}$ are class functions.
(b) *For any pair of functions ψ_1, ψ_2 on $G(L/K)$, define

$$\langle \psi_1, \psi_2 \rangle = \frac{1}{g_0(L/K)} \sum_{\sigma \in G(L/K)} \psi_1(\sigma)\psi_2(\sigma^{-1}).$$

If χ is the character of an irreducible representation of $G(L/K)$, prove that $\langle \chi, \mathrm{Ar}_{L/K} \rangle$ and $\langle \chi, \mathrm{Sw}_{L/K} \rangle$ are non-negative integers.
(c) Replace K by its maximal unramified extension, and let E/K be an elliptic curve with integral j-invariant. Let L/K be a finite Galois extension such that E has good reduction over L. Further let χ_E be the character of the representation of $G(L/K)$ on $T_\ell(E)$. Prove that χ_E takes values in $\mathbb{Z}$ and is independent of ℓ.
(d) *Continuing with the assumptions from (c), prove that

$$\delta(E/K) = \langle \mathrm{Sw}_{L/K}, \chi_E \rangle \qquad \text{and} \qquad f(E/K) = \langle \mathrm{Ar}_{L/K}, \chi_E \rangle.$$

Deduce that $\delta(E/K)$ and $f(E/K)$ are integers that are independent of ℓ. This proves (10.2c) in the case that E has potential good reduction.
(e) If E/K has non-integral j-invariant, prove that $\delta(E/K)$ and $f(E/K)$ are integers and are independent of ℓ, thus completing the proof of (10.2c). (*Hint.* For (e), use the isomorphism $E(\bar{K}) \cong \bar{K}^*/q^{\mathbb{Z}}$ described in (V.5.3) and exercise 5.11.)

4.47. Let $K/\mathbb{Q}_p$ be a p-adic field with $p \geq 5$, let v_K be the normalized valuation on K, and let E/K be an elliptic curve. Prove that there exists a minimal Weierstrass equation for E/K with $a_1 = a_2 = a_3 = 0$, and with a_4, a_6, and Δ as described in the following table:

Type	I_0	I_n	II	III	IV	I_0^*	I_n^*	IV*	III*	II*
$v_K(a_4)$		$= 0$	≥ 1	$= 1$	≥ 2	$= 2$	$= 2$	≥ 3	$= 3$	≥ 4
$v_K(a_6)$		$= 0$	$= 1$	≥ 2	$= 2$	$= 3$	$= 3$	$= 4$	≥ 5	$= 5$
$v_K(\Delta)$	$= 0$	$= n$	$= 2$	$= 3$	$= 4$	$= 6$	$= n+6$	$= 8$	$= 9$	$= 10$

4.48. Let $K/\mathbb{Q}_3$ be a 3-adic field with normalized valuation v_K, and let E/K be an elliptic curve. Prove that there exists a minimal Weierstrass equation for E/K with $a_1 = a_3 = 0$, and with a_2, a_4, a_6, and Δ as described in the following table:

Type	I_0	I_n	II	III	IV	I_0^*	I_n^*	IV*	III*	II*
$v_K(a_2)$		$= 0$	≥ 1	≥ 1	≥ 1	≥ 1	$= 1$	≥ 2	≥ 2	≥ 2
$v_K(a_4)$		≥ 1	≥ 1	$= 1$	≥ 2	≥ 2	$\geq \left[\frac{n+5}{2}\right]$	≥ 3	$= 3$	≥ 4
$v_K(a_6)$		≥ 1	$= 1$	≥ 2	$= 2$	≥ 3	$\geq n+3$	$= 4$	≥ 5	$= 5$
$v_K(\Delta)$	$= 0$	$= n$	≥ 3	$= 3$	≥ 5	$= 6$	$= n+6$	≥ 9	$= 9$	≥ 11

4.49. Let $K/\mathbb{Q}_p$ be a p-adic field with $p \geq 3$, let v_K be the normalized valuation on K, and let E/K be an elliptic curve with Type I_n^* reduction.
(a) If $n = 0$, prove that $v_K(j(E)) \geq 0$ and $v_K(\mathcal{D}_{E/K}) = 6$.
(b) If $n \geq 1$, prove that $v_K(j(E)) = -n$ and $v_K(\mathcal{D}_{E/K}) = n + 6$.
(c) Let L/K be a tamely ramified extension with ramification degree $e = e(L/K)$. Prove that E/L has Type I_{ne} reduction if $e \equiv 0 \pmod 2$, and E/L has Type I_{ne}^* reduction if $e \equiv 1 \pmod 2$.
(d) Give an example of an elliptic curve $E/\mathbb{Q}_2$ with Type I_n^* reduction satisfying $n \geq 1$ and $v_2(j(E)) \geq 0$. This shows that (b) is not true for $p = 2$. What is the largest possible value of n in this situation?

4.50. Let $K/\mathbb{Q}_3$ be a 3-adic field with normalized valuation v_K, and let L/K be a tamely ramified extension with normalized valuation v_L. Let $e = e(L/K)$ be the relative ramification degree, so $v_L = ev_K$. Let E/K be an elliptic curve whose reduction type is one of II, IV, IV*, or II*.
(a) Prove that the reduction type of E/L is given by the following table:

Type(E/K)	$e \equiv 1$ (6)	$e \equiv 2$ (6)	$e \equiv 4$ (6)	$e \equiv 5$ (6)
II	II	IV	IV*	II*
IV	IV	IV*	IV	IV*
IV*	IV*	IV	IV*	IV
II*	II*	IV*	IV	II

(b) Let $\mathcal{D}_{E/K}$ and $\mathcal{D}_{E/L}$ be the minimal discriminants of E/K and E/L respectively. Prove that the value of the difference

$$e(L/K)v_K(\mathcal{D}_{E/K}) - v_L(\mathcal{D}_{E/L})$$

is given by the following table:

Type(E/K)	$e \equiv 1$ (6)	$e \equiv 2$ (6)	$e \equiv 4$ (6)	$e \equiv 5$ (6)
II	$2e - 2$	$2e - 4$	$2e - 8$	$2e - 10$
IV	$4e - 4$	$4e - 8$	$4e - 4$	$4e - 8$
IV*	$8e - 8$	$8e - 4$	$8e - 8$	$8e - 4$
II*	$10e - 10$	$10e - 8$	$10e - 4$	$10e - 2$

4.51. This exercise illustrates how wild ramification can cause the reduction type of an elliptic curve to change in an irregular fashion. It may be compared with the previous exercise, which dealt with the tamely ramified case.

Let $K/\mathbb{Q}_3$ be a 3-adic field with normalized valuation v_K, and let E/K be an elliptic curve given by a Weierstrass equation

$$E : y^2 = x^3 + a_6 \qquad \text{with } v_K(a_6) = 1.$$

Let L/K be a ramified extension of degree 3, so L/K is wildly ramified.
(a) Prove that E/K has Type II reduction.
(b) If $v_K(3) = 1$ (i.e., if $K/\mathbb{Q}_3$ is unramified), prove that E/L has Type III* reduction.
(c) If $v_K(3) = 2$, prove that E/L has Type I_0 (i.e., good) reduction.
(d) If $v_K(3) = 3$, prove that E/L has Type III reduction.
(e) Try to find a general formula for the reduction type of E/L. Does the reduction type of E/L depend only on $v_K(3)$?

4.52. We continue with the notation from the previous exercise, so $K/\mathbb{Q}_3$ is a 3-adic field and E/K is an elliptic curve

$$E : y^2 = x^3 + a_6 \qquad \text{with } v_K(a_6) = 1.$$

Prove that the conductor exponent of E/K is $f(E/K) = 2 + 3v_K(3)$. Notice that (10.4) says that this is the largest allowable conductor exponent for an elliptic curve defined over a 3-adic field.

4.53. Let $K/\mathbb{Q}_2$ be a 2-adic field with normalized valuation v_K, and let E/K be an elliptic curve given by a Weierstrass equation

$$E : y^2 + 2xy = x^3 - x^2 + a_4x \qquad \text{with } v_K(a_4) = 1.$$

(a) Prove that the equation given for E is a minimal Weierstrass equation and that $v_K(\mathcal{D}_{E/K}) = 6v_K(2) + 3$.
(b) Prove that E has Type III reduction and that the special fiber of E has two components.
(c) Prove that the conductor exponent of E/K is $f(E/K) = 2 + 6v_K(2)$. Notice that (10.4) says that this is the largest allowable conductor exponent for an elliptic curve defined over a 2-adic field.

4.54. This exercise asks you to verify two cases of Ogg's formula (11.1) that were not completed in §11. Let $K/\mathbb{Q}_3$ be a 3-adic field, let E/K be an elliptic curve, and let $L = K(E[2])$.
(a) If E/K has Type III* reduction, prove directly that L/K is a tamely ramified extension. Use this to verify Ogg's formula for E/K.
(b) Let $M = K(\sqrt{\Delta})$, and suppose that E/M has Type IV* reduction. Give a direct proof of Ogg's formula in this situation. (*Hint.* Mimic the proof for Type IV reduction given in §11.)

CHAPTER V

Elliptic Curves over Complete Fields

Every elliptic curve $E/\mathbb{C}$ admits an isomorphism $\mathbb{C}^*/q^{\mathbb{Z}} \cong E(\mathbb{C})$ by complex analytic functions, and we have seen amply demonstrated in Chapters I and II the importance of such uniformizations. In this chapter we are going to study uniformizations over other complete fields such as $\mathbb{R}$ and finite extensions $K/\mathbb{Q}_p$. We begin in §1 with a brief review of the relevant formulas over $\mathbb{C}$, and then in §2 we use the complex uniformization to investigate elliptic curves over $\mathbb{R}$.

We next turn to elliptic curves defined over p-adic fields $K/\mathbb{Q}_p$. Tate [9] has shown that for every $q \in K^*$ with $|q| < 1$ there is an elliptic curve E_q/K and a p-adic analytic isomorphism $K^*/q^{\mathbb{Z}} \cong E_q(K)$. In §3 we will describe the Tate curve E_q and prove all of its main properties except for the surjectivity of the map $K^* \to E_q(K)$, which we reserve for §4. Tate has also shown that every elliptic curve E/K with non-integral j-invariant is isomorphic, possibly over a quadratic extension of K, to some E_q. We will prove this result and describe the necessary twisting in §5, and then in §6 we will give some applications, including Serre's proof that an elliptic curve with complex multiplication has integral j-invariant.

§1. Elliptic Curves over $\mathbb{C}$

We have already discussed elliptic curves and elliptic functions over $\mathbb{C}$ in some detail; see [AEC VI] and (I §§5–8). The purpose of this section is to gather and rewrite in a convenient form the formulas we will use later in this chapter when we study elliptic curves over $\mathbb{R}$ and over p-adic fields.

Let $E/\mathbb{C}$ be the elliptic curve corresponding to the normalized lattice

$$\Lambda_\tau = \mathbb{Z}\tau + \mathbb{Z} \quad \text{for some } \tau \in \mathbf{H}.$$

We know that $E(\mathbb{C}) \cong \mathbb{C}/\Lambda_\tau$, the isomorphism being given in terms of the Weierstrass $\wp$-function and its derivative. As in (I §6), it is convenient to let

$$u = e^{2\pi i z}, \quad q = e^{2\pi i \tau}, \quad \text{and} \quad q^{\mathbb{Z}} = \{q^k : k \in \mathbb{Z}\}.$$

Note that since $\text{Im}(\tau) > 0$, we have $|q| < 1$. There is a complex analytic isomorphism

$$\mathbb{C}/\Lambda_\tau \xrightarrow{\sim} \mathbb{C}^*/q^{\mathbb{Z}}, \quad z \longmapsto u = e^{2\pi i z},$$

and we can use the q-expansions from (I §§6,7) to explicitly describe the isomorphism $E(\mathbb{C}) \cong \mathbb{C}^*/q^{\mathbb{Z}}$.

The elliptic curve E has the Weierstrass equation

$$E : y^2 = 4x^3 - g_2(\tau)x - g_3(\tau).$$

From (I.7.3.2) we have

$$\frac{1}{(2\pi i)^4} g_2(\tau) = \frac{1}{12}\big[1 + 240 s_3(q)\big],$$

$$\frac{1}{(2\pi i)^6} g_3(\tau) = \frac{1}{216}\big[-1 + 504 s_5(q)\big],$$

where in general

$$s_k(q) = \sum_{n \geq 1} \sigma_k(n) q^n = \sum_{n \geq 1} \frac{n^k q^n}{1 - q^n}.$$

(For the second equality, see exercise 5.1. Here $\sigma_k(n) = \sum_{d|n} d^k$ as usual.) We've collected the powers of $2\pi i$ as indicated to make it easier to eliminate them.

Next, the isomorphism

$$\begin{array}{ccc} \mathbb{C}^*/q^{\mathbb{Z}} & \longrightarrow & E(\mathbb{C}) \\ u & \longmapsto & \big(\wp(u,q), \wp'(u,q)\big) \end{array}$$

is given by the power series described in (I.6.2) and (I.6.2.1):

$$\frac{1}{(2\pi i)^2} \wp(u,q) = \sum_{n \in \mathbb{Z}} \frac{q^n u}{(1 - q^n u)^2} + \frac{1}{12} - 2 s_1(q),$$

$$\frac{1}{(2\pi i)^3} \wp'(u,q) = \sum_{n \in \mathbb{Z}} \frac{q^n u(1 + q^n u)}{(1 - q^n u)^3}.$$

(Note that $\wp'$ is the derivative of $\wp$ with respect to z, where $u = e^{2\pi i z}$.)

Jacobi's formula (I.8.1) says that the discriminant of the Weierstrass equation for E has the product expansion

$$\Delta(\tau) = g_2(\tau)^3 - 27 g_3(\tau)^2 = (2\pi i)^{12} q \prod_{n \geq 1} (1 - q^n)^{24},$$

and the j-invariant of E is given by the series (I.7.4)

$$j(q) = \frac{1}{q} + 744 + 196884q + \cdots = \frac{1}{q} + \sum_{n \geq 0} c(n)q^n \quad \text{with } c(n) \in \mathbb{Z}.$$

It is convenient to make a change of variables, partially to remove the powers of $2\pi i$ and partially to eliminate the powers of 2 and 3 appearing in the denominators of the series for g_2, g_3, and $\wp$. Thus we let

$$\frac{1}{(2\pi i)^2} x = x' + \frac{1}{12},$$

$$\frac{1}{(2\pi i)^3} y = 2y' + x',$$

which gives the new Weierstrass equation

$$y'^2 + x'y' = x'^3 + a_4 x' + a_6$$

with

$$a_4 = -\frac{1}{4} \cdot \frac{1}{(2\pi i)^4} g_2(\tau) + \frac{1}{48},$$

$$a_6 = -\frac{1}{4} \cdot \frac{1}{(2\pi i)^6} g_3(\tau) - \frac{1}{48} \cdot \frac{1}{(2\pi i)^4} g_2(\tau) + \frac{1}{1728}.$$

Now using the series for g_2, g_3, $\wp$, $\wp'$ and doing a little algebra, we find we have proven virtually all of the following result.

Theorem 1.1. *For $u, q \in \mathbb{C}$ with $|q| < 1$, define quantities*

$$s_k(q) = \sum_{n \geq 1} \sigma_k(n)q^n = \sum_{n \geq 1} \frac{n^k q^n}{1 - q^n},$$

$$a_4(q) = -5s_3(q), \qquad a_6(q) = -\frac{5s_3(q) + 7s_5(q)}{12},$$

$$X(u, q) = \sum_{n \in \mathbb{Z}} \frac{q^n u}{(1 - q^n u)^2} - 2s_1(q),$$

$$Y(u, q) = \sum_{n \in \mathbb{Z}} \frac{(q^n u)^2}{(1 - q^n u)^3} + s_1(q),$$

$$E_q : y^2 + xy = x^3 + a_4(q)x + a_6(q).$$

(a) *E_q is an elliptic curve, and X and Y define a complex analytic isomorphism*

$$\phi : \quad \mathbb{C}^*/q^{\mathbb{Z}} \quad \longrightarrow \qquad E_q(\mathbb{C})$$

$$u \quad \longmapsto \quad \begin{cases} \big(X(u, q), Y(u, q)\big) & \text{if } u \notin q^{\mathbb{Z}}, \\ O & \text{if } u \in q^{\mathbb{Z}}. \end{cases}$$

(b) *Written as power series in q, both $a_4(q)$ and $a_6(q)$ have integer coefficients; that is, $a_4(q), a_6(q) \in \mathbb{Z}[\![q]\!]$.*

(c) *The discriminant and j-invariant of E_q are given by the formulas*

$$\Delta(q) = -a_6 + a_4^2 + 72a_4a_6 - 64a_4^3 - 432a_6^2$$

$$= q \prod_{n \geq 1} (1 - q^n)^{24} \in \mathbb{Z}[\![q]\!],$$

$$j(q) = \frac{1}{q} + 744 + 196884q + \cdots$$

$$= \frac{1}{q} + \sum_{n \geq 0} c(n)q^n \in \frac{1}{q} + \mathbb{Z}[\![q]\!].$$

(d) *For every elliptic curve $E/\mathbb{C}$ there is a $q \in \mathbb{C}^*$ with $|q| < 1$ such that E is isomorphic to E_q.*

PROOF. The discussion given above has proven all of (a), (b), and (c) except for the minor point that the power series for $a_6(q)$ has integer coefficients. Since

$$a_6(q) = -\frac{5s_3(q) + 7s_5(q)}{12} = -\sum_{n \geq 1} \frac{5\sigma_3(q) + 7\sigma_5(q)}{12} q^n$$

$$= -\sum_{n \geq 1} \left(\sum_{d \mid n} \frac{5d^3 + 7d^5}{12} \right) q^n,$$

it suffices to observe that

$$5d^3 + 7d^5 \equiv 0 \,(\mathrm{mod}\ 12) \qquad \text{for all } d \in \mathbb{Z}.$$

(Notice that we used this same fact in the proof of (I.7.4a).) This completes the proof of (c).

Finally, to prove (d), we note that the uniformization theorem (I.4.4) says that every elliptic curve $E/\mathbb{C}$ is isomorphic to $\mathbb{C}/\Lambda$ for some lattice Λ. Then the change of variables used above transforms the Weierstrass equation for E into an E_q.

$\square$

Remark 1.2. It is sometimes convenient to rewrite $s_1(q)$ in the alternative form

$$s_1(q) = \sum_{n \geq 1} \frac{q^n}{(1 - q^n)^2}.$$

To check that these two expressions for $s_1(q)$ are the same, we substitute $T = q^n$ into

$$\frac{T}{(1 - T)^2} = T\frac{d}{dT}\left(\frac{1}{1 - T}\right) = T\frac{d}{dT}\sum_{m \geq 0} T^m = \sum_{m \geq 1} mT^m$$

and sum over $n \geq 1$, which yields

$$\sum_{n \geq 1} \frac{q^n}{(1-q^n)^2} = \sum_{n \geq 1} \sum_{m \geq 1} m q^{nm} = \sum_{m \geq 1} m \sum_{n \geq 1} q^{mn} = \sum_{m \geq 1} \frac{m q^m}{1 - q^m}.$$

For later reference, it will be helpful to rewrite the formulas

$$\wp(z) - \wp(a) = -\frac{\sigma(z+a)\sigma(z-a)}{\sigma(z)^2 \sigma(a)^2} \quad \text{and} \quad \wp'(z) = -\frac{\sigma(2z)}{\sigma(z)^4}$$

from (I.5.6) in terms of $X(u,q)$ and $Y(u,q)$. For this purpose, we introduce a normalized theta function.

Proposition 1.3. *Define a normalized theta function $\theta(u,q)$ by the formula*

$$\theta(u,q) = (1-u) \prod_{n \geq 1} \frac{(1-q^n u)(1-q^n u^{-1})}{(1-q^n)^2}.$$

(a) $\theta(u,q)$ *converges for all $u, q \in \mathbb{C}^*$ with $|q| < 1$ and satisfies the functional equation*

$$\theta(qu, q) = -\frac{1}{u}\theta(u,q).$$

(b) *θ is related to the Weierstrass σ function by the formula*

$$\sigma(u,q) = -\frac{1}{2\pi i} e^{\frac{1}{2}\eta(1)z^2} e^{-\pi i z} \theta(u,q),$$

where $u = e^{2\pi i z}$, $q = e^{2\pi i \tau}$, and $\eta(1)$ is the quasi-period associated to the period 1 in the lattice $\mathbb{Z}\tau + \mathbb{Z}$.

(c) *θ is related to the functions $X(u,q)$ and $Y(u,q)$ described in (1.1) by the formulas*

(i) $$X(u_1, q) - X(u_2, q) = -\frac{u_2 \theta(u_1 u_2, q)\theta(u_1 u_2^{-1}, q)}{\theta(u_1, q)^2 \theta(u_2, q)^2},$$

(ii) $$2Y(u,q) + X(u,q) = -\frac{u\theta(u^2, q)}{\theta(u,q)^4}.$$

PROOF. (a) Since $|q^n u| < 1$ and $|q^n u^{-1}| < 1$ for all sufficiently large n, it is clear that the product defining θ converges. Then replacing u by qu and renumbering the factors in the product gives

$$\theta(qu, q) = (1-qu) \prod_{n \geq 1} \frac{(1-q^{n+1}u)(1-q^{n-1}u^{-1})}{(1-q^n)^2}$$

$$= (1-u^{-1}) \prod_{n \geq 1} \frac{(1-q^n u)(1-q^n u^{-1})}{(1-q^n)^2} = \frac{-1}{u}\theta(u,q).$$

(b) This is immediate from the definition of θ and the product formula for σ given in (I.6.4).

(c) From above, X and Y are related to $\wp$ and $\wp'$ by the formulas

$$X = \frac{1}{(2\pi i)^2}\wp - \frac{1}{12} \quad \text{and} \quad 2Y + X = \frac{1}{(2\pi i)^3}\wp'.$$

Hence writing $u_1 = e^{2\pi i z_1}$ and $u_2 = e^{2\pi i z_2}$, we find (dropping the q from our notation)

$$X(u_1) - X(u_2) = \frac{1}{(2\pi i)^2}\left(\wp(u_1) - \wp(u_2)\right)$$

$$= -\frac{1}{(2\pi i)^2}\frac{\sigma(u_1 u_2)\sigma(u_1 u_2^{-1})}{\sigma(u_1)^2\sigma(u_2)^2} \quad \text{from (I.5.6a)}$$

$$= -e^{\frac{1}{2}\eta(1)((z_1+z_2)^2+(z_1-z_2)^2-2z_1^2-2z_2^2)}$$

$$\cdot e^{-\pi i((z_1+z_2)+(z_1-z_2)-2z_1-2z_2)} \cdot \frac{\theta(u_1 u_2)\theta(u_1 u_2^{-1})}{\theta(u_1)^2\theta(u_2)^2}$$

$$\text{from (b)}$$

$$= -u_2\frac{\theta(u_1 u_2)\theta(u_1 u_2^{-1})}{\theta(u_1)^2\theta(u_2)^2}.$$

This proves (i), and (ii) is proven by the similar calculation

$$2Y(u) + X(u) = \frac{1}{(2\pi i)^3}\wp'(u)$$

$$= -\frac{1}{(2\pi i)^3}\frac{\sigma(u^2)}{\sigma(u)^4} \quad \text{from (I.5.6b)}$$

$$= -e^{\frac{1}{2}\eta(1)((2z)^2-4z^2)} \cdot e^{-\pi i(2z-4z)}\frac{\theta(u^2)}{\theta(u)^4} \quad \text{from (b)}$$

$$= -u\frac{\theta(u^2)}{\theta(u)^4}.$$

$\square$

§2. Elliptic Curves over $\mathbb{R}$

The uniformization theorem (I.4.4) says that every elliptic curve defined over $\mathbb{C}$ is analytically isomorphic to $\mathbb{C}^*/q^{\mathbb{Z}}$ for some $q = e^{2\pi i\tau}$. Since an elliptic curve defined over $\mathbb{R}$ is automatically defined over $\mathbb{C}$, it has such a model. We begin by describing a set of τ's which classifies elliptic curves over $\mathbb{R}$ up to $\mathbb{C}$-isomorphism.

Proposition 2.1. *Let $E/\mathbb{R}$ be an elliptic curve. Then there exists a unique τ in the set*

$$\mathcal{C} = \left\{ it : t \geq 1 \right\} \cup \left\{ e^{i\theta} : \frac{\pi}{3} \leq \theta \leq \frac{\pi}{2} \right\} \cup \left\{ \frac{1}{2} + it : t \geq \frac{\sqrt{3}}{2} \right\}$$

such that

$$j(\tau) = j(E).$$

(The set $\mathcal{C}$ is illustrated in Figure 5.1.)

PROOF. First we check that $j(\mathcal{C}) \subset \mathbb{R}$. If $\tau = it$ or $\tau = \frac{1}{2} + it$ with $t \in \mathbb{R}$, then $q \in \mathbb{R}$, so the q-expansion (I.7.4b)

$$j(\tau) = q^{-1} + \sum c(n)q^n, \qquad c(n) \in \mathbb{Z},$$

shows that $j(\tau) \in \mathbb{R}$. Next, for any τ we have

$$\bar{q} = e^{-2\pi i \bar{\tau}} = e^{-2\pi i |\tau|^2/\tau};$$

so in general

$$\overline{j(\tau)} = j\left(-\frac{|\tau|^2}{\tau} \right).$$

Hence for $\tau = e^{i\theta}$ we have

$$\overline{j(e^{i\theta})} = j\left(-\frac{1}{e^{i\theta}} \right) = j\left(\begin{pmatrix} 0 & -1 \\ 1 & 0 \end{pmatrix} e^{i\theta} \right) = j(e^{i\theta}), \quad \text{so } j(e^{i\theta}) \in \mathbb{R}.$$

This proves that $j(\mathcal{C}) \subset \mathbb{R}$.

Next we observe that

$$\lim_{t \to \infty} j(it) \qquad = \lim_{q \to 0^+} q^{-1} + \sum c(n)q^n = +\infty,$$

$$\lim_{t \to \infty} j\left(\frac{1}{2} + it \right) = \lim_{q \to 0^-} q^{-1} + \sum c(n)q^n = -\infty.$$

By continuity, we conclude that $j(\mathcal{C}) = \mathbb{R}$. (Note that $j : \mathcal{C} \to \mathbb{R}$ is continuous, since $j : \mathbf{H} \to \mathbb{C}$ is holomorphic.) Finally, (I.1.5b) and (I.4.1) imply that $j : \mathcal{C} \to \mathbb{R}$ is injective, which concludes the proof that the map $j : \mathcal{C} \to \mathbb{R}$ is a bijection. □

Proposition 2.1 completely describes all $\mathbb{C}$-isomorphism classes of elliptic curves defined over $\mathbb{R}$. However, since $\mathbb{R}$ is not algebraically closed, it is possible to have more than one $\mathbb{R}$-isomorphism class in each $\mathbb{C}$-isomorphism class. For a given $E/\mathbb{R}$, these other curves are called the *twists of E* (see [AEC X §5]). Our next result classifies these twists.

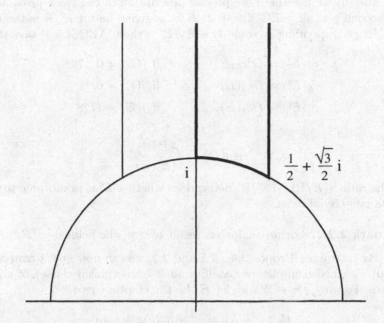

The Set $\mathcal{C}$ for Which $j(\tau) \in \mathbb{R}$

Figure 5.1

Proposition 2.2. (a) *Let $E/\mathbb{R}$ be an elliptic curve. Then the $\mathbb{C}$-isomorphism class of E contains exactly two $\mathbb{R}$-isomorphism classes. (In the notation of [AEC X §5], $\mathrm{Twist}\big((E, O)/\mathbb{R}\big) \cong \{\pm 1\}$.)*
(b) *More precisely, define an invariant $\gamma(E/\mathbb{R}) \in \{\pm 1\}$ by the rule*

$$\gamma(E/\mathbb{R}) = \begin{cases} \mathrm{sign}(c_6), & \text{if } j \neq 1728 \ (\text{i.e., if } c_6 \neq 0), \\ \mathrm{sign}(c_4), & \text{if } j = 1728 \ (\text{i.e., if } c_6 = 0). \end{cases}$$

(Here c_4 and c_6 are the usual quantities associated to some Weierstrass equation for $E/\mathbb{R}$.) Let $E/\mathbb{R}$ and $E'/\mathbb{R}$ be elliptic curves. Then

$$E \cong E' \text{ over } \mathbb{R} \quad \Longleftrightarrow \quad j(E) = j(E') \quad \text{and} \quad \gamma(E/\mathbb{R}) = \gamma(E'/\mathbb{R}).$$

PROOF. From [AEC X.5.4], the twists of E are in one-to-one correspondence with the elements of the group

$$\mathbb{R}^*/\mathbb{R}^{*n}, \qquad \text{where } n = \#\,\mathrm{Aut}(E) \in \{2, 4, 6\}.$$

Since n is even, $\mathbb{R}^*/\mathbb{R}^{*n} \cong \{\pm 1\}$ has two elements. This completes the proof of (a).

For (b) we use the more precise description of the twist provided by the second part of [AEC X.5.4]. If $E/\mathbb{R}$ is given, and if $E'/\mathbb{R}$ is the twist of $E/\mathbb{R}$ corresponding to some $D \in \mathbb{R}^*/\mathbb{R}^{*n}$, then [AEC X.5.4] says that

$$
\begin{aligned}
c_6(E') &= D^3 c_6(E) &&\text{if } j(E) \neq 0, 1728, \\
c_6(E') &= D c_6(E) &&\text{if } j(E) = 0, \\
c_4(E') &= D c_4(E) &&\text{if } j(E) = 1728.
\end{aligned}
$$

Hence in all cases

$$
\mathrm{sign}(D) = \frac{\gamma(E/\mathbb{R})}{\gamma(E'/\mathbb{R})},
$$

so the ratio $\gamma(E/\mathbb{R})/\gamma(E'/\mathbb{R})$ determines whether E' is isomorphic to E or to its non-trivial twist. $\qquad\square$

Remark 2.2.1. For an analogous result over p-adic fields, see (5.2).

By combining Propositions 2.1 and 2.2, we can now give a convenient set of q's which completely classifies all $\mathbb{R}$-isomorphism classes of elliptic curves. For any $q = e^{2\pi i \tau}$, we let E_q be the elliptic curve

$$
E_q : y^2 + xy = x^3 + a_4(q)x + a_6(q),
$$

where $a_4(q)$ and $a_6(q)$ are the power series described in (1.1), and we let

$$
\phi : \mathbb{C}^*/q^{\mathbb{Z}} \xrightarrow{\sim} E_q(\mathbb{C}), \qquad \phi(u) = (X(u, q), Y(u, q)),
$$

be the $\mathbb{C}$-analytic isomorphism from (1.1).

Theorem 2.3. *Let $E/\mathbb{R}$ be an elliptic curve.*
(a) *There is a unique $q \in \mathbb{R}$ with $0 < |q| < 1$ such that*

$$
E \cong_{/\mathbb{R}} E_q
$$

(i.e., E is $\mathbb{R}$-isomorphic to E_q).
(b) *Composing the isomorphism from (a) with the map ϕ described above, we obtain an isomorphism*

$$
\psi : \mathbb{C}^*/q^{\mathbb{Z}} \longrightarrow E(\mathbb{C})
$$

which commutes with complex conjugation, that is, ψ is defined over $\mathbb{R}$. In particular,

$$
\psi : \mathbb{R}^*/q^{\mathbb{Z}} \longrightarrow E(\mathbb{R})
$$

is an $\mathbb{R}$-analytic isomorphism.

PROOF. Note first that if $q \in \mathbb{R}$, then the series (1.1) imply that $a_4(q)$ and $a_6(q)$ are in $\mathbb{R}$, so E_q is defined over $\mathbb{R}$. We want to start by using (2.1) to find a τ with $j(\tau) = j(E)$. As noted during the proof of (2.1), the points $\tau = it$ and $\tau = \frac{1}{2} + it$ give real values of q. However, for $\tau = e^{i\theta}$ we do not have $q \in \mathbb{R}$. So we use the transformation

$$\alpha(\tau) = \frac{-1}{\tau - 1} \quad \text{which satisfies} \quad \alpha(e^{i\theta}) = \frac{1}{2} + \frac{i}{2}\cot\frac{\theta}{2}.$$

Thus $\alpha \in \Gamma(1)$ yields a bijection

$$\alpha : \left\{ e^{i\theta} : \frac{\pi}{3} \le \theta \le \frac{\pi}{2} \right\} \xrightarrow{\sim} \left\{ \frac{1}{2} + it : \frac{1}{2} \le t \le \frac{\sqrt{3}}{2} \right\}.$$

Since $j(\alpha\tau) = j(\tau)$, we conclude from (2.1) that there is a unique τ in the set

$$\left\{ it : t \ge 1 \right\} \cup \left\{ \frac{1}{2} + it : t > \frac{1}{2} \right\}$$

such that $j(\tau) = j(E)$. Note that in the second set we do not allow $t = \frac{1}{2}$, since this would give the duplicate value

$$j\left(\frac{1}{2} + \frac{1}{2}i \right) = j(\alpha(i)) = j(i).$$

The set of $q = e^{2\pi i \tau} \in \mathbb{R}$ with $0 < |q| < 1$ corresponds bijectively with the τ's in the set

$$\mathcal{T} = \left\{ it : t > 0 \right\} \cup \left\{ \frac{1}{2} + it : t > 0 \right\}.$$

Further, the transformations

$$S\tau = -\frac{1}{\tau}, \qquad \beta\tau = \frac{\tau - 1}{2\tau - 1}, \qquad \alpha\tau = \frac{-1}{\tau - 1}$$

give identifications

$$S : \{ it : t > 1 \} \xrightarrow{\sim} \{ it : t < 1 \},$$

$$\beta : \left\{ \frac{1}{2} + it : t > \frac{1}{2} \right\} \xrightarrow{\sim} \left\{ \frac{1}{2} + it : t < \frac{1}{2} \right\},$$

$$\alpha : i \longmapsto \frac{1}{2} + \frac{1}{2}i.$$

(See Figure 5.2.)

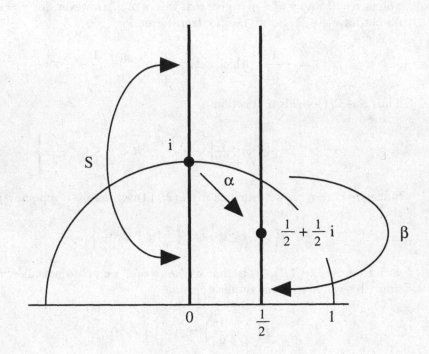

A Set of τ Giving All $q \in \mathbb{R}$ with $0 < |q| < 1$
Figure 5.2

Since $j(\gamma\tau) = j(\tau)$ for any $\gamma \in \mathrm{SL}_2(\mathbb{Z})$, we see from above that the map

$$j : \mathcal{T} \longrightarrow \mathbb{R}$$

is exactly two-to-one. From (2.2a) there are exactly two $\mathbb{R}$-isomorphism classes of elliptic curves with a given j-invariant. Hence to complete the proof of (a) we must check that if $\tau, \tau' \in \mathcal{T}$ are distinct points with $j(\tau) = j(\tau')$, then E_q and $E_{q'}$ are non-trivial twists of one another. To do this we will use (2.2b).

The change-of-variable formulas for Weierstrass equations [AEC III §1] imply that

$$c_4(q) = u^4(12g_2(\tau)) \qquad \text{and} \qquad c_6(q) = u^6(216g_3(\tau))$$

for some $u \in \mathbb{C}^*$. (In fact, our explicit formulas imply that $u = (2\pi i)^{-1}$, but for our purposes it suffices to know that u does not depend on τ.) Now suppose that

$$\tau' = \frac{a\tau + b}{c\tau + d} \qquad \text{for some } \begin{pmatrix} a & b \\ c & d \end{pmatrix} \in \mathrm{SL}_2(\mathbb{Z}).$$

Then using the γ-invariant defined in (2.2b) and the fact that g_2 and g_3 are modular forms, we find that

$$\frac{\gamma(E_{q'}/\mathbb{R})}{\gamma(E_q/\mathbb{R})} = \text{sign}\left(\frac{c_6(q')}{c_6(q)}\right) = \text{sign}\left(\frac{g_3(\tau')}{g_3(\tau)}\right) = \text{sign}\{(c\tau + d)^6\}$$
$$\text{if } j(\tau) \neq 1728.$$

Similarly,

$$\frac{\gamma(E_{q'}/\mathbb{R})}{\gamma(E_q/\mathbb{R})} = \text{sign}\left(\frac{c_4(q')}{c_4(q)}\right) = \text{sign}\left(\frac{g_2(\tau')}{g_2(\tau)}\right) = \text{sign}\{(c\tau + d)^4\}$$
$$\text{if } j(\tau) = 1728.$$

We must show that for the τ's and τ''s described above, all of these signs are -1. This requires checking several cases.

Case I: $\tau = it, t > 1.$ $\tau' = S\tau = -\dfrac{1}{\tau}.$

In this case $j(\tau) \neq 1728$, and

$$(c\tau + d)^6 = (it)^6 = -t^6 < 0.$$

Case II: $\tau = \dfrac{1}{2} + it, t > 1.$ $\tau' = \beta\tau = \dfrac{\tau - 1}{2\tau - 1}.$

Again we are in a case in which $j(\tau) \neq 1728$, and

$$(c\tau + d)^6 = (2\tau - 1)^6 = (2it)^6 = -64t^6 < 0.$$

Case III: $\tau = i.$ $\tau' = \alpha\tau = \dfrac{-1}{\tau - 1} = \dfrac{1}{2} + \dfrac{1}{2}i.$

This is the case that $j(\tau) = 1728$, and

$$(c\tau + d)^4 = (\tau - 1)^4 = (i - 1)^4 = -4 < 0.$$

Hence in all cases E_q and $E_{q'}$ are distinct twists, which completes the proof of (a).

(b) The series for $X(u, q)$ and $Y(u, q)$ show that if $q \in \mathbb{R}$, then the map

$$\phi : \mathbb{C}^*/q^{\mathbb{Z}} \xrightarrow{\sim} E_q(\mathbb{C}), \qquad \phi(u) = \big(X(u, q), Y(u, q)\big)$$

commutes with complex conjugation. Since the isomorphism $E \cong E_q$ in (a) is defined over $\mathbb{R}$, it follows that the composition $\psi : \mathbb{C}^*/q^{\mathbb{Z}} \to E(\mathbb{C})$ is also defined over $\mathbb{R}$. This proves the first half of (b), and the second follows by taking $G_{\mathbb{C}/\mathbb{R}}$-invariants of the exact sequence

$$1 \longrightarrow q^{\mathbb{Z}} \longrightarrow \mathbb{C}^* \longrightarrow E(\mathbb{C}) \longrightarrow 0,$$

so we get $\mathbb{R}^*/q^{\mathbb{Z}} \xrightarrow{\sim} E(\mathbb{R})$. (Note that $H^1(G_{\mathbb{C}/\mathbb{R}}, q^{\mathbb{Z}}) = \text{Hom}(G_{\mathbb{C}/\mathbb{R}}, \mathbb{Z}) = 0$.) Finally, it is clear that this isomorphism is $\mathbb{R}$-analytic, since we know that ψ is $\mathbb{C}$-analytic and is given by power series with coefficients in $\mathbb{R}$.

$\square$

An elliptic curve over $\mathbb{R}$ has either one or two components. We can use (2.3) to give a criterion to determine which case holds.

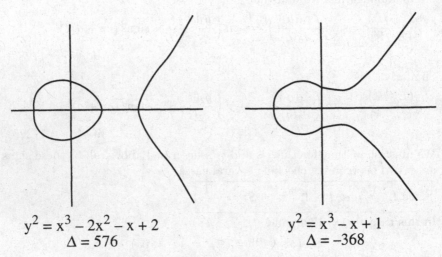

$$y^2 = x^3 - 2x^2 - x + 2 \qquad\qquad y^2 = x^3 - x + 1$$
$$\Delta = 576 \qquad\qquad\qquad \Delta = -368$$

Elliptic Curves over $\mathbb{R}$ with One and Two Components

Figure 5.3

Corollary 2.3.1. *Let $E(\mathbb{R})$ be an elliptic curve, and let $\Delta(E)$ be the discriminant of some Weierstrass equation for $E/\mathbb{R}$. Then there is an isomorphism of real Lie groups*

$$E(\mathbb{R}) \cong \begin{cases} \mathbb{R}/\mathbb{Z}, & \text{if } \Delta(E) < 0, \\ (\mathbb{R}/\mathbb{Z}) \times (\mathbb{Z}/2\mathbb{Z}), & \text{if } \Delta(E) > 0. \end{cases}$$

PROOF. Fix an isomorphism $E \cong_{/\mathbb{R}} E_q$ as in (2.3a). Then

$$u^{12}\Delta(E) = \Delta(E_q) = q \prod_{n \geq 1} (1 - q^n)^{24}$$

for some $u \in \mathbb{R}$, so

$$\operatorname{sign} \Delta(E) = \operatorname{sign} \Delta(E_q) = \operatorname{sign} q.$$

Now (2.3b) says that $E(\mathbb{R}) \cong E_q(\mathbb{R}) \cong \mathbb{R}^*/q^{\mathbb{Z}}$, so the following isomorphisms complete the proof of Corollary 2.3.1:

$$\mathbb{R}^*/q^{\mathbb{Z}} \xrightarrow{\sim} \mathbb{R}/\mathbb{Z}, \quad u \longmapsto \frac{1}{2}\left(\frac{\log|u|}{\log|q|} - \operatorname{sign}(u) + 1\right) (\operatorname{mod}\mathbb{Z}), \quad \text{if } q < 0,$$

$$\mathbb{R}^*/q^{\mathbb{Z}} \xrightarrow{\sim} (\mathbb{R}/\mathbb{Z}) \times \{\pm 1\}, \quad u \longmapsto \left(\frac{\log|u|}{\log q}(\operatorname{mod}\mathbb{Z}), \operatorname{sign}(u)\right), \quad \text{if } q > 0.$$

$\square$

We conclude our discussion of elliptic curves over $\mathbb{R}$ by describing the Weil-Châtelet group $WC(E/\mathbb{R})$. (See [AEC X §3] for basic facts about the Weil-Châtelet group.) As in (2.2), the fact that $G_{\mathbb{C}/\mathbb{R}}$ is so small leads to a very simple answer.

Theorem 2.4. *Let $E/\mathbb{R}$ be an elliptic curve, and let $\Delta(E)$ be the discriminant of some Weierstrass equation for $E/\mathbb{R}$. Then*

$$WC(E/\mathbb{R}) \cong \begin{cases} 0 & \text{if } \Delta(E) < 0, \\ \mathbb{Z}/2\mathbb{Z} & \text{if } \Delta(E) > 0. \end{cases}$$

PROOF. From [AEC X.3.6] there is an isomorphism

$$WC(E/\mathbb{R}) \cong H^1\big(G_{\mathbb{C}/\mathbb{R}}, E(\mathbb{C})\big).$$

Choose a $q \in \mathbb{R}$ and an $\mathbb{R}$-isomorphism $E \cong_{/\mathbb{R}} E_q$ as in (2.3). Then

$$E(\mathbb{C}) \cong \mathbb{C}^*/q^{\mathbb{Z}} \qquad \text{as } G_{\mathbb{C}/\mathbb{R}}\text{-modules},$$

so we have an exact sequence

$$0 \longrightarrow q^{\mathbb{Z}} \longrightarrow \mathbb{C}^* \longrightarrow E(\mathbb{C}) \longrightarrow 0$$

of $G_{\mathbb{C}/\mathbb{R}}$-modules. Since

$$H^1\big(G_{\mathbb{C}/\mathbb{R}}, \mathbb{C}^*\big) = 0$$

from Hilbert's Theorem 90 (or by an easy direct calculation), the long exact sequence in $G_{\mathbb{C}/\mathbb{R}}$-cohomology gives

$$0 \longrightarrow WC(E/\mathbb{R}) \longrightarrow H^2\big(G_{\mathbb{C}/\mathbb{R}}, q^{\mathbb{Z}}\big) \longrightarrow H^2\big(G_{\mathbb{C}/\mathbb{R}}, \mathbb{C}^*\big).$$

Now $G_{\mathbb{C}/\mathbb{R}} = \{1, \sigma\}$ is cyclic of order 2, so for any $G_{\mathbb{C}/\mathbb{R}}$-module M (written multiplicatively) we have

$$H^2\big(G_{\mathbb{C}/\mathbb{R}}, M\big) \cong \frac{\{x \in M \ : \ \sigma(x) = x\}}{\{x \cdot \sigma(x) \ : \ x \in M\}}.$$

(This is a special case of a general formula for the cohomology of cyclic groups. See exercise 5.2.) Since $G_{\mathbb{C}/\mathbb{R}}$ acts trivially on $q^{\mathbb{Z}}$, we find

$$H^2\big(G_{\mathbb{C}/\mathbb{R}}, q^{\mathbb{Z}}\big) \cong q^{\mathbb{Z}}/q^{2\mathbb{Z}},$$

$$H^2\big(G_{\mathbb{C}/\mathbb{R}}, \mathbb{C}^*\big) \cong \frac{\{u \in \mathbb{C}^* \ : \ \bar{u} = u\}}{\{|u|^2 \ : \ u \in \mathbb{C}^*\}} \cong \mathbb{R}^*/\mathbb{R}^{*2}.$$

Hence we finally obtain an exact sequence

$$0 \longrightarrow \mathrm{WC}(E/\mathbb{R}) \longrightarrow q^{\mathbb{Z}}/q^{2\mathbb{Z}} \longrightarrow \mathbb{R}^*/\mathbb{R}^{*2},$$

where the right-hand map is induced by the natural inclusion $q^{\mathbb{Z}} \hookrightarrow \mathbb{R}^*$.
From this exact sequence we immediately conclude that

$$\mathrm{WC}(E/\mathbb{R}) \cong \begin{cases} 0 & \text{if } q < 0, \\ \mathbb{Z}/2\mathbb{Z} & \text{if } q > 0. \end{cases}$$

But as we observed during the proof of (2.3.1),

$$u^{12}\Delta(E) = \Delta(q) = q \prod_{n \geq 1} (1 - q^n)^{24} \qquad \text{for some } u \in \mathbb{R},$$

so $\mathrm{sign}\,\Delta(E) = \mathrm{sign}\,q$. This completes the proof of Theorem 2.4. $\square$

§3. The Tate Curve

We have seen that every elliptic curve defined over the complex numbers has a parametrization $\mathbb{C}/\Lambda$ for some lattice $\Lambda \subset \mathbb{C}$. Suppose we replace $\mathbb{C}$ by $\mathbb{Q}_p$ and endeavor to parametrize an elliptic curve $E/\mathbb{Q}_p$ by a group of the form $\mathbb{Q}_p/\Lambda$. Unfortunately, this approach immediately fails, because $\mathbb{Q}_p$ has no non-trivial lattices. Indeed, if $\Lambda \subset \mathbb{Q}_p$ is any non-zero subgroup and $0 \neq t \in \Lambda$, then

$$p^n t \in \Lambda \text{ for all } n \geq 0 \quad \text{and} \quad \lim_{n \to \infty} p^n t = 0,$$

so 0 is an accumulation point of Λ. Hence $\mathbb{Q}_p$ contains no discrete subgroups other than 0.

Tate's idea is to first exponentiate, which leads to the alternative description $\mathbb{C}^*/q^{\mathbb{Z}}$ for elliptic curves over $\mathbb{C}$. Now the analogous situation over $\mathbb{Q}_p$ is much more promising, since $\mathbb{Q}_p^*$ has lots of discrete subgroups. For example, any $q \in \mathbb{Q}_p^*$ with $|q| < 1$ defines the discrete subgroup

$$q^{\mathbb{Z}} = \{q^n : n \in \mathbb{Z}\} \subset \mathbb{Q}_p^*.$$

Further, the series described in (1.1) will converge in $\mathbb{Q}_p$ and give a p-adic analytic isomorphism of the quotient $\mathbb{Q}_p^*/q^{\mathbb{Z}}$ with a certain elliptic curve E_q.

The situation is nicely summarized by the following picture (taken from Robert [1, II §5]).

Complex case	p-adic case
$\mathbb{C}/\Lambda$	no p-adic analogue
$\Big\downarrow$ exponential map $e^{2\pi i z}$	no exponential available
$\mathbb{C}^*/q^{\mathbb{Z}}$	$\mathbb{Q}_p^*/q^{\mathbb{Z}}$: p-adic elliptic curve.

More generally, we can work over any p-adic field K, by which we mean a finite extension $K/\mathbb{Q}_p$. All of these facts (and more) are contained in the next theorem, the proof of which will keep us busy for the next two sections.

Theorem 3.1.   (Tate) *Let K be a p-adic field with absolute value $|\cdot|$, let $q \in K^*$ satisfy $|q| < 1$, and let*

$$s_k(q) = \sum_{n \geq 1} \frac{n^k q^n}{1 - q^n}, \qquad a_4(q) = -5s_3(q), \qquad a_6(q) = -\frac{5s_3(q) + 7s_5(q)}{12}$$

be the series described in (1.1).
(a) *The series $a_4(q)$ and $a_6(q)$ converge in K. Define the Tate curve E_q by the equation*

$$E_q : y^2 + xy = x^3 + a_4(q)x + a_6(q).$$

(b) *The Tate curve is an elliptic curve defined over K with discriminant*

$$\Delta = q \prod_{n \geq 1} (1 - q^n)^{24}$$

and j-invariant

$$j(E_q) = \frac{1}{q} + 744 + 196884q + \cdots = \frac{1}{q} + \sum_{n \geq 0} c(n)q^n,$$

where the $c(n)$'s are the integers described in (1.1).
(c) *The series*

$$X(u, q) = \sum_{n \in \mathbb{Z}} \frac{q^n u}{(1 - q^n u)^2} - 2s_1(q),$$

$$Y(u, q) = \sum_{n \in \mathbb{Z}} \frac{(q^n u)^2}{(1 - q^n u)^3} + s_1(q),$$

converge for all $u \in \bar{K}$, $u \notin q^{\mathbb{Z}}$. They define a surjective homomorphism

$$\phi : \quad \bar{K}^* \quad \xrightarrow{\sim} \quad E_q(\bar{K})$$
$$u \quad \longmapsto \quad \begin{cases} \big(X(u,q), Y(u,q)\big) & \text{if } u \notin q^{\mathbb{Z}}, \\ O & \text{if } u \in q^{\mathbb{Z}}. \end{cases}$$

The kernel of ϕ is $q^{\mathbb{Z}}$.

(d) The map ϕ in (c) is compatible with the action of the Galois group $G_{\bar{K}/K}$ in the sense that

$$\phi(u^\sigma) = \phi(u)^\sigma \quad \text{for all } u \in \bar{K}^*, \ \sigma \in G_{\bar{K}/K}.$$

In particular, for any algebraic extension L/K, ϕ induces an isomorphism

$$\phi : L^*/q^{\mathbb{Z}} \xrightarrow{\sim} E_q(L).$$

Remark 3.1.1. The p-adic uniformization described in Theorem 3.1 is especially useful for arithmetic applications because it is compatible with the action of Galois as described in (3.1d). Note that a complex uniformization $\mathbb{C}/\Lambda \to E(\mathbb{C})$ or $\mathbb{C}^*/q^{\mathbb{Z}} \to E(\mathbb{C})$ does not have this compatibility (except relative to $G_{\mathbb{C}/\mathbb{R}}$), since in general one cannot apply an element of Galois to the value of a convergent series by applying it to each term of the series. (See exercise 5.8.)

Remark 3.1.2. Theorem 3.1 is actually true for any field K that is complete with respect to a non-archimedean absolute value. The only time we will use the fact that K is a finite extension of $\mathbb{Q}_p$ will be in the proof that the map ϕ in (3.1c) is surjective. (In fact, we will really only need the fact that the absolute value is discrete, so our proof actually is valid somewhat more generally, for example over the completion of $\mathbb{Q}_p^{\mathrm{nr}}$.) For a proof of Theorem 3.1 in the most general setting, using p-adic analytic methods, see Roquette [1].

PROOF. (a) From (1.1b), the series defining a_4 and a_6 are in $\mathbb{Z}[\![q]\!]$, so they will converge in K for any value of $q \in K$ satisfying $|q| < 1$.

(b) The discriminant of E_q is

$$\Delta(q) = -a_6 + a_4^2 + 72a_4a_6 - 64a_4^3 - 432a_6^2.$$

Substituting in the series for $a_4(q)$ and $a_6(q)$, we find the usual power series

$$\Delta(q) = q - 24q^2 + 252q^3 + \cdots \equiv q \pmod{q^2}.$$

Hence $|\Delta(q)| = |q| \neq 0$, so E_q is a non-singular elliptic curve.

Next we observe from (1.1c) that the identity

$$\Delta(q) = q \prod_{n \geq 1} (1 - q^n)^{24} \quad .$$

holds for all $q \in \mathbb{C}$ with $|q|_\infty < 1$, where for the moment we write $| \cdot |$ for the usual absolute value on $\mathbb{C}$. It follows that this identity is true as an identity of formal power series in $\mathbb{Z}[\![q]\!]$. Hence it remains true when we take q to be an element of absolute value less than 1 in any field that is complete with respect to a non-archimedean absolute value.

Finally, the formula

$$
\begin{aligned}
j(q) &= \frac{\left(1 + 48a_4(q)\right)^3}{\Delta(q)} \\
&= \frac{1 + 240q + 2160q^2 + \cdots}{q - 24q^2 + 252q^3 + \cdots} \\
&= \frac{1}{q}(1 + 744q + 196884q^2 + \cdots)
\end{aligned}
$$

holds in the non-archimedean case, since it is obtained formally by taking the quotient of the appropriate power series.

(c) We begin by rewriting the series for X and Y as follows, where we've used the alternative expression (1.2) for $s_1(q)$:

$$
\begin{aligned}
X(u,q) &= \frac{u}{(1-u)^2} + \sum_{n \geq 1} \left(\frac{q^n u}{(1 - q^n u)^2} + \frac{q^{-n} u}{(1 - q^{-n} u)^2} - 2\frac{q^n}{(1 - q^n)^2} \right) \\
&= \frac{1}{u + u^{-1} - 2} + \sum_{n \geq 1} \left(\frac{q^n u}{(1 - q^n u)^2} + \frac{q^n u^{-1}}{(1 - q^n u^{-1})^2} - 2\frac{q^n}{(1 - q^n)^2} \right),
\end{aligned}
$$

$$
\begin{aligned}
Y(u,q) &= \frac{u^2}{(1-u)^3} + \sum_{n \geq 1} \left(\frac{(q^n u)^2}{(1 - q^n u)^3} + \frac{(q^{-n} u)^2}{(1 - q^{-n} u)^3} + \frac{q^n}{(1 - q^n)^2} \right) \\
&= \frac{u^2}{(1-u)^3} + \sum_{n \geq 1} \left(\frac{(q^n u)^2}{(1 - q^n u)^3} - \frac{q^n u^{-1}}{(1 - q^n u^{-1})^3} + \frac{q^n}{(1 - q^n)^2} \right).
\end{aligned}
$$

These expressions show immediately that $X(u,q)$ and $Y(u,q)$ converge for all $u \in \bar{K}^* \smallsetminus q^{\mathbb{Z}}$. (Note that although $\bar{K}$ itself is not complete, every term in these series is in the field $K(u) = \mathbb{Q}_p(u, q)$, which is a finite extension of $\mathbb{Q}_p$; so we are really working in the complete field $K(u)$. Similar comments will apply below whenever we speak of substituting elements of $\bar{K}$ into a series.) The functional equations

$$X(qu, q) = X(u, q) = X(u^{-1}, q)$$

are now obvious, the first equality from the original series for X and the second from the rearranged series we just gave. A little algebra gives similar functional equations for Y,

$$Y(qu, q) = Y(u, q) \quad \text{and} \quad Y(u^{-1}, q) = -Y(u, q) - X(u, q).$$

If we restrict u to the range $|q| < |u| < |q|^{-1}$, we have $|q^n u| < 1$ and $|q^n u^{-1}| < 1$ for all positive integers n. So we can use the expansion

$$\sum_{n \geq 1} \frac{q^n T}{(1 - q^n T)^2} = \sum_{n \geq 1} \sum_{m \geq 1} m(q^n T)^m = \sum_{d \geq 1} \Big(\sum_{m \mid d} m T^m \Big) q^d$$

with T equal successively to u, u^{-1}, and 1 to rewrite the series for X as

$$X(u, q) = \frac{u}{(1 - u)^2} + \sum_{d \geq 1} \Big(\sum_{m \mid d} m(u^m + u^{-m} - 2) \Big) q^d \in \mathbb{Q}(u)[\![q]\!],$$

$$\text{valid for } |q| < |u| < |q|^{-1}.$$

A similar calculation allows us to write Y as a power series in q with coefficients in $\mathbb{Q}(u)$:

$$Y(u, q) = \frac{u^2}{(1 - u)^3} + \sum_{d \geq 1} \Big(\sum_{m \mid d} \Big\{ \frac{(m - 1)m}{2} u^m - \frac{m(m + 1)}{2} u^{-m} + m \Big\} \Big) q^d$$

$$\in \mathbb{Q}(u)[\![q]\!], \quad \text{valid for } |q| < |u| < |q|^{-1}.$$

We begin our proof of (c) by showing that the image of the map ϕ is contained in the curve E_q given by the Weierstrass equation

$$E_q : y^2 + xy = x^3 + a_4(q)x + a_6(q).$$

This amounts to showing that when we substitute the series $X(u, q)$ and $Y(u, q)$ for x and y in this equation, we get an identity valid for all $u \in \bar{K}^* \smallsetminus q^{\mathbb{Z}}$. By the periodicity of X and Y, it is enough to consider values of u such that $|q| < |u| \leq 1$ and $u \neq 1$. In this range we can use the above formulas which express X and Y as power series in q with coefficients that are rational functions of u. Thus we will be done if we can show that the equation

$$Y(u, q)^2 + X(u, q)Y(u, q) = X(u, q)^3 + a_4(q)X(u, q) + a_6(q)$$

is valid as a formal identity in the ring of formal power series in q with coefficients which are rational functions of the indeterminate u. In other words, we want to verify that this identity holds in the ring $\mathbb{Q}(u)[\![q]\!]$.

From (1.1), we know that this equation is true numerically if we substitute any pair of complex numbers $u, q \in \mathbb{C}$ in the domain of convergence $|q|_\infty < |u|_\infty < |q|_\infty^{-1}$, $u \neq 1$. If we fix some u with $|q|_\infty < |u|_\infty < 1$ and let q vary, we conclude that the resulting power series in q with complex coefficients are equal coefficient-wise. Then letting u vary, we deduce that the coefficients are formally equal as rational functions of u. Hence we have an equality of formal power series in $\mathbb{Q}(u)[\![q]\!]$.

Next we prove that ϕ is a homomorphism. Given $u_1, u_2 \in \bar{K}^*$, we put $u_3 = u_1 u_2$ and must prove that

$$P_3 = P_1 + P_2 \qquad \text{where } P_i = \phi(u_i), \ i = 1, 2, 3.$$

In view of the periodicity $\phi(qu) = \phi(u)$, we may restrict consideration to values of u_1 and u_2 in the ranges

$$|q| < |u_1| \leq 1 \text{ and } 1 \leq |u_2| < |q|^{-1}, \text{ which means that } |q| < |u_3| < |q|^{-1}.$$

Then all three u_i are within the domain of convergence of the power series expressions for $X, Y \in \mathbb{Q}(u)[\![q]\!]$ described above.

Since $\phi(1) = O$ by definition, the relation $P_3 = P_1 + P_2$ holds trivially if $u_1 = 1$ or $u_2 = 1$. Using the functional equations for $X(u^{-1}, q)$ and $Y(u^{-1}, q)$ and the fact that $P_1 + P_2 = O$ if and only if $x_1 = x_2$ and $y_1 + y_2 = -x_1$, it is also not hard to verify that $P_3 = P_1 + P_2$ in the case that $u_1 u_2 = 1$. So we are reduced to the case that P_1, P_2, and P_3 are all different from O. We write $P_i = (x_i, y_i)$; that is, we set $x_i = X(u_i, q)$ and $y_i = Y(u_i, q)$ for $i = 1, 2, 3$.

Suppose first that $x_1 \neq x_2$. Then writing out the addition law on E_q, we see from [AEC, III.2.3] that the relation $P_1 + P_2 = P_3$ is equivalent to the two identities

$$(x_2 - x_1)^2 x_3 = (y_2 - y_1)^2 + (y_2 - y_1)(x_2 - x_1) - (x_2 - x_1)^2(x_1 + x_2),$$
$$(x_2 - x_1)y_3 = -\big((y_2 - y_1) + (x_2 - x_1)\big)x_3 - (y_1 x_2 - y_2 x_1).$$

Now we can argue as above that (1.1) implies that these identities hold for all complex numbers u_1, u_2, q in the specified ranges. Hence they are identities in the ring $\mathbb{Q}(u_1, u_2)[\![q]\!]$ of formal power series in q with coefficients that are rational functions of u_1 and u_2, and so are true for $u_1, u_2, q \in \bar{K}$.

To deal with the remaining case $x_1 = x_2$, we could use the duplication formula, or we could invoke a p-adic continuity argument, but perhaps the simplest solution is to observe that $x_1 = x_2$ if and only if $P_1 = \pm P_2$, and then use the following lemma.

Lemma 3.1.2. *Let ϕ be a map of a (multiplicative) group into an (additive) group which takes on an infinite number of distinct values and satisfies the identity*

$$\phi(u_1 u_2) = \phi(u_1) + \phi(u_2) \qquad \text{whenever } \phi(u_1) \neq \pm\phi(u_2).$$

Then ϕ is a homomorphism.

PROOF. Given any u_1 and u_2, the fact that ϕ takes on infinitely many distinct values means that we can choose a u such that

$$\phi(u) \neq \pm\phi(u_1), \qquad \phi(u) \neq -\phi(u_1) \pm \phi(u_2), \qquad \phi(u) \neq \pm\phi(u_1 u_2).$$

Then $\phi(uu_1) = \phi(u) + \phi(u_1) \neq \pm\phi(u_2)$, so

$$\phi(u) + \phi(u_1) + \phi(u_2) = \phi(uu_1) + \phi(u_2) = \phi(uu_1 u_2) = \phi(u) + \phi(u_1 u_2).$$

Canceling $\phi(u)$ gives $\phi(u_1)+\phi(u_2) = \phi(u_1 u_2)$, valid for all u_1 and u_2, which shows that ϕ is a homomorphism. $\qquad\qquad\square$

To finish the proof that our $\phi : \bar{K}^* \to E_q(\bar{K})$ is a homomorphism, we need merely observe that ϕ certainly takes on infinitely many distinct values. For example, the series for $X(u, q)$ shows that for any $t \in K$ with $|t| < 1$, we have $\big|X(1 + t, q)\big| = |t|^{-2}$. Hence (3.1.2) applies in our case.

So we now know that ϕ is a homomorphism of $\bar{K}^*$ into $E_q(\bar{K})$. That the kernel of ϕ is $q^{\mathbb{Z}}$ is apparent from its very definition. It remains to prove that ϕ is surjective. This is the hardest part of the proof, which we will leave to the next section.

(d) As noted above, the series for $X(u, q)$ and $Y(u, q)$ converge in the complete field $K(u)$, so it really suffices to prove (d) for $\sigma \in G_{L/K}$, where L is any finite Galois extension of K containing $K(u)$. Any such σ maps the maximal ideal of the ring of integers of L to itself, so σ will preserve the absolute value on L:

$$|\alpha^\sigma| = |\alpha| \quad \text{for all } \sigma \in G_{L/K} \text{ and all } \alpha \in L.$$

It follows easily from this that if $\sum \alpha_i$ is a convergent series with $\alpha_i \in L$, then $\left(\sum \alpha_i\right)^\sigma = \sum \alpha_i^\sigma$. (See exercise 5.8.) Applying this to the series for $X(u, q)$ and $Y(u, q)$, we deduce that $\phi(u)^\sigma = \phi(u^\sigma)$. This proves the first part of (d).

For the second part, we use (c) to produce the exact sequence

$$1 \longrightarrow q^{\mathbb{Z}} \longrightarrow \bar{K}^* \xrightarrow{\ \phi\ } E_q(\bar{K}) \longrightarrow 0.$$

We now know that the maps in this exact sequence commute with the action of $G_{\bar{K}/K}$. Hence for any algebraic extension L/K, we can take $G_{\bar{K}/L}$ invariants of this short exact sequence to obtain the exact sequence

$$1 \longrightarrow q^{\mathbb{Z}} \longrightarrow L^* \xrightarrow{\ \phi\ } E_q(L).$$

To obtain surjectivity on the right, we observe that it suffices to prove surjectivity in the case that L is a finite extension of $\mathbb{Q}_p$; we will prove exactly this fact in the next section. Alternatively, we may observe that the next term in this last sequence is the cohomology group $H^1(G_{\bar{K}/L}, q^{\mathbb{Z}})$. Since $q \in K$, the action of $G_{\bar{K}/L}$ on $q^{\mathbb{Z}}$ is trivial, so this is just the group of continuous homomorphisms from the profinite group $G_{\bar{K}/L}$ to the discrete group $q^{\mathbb{Z}} \cong \mathbb{Z}$. The only such homomorphism is the trivial one, so $H^1(G_{\bar{K}/L}, q^{\mathbb{Z}}) = 0$, which proves that $L^* \to E_q(L)$ is surjective. This completes the proof of (d). $\qquad\qquad\square$

Before resuming the proof of Theorem 3.1 in the next section, we briefly pause to repeat Proposition 1.3 in the context of p-adic theta functions. These formulas will be used in our study of local height functions in Chapter 6.

Proposition 3.2. *Define a function $\theta(u, q)$ by the formula*

$$\theta(u, q) = (1 - u) \prod_{n \geq 1} \frac{(1 - q^n u)(1 - q^n u^{-1})}{(1 - q^n)^2}.$$

(a) *$\theta(u, q)$ converges for all $u, q \in \bar{\mathbb{Q}}_p^*$ with $|q| < 1$ and satisfies the functional equation*

$$\theta(qu, q) = -\frac{1}{u}\theta(u, q).$$

(b) *θ is related to the functions $X(u, q)$ and $Y(u, q)$ described in (3.1c) by the formulas*

$$\text{(i)} \qquad X(u_1, q) - X(u_2, q) = -\frac{u_2\theta(u_1 u_2, q)\theta(u_1 u_2^{-1}, q)}{\theta(u_1, q)^2\theta(u_2, q)^2},$$

$$\text{(ii)} \qquad 2Y(u, q) + X(u, q) = -\frac{u\theta(u^2, q)}{\theta(u, q)^4}.$$

PROOF. The convergence of the infinite product defining θ is clear, and the functional equation follows formally by substituting qu for u and renumbering. This proves (a). Next we observe from (1.3) that the two formulas in (b) are valid over $\mathbb{C}$. Now an argument similar to that used to prove (3.1c) shows that they are valid over K. We will leave the details to the reader. $\qquad\qquad\square$

§4. The Tate Map Is Surjective

The final step in the proof of (3.1), which we postponed from the previous section, is to show that the map

$$\phi : \bar{K}^* \longrightarrow E_q(\bar{K}), \qquad \phi(u) = \big(X(u,q), Y(u,q)\big),$$

is surjective. One approach is to reprove in the p-adic case some classical results from complex analysis concerning Laurent series. In particular, one proves Schnirelmann's Theorem that a Laurent series $f(X)$ which converges for all $X \neq 0$ can be written as a convergent product

$$f(X) = cX^k \prod_{|\alpha|<1} \left(1 - \frac{\alpha}{X}\right) \prod_{|\alpha|\geq 1} \left(1 - \frac{X}{\alpha}\right),$$

where the product is over all roots α of f. Then one constructs the field of p-adic meromorphic functions on $\bar{K}^*/q^{\mathbb{Z}}$ and shows using Riemann-Roch that it is a field of genus 1 over K. This leads to an isomorphism with some elliptic curve, and after some work with the classical power series for $\wp$ and $\wp'$, one deduces that the elliptic curve is indeed the curve we have denoted E_q. For details of this line of proof, see Robert [1], Roquette [1], and Tate [9]. We will take a more computational, geometrically inspired approach. However, we should note that the theory of p-adic analytic functions has many important applications in modern arithmetic geometry. The reader might consult Bosch-Günter-Remmert [1] for a thorough introduction to this subject which is called rigid analysis.

In order to prove that $\phi : \bar{K}^* \to E_q(\bar{K})$ is surjective, we need to show that for any given point $P \in E_q(\bar{K})$ there is some $u \in \bar{K}^*$ with $\phi(u) = P$. But the point P will be defined over some finite extension of K, so it suffices to prove that $\phi : L^* \to E_q(L)$ is surjective for all finite extensions L/K. In fact, this is even stronger than the original statement of (3.1c), although it is precisely the result we needed to complete the proof of (3.1d). For notational simplicity, we will write K in place of L, so we are reduced to showing that for any finite extension $K/\mathbb{Q}_p$ and any $q \in K$ with $|q| < 1$, the map $\phi : K^* \to E_q(K)$ described in (3.1) is surjective.

We also set the following notation which we will use for the remainder of this section:

R the ring of integers of K,

$\mathfrak{M}$ the maximal ideal of R,

π a uniformizer for R, $\mathfrak{M} = \pi R$,

k the residue field of R, $k = R/\mathfrak{M}$,

ord_v the normalized valuation $\mathrm{ord}_v : K^* \twoheadrightarrow \mathbb{Z}$ on K.

The group $E_q(K)$ admits the usual filtration (see [AEC, VII §2])

$$E_q(K) \supset E_{q,0}(K) \supset E_{q,1}(K),$$

where

$$E_{q,0}(K) = \big\{ P \in E_q(K) : \tilde{P} \in \tilde{E}_{q,ns}(k) \big\},$$
$$E_{q,1}(K) = \big\{ P \in E_q(K) : \tilde{P} = \tilde{O} \big\}.$$

Here $\tilde{E}_q/k$ is the reduction of E_q modulo $\mathfrak{M}$, and $\tilde{E}_{q,ns}$ are the non-singular points on $\tilde{E}_q$. From [AEC, VII.2.1] and [AEC, VII.2.2] we have isomorphisms

$$E_{q,0}(K)/E_{q,1}(K) \cong \tilde{E}_{q,ns}(k) \quad \text{and} \quad E_{q,1}(K) \cong \hat{E}_q(\mathfrak{M})$$
$$P \mapsto \tilde{P} \qquad\qquad P = (x,y) \mapsto -\frac{x}{y},$$

where $\hat{E}$ is the formal group of E [AEC, IV §1].

Similarly, the quotient group $K^*/q^{\mathbb{Z}}$ has a natural filtration

$$K^*/q^{\mathbb{Z}} \supset R^* \supset R_1^*,$$

where

$$R_1^* = \big\{ u \in R : u \equiv 1 \,(\mathrm{mod}\,\mathfrak{M}) \big\}$$

is the group of 1-units in R. There are also isomorphisms

$$R^*/R_1^* \cong k^* \quad \text{and} \quad R_1^* = \hat{\mathbb{G}}_m(\mathfrak{M})$$
$$a \mapsto \tilde{a} \qquad\qquad u \mapsto 1 - u,$$

where $\hat{\mathbb{G}}_m$ is the formal multiplicative group [AEC, VI.2.2.2]. We are going to prove not only that the map $\phi : K^*/q^{\mathbb{Z}} \to E_q(K)$ is an isomorphism but that it respects the filtrations we just described.

We begin with the formal groups. First, from the formula for $X(u,q)$, it is clear that

$$u \equiv 1 \,(\mathrm{mod}\,\mathfrak{M}) \implies \mathrm{ord}_v\big(X(u,q)\big) < 0,$$

since only the term $\dfrac{u}{(1-u)^2}$ will be non-integral. This proves that

$$\phi(R_1^*) \subset E_{q,1}(K).$$

Next we show that this inclusion is an equality.

Using the isomorphisms described above, we look at the map

$$\hat{\mathbb{G}}_m(\mathfrak{M}) \xrightarrow{\;\cong\;} R_1^* \xrightarrow{\;\phi\;} E_{q,1}(K) \xrightarrow{\;\cong\;} \hat{E}_q(\mathfrak{M})$$
$$t \xrightarrow{\hspace{6cm}} -\frac{X(1+t,q)}{Y(1+t,q)}.$$

Note that as sets, $\hat{\mathbb{G}}_m(\mathfrak{M})$ and $\hat{E}_q(\mathfrak{M})$ are just the set $\mathfrak{M}$; they merely have different group structures attached to them. If we substitute $u = 1 + t$ into the series for $X(u, q)$ and $Y(u, q)$ and expand as Laurent series in t, we find that

$$X(1+t, q) = t^{-2}\Big(1 + \sum_{m \geq 1} \alpha_m t^m\Big) \quad \text{and} \quad Y(1+t, q) = t^{-3}\Big(1 + \sum_{m \geq 1} \beta_m t^m\Big),$$

with coefficients $\alpha_m, \beta_m \in R$. By taking the ratio of X and Y, we are reduced to showing that if $\gamma_1, \gamma_2, \ldots \in R$, then the map

$$\psi : \mathfrak{M} \longrightarrow \mathfrak{M}, \qquad t \longmapsto t\Big(1 + \sum_{m \geq 1} \gamma_m t^m\Big)$$

is surjective. This follows immediately from [AEC, IV.2.4], which asserts the existence of a power series $\lambda(T) \in R[\![T]\!]$ satisfying $\psi\big(\lambda(T)\big) = T$ (i.e., ψ is surjective, since for any $w \in \mathfrak{M}$ we have $\lambda(w) \in \mathfrak{M}$ and $\psi\big(\lambda(w)\big) = w$.)

Next we look at the behavior of ϕ on R^*. If we take the series for $X(u, q)$ and reduce it modulo $\mathfrak{M}$, we see that

$$X(u, q) \equiv \frac{u}{(1-u)^2} \not\equiv 0 \pmod{\mathfrak{M}} \qquad \text{for all } u \in R^*,$$

so $\phi(R^*) \subset E_{q,0}(K)$. Since $\phi(R_1^*) = E_{q,1}(K)$ from above, we get a well-defined injective homomorphism on the quotient groups

$$k^* \;\cong\; R^*/R_1^* \;\xrightarrow{\;\phi\;}\; E_{q,0}(K)/E_{q,1}(K) \;\cong\; \tilde{E}_{q,ns}(k)$$
$$u \longmapsto \hspace{6.5cm} \Big(\frac{u}{(1-u)^2}, \frac{u^2}{(1-u)^3}\Big).$$

This map $k^* \to \tilde{E}_{q,ns}(k)$ is clearly surjective, the inverse being

$$(x, y) \longmapsto \frac{y^2}{x^3},$$

so the map on the quotient groups is an isomorphism. (Note that $\tilde{E}_q$ has the equation $y^2 + xy = x^3$.) Then the commutative diagram

$$
\begin{array}{ccccccccc}
1 & \longrightarrow & R_1^* & \longrightarrow & R^* & \longrightarrow & k^* & \longrightarrow & 1 \\
 & & \downarrow{\wr} & & \downarrow{\phi} & & \downarrow{\wr} & & \\
0 & \longrightarrow & E_{q,1}(K) & \longrightarrow & E_{q,0} & \longrightarrow & \tilde{E}_{q,ns}(k) & \longrightarrow & 0
\end{array}
$$

implies that the map $\phi : R^* \longrightarrow E_{q,0}(K)$ is an isomorphism.

We are left to show that the injective homomorphism

$$\phi : K^*/R^* q^{\mathbb{Z}} \longrightarrow E_q(K)/E_{q,0}(K)$$

is surjective. The group on the left is easy to describe, since the map

$$K^*/R^* q^{\mathbb{Z}} \longrightarrow \mathbb{Z}/\mathrm{ord}_v(q)\mathbb{Z}$$
$$u \longmapsto \mathrm{ord}_v(u)$$

is clearly an isomorphism. So the following proposition will complete the proof of (3.1).

Proposition 4.1.

$$\#E_q(K)/E_{q,0}(K) \le \mathrm{ord}_v(q).$$

To prove this estimate, we will use geometry to divide $E_q(K)$ into several subsets, and then we will show that these subsets actually correspond to the cosets of $E_{q,0}(K)$ inside $E(K)$. We start with an elementary characterization of the points in $E_{q,0}(K)$.

Lemma 4.1.1. *Let* $P = (x, y) \in E_q(K)$. *The following are equivalent:*
(i) $P \in E_{q,0}(K)$,
(ii) $|x| \ge 1$,
(iii) $|y| \ge 1$.

PROOF. Taking partial derivatives of the equation for E_q, we observe that

$$
\begin{aligned}
P \in E_{q,0}(K) &\iff |y - 3x^2 - a_4| \ge 1 \quad \text{or} \quad |2y + x| \ge 1 \\
&\iff \max\{|y - 3x^2|, |2y + x|\} \ge 1 \quad \text{since } |a_4| = |q| < 1 \\
&\implies \max\{|x|, |y|\} \ge 1.
\end{aligned}
$$

Suppose first that $|x| \ge 1 > |y|$. Then

$$|x|^3 = |y^2 + xy - a_4x - a_6| \le \max\{|y|^2, |xy|, |a_4x|, |a_6|\} < \max\{1, |x|\}.$$

This strict inequality is a contradiction, so $|x| \ge 1$ implies $|y| \ge 1$. Similarly, the assumption $|y| \ge 1 > |x|$ gives the contradiction

$$|y|^2 = |x^3 + a_4x + a_6 - xy| < \max\{1, |y|\},$$

so $|y| \ge 1$ implies $|x| \ge 1$. This proves that

$$\max\{|x|, |y|\} \ge 1 \implies |x| \ge 1 \text{ and } |y| \ge 1,$$

so (i) implies (ii) and (iii).

Conversely, if $P \notin E_{q,0}(K)$, then $P = (x, y)$ reduces to the singular point $(0, 0)$ of $\tilde{E}_q(k)$, so $|x| < 1$ and $|y| < 1$. Hence either of (ii) or (iii) implies (i), which completes the proof of the lemma.   □

Next we use similar criteria to partition the points of $E_q(K)$ that do not lie in $E_{q,0}(K)$.

Lemma 4.1.2. *Let* $P = (x, y) \in E_q(K) \smallsetminus E_{q,0}(K)$. *Then exactly one of the following three conditions is true:*
(i) $1 > |y| > |x + y|$, *in which case* $|y| > |q|^{\frac{1}{2}}$,
(ii) $1 > |x + y| > |y|$, *in which case* $|x + y| > |q|^{\frac{1}{2}}$,

(iii) $|y| = |x + y| = |q|^{\frac{1}{2}}$.

(Note that (iii) can only occur if $\mathrm{ord}_v(q)$ is even.)

PROOF. Let

$$n = \min\{\mathrm{ord}_v\, x, \mathrm{ord}_v\, y\} \qquad \text{and} \qquad N = \mathrm{ord}_v\, q.$$

Dividing the equation for E_q by π^{2n}, we obtain the equation

$$y_n^2 + x_n y_n = \pi^n x_n^3 + \pi^{-n} a_4(q) x_n + \pi^{-2n} a_6(q),$$

where

$$x_n = \pi^{-n} x \in R \qquad \text{and} \qquad y_n = \pi^{-n} y \in R.$$

(In fancy terminology, we've blown up the scheme $E_q/\mathrm{Spec}(R)$ to find the affine subscheme on which P lies.) Since $|a_4(q)| = |a_6(q)| = |q|$, it is immediate from this equation that $\pi^{-2n} a_6(q) \in R$, so

$$1 \le n \le \frac{1}{2}\,\mathrm{ord}_v\, a_6(q) = \frac{1}{2}\,\mathrm{ord}_v(q) = \frac{1}{2}N.$$

(The fact that $1 \le n$ comes from (4.1.1).)

We now consider two cases. First, if $n < \frac{1}{2}N$, then reducing the above equation modulo π gives

$$y_n^2 + x_n y_n \equiv 0 \,(\mathrm{mod}\ \pi).$$

This means that either

$$y_n \equiv 0 \,(\mathrm{mod}\ \pi), \quad \text{or} \quad y_n + x_n \equiv 0 \,(\mathrm{mod}\ \pi), \quad \text{or both.}$$

But they cannot both be zero, since otherwise $x_n \equiv y_n \equiv 0 \,(\mathrm{mod}\ \pi)$, which would contradict the definition of n. Hence one of the following two assertions is true:

(i) $y_n \not\equiv 0 \,(\mathrm{mod}\ \pi)$ and $y_n + x_n \equiv 0 \,(\mathrm{mod}\ \pi)$,
(ii) $y_n \equiv 0 \,(\mathrm{mod}\ \pi)$ and $y_n + x_n \not\equiv 0 \,(\mathrm{mod}\ \pi)$.

These correspond to (i) and (ii) in the statement of the lemma. For example, for (i) we find that

$$|y| = |\pi^n y_n| = |\pi|^n > |q|^{\frac{1}{2}} \qquad \text{and} \qquad |y+x| = \left|\pi^n(y_n + x_n)\right| < |\pi|^n = |y|.$$

It remains to deal with the case $n = \frac{1}{2}N$. Since $a_6(q) = -q + \cdots$, our equation becomes

$$y_n^2 + x_n y_n \equiv (-q/\pi^{2n}) \equiv 0 \,(\mathrm{mod}\ \pi).$$

Hence $|y_n| = |y_n + x_n| = 1$, which implies that

$$|y| = |y + x| = |\pi|^n = |\pi|^{\frac{1}{2}N} = |q|^{\frac{1}{2}}.$$

□

Lemmas (4.1.1) and (4.1.2) allow us to divide $E_q(K)$ into the following subsets:

$$
\begin{aligned}
E_{q,0}(K) &= \left\{ (x,y) \in E_q(K) : |x| \geq 1 \text{ or } |y| \geq 1 \right\}, \\
U_n &= \left\{ (x,y) \in E_q(K) : |\pi|^n = |y| > |x+y| \right\}, \\
V_n &= \left\{ (x,y) \in E_q(K) : |\pi|^n = |x+y| > |y| \right\}, \\
W &= \left\{ (x,y) \in E_q(K) : |y| = |x+y| = |q|^{\frac{1}{2}} \right\}.
\end{aligned}
$$

Notice that (4.1.2) says that U_n and V_n are empty unless $n < \frac{1}{2} \operatorname{ord}_v q$, so $E_q(K)$ can be written as the union

$$E_q(K) = E_{q,0}(K) \cup W \cup \bigcup_{1 \leq n < \frac{1}{2} \operatorname{ord}_v q} (U_n \cup V_n).$$

Further, if $\operatorname{ord}_v q$ is odd, then $W = \emptyset$. So we have partitioned $E_q(K)$ into (at most) $\operatorname{ord}_v(q)$ pieces. The final step in the proof of Proposition 4.1, which will also complete the proof of Theorem 3.1, is to show that these subsets are the cosets of $E_{q,0}(K)$ in $E_q(K)$. More precisely, it will suffice to show that two points in the same subset are in the same coset, since this will imply that the number of cosets is no larger than $\operatorname{ord}_v(q)$. This is exactly what we do in the following lemma.

Remark 4.1.3. A more intrinsic explanation for the above decomposition of $E_q(K)$ is that the subsets U_n, V_n, and W are neighborhoods of the non-identity components of the special fiber of the Néron model of E_q over $\operatorname{Spec}(R)$. This decomposition may be compared with the description of special fibers of Type I_n in (IV §8), Tate's algorithm (IV.9.1), and the discussion in (IV.9.6).

Lemma 4.1.4. *Let $P, P' \in E_q(K)$ be points satisfying any one of the following conditions:*

(i) $P, P' \in U_n$;   *(ii)* $P, P' \in V_n$;   *(iii)* $P, P' \in W$.

Then $P - P' \in E_{q,0}(K)$.

PROOF. The proof of this lemma is completely elementary, although somewhat computationally involved. We merely have to combine the geometric description of U_n, V_n, and W with the algebraic formulas giving the group law on E_q.

If $P = P'$, there is nothing to prove. Assume for now that also $P \neq -P'$. (We'll deal with $P = -P'$ at the end.) Then writing $P = (x, y)$

and $P' = (x', y')$, the addition law on E_q and a little algebra yield the formula

$$x(P - P') = \frac{(y + y' + x')(y + x + y')}{(x - x')^2} - x - x'.$$

In all three cases, (i), (ii), and (iii), we have $|x| < 1$ and $|x'| < 1$, so (4.1.1) and the formula for $x(P - P')$ gives

$$P - P' \in E_{q,0}(K) \iff |x(P - P')| \geq 1$$
$$\iff |y + y' + x'| \cdot |y + x + y'| \geq |x - x'|^2.$$

Suppose first that $P, P' \in U_n$. Then

$$|x| = |y| = |\pi|^n \quad \text{and} \quad |x'| = |y'| = |\pi|^n, \quad \text{so} \ |x - x'| \leq |\pi|^n.$$

On the other hand, since $|y| = |\pi|^n$ and $|y' + x'| < |\pi|^n$, we get

$$|y + y' + x'| = |\pi|^n;$$

and similarly $|y'| = |\pi|^n$ and $|y + x| < |\pi|^n$, so

$$|y + x + y'| = |\pi|^n.$$

Therefore

$$|y + y' + x'| \cdot |y + x + y'| = |\pi|^{2n} \geq |x - x'|^2, \quad \text{so} \ P - P' \in E_{q,0}(K).$$

This proves (i).

The proof of (ii) can be done in a similar fashion, but it is even easier to observe that

$$P \in U_n \iff -P \in V_n.$$

This follows from the formula $-(x, y) = (x, -y - x)$. Hence (i) implies (ii). Further, since U_n and V_n are disjoint, we see that $P \neq -P'$ in cases (i) and (ii).

We turn now to case (iii), which is the most difficult. We claim that in this case we have

$$|y + x + y'| = |y + y' + x'|.$$

To see this, we note that

$$|x| \leq \max\{|x + y|, |y|\} = |q|^{\frac{1}{2}} \quad \text{and} \quad |x'| \leq \max\{|x' + y'|, |y'|\} = |q|^{\frac{1}{2}};$$

then we compute

$$
\begin{aligned}
\big|(y + x + y')y &- (y + y' + x')y'\big| \\
&= \big|(y^2 + xy) - (y'^2 + x'y')\big| \\
&= \big|(x^3 + a_4 x + a_6) - (x'^3 + a_4 x' + a_6)\big| \\
&= |x - x'| \cdot |x^2 - xx' + x'^2 + a_4| \\
&\le |x - x'||q| \quad \text{since } |x|, |x'| \le |q|^{\frac{1}{2}}, \ |a_4| = |q| \\
&= \big|(y + x + y') - (y + y' + x')\big| \cdot |q| \\
&\le \max\{|y + x + y'|, |y + y' + x'|\} \cdot |q| \\
&= \max\{|(y + x + y')y|, |(y + y' + x')y'|\} \cdot |q|^{\frac{1}{2}} \\
&\qquad\qquad\qquad\qquad \text{since } |y| = |y'| = |q|^{\frac{1}{2}}, \\
&< \max\{|(y + x + y')y|, |(y + y' + x')y'|\} \\
&\qquad\qquad\qquad\qquad\qquad \text{since } |q| < 1.
\end{aligned}
$$

The only way that this strict inequality can possibly be true is if

$$
\big|(y + x + y')y\big| = \big|(y + y' + x')y'\big|.
$$

Further, we know that $|y| = |y'|$, so we have proven our claim

$$
|y + x + y'| = |y + y' + x'|.
$$

Using this equality, we compute

$$
\begin{aligned}
|x - x'|^2 &= \big|(y + x + y') - (y + y' + x')\big|^2 \\
&\le \max\{|y + x + y'|, |y + y' + x'|\}^2 \\
&= |y + x + y'| \cdot |y + y' + x'|.
\end{aligned}
$$

From above, this inequality implies that $P - P' \in E_{q,0}(K)$.

Finally, we must deal with the case $P' = -P \in W$, so $P - P' = 2P$. We could argue by continuity, but here is a direct argument using the duplication formula, which on E_q reads

$$
x(P - P') = x(2P) = \frac{x^4 - 2a_4 x^2 - 8a_6 x + a_4^2 - a_6}{4x^3 + x^2 + 4a_4 x + 4a_6} = \frac{f(x)}{g(x)}.
$$

Here $f(x)$ and $g(x)$ are the indicated polynomials. From general principles, one knows that $f(x)$ and $g(x)$ are relatively prime in $K[x]$. More precisely, if we let

$$
F(x) = 48x^2 + 8x + 64a_4 - 1 \quad \text{and} \quad G(x) = 12x^3 - x^2 - 20a_4 x + 2a_4 - 108a_6,
$$

then a little algebra suffices to verify the relation

$$f(x)F(x) - g(x)G(x) = \Delta,$$

where Δ is the discriminant of the Weierstrass equation for E_q. Substituting $f(x) = x(2P)g(x)$ into this relation gives

$$g(x)\{x(2P)F(x) - G(x)\} = \Delta.$$

We are assuming that $P \in W$, so

$$\left|g(x)\right| = |2y + x|^2 \leq \max\{|y|, |y + x|\}^2 = |q|,$$
$$\left|G(x)\right| = |12x^3 - x^2 - 20a_4x + 2a_4 - 108a_6| \leq \max\{|x|^2, |a_4|, |a_6|\} = |q|,$$
$$\left|F(x)\right| = |48x^2 + 8x + 64a_4 - 1| = 1,$$
$$\left|\Delta\right| = |q - 24q^2 + 252q^3 - \cdots| = |q|.$$

Hence

$$1 \leq \left|\frac{\Delta}{g(x)}\right| = \left|x(2P)F(x) - G(x)\right|$$
$$\leq \max\{\left|x(2P)\right| \cdot \left|F(x)\right|, \left|G(x)\right|\}$$
$$\leq \max\{\left|x(2P)\right|, |q|\}.$$

Since $|q| < 1$, it follows that $\left|x(2P)\right| \geq 1$, so $2P \in E_{q,0}(K)$ from (4.1.1). This completes the proof of Lemma 4.1.4, and with it the proofs of Proposition 4.1 and Theorem 3.1.

$\square$

§5. Elliptic Curves over p-adic Fields

In the previous two sections we have shown that for any p-adic field $K/\mathbb{Q}_p$ and any $q \in K^*$ with $|q| < 1$, the quotient group $\bar{K}^*/q^{\mathbb{Z}}$ is (analytically) isomorphic to an elliptic curve $E_q(\bar{K})$. In the analogous situation over the complex numbers, we know (1.1d) that every elliptic curve $E/\mathbb{C}$ is isomorphic to E_q for some $q \in \mathbb{C}^*$. However, in the p-adic case we have

$$\left|j(E_q)\right| = \left|\frac{1}{q} + 744 + 196884q + \cdots\right| = \frac{1}{|q|} > 1,$$

so it is clear that not every elliptic curve over K can be isomorphic (over $\bar{K}$) to an E_q. A necessary condition is $\left|j(E)\right| > 1$. We begin by showing that this condition is also sufficient.

Lemma 5.1. *Let $\alpha \in \bar{\mathbb{Q}}_p$ be an element with $|\alpha| > 1$. Then there is a unique $q \in \bar{\mathbb{Q}}_p^*$ with $|q| < 1$ such that $j(E_q) = \alpha$. This value of q lies in $\mathbb{Q}_p(\alpha)$.*

PROOF. The j-invariant of E_q is given by the series (3.1), which we write as

$$j(q) = \frac{1 + 744q + 196884q^2 + \cdots}{q}.$$

The reciprocal of this series, which we will call $f(q)$, is given by the formula

$$f(q) = \frac{1}{j(q)} = \frac{q}{1 + 744q + 196884q^2 + \cdots}$$
$$= q - 744q^2 + 356652q^3 - \cdots \in \mathbb{Z}[\![q]\!].$$

Applying [AEC, IV.2.4] to the series f, we get a series $g(q) = q + \cdots \in \mathbb{Z}[\![q]\!]$ such that $g(f(q)) = q$ as formal power series in $\mathbb{Z}[\![q]\!]$. Since g has integer coefficients and leading term q, it will converge if we evaluate it at any element $\beta \in \bar{\mathbb{Q}}_p$ of absolute value less than 1 and will satisfy $|g(\beta)| = |\beta|$. In particular, since $|\alpha| > 1$, we find that

$$q = g\left(\frac{1}{\alpha}\right) \in \mathbb{Q}_p(\alpha)$$

satisfies

$$0 < |q| = \left|\frac{1}{\alpha}\right| < 1 \quad \text{and} \quad \frac{1}{j(q)} = f(q) = f\left(g\left(\frac{1}{\alpha}\right)\right) = \frac{1}{\alpha}.$$

Hence $j(q) = \alpha$ as desired. This proves the existence part of (5.1).

To prove uniqueness, suppose that $j(q) = j(q')$ with $|q| < 1$ and $|q'| < 1$. Then $f(q) = f(q')$, so

$$0 = |f(q) - f(q')|$$
$$= |q - q'| \cdot |1 - 744(q + q') + 356652(q^2 + qq' + q'^2) + \cdots|$$
$$= |q - q'|.$$

Therefore $q = q'$. $\qquad\square$

Before proving the p-adic uniformization theorem, we describe an invariant which is useful for studying the twists of a curve.

Lemma 5.2. *Let E/K be an elliptic curve defined over a field of characteristic not equal to 2 or 3, and choose a Weierstrass equation*

$$y^2 + a_1 xy + a_3 y = x^3 + a_2 x^2 + a_4 x + a_6$$

for E/K. Let c_4 and c_6 be the usual quantities [AEC, III §1] associated to this equation. Assuming that $j(E) \neq 0, 1728$, we define

$$\gamma(E/K) = -c_4/c_6 \in K^*/K^{*2}.$$

(The reason for the negative sign will become apparent later when we prove that $\gamma(E_q/K) = 1$.)

(a) $\gamma(E/K)$ is well-defined as an element of K^*/K^{*2}, independent of the choice of Weierstrass equation for E/K.

(b) Let E'/K be another elliptic curve with $j(E') \neq 0, 1728$. Then E and E' are isomorphic over K if and only if

$$j(E) = j(E') \qquad \text{and} \qquad \gamma(E/K) = \gamma(E'/K).$$

(c) Let E/K and E'/K be elliptic curves with $j(E') = j(E) \neq 0, 1728$, and suppose that $\gamma(E/K) \neq \gamma(E'/K)$, so

$$L = K\left(\sqrt{\frac{\gamma(E/K)}{\gamma(E'/K)}}\right)$$

is a quadratic extension of K. Let

$$\chi : G_{\bar{K}/K} \longrightarrow G_{L/K} \longrightarrow \{\pm 1\}$$

be the quadratic character associated to L/K. Then there is an isomorphism

$$\psi : E \to E'$$

with the property that

$$\psi(P^\sigma) = \chi(\sigma)\psi(P) \quad \text{for all } \sigma \in G_{\bar{K}/K} \text{ and all } P \in E(\bar{K}).$$

PROOF. (a) The condition $j(E) \neq 0, 1728$ is equivalent to $c_4 \neq 0$ and $c_6 \neq 0$, so $\gamma(E/K)$ exists. If we choose a new Weierstrass equation for E/K, then the new c_4 and c_6 are related to the old ones by the formulas $u^4 c_4' = c_4$ and $u^6 c_6' = c_6$ for some $u \in \bar{K}^*$. (See [AEC, III Table 1.2]. The fact that $u \in K$ follows from [AEC, III.3.1].) Hence

$$\frac{c_4'}{c_6'} = u^2 \frac{c_4}{c_6} \equiv \frac{c_4}{c_6} \ (\text{mod } K^{*2}),$$

which proves that $\gamma(E/K)$ is independent of the chosen Weierstrass equation.

(b) If E and E' are isomorphic over K, then [AEC, III.1.4b] asserts that $j(E) = j(E')$. Further, since the Weierstrass equations for E and E' are Weierstrass equations for the same elliptic curve over K, it follows from (a) that $\gamma(E/K) = \gamma(E'/K)$.

Conversely, suppose that $j(E) = j(E')$ and $\gamma(E/K) = \gamma(E'/K)$. Since the characteristic of K is not 2 or 3, we can find Weierstrass equations for E and E' over K of the form

$$E : y^2 = x^3 + Ax + B, \qquad E' : y^2 = x^3 + A'x + B',$$

with $A, B, A', B' \in K$. The fact that $j(E) = j(E') \neq 0, 1728$ implies that

$$\frac{A^3}{B^2} = \frac{A'^3}{B'^2}, \quad \text{since } j(E) = 1728\frac{4A^3}{4A^3 + 27B^2}.$$

Similarly, since $c_4 = -48A$ and $c_6 = -864B$, our assumption $\gamma(E/K) = \gamma(E'/K)$ means that

$$\frac{2A}{B} \equiv \frac{c_4}{c_6} \equiv -\gamma(E/K) \equiv -\gamma(E'/K) \equiv \frac{c_4'}{c_6'} \equiv \frac{2A'}{B'} \pmod{K^{*2}},$$

so there is some $t \in K^*$ such that $AB' = t^2 A'B$. Using these relations between A, B and A', B', it is now easy to check that the map

$$E \longrightarrow E', \qquad (x, y) \longmapsto (t^2 x, t^3 y)$$

is a K-isomorphism.

(c) We take models for E/K and E'/K as in (b), and again the assumption $j(E) = j(E') \neq 0, 1728$ implies that $A^3 B'^2 = A'^2 B^3$. Next we let

$$t = \sqrt{\frac{c_4 c_6'}{c_4' c_6}} = \sqrt{\frac{AB'}{A'B}}, \quad \text{so} \quad t^2 \equiv \frac{\gamma(E/K)}{\gamma(E'/K)} \pmod{K^{*2}}.$$

Since $\gamma(E/K) \neq \gamma(E'/K)$, we know that $L = K(t)$ is a quadratic extension of K; and as in (b), the map

$$\psi : E \longrightarrow E', \qquad (x, y) \longmapsto (t^2 x, t^3 y)$$

is easily seen to be an isomorphism. Finally, for any $\sigma \in G_{\bar{K}/K}$, we know that $t^\sigma = \chi(\sigma)t$. So for $P = (x, y) \in E(\bar{K})$ we have

$$\begin{aligned}
\psi(P)^\sigma = \psi(x, y)^\sigma &= (t^2 x, t^3 y)^\sigma = \left(\chi(\sigma)^2 t^2 x^\sigma, \chi(\sigma)^3 t^3 y^\sigma\right) \\
&= \left(t^2 x^\sigma, \chi(\sigma) t^3 y^\sigma\right) = \chi(\sigma)(t^2 x^\sigma, t^3 y^\sigma) = \chi(\sigma)\psi(P^\sigma).
\end{aligned}$$
$\square$

We are now ready to prove Tate's p-adic uniformization theorem, which applies to all curves whose j-invariant has absolute value greater than 1.

Theorem 5.3. (Tate) *Let K be a p-adic field, let E/K be an elliptic curve with $|j(E)| > 1$, and let $\gamma(E/K) \in K^*/K^{*2}$ be the invariant defined in (5.2).*

(a) *There is a unique $q \in \bar{K}^*$ with $|q| < 1$ such that E is isomorphic over $\bar{K}$ to the Tate curve E_q. Further, this value of q lies in K.*

(b) *Let q be chosen as in (a). Then the following three conditions are equivalent:*

    (i)   *E is isomorphic to E_q over K.*

    (ii)   *$\gamma(E/K) = 1$.*

    (iii) *E has split multiplicative reduction.*

PROOF. (a) From (5.1) there is a unique $q \in \bar{K}^*$ with $|q| < 1$ such that $j(E_q) = j(E)$. This implies [AEC, III.1.4b] that E_q is isomorphic to E over $\bar{K}$, which completes the proof of (a).

(b) From (5.2) we know that E is isomorphic to E_q over K if and only if $j(E) = j(E_q)$ and $\gamma(E/K) = \gamma(E_q/K)$. So in order to prove that (i) and (ii) are equivalent, we must show that $\gamma(E_q/K) = 1$. Using (3.1) we find that the c_4 and c_6 values associated to the Tate curve

$$E_q : y^2 + xy = x^3 + a_4(q)x + a_6(q)$$

are

$$c_4(q) = 1 - 48a_4(q) = 1 + 240s_3(q),$$
$$c_6(q) = -1 + 72a_4(q) - 864a_6(q) = -1 + 504s_5(q).$$

So the γ-invariant of E_q/K equals

$$\gamma(E_q/K) \equiv -\frac{c_4(q)}{c_6(q)} \equiv \frac{1 + 240s_3(q)}{1 - 504s_5(q)} \pmod{K^{*2}}.$$

To see that $\gamma(E_q/K)$ is a square, we use the following elementary calculation which implies that $c_4(q)$ and $-c_6(q)$ are themselves squares in K.

Lemma 5.3.1. *Let $\alpha \in K$ with $|\alpha| < 1$. Then $1 + 4\alpha$ is a square in K.*

PROOF. We first observe that the binomial coefficient

$$\binom{-1/2}{n} = \frac{\left(-\frac{1}{2}\right)\left(-\frac{3}{2}\right)\left(-\frac{5}{2}\right)\cdots\left(-\frac{2n-1}{2}\right)}{n!} = \frac{(-1)^n}{4^n}\binom{2n}{n}$$

is an integer divided by 4^n. Hence the coefficients of the series

$$(1 + 4\alpha)^{-\frac{1}{2}} = \sum_{n=0}^{\infty} \binom{-1/2}{n}(4\alpha)^n = \sum_{n=0}^{\infty}(-1)^n\binom{2n}{n}\alpha^n$$

are integers, so the series converges in K. Therefore $(1+4\alpha)^{-1}$ is a square in K, so the same is true of $1+4\alpha$. This completes the proof of the lemma, and with it the fact that (i) and (ii) are equivalent. $\qquad\qquad\square$

Next we note that since $|a_4(q)| = |a_6(q)| = |q| < 1$, the equation of the reduced curve $\tilde{E}_q$ is

$$\tilde{E}_q : y^2 + xy = x^3,$$

which clearly has split multiplicative reduction. This shows that (i) implies (iii).

Conversely, suppose that E has split multiplicative reduction. We will show that $\gamma(E/K) = 1$, which will prove that (iii) implies (ii). Take a minimal Weierstrass equation for E,

$$E : y^2 + a_1 xy + a_3 y = x^3 + a_2 x^2 + a_4 x + a_6.$$

Making a linear change of variables, we may assume that the singular point modulo $\mathfrak{M}$ is the point $(0,0)$, where as usual we write $\mathfrak{M}$ for the maximal ideal of the ring of integers of K. Then the fact that $(0,0)$ is on the curve and singular modulo $\mathfrak{M}$ implies that

$$a_3 \equiv a_4 \equiv a_6 \equiv 0 \,(\mathrm{mod}\ \mathfrak{M}),$$

and hence that

$$b_4 = a_1 a_3 + 2a_4 \equiv 0 \,(\mathrm{mod}\ \mathfrak{M}) \quad \text{and} \quad c_4 = b_2^2 - 24b_4 \equiv b_2^2 \,(\mathrm{mod}\ \mathfrak{M}).$$

From [AEC, VII.5.1b], the fact that E has multiplicative reduction implies $c_4 \not\equiv 0 \,(\mathrm{mod}\ \mathfrak{M})$, so we see that $b_2 \not\equiv 0 \,(\mathrm{mod}\ \mathfrak{M})$. It follows that b_2 is a unit (i.e., $|b_2| = 1$). Hence

$$\gamma(E/K) = -\frac{c_4}{c_6} = \frac{1}{b_2} \cdot \left(\frac{1 - 24\dfrac{b_4}{b_2^2}}{1 - 36\dfrac{b_4}{b_2^2} + 216\dfrac{b_6}{b_2^3}} \right) \quad (\mathrm{mod}\ K^{*2}).$$

Applying (5.3.1) to the numerator and denominator of the bracketed fraction on the right-hand side of this equation, we find that

$$\gamma(E/K) \equiv \frac{1}{b_2} \equiv b_2 \quad (\mathrm{mod}\ K^{*2}).$$

It remains to show that if the multiplicative reduction of E is split, then b_2 is a square in K^*.

Note that the reduction of E is

$$\tilde{E} : y^2 + \tilde{a}_1 xy = x^3 + \tilde{a}_2 x^2.$$

We factor the polynomial

$$y^2 + \tilde{a}_1 xy - \tilde{a}_2 x^2 = (y - \tilde{\alpha}x)(y - \tilde{\beta}x).$$

The fact that E has multiplicative reduction means that $\tilde{E}$ has a node, so $\tilde{\alpha} \neq \tilde{\beta}$; and the fact that the reduction is split means that $\tilde{\alpha}$ and $\tilde{\beta}$ are actually in the residue field of K, rather than in a quadratic extension. (See [AEC, III §1, VII §5].) It follows from Hensel's lemma applied to the polynomial $T^2 + a_1 T - a_2$ that $\tilde{\alpha}$ and $\tilde{\beta}$ lift uniquely to elements $\alpha, \beta \in K$ such that

$$y^2 + a_1 xy - a_2 x^2 = (y - \alpha x)(y - \beta x).$$

Hence

$$b_2 = a_1^2 + 4a_2 = (-\alpha - \beta)^2 + 4(-\alpha\beta) = (\alpha - \beta)^2 \in K^{*2},$$

so $\gamma(E/K) \equiv b_2 \equiv 1 \pmod{K^{*2}}$.

We have now proven (ii) $\Longleftrightarrow$ (i) $\Longrightarrow$ (iii) $\Longrightarrow$ (ii), which completes the proof of Theorem 5.3. $\qquad\square$

Suppose that we have an elliptic curve E/K as in Theorem 5.3 with invariant $\gamma(E/K) \neq 1$. If we let $L = K\left(\sqrt{\gamma(E/K)}\right)$, which is well-defined, since $\gamma(E/K)$ is defined up to squares in K, then it is clear that $\gamma(E/L) = 1$. Applying (5.3) to E/L, we find that E is isomorphic to E_q over L, so

$$E(L) \cong E_q(L) \cong L^*/q^{\mathbb{Z}}.$$

We will now describe $E(K)$ in terms of this identification.

Corollary 5.4. *With notation as in the preceding paragraph,*

$$E(K) \cong \left\{ u \in L^*/q^{\mathbb{Z}} : \mathrm{N}_K^L(u) \in q^{\mathbb{Z}}/q^{2\mathbb{Z}} \right\}.$$

PROOF. First we observe that the norm map N_K^L is a homomorphism

$$\mathrm{N}_K^L : L^*/q^{\mathbb{Z}} \longrightarrow K^*/q^{2\mathbb{Z}},$$

so $\mathrm{N}_K^L(u)$ is well-defined modulo $q^{2\mathbb{Z}}$. Applying (5.2c) to E and E_q, there is an isomorphism

$$\psi : E_q(\bar{K}) \longrightarrow E(\bar{K})$$

satisfying $\psi(P^\sigma) = \chi(\sigma)\psi(P)^\sigma$ for all $\sigma \in G_{\bar{K}/K}$, where $\chi : G_{\bar{K}/K} \to G_{L/K} \to \{\pm 1\}$ is the quadratic character associated to L/K. On the other hand, the isomorphism $\phi : \bar{K}^*/q^{\mathbb{Z}} \to E_q(\bar{K})$ is defined over K, which means that $\phi(P^\sigma) = \phi(P)^\sigma$. We look at the composition

$$L^*/q^{\mathbb{Z}} \xrightarrow{\phi} E_q(L) \xrightarrow{\psi} E(L),$$

which we know from above is an isomorphism of groups. Let $\tau \in G_{\bar{K}/K}$ be an element with $\chi(\tau) = -1$, so τ represents the non-trivial element in $G_{L/K}$. Then for any $u \in L^*$,

$$
\begin{aligned}
(\psi \circ \phi)(u) \in E(K) &\iff \psi\big(\phi(u)\big)^\tau = \psi\big(\phi(u)\big) \\
&\iff -\psi\big(\phi(u^\tau)\big) = \psi\big(\phi(u)\big) \quad \text{since } \chi(\tau) = -1 \\
&\iff \psi\big(\phi(u^{-\tau})\big) = \psi\big(\phi(u)\big) \\
&\qquad \text{since } -\psi(P) = \psi(-P) \text{ and } -\phi(u) = \phi(u^{-1}) \\
&\iff u^{-\tau} \equiv u \pmod{q^{\mathbb{Z}}} \\
&\qquad\qquad\qquad\quad \text{since } \phi \text{ and } \psi \text{ are isomorphisms} \\
&\iff u^{1+\tau} \in q^{\mathbb{Z}}.
\end{aligned}
$$

Since $u^{1+\tau} = \mathrm{N}_K^L(u)$, this completes the proof of the corollary. $\qquad\square$

§6. Some Applications of p-adic Uniformization

As we have seen amply demonstrated, the arithmetic properties of the torsion points on an elliptic curve are of fundamental importance. In the case that the curve has a p-adic uniformization, $E(\bar{K}) \cong \bar{K}^*/q^{\mathbb{Z}}$, it is easy to describe the torsion subgroup of E. Further, since the p-adic uniformization commutes with the action of $G_{\bar{K}/K}$, it is similarly easy to describe the action of $G_{\bar{K}/K}$ on the torsion subgroup of E. We will not prove the most general theorem in this direction but will be content with the following fundamental result. (See also exercise 5.13.)

Proposition 6.1. Let K be a p-adic field with normalized valuation ord_v, let E/K be an elliptic curve with $|j(E)| > 1$, and let $\ell \geq 3$ be a prime not dividing $\mathrm{ord}_v j(E)$. Then there is an element σ in the inertia subgroup of $G_{\bar{K}/K}$ which acts on the ℓ-torsion subgroup $E[\ell]$ of E via a matrix of the form $\left(\begin{smallmatrix} 1 & 1 \\ 0 & 1 \end{smallmatrix}\right)$. In other words, there is a basis $P_1, P_2 \in E[\ell]$ such that

$$
P_1^\sigma = P_1 \qquad \text{and} \qquad P_2^\sigma = P_1 + P_2.
$$

(One sometimes says that σ acts as a transvection on $E[\ell]$.)

Remark 6.1.1. Recall that there is an ℓ-adic representation [AEC, III §7]

$$
\rho_\ell : G_{\bar{K}/K} \longrightarrow \mathrm{Aut}\big(T_\ell(E)\big).
$$

Since $E[\ell] \cong T_\ell(E)/\ell T_\ell(E)$, another way to state (6.1) is that relative to an appropriate basis, there is a $\sigma \in G_{\bar{K}/K}$ satisfying

$$\rho_\ell(\sigma) \equiv \begin{pmatrix} 1 & 1 \\ 0 & 1 \end{pmatrix} \pmod{\ell}.$$

Remark 6.1.2. It is also worth pointing out that the proof of Proposition 6.1 does not need the full strength of Theorem 3.1. Specifically, we only need to know that the map $\phi : \bar{K}^*/q^{\mathbb{Z}} \to E_q(\bar{K})$ is an injective homomorphism; we do not need to know that it is surjective. The reason injectivity suffices is that we are really only interested in the torsion subgroup of E_q, and a simple count shows that the there are m^2 points of order m in $\bar{K}^*/q^{\mathbb{Z}}$, so we get (essentially for free) that ϕ is an isomorphism on torsion.

PROOF. First we observe that if L/K is a finite extension of degree prime to ℓ, and if (6.1) is true for E/L, then it is true for E/K. This follows from the equality $\mathrm{ord}_w\, j(E) = e_{w/v}\, \mathrm{ord}_v\, j(E)$, where w is the extension of v to L, and the ramification index $e_{w/v}$ is prime to ℓ, since it divides $[L : K]$. Hence ℓ will not divide $\mathrm{ord}_w\, j(E)$, so there is a $\sigma \in G_{\bar{K}/L} \subset G_{\bar{K}/K}$ that acts as a transvection on $E[\ell]$.

From (5.3b) we know that E is isomorphic to a (unique) Tate curve E_q over an (at most) quadratic extension of K. So replacing K by this extension, it suffices to prove (6.1) for E_q, where $q \in K^*$. Similarly, we may assume that K contains a primitive ℓ^{th}-root of unity ζ, since the degree of $K(\zeta)/K$ divides $\ell - 1$, so the degree is prime to ℓ.

Let $Q = q^{\frac{1}{\ell}} \in \bar{K}$ be a fixed ℓ^{th}-root of q. Since $\mathrm{ord}_v\, j(E_q) = -\,\mathrm{ord}_v\, q$ is not divisible by ℓ, the Kummer extension $K(Q)/K$ is totally ramified of degree ℓ. Hence there exists a σ in the inertia subgroup of $G_{\bar{K}/K}$ such that $Q^\sigma = \zeta Q$. We claim that this is the desired σ; it remains to pick the right basis for $E_q[\ell]$.

To do this, we use the p-adic uniformization (3.1)

$$\phi : \bar{K}^*/q^{\mathbb{Z}} \xrightarrow{\sim} E_q(\bar{K}).$$

With this identification we clearly have

$$\phi : (\zeta^{\mathbb{Z}} \cdot Q^{\mathbb{Z}})/q^{\mathbb{Z}} \xrightarrow{\sim} E_q[\ell].$$

Further, the p-adic uniformization map ϕ commutes with the action of Galois (3.1d) (i.e., $\phi(P^\sigma) = \phi(P)^\sigma$), so the action of $G_{\bar{K}/K}$ on $E_q[\ell]$ is the same as its action on the quotient group $(\zeta^{\mathbb{Z}} \cdot Q^{\mathbb{Z}})/q^{\mathbb{Z}}$. As our basis for $E_q[\ell]$, we take the elements $P_1 = \phi(\zeta)$ and $P_2 = \phi(Q)$. Then

$$P_1^\sigma = \phi(\zeta)^\sigma = \phi(\zeta^\sigma) = \phi(\zeta) = P_1,$$
$$P_2^\sigma = \phi(Q)^\sigma = \phi(Q^\sigma) = \phi(\zeta Q) = \phi(\zeta) + \phi(Q) = P_1 + P_2. \qquad \square$$

We next observe that Proposition 6.1 remains true for certain elliptic curves over number fields.

Corollary 6.2.   *Let $K/\mathbb{Q}$ be a number field, let E/K be an elliptic curve, and assume that the j-invariant of E is not in the ring of integers of K. Then for all but finitely many primes ℓ, the image of the ℓ-adic representation $\rho_\ell : G_{\bar{K}/K} \to \operatorname{Aut}\big(T_\ell(E)\big)$ contains an element satisfying*

$$\rho_\ell(\sigma) \equiv \begin{pmatrix} 1 & 1 \\ 0 & 1 \end{pmatrix} \pmod{\ell}$$

relative to a suitable basis for $T_\ell(E)/\ell T_\ell(E) = E[\ell]$.

PROOF. Let v be a (finite) place of K for which $j(E)$ is non-integral, so $\big|j(E)\big|_v > 1$. Let $G_{\bar{K}_v/K_v} \subset G_{\bar{K}/K}$ be the decomposition group of v for the extension of v to $\bar{K}$ corresponding to some embedding $\bar{K} \hookrightarrow \bar{K}_v$. Now (6.1) gives an element $\sigma \in G_{\bar{K}_v/K_v}$ which acts like $\begin{pmatrix} 1 & 1 \\ 0 & 1 \end{pmatrix}$ on $E[\ell]$. But with our identifications, $\sigma \in G_{\bar{K}/K}$ and $E[\ell] \subset E(\bar{K}) \subset E(\bar{K}_v)$, which gives the desired result. $\square$

It is a legitimate question to ask why one should care that $\operatorname{Gal}(\bar{K}/K)$ contains an element that acts on $E[\ell]$ as a transvection. One answer is that this puts severe constraints on the allowable maps between such elliptic curves. For example, we will now give Serre's p-adic proof that an elliptic curve with complex multiplication has integral j-invariant. (For alternative proofs of this important fact, see [AEC, exercise 7.10] and (II §6).)

Theorem 6.3.   *Let $K/\mathbb{Q}$ be a number field, and let E/K be an elliptic curve whose j-invariant $j(E)$ is not in the ring of integers of K. Then $\operatorname{End}(E) = \mathbb{Z}$.*

PROOF. (Serre) We begin by recalling that there is a representation of the endomorphism ring of E [AEC, III §7],

$$\operatorname{End}(E) \longrightarrow \operatorname{End}\big(T_\ell(E)\big), \qquad \psi \longmapsto \psi_\ell.$$

Further, we proved in [AEC, V.2.3] that for any $\psi \in \operatorname{End}(E)$, this representation can be used to compute the degree of ψ via the formula

$$\deg(\psi) = \det(\psi_\ell).$$

(This result appears in [AEC] in the chapter on elliptic curves over finite fields, but the proof depends only on the non-degeneracy of the Weil pairing, which is valid in general.)

Let $\psi \in \operatorname{End}(E)$ be an isogeny. Taking a finite extension of K if necessary, we may assume that ψ is defined over K. This means that

$$\psi(P^\sigma) = \psi(P)^\sigma \qquad \text{for all } \psi \in G_{\bar{K}/K} \text{ and all } P \in E(\bar{K}).$$

We need to show that $\psi \in \mathbf{Z}$.
  Let
$$m = \deg(1 + \psi) - \deg(\psi) - 1.$$
Notice that if we knew that ψ was in $\mathbf{Z}$, then m would equal 2ψ. So we will try to show that $m = 2\psi$ by showing that the degree of $m - 2\psi$ is 0.

  Using (6.2), choose a "large" prime ℓ, an element $\sigma \in G_{\bar{K}/K}$, and an ordered basis $\{P_1, P_2\}$ for $E[\ell]$ so that relative to this basis,

$$\rho_\ell(\sigma) \equiv \begin{pmatrix} 1 & 1 \\ 0 & 1 \end{pmatrix} \quad (\text{mod } \ell).$$

(We will see below that any ℓ larger than $\deg(m - 2\psi)$ will suffice.) Looking at the action of ψ on $E[\ell]$, we find that ψ is represented by a matrix

$$\psi_\ell \equiv \begin{pmatrix} a & b \\ c & d \end{pmatrix} \quad (\text{mod } \ell)$$

for some $a, b, c, d \in \mathbf{Z}/\ell\mathbf{Z}$. In other words, $\psi(P_1) = aP_1 + cP_2$ and $\psi(P_2) = bP_1 + dP_2$. Now, since ψ and σ commute in their action on $E(\bar{K})$, it follows that their matrices commute in $\text{End}(E[\ell]) \cong \text{GL}_2(\mathbf{Z}/\ell\mathbf{Z})$:

$$\begin{pmatrix} 1 & 1 \\ 0 & 1 \end{pmatrix} \begin{pmatrix} a & b \\ c & d \end{pmatrix} \equiv \begin{pmatrix} a & b \\ c & d \end{pmatrix} \begin{pmatrix} 1 & 1 \\ 0 & 1 \end{pmatrix} \quad (\text{mod } \ell).$$

Multiplying this out, we find that $a = d$ and $c = 0$, so

$$\psi_\ell \equiv \begin{pmatrix} a & b \\ 0 & a \end{pmatrix} \quad (\text{mod } \ell).$$

Next we determine the relationship between a and m:

$$\begin{aligned}
m &= \deg(1 + \psi) - \deg(\psi) - 1 \quad \text{by definition of } m \\
&= \det(1 + \psi_\ell) - \det(\psi_\ell) - 1 \quad \text{from [AEC, V.2.3]} \\
&\equiv \det \begin{pmatrix} 1 + a & b \\ 0 & 1 + a \end{pmatrix} - \det \begin{pmatrix} a & b \\ 0 & a \end{pmatrix} - 1 \quad (\text{mod } \ell) \\
&\equiv 2a \quad (\text{mod } \ell).
\end{aligned}$$

This now allows us to compute the degree of $m - 2\psi$, at least modulo ℓ.

$$\begin{aligned}
\deg(m - 2\psi) &= \det(m - 2\psi_\ell) \quad \text{from [AEC, V.2.3] again} \\
&\equiv \det \left[\begin{pmatrix} m & 0 \\ 0 & m \end{pmatrix} - 2 \begin{pmatrix} a & b \\ 0 & a \end{pmatrix} \right] \quad (\text{mod } \ell) \\
&\equiv 0 \quad (\text{mod } \ell) \quad \text{since } m \equiv 2a \, (\text{mod } \ell) \text{ from above.}
\end{aligned}$$

  We have now proven that $\deg(m - 2\psi) \equiv 0 \, (\text{mod } \ell)$ for all primes ℓ such that (6.2) is true, which means that $\deg(m - 2\psi) \equiv 0 \, (\text{mod } \ell)$ for all but finitely many ℓ's. Hence $\deg(m - 2\psi) = 0$, so $m = 2\psi$. But every endomorphism is integral over $\mathbf{Z}$ [AEC, III.9.4], so m must be even and $\psi \in \mathbf{Z}$. This completes the proof that $\text{End}(E) = \mathbf{Z}$. $\qquad \square$

EXERCISES

5.1. Let $q \in \mathbb{C}$, $|q| < 1$, and let $k \in \mathbb{R}$. Prove that

$$s_k(q) = \sum_{n \geq 1} \sigma_k(n) q^n = \sum_{n \geq 1} \frac{n^k q^n}{1 - q^n},$$

where as usual $\sigma_k(n) = \sum_{d \mid n} d^k$.

5.2. Let G be a cyclic group of order n, let σ be a generator for G, and let M be a G-module.
  (a) Prove that

$$H^2(G, M) \cong \frac{\{x \in M \; : \; x - \sigma x = 0\}}{\{x + \sigma x + \cdots + \sigma^{n-1} x \; : \; x \in M\}}.$$

(This piece of elementary group cohomology is used in the proof of (2.4).)
  (b) Prove that

$$H^1(G, M) \cong \frac{\{x \in M \; : \; x + \sigma x + \cdots + \sigma^{n-1} x = 0\}}{\{x - \sigma x \; : \; x \in M\}}.$$

Use this directly to show that for $q \in \mathbb{R}^*$, the Weil-Châtelet group

$$\mathrm{WC}(E_q / \mathbb{R}) \cong H^1 \left(G_{\mathbb{C}/\mathbb{R}}, \mathbb{C}^* / q^{\mathbb{Z}} \right)$$

has order 1 (respectively 2) if $q < 0$ (respectively $q > 0$.)

5.3. Let $E/\mathbb{R}$ be an elliptic curve, let $\Delta(E)$ be the discriminant of a Weierstrass equation for $E/\mathbb{R}$, and let m be an *even* integer.
  (a) Prove that

$$E(\mathbb{R})[m] \cong \begin{cases} \mathbb{Z}/m\mathbb{Z}, & \text{if } \Delta(E) < 0, \\ (\mathbb{Z}/2\mathbb{Z}) \times (\mathbb{Z}/m\mathbb{Z}), & \text{if } \Delta(E) > 0. \end{cases}$$

  (b) Consider the Kummer sequence

$$0 \longrightarrow E(\mathbb{R})/mE(\mathbb{R}) \longrightarrow H^1(G_{\mathbb{C}/\mathbb{R}}, E[m]) \longrightarrow \mathrm{WC}(E/\mathbb{R}) \longrightarrow 0.$$

If $\Delta(E) < 0$, prove that all three terms are 0. If $\Delta(E) > 0$, prove that the sequence is

$$0 \longrightarrow \mathbb{Z}/2\mathbb{Z} \longrightarrow \mathbb{Z}/2\mathbb{Z} \times \mathbb{Z}/2\mathbb{Z} \longrightarrow \mathbb{Z}/2\mathbb{Z} \longrightarrow 0.$$

5.4.   (a) Let $a, b \in \mathbb{R}$, and let $E/\mathbb{R}$ be the elliptic curve

$$E : y^2 = x^3 + ax^2 + bx.$$

If $\Delta(E) > 0$, so in particular if $b < 0$, prove that $\mathrm{WC}(E/\mathbb{R})$ has order 2, and its non-trivial element is represented by the homogeneous space

$$C : w^2 = 4bz^2 - (1 + az^2)^2.$$

(*Hint.* See [AEC X.3.7].)
(b) Let $E/\mathbb{R}$ be an elliptic curve with $\Delta(E) > 0$. Exercise 5.3(a) says that $E[2] \subset E(\mathbb{R})$, so we can factor

$$4x^3 + b_2 x^2 + 2b_4 x + b_6 = 4(x - e_1)(x - e_2)(x - e_3)$$

$$\text{with } e_1 < e_2 < e_3.$$

Prove that the non-trivial element of $\mathrm{WC}(E/\mathbb{R})$ is represented by the homogeneous space

$$C : -w^2 = 1 + 2(2e_2 - e_1 - e_3)z^2 + (e_1 - e_3)^2 z^4.$$

5.5.   Let $E/\mathbb{R}$ be an elliptic curve, and choose $q \in \mathbb{R}$, $0 < |q| < 1$, so that $j(E) = j(E_q)$. Suppose, however, that E is *not* $\mathbb{R}$-isomorphic to E_q. Then if we consider the isomorphisms

$$\mathbb{C}^*/q^{\mathbb{Z}} \xrightarrow{\sim} E_q(\mathbb{C}) \xrightarrow{\sim} E(\mathbb{C}),$$

the second map will not be defined over $\mathbb{R}$. Prove that with this identification,

$$E(\mathbb{R}) \cong \{u \in \mathbb{C}^*/q^{\mathbb{Z}} : |u|^2 \in q^{\mathbb{Z}}/q^{2\mathbb{Z}}\}.$$

5.6.   For this problem we will write $E\langle\tau\rangle$ for E_q with $q = e^{2\pi i \tau}$, and for a given $E/\mathbb{R}$ we write E^χ for the non-trivial twist of E. (See Proposition 2.2(a).)
(a) Prove that

$$E\langle it\rangle^\chi \cong_{/\mathbb{R}} E\left\langle \frac{i}{t} \right\rangle, \qquad \text{for all } t > 0,\, t \neq 1,$$

$$E\left\langle \frac{1}{2} + \frac{it}{2} \right\rangle^\chi \cong_{/\mathbb{R}} E\left\langle \frac{1}{2} + \frac{i}{2t} \right\rangle, \qquad \text{for all } t > 0,\, t \neq 1,$$

$$E\langle i\rangle^\chi \cong_{/\mathbb{R}} E\left\langle \frac{1}{2} + \frac{i}{2} \right\rangle.$$

(b) Fix $E/\mathbb{R}$ with $j(E) \neq 1728$. From (2.3) there is a unique $t > 0$ so that

$$E \cong_{/\mathbb{R}} \begin{cases} E\langle it\rangle, & \text{if } \Delta(E) > 0, \\ E\left\langle \dfrac{1}{2} + \dfrac{it}{2} \right\rangle, & \text{if } \Delta(E) < 0. \end{cases}$$

Let $\gamma(E/\mathbb{R}) = \mathrm{sign}\, c_6(E)$ be the invariant defined in (2.2b). Prove that

$$\gamma(E/\mathbb{R}) = \mathrm{sign}(1 - t).$$

5.7. Let $K/\mathbb{Q}_p$ be a finite extension with ring of integers R and normalized valuation ord_v. Let $q \in K^*$ satisfy $|q| < 1$, let E_q be the corresponding Tate curve, and let $\phi : K^* \to E_q(K)$ be the homomorphism described in (3.1). Prove that for every $r \geq 1$, ϕ induces an isomorphism

$$\phi : R_r^* \longrightarrow E_{q,r}(K),$$

where

$$R_r^* = \{u \in K^* : \text{ord}_v(u-1) \geq r\},$$
$$E_{q,r}(K) = \{(x,y) \in E_q(K) : \text{ord}_v(x) \leq -2r\} \cup \{O\}.$$

5.8. (a) Let $L/K/\mathbb{Q}_p$ be a finite tower of fields with L Galois over K, let $\alpha_i \in L$ be a sequence of elements such that the series $\sum \alpha_i$ converges, and let $\sigma \in G_{L/K}$. Prove that

$$\left(\sum_{i=1}^{\infty} \alpha_i\right)^{\sigma} = \sum_{i=1}^{\infty} \alpha_i^{\sigma}.$$

(b) Show that (a) is true if we replace $\mathbb{Q}_p$ by $\mathbb{R}$ and take $K = \mathbb{R}$ and $L = \mathbb{C}$.

(c) Find a sequence of elements $\alpha_i \in \mathbb{Q}$ such that $\sum \alpha_i = \sqrt{2}$, and deduce that there is an element $\sigma \in G_{\bar{\mathbb{Q}}/\mathbb{Q}}$ such that

$$\left(\sum_{i=1}^{\infty} \alpha_i\right)^{\sigma} \neq \sum_{i=1}^{\infty} \alpha_i^{\sigma}.$$

Thus (a) is not true if we replace $\mathbb{Q}_p$ by $\mathbb{Q}$.

5.9. Fill in the details needed to rigorously prove the formulas (i) and (ii) in Proposition 3.2(b).

5.10. Let K be a p-adic field, let $q, q' \in K^*$ satisfy $|q| < 1$ and $|q'| < 1$, and let E_q and $E_{q'}$ be the corresponding Tate curves.

(a) If E_q and $E_{q'}$ are isogenous, prove that there are positive integers m, n such that $q^m = q'^n$.

(b) *Conversely, if $q^m = q'^n$ for some integers $m, n \geq 1$, prove that E_q and $E_{q'}$ are isogenous. (Clearly, there are homomorphisms from $\bar{K}^*/q^{\mathbb{Z}}$ to $\bar{K}^*/q'^{\mathbb{Z}}$, for example $u \to u^m$. What is unclear is that the corresponding homomorphisms $E_q \to E_{q'}$ are given by rational functions, rather than by power series.)

5.11. Let K be a p-adic field, and let E/K be an elliptic curve such that $|j(E)| > 1$. Let $\gamma(E/K)$ be the invariant described in (5.2), and consider the field $L = K\left(\sqrt{\gamma(E/K)}\right)$.

(a) Prove E/K has split multiplicative reduction if and only if $L = K$.

(b) Prove E/K has non-split multiplicative reduction if and only if L/K is unramified of degree 2.

(c) Prove E/K has additive reduction if and only if L/K is ramified of degree 2.

(d) Let $E_0(K)$ be the group of points whose reduction is non-singular. Describe the quotient group $E(K)/E_0(K)$ in each of the three cases (a), (b), and (c).

5.12. (a) Prove that Lemma 5.2 is still true if K has characteristic 3.

(b) Show that if K has characteristic 2 and E/K is an elliptic curve with $j(E) \neq 0$, then $\gamma(E/K)$ is always equal to 1. Hence Lemma 5.2 is not true in characteristic 2.

5.13. Let K be a p-adic field, and let E/K be an elliptic curve with split multiplicative reduction.

(a) Prove that for each prime $\ell \neq p$ there is an exact sequence of $G_{\bar{K}/K}$-modules

$$1 \longrightarrow T_\ell(\boldsymbol{\mu}) \longrightarrow T_\ell(E) \longrightarrow \mathbb{Z}_\ell \longrightarrow 0,$$

where $T_\ell(\boldsymbol{\mu})$ is the Tate module of K (see [AEC, III.7.3]) and $G_{\bar{K}/K}$ acts trivially on $\mathbb{Z}_\ell$.

(b) Prove that there is a basis for $T_\ell(E)$ so that the image of the inertia group of $\bar{K}/K$ in $\mathrm{Aut}(T_\ell(E)) \cong \mathrm{GL}_2(\mathbb{Z}_\ell)$ is equal to

$$\left\{ \begin{pmatrix} 1 & b \\ 0 & 1 \end{pmatrix} \in \mathrm{GL}_2(\mathbb{Z}_\ell) \, : \, \mathrm{ord}_\ell(b) \geq \mathrm{ord}_\ell(v_K(j_E)) \right\}.$$

(c) Prove that the exact sequence in (a), considered as a sequence of $G_{\bar{K}/K}$-modules, does not split.

5.14. Let E_q/K be a Tate curve over a p-adic field K.

(a) Prove that there is an exact sequence

$$0 \longrightarrow \mathrm{WC}(E_q/K) \longrightarrow H^2\left(G_{\bar{K}/K}, q^\mathbb{Z}\right) \longrightarrow H^2\left(G_{\bar{K}/K}, \bar{K}^*\right).$$

(b) Prove that

$$H^2\left(G_{\bar{K}/K}, q^\mathbb{Z}\right) \cong \mathrm{Hom}\left(G_{\bar{K}/K}, \mathbb{Q}/\mathbb{Z}\right) \cong \mathrm{Hom}\left(\bar{K}^*, \mathbb{Q}/\mathbb{Z}\right).$$

(*Hint.* For the second isomorphism, use local class field theory.)

(c) It is well known from local class field theory Serre [4] that there is an isomorphism $H^2(G_{\bar{K}/K}, \bar{K}^*) = \mathrm{Br}(K) \cong \mathbb{Q}/\mathbb{Z}$. Prove that the map

$$\mathrm{Hom}\left(\bar{K}^*, \mathbb{Q}/\mathbb{Z}\right) \cong H^2\left(G_{\bar{K}/K}, q^\mathbb{Z}\right) \longrightarrow H^2\left(G_{\bar{K}/K}, \bar{K}^*\right) \cong \mathbb{Q}/\mathbb{Z}$$

obtained by composing this isomorphism with the maps from (a) and (b) is given by the rule $f \longmapsto f(q)$.

(d) Deduce that

$$WC(E_q/K) \cong \mathrm{Hom}\left(E_q(K), \mathbb{Q}/\mathbb{Z}\right).$$

In other words, the Weil-Châtelet group of E_q/K is dual to the group of rational points $E_q(K)$.

(e) *More generally, prove that $WC(E/K) \cong \mathrm{Hom}(E(K), \mathbb{Q}/\mathbb{Z})$ for any elliptic curve E/K satisfying $|j(E)| > 1$.

(In fact, it is true that $WC(E/K) \cong \mathrm{Hom}(E(K), \mathbb{Q}/\mathbb{Z})$ for all elliptic curves over p-adic fields, not just those with non-integral j-invariant. The proof of this result, which is due to Tate [5,6], requires different methods than those used in this chapter. See also Milne [1])

5.15. Let $K/\mathbb{Q}_p$ be a p-adic field, let E/K be an elliptic curve, let $N \geq 5$ be a prime not equal to p, and suppose that there is a point $P \in E(K)$ of exact order N.

(a) Prove that E has either good or multiplicative reduction.

(b) Let $E \to E'$ be an isogeny of elliptic curves whose kernel is the cyclic subgroup generated by P (i.e., $E' = E/\mathbb{Z}P$). Prove that

$$v_K(\mathcal{D}_{E/K}) + v_K(\mathcal{D}_{E'/K}) \equiv 0 \pmod{N+1}.$$

Here $v_K : K^* \to \mathbb{Z}$ is the normalized valuation on K, and $\mathcal{D}_{E/K}$ and $\mathcal{D}_{E'/K}$ are the minimal discriminants of E/K and E'/K respectively. (*Hint.* Take Tate models E_q and $E_{q'}$ and look at the isogeny $E_q \to E_{q'}$.)

5.16. Let E/K be an elliptic curve defined over a number field, let N be a prime, and suppose that there is a point $P \in E(K)$ of exact order N. Use the previous exercise and Szpiro's conjecture (IV.10.6) to prove that N is bounded by a constant that depends only on the field K. This approach to proving the boundedness conjecture [AEC, VIII.7.7] is due to Frey.

CHAPTER VI

Local Height Functions

The canonical height function

$$\hat{h} : E(\bar{K}) \longrightarrow [0, \infty)$$

is a quadratic form whose value at a point P measures the arithmetic complexity of P. The importance of the canonical height stems from the fact that it relates the geometrically defined group law to the arithmetic properties of the algebraic points on E. See [AEC VIII, §9] for details.

Recall that the ordinary height of a non-zero point $P \in E(K)$ (relative to the function x) is defined as a sum of local terms, one for each absolute value. Thus

$$h(P) = \frac{1}{2[K:\mathbb{Q}]} \sum_{v \in M_K} n_v \max\{-v(x(P)), 0\}.$$

The canonical height is then the limiting value of these ordinary heights,

$$\hat{h}(P) = \lim_{n \to \infty} \frac{1}{n^2} h([n]P).$$

It is natural to ask whether the canonical height itself can be naturally decomposed as a sum of quadratic forms, one for each absolute value in K. The answer is "no," but Néron and Tate have shown that there is a decomposition into local functions which are almost quadratic. Precisely, they show that for each $v \in M_K$ there is an almost quadratic function

$$\lambda_v : E(K_v) \smallsetminus \{O\} \longrightarrow \mathbb{R}$$

such that

$$\hat{h}(P) = \frac{1}{[K:\mathbb{Q}]} \sum_{v \in M_K} n_v \lambda_v(P) \qquad \text{for all } P \in E(K) \smallsetminus \{O\}.$$

In §1 of this chapter we will use an averaging argument to prove the existence of local height functions for all absolute values $v \in M_K$, and in §2 we will prove that the canonical height is equal to the sum of the local heights. It is also of interest to have explicit formulas for the local height functions, so we will give such formulas for archimedean absolute values in §3 and for non-archimedean absolute values in §4. For further information about local height functions, see for example Lang [3] and Zimmer [2].

§1. Existence of Local Height Functions

Let K be a field and let $|\cdot|_v$ be an absolute value on K. The absolute value can be used to define a topology on K in the usual way: a basis of open neighborhoods around an element $\alpha \in K$ is the collection of (open) balls

$$U_\varepsilon = \{\beta \in K : |\beta - \alpha|_v < \varepsilon\}, \qquad \text{all } \varepsilon > 0.$$

Let E/K be an elliptic curve. In a similar way we can define a topology on $E(K)$.

Definition. The *v-adic topology on $E(K)$* is defined as follows. For a point $P_0 = (x_0, y_0) \in E(K)$, a basis of open neighborhoods of P_0 consists of the sets

$$U_\varepsilon = \{(x, y) \in E(K) : |x - x_0|_v < \varepsilon \text{ and } |y - y_0| < \varepsilon\}, \qquad \text{all } \varepsilon > 0.$$

For the point $O \in E(K)$ at infinity we take as a basis the open neighborhoods

$$U_\varepsilon = \{(x, y) \in E(K) : |x|_v > \varepsilon^{-1}\} \cup \{O\}, \qquad \text{all } \varepsilon > 0.$$

(Notice that for the neighborhoods of O there is no need to require both $|x|_v$ and $|y|_v$ to be large. The Weierstrass equation ensures that they simultaneously go to ∞. For an alternative definition of the v-adic topology, see exercise 6.1.)

In this section we will prove the existence of almost quadratic local height functions. These will be certain continuous functions

$$E(K) \smallsetminus \{O\} \longrightarrow \mathbb{R},$$

where $E(K) \smallsetminus \{O\}$ is given the v-adic topology induced from $E(K)$, and $\mathbb{R}$ is given its usual topology. The following formulation is due to Tate.

Theorem 1.1. (Néron, Tate) *Let K be a field which is complete with respect to an absolute value $|\cdot|_v$, and let*

$$v(\cdot) = -\log|\cdot|_v$$

be the corresponding additive absolute value. Let E/K be an elliptic curve. Choose a Weierstrass equation for E/K,

$$E : y^2 + a_1 xy + a_3 y = x^3 + a_2 x^2 + a_4 x + a_6,$$

and let Δ be the discriminant of this equation.
(a) There exists a unique function

$$\lambda : E(K) \smallsetminus \{O\} \longrightarrow \mathbb{R}$$

with the following three properties:

(i) λ is continuous on $E(K) \smallsetminus \{O\}$ and is bounded on the complement of any v-adic neighborhood of O.

(ii) The limit
$$\lim_{P \xrightarrow{v-\text{adic}} O} \left\{ \lambda(P) + \tfrac{1}{2}v\big(x(P)\big) \right\}$$

exists.

(iii) For all $P \in E(K)$ with $[2]P \neq O$,

$$\lambda([2]P) = 4\lambda(P) + v\big((2y + a_1 x + a_3)(P)\big) - \tfrac{1}{4}v(\Delta).$$

(b) λ is independent of the choice of Weierstrass equation for E/K.

(c) Let L/K be a finite extension and w the extension of v to L. Then (with the obvious notation)

$$\lambda_w(P) = \lambda_v(P) \qquad \text{for all } P \in E(K) \smallsetminus \{O\}.$$

Definition. The function λ described in Theorem 1.1 is called the *(local) Néron height function on E associated to v*.

Remark 1.1.1. For other properties of λ which are equivalent to (iii), see exercises 6.3 and 6.4.

PROOF (of Theorem 1.1). (a) We begin with uniqueness. Let

$$\lambda, \lambda' : E(K) \smallsetminus \{O\} \longrightarrow \mathbb{R}$$

be two functions satisfying (i), (ii), and(iii); and let $\Lambda = \lambda - \lambda'$ be their difference. From (ii) we see that the limit

$$\lim_{P \to O} \Lambda(P)$$

exists; so if we define $\Lambda(O)$ to be this limiting value, then (i) implies that

$$\Lambda : E(K) \longrightarrow \mathbb{R}$$

is a continuous bounded function on all of $E(K)$.

Next we observe from (iii) that $\Lambda([2]P) = 4\Lambda(P)$ provided $[2]P \neq O$. But the points satisfying $[2]P = O$ form a discrete subset of $E(K)$ (notice there are at most four such points), so by continuity $\Lambda([2]P) = 4\Lambda(P)$ holds for all P. Iterating this relation N times and dividing by 4^N gives

$$\Lambda(P) = \frac{1}{4^N}\Lambda(2^N P), \quad \text{valid for all } P \in E(K) \text{ and all } N \geq 1.$$

Since Λ is bounded, we may let $N \to \infty$ to deduce that $\Lambda(P) = 0$. Hence $\lambda = \lambda'$, which proves uniqueness.

Before proving the existence of λ, which is more complicated, we will prove (b) and (c).

(b) It is clear that conditions (i) and (ii) are independent of the choice of Weierstrass equation. Since the quantity

$$\frac{(2y + a_1 x + a_3)^4}{\Delta}$$

is invariant under change of coordinates (see [AEC, III §1]), we see that (iii) is likewise independent. Hence λ, if it exists at all, does not depend on the Weierstrass equation.

(c) Since λ_w satisfies conditions (i), (ii), and (iii) for $E(L)$ and w, and since w restricted to K equals v, we see that λ_w satisfies (i), (ii), and (iii) for $E(K)$ and v. By the uniqueness already proven, the restriction of λ_w to $E(K)$ equals λ_v.

We turn now to the proof of existence. Property (ii) says that λ should look like $\frac{1}{2} v \circ x^{-1}$, at least close to O. Of course, this is no good for points with $x(P)$ close to 0, since away from O, λ is supposed to be bounded. So as a first guess for λ we might try the function

$$\lambda_1(P) = \frac{1}{2} \max\{v(x(P)^{-1}), 0\}.$$

It turns out that λ_1 almost satisfies property (iii). We first need to make this precise, and then we will modify λ_1 to produce λ.

Let

$$\phi(x) = x^4 - b_4 x^2 - 2b_6 x - b_8,$$

$$\psi(x) = 4x^3 + b_2 x^2 + 2b_4 x + b_6 = (2y + a_1 x + a_3)^2$$

be the usual functions on E, so the duplication formula [AEC III.2.3(d)] reads

$$x(2P) = \frac{\phi(P)}{\psi(P)}.$$

Define a function

$$f(P) = \lambda_1([2]P) - 4\lambda_1(P) - \frac{1}{2} v(\psi(P)) + \frac{1}{4} v(\Delta)$$

$$\text{for all } P \in E(K) \text{ with } [2]P \neq O.$$

Notice that if λ_1 were to satisfy (iii), then f would be identically 0. We are going to show that f is bounded.

Using the definition of λ_1 and the duplication formula, we can rewrite f as

$$f(P) = \frac{1}{2} \log \left(\frac{\max\{|\phi(P)|_v, |\psi(P)|_v\}}{\max\{|x(P)|_v^4, 1\}} \right) + \frac{1}{4} v(\Delta).$$

(Recall that $v(t) = -\log|t|_v$.) A priori, f is defined at all points $P \in E(K)$ with $[2]P \neq O$. The following crucial result shows that more is true.

Lemma 1.2. *With notation as above, f extends to a bounded continuous function on all of $E(K)$.*

PROOF. Let

$$F(P) = \frac{\max\{|\phi(P)|_v, |\psi(P)|_v\}}{\max\{|x(P)|_v^4, 1\}},$$

so $f = \frac{1}{2}\log(F) + \frac{1}{4}v(\Delta)$. Clearly, F is continuous on $E(K) \smallsetminus \{O\}$. Further, as $P \to O$, we find

$$\lim_{P \to O} F(P) = \lim_{|x(P)|_v \to \infty} \frac{\max\{|\phi(P)|_v, |\psi(P)|_v\}}{|x(P)|_v^4} = 1.$$

Hence F extends to a continuous, bounded function on $E(K)$.

Since the limit equals 1, we also see that there is a constant $c_1 > 0$, depending on the chosen Weierstrass equation, such that

$$|x(P)|_v \geq c_1 \implies F(P) \geq \frac{1}{2}.$$

So in order to prove that $\log(F)$ is continuous and bounded on $E(K)$, we are reduced to showing that F is bounded away from 0 on the set $|x|_v \leq c_1$. From the definition of F, we must show that there is a constant $c_2 > 0$ so that

$$\max\{|\phi(P)|_v, |\psi(P)|_v\} \geq c_2 \qquad \text{for all } P \in E(K) \text{ with } |x(P)|_v \leq c_1.$$

The polynomials $\phi(x)$ and $\psi(x)$ are relatively prime in $K[x]$. We give two quick proofs of this fact.

Proof 1 (theoretical): The map $[2] : E \to E$ has degree 4 [AEC III.6.2(d)]. From the commutative diagram

$$
\begin{array}{ccc}
E & \xrightarrow{\;[2]\;} & E \\
\downarrow{\scriptstyle x} & & \downarrow{\scriptstyle x} \\
\mathbb{P}^1 & \xrightarrow{\;\phi/\psi\;} & \mathbb{P}^1
\end{array}
$$

the rational map ϕ/ψ has degree 4, so $\phi(x)$ and $\psi(x)$ have no common roots.

Proof 2 (computational): An explicit computation shows that

$$\text{Resultant}(\phi(x), \psi(x)) = \Delta^2,$$

where Δ is the discriminant of the given Weierstrass equation. Since $\Delta \neq 0$, we see that $\phi(x)$ and $\psi(x)$ have no common roots.

Therefore we can find polynomials $\Phi, \Psi \in K[x]$ satisfying

$$\phi(x)\Phi(x) + \psi(x)\Psi(x) = 1.$$

Evaluating this identity at $x = x(P)$ and using the triangle inequality yields the desired result:

$$\begin{aligned}
1 &\leq \left|\phi(x)\Phi(x)\right|_v + \left|\psi(x)\Psi(x)\right|_v \\
&\leq 2\max\{\left|\phi(x)\right|_v, \left|\psi(x)\right|_v\} \cdot \max\{\left|\Phi(x)\right|_v, \left|\Psi(x)\right|_v\} \\
&\leq c_2^{-1} \max\{\left|\phi(x)\right|_v, \left|\psi(x)\right|_v\} \qquad \text{for } |x|_v \leq c_1.
\end{aligned}$$

$\square$

According to Lemma 1.2, the "naive" local height function λ_1 satisfies condition (iii) up to a bounded function. The next proposition shows how to decompose such a bounded function into a difference of two functions. Then these new functions will be used to modify λ_1 so as to make (iii) hold exactly.

Proposition 1.3. (Tate) *Let*

$$f : E(K) \longrightarrow \mathbb{R}$$

be any bounded continuous function. Then there exists a unique bounded continuous function

$$\mu : E(K) \longrightarrow \mathbb{R}$$

such that

$$f(P) = 4\mu(P) - \mu([2]P) \qquad \text{for all } P \in E(K).$$

PROOF. If the function μ exists, then we can use its defining relation N times to compute

$$\begin{aligned}
\mu(P) &= \frac{1}{4}f(P) + \frac{1}{4}\mu([2]P) \\
&= \frac{1}{4}f(P) + \frac{1}{16}f([2]P) + \frac{1}{16}\mu([4]P) \\
&\ \ \vdots \\
&= \sum_{n=0}^{N-1} \frac{1}{4^{n+1}} f([2^n]P) + \frac{1}{4^N}\mu([2^N]P).
\end{aligned}$$

Since μ is supposed to be bounded, if we let $N \to \infty$, then the last term should disappear. So we *define* μ by the formula

$$\mu(P) = \sum_{n=0}^{\infty} \frac{1}{4^{n+1}} f([2^n]P)$$

and verify that it has the required properties. (Notice that if μ exists, it must be given by this formula, so we get uniqueness for free.)

First, since f is a bounded function, it is clear that the series is absolutely convergent, so μ is well-defined. But more is true. Each of the functions

$$f \circ [2^n] : E(K) \longrightarrow \mathbb{R}$$

is bounded and continuous. (It is easy to check that the multiplication maps $[m] : E(K) \to E(K)$ are continuous for the v-adic topology.) It follows that the series defining μ gives a bounded continuous function on $E(K)$. Finally, using Tate's telescoping series trick,

$$4\mu(P) - \mu([2]P) = \sum_{n=0}^{\infty} \frac{1}{4^n} f([2^n]P) - \sum_{n=0}^{\infty} \frac{1}{4^{n+1}} f([2^{n+1}]P) = f(P).$$

$\square$

We now have all the tools needed to prove the existence of λ and so complete the proof of Theorem 1.1. As above, let

$$\lambda_1(P) = \frac{1}{2} \max\{v(x(P)^{-1}), 0\},$$

$$f(P) = \lambda_1([2]P) - 4\lambda_1(P) - v((2y + a_1 x + a_3)(P)) + \frac{1}{4} v(\Delta).$$

From Lemma 1.2, f extends to a bounded continuous function (also denoted f) on all of $E(K)$. Then Proposition 1.3 gives a bounded continuous function $\mu : E(K) \to \mathbb{R}$ satisfying

$$f(P) = 4\mu(P) - \mu([2]P) \qquad \text{for all } P \in E(K).$$

Define

$$\lambda(P) = \lambda_1(P) + \mu(P).$$

We now verify that λ satisfies properties (i), (ii), and (iii) of Theorem 1.1.

(i) By inspection, λ_1 is continuous on $E(K) \smallsetminus \{O\}$ and is bounded on the complement of any v-adic neighborhood of O. Since μ is continuous and bounded on all of $E(K)$, λ satisfies (i).

(ii) If P is v-adically close to O, then $\lambda_1(P)$ equals $-\frac{1}{2}v(x(P))$. We compute

$$\lim_{P \to O}\left\{\lambda(P) + \frac{1}{2}v(x(P))\right\} = \lim_{P \to O}\left\{\lambda_1(P) + \mu(P) + \frac{1}{2}v(x(P))\right\}$$

$$= \lim_{P \to O} \mu(P) = \mu(O).$$

Hence λ satisfies (ii).

(iii) Using the formulas defining and relating λ, λ_1, μ, and f, we find

$$
\begin{aligned}
\lambda([2]P) &= \lambda_1([2]P) + \mu([2]P) \\
&= \lambda_1([2]P) - f(P) + 4\mu(P) \\
&= 4\lambda_1(P) + v\big((2y + a_1 x + a_3)(P)\big) - \tfrac{1}{4}v(\Delta) + 4\mu(P) \\
&= 4\lambda(P) + v\big((2y + a_1 x + a_3)(P)\big) - \tfrac{1}{4}v(\Delta).
\end{aligned}
$$

This proves that λ verifies (iii) and completes the proof of Theorem 1.1.

$\square$

§2. Local Decomposition of the Canonical Height

The canonical height [AEC VIII §9]

$$
\hat{h} : E(K) \longrightarrow [0, \infty)
$$

is a quadratic form defined in terms of the arithmetic of $E(K)$. We now show that $\hat{h}$ can be decomposed as a sum of local height functions.

Theorem 2.1. *Let K be a number field, M_K the standard set of absolute values on K, and $n_v = [K_v : \mathbb{Q}_v]$ the local degree of $v \in M_K$. (See [AEC VIII §5] for a description of M_K.) Let E/K be an elliptic curve, and for each $v \in M_K$, let $\lambda_v : E(K_v) \smallsetminus \{O\} \to \mathbb{R}$ be the local Néron height function associated to v as described in (1.1). Then*

$$
\hat{h}(P) = \frac{1}{[K : \mathbb{Q}]} \sum_{v \in M_K} n_v \lambda_v(P) \qquad \text{for all } P \in E(K) \smallsetminus \{O\}.
$$

In order to prove Theorem 2.1, we will use the defining properties of λ (especially Theorem 1.1(iii)), together with the following fact, which we will prove later.

Lemma 2.2. *There is a finite set of absolute values $S \subset M_K$ so that for all $v \notin S$,*

$$
\lambda_v(P) = \frac{1}{2} \max\{v(x(P)^{-1}), 0\} \qquad \text{for all } P \in E(K_v) \smallsetminus \{O\}.
$$

PROOF. This lemma says that for almost all absolute values, the naive local height $\frac{1}{2}\max\{v(x^{-1}), 0\}$ actually satisfies the quadratic property (iii) of Theorem 1.1. We will postpone the proof of (2.2) until §4, where we will prove the more precise result (4.1) that $\lambda_v = \frac{1}{2}\max\{v(x^{-1}),0\}$ for all finite places v such that the given Weierstrass equation has good reduction. □

PROOF (of Theorem 2.1). Let S be the set described in (2.2). Define a function

$$L : E(K) \smallsetminus \{O\} \longrightarrow \mathbb{R}, \quad L(P) = \frac{1}{[K:\mathbb{Q}]} \sum_{v \in M_K} n_v \lambda_v(P).$$

For any given $P \in E(K) \smallsetminus \{O\}$, (2.2) implies that

$$\lambda_v(P) = 0 \qquad \text{if } v \notin S \text{ and } v(x(P)) \geq 0.$$

Hence the sum $\sum n_v \lambda_v(P)$ has only finitely many non-zero terms, so $L(P)$ is well-defined.

Next we compare $L(P)$ with $h(x(P))$. From Theorem 1.1(i),(ii), for each $v \in M_K$ there is a constant c_v so that

$$-c_v \leq \lambda_v(P) - \frac{1}{2}\max\{v(x(P)^{-1}),0\} \leq c_v \qquad \text{for all } P \in E(K_v) \smallsetminus \{O\},$$

and (2.2) allows us to take $c_v = 0$ for all $v \notin S$. Now multiply by n_v, sum over $v \in M_K$, and divide by $[K:\mathbb{Q}]$. This gives

$$-c \leq L(P) - \frac{1}{2}h(x(P)) \leq c \qquad \text{for all } P \in E(K) \smallsetminus \{O\},$$

where

$$c = \frac{1}{[K:\mathbb{Q}]} \sum_{v \in S} n_v c_v$$

is finite and independent of P. In other words, if we set $L(O) = 0$, then

$$L(P) = \frac{1}{2}h(x(P)) + O(1) \qquad \text{for all } P \in E(K).$$

Finally, we verify the quadratic nature of L. Let $P \in E(K)$ be a point with $[2]P \neq O$. Then

$$L([2]P) = \frac{1}{[K:\mathbb{Q}]} \sum_{v \in M_K} n_v \lambda_v([2]P)$$

$$= \frac{1}{[K:\mathbb{Q}]} \sum_{v \in M_K} n_v \Big\{ 4\lambda_v(P) + v\big((2y + a_1 x + a_3)(P)\big) + v(\Delta) \Big\}$$

$$\text{by Theorem 1.1(iii)}$$

$$= \frac{4}{[K:\mathbb{Q}]} \sum_{v \in M_K} n_v \lambda_v(P) \quad \text{product formula [AEC VIII.5.3]}$$

$$= 4L(P).$$

Since we have defined $L(O) = 0$, the relation $L([2]P) = 4L(P)$ holds for $P = O$, too. We must also verify it for $P \in E[2]$, $P \neq O$. The quickest way to do this is to use the triplication formula for λ (exercise 6.4e),

$$\lambda_v([3]Q) = 9\lambda_v(Q) + v\big((3x^4 + b_2x^3 + 3b_4x^2 + 3b_6x + b_8)(Q)\big) - \frac{2}{3}v(\Delta)$$
$$\text{for all } Q \in E(K_v) \text{ with } [3]Q \neq O.$$

Summing over $v \in M_K$ as above gives

$$L([3]Q) = 9L(Q) \qquad \text{for all } Q \in E(K) \text{ with } [3]Q \neq O.$$

In particular, if $P \neq O$ and $[2]P = O$, then

$$L(P) = L([3]P) = 9L(P), \quad \text{so } L(P) = 0.$$

Hence

$$L([2]P) = L(O) = 0 = L(P) = 4L(P).$$

We have now proven the two relations

$$L(P) = \frac{1}{2}h\big(x(P)\big) + O(1) \quad \text{and} \quad L([2]P) = 4L(P) \quad \text{for all } P \in E(K).$$

The canonical height $\hat{h}$ also satisfies these relations [AEC VIII.9.3]. It follows that the difference $F = L - \hat{h}$ is bounded and satisfies $F([2]P) = 4F(P)$. Hence

$$F(P) = \frac{1}{4^N}F\big([2^N]P\big) \xrightarrow[N \to \infty]{} 0, \qquad \text{so } F(P) = 0 \text{ for all } P \in E(K).$$

Therefore $L = \hat{h}$. $\qquad\qquad\qquad\qquad\qquad\qquad\qquad\qquad\qquad\qquad\qquad$ □

§3. Archimedean Absolute Values — Explicit Formulas

Let K be a field which is complete with respect to an archimedean absolute value $|\cdot|_v$, and let E/K be an elliptic curve. Then K is isomorphic to either $\mathbb{R}$ or $\mathbb{C}$, and $|\cdot|_v$ corresponds to some power of the usual absolute value. Thus in order to compute the local height function λ over $E(K)$, it suffices to consider the case that $K = \mathbb{C}$, so in this section we will derive explicit formulas for the local height for elliptic curves over the complex numbers.

Recall that an elliptic curve $E/\mathbb{C}$ has an analytic parametrization

$$\mathbb{C}/\Lambda \longrightarrow E(\mathbb{C}), \qquad z \longmapsto \left(\wp(z;\Lambda), \wp'(z;\Lambda)\right),$$

with Weierstrass equation

$$E : (\wp')^2 = 4\wp^3 - g_2(\Lambda)\wp - g_3(\Lambda)$$

having discriminant

$$\Delta(\Lambda) = g_2(\Lambda)^3 - 27g_3(\Lambda)^2.$$

(See [AEC VI.3.6] and (I.4.4).) We put this in standard Weierstrass form by the substitution

$$x = \wp(z), \qquad y = \frac{1}{2}\wp'(z),$$

yielding the equation

$$E : y^2 = x^3 - \frac{1}{4}g_2(\Lambda)x - \frac{1}{4}g_3(\Lambda) \qquad \text{with discriminant } \Delta = \Delta(\Lambda).$$

Recall also the Weierstrass σ-function (I.5.4)

$$\sigma(z) = \sigma(z;\Lambda) = \prod_{\substack{\omega \in \Lambda \\ \omega \neq 0}} \left(1 - \frac{z}{\omega}\right) e^{\frac{z}{\omega} + \frac{1}{2}\left(\frac{z}{\omega}\right)^2},$$

which has a simple zero at each lattice point and satisfies the transformation formula

$$\sigma(z + \omega) = \psi(\omega)e^{\eta(\omega)\left(z + \frac{1}{2}\omega\right)}\sigma(z) \qquad \text{for all } z \in \mathbb{C}, \omega \in \Lambda.$$

Here $\psi : \Lambda \to \{\pm 1\}$ is the map with $\psi(\omega) = 1$ if and only if $\omega \in 2\Lambda$, and $\eta : \Lambda \to \mathbb{C}$ is the quasi-period homomorphism.

We have seen (I.5.6b) that there is a factorization

$$\wp'(z) = -\frac{\sigma(2z)}{\sigma(z)^4}.$$

Applying $\log|\cdot|$ yields

$$\log|\sigma(2z)| = 4\log|\sigma(z)| + \log|\wp'(z)|.$$

Since $\wp'(z) = 2y$, comparison of this equation with Theorem 1.1(iii) (and the fact that the local height has a pole at O) suggests that the local height function on $E(\mathbb{C}) = \mathbb{C}/\Lambda$ should look like $-\log|\sigma(z)|$. Unfortunately, the transformation formula for $\sigma(z)$ shows that $|\sigma(z)|$ is not invariant under translation by Λ, so $-\log|\sigma(z)|$ is not well-defined on $\mathbb{C}/\Lambda$. The next proposition explains how to modify $\sigma(z)$ to obtain an invariant function.

Proposition 3.1. *Let $\Lambda \subset \mathbb{C}$ be a lattice. Extend the quasi-period map $\eta : \Lambda \to \mathbb{C}$ linearly to obtain an $\mathbb{R}$-linear homomorphism (also denoted η)*

$$\eta : \mathbb{C} \cong \Lambda \otimes_{\mathbb{Z}} \mathbb{R} \longrightarrow \mathbb{C}.$$

(a) *For all $z, w \in \mathbb{C}$, the quantity*

$$z\eta(w) - w\eta(z)$$

is purely imaginary.
(b) *Define a function*

$$F(z) = e^{-\frac{1}{2}z\eta(z)}\sigma(z).$$

Then

$$F(z + \omega) = \psi(\omega)e^{\frac{1}{2}(z\eta(\omega) - \omega\eta(z))}F(z) \qquad \text{for all } z \in \mathbb{C}, \, \omega \in \Lambda.$$

(Note that $F(z)$ is not holomorphic, because $\eta(z)$ is only $\mathbb{R}$-linear.)
(c) *The function $|F(z)|$ is a well-defined function on $\mathbb{C}/\Lambda$ and is real-analytic and non-vanishing away from 0.*

PROOF. (a) Choose a basis ω_1, ω_2 for Λ with $\text{Im}(\omega_1/\omega_2) > 0$. Legendre's relation (I.5.2d) says that

$$\omega_1\eta(\omega_2) - \omega_2\eta(\omega_1) = 2\pi i.$$

Write

$$z = a\omega_1 + b\omega_2, \quad w = c\omega_1 + d\omega_2, \qquad \text{with } a, b, c, d \in \mathbb{R}.$$

Then the $\mathbb{R}$-linearity of η and Legendre's relation give

$$z\eta(w) - w\eta(z) = (ad - bc)\big(\omega_1\eta(\omega_2) - \omega_2\eta(\omega_1)\big)$$
$$= (ad - bc)2\pi i.$$

(b) Using the transformation formula (I.5.4) for $\sigma(z)$ stated above, we compute

$$F(z + \omega) = e^{-\frac{1}{2}(z+\omega)\eta(z+\omega)}\sigma(z + \omega)$$
$$= e^{-\frac{1}{2}(z+\omega)\eta(z+\omega)}\psi(\omega)e^{\eta(\omega)(z+\frac{1}{2}\omega)}\sigma(z)$$
$$= \psi(\omega)e^{\frac{1}{2}(z\eta(\omega) - \omega\eta(z))}F(z).$$

(c) Since $\psi(\omega) = \pm 1$, and since (a) implies that

$$\left| e^{\frac{1}{2}(z\eta(\omega) - \omega\eta(z))} \right| = 1,$$

it follows from (b) that $|F(z)|$ is well-defined on $\mathbb{C}/\Lambda$. Further, $F(z)$ is clearly real-analytic on $\mathbb{C}$ and vanishes only at points of Λ, so $|F(z)|$ is real-analytic and non-vanishing on $\mathbb{C}\backslash\Lambda$. $\qquad\qquad\qquad\qquad\qquad\square$

Theorem 3.2. *Let $E/\mathbb{C}$ be an elliptic curve with period lattice Λ. Then the Néron local height function*

$$\lambda : E(\mathbb{C}) \smallsetminus \{O\} \longrightarrow \mathbb{R}$$

is given by the formula

$$\lambda(z) = -\log \left| e^{-\frac{1}{2}z\eta(z)} \sigma(z)\Delta(\Lambda)^{\frac{1}{12}} \right|$$
$$= \frac{1}{2}\operatorname{Re}\big(z\eta(z)\big) - \log\big|\sigma(z)\big| - \frac{1}{12}\log\big|\Delta(\Lambda)\big|,$$

where $\eta : \mathbb{C} \to \mathbb{C}$ is the extension of the quasi-period map described above in (3.1).

PROOF. Let $\lambda(z)$ be the indicated function. We must verify that λ satisfies properties (i), (ii), and (iii) of Theorem 1.1. First, (3.1c) ensures that λ is well-defined on $E(\mathbb{C}) \smallsetminus \{O\}$. Further, since $\sigma(z)$ is holomorphic on $\mathbb{C}$ and non-vanishing on $\mathbb{C} \smallsetminus \Lambda$, it is clear that $\lambda(z)$ is actually a real-analytic function on $E(\mathbb{C}) \smallsetminus \{O\}$. Hence it satisfies property (i) of Theorem 1.1.

Next we observe that the limit

$$\lim_{z\to 0} \lambda(z) + \frac{1}{2}v\big(\wp(z)\big)$$
$$= \lim_{z\to 0} \left\{ \frac{1}{2}\operatorname{Re}\big(z\eta(z)\big) - \frac{1}{2}\log\big|\sigma(z)^2\wp(z)\big| - \frac{1}{12}\log\big|\Delta(\Lambda)\big| \right\}$$

exists, since $\sigma(z)$ has a simple zero and $\wp(z)$ has a double pole at $z = 0$. (In fact, the limit equals $\frac{1}{12}\log|\Delta(\Lambda)|$.) This verifies property (ii).

Finally, we must check property (iii). From (I.5.6b) we have

$$\log\big|\sigma(2z)\big| = 4\log\big|\sigma(z)\big| + \log\big|\wp'(z)\big|,$$

and the linearity of $\eta(z)$ gives

$$\operatorname{Re}\big(2z\eta(2z)\big) = 4\operatorname{Re}\big(z\eta(z)\big).$$

Subtracting the first equation from half the second, and then subtracting $\frac{1}{12}\log|\Delta(\Lambda)|$ from both sides, we obtain the desired relation

$$\lambda(2z) = 4\lambda(z) - \log\big|\wp'(z)\big| + \frac{1}{4}\log\big|\Delta(\Lambda)\big|.$$

(Note that $\wp'(z) = 2y$ and $\Delta(\Lambda) = \Delta$ for the Weierstrass equation $y^2 = x^3 - \frac{1}{4}g_2 x - \frac{1}{4}g_3$.)

This proves that λ has properties (i), (ii),and (iii) of Theorem 1.1, so λ is the Néron local height function on $E(\mathbb{C})$. $\square$

Remark 3.2.1. In proving Theorem 3.2, we verified directly that the function

$$- \log \left| e^{-\frac{1}{2} z \eta(z)} \sigma(z) \Delta(\Lambda)^{\frac{1}{12}} \right|$$

has properties (i), (ii), and (iii). This gives an alternative proof of existence for Theorem 1.1 in the case of archimedean absolute values.

Corollary 3.3. *The local height function*

$$\lambda : E(\mathbb{C}) \smallsetminus \{O\} \longrightarrow \mathbb{R}$$

satisfies the quasi-parallelogram law

$$\lambda(P + Q) + \lambda(P - Q) = 2\lambda(P) + 2\lambda(Q) + v\big(x(P) - x(Q)\big) - \frac{1}{6} v(\Delta)$$

$$\text{for all } P, Q \in E(\mathbb{C}) \text{ with } P, Q, P \pm Q \neq O.$$

(Note that the quantity $(x(P) - x(Q))^6 / \Delta$ is well-defined, independent of the choice of a particular Weierstrass model for E.)

PROOF. The Weierstrass $\wp$-function has the factorization (I.5.6a)

$$\wp(z) - \wp(w) = - \frac{\sigma(z + w)\sigma(z - w)}{\sigma(z)^2 \sigma(w)^2} \qquad \text{for all } z, w \in \mathbb{C}.$$

Hence

$$- \log|\sigma(z+w)| - \log|\sigma(z-w)| = -2\log|\sigma(z)| - 2\log|\sigma(w)| - \log|\wp(z) - \wp(w)|.$$

Next, the linearity of $\eta(z)$ immediately implies

$$(z + w)\eta(z + w) + (z - w)\eta(z - w) = 2z\eta(z) + 2w\eta(w).$$

Apply $\frac{1}{2}\,\mathrm{Re}(\,\cdot\,)$ to this last equation and add it to the previous one. Comparison with the formula (3.2) for $\lambda(z)$ yields

$$\lambda(z + w) + \lambda(z - w) = 2\lambda(z) + 2\lambda(w) - \log|\wp(z) - \wp(w)| + \frac{1}{6}\log|\Delta(\Lambda)|,$$

which is exactly the desired identity. $\square$

It is often convenient to use the Fourier expansions for $\sigma(z)$ and $\Delta(\tau)$ to rewrite the formula for the local height $\lambda(z)$.

Theorem 3.4. *Let $E/\mathbb{C}$ be an elliptic curve with lattice $\mathbb{Z}\tau + \mathbb{Z}$ normalized so that $\mathrm{Im}(\tau) > 0$. As usual, let*

$$u = e^{2\pi i z} \quad \text{and} \quad q = e^{2\pi i \tau},$$

and identify

$$E(\mathbb{C}) \;\cong\; \mathbb{C}/(\mathbb{Z}\tau + \mathbb{Z}) \;\cong\; \mathbb{C}^*/q^{\mathbb{Z}}$$
$$z \longmapsto u.$$

(See (I §6) and (V §1).) Then the local height function

$$\lambda : E(\mathbb{C}) \smallsetminus \{O\} \longrightarrow \mathbb{R}$$

is given by the formula

$$\lambda(z) = -\frac{1}{2}B_2\left(\frac{\mathrm{Im}\,z}{\mathrm{Im}\,\tau}\right)\log|q| - \log|1-u| - \sum_{n\geq 1}\log\left|(1-q^n u)(1-q^n u^{-1})\right|,$$

where

$$B_2(T) = T^2 - T + \frac{1}{6}$$

is the second Bernoulli polynomial.

Remark 3.4.1. The formula (3.4) for the local height $\lambda(z)$ is sometimes rewritten using the equivalent quantities

$$\frac{\mathrm{Im}\,z}{\mathrm{Im}\,\tau} = \frac{\log|u|}{\log|q|} = \frac{v(u)}{v(q)}.$$

PROOF. The Weierstrass σ-function has the product expansion (I.6.4)

$$\sigma(z) = -\frac{1}{2\pi i}e^{\frac{1}{2}\eta(1)z^2}e^{-\pi i z}(1-u)\prod_{n\geq 1}\frac{(1-q^n u)(1-q^n u^{-1})}{(1-q^n)^2}.$$

The modular discriminant function has the product expansion (I.8.1)

$$\Delta(\tau) = (2\pi)^{12}q\prod_{n\geq 1}(1-q^n)^{24}.$$

Hence

$$\left|e^{-\frac{1}{2}z\eta(z)}\sigma(z)\Delta(\tau)^{\frac{1}{12}}\right|$$
$$= \left|e^{\frac{1}{2}z(\eta(1)z - \eta(z) - 2\pi i)}q^{\frac{1}{12}}(1-u)\prod_{n\geq 1}(1-q^n u)(1-q^n u^{-1})\right|.$$

To simplify the exponential, we use Legendre's relation (I.5.2d), which in the case of a normalized lattice $\mathbb{Z}\tau + \mathbb{Z}$ says

$$\tau\eta(1) - \eta(\tau) = 2\pi i.$$

Writing

$$z = a\tau + b \qquad \text{with } a, b \in \mathbb{R},$$

we find

$$\eta(1)z - \eta(z) - 2\pi i = a\big(\tau\eta(1) - \eta(\tau)\big) - 2\pi i = 2\pi i(a - 1).$$

Hence

$$e^{\frac{1}{2}z(\eta(1)z - \eta(z) - 2\pi i)}q^{\frac{1}{12}} = e^{\frac{1}{2}(a\tau + b)\cdot 2\pi i(a-1)} \cdot e^{\frac{1}{12}\cdot 2\pi i\tau}$$

$$= e^{(a^2 - a + \frac{1}{6})\pi i\tau} \cdot e^{(a-1)b\pi i},$$

so

$$\left| e^{\frac{1}{2}z(\eta(1)z - \eta(z) - 2\pi i)}q^{\frac{1}{12}} \right| = \left| q^{\frac{1}{2}(a^2 - a + \frac{1}{6})} \right|.$$

Substituting this in above yields the formula

$$\left| e^{-\frac{1}{2}z\eta(z)}\sigma(z)\Delta(\tau)^{\frac{1}{12}} \right| = \left| q^{\frac{1}{2}(a^2 - a + \frac{1}{6})}(1 - u) \prod_{n \geq 1}(1 - q^n u)(1 - q^n u^{-1}) \right|.$$

Theorem 3.2 says that applying $-\log(\cdot)$ to the left-hand side gives the local height $\lambda(z)$. Since $\mathrm{Im}(z) = \mathrm{Im}(a\tau + b) = a\,\mathrm{Im}(\tau)$, this completes the proof of Theorem 3.4. $\qquad\qquad\qquad\qquad\qquad\qquad\qquad\qquad\qquad\qquad\square$

§4. Non-Archimedean Absolute Values — Explicit Formulas

Let K be a field with absolute value v, and let E/K be an elliptic curve. If the absolute value v on K is non-archimedean, then we can talk about the reduction of E modulo v. More precisely, fix a Weierstrass equation for E with v-integral coefficients (i.e. $v(a_i) \geq 0$). We consider the reduction $\tilde{E}$ of that Weierstrass equation modulo the maximal ideal of the local ring $R = \{\alpha \in K : v(\alpha) \geq 0\}$. The reduced curve $\tilde{E}$ may be singular. Define a subset $E_0(K)$ of $E(K)$ by

$$E_0(K) = \{P \in E(K) : \tilde{P} \text{ is a smooth point of } \tilde{E}\}.$$

In particular, if the reduced equation is smooth (E has good reduction), then $E_0(K) = E(K)$. (See [AEC VII], which discusses in some detail the case that v is a discrete valuation.)

We now show that for points in $E_0(K)$ the local height is given by a simple formula.

Theorem 4.1.   *Let K be a field complete with respect to a non-ar-chimedean absolute value v, let E/K be an elliptic curve, and choose a Weierstrass equation for E with v-integral coefficients,*

$$E : y^2 + a_1 xy + a_3 y = x^3 + a_2 x^2 + a_4 x + a_6.$$

Let Δ be the discriminant of this equation. Then the Néron local height function $\lambda : E(K) \smallsetminus \{O\} \to \mathbb{R}$ is given by the formula

$$\lambda(P) = \frac{1}{2} \max\{v(x(P)^{-1}), 0\} + \frac{1}{12} v(\Delta) \qquad \text{for all } P \in E_0(K).$$

Remark 4.1.1. If E has good reduction, then we can find a Weierstrass equation for E/K with $E_0(K) = E(K)$ and $v(\Delta) = 0$. In this situation, the proof of (4.1) will show that the function

$$\frac{1}{2} \max\{v(x(P)^{-1}), 0\}$$

has properties (i), (ii), and (iii) of Theorem 1.1. This provides an alternative proof of the existence of λ in this case. Further, since λ is invariant under finite extension of the field K (1.1c) and is independent of the choice of Weierstrass equation (1.1b), we actually obtain an existence proof whenever E has potential good reduction.

Remark 4.1.2. The local height λ is independent of the choice of Weierstrass equation. But the formula for λ given in Theorem 4.1 does not appear to be independent of this choice. For example, the change of coordinates $x = u^{-2} x'$, $y = u^{-3} y'$ will alter the formula in Theorem 4.1 to

$$\frac{1}{2} \max\{v(x'(P)^{-1}), 0\} + \frac{1}{12} v(\Delta')$$
$$= \frac{1}{2} \max\{v(u^{-2} x(P)^{-1}), 0\} + \frac{1}{12} v(u^{-12} \Delta).$$

However, this new formula for $\lambda(P)$ is valid only for points in $E_0'(K)$, where $E_0'(K)$ is defined using the equation with coordinates (x', y'). One can verify that the two formulas agree on the intersection $E_0(K) \cap E_0'(K)$, as they should.

PROOF (of Theorem 4.1). Let

$$\lambda_1(P) = \frac{1}{2} \max\{v(x(P)^{-1}), 0\} + \frac{1}{12} v(\Delta).$$

By inspection, λ_1 satisfies conditions (i) and (ii) of Theorem 1.1. We next verify condition (iii) for points in $E_0(K) \smallsetminus \{O\}$.

As usual, let

$$\phi(x) = x^4 - b_4 x^2 - 2b_6 x - b_8,$$
$$\psi(x) = 4x^3 + b_2 x^2 + 2b_4 x + b_6 = (2y + a_1 x + a_3)^2$$

be the functions appearing in the duplication formula

$$x([2]P) = \frac{\phi(P)}{\psi(P)}.$$

Then the equation

$$\lambda_1([2]P) = 4\lambda_1(P) + v\big((2y + a_1 x + a_3)(P)\big) - \frac{1}{4}v(\Delta)$$

to be verified is equivalent (after some algebra) to

$$\min\{v(\phi(P)), v(\psi(P))\} = \min\{4v(x(P)), 0\}.$$

First suppose that $v(x(P)) < 0$. Then the (non-archimedean) triangle inequality yields

$$v(\phi(P)) = 4v(x(P)) \quad \text{and} \quad v(\psi(P)) \geq v(4x(P)^3) > 4v(x(P)),$$

so the desired relation is true. We are left to prove

$$P \in E_0(K) \quad \text{and} \quad v(x(P)) \geq 0 \Longrightarrow \min\{v(\phi(P)), v(\psi(P))\} = 0.$$

To prove this, we must express the condition $P \in E_0(K)$ in terms of $\phi(P)$ and $\psi(P)$. Let

$$F(X, Y) = Y^2 + a_1 XY + a_3 Y - X^3 - a_2 X^2 - a_4 X - a_6$$

be the polynomial defining E. Recall that a point $(x_0, y_0) \in E$ is singular if and only if

$$F_X(x_0, y_0) = F_Y(x_0, y_0) = 0.$$

(See [AEC I.1.5]. The subscripts denote partial derivatives.) Now $E_0(K)$ consists of all points whose reduction modulo v is non-singular on the reduced curve $\tilde{E}$. So we see that

$$P \in E_0(K) \Longrightarrow v(F_X(P)) \leq 0 \quad \text{or} \quad v(F_Y(P)) \leq 0.$$

We also recall the addition formula [AEC III.2.3(c)], which says that

$$x([2]P) = m^2 + a_1 m - a_2 - 2x(P),$$

where m is the slope of the tangent line to E at P. (Note that $m \neq \infty$ since $[2]P \neq 0$.) Thus

$$m = \frac{F_X(P)}{F_Y(P)},$$

and so we find

$$x([2]P) = \frac{F_X(P)^2 + G(P)F_Y(P)}{F_Y(P)^2}$$

for the polynomial

$$G = a_1 F_X - (a_2 + 2X)F_Y \in R[X, Y] \subset K(E).$$

(N.B. G has coefficients in the valuation ring R, since by assumption the coefficients of F are v-integral.) Thus

$$\phi = F_X^2 + GF_Y \qquad \text{and} \qquad \psi = F_Y^2.$$

Now let $P \in E_0(K)$ satisfy $v(x(P)) \geq 0$. Then

$$0 \leq \min\{v(\phi(P)), v(\psi(P))\} \qquad \text{since } v(x(P)) \geq 0$$

$$= \min\{v(F_X(P)^2 + G(P)F_Y(P)), v(F_Y(P)^2)\}$$

$$= 0 \qquad \text{since either } v(F_X(P)) \leq 0 \text{ or } v(F_Y(P)) \leq 0.$$

This completes the proof that

$$\lambda_1([2]P) = 4\lambda_1(P) + v((2y + a_1 x + a_3)(P)) - \frac{1}{4}v(\Delta)$$

$$\text{for all } P \in E_0(K) \smallsetminus \{O\}.$$

We have now shown that λ_1 satisfies conditions (i), (ii), and (iii) of Theorem 1.1 for all points in $E_0(K) \smallsetminus \{O\}$. If $E_0(K) = E(K)$, that is, if E has good reduction and we take a minimal Weierstrass equation for E, then the uniqueness assertion of Theorem 1.1 implies that $\lambda = \lambda_1$. However, even if the Weierstrass equation for E has singular reduction, the proof of uniqueness in Theorem 1.1 works for the subgroup $E_0(K)$. A brief sketch follows.

From (i) and (ii), the difference $\Lambda = \lambda - \lambda_1$ extends to a bounded, continuous function on all of $E_0(K)$ (in fact, on all of $E(K)$). Further, from (iii) it satisfies $\Lambda([2]P) = 4\Lambda(P)$ for all $P \in E_0(K)$. (By continuity, this holds even when $[2]P = 0$.) Then

$$\Lambda(P) = \frac{1}{4^N}\Lambda([2^N]P) \xrightarrow[N \to \infty]{} 0, \qquad \text{so } \Lambda(P) = 0 \text{ for all } P \in E_0(K). \quad \square$$

It remains to find an explicit formula for the local height in the case that E has bad reduction at v and P is not in $E_0(K)$. Since the local height is invariant under finite extension of K (1.1c), it suffices in principle to consider the case that E has split multiplicative reduction at v. For computational purposes, however, it is often more convenient to work directly over K. Explicit formulas for λ in the case of additive reduction are given in exercises 6.7 and 6.8.

Theorem 4.2. *Let K be a p-adic field (i.e., a finite extension of $\mathbb{Q}_p$) with absolute value $v = -\log| \cdot |_v$, let $q \in K^*$ satisfy $|q|_v < 1$, and let E_q/K be the Tate curve (V.3.4) with its parametrization*

$$\phi : K^*/q^{\mathbb{Z}} \xrightarrow{\sim} E_q(K).$$

(a) *The Néron local height function*

$$\lambda \circ \phi : E_q(K) \smallsetminus \{O\} \longrightarrow \mathbb{R}$$

is given by the formula

$$\lambda(\phi(u)) = \frac{1}{2}B_2\left(\frac{v(u)}{v(q)}\right)v(q) + v(1-u) + \sum_{n \geq 1} v\big((1-q^nu)(1-q^nu^{-1})\big).$$

(b) *If we choose u (by periodicity) to satisfy*

$$0 \leq v(u) < v(q),$$

then

$$\lambda(\phi(u)) = \begin{cases} \dfrac{1}{2}B_2\left(\dfrac{v(u)}{v(q)}\right)v(q), & \text{if } 0 < v(u) < v(q), \\[2ex] v(1-u) + \dfrac{1}{12}v(q), & \text{if } v(u) = 0. \end{cases}$$

(Note that for a Tate curve, $v(q) = -v\big(j(E_q)\big) = v(\Delta(q))$.)

PROOF. (a) The Tate parametrization (V.3.4) is a v-adic analytic map from $K^*/q^{\mathbb{Z}}$ to the elliptic curve with Weierstrass equation

$$E_q : y^2 + xy = x^3 + a_4(q)x + a_6(q)$$

defined by

$$\phi : K^*/q^{\mathbb{Z}} \xrightarrow{\sim} E_q(K), \qquad \phi(u) = \big(X(u), Y(u)\big).$$

(For the series defining $a_4(q)$, $a_6(q)$, $X(u)$, and $Y(u)$, see (V.3.1).) The discriminant of the Weierstrass equation for E_q has the product expansion (V.3.1b)

$$\Delta(q) = q \prod_{n \geq 1}(1-q^n)^{24}.$$

Recall also the v-adic θ-function (V.3.2)

$$\theta(u) = (1-u)\prod_{n \geq 1} \frac{(1-q^nu)(1-q^nu^{-1})}{(1-q^n)^2}$$

and the factorization (V.3.2b)

$$2Y(u) + X(u) = -\frac{u\theta(u^2)}{\theta(u)^4}.$$

Applying v to this relation and doing a little algebra, we find that

$$\left\{v\big(\theta(u^2)\big) - \frac{1}{2}v(u^2)\right\} = 4\left\{v\big(\theta(u)\big) - \frac{1}{2}v(u)\right\} + v\big(2Y(u) + X(u)\big).$$

This suggests that $v\big(\theta(u)\big) - \frac{1}{2}v(u) - \frac{1}{12}v(\Delta)$ would be a good candidate for λ, since it has property (iii), but unfortunately it is not invariant under the transformation $u \mapsto qu$. So we make a slight alteration to obtain an invariant function.

Define

$$\lambda(u) = \frac{1}{2}B_2\left(\frac{v(u)}{v(q)}\right)v(q) + v\big(\theta(u)\big).$$

We will show that λ is the Néron local height function by verifying that it satisfies properties (i), (ii), and (iii) of Theorem 1.1.

First, using the identities

$$\frac{v(qu)}{v(q)} = \frac{v(u)}{v(q)} + 1, \qquad B_2(T+1) = B_2(T) + 2T, \qquad \theta(qu) = -\frac{1}{u}\theta(u)$$

(the last is (V.3.2a)), it is easy to check that $\lambda(qu) = \lambda(u)$; so λ is well-defined on

$$E_q(K) \smallsetminus \{O\} \cong (K^*/q^{\mathbb{Z}}) \smallsetminus \{1\}.$$

Next, the product defining θ is absolutely convergent and non-zero away from $q^{\mathbb{Z}}$. Hence λ is continuous (in fact, v-adically analytic) on $E_q(K) \smallsetminus \{O\}$ and bounded on the complement of any neighborhood of O. This verifies (i).

To check (ii) we compute

$$\lim_{u\to 1}\left\{\lambda(u) + \frac{1}{2}v\big(X(u)\big)\right\} = \frac{1}{2}B_2(0)v(q) + \frac{1}{2}\lim_{u\to 1}v\big(\theta(u)^2 X(u)\big)$$

$$= \frac{1}{2}B_2(0)v(q) + \frac{1}{2}\lim_{u\to 1}v\big((1-u)^2 X(u)\big).$$

The series for $X(u)$ given in (V.3.1) is

$$X(u) = \sum_{n\in\mathbb{Z}}\frac{q^n u}{(1-q^n u)^2} - 2s_1(q)$$

$$= \frac{u}{(1-u)^2} + \sum_{n\geq 1}\left\{\frac{q^n u}{(1-q^n u)^2} + \frac{q^n u^{-1}}{(1-q^n u^{-1})^2}\right\} - 2s_1(q),$$

so we see that the pole at $u = 1$ comes only from the $n = 0$ term. Hence

$$\lim_{u \to 1} (1 - u)^2 X(u) = 1,$$

which proves that the above limit exists. (In fact, the limit is $\frac{1}{12} v(q)$.)

Finally, we verify the duplication formula (iii). Note first that for all $n \geq 1$ we have $v(1 - q^n) = 0$, so

$$v\big(\Delta(q)\big) = v\Big(q \prod_{n \geq 1} (1 - q^n)^{24}\Big) = v(q).$$

Next we add the formula

$$\Big(v(\theta(u^2)) - \frac{1}{2} v(u^2)\Big) = 4\Big(v(\theta(u)) - \frac{1}{2} v(u)\Big) + v\big(2Y(u) + X(u)\big)$$

obtained above to the identity

$$\left(\frac{1}{2}\left(\frac{v(u^2)}{v(q)}\right)^2 + \frac{1}{12}\right) v(q) = 4\left(\frac{1}{2}\left(\frac{v(u)}{v(q)}\right)^2 + \frac{1}{12}\right) v(q) - \frac{1}{4} v(q).$$

Since $v(q) = v\big(\Delta(q)\big)$, this gives property (iii):

$$\lambda(u^2) = 4\lambda(u) + v\big(2Y(u) + X(u)\big) - \frac{1}{4} v\big(\Delta(q)\big).$$

We have now shown that $\lambda(u)$ satisfies properties (i), (ii), and (iii) of Theorem 1.1, so it is the Néron local height function.

(b) Since $0 \leq v(u) < v(q)$, we have

$$v(1 - q^n u) = v(1 - q^n u^{-1}) = 0 \qquad \text{for all } n \geq 1.$$

So the formula in (a) becomes

$$\lambda(u) = \frac{1}{2} B_2\left(\frac{v(u)}{v(q)}\right) v(q) + v(1 - u).$$

If in addition $v(u) > 0$, then $v(1 - u) = 0$, which gives the first expression. Similarly, if $v(u) = 0$, then the second expression is a consequence of $B_2(0) = 1/6$. □

Remark 4.2.1. The proof of (4.2) shows directly that the function described in (4.2a) satisfies conditions (i), (ii), and (iii) of Theorem 1.1, so we obtain an independent proof of the existence of the Néron local height for Tate curves. We have now given proofs of the existence of the Néron local height, independent of the proof in §1, in the following three cases:

(A) $K = \mathbb{C}$, see (3.2.1);

(B) $K/\mathbb{Q}_p$ and E/K has good reduction, see (4.1.1);

(C) $K/\mathbb{Q}_p$ and $E = E_q$ is a Tate curve, see (4.2).

But if K is the completion of a number field with respect to some absolute value, and E/K is an elliptic curve, then we can find a finite extension L/K so that E/L falls into one of the three cases (A), (B), or (C). This follows from [AEC, VII.5.5] and (V.5.3). Now using the elementary fact (1.1c) that the local height is invariant under field extension, we obtain an independent proof of the existence of λ for all E/K.

EXERCISES

6.1. Let K be a field, $| \cdot |_v$ an absolute value on K, and E/K an elliptic curve. For any rational function $f \in K(E)$, let

$$U_f(K) = \{P \in E(K) : f \text{ is defined at } P\}.$$

(a) Prove that the map

$$U_f(K) \longrightarrow \mathbb{R}, \qquad P \longmapsto |f(P)|_v$$

is continuous. (Here $U_f(K)$ inherits the v-adic topology from $E(K)$, and $\mathbb{R}$ is given the usual topology.)

(b) Prove that the topology on $E(K)$ described in §1 is the weakest topology (i.e., the topology containing the fewest open sets) such that the maps in (a) are continuous for every rational function $f \in K(E)$.

6.2. Let K be a field, $| \cdot |_v$ an absolute value on K, and E/K an elliptic curve. If K is locally compact, prove that $E(K)$ is compact. In particular, $E(K)$ is compact if $K = \mathbb{R}$, $K = \mathbb{C}$, or K is a finite extension of $\mathbb{Q}_p$.

6.3. Let K be the completion of a number field with respect to some absolute value, and let E/K be an elliptic curve. Prove that for all $P, Q \in E(K)$ with $P, Q, P \pm Q \neq O$, the Néron local height λ satisfies the *quasi-parallelogram law*

$$\lambda(P + Q) + \lambda(P - Q) = 2\lambda(P) + 2\lambda(Q) + v(x(P) - x(Q)) - \tfrac{1}{6}v(\Delta).$$

(*Hint.* We already proved this for $K = \mathbb{C}$ in Corollary 3.3. Going to an extension field, it suffices to prove the result when K is a finite extension of $\mathbb{Q}_p$ and E has either good or split multiplicative reduction. For the former, the formulas for $x_3 + x_4$ and $x_3 x_4$ in the proof of [AEC, VIII.6.2] may prove useful, at least if $p \neq 2, 3$; whereas for the latter, you can use the Tate curve together with (V.3.2b) and (4.2) to mimic the proof of Corollary 3.3.)

6.4.   Let K and E be as in the previous exercise, and fix a Weierstrass equation

$$E : y^2 + a_1 xy + a_3 y = x^3 + a_2 x^2 + a_4 x + a_6.$$

For each integer m define a function

$$F_m(x) = m^2 \prod_{\substack{T \in E[m] \\ T \neq O}} (x - x(T)) \in K(E).$$

(a) Prove that

$$\mathrm{div}(F_m) = 2 \Big(\sum_{T \in E[m]} (T) \Big) - 2m^2 (O).$$

(b) Prove that

$$F_2 = 4x^3 + b_2 x^2 + 2b_4 x + b_6 = (2y + a_1 x + a_3)^2,$$
$$F_3 = (3x^4 + b_2 x^3 + 3b_4 x^2 + 3b_6 x + b_8)^2,$$
$$F_4 = F_2 (2x^6 + b_2 x^5 + 5b_4 x^4 + 10b_6 x^3$$
$$+ 10b_8 x^2 + (b_2 b_8 - b_4 b_6)x + b_4 b_8 - b_6^2)^2.$$

Generally, show that there exist functions $\psi_m \in K(x, y) = K(E)$ satisfying $F_m = \psi_m^2$. (ψ_m is the mth-division polynomial; see [AEC, exercise 3.7].)
(c) Let Δ be the discriminant of the given equation. Prove that the function

$$\frac{F_m(x)^6}{\Delta^{m^2 - 1}}$$

is independent of the Weierstrass equation.
(d) Prove the recurrence formula

$$F_{m+1} F_{m-1} = (x \circ [m] - x) F_m^2, \qquad \text{for all } m \geq 2.$$

By convention, we set $F_1(x) = 1$. (*Hint.* Compare divisors. Then to find the constant, let $P \to O$.)
(e) Prove that

$$\lambda([m]P) = m^2 \lambda(P) + \frac{1}{2} v(F_m(P)) - \frac{m^2 - 1}{12} v(\Delta),$$

$$\text{for all } P \in E(K) \text{ with } [m]P \neq O.$$

6.5.   Let $E/\mathbb{C}$ be an elliptic curve with normalized lattice $\mathbb{Z}\tau + \mathbb{Z}$, and let $\lambda(z) = \lambda(x + iy)$ be the local height function on $E(\mathbb{C}) \smallsetminus \{O\}$.
(a) Prove that

$$\int_{E(\mathbb{C})} \lambda(z) \, dx \, dy = 0.$$

(Note that this is an improper integral, since $\lambda(z)$ blows up at $z = 0$. Be sure to check that the integral converges.)
(b) Prove that $\lambda(z)$ is a solution of the differential equation

$$\left(\frac{\partial^2}{\partial x^2} + \frac{\partial^2}{\partial y^2} \right) \lambda(z) = \frac{2\pi}{\mathrm{Im}\, \tau}.$$

(This exercise, together with Theorem 1.1(ii), says that $\lambda(z)$ is the *Green's function* on $E(\mathbb{C})$ for the divisor (O).)

6.6. Let K be a field with absolute value v and let E/K be an elliptic curve. Fix a Weierstrass equation for E,

$$E : y^2 = x^3 + Ax + B,$$

with discriminant and j-invariant

$$\Delta = -16(4A^3 + 27B^2) \qquad \text{and} \qquad j = -(48A)^3/\Delta.$$

(a) If v is non-archimedean, and A and B are v-integral, prove that for all $P \in E(K_v) \smallsetminus \{O\}$,

$$-\frac{1}{24}\max\{v(j(E)^{-1}), 0\} \le \lambda(P) - \frac{1}{2}\max\{v(x(P)^{-1}), 0\} \le \frac{1}{12}v(\Delta).$$

(*Hint.* Use the explicit formulas (4.1) and (4.2).)

(b) If v is archimedean, prove that for all $P \in E(K_v) \smallsetminus \{O\}$,

$$\left| \lambda(P) - \frac{1}{2}\max\left\{ v\left(\frac{\Delta^{1/6}}{x(P)} \right), 0 \right\} \right| \le \frac{1}{8}\max\{v(j(E)^{-1}), 0\} + v(3).$$

(c) Now suppose that K is a number field, and A and B are in the ring of integers of K. Prove that for all $P \in E(K)$,

$$\left| \hat{h}(P) - \frac{1}{2}h(x(P)) \right| \le \frac{1}{8}h(j) + \frac{1}{12}h(\Delta) + \log 3.$$

(The constants in (b) and (c) are certainly not best possible. See if you can improve them.)

6.7. The next three exercises give formulas for the local height which are especially well suited for numerical computations. We begin with the non-archimedean case.

Let K be a field complete with respect to a discrete valuation v, let E/K be an elliptic curve, and fix a *minimal* Weierstrass equation for E,

$$y^2 + a_1xy + a_3y = x^3 + a_2x^2 + a_4x + a_6.$$

Let Δ be the discriminant of this equation, and let $P \in E(K)$.

(a) Prove that $P \in E_0(K)$ if and only if either

$$v((3x^2 + 2a_2x + a_4 - a_1y)(P)) \le 0, \quad \text{or} \quad v((2y + a_1x + a_3)(P)) \le 0.$$

Note that if $P \in E_0(K)$, then (4.1) says that

$$\lambda(P) = \frac{1}{2}\max\{v(x(P)^{-1}), 0\} + \frac{1}{12}v(\Delta).$$

(b) Assume that $v(\Delta) > 0$ and $v(c_4) = 0$. (This means that E has multiplicative reduction at v; see [AEC VII.5.1b].) Let

$$\alpha(P) = \min\left\{ \frac{v((2y + a_1x + a_3)(P))}{v(\Delta)}, \frac{1}{2} \right\}.$$

Prove that if $P \notin E_0(K)$, then the local height of P is given by the formula

$$\lambda(P) = \frac{1}{2}B_2(\alpha(P))v(\Delta).$$

6.8. *Let K, v, E and Δ be as in the previous exercise. Suppose that E has additive reduction (i.e., $v(\Delta) > 0$ and $v(c_4) > 0$.) Let $P \in E(K)$ with $P \notin E_0(K)$. Let F_2 and F_3 be the polynomials defined in exercise 6.4b. Prove that

$$\lambda(P) = \begin{cases} -\dfrac{1}{6}v(F_2(P)) + \dfrac{1}{12}v(\Delta) & \text{if } v(F_3(P)) \geq 3v(F_2(P)), \\[2ex] -\dfrac{1}{16}v(F_3(P)) + \dfrac{1}{12}v(\Delta) & \text{otherwise.} \end{cases}$$

6.9. Let $E/\mathbb{R}$ be an elliptic curve given by the usual Weierstrass equation

$$y^2 + a_1 xy + a_3 y = x^3 + a_2 x^2 + a_4 x + a_6,$$

and suppose that $x(P) \neq 0$ for all $P \in E(\mathbb{R})$. (Note that one can always achieve this condition by making a shift $x = x' + r$ for sufficiently large r.) Then the functions

$$t = \frac{1}{x}, \quad w = 4t + b_2 t^2 + 2b_4 t^3 + b_6 t^4, \quad z = 1 - b_4 t^2 - 2b_6 t^3 - b_8 t^4,$$

are well-defined for all points in $E(\mathbb{R})$.
(a) Prove that
$$t([2]P) = \frac{w(P)}{z(P)}.$$

This gives a convenient recursive formula for computing $z([2^n]P)$.
(b) Prove that there are constants $c_1, c_2 > 0$, depending on the Weierstrass equation, so that

$$c_1 \leq z(P) \leq c_2 \qquad \text{for all } P \in E(\mathbb{R}).$$

Conclude that the series

$$\sum_{n=0}^{\infty} \frac{1}{4^n} \log|z([2^n]P)|$$

is absolutely convergent for all $P \in E(\mathbb{R})$.
(c) Prove that the local height function on $E(\mathbb{R}) \smallsetminus \{O\}$ is given by the formula

$$\lambda(P) = \frac{1}{2}\log|x(P)| - \frac{1}{12}\log|\Delta| + \frac{1}{8}\sum_{n=0}^{\infty}\frac{1}{4^n}\log|z([2^n]P)|.$$

(d) Give a modification of the series in (c) which converges to give the local height function on $E(\mathbb{C}) \smallsetminus \{O\}$. (Note that in this case there will always be points with $x(P) = 0$.)

6.10. Using the previous three exercises, compute the canonical height $\hat{h}(P)$ of the indicated point on each of the following curves.

(a) $\qquad y^2 + y = x^3 - x,$ $\qquad\qquad\qquad\qquad P = (0,0),$

(b) $\qquad y^2 + y = x^3 - x^2,$ $\qquad\qquad\qquad\qquad P = (0,0),$

(c) $\quad y^2 + xy + y = x^3 - x^2 - 48x + 147,$ $\qquad P = (13,33),$

(d) $\quad y^2 + xy + y = x^3 + x^2 - 1001x + 12375,$ $\quad P = (45,224).$

6.11. Define the *periodic second Bernoulli polynomial* $\mathbf{B}_2(t)$ by

$$\mathbf{B}_2(t) = \left(t - [t]\right)^2 - \left(t - [t]\right) + \frac{1}{6}$$

(i.e., $\mathbf{B}_2$ equals B_2 on the interval $0 \le t < 1$, and is extended periodically modulo 1 to all of $\mathbb{R}$).

(a) Prove that $\mathbf{B}_2(t)$ has the Fourier expansion

$$\mathbf{B}_2(t) = \frac{1}{2\pi^2} \sum_{k \in \mathbb{Z},\, k \ne 0} \frac{e^{2\pi i k t}}{k^2}.$$

(b) Let $t_1, \ldots, t_N \in \mathbb{R}$. Prove that

$$\sum_{\substack{1 \le i,j \le N \\ i \ne j}} \mathbf{B}_2(t_i - t_j) \ge -\frac{N}{6}.$$

(c) Let K be a complete field with discrete valuation v, let E/K be an elliptic curve, and let Δ be the discriminant of a minimal Weierstrass equation for E at v. Prove that for any collection of distinct points $P_1, \ldots, P_N \in E(K)$,

$$\sum_{\substack{1 \le i,j \le N \\ i \ne j}} \lambda(P_i - P_j) \ge -\frac{N}{12} v(\Delta).$$

(*Hint.* Since λ is invariant under finite extension of K, it suffices to consider the two cases of good and split multiplicative reduction.)

(d) Let K be complete with respect to a discrete valuation v, and suppose that E has split multiplicative reduction. Choose a parametrization $E(K) \cong K^*/q^{\mathbb{Z}}$, and let $P \in E[N+1]$ correspond to $q^{1/(N+1)}$. (Take any root in $\bar{K}$.) For each i, $1 \le i \le N$, let $P_i = [i]P$. Prove that

$$\sum_{\substack{1 \le i,j \le N \\ i \ne j}} \lambda(P_i - P_j) = -\frac{N}{12} v(\Delta) \left(1 - \frac{2}{N+1}\right).$$

Thus the estimate in (c) is essentially best possible.

(e) *Let $E/\mathbb{C}$ be an elliptic curve. Prove that there is a constant $c = c(E)$ so that for any set of distinct points $P_1, \ldots, P_N \in E(\mathbb{C})$,

$$\sum_{\substack{1 \le i,j \le N \\ i \ne j}} \lambda(P_i - P_j) \ge -\frac{1}{2} N \log N - c(E) N.$$

(This is the archimedean analogue of (c). It is quite difficult.)

APPENDIX A

Some Useful Tables

§1. Bernoulli Numbers and $\zeta(2k)$

Values of the Riemann ζ-Function at Even Integers, $\zeta(s) = \sum_{n \geq 1} \dfrac{1}{n^s}$.

$$\zeta(2) = \frac{\pi^2}{2 \cdot 3} \qquad \zeta(4) = \frac{\pi^4}{2 \cdot 3^2 \cdot 5}$$

$$\zeta(6) = \frac{\pi^6}{3^3 \cdot 5 \cdot 7} \qquad \zeta(8) = \frac{\pi^8}{2 \cdot 3^3 \cdot 5^2 \cdot 7}$$

$$\zeta(10) = \frac{\pi^{10}}{3^5 \cdot 5 \cdot 7 \cdot 11} \qquad \zeta(12) = \frac{691\pi^{12}}{3^6 \cdot 5^3 \cdot 7^2 \cdot 11 \cdot 13}$$

$$\zeta(14) = \frac{2\pi^{14}}{3^6 \cdot 5^2 \cdot 7 \cdot 11 \cdot 13}$$

Bernoulli Numbers $\dfrac{x}{e^x - 1} = \displaystyle\sum_{k=0}^{\infty} B_k \frac{x^k}{k!}$

$$B_2 = \frac{1}{6} \qquad B_4 = -\frac{1}{30} \qquad B_6 = \frac{1}{42}$$

$$B_8 = -\frac{1}{30} \qquad B_{10} = \frac{5}{66} \qquad B_{12} = -\frac{691}{2730}$$

$$B_{14} = \frac{7}{6} \qquad B_{16} = -\frac{3617}{510} \qquad B_{18} = \frac{43867}{798}$$

$$B_{20} = -\frac{283 \cdot 617}{330} \qquad B_{22} = \frac{11 \cdot 131 \cdot 593}{138} \qquad B_{24} = -\frac{103 \cdot 2294797}{2730}$$

$$B_{26} = \frac{13 \cdot 657931}{6} \qquad B_{28} = -\frac{7 \cdot 9349 \cdot 362903}{870}$$

§2. Fourier Coefficients of $\Delta(\tau)$ and $j(\tau)$

Fourier Coefficients of $(2\pi)^{-12}\Delta(\tau) = \sum_{n\geq 0} \tau(n)q^n = q\prod_{n\geq 1}(1-q^n)^{24}$.

$\tau(1) =$	1	$\tau(2) =$	-24	$\tau(3) =$	252
$\tau(4) =$	-1472	$\tau(5) =$	4830	$\tau(6) =$	-6048
$\tau(7) =$	-16744	$\tau(8) =$	84480	$\tau(9) =$	-113643
$\tau(10) =$	-115920	$\tau(11) =$	534612	$\tau(12) =$	-370944

The function $\tau(n)$ is called the Ramanujan τ function. For values of $\tau(n)$ with $n \leq 300$, see Lehmer [1].

Fourier Coefficients of $j(\tau) = \dfrac{1}{q} + \sum_{n\geq 0} c(n)q^n$.

$c(0) =$	744	$c(1) =$	196884
$c(2) =$	21493760	$c(3) =$	864299970
$c(4) =$	20245856256	$c(5) =$	333202640600
$c(6) =$	4252023300096	$c(7) =$	44656994071935
$c(8) =$	401490886656000		

Inversion of Series for j Function, $q = \sum_{n\geq 1} d(n)\dfrac{1}{j^n}$.

$d(1) =$	1	$d(2) =$	744
$d(3) =$	750420	$d(4) =$	872769632
$d(5) =$	1102652742882	$d(6) =$	1470561136292880
$d(7) =$	2037518752496883080	$d(8) =$	2904264865530359889600

§3. Elliptic Curves over $\mathbb{Q}$ with Complex Multiplication

In this section we describe all elliptic curves defined over $\mathbb{Q}$ with complex multiplication by an order $R = \mathbb{Z} + fR_K$ of conductor f in a quadratic imaginary field $K = \mathbb{Q}\left(\sqrt{-D}\right)$ of discriminant $-D$. The first table gives the j-invariant for each such order. The second table gives a representative elliptic curve E over $\mathbb{Q}$ with the specified j, together with the minimal discriminant Δ_E and conductor N_E of E. Those curves possessing endomorphisms of degree 2 are discussed in (II.2.3.1).

Discriminant $-D$ of K	Conductor f of R	j-invariant of E
-3	1	0
	2	$2^4 3^3 5^3$
	3	$-2^{15} 3 \cdot 5^3$
-4	1	$2^6 3^3$
	2	$2^3 3^3 11^3$
-7	1	$-3^3 5^3$
	2	$3^3 5^3 17^3$
-8	1	$2^6 5^3$
-11	1	-2^{15}
-19	1	$-2^{15} 3^3$
-43	1	$-2^{18} 3^3 5^3$
-67	1	$-2^{15} 3^3 5^3 11^3$
-163	1	$-2^{18} 3^3 5^3 23^3 29^3$

$-D$	f	Minimal Weierstrass equation of E over $\mathbb{Q}$	Δ_E	N_E
-3	1	$y^2 + y = x^3$	3^3	3^3
	2	$y^2 = x^3 - 15x + 22$	$2^8 3^3$	$2^2 3^3$
	3	$y^2 + y = x^3 - 30x + 63$	3^5	3^3
-4	1	$y^2 = x^3 + x$	2^6	2^6
	2	$y^2 = x^3 - 11x + 14$	2^9	2^5
-7	1	$y^2 + xy = x^3 - x^2 - 2x - 1$	7^3	7^2
	2	$y^2 = x^3 - 595x + 5586$	$2^{12} 7^3$	$2^4 7^2$
-8	1	$y^2 = x^3 + 4x^2 + 2x$	2^9	2^8
-11	1	$y^2 + y = x^3 - x^2 - 7x + 10$	11^3	11^2
-19	1	$y^2 + y = x^3 - 38x + 90$	19^3	19^2
-43	1	$y^2 + y = x^3 - 860x + 9707$	43^3	43^2
-67	1	$y^2 + y = x^3 - 7370x + 243528$	67^3	67^2
-163	1	$y^2 + y = x^3 - 2174420x + 1234136692$	163^3	163^2

Notes on Exercises

Many of the exercises in this book are standard results which were not included in the text due to lack of space, whereas others are special cases of results which appear in the literature. The following list thus serves two purposes. First, it is an attempt by the author to give credit for the theorems which appear in the exercises, and second, it will aid the reader who wishes to delve more deeply into some aspect of the theory. However, since any attempt to assign credit is bound to be incomplete in some respects, the author herewith tenders his apologies to anyone who feels that they have been slighted.

Except for an occasional computational problem, we have not included solutions (nor even hints). Indeed, since it is hoped that this book will lead the student on into the realm of active mathematics, the benefits of working without aid clearly outweigh any advantage that might be gained by having solutions readily available.

CHAPTER I

(1.1) For an elementary proof, see Alperin [1].

(1.5) (e) Let $h(-D)$ denote the class number of $\mathbb{Q}\left(\sqrt{-D}\right)$. Then $h(-3) = 1$, $h(-5) = 2$, $h(-23) = 3$, $h(-29) = 6$, $h(-47) = 5$.

(1.10) See Serre [3, VII, §3.2, Cor. 2].

(1.11) A similar argument is given in Serre [3, VII, Thm. 3] and Apostol [1, Thm. 2.4].

(1.13) See de Shalit [1, II §2, equation (4)]. See also Weil [1, Ch. III, IV].

(1.14) Proven in Stark [2] using Kronecker's limit formula. Alternatively, one can compare poles and zeros to see that the ratio is constant, and then let $z \to 0$ to find the constant.

(1.16) Answer: $s(2,y) = (y-1)(y-5)/24$ if y is odd, and $s(2,y) = (y^2+5)/24y$ if y is even.

(1.19) (a) See Shimura [1, exercise 3.27]. (b,c) See Shimura [1, Thm. 3.24].

(1.20) (a,b) See Shimura [1, exercise 2.8].

(1.22) See Lang [2, Ch. III §4], Shimura [1, Ch. 3, §§4,5], or Ogg [1].

(1.24) See Serre [3, Ch. VII §4.3].

(1.25) See Apostol [1, Thm. 6.16].

(1.26) This is due to Hecke. See Apostol [1, Thm. 7.20].

(1.29) (c) This is due to Hecke. See Ogg [1, Ch. I, Thm. 1, p. I-5].

CHAPTER II

(2.1) Write $R_K = \mathbb{Z} + \mathbb{Z}\tau$, and for any $\alpha \in R$, write $\alpha = a_\alpha + b_\alpha\tau$ with $a_\alpha \in \mathbb{Z}$ and $b_\alpha \in \mathbb{Z}$. Then $f = \min\{b_\alpha : \alpha \in R, b_\alpha > 0\}$.

(2.2) This is due to Hurwitz [1]. A nice exposition of the proof is given in the appendix to Rosen [1].

(2.9) (a) See Shimura [1, (5.4.3), p. 124]. (b) See Shimura [1, exercise 5.8, p. 124].

(2.12) (b) This exercise was suggested by David Rohrlich.

(2.13) The minimal polynomial of β is $27x^8 + 72x^4 - 16$.

(2.14) $K_2 = L_2 = K$, $K_3 = K$, $L_3 = K(\sqrt[3]{4})$, $K_4 = K(\sqrt{3})$, and $L_4 = K(\sqrt{-9 + 6\sqrt{3}})$. $\mathrm{Gal}(L_4/K)$ is cyclic of order 4. In computing L_4, the identity $-10 - 6\sqrt{3} = (-1 - \sqrt{3})^3$ is useful.

(2.17) See Lang [1, Ch. 5, §1].

(2.18) (a,b,c,d) See Lang [1, Ch. 5, §2], Shimura [1, Ch. 4.6]. (f) See P. Cohen [1] and Silverman [4].

(2.19) See Lang [1, Ch. 5, §3].

(2.20) See Lang [1, Ch. 5, §2].

(2.21) See Lang [1, Ch. 5, Thm. 2].

(2.25) See Gross [1, Lemma 9.2.5].

(2.30) See Lang [1, Ch. 10, §4, Thm. 10]. For the general case of abelian varieties, see Shimura [1, Thm. 7.46 and Prop. 7.47].

(2.33) See Ireland-Rosen [1, Ch. 18, §4].

CHAPTER III

(3.4) The Mordell-Weil theorem for abelian varieties over finitely generated fields is due to Néron. See, for example, Lang [4, Ch. 6, Thm. 1].

(3.11) (c) See also Chapter VI text and exercises for an approach using local height functions which give a better estimate in (c). One can then use (c) to prove (a) and (b).

(3.12) $[0, 1, 0]$, $[0, \pm T, 1]$, $[T, 0, 1]$, $[1 - T, \pm(1 - 2T), 1]$, $[\sqrt{2}, \pm(2T - 1), 2\sqrt{2}]$.

(3.14) See Mumford [1, §6, Lemma on p. 56].

(3.15) In general, the Mordell-Weil group of an abelian variety is finitely generated in the following two cases: (i) (Néron) if the field of definition is finitely generated over $\mathbb{Q}$. (ii) (Lang-Néron) if one takes the quotient by the subgroup of points defined over the constant field. For details, see Lang [4, Ch. 6, theorems 1 and 2].

(3.16) This version of (III.11.3.1) and (III.11.4) for split elliptic surfaces is due to Dem'janenko [1], with a generalization to abelian varieties due to Manin [1]. The example in (d) is due to Dem'janenko.

(3.17) See Kuwata [1,2].

(3.20) (a) 2. (b) 3. (c) 2. (d) 4.

(3.21) This is a special case of Zariski's Main Theorem, see Hartshorne [1, V.5.2, III.11.3, exercise III.11.4].

(3.23) This is a special case of the general fact that algebraically equivalent divisors are numerically equivalent, see Hartshorne [1, exercise V.1.7].

(3.24) (b) $\det(I_{00}) = n$. (c) $a_i = i(n-k)/n$ if $1 \le i \le k$, and $a_i = k(n-i)/n$ if $k < i < n$. See, for example, Cox and Zucker [1, Table 1.14].

(3.31) See Lang [4, Ch. 4, Prop. 5.2].

(3.32) See Lang [4, Ch. 4, Prop. 3.3 and Cor. 3.4].

(3.34) This estimate is due to Tate [4], see also Lang [4, 12 Cor. 5.4].

(3.36) (b) This is due to Kodaira, see Shioda [3, Prop. 2.8]. It has been frequently rediscovered, see for example Hindry-Silverman [2, Thm. 5.1] and Szpiro [1].

(3.37) This result is due to Hindry and Silverman [2].

(3.38) For an explicit construction of the Jacobian variety of a hyperelliptic curve, see Mumford [3, Ch. IIIa]. In particular, $\mathrm{Pic}(C)[2]$ is described in Mumford [3, Ch. IIIa, Lemma 2.4 and Cor. 2.11].

(3.40) See Shioda [3, Prop. 1.6].

CHAPTER IV

(4.5) (a) See Matsumura [1, corollary to Thm. 45 (18.G)].

(4.7) See Bosch-Lütkebohmert-Raynaud [1, §2.4, Prop. 8].

(4.10) See Shatz [1, §2] or Waterhouse [1, Ch. 1 and 2].

(4.15) (a) This is a restatement of Lemma IV.9.5.

(4.16) $\mathcal{C}_7 : (y^2 - x^3)(x - 1)^3(y - 2x) = 0$.

(4.19) See Artin [1, §0].

(4.24) See Bosch-Lütkebohmert-Raynaud [1, Ch. 7, Prop. 6].

(4.25) Combine (IV.5.3) and (IV.9.1). For a more intrinsic proof for general group schemes, see Artin [1, Lemma 1.16].

(4.26) See Greenberg [1, §3, Lemma 2] for a multi-variable version.

(4.27) (b) See Bosch-Lütkebohmert-Raynaud [1, 2.2, Prop. 7]. (c) See Bosch-Lütkebohmert-Raynaud [1, 2.3, Prop. 5].

(4.29) See Milne [3, §7] or Weil [3].

(4.30) See Artin [1, Cor. 1.6].

(4.31) (c) The special fiber has five components, three of multiplicity 1 and two of multiplicity 2. This is a fiber of Type IX-1 in the classification of Namikawa and Ueno [1]. (d) The special fiber has four components, one of multiplicity 1, two of multiplicity 2, and one of multiplicity 3. This is a fiber of Type VIII-3 in the classification of Namikawa and Ueno [1].

(4.32) (c) Raynaud [1] has proven a general result which allows one to compute the group of components on the Néron model of the Jacobian of a curve in terms of the incidence matrix of the special fiber of a minimal proper regular model of the curve.

(4.35) This exercise is taken from Tate [2, end of §6].

(4.36) (a,b) This is an unpublished result of Mestre; see Kraus [1]. (c) This is due to Kraus [1].

(4.37) See Néron [1, §III].

(4.39) The proof is essentially the same as the proof of [AEC, IV.6.1]. See also exercises 2.22 and 2.23.

(4.43) (a) $\{1\}$. (b) $\mathbb{Z}/2\mathbb{Z}$. (c) H_8.

(4.44) This follows easily from [AEC, IX.6.1].

(4.45) See Serre [4, Ch. IV, Section 1].

(4.46) (b) See Serre [4, Ch. VI, Thm. 1']. (d) See Ogg [2], Serre-Tate [1, §3], and Serre [7, Ch. 19]. (e) This is due to Ogg [2].

CHAPTER V

(5.2) See Serre [4, VIII §4].

(5.13) This is due to Serre [1, (A.1.2), pp. IV-31,2.].

(5.14) This approach to Tate's theorem for elliptic curves with non-integral j is due to Shatz [1].

CHAPTER VI

(6.5) See Lang [6, II §5].

(6.6) (a) This is due to Tate; see Lang [3, III, Thm. 4.5]. (b,c) See Silverman [3].

(6.7) See Silverman [2, Lemma 5.1].

(6.8) See Silverman [2, Thm. 5.2].

(6.9) The series (c) is due to Tate, and the extension to $E(\mathbb{C})$ is due to Silverman. For proofs with error estimates, see Silverman [2].

(6.10) Solutions: (a) $0.02555\ldots$ (b) $\hat{h}(P) = 0$, P is a 5-torsion point, (c) $0.01028\ldots$, this point has very small height, (d) $0.01049\ldots$, this point has very small ratio $\hat{h}(P)/\log(\Delta)$.

(6.11) (b) This is due to Blanksby and Montgomery; see Hindry-Silverman [1]. (c) Hindry-Silverman [1]. (e) This is due to Elkies, see Lang [6, VI §5]. One can give the explicit lower bound

$$-\frac{1}{2}N\log N - \frac{1}{12}N\max\{\log|j(E)|, 0\} - 3.64N,$$

where the 3.64 is not best possible.

References

[AEC]   *The Arithmetic of Elliptic Curves,* GTM 106, J.H. Silverman, Springer-Verlag, New York, 1986.

Abhyankar, S.S.
   [1]   Resolution of singularities of arithmetical surfaces. In *Arithmetical Algebraic Geometry,* Harper and Row, New York, 1965.
   [2]   Resolution of singularities of algebraic surfaces. In *Algebraic Geometry,* Oxford University Press, London, 1969, 1–11.

Ahlfors, L.
   [1]   *Complex Analysis: An Introduction to the Theory of Analytic Functions of One Complex Variable,* 3rd ed., McGraw-Hill, New York, 1979.

Alling, N.
   [1]   *Real Elliptic Curves,* North-Holland, New York, 1981.

Alperin, R.
   [1] $PSL_2(\mathbb{Z}) = \mathbb{Z}_2 * \mathbb{Z}_3$. *Amer. Math. Monthly* **100** (1993), 385–386.

Altman, A. and Kleiman, S.
   [1]   *Introduction to Grothedieck Duality Theory,* Lect. Notes in Math. 146, Springer-Verlag, Berlin, 1970.

Apostol, T.
   [1]   *Modular Functions and Dirichlet Series in Number Theory,* GTM 41, Springer-Verlag, New York, 1976.

Arakelov, S.
   [1]   Intersection theory of divisors on an arithmetic surface. *Izv. Akad. Nauk SSR Ser. Mat.* **38** (1974), *AMS Transl.* **8** (1974), 1167–1180.

Artin, M.
   [1]   Néron models. In *Arithmetic Geometry,* G. Cornell and J. Silverman, eds., Springer-Verlag, New York, 1986, 213–230.
   [2]   Lipman's proof of resolution of singularities for surfaces. In *Arithmetic Geometry,* G. Cornell and J. Silverman, eds., Springer-Verlag, New York, 1986, 268–287.

Atiyah, M.F. and MacDonald, I.G.
   [1]   *Introduction to Commutative Algebra,* Addison-Wesley, Reading, MA, 1969.

Baker, A.
[1]   *Transcendental Number Theory,* Cambridge Univ. Press, Cambridge, 1975.

Beauville, A.
[1]   *Complex Algebraic Surfaces,* London Math. Soc. Lect. Note 68, Cambridge Univ. Press, Cambridge, 1983.

Birch, B.J. and Kuyk, W., eds.
[1]   *Modular Functions of One Variable IV,* Lect. Notes in Math. 476, Springer-Verlag, Berlin, 1975.

Birch, B.J. and Swinnerton-Dyer, H.P.F.
[1]   Elliptic curves and modular functions. In *Modular Functions of One Variable IV,* Lect. Notes in Math. 476, B.J. Birch and W. Kuyk, eds., Springer-Verlag, Berlin, 1975, 2–32.

Borel, A., Chowla, S., Herz, C.S., Iwasawa, K. and Serre, J.-P.
[1]   *Seminar on Complex Multiplication,* Lect. Notes in Math. 21, Springer-Verlag, Berlin, 1966.

Bosch, S., Günter, U. and Remmert, R.
[1]   *Non-Archimedean Analysis: A Systematic Approach to Rigid Analytic Geometry,* Springer-Verlag, Berlin, 1984.

Bosch, S., Lütkebohmert, W. and Raynaud, M.
[1]   *Néron Models,* Springer-Verlag, Berlin, 1990.

Brumer, A. and Kramer, K.
[1]   The conductor of an abelian variety. *Compositio Math.* (1994), to appear.

Cassou-Noguès, Ph. and Taylor M.J.
[1]   *Elliptic Functions and Rings of Integers,* Birkhäuser, Boston, 1987.

Chinburg, T.
[1]   An introduction to Arakelov intersection theory. In *Arithmetic Geometry,* G. Cornell and J. Silverman, eds., Springer-Verlag, New York, 1986, 291–307.
[2]   Minimal models for curves over Dedekind rings. In *Arithmetic Geometry,* G. Cornell and J. Silverman, eds., Springer-Verlag, New York, 1986, 309–326.

Coates, J.
[1]   Elliptic curves and Iwasawa theory. In *Modular Forms,* R.A. Rankin, ed., Ellis Horwood Ltd., Chichester, 1984, 51–74.

Cohen, H.
[1]   *A Course in Computational Algebraic Number Theory,* Springer-Verlag, Berlin, 1993.

Cohen, P.
[1]   On the coefficients of the transformation polynomials for the elliptic modular function. *Math. Proc. Camb. Philos. Soc.* **95** (1984), 389–402.

Conway, J.H.
[1]   Monsters and moonshine. *Math. Intelligencer* **2** (1979/80), 165–171.

Conway, J.H. and Norton, S.
[1]   Monstrous moonshine. *Bull. London Math. Soc.* **11** (1979), 308–339.

Cox, D. and Zucker, S.
  [1]   Intersection numbers of sections of elliptic surfaces. *Invent. Math.* **53**
         (1979), 1–44.
Cremona, J.E.
  [1]   *Algorithms for Modular Elliptic Curves,* Cambridge Univ. Press, Cam-
         bridge, 1992.
Deligne, P.
  [1]   Formes modulaires et représentations *l*-adic.   In *Sém. Bourbaki,* 21e
         année, 1968/69, no. 355, Lect. Notes in Math. 179, ˙Springer-Verlag,
         Berlin, 1971, 139–172.
  [2]   La conjecture de Weil I. *Publ. Math. IHES* **43** (1974), 273–307.
Dem'janenko, V.A.
  [1]   Rational points of a class of algebraic curves. *Izv. Akad. Nauk SSSR
         Ser. Math.* **30** (1966), 1373–1396. *AMS Transl.* **66** (1968), 246–272.
  [2]   An estimate of the remainder term in Tate's formula. *Mat. Zametki* **3**
         (1968), 271–278. *Math. Notes* **3** (1968), 173–177.
de Shalit, E.
  [1]   *Iwasawa theory of elliptic curves with complex multiplication,* Academic
         Press, Boston, 1987.
Eisenbud, D. and Harris, J.
  [1]   *Schemes: The language of modern algebraic geometry,*   Wadsworth &
         Brooks/Cole, Pacific Grove, CA, 1992.
Elkies, N.
  [1]   The existence of infinitely many supersingular ˙primes for every elliptic
         curve over $\mathbb{Q}$. *Invent. Math.* **89** (1987), 561–567.
Faltings, G.
  [1]   Calculus on arithmetic surfaces. *Ann. Math.* **119** (1984), 387–424.
Fermigier, S.
  [1]   Un exemple de courbe elliptique definie sur $\mathbb{Q}$ de rang $\geq$ 19. *CRAS
         Serie 1* **315** (1992), 719–722.
Greenberg, M.
  [1]   Rational points in Henselian discrete valuation rings. *Publ. Math. IHES*
         31 (1966), 563–567.
Griffiths, P. and Harris, J.
  [1]   *Principles of Algebraic Geometry,* Wiley, New York, 1978.
Gross, B.
  [1]   *Arithmetic on Elliptic Curves,* Lect. Notes in Math. 776, Springer-Ver-
         lag, Berlin, 1980.
Gross, B., Kohnen, W. and Zagier, D.
  [1]   Heegner points and derivatives of *L*-series. *Math. Ann.* **278** (1987),
         497–562.
Gross, B.H. and Zagier, D.B.
  [1]   Heegner points and derivatives of *L*-series. *Invent. Math.* **84** (1986),
         225–320.
Grosswald, E. and Rachmacher, H.
  [1]   *Dedekind Sums,* Carus Math. Monograph, Math. Assoc. of America,
         Providence, RI, 1972.

Harris, J.
[1]   *Algebraic Geometry: A First Course,* Springer-Verlag, New York, 1992.
Hartshorne, R.
[1]   *Algebraic Geometry,* Springer-Verlag, New York, 1977.
Heegner, K.
[1]   Diophantische Analysis und Modulfunktionen. *Math. Zeit.* **56** (1952), 227–253.
Hindry, M. and Silverman, J.H.
[1]   *Introduction to Diophantine Geometry,* in preparation.
[2]   The canonical height and integral points on elliptic curves. *Invent. Math.* **93** (1988), 419–450.
Hurwitz, A.
[1]   Über die Entwicklungskoeffizienten der lemniscatischen Funktionen. In *Mathematische Werke,* vol. 2, Birkhäuser, Basel, 1962, 342–373.
Ireland, K. and Rosen, M.
[1]   *A Classical Introduction to Modern Number Theory,* GTM 84, Springer-Verlag, New York, 1982.
Katz, N. and Mazur, B.
[1]   *Arithmetic Moduli of Elliptic Curves,* Princeton Univ. Press, Princeton, NJ, 1985.
Kenku, M.A.
[1]   On the number of $\mathbb{Q}$-isomorphism classes of elliptic curves in each $\mathbb{Q}$-isogeny class. *J. Number Theory* **15** (1982), 199–202.
Knapp, A.
[1]   *Elliptic Curves,* Math. Notes 40, Princeton Univ. Press, Princeton, NJ, 1992.
Koblitz, N.
[1]   *Introduction to Elliptic Curves and Modular Forms,* Springer-Verlag, New York, 1984.
Kodaira, K.
[1]   On the structure of compact complex analytic surfaces I, II. *Amer. J. Math.* **86** (1964), 751–798; **88** (1966), 682–721.
Kolyvagin, V.A.
[1]   Finiteness of $E(\mathbb{Q})$ and $\mathrm{III}(\mathbb{Q})$ for a class of Weil curves. *Math. USSR Izv.* **32** (1989), 523–542.
Kraus, A.
[1]   Quelques remarques a propos des invariants c_4, c_6, et Δ d'une courbe elliptique. *Acta Arith.* **54** (1989), 75–80.
Kubert, D. and Lang, S.
[1]   *Modular Units,* Springer-Verlag, New York, 1981.
Kuwata, M.
[1]   The field of definition of the Mordell-Weil group of an elliptic curve over a function field. *Compos. Math.* **76** (1990), 399–406.
[2]   Ramified primes in the field of definition for the Mordell-Weil group of an elliptic surface. *Proc. AMS* **116** (1992), 955–959.

Lang, S.
[1]   *Elliptic Functions,* GTM 112, Springer-Verlag, New York, 1987.
[2]   *Introduction to Modular Forms,* Springer-Verlag, Berlin, 1976.
[3]   *Elliptic Curves: Diophantine Analysis,* Springer-Verlag, Berlin, 1978.
[4]   *Fundamentals of Diophantine Geometry,* Springer-Verlag, New York, 1983.
[5]   *Algebraic Number Theory,* GTM 111, Springer-Verlag, New York, 1986.
[6]   *Introduction to Arakelov Theory,* Springer-Verlag, New York, 1988.
[7]   *Algebra,* 2nd edition, Addison-Wesley, Menlo Park, CA, 1984.
[8]   Integral points on curves. *Publ. Math. IHES* **6** (1960), 27–43.

Laska, M.
[1]   An algorithm for finding a minimal Weierstrass equation for an elliptic curve. *Math. Comp.* **38** (1982), 257–260.

Lehmer, D.H.
[1]   Ramanujan's function $\tau(n)$. *Duke Math. J.* **10** (1943), 483–492.

Lehner, J.
[1]   Divisibility properties of the Fourier coefficients of the modular invariant $j(\tau)$. *Amer. J. Math.* **71** (1949), 136–148.
[2]   Further congruence properties for the Fourier coefficients of the modular invariant $j(\tau)$. *Amer. J. Math.* **71** (1949), 373–386.

Lenstra, H.W.
[1]   Factoring integers with elliptic curves. *Ann. of Math.* **126** (1987), 649–673.

Lichtenbaum, S.
[1]   Curves over discrete valuation rings. *Amer. J. Math.* **90** (1968), 380–403.

Lipman, J.
[1]   Rational singularities with applications to algebraic surfaces and unique factorization. *Publ. Math. IHES* **38** (1970), 195–279.
[2]   Desingularization of two-dimensional schemes. *Ann. Math.* **107** (1978), 151–207.

Liu, Q.
[1]   Formule d'Ogg d'aprés Saito. preprint.

Lockhart, P., Rosen, M. and Silverman, J.
[1]   An upper bound for the conductor of an abelian variety. *J. Alg. Geo.* **2** (1993), 569–601.

Manin, Ju.
[1]   The Tate height of points on an abelian variety. Its variants and applications. *Transl. AMS* **59** (1966), 82–110.
[2]   The p-torsion of elliptic curves is uniformly bounded. *Izv. Akad. Nauk SSSR* **33** (1969), 433–438.
[3]   Rational points on an algebraic curve over function fields. *Transl. AMS* **50** (1966), 189–234.

Mason, R.C.
[1]   The hyperelliptic equation over function fields. *Math. Proc. Camb. Philos. Soc.* **93** (1983), 219–230.

Matsumura, H.
[1]   *Commutative Algebra* 2nd ed., Benjamin/Cummings, Reading, MA, 1980.

Mazur, B.

[1]   Modular curves and the Eisenstein ideal. *Publ. Math. IHES* **47** (1977), 33–186.

[2]   Rational isogenies of prime degree. *Invent. Math.* **44** (1978), 129–162.

Mestre, J.-F.

[1]   Formules explicites et minorations de conducteurs de variétiés algébriques. *Compos. Math.* **58** (1986), 209–232.

[2]   Courbes elliptiques de rang ≥ 11 sur $\mathbb{Q}(T)$, Courbes elliptiques de rang ≥ 12 sur $\mathbb{Q}(T)$, Un example de courbes elliptiques sur $\mathbb{Q}$ de rang ≥ 15. *C.R. Acad. Sci. Paris* **313** (1991), 139–142; **313** (1991), 171–174; **314** (1992), 453–455.

Milne, J.S.

[1]   *Arithmetic Duality Theorems,* Academic Press, Boston, 1986.

[2]   Abelian varieties. In *Arithmetic Geometry,* G. Cornell and J. Silverman, eds., Springer-Verlag, New York, 1986, 103–150.

[3]   Jacobian varieties. In *Arithmetic Geometry,* G. Cornell and J. Silverman, eds., Springer-Verlag, New York, 1986, 167–212.

[4]   *Étale Cohomology,* Princeton University Press, Princeton, NJ, 1980.

Miranda, R.

[1]   *The basic theory of elliptic surfaces,* Dottorato di Ricerca in Matematica, ETS Editrice, Pisa, 1989.

Mumford, D.

[1]   *Abelian Varieties,* Oxford Univ. Press, Oxford, 1974.

[2]   *Curves and their Jacobians,* The University of Michigan Press, Ann Arbor, MI, 1975.

[3]   *Tata Lectures on Theta II,* Prog. in Math. 43, Birkhäuser, Basel, 1984.

Mumford, D. and Suominen, K.

[1]   Introduction to the theory of moduli. In *Algebraic Geometry, Oslo 1970,* F. Oort, ed., Wolters-Noordhoff, Groningen, 1972, .

Nagao, K.

[1]   An example of an elliptic curve over $\mathbb{Q}$ with rank ≥ 20. *Proc. Japan Acad.* **69** (1993), 291–293.

[2]   An example of an elliptic curve over $\mathbb{Q}(T)$ with rank ≥ 13. *Proc.Japan Acad.* **70** (1994), 152–153.

Nagao, K. and Kouya, T.

[1]   An example of an elliptic curve over $\mathbb{Q}$ with rank ≥ 21. *Proc. Japan Acad.* **70** (1994), 104–105.

Namikawa, Y. and Ueno, K.

[1]   The complete classification of fibres in pencils of curves of genus two. *Manuscripta Math.* **9** (1973), 143–186.

Néron, A.

[1]   Modèles minimaux des variétés abéliennes sur les corps locaux et globaux. *Publ. Math. IHES* **21** (1964), 361–482.

[2]   Quasi-fonctions et hauteurs sur les variétés abéliennes. *Ann. of Math.* **82** (1965), 249–331.

[3]   Problèmes arithmétiques et géométriques rattachés a la notion de rang d'une courbe algébrique dans un corps. *Bull. Soc. Math. France* **80** (1952), 101–166.

Neukirch, J.
[1]   *Class Field Theory,* Grund. der Math. Wiss. 280,   Springer-Verlag, Berlin, 1986.

Ogg, A.
[1]   *Modular Forms and Dirichlet Series,* Benjamin, New York, 1969.
[2]   Elliptic curves and wild ramification. *Amer. J. Math.* **89** (1967), 1–21.
[3]   On pencils of curves of genus two. *Topology* **5** (1966), 355–362.

Perrin-Riou, B.
[1]   Arithmétique des courbes elliptiques et théorie d'Iwasawa. *Mém. Soc. Math. France* **17** (1984), 1–130.

Petersson, H.
[1]   Über die Entwicklungskoeffizienten der automorphen formen.   *Acta Math.* **58** (1932), 169–215.
[2]   Über eine Metrisierung der ganzen Modulformen. *Jber. Deutsche Math.* **49** (1939), 49–75.

Raynaud, M.
[1]   Spécialisation du foncteur de Picard. *Publ. Math. IHES* **38** (1970), 27–76.

Robert, A.
[1]   *Elliptic Curves,* Lect. Notes in Math. 326, Springer-Verlag, Berlin, 1973.

Roquette, P.
[1]   *Analytic Theory of Elliptic Functions Over Local Fields,* Vandenhoeck & Ruprecht, Göttingen, 1970.

Rosen, M.
[1]   Abel's theorem on the lemniscate. *Amer. Math. Monthly* **88** (1981), 387–395.

Rubin, K.
[1]   Tate-Shafarevich groups and *L*-functions of elliptic curves with complex multiplication. *Invent. Math.* **89** (1987), 527–560.

Saito, T.
[1]   Conductor, discriminant, and the Noether formula of arithmetic surfaces. *Duke Math. J.* **57** (1988), 151–173.

Schmidt, W.
[1]   Thue's equation over function fields. *Aust. Math. Soc. Gaz.* **25** (1978), 385–422.

Schneider, Th.
[1]   *Introduction aux Nombres Transcendents,* Grund. der Math. Wiss. 81, Springer-Verlag, Berlin, 1957.

Serre, J.-P.
[1]   *Abelian ℓ-adic Representations and Elliptic Curves,* Benjamin, New York, 1968.
[2]   Propriétés galoisiennes des points d'ordre fini des courbes elliptiques. *Invent. Math.* **15** (1972), 259–331.
[3]   *A Course in Arithmetic,* GTM 7, Springer-Verlag, New York, 1973.
[4]   *Local Fields,* transl. by M. Greenberg, Springer-Verlag, New York, 1979.
[5]   Local class field theory. In *Algebraic Number Theory,* J.W.S. Cassels and A. Fröhlich, eds., Academic Press, London, 1967, 129–162.
[6]   Complex multiplication. In *Algebraic Number Theory,* J.W.S. Cassels and A. Fröhlich, eds., Academic Press, London, 1967, 292–296.

Serre, J.-P. (continued)

[7]   *Linear Representations of Finite Groups,* GTM 42, Springer-Verlag, New York, 1977.

Serre, J.-P. and Tate, J.

[1]   Good reduction of abelian varieties. *Ann. Math.* **68** (1968), 492–517.

Shafarevich, I.R.

[1]   *Basic Algebraic Geometry,* Springer-Verlag, New York, 1977.

[2]   *Lectures on Minimal Models,* Tata Institute, Bombay, 1966.

Shatz, S.

[1]   The cohomology of certain elliptic curves over local and quasi-local fields. *Illinois J. Math.* **11** (1967), 234–241.

[2]   Group schemes, formal groups, and p-divisible groups. In *Arithmetic Geometry,* G. Cornell and J. Silverman, eds., Springer-Verlag, New York, 1986, 29–78.

Shimura, G.

[1]   *Introduction to the Arithmetic Theory of Automorphic Forms,* Princeton Univ. Press, Princeton, NJ, 1971.

Shimura, G. and Taniyama, Y.

[1]   *Complex Multiplication of Abelian Varieties and its Application to Number Theory,* Publ. Math. Soc. Japan 6, 1961.

Shioda, T.

[1]   The Galois representation of type E_8 arising from certain Mordell-Weil groups. *Proc. Japan. Acad.* **65** (1989), 195–197.

[2]   On Mordell-Weil lattices. *Univ. Sancti Pauli* **39** (1990), 211–240.

[3]   On elliptic modular surfaces. *J. Math. Soc. Japan* **24** (1972), 20–59.

[4]   Mordell-Weil lattices and Galois representations I, II, III. *Proc. Japan Acad.* **65** (1989), 268–271, 296–299, 300–303.

[5]   Mordell-Weil lattices and sphere packings. *Amer. J. Math.* **113** (1991), 931–948.

[6]   Mordell-Weil lattices of type E_8 and deformation of singularities. *Lect. Notes Math.* **1468** (1991), 177–202.

[7]   Construction of elliptic curves with high rank via the invariants of the Weyl groups. *J. Math. Soc. Japan* **43** (1991), 673–719.

Siegel, C.L.

[1]   A simple proof that $\eta(-1/\tau) = \eta(\tau)\sqrt{\tau/i}$. *Mathematika* **1** (1954), 4.

Silverman, J.H.

[AEC]   *The Arithmetic of Elliptic Curves,* GTM 106, Springer-Verlag, New York, 1986.

[1]   Heights and the specialization map for families of abelian varieties. *J. Reine Angew. Math.* **342** (1983), 197–211.

[2]   Computing heights on elliptic curves. *Math. Comp.* **51** (1988), 339–358.

[3]   The difference between the Weil height and the canonical height on elliptic curves. *Math. Comp.* **55** (1990), 723–743.

[4]   Hecke points on modular curves. *Duke Math. J.* **60** (1990), 401–423.

[5]   Variation of the canonical height on elliptic surfaces I. *J. Reine Angew. Math.* **426** (1992), 151–178.

[6]   Variation of the canonical height on elliptic surfaces II, III. *J. Number Theory* (1994), to appear.

Silverman, J.H. (continued)
[7]   The Néron-Tate Height on Elliptic Curves. Ph.D. thesis, Harvard, 1981.
[8]   The S-unit equation over function fields. *Math. Proc. Camb. Philos. Soc.* **95** (1984), 3–4.

Stark, H.
[1]   A complete determination of the complex quadratic fields of class-number one. *Michigan Math. J.* **14** (1967), 1–27.
[2]   The Coates-Wiles theorem revisited. In *Number Theory Related to Fermat's Last Theorem*, N. Koblitz, ed., Birkhäuser, Boston, 1982, 349–362.

Szpiro, L.
[1]   Séminaire sur les pinceaux de courbes de genre au moins deux. *Astérisque* **86** (1981), 44–78.

Tate, J.
[1]   The arithmetic of elliptic curves. *Invent. Math.* **23** (1974), 171–206.
[2]   Algorithm for determining the type of a singular fiber in an elliptic pencil. In *Modular Functions of One Variable IV*, Lect. Notes in Math. 476, B.J. Birch and W. Kuyk, eds., Springer-Verlag, Berlin, 1975, 33–52.
[3]   Letter to J.-P. Serre (1979), unpublished.
[4]   Variation of the canonical height of a point depending on a parameter. *Amer. J. Math.* **105** (1983), 287–294.
[5]   WC-groups over $\mathfrak{p}$-adic fields. *Sém. Bourb.* Exposé **156** (1957/58), 13pp.
[6]   Duality thoerems in Galois cohomology. *Proc. Intern. Congress Math.*, Stockholm, 1962, 234–241.
[7]   Global class field theory. In *Algebraic Number Theory*, J.W.S. Cassels and A. Fröhlich, eds., Academic Press, London, 1967, 163–203.
[8]   Fourier analysis in number fields and Hecke's zeta-functions. In *Algebraic Number Theory*, J.W.S. Cassels and A. Fröhlich, eds., Academic Press, London, 1967, 305–347.
[9]   A review of non-archimedean elliptic functions. In *Elliptic Curves, Modular Forms, & Fermat's Last Theorem*, J. Coates and S.T. Yau, eds., International Press, Boston, 1995, 162–184.

Vlăduţ, S.G.
[1]   *Kronecker's Jugendtraum and modular functions*, Gordon and Breach, New York, 1991.

Vojta, P.
[1]   *Diophantine approximations and value distribution theory*, Lect. Notes in Math. 1239, Springer-Verlag, Berlin, 1987.

Voloch, J.F.
[1]   Siegel's theorem for complex function fields. *Proc. AMS* **121** (1994), 1307–1308.

Waldschmidt, M.
[1]   Nombres transcendent et groupes algébriques, *Astérisque* **69–70**, 1979.

Waterhouse, W.
[1]   *Introduction to Affine Group Schemes*, GTM 66, Springer-Verlag, New York, 1979.

Weil, A.
[1]   *Elliptic Functions According to Eisenstein and Kronecker,* Springer-Verlag, Berlin, 1976.
[2]   Arithmetic on algebraic varieties. *Ann. Math.* **53** (1951), 412–444.
[3]   *Variétés Abéliennes et Courbes Algébriques,* Hermann, Paris, 1948.

Zimmer, H.
[1]   On the difference of the Weil height and the Néron-Tate height. *Math. Z.* **147** (1976), 35–51.
[2]   Quasifunctions on elliptic curves over local fields. *J. reine angew. Math.* **307/308** (1979), 221–246; Corrections and remarks concerning quasifunctions on elliptic curves. *J. reine angew. Math.* **343** (1983), 203–211.

List of Notation

E_Λ	the elliptic curve $y^2 = 4x^3 - g_2(\Lambda)x - g_3(\Lambda)$, 6
$\wp(z; \Lambda)$	Weierstrass $\wp$ function relative to the lattice Λ, 6
$\mathcal{L}$	the set of lattices in $\mathbb{C}$, 6
$\mathbf{H}$	the upper half plane $\{\tau \in \mathbb{C} : \mathrm{Im}(\tau) > 0\}$, 7
$SL_2(\mathbb{Z})$	special linear group over $\mathbb{Z}$, 8
Λ_τ	the lattice $\mathbb{Z}\tau + \mathbb{Z}$, 9
$\Gamma(1)$	the modular group, 10
S	the element $\begin{pmatrix} 0 & -1 \\ 1 & 0 \end{pmatrix} \in \Gamma(1)$, 10
T	the element $\begin{pmatrix} 1 & 1 \\ 0 & 1 \end{pmatrix} \in \Gamma(1)$, 10
$\mathcal{F}$	fundamental domain for action of $SL_2(\mathbb{Z})$ on $\mathbf{H}$, 10
$I(\tau)$	the stabilizer of τ, 11
$X(1)$	modular curve, 14
$\mathbf{H}^*$	extended upper half plane, 14
$Y(1)$	affine modular curve, 14
$X(1)$	projective modular curve, 14
$I(\tau_1, \tau_2)$	transformations sending τ_1 to τ_2, 17
$I(U_1, U_2)$	transformations γ with $\gamma U_1 \cap U_2 \neq \emptyset$, 17
ψ_x	local parameter for $X(1)$, 20
j	j-invariant map $X(1) \to \mathbb{P}^1(\mathbb{C})$, 23
$\mathrm{ord}_\infty(f)$	order at ∞ of a modular function, 24
$f(\infty)$	value at ∞ of a modular function, 24
$G_{2k}(\Lambda)$	Eisenstein series, 25
$G_{2k}(\tau)$	Eisenstein series, 25
g_2	equals $60G_4$, 26
g_3	equals $140G_6$, 26
$\Delta(\tau)$	the modular discriminant, 26
Ω_X	space of differential 1-forms on X, 27
Ω_X^k	space of meromorphic k-forms on X, 27
$\mathrm{ord}_x(\omega)$	order of a differential form, 27
ω_f	differential form on $X(1)$ attached to modular form f, 28
M_{2k}	space of modular forms of weight $2k$, 31
M_{2k}^0	space of cusp forms of weight $2k$, 31
$\mathcal{ELL}_\mathbb{C}$	isomorphism classes of elliptic curves over $\mathbb{C}$, 36
$\zeta(z; \Lambda)$	Weierstrass ζ function for the lattice Λ, 39

η	quasi-period homomorphism, 41	
$\sigma(z;\Lambda)$	Weierstrass σ function for the lattice Λ, 44	
ψ	character describing periodicity of Weierstrass σ function, 44	
Λ_τ	normalized lattice $\mathbb{Z}\tau + \mathbb{Z}$, 47	
$\wp(z;\tau)$	Weierstrass $\wp$ function for the normalized lattice Λ_τ, 47	
$\zeta(z;\tau)$	Weierstrass ζ function for the normalized lattice Λ_τ, 47	
$\sigma(z;\tau)$	Weierstrass σ function for the normalized lattice Λ_τ, 47	
$u = u_z$	$= e^{2\pi i z}$, 47	
$q = q_\tau$	$= e^{2\pi i \tau}$, 47	
$q^{\mathbb{Z}}$	the cyclic group generated by q, 47	
$F(u;q)$	an elliptic function, 49	
$\sigma_k(n)$	sum of kth powers of divisors of n, 55	
B_k	Bernoulli number, 57	
$E_{2k}(\tau)$	normalized Eisenstein series, 58	
$\tau(n)$	Ramanujan τ function, 59	
$c(n)$	Fourier coefficients of the modular j function, 59	
$\eta(\tau)$	Dedekind η function, 65	
$\Phi(\gamma)$	integer in Dedekind η function transformation formula, 66	
$s(x,y)$	Dedekind sum, 66	
$\text{Div}(S)$	divisor group of a set, 68	
$T(n)$	n^{th} Hecke operator, 68	
R_λ	homothety operator, 68	
$\Lambda' \overset{n}{\subset} \Lambda$	Λ' is a sublattice of Λ of index n, 69	
$M_2(\mathbb{Z})$	ring of 2×2 matrices with integral coefficients, 72	
$\alpha(\Lambda)$	the lattice $\mathbb{Z}(a\omega_1 + b\omega_2) + \mathbb{Z}(c\omega_1 + d\omega_2)$, 72	
$\mathcal{D}_n$	integer matrices of determinant n, 72	
$\mathcal{S}_n$	special integer matrices of determinant n, 72	
F_f	lattice function associated to the modular function f, 74	
f_F	modular function associated to the lattice function F, 74	
$T_{2k}(n)$	n^{th} Hecke operator on space of modular functions, 76	
$\gamma(m)$	m^{th} Fourier coefficient of $T_{2k}(n)f$, 76	
$L(f,s)$	formal Dirichlet series attached to the power series f, 80	
$R(f,s)$	normalized L-series attached to f, 83	
$\Gamma(s)$	gamma function, 83	
$\Gamma(N)$	a congruence subgroup of $\Gamma(1)$, 86	
$\Gamma_0(N)$	a congruence subgroup of $\Gamma(1)$, 87	
$\Gamma_1(N)$	a congruence subgroup of $\Gamma(1)$, 87	
$\mathcal{L}(D)$	linear series attached to the divisor D, 87	
e_m	Weil pairing, 89	
$\mu(\alpha,\tau)$	$= c\tau + d$, where $\alpha = \begin{pmatrix} a & b \\ c & d \end{pmatrix}$, 91	
$f|[\alpha]_{2k}$	$= (\det \alpha)^k \mu(\alpha,\tau)^{-2k} f \circ \alpha$, 91	
$\langle f, g \rangle$	Petersson inner product of f and g, 92	
$\Gamma(s,x)$	incomplete gamma function, 93	
$g(\chi)$	Gauss sum associated to χ, 93	
$f(\chi,\tau)$	twist of the cusp form f by the character χ, 93	
$L(f,\chi,s)$	twist of the L-series $L(f,s)$ by the character χ, 93	
R_K	ring of integers (maximal order) of K, 96	
E_Λ	the elliptic curve isomorphic to $\mathbb{C}/\Lambda$, 97	

$[\cdot]$	normalized isomorphism $R \to \mathrm{End}(E)$, 97	
$\mathcal{ELL}(R)$	elliptic curve, with endomorphism ring R, 98	
$\mathcal{CL}(R_K)$	ideal class group of R_K, 99	
$\bar{\mathfrak{a}}$	the ideal class of $\mathfrak{a}$, 99	
$\mathfrak{a}\Lambda$	the product of the ideal $\mathfrak{a}$ and lattice Λ, 99	
$\bar{\mathfrak{a}} * E_\Lambda$	action of ideal class $\bar{\mathfrak{a}}$ on elliptic curve E_Λ, 99	
$E[\mathfrak{a}]$	group of $\mathfrak{a}$-torsion points, 102	
h_K	class number of K, 107	
F	complex multiplication map $\mathrm{Gal}(\bar{K}/K) \to \mathcal{CL}(R_K)$, 112	
$\sigma_{\mathfrak{p}}$	Frobenius element in $\mathrm{Gal}(L/K)$, 116	
$I(\mathfrak{c})$	fractional ideal prime to $\mathfrak{c}$, 116	
$P(\mathfrak{c})$	principal ideals congruent to 1 modulo $\mathfrak{c}$, 117	
$K_{\mathfrak{c}}$	ray class field of K modulo $\mathfrak{c}$, 117	
H, H_K	Hilbert class field of K, 118	
K_v	completion of K at v, 119	
R_v	ring of integers of K_v, or K_v if v is archimedean, 119	
$\mathbf{A}_K^*$	idele group of K, 119	
$K_{\mathfrak{p}}$	completion of K at $\mathfrak{p}$, 119	
$R_{\mathfrak{p}}$	ring of integers of $K_{\mathfrak{p}}$, 119	
$\mathrm{ord}_{\mathfrak{p}}$	normalized valuation on $K_{\mathfrak{p}}$ and $R_{\mathfrak{p}}$, 119	
(s)	ideal of the idele s, 119	
$U_{\mathfrak{c}}$	an open subgroup of the idele group $\mathbf{A}_K^*$, 119	
N_K^L	norm map on idele groups, 119	
K^{ab}	maximal abelian extension of K, 120	
$[\cdot, K]$	the reciprocity map for K, 120	
h	Weber function $E \to E/\mathrm{Aut}(E) \equiv \mathbb{P}^1$, 134	
$F_n(X)$	the polynomial $\prod_{\alpha \in \mathcal{S}_n}(X - j \circ \alpha)$, 144	
$s_m(\tau)$	the coefficients of the polynomial $F_n(X)$, 144	
$F_n(Y, X)$	the modular polynomial $F_n(j, X) = \prod_{\alpha \in \mathcal{S}_n}(X - j \circ \alpha)$, 146	
$H_n(X)$	the modular polynomial $F_n(X, X)$, 146	
$M[\mathfrak{p}^\infty]$	$\mathfrak{p}$-primary component of an R_K-module M, 157	
$\psi_{E/L}$	the Grössencharacter of a CM elliptic curve E/L, 168	
$L_{\mathfrak{P}}(E/L, T)$	local L-series of E at $\mathfrak{p}$, 171	
$L(s, \psi)$	Hecke L-series attached to the Grössencharacter ψ, 173	
$\Lambda(E/L), s)$	modified L-series of E/L, 176	
$w_{E/L}$	sign of the functional equation of E/L, 176	
$\mathcal{D}_n^*$	primitive integer matrices of determinant n, 181	
$\mathcal{S}_n^*$	primitive special integer matrices of determinant n, 181	
$\Phi_n(X)$	the modular polynomial of order n, 181	
$Z(E/\mathbb{F}_{\mathfrak{P}}, T)$	zeta function of the elliptic curve $E/\mathbb{F}_{\mathfrak{P}}$, 183	
$\zeta_L(s)$	zeta function of the field L, 183	
$\zeta(E/L, s)$	zeta function of the elliptic curve E/L, 183	
$T_{\mathfrak{p}}(E)$	$\mathfrak{p}$-adic Tate module, 186	
$K(S, 2)$	subgroup of K^*/K^{*2}, 194	
$K(S, m)$	subgroup of K^*/K^{*m}, 195	
$\mathrm{Pic}^0(C)$	group of divisor classes of degree zero, 197	
$\mathrm{Jac}(C)$	Jacobian variety of C, 197	
$C^{(g)}$	symmetric product of a curve, $= C^g/S_g$, 197	

$i_{L/K}$ index function on the Galois group of local fields, 405

Ar   the Artin character, 405

Sw   the Swan character, 405

$\sigma_k(n)$ the sum $\sum_{d|n} d^k$, 409

E_q elliptic curve over $\mathbb{C}$ with $j(E_q) = j(q)$, 410

$s_k(q)$ the series $\sum_{n \geq 1} \sigma_k(n) q^n$, 410, 423

$a_4(q), a_6(q)$ Weierstrass coefficients of E_q, 410, 423

$X(u,q), Y(u,q)$ series giving parametrization of E_q, 410, 423

$\theta(u,q)$ theta function, 412

$\gamma(E/\mathbb{R})$ γ-invariant for an elliptic curve defined over $\mathbb{R}$, 414

E_q Tate curve over p-adic field, 423

$\hat{E}$ formal group of an elliptic curve, 431

$\mathbb{G}_m$ formal multiplicative group, 431

$\gamma(E/K)$ γ-invariant for an elliptic curve defined over K, 439

$E_{q,r}$ filtration of Tate curve E_q, 450

$\hat{h}$ the canonical height on an elliptic curve, 454

$|\cdot|_v$ absolute value on the field K, 455

λ the local Néron height function on E associated to v, 455

M_K the standard set of absolute values on K, 461

n_v $= [K_v : \mathbb{Q}_v]$, the local degree of $v \in M_K$, 461

λ_v local height associated to v, 461

B_2 the second Bernoulli polynomial $T^2 - T + \frac{1}{6}$, 468

$\tilde{E}$ the reduction of E modulo v, 469

$E_0(K)$ points in $E(K)$ with non-singular reduction, 469

ψ_m division polynomial, 477

Index

Modular curve $X(1)$ (*continued*)
  is connected, 21
  j-function on, 23, 34, 36
  local parameter, 20, 21, 29, 30, 85
  meromorphic functions on, 23, 35
  open cover of, 20
  order of a differential form, 28
  projection is open map, 19
  topology on, 16, 18, 19
Modular curve $X_1(11)$, 190, 279
Modular discriminant
  See Discriminant
Modular form, 25, 91 *See also* Modular
  function, cusp form
  action of Hecke operator, 76
  dimension of space of, 31, 87
  Eisenstein series is a, 25, 31, 32
  Fourier coefficients of, 82, 92
  L-series attached to, 93
  size of Fourier coefficients, 82, 92
  space of (M_{2k}), 31, 87
  space of is generated by G_4, G_6, 88
  spaces with dimension one, 33
Modular function, 24, 91
  See also j-invariant; modular form;
    cusp form
  j is a, 34
  action of Hecke operator, 76, 79, 91
  associated differential form, 28, 31, 87,
    91
  associated lattice function, 74
  eigenfunction for Hecke operator, 77
  formula for order, 30
  Fourier series, 24, 76, 145
  holomorphic at ∞, 24
  meromorphic at ∞, 24
  of weight zero, 34, 35, 182
  order at ∞, 24
  value at ∞, 24, 32
  weight of, 24
  why even weight, 24
Modular group $\Gamma(1)$, 6–14
  See also Action of $SL_2(\mathbb{Z})$ on **H**
  generated by S and T, 14, 24, 62, 85
  is a free product, 14, 85
Modular polynomial
  examples, 148
  F_n, 144, 146, 181, 182
  Φ_n, 181, 182
  H_n, 144, 146
  size of coefficients, 181
Monster group, 61
Montgomery, H., 487
Mordell, L.J., 61, 79
Mordell conjecture, 271
Mordell-Weil group
  Euclidean structure, 247
  lattice structure, 254
Mordell-Weil theorem
  bound for rank, 279
  for abelian varieties, 485

  for elliptic surfaces, 276
  for function fields, 230, 279
  for split elliptic surfaces, 231, 281
  ineffective, 277
  over finitely generated fields, 279
  relative, 231, 281
  weak, 191–195, 230
Morphism
  birational, 235
  diagonal, 300, 309, 327
  divisor associated to, 258
  étale, 306
  flat, 304, 397
  graph of, 300
  of S-schemes, 297
  proper, 305
  separated, 305
  smooth, 304, 305, 306, 320
  universally closed, 305
Multiple fibers, 203
Multiplication map
  continuous for v-adic topology, 460
  on elliptic curve, 310
  on group scheme, 310, 398
Multiplication by an idele, 159, 170
Multiplication-by-n map on an elliptic
  surface, 266
Multiplication-by-x map on $\mathbb{Q}/\mathbb{Z}$, 152,
  153
Multiplicative group, 128, 151, 398, 399
  exponential map, 151
  formal, 149, 431
  over a field, 291, 293, 398
  over R, 308
  over S, 308
  over $\mathbb{Z}$, 308
  rational points of, 292
  S-valued points, 397
  Tate module of, 382
  torsion subgroup, 151
Multiplicative reduction, 287, 369, 388,
  399
  conductor, 381
  Néron model, 403
  non-split, 171, 378
  split, 171, 362, 377, 379
  torsion for, 453
Multiplicity of fibral components, 313

Nagao, K., 272
Nakai-Moishezon criterion, 258, 261, 283,
  284
Nakayama's lemma, 103
Namikawa, Y., 355, 486
Natural map, 197, 198
Negation map, on an elliptic curve, 325
Néron, A., 231, 265, 272, 352, 454, 455,
  485
Néron height function
  See Local height function

Graduate Texts in Mathematics

(continued from page ii)